Polarization Measurement and Control in Optical Fiber Communication and Sensor Systems

Polarization Measurement and Control in Optical Fiber Communication and Sensor Systems

X. Steve Yao
Xiaojun (James) Chen

This edition first published 2023

Registered Office
John Wiley & Sons, Inc., 111 River Street, Hoboken, NJ 07030, USA

Editorial Office
John Wiley & Sons, Inc., 111 River Street, Hoboken, NJ 07030, USA

For details of our global editorial offices, customer services, and more information about Wiley products visit us at www.wiley.com.

Wiley also publishes its books in a variety of electronic formats and by print-on-demand. Some content that appears in standard print versions of this book may not be available in other formats.

Library of Congress Cataloging-in-Publication Data applied for:

Hardback ISBN: 9781119758471

Cover design by Wiley
Cover image: Courtesy of X. Steve Yao

Set in 9.5/12.5pt STIXTwoText by Straive, Chennai, India

To the memory of my father, Dunli Yao, whose lasting faith continues to inspire.

Contents

Author Biographies

X. Steve Yao is the founder of PolaLight Consulting LLC in Las Vegas, Nevada, and was the founder and Chief Technology Officer of General Photonics Corporation (now part of Luna Innovations) in Chino, California, dedicated to the design and engineering of polarization control and measurement products for over 25 years. He is also the founding director of the Photonics Information Innovation Center (PIIC) at Hebei University (his alma mater) in China. With over 100 journal publications and 80 US patents, Dr. Yao is a fellow of both IEEE and Optica, and holds a PhD degree in Electrical Engineering from the University of Southern California, USA.

Xiaojun (James) Chen is the founder and Chief Technology Officer of In-line Photonics Inc. in San Gabriel, California, and was the Chief Scientist of General Photonics Corporation (now part of Luna Innovations) in Chino, California, dedicated to the design and engineering of polarization control and measurement products for over 20 years. Dr. Chen holds a PhD degree in Condense Matter Physics from Nankai University, China.

Preface

Polarization is one of the seven fundamental parameters that define the properties of light waves, along with intensity, wavelength, phase, direction, speed, and coherence. Unlike the other parameters, polarization is a multidimensional problem; it requires at least a three-dimensional sphere, known as the Poincaré Sphere, to describe its behavior, which becomes more complicated when the light interacts with anisotropic materials.

Étienne-Louis Malus is considered by many to be the father of polarization because he was the first to systematically study the polarization properties of reflected light, publishing his findings on the subject in 1809. He also published his theory of double refraction of light in crystals in 1810, although the phenomenon of double refraction was first observed by Vikings and by Thomas Bartholin. Throughout most of the 250-year history of polarization research, studies have been limited to observations in free space or in short lengths of optical media that are stable over time.

The polarization behavior of light in an optical fiber is quite different from its behavior in free space or in a short span of a bulk optical medium, although the fundamental physics is the same. It is therefore not straightforward to apply the knowledge of polarization commonly covered in classical physics, which mostly focuses on light traveling in free space or in crystals, to fully understand the polarization characteristics of light in optical fibers for two main reasons. First, unlike in free space or in an optical crystal, the birefringence distribution over a long segment of single mode fiber is random and changes over time, which causes rapid polarization changes both along the fiber and over time. Second, most physics textbooks assume monochromatic light, whereas in optical fiber communication systems, the optical signals have finite bandwidths which increase with the data transmission rate, resulting in much more complicated polarization behavior when such broadband signals travel in a single mode fiber whose birefringence varies over distance and time.

The peculiar polarization behavior of light in optical fibers was first observed and studied in 1972 by researchers at Corning (F. Kapron, N. Borrelli, and D. Keck) and at Bell Labs (W. Schosser), when optical fibers with a loss less than 16 dB/km were first achieved. At that time, the adaptation of a coherent detection approach from radio communication to optical fiber communication was assumed, and therefore the studies of polarization behavior were actively pursued. Soon after, the simplest on–off key (OOK) modulation format was found to be adequate for optical fiber communications, especially after the adoption of Erbium doped fiber amplifiers (EDFA); the motivation for understanding polarization behavior in optical fibers in communication systems diminished because OOK at low data rates is essentially polarization independent. The topic remained largely neglected in optical fiber communication systems until around the year 2000, when data rates in new communications systems climbed beyond 10 Gbps, a rate at which the effects of polarization related issues such as polarization mode dispersion and polarization dependent loss could no longer be ignored. In contrast, polarization studies for optical fiber sensor systems have always remained active.

This book is based on the authors' experience over more than 20 years at General Photonics Corporation (now part of Luna Innovations) in designing and building devices and instruments for the control and measurement of polarization for optical fiber communication and sensing systems. It is intended to be a reference book to enable scientists, engineers, and graduate students to obtain a quick grasp of polarization issues in optical fiber systems, with detailed descriptions of practical devices or configurations for the control and measurement of polarization related parameters. The book includes a brief history of polarization (Chapter 1) with a drawing to show the historical milestones of human beings in understanding the polarization of light. Unlike many other books on polarization, which start with Maxwell equations and heavy mathematics for polarization analysis, this book starts the subject by introducing the basics of the polarization of light (Chapter 2) with minimal mathematics, followed by descriptions of polarization effects unique to optical fibers (Chapter 3), again without involving heavy mathematics. Instead, all mathematical methods for polarization analysis and related formulae are put into Chapter 4 for interested readers. Many of the formulae are directly used in subsequent chapters whenever applicable. In fact, Chapter 4 includes detailed derivations of almost all major polarization-related formulae or conclusions, providing a "one-stop shop" where interested readers can find not only the equations themselves, but also where they came from, without having to search for answers scattered over dozens of publications.

In Chapter 5, we review and summarize the properties of certain polarization phenomena in common anisotropic materials, such as the double refraction, optical activity, linear electro-optic effect, and photo-elastic effect, which are important for the discussion of polarization control in Chapter 6 that focuses on polarization management components and devices. Chapter 7 covers modules and instrument for active polarization management, focusing on the integration of components and devices introduced in Chapter 6 with electronics and algorithms for common polarization management applications.

Chapter 8 provides detailed coverage of the important topic of polarization measurement, including discussion of both Stokes polarimeters, which can measure the polarization properties of the light itself, such as its state of polarization, degree of polarization, and polarization extinction ratio; and Mueller matrix polarimeters, which can determine not only the polarization properties of the light, but also the polarization properties of optical media through which the light passes, including polarization mode dispersion and polarization dependent loss.

The polarization generators and analyzers described in Chapter 8 are traditional analog devices, relying on mechanical rotations and analog modulations with relatively poor repeatability and accuracy due to mechanical wear and tear and varied environmental conditions. To overcome these issues, we developed binary polarization generators and analyzers utilizing binary magneto-optic rotators, with the specifics covered in Chapter 9. The binary polarization generation and analysis techniques pioneered by the authors of this book have never been discussed in any other books, which in our opinion represent a major advancement on the topic of polarization measurements.

The polarization analysis systems described in Chapters 8 and 9 can only measure the cumulative polarization effect of an optical medium (such as a fiber) on a light wave after transmission through the medium. Sometimes, it is important to know the position resolved polarization properties of the light wave in the medium. Chapter 10 focuses on such distributed polarization analysis techniques and their applications, mostly based on the authors' own research on this topic over the past 10 years.

Finally, in Chapter 11, we describe techniques for using polarization analysis to enable several new applications, including fast optical frequency detection, high accuracy rotation detection (polarimetric fiber optic gyroscope, P-FOG), and high accuracy electrical current and magnetic field sensing, which summarize the authors' own research in these exciting directions over the past 15 years.

We have tried not to involve more advanced mathematics, such as spin-vector calculus and group theory, in the polarization analysis in order to make it comprehensible to most readers with engineering backgrounds. Only the more well-known Jones matrix calculus and Mueller matrix calculus are introduced in this book. We recognize that the spin-vector calculus may be more elegant mathematically in treating certain polarization problems, and is easier to grasp for people with strong backgrounds in quantum mechanics. Interested readers are encouraged to read the book *Polarization Optics in Telecommunications* by Jay N. Damask (2005) on the spin-vector calculus.

It is often difficult to visualize how a state of polarization (SOP) evolves on the Poincaré Sphere in response to the variations of linear or circular birefringence; we therefore include a SOP trace visualization software called PolaTrace™ written in Python to help readers to view SOP variation traces on the Poincaré Sphere when the magnitudes and/or the orientation angles of the birefringence in one or more locations in the optical path are changed. With the PolaTrace, one is able to see how the SOP can be manipulated with either a single polarization component, or multiple polarization components cascaded in series, such that polarization variations in an optical fiber discussed in Chapter 3 and various polarization controlling devices described in Chapters 4, 6, and 7 can be more easily understood. The owners of this book have the privilege to download the Python source codes at https://github.com/PolarizationInFiber/PolaTrace.

X. Steve Yao would like to dedicate this book to his parents, Dunli Yao and Xianrong Fu, for their constant love and unconditional support. He is grateful to Professors Jack Feinberg and Robert Hellworth at the University of Southern California (USC) for their excellent guidance in the graduate school, Professor Alan Willner of USC for his collaboration and support in the past 20 years (many of the joint research results are included in this book), and Mr. Eric Udd of Columbia Gorge Research for his encouragement for writing this book. His appreciation also goes to his students, Xiaosong Ma and Penghui Yao at Hebei University and Tianjin University, respectively, for their help in preparing some of the figures used in the book.

Xiaojun (James) Chen would like to dedicate this book to his lovely wife, Mengyan, for her understanding and care, and his amazing daughter, Jenny, for filling smiles in his life. He would also like to thank Steve for the opportunity to join General Photonics in 2004, and for his trust and support throughout the years while working at this highly innovative company.

Both of us are grateful to our formal colleagues at General Photonics for their help in turning many of our ideas on paper into real products (many of which are described in this book) and Susan Wey for editing most of the technical support documents of the company over the years (some of which are adapted in this book), including her help in editing this preface. Finally, we would like to express our sincere appreciations to Wiley editor Brett Kurzman, managing editor Sarah Lemore, cover editor Becky Cowan, and content refinement specialist Judit Anbu Hena, whose hard work and support make this book possible.

Las Vegas, NV 89109, USA
San Gabriel, CA 91775, USA
August 2022

X. Steve Yao
Xiaojun (James) Chen

1

History of Light and Polarization

1.1 Early History of Light

Light is the most important element to the life on Earth, particularly to the existence of human beings. Light not only enables us to "see" for receiving visual information from our surroundings and communicating with others via gestures and expressions, but also provides the energy for living things to grow. The main source of light on Earth is the Sun, which provides the energy for the plants to grow via a process called photosynthesis (Morton 2009). While growing, the plants create sugars and carbohydrates (Stichler 2002), which release energy into the living things that digest them. Buried dead plants can also be turned into coal and fossil fuels by nature as a different form of energy. Another important source of light for humans has been fire, from ancient campfires using woods to last century's kerosene lamps using fossil fuels, which also originated from the Sun light. With the development of electric lights and power systems, electric lighting has effectively replaced firelight. Interestingly, the origin of the electrical energy is still the Sun light because it is produced by burning the coal or fossil oil, or by the direct conversion using solar panels via photovoltaic effect. Some species of animals generate their own light via a process called bioluminescence. For example, fireflies use light to locate mates, and vampire squids use it to hide themselves from prey.

The most recent form of man-made light is laser beams, which are being widely utilized in human societies today, such as in optic fiber communication networks (the backbone of internet), printing (laser printers), navigation and guidance (optic gyroscopes and Lidar), medicine (laser diagnosis, surgical and treatment equipment), sensor (facial recognition in smart phones and temperature, pressure, and other sensors in industrial, medical, and defense equipment), and manufacturing (laser cutting and laser additive manufacturing), just to name a few. These laser applications have dramatically improved the quality of our lives behind scenes and many of us may not notice them. New laser based developments and applications are emerging almost every day to continue better our lives. That is why twenty-first century is considered the century of optics by many, just as twentieth century was regarded as the century of electronics.

Humans have been trying to understand and utilize light since very early stage of our civilization. The early studies of light may likely be initialed from trying to understand vision because ancient men started to wonder why they could see, while the early man-made optical devices may be for harvesting sun light. For example, the 3000-year old Nimrud lens made with a quartz crystal unearthed in modern-day Iraq was likely a sun-light concentrator for starting fire. Bronze concave mirrors dating back around West Zhou period (c. 1046–771 BCE) unearthed in Shan'Xi, China, from 1972 to 1995 are also believed for concentrating sun light for fire-starting. In fact, it was recorded in

Polarization Measurement and Control in Optical Fiber Communication and Sensor Systems, First Edition.
X. Steve Yao and Xiaojun (James) Chen.

a book called *Etiquette of Zhou* that the government had a dedicated official for using the concave mirror to start fire for religious rituals or ceremonies.

As early as sixth to fifth BCE, **Ancient Indian** developed some theories on light and believed that light rays were a stream of high velocity of *tejas* (fire) particles. In **Ancient China**, **Mozi** (c. 468–376 BCE), a renowned Chinese scientist and thinker during Warring States Period (c. 476–221 BCE) summarized 16 rules (eight each in two separate articles) regarding shadows, pin-hole images, and mirror images in his book *Mojing* (*Mo's Articles*) around 388 BCE, describing the relationships between the shadows and the light source positions, the relationship between the images and pin-hole positions, as well as the relationships between the image properties (size, location, and direction [inverted or upright]) and different types of mirrors (flat, concave, and convex).

Euclid of Alexandria (c. 325–265 BCE) is the first known author of a treatise on geometrical optics. His book, known as *Euclid's Optics*, influenced the work of later Greek, Islamic, and Western European Renaissance scientists and artists. In his book, Euclid observed that "things seen under a greater angle appear greater, and those under a lesser angle less, while those under equal angles appear equal." In the 36 propositions that follow, Euclid relates the apparent size of an object to its distance from the eye and investigates the apparent shapes of cylinders and cones when viewed from different angles. While Euclid had limited his analysis to simple direct vision, **Hero of Alexandria** (c. 10–70 CE) extended the principles of geometrical optics to consider problems of reflection (catoptrics) and demonstrated the equality of the angle of incidence and reflection on the grounds that this is the shortest path from the object to the observer. On this basis, he was able to define the fixed relation between an object and its image in a plane mirror. Specifically, the image appears to be as far behind the mirror as the object really is in front of the mirror. **Claudius Ptolemy** (c. 323–283 BCE), in his book *Optics* (known as *Ptolemy's Optics*), undertook studies of reflection and refraction of light on flat surfaces in the second century BCE. Both Euclid and Ptolemy believed that sight worked by the eyes emitting rays of light.

Ibn Sahl (c. 940–1000 CE), a Persian mathematician and scientist, is known to have compiled a commentary on *Ptolemy's Optics* and wrote an optical treaties around 984. He studied the optical properties of curved mirrors and lenses, and discovered the law of refraction which was mathematically equivalent to Snell's law. He used his law of refraction to compute the shapes of lenses and mirrors that focus light at a single point on the axis. **Hasan Ibn al-Haytham** (c. 965–1040), an Arab mathematician, astronomer, and physicist, often referred to as "the father of modern optics," studied the characteristics of light and the mechanism/process of vision, produced a comprehensive and systematic analysis of Greek optical theories, and wrote the influential *Book of Optics*. He was against Euclid and Ptolemy's opinion on eyes emitting light rays and insisted that vision occurs because rays enter the eyes and considered these rays as the forms of light and color. He then analyzed these rays according to the principles of geometrical optics and carried out various experiments with lenses, mirrors, reflections, and refractions to verify his analysis. He was an early proponent of the concept that a hypothesis must be supported by experiments based on confirmable procedures or mathematical evidence – an early pioneer in scientific method five centuries before Renaissance scientists. In the Far East, **Shen Kuo** (c. 960–1279), a Chinese polymathic scientist, was the first to have detailed description of camera obscura or pin-hole imaging in his book *Dreams Pool Essays* in 1088. In the late thirteenth and early fourteenth centuries, **Qutb-al-Din al-Shirazi** (1236–1311) and his student **Kamal al-Din al-Farisi** (1260–1320) continued the work of Ibn al-Haytham, and they were among the first to give the correct explanations for the rainbow phenomenon. al-Farisi published his findings in his book, *The Revision of Optics (Kitab Tanqih al-Manazir)*, for refining al-Haytham's *Book of Optics*, which signified the beginning for humans to understand another important aspect of light: the color.

The English bishop, **Robert Grosseteste** (c. 1175–1253), may be the most influential figure in the understanding of light during the medieval Europe. He wrote on a wide range of scientific topics on the time of the origin and tended to apply mathematics and the Platonic metaphor of light in many of his writings. His works on light are more philosophical and have been credited with discussing light from four different perspectives: an epistemology of light, a metaphysics or cosmogony of light, an etiology of light, and a theology of light (Lindberg 1976). Grosseteste was the first of the Scholastics to fully understand Aristotle's vision of the dual path of scientific reasoning: generalizing from particular observations into a universal law, and then back again from universal laws to prediction of particulars. He had read several important works translated from Greek via Arabic and produced important work in optics of his own. In *De Iride* he wrote: "This part of optics, when well understood, shows us how we may make things a very long distance off appear as if placed very close, and large near things appear very small, and how we may make small things placed at a distance appear any size we want, so that it may be possible for us to read the smallest letters at incredible distances, or to count sand, or seed, or any sort of minute objects." Grosseteste is now believed to have had a very modern understanding of color, as described in his treatise *De Luce* (*On Light*) written in about 1225. In *De Luce*, Grosseteste also explored the nature of matter and the cosmos, and described the birth of the Universe in an explosion, four centuries before Isaac Newton proposed gravity and seven centuries before the Big Bang theory. Other important figures in understanding of light during the medieval Europe include **Roger Bacon** (c. 1214–1294), **John Pecham** (died 1292), **Witelo** (born c. 1230, died between 1280 and 1314), and **Theodoric Freiberg** (c. 1250–1310). Bacon wrote three books on light (*Perspectiva*, the *De multiplicatione specierum*, and the *De speculis comburentibus*) and cited a wide range of translated optical and philosophical works, including those of Euclid, Ptolemy, al-Haytham, and Aristotle. Pecham built on the work of Bacon, Grosseteste, and a diverse range of earlier writers to produce what became the most widely used textbook on Optics of the Middle Ages, the *Perspectiva communis*. His book centered on the question of vision, on how we see, rather than on the nature of light and color. Witelo drew on the extensive body of optical works translated from Greek and Arabic to produce a massive presentation of the subject titled the *Perspectiva*. Freiberg was among the first in Europe to provide the correct scientific explanation for the rainbow phenomenon. Although his work was independent from that of al-Farisi mentioned earlier, both authors relied on the *Book of Optics* by Ibn al-Haytham.

1.2 History of Polarization

Polarization is one of seven fundamental properties of light waves, together with the propagation direction, wavelength, intensity, speed, phase, polarization, and coherence. The early studies of light mostly concerned with the direction, color (wavelength), and brightness (intensity) because they can be directly observed with human naked eyes. For example, the reflection and refraction in the works of Euclid, Ptolemy, Ibn Sahl, and Ibn al-Haytham basically describes how the direction of a light ray changes upon incident on an object. The word "ray" basically implies the involvement of the direction. Similarly, the studies on color were the direct results of humans' observation of rainbows and other colorful objects, such as flowers, leaves, and sky. On the other hand, the other fundamental properties of light, the speed, phase, polarization, and coherence, could not be observed directly and therefore were not studied until much later. In order not to defocus our attention to the subject of this book, the polarization, we will not go into the details of the other properties of light, but concentrate on humans' understanding of polarization as it evolved in the history.

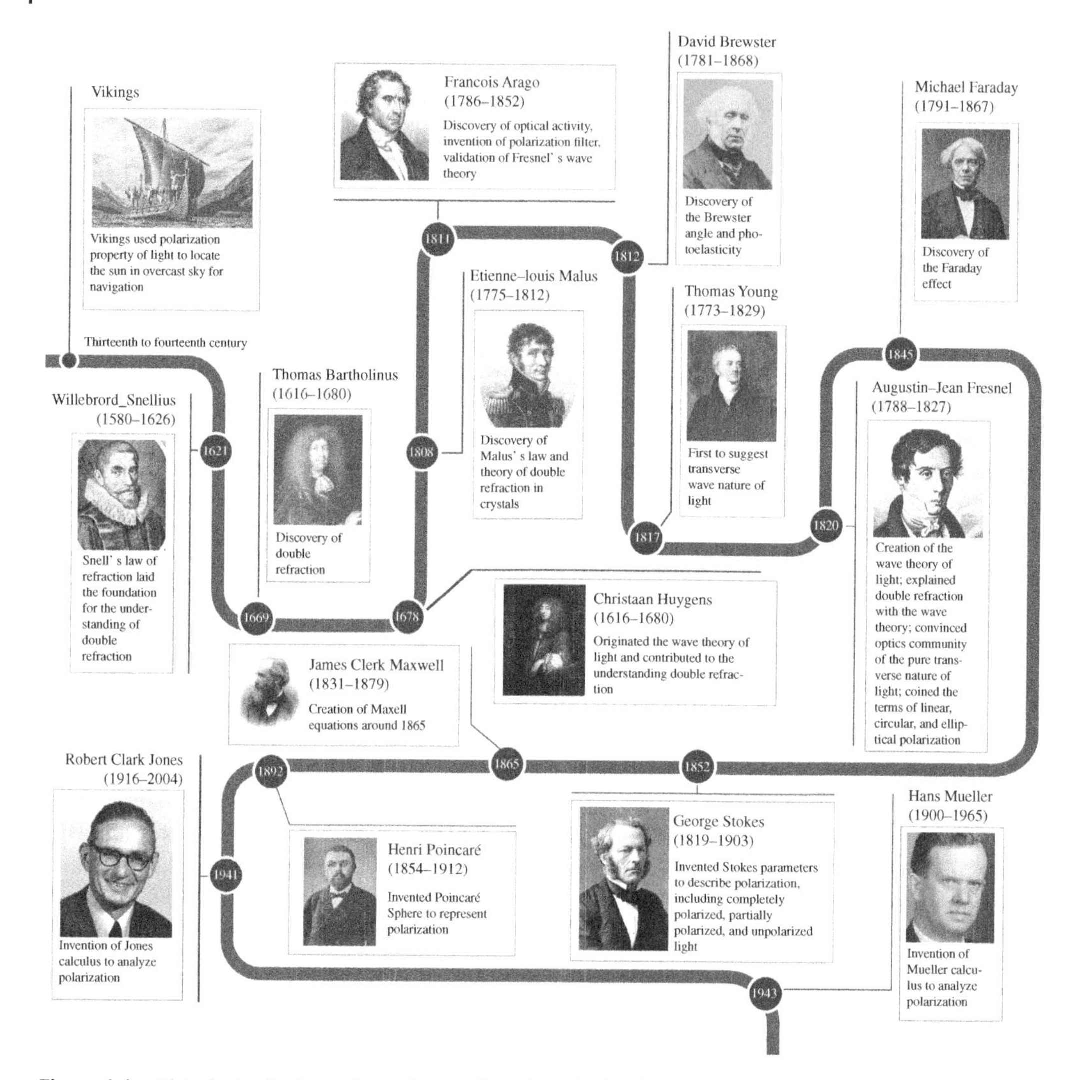

Figure 1.1 Historical milestones in understanding the polarization of light. Sources: Nejron Photo/Adobe Stock; Unidentified painter/Wikimedia Commons/Public domain; Unknown author/Wikimedia Commons/Public domain; Henry Perronet Briggs/Wikimedia Commons/Public_Domain; Unknown author/Wikimedia Commons/Public domain; World History Archive/Alamy Stock Photo; ART Collection/Alamy Stock Photo; Photo Researchers/Science History Images/Alamy Stock PhotoUnknown author/Wikimedia Commons/Public domain; John Simon Guggenheim Memorial Foundation.

Figure 1.1 is the chronicles of polarization, showing the important milestones for the understanding of polarization by humankind in the history, manifested by the fundamental discoveries or contributions made by various individuals or group, which had profound impacts on the subject matter. **Vikings** were probably the earliest men who noticed the polarization property of light. In around thirteenth to fourteenth century, they used polarization property of light to locate the Sun in overcast sky for navigation with the aid of Sunstone or Iceland spar (calcite). In 1621, **Willebrord Snellius** (13 June 1580–30 October 1626), a Dutch physicist, astronomer, and mathematician, known in the English-speaking world as Snell, published the Snell law of refraction, which was mathematically equivalent to Ibn Sahl's law of refraction published around 984. Unfortunately, Ibn Sahl's law of refraction was not recognized earlier probably because it was not singled out as a law, but buried in the calculations of lenses' focusing effect in Ibn Sahl's writings. Although Snell's law was not directly related to polarization, it was important for the understanding double refraction, which **Thomas Bartholin** (20 October 1616–4 December 1680), Danish physician, mathematician, and physicist, discovered and described in 1669.

In 1678, **Christiaan Huygens** (14 April 1629–8 July 1695), a Dutch physicist, proposed the wave theory of light: "Treatise on Light" and described it in 1690, which was regarded as the first mathematical theory of light. The theory was rejected in favor of Newton's corpuscular theory of light, until adopted by Augustin-Jean Fresnel in 1818. Now the theory is known as Huygens–Fresnel principle.

Many people considered **Étienne-Louis Malus** (23 July 1775–24 February 1812), a French physicist, as the father of polarization. His mathematical work was almost entirely concerned with the study of light. He conducted experiments to verify **Christiaan Huygens**' theories of light and rewrote the theory in analytical form. His discovery of the polarization of light by reflection was published in 1809 and his theory of double refraction of light in crystals, in 1810. Malus attempted to identify the relationship between the polarizing angle of reflection that he had discovered and the refractive index of the reflecting material. While he deduced the correct relation for water, he was unable to do so for glasses due to the low quality of materials available to him (most glasses at that time showed a variation in refractive index between the surface and the interior of the glass). It was not until 1815 that Sir **David Brewster** was able to experiment with higher quality glasses and correctly formulated what is known as Brewster's law. This law was later explained theoretically by **Augustin Fresnel**, as a special case of his Fresnel equation. **Malus** is probably best remembered for Malus's law, giving the resultant intensity, when a polarizer is placed in the path of an incident beam. His name is one of the 72 names inscribed on the Eiffel tower.

Francois Jean Dominique Arago (26 February 1786–2 October 1853), a French mathematician and physicist, discovered optical activity in 1811, invented polariscope and polarization filter in 1812, discovered polarized light from the tail of the Great Comet of 1819, and experimentally validated Fresnel's wave theory of light. Nowadays, polarization filters are used extensively in photography to cut out unwanted reflections or to enhance reflection. Interestingly, Arago was one of five judges in the award committee to judge the entries for the biannual physics *Grand Prix* to be awarded in 1819 and eventually wrote the committee's report. The other judges were **Laplace**, **Biot**, **Gay-Lussac**, and **Poisson**, mostly corpuscularists. With the encouragement from **Arago**, **Fresnel** submitted his prize entry titled "Memoir on the diffraction of light" on 29 July 1818, describing using wave theory of light to analyze diffraction. In order to disapprove Fresnel's theory, **Poisson**, a strong supporter of Newton's corpuscular theory of light, exploiting a scenario in which Fresnel's theory gave easy integrals, predicted that if a circular obstacle were illuminated by a point-source, there should be (according to the theory) a bright spot in the center of the shadow, illuminated as brightly as the exterior. Arago, undeterred, assembled an experiment with an obstacle 2 mm in diameter and observed indeed in the center of the shadow a bright spot which was later called Poisson's spot. Arago's verification of Poisson's counter-intuitive prediction passed into folklore as if it had decided the prize and was important to convince the scientific community to accept Fresnel's wave theory of light finally published in 1820.

Jean-Baptiste Biot (21 April 1774–3 February 1862), French physicist, astronomer, and mathematician, discovered optical activity in organic materials in 1815, laid foundation for liquid crystal display (LCD) display. In 1812, Biot turned his attention to the study of optics, particularly the polarization of light. Prior to the nineteenth century, light was believed to consist of discrete packets called corpuscles proposed and advocated by Newton. During the early nineteenth century, many scientists began to disregard Newton's corpuscular theory in favor of the wave theory of light. Biot began his work on polarization to show that the results he was obtaining could appear only if light were made of corpuscles. In 1815 he demonstrated that "polarized light, when passing through an organic substance, could be rotated clockwise or counterclockwise, dependent upon the optical axis of the material." His work in chromatic polarization and rotary polarization greatly advanced the field of optics, although it was later shown that his findings could also be obtained using the wave theory of light. Biot's work on the polarization of light has led to many breakthroughs in the field of optics. LCDs, such as television and computer screens, use light that is polarized by a filter as it enters the liquid crystal, to allow the liquid crystal to modulate the intensity of the transmitted light. This happens as the liquid crystal's polarization varies in response to an electric control signal applied across it.

Thomas Young (13 June 1773–10 May 1829), a British polymath, known for his double-slit experiment, experimentally demonstrated wave nature of light and laid the foundation of Fresnel's wave theory of light. He was also the first person to suggest to Arago that the light waves were pure transverse in order to explain Arago's experimental results, which could not be explained with Fresnel's wave theory when light wave was believed to be longitudinal as **Huygens** assumed. In 1807, he invented the name "index of refraction," which was called by Newton as "proportion of the sines of incidence and refraction" and by Hauksbee as the "ration of refraction."

David Brewster (11 December 1781–10 February 1868), a Scottish scientist, discovered Brewster angle, which is one of the most important quantities related to polarization. He is principally remembered for his experimental work in physical optics, mostly concerned with the study of the polarization of light, including the discovery of Brewster angle. He studied the birefringence of crystals under compression and discovered photoelasticity in 1815, thereby creating the field of optical mineralogy. For this work, William Whewell dubbed him the "father of modern experimental optics" and "the Johannes Kepler of optics."

Augustin-Jean Fresnel (10 May 1788–14 July 1827) perhaps is the most important figure in the history of polarization, as well as in optics. Fresnel was a French civil engineer and physicist whose wave theory of light published in 1820 completely won over Newton's corpuscular theory. By supposing that light waves are purely transverse as suggested by Thomas Young, he explained the nature of polarization, the mechanism of chromatic polarization, and the transmission and reflection coefficients at the interface between two transparent isotopic media. By generalizing the direction–speed–polarization relation for calcite, he accounted for the directions and polarizations of the refracted rays in double-refractive crystals of the *biaxial* class (those for which Huygens' secondary wave-fronts are not axisymmetric). The period between the first publication of his pure-transverse-wave hypothesis and the submission of his first correct solution to the biaxial problem was less than a year. Later, he coined the terms linear polarization, circular polarization, and elliptical polarization, explained how optical rotation could be understood as a difference in propagation speeds for the two directions of circular polarization, and (by allowing the reflection coefficient to be complex) accounted for the change in polarization due to total internal reflection (TIR), as exploited in the Fresnel rhomb. Not only Fresnel correctly predicted the pure transverse-wave nature of light in his wave theory of light in 1820, but he also correctly described "direct" (unpolarized) light wave as "*the rapid succession of systems of waves polarized in all directions*" in 1821, and gave the modern explanation of chromatic polarization. According to this new view, he wrote, "*the act of polarization consists not in creating transverse motions, but in decomposing them in two fixed, mutually perpendicular directions, and in separating the two components.*" Defenders of the established corpuscular theory could not match Fresnel's quantitative explanations of so many phenomena on so few assumptions.

Despite the great success, there still remained a last weakness with the wave theory of light: the light waves, like sound waves, would need a medium for transmission and there was still no experimental proof that the hypothetical substance *luminiferous aether* proposed by Huygens in

1678 existed. This issue was not resolved until the wave theory of light was subsumed by Maxwell's electromagnetic theory in the 1860s. In the period between Fresnel's unification of physical optics and Maxwell's wider unification, a contemporary authority, Humphrey Lloyd, described Fresnel's transverse-wave theory as "the noblest fabric which has ever adorned the domain of physical science, Newton's system of the universe alone excepted."

An important development in the understanding of light is due to **Michael Faraday** (22 September 1791–25 August 1867), an English scientist, who discovered Faraday Effect in 1845 describing polarization rotation in certain media caused by a magnetic field. Nowadays the Faraday Effect is successfully used for sensing magnetic field and electrical current with light, particularly with optical fibers. Perhaps more importantly, it inspired **James Clerk Maxwell** (13 June 1831–5 November 1879), a Scottish scientist, to realize the linkage between the magnetic field and light, and led to the creation of Maxwell's electromagnetic theory (known as Maxwell Equations) between 1860 and 1871, the second great unification in physics unifying electricity, magnetism, and light, after the first one realized by **Isaac Newton** (unification of gravity). Maxwell's equations firmly establish that light is alternating electrical and magnetic fields of transverse waves, a view well accepted in the scientific community today. The last weakness of wave theory of light was finally overcome by Maxwell's electromagnetic theory of light: no medium is required for the light to transmit because the oscillating electrical field induces magnetic field, and vice versa as the light propagates in space or vacuum.

Note that Fresnel's wave theory was an amplitude description of light, completely successful in describing polarized light. However, the interference experiments by Fresnel and Arago were carried out with unpolarized or partially polarized light. Although Fresnel, together with Arago, summarized their interference experiment results as the four interference laws (Fresnel–Arago interference laws), unfortunately he was not able to mathematically describe the unpolarized light or to provide the mathematical statement of the interference laws until his early death in 1827. In the next 35 years, the mathematical description of unpolarized light and the Fresnel and Arago interference laws still remained unsolved.

In 1852, **George Gabriel Stokes** (13 August 1819–1 February 1903), an Irish physicist and mathematician, published a remarkable paper titled "On the composition and resolution of streams of polarized light from different sources," reporting a novel approach for mathematically describing not only the polarized light, but also partially and totally unpolarized light. He abandoned his predecessors' fruitless attempts to describe unpolarized light with amplitudes; instead, he used four measured intensities, or observables, known as Stokes parameters to define the polarization properties of light. With this approach, the mathematical statements of unpolarized light, partially polarized light, and completely polarized light, as well as Fresnel–Arago interference laws could be written within the framework of the wave theory of light. **Stokes** practically completed the unfinished task of Fresnel on the wave theory of light, but unfortunately his great paper was practically forgotten for nearly a century till it was re-discovered by the Nobel laureate, **Subrahmanyan Chandrasekhar** (19 October 1910–21 August 1995) who used Stokes parameters to describe the effect of polarized light in his study of radiative transfer published in 1950. Nowadays, Stokes parameters are widely used to describe the polarizations of both coherent and incoherent lights with the introduction of Mueller matrix calculus by **Han Mueller** in 1943.

Another important development in polarization optics is the **Poincaré Sphere**, the three-dimensional spherical volume with which any state of polarization and degree of polarization can be represented was credited to **Jules Henri Poincaré** (29 April 1854–17 July 1912), a French polymath. Poincaré made contributions in many areas of mathematics as well as astronomy and physics. The Poincaré sphere representation was first introduced in 1892 in Poincaré's book

Théorie Mathématique de la Lumière. Poincaré apparently was not aware of the Stokes parameters because he did not present the sphere as being generated using the Stokes parameters as values along orthogonal axes in a Cartesian coordinate system, a representation we often used today.

Robert Clark Jones (30 June 1916–26 April 2004), a researcher at Land Corporation and a Harvard professor, invented Jones vector and calculus to calculate polarization states after passing through optical elements in 1941. The work was published in a series of eight papers in the *Journal of the Optical Society of America* from 1941 to 1956. Nowadays Jones vector and calculus, sometimes referred to as Jones matrix formulation, are widely used in scientific research and engineering designs.

Hans Mueller (27 October 1900–10 June 1965), a Swiss–American physicist at the Massachusetts Institute of Technology (MIT), invented Mueller matrix calculus to manipulate Stokes vectors in 1943. Born in Switzerland and obtained his degrees from the Eidgenössische Technische Hochschule in Zurich, he came to MIT in 1925 and remained there for the next 40 years. Mueller matrix calculus is also widely used around the world today, especially when dealing with partially polarized or partially coherent light.

1.3 History of Polarization in Optical Fibers and Waveguides

1.3.1 The History of Optical Fiber

The history of polarization in optical fibers and waveguides could not even start without the history of guided light, which relies on a phenomenon called total internal reflection (TIR). The discovery of TIR was generally attributed to Johannes Kepler (27 December 1571–15 November 1630), a German astronomer, mathematician, and astrologer, who published his findings in his *Dioptrice* in 1611. Although Kepler did not find Snell's law of refraction (which was discovered 10 years later by **Willebrord Snellius)**, from his extensive experimental studies on refraction between air-to-glass and glass-to-air, he concluded that when a ray is incident from glass-to-air at an angle beyond 42°, it could only be totally *reflected* (Total internal reflection – Wikipedia). This 42° angle was later termed as the critical angle of glass.

The first time that light was ever guided by the TIR was demonstrated in 1841 by Swiss physicist **Daniel Colladon** (15 December 1802–30 June 1893) in his attempts to show the speed of sound in water jets to the audience. In order to visualize the sound wave in water jet, he collected sunlight and directed into the water jet. The light was trapped by the TIR in the water jet and curved with it. A sound wave traveled in the water jet would disturb the TIR especially at the curved sections and cause glows to be seen. Colladon reported this experiment in 1842 to a wider audience in the *Comptes Rendus*, the French Academy of Sciences' journal. He later taught the tricks to **Louis Jules Duboscq**, a famous Paris instrument maker hired by Paris Opera, who then deployed them for special effect stage lighting, first appeared on stage in "Elias et Mysis" in 1853 as background fountains and later in Charles Gounod's opera "Faust" in 1859 to show the devil making a stream of fire flash from a wine barrel. With the successes on the stage, Duboscq published a catalog of luminous fountains aimed at the upper class, which even included rotating color filters to change colors, with list prices from 50 to 1000 French francs. In the International Health Exhibition held in 1884 in London, eight London water companies jointly built a huge luminous fountain to show their sparkling clean water. The fountain was designed and operated by **Sir Francis Bolton** with color changing filters and earned "unqualified admiration" remarks from *Nature*. More refined luminous fountains were later exhibited in Royal Jubilee Exhibition in Manchester in 1887 and in Universal Exhibition in Paris in 1889.

Around 1842, a French Physicist **Jacques Babinet** (5 March 1794–2 October 1872), the inventor of Babinet compensator, independently demonstrated light guiding in Paris by the TIR in water stream by focusing candlelight on the bottom of a bottle and pouring the water from the mouth of the bottle. In his article in *Comptes Rendus* per Arago's invitation to describe his experiments, he first reported light guiding in a curved glass rod to deliver illumination to the inside of a mouth. In 1854, a man named **John Tyndall** (2 August 1820–4 December 1893) demonstrated light guiding in water jets at the suggestion of Michael Faraday that made the topic more common knowledge. Tyndall also wrote about the property of TIR in an introductory book about the nature of light in 1870. In 1888, **Dr. Roth** and **Prof. Reuss** of Vienna used bent glass rods to illuminate body cavities for dentistry and surgery and realized Babinet's suggestion in practice. In 1889, **David D. Smith** of Indianapolis applied for patent on bent glass rod as a surgical lamp. In October of 1926, **John Logie Baird** applied for a patent on transferring images through glass rods and hollow tubes, and in December of 1926, **Clarence Hansell** proposed a fiber-optic imaging bundle in his notebook at the RCA Rocky Point Laboratory on Long Island, and received American and British patents later. Then in the 1930s **Heinrich Lamm** used the technology developed by Hansell and Baird to conduct internal medical examinations.

Circa 1949, **Holger Møller Hansen** in Denmark and **Abraham van Heel** in the Netherlands started to investigate image transmission in glass fiber bundles, and in 1951, Hansen filed a Danish patent application to propose applying a cladding for glass and plastic fibers with a transparent low index material. In October 1951, **Brian O'Brien** of the University of Rochester suggested to van Heel to apply a transparent cladding for improving transmission of fibers in his imaging bundle. On 12 June 1953, Dutch-language weekly *De Ingenieur* published van Heel's first report of clad fiber. On 2 January 1954, *Nature* published van Heel's paper on clad fiber bundles (van Heel 1954), and **Harold Hopkins and Narinder Singh Kapany**'s paper on bundles of unclad fibers for image transmission next to each other (Hopkins and Kapany 1954, 1955). In 1956 **Basil Hirschowitz**, **C. Wilbur Peters**, and **Lawrence E. Curtiss** patented the first fiber optic partially flexible gastroscope, which was licensed to American Cystoscope Manufacturers Inc. in 1957. In early 1958, **Will Hicks** of American Optical (AO) Company developed practical fiber-optic faceplates for military imaging systems and later in the same year he, together with **Paul Kiritsy** and **Chet Thompson**, left American Optical to form the first fiber-optic company, Mosaic Fabrications in Southbridge, Massachusetts, the same town where American Optical is located.

In a job interview at American Optical (AO) in 1958, **Elias Snitzer** (27 February 1925–21 May 2012) was shown with a picture of the end of fiber-optic faceplate and he immediately recognized waveguide modes because he was very familiar with the mode patterns in dielectric microwave waveguides when he consulted with a firm called High Voltage Engineering on a proposal for Stanford Linear Accelerator to calculate wave-guide mode propagation in linear accelerator type applications. In 1958, working with Hicks, AO produced the first single mode fiber and **Elias Snitzer**, after hired by AO, started to develop the dielectric waveguide theory using the Maxwell equations to describe waveguide modes in thin optical fibers with cores on the order of wavelengths, because the geometrical optic treatment of optical fibers was no longer sufficient for such small glass core fibers. He first reported his theory (Snitzer 1959a, paper TB36 by himself) and related experiment (Snitzer et al. 1959, paper TB37 with his colleagues **H. Osterberg**, **M. Polanyi, and W. Hicks**) at Optical Society of America's (OSA) annual meeting in October 1959. He even developed the well-known coupled mode equations to describe optical coupling between two parallel waveguides and reported the work in the same conference (Snitzer 1959b, paper TB38). **Snitzer** recognized the importance of single-mode transmission and applied for a patent with Hicks in 1960. In 16 May 1960, **Theodore Maiman** demonstrated the first laser at

Hughes Research Laboratories in Malibu, California. In October 1964, **Charles Koester** and **Elias Snitzer** described the first optical amplifier using neodymium-doped glass. In 1965 **Manfred Börner** demonstrated the very first fiber-optic data transmission system, and one year later, in 1966, he patented this new innovation that should be considered as a serious alternative to copper wires.

Charles K. Kao (4 November 1933–23 September 2018) is perhaps the most critical figure in advocating optical fiber for communication and was recognized by being awarded with a Nobel Prize in Physics in 2009. In December 1964, Kao took over Standard Telecommunication Laboratories' (STL) optical communication program when **Antoni E. Karbowiak** left to become chair of electrical engineering at the University of New South Wales. Kao and **George Hockham** soon abandoned thin-film waveguide in favor of single mode clad optical fiber. In the fall of 1965, Kao concluded that the fundamental limit on glass transparency was below 20 dB/km, which would be practical for communications and Hockham calculated that the radiation loss in clad fibers should sufficiently low. They prepared a paper proposing fiber-optic communications. In January 1966, Kao announced to the Institution of Electrical Engineers in London that glass fibers could be made with loss below 20 dB/km for communications, and in July 1966, Kao and Hockham published the paper proposing optical fibers for communications with convincing evidence in *Proceedings of the Institution of Electrical Engineers* (Kao and Hockham 1966). In their theory they showed the attenuation (light loss) could go below 20 dB/km if the correct materials were used. However, at the time of this determination, optical fibers commonly exhibited light loss as high as 1000 dB/km and even more. In 1968, Kao and **M.W. Jones** measured intrinsic loss of bulk fused silica at 4 dB/km, the first evidence of ultra-transparent glass, prompting Bell Labs to seriously consider fiber optics.

In the summer of 1970, **Robert D. Maurer, Donald Keck, Peter C. Schultz, and Frank Zimar** produced an optical fiber with 16 dB/km attenuation at 633 nm by doping titanium into fiber core and two years later in 1972, they were able to produce multimode germanium-doped fiber with attenuation levels as low as 4 dB/km, with much greater strength than titanium-doped fiber. Around 1970–1971, **Dick Dyott** at British Post Office (Dyott and Stern 1971) and **Felix Kapron** of Corning (Kapron and Keck 1971) separately discovered that pulse spreading caused by dispersion was lowest at 1.2–1.3 μm, and in 1965, **Dave Payne** and **Alex Gambling** at University of Southampton confirmed that pulse spreading should be zero at 1.27 μm. In early 1976, **Masaharu Horiguchi** and **Hiroshi Osanai** of Japan made first fibers with a low loss of 0.47 dB/km at 1.2 μm. In the summer of 1976, **Horiguchi** and **Osanai** of Japan discover third fiber-optic transmission window at 1.55 μm. Late in 1976, **J. Jim Hsieh** made the first indium-gallium arsenide phosphide (InGaAsP) lasers emitting continuously at 1.25 μm, providing the light source required for the 1.2–1.3 μm communication window. In late 1978, **NTT Ibaraki lab** made single-mode fiber with record 0.2 dB/km loss at 1.55 μm. Such low levels of attenuation and dispersion made optical fiber a very practical choice for communications and triggered a series chain effects in the commercialization of the technology. For example, **Bell Labs** publicly committed to single-mode 1.3-μm technology for the first transatlantic fiber-optic cable, TAT-8 in 1980; **ITT** signed consent agreement to pay **Corning** and licensed Corning communication fiber patents in the same year; **British Telecom** transmitted 140 Mb/s data through 49 km of single-mode fiber at 1.3 μm in 1981; **MCI** leased right of way to install single-mode fiber from New York to Washington operating at 400 Mb/s at 1.3 μm in December 1982. After these notable discoveries and commercialization efforts described earlier, little bits of research here and there have accumulated into the modern day optical fiber we know and love.

1.3.2 History of Polarization in Optical Fibers

The first mention of polarization in optical fibers was perhaps by **W. Hicks** representing both Mosaic Fabrication and American Optical, and **R.J. Potter** of University of Rochester, in their attempts to develop an electromagnetic theory of optical fiber reported at OSA's annual meeting in October 1959 (Hicks and Potter 1959, paper FB39). In the abstract, they outlined their presentation to include "a few comments about the splitting of momentum levels due to polarization, a discussion of bent fibers and cones and cross coupling between modes…" Interestingly, Potter and Hicks were also working on the geometrical theory of optical fibers, as reported in the related work at the same conference (Potter and Hicks 1959, paper FB38). Apparently, Hicks was very familiar with the theoretical work of **Elias Snitzer**, and had collaborated with Snitzer on the experiments (making the fiber for Snitzer) to observe the modes predicted by Snitzer's electromagnetic theory (Snitzer 1959a, paper TB37). Hicks and Potter might have realized the limitations of their geometrical theory and started to develop their own theory with the similar electromagnetic wave approach. Clearly, Snitzer was much more advanced in the efforts because he presented three reports (Snitzer 1959a, 1959b; Snitzer et al. 1959, papers TB36, TB38, TB37) in the same conference describing his theoretical and experimental work on the electromagnetic wave description of waveguide modes in optical fibers with much more solid results.

Elias Snitzer is the first to report solid theoretical and experimental studies of the polarization of light in optical fibers, particularly associated with waveguide modes. In May 1961, **Snitzer** published his detailed electromagnetic theory of optical fibers, titled "Cylindrical dielectric waveguide modes," in the *Journal of the Optical Society of America* (Snitzer 1961). In the paper, he started with Maxwell Equations to derive expressions for different waveguide modes in optical fibers and plotted electrical and magnetic fields inside the fiber, which was the first description of polarization in optical fibers. In the same issue of the journal, together with his colleague **Harold Osterberg**, Snitzer also published his experimental results titled "Observed dielectric waveguide modes in the visible spectrum" to verify his theoretical findings (Snitzer and Osterberg 1961). This paper shows that he is the first to measure the polarization of different modes in the fiber. At the end of the paper, Snitzer acknowledged Hicks for helpful discussions and for providing the special fibers in the study.

In January 1971, **L.G. Cohen** of Bell Labs published a paper in *The Bell System Technical Journal* titled "Measured attenuation and depolarization of light transmitted along glass fibers" describing his measurements of Stokes parameters and depolarization of a linearly polarized input light after passing through different multimode optical fibers (Cohen 1971). He observed significant depolarization due to light coupling into multiple waveguide modes and attributed the cause of the circularly polarized components to the cross-sectional ellipticity caused by squeezing during the fiber drawing process. In February 1972, **Felix Kapron**, **Nicholas Borrelli**, and **Donald Keck** of Corning published their detailed theoretical and experimental studies titled "Birefringence in dielectric optical waveguides" (Kapron et al. 1972) to describe polarization evolution in a single mode optical fiber under the influence of birefringence at 633 nm (He–Ne laser wavelength). They described polarization having some anomalous behavior which could be explained in terms of birefringence varying locally along the fiber, observed a depolarization about 1% per km, and estimated an upper information rate of over **1000** Gb/s due to the birefringence. They finally concluded that because of the rotating fast-axis effect along the fiber, the overall phase retardation would not increase indefinitely with length, and that the guide dispersion rather than birefringence would limit the data rates. Concurrently, **W.O. Schlosser** of Bell Labs published his theoretical study on the data rate limitation imposed by birefringence in a paper titled "Delay distortion in weakly guiding optical fibers due to elliptic deformation of the boundary" in February 1972's issue of *The Bell System Technical Journals* (Schlosser 1972). This birefringence on the data rate limit is in fact the polarization mode dispersion (PMD) as we know today. Although the manuscript submission date of Schlosser's paper was two months later than that of Kapron's (13 September 1971 vs. 7 July 1971), we believe it is fair to credit the discovery of PMD effect to Kapron, Borrelli, Keck, and Schlosser. Note that Kapron's conclusion that "the overall phase retardation would not increase indefinitely with length" is not exactly accurate. It is more proper to say that "overall phase retardation would not increase linearly with length, but rather increase with the square root of length."

The first report on light induced birefringence in a single mode optical fiber via optical Kerr effect was first reported by **R.H. Stolen** and **A. Ashkin** in 1973 in a paper titled "Optical Kerr effect in glass waveguide" (Stolen and Ashkin 1973), in which a single mode fiber with a length of 0.58 m was used. A pulsed Xe laser at 535.3 nm was used to induce the birefringence change while a He–Ne laser at 632.8 nm was used to sense polarization rotation caused by the induced birefringence.

As the optical fibers became generally available, some researchers started to turn their attentions to using the optical fiber for sensing in which understanding the polarization properties was critical. In October 1975, **A. Rapp** and **H. Harms** from Siemens AG first reported the polarization maintaining capability of an optical fiber in their paper titled "Polarization optics of index-gradient optical waveguide fibers" (Rapp and Harms 1975) by launching a linear polarization into the fast axis of a graded-index fiber having a fairly large linear birefringence that resulted from mechanical stress created during fabrication. They found that the exit polarization can be rotated with the fiber output end. They are also the first to visually observe the polarization evolution along the fiber by looking at the Rayleigh scattering of light from the side perpendicular to light's propagation direction inside the fiber, although they might have mistakenly attributed the scattering as the Tyndall scattering. In particular, light scattering in the glass fiber is preferentially normal to the polarization direction and no scattering can be seen by looking into the direction of linear polarization. Therefore by launching the input polarization 45° from the fast axis, the polarization would evolve from linear to elliptical, circular, elliptical, and finally back to linear along the fiber periodically, and the Rayleigh scattering of He–Ne red light viewed from the side would show such periodic change in intensity. In March 1976, the same group in Siemens, **H. Harms**, **A. Rapp**, and **K. Kemper**, first reported the magneto-optic properties in an optical fiber (Harms et al. 1976) and presented the theory to describe the polarization evolution of light under the influence of both Faraday Effect and linear birefringence inside fiber.

In 1976, **V. Vali** and **R.W. Shorthill** (Vali and Shorthill 1976) demonstrated the first optic fiber Sagnac interferometer for rotation sensing (the first fiber-optic gyroscope), pointing out the need for maintaining polarization in optical fibers for optical fiber interferometers. In October 1976, **R.A. Steinberg** and **T.G. Giallorenzi** first reported the performance limitations of waveguide switches and modulators imposed by polarization (Steinberg and Giallorenzi 1976), further elevated the need for polarization maintaining fibers. One approach was to make fibers with extremely low birefringence. In October 1978, a research group at Siemens published the results of their efforts in making an ultra-low birefringence fiber in their efforts to maintain polarization in optical fibers for current sensing and fiber-optic gyro applications (Schneider et al. 1978) with a linear retardation about 10°/m, five times better than the lowest reported previously. A year later, a University of Southampton group, **S.R. Norman**, **D.N. Payne**, and **M.J. Adams**, in England achieved a low birefringence fiber with a retardation of 2.6°/m (Norman et al. 1979).

Another approach for maintaining the polarization was to make high birefringence fibers following **A. Rapp and H. Harms**' initial demonstration in 1975. From 1978 to 1982, several research groups were quite active in making such polarization maintaining (PM) fibers. The Bell Labs group included **Ivan P. Kaminow**, **Roger H. Stolen**, **V. Ramaswamy**, **R.D. Standley**, **D. Sze**, **W.G. French**, **J.R. Simpson**, **H.M. Presby**, **P. Kaiser**, and **W. Pleibel**, investigated different ways to introduce birefringence, including using non-circular fiber cores (dumb-bell and elliptical cores), non-circular claddings, and anisotropic strains. In March 1978, **Ramaswamy** and **French** first concluded that the birefringence produced from non-circular shape of the core was not sufficient to maintain the state of polarization when the fiber was subject to bend, twist, and pressure and anisotropic strain to increase the birefringence was suggested (Ramaswamy and French 1978), pointing to new directions for PM fiber development. Six months later, Stolen, Ramaswamy, Kaiser, and Pleibel reported a first PM fiber using non-circular cladding to cause the anisotropic strain and hence increase the birefringence (Stolen et al. 1978). In February 1979, **Kaminow** and **Ramaswamy** first theoretically predicted that non-circular cladding fibers could be fabricated from the usual low-loss high-silica glasses to have a strain birefringence as large as 5×10^{-4} resulting from the large differential thermal expansion between a silica outer jacket and a cladding–core region of silica doped with boron, germanium, or phosphorous (Kaminow and Ramaswamy 1979). In a short period of seven months, **Kaminow**, **Simpson**, **and Presby** successfully achieved a modal birefringence of 3×10^{-4} using the differential thermal expansion in a germanosilicate fiber with a circular core but the non-circular exposed-cladding. Measurements of birefringence agree with earlier calculations (Kaminow, Simpson, and Presby 1979). These studies paved way to the rapid development of PM fibers in the years followed.

Other groups included **R.B. Dyott**, **J.R. Cozens**, and **D.G. Morris** from Andrews Corporation of USA and Imperial College of England, **Luc B. Jeunhomme** and **M. Monerie** of CNET Lab, France, **S.R. Rengarajan** and **J.E. Lewis** of University of New Brunswick, Canada, **T. Hosaka**, **K. Okamoto**, **T. Miya**, **Y. Sasaki**, and **T. Edahiro** of NTT, Japan, and **T. Katsuyama**, **H. Matsumura**, and **T. Suganuma** of Hitachi, Japan. The Dyott group concentrated on using elliptical cores or elliptical cladding to introduce birefringence inside the fiber core, while Jeunhomme's group first proposed (Jeunhomme and Monerie 1980) circular core polarization maintaining fiber. After realizing from the stress analysis that the contribution of the stress-induced strain birefringence to modal birefringence was much larger than that of the geometrical anisotropy, the NTT group led by Hosaka started to investigate introducing birefringence using strain rods and reported the results in 1981 (Hosaka et al. 1981), the first PANDA fiber as we know it today.

As single mode fiber became generally available, many researchers also started to explore using fibers for sensing. Inevitably, understanding polarization variations in optical fibers became increasingly important. In January 1978, **A.M. Smith** of Central Electricity Research Laboratories in England reported the first use of a single mode fiber for current sensing (Smith 1978) and his study on the Faraday Effect induced polarization change in the fiber. Two years later, he also published the detailed studies on pressure induced birefringence (Smith 1980a) and on bend and twist induced birefringence in single mode fibers (Smith 1980b).

In **May 1979**, **G.B. Hocker** of Honeywell first proposed using optical fiber interferometers for temperature and pressure sensing (Hocker 1979).

The Max-Plant group led by **R. Ulrich**, with members including **A. Simon**, **M. Johnson**, **S.C. Rashleigh**, **W. Eickhoff**, and **A. Kumar**, was most active and productive in polarization related research in optical fibers from 1977 to 1981. They mostly focused on understanding polarization changes induced by bend, pressure, twist, magnetic field, and how to control or manage the state of polarization. Their studies on the fiber bend and twist induced birefringence in a single mode fiber were reported a few months earlier than those by A. Smith. It is important to point out that Rashleigh and Ulrich were the first one to coin the term polarization mode dispersion (PMD) to describe the relative group delay between two orthogonal polarizations caused by birefringence in a paper titled "Polarization mode dispersion in single-mode fibers" (Rashleigh and Ulrich 1978). They also demonstrated the first passive PMD compensator by using a ±68° double twist at the center of the fiber, as reported in the same paper. It is also important to point out that M. Johnson made the first polarization controller using electro-magnet based fiber squeezers in 1978 (Johnson 1979) and Ulrich was the first to stabilize the output state of polarization in an optical fiber using such a polarization controller and a polarization detector in a feedback loop in 1979 (Ulrich 1979). In 1981, Kumar and Ulrich published their study on birefringence of optical fiber pressed into a V-groove, which concluded that in the absence of friction, the induced birefringence could be zero for a V-groove angle of 60° (Kumar and Ulrich 1981). Note that Rashleigh later joined US Navy Research Laboratory to develop fiber-optic sensors.

There have been tremendous advancements of polarization optics in optical fibers every year by many scientists and engineers working in the field since 1981. The history for this booming period is not covered in this book and may be covered in the second edition of the book.

1.3.3 Chronicles of Polarization Optics in Optical Fibers from 1959 to 1981

October 1959: **Elias Snitzer, R.J. Potter, and W. Hicks** reported initial investigation of polarization related issues in optical fibers at 1959 OSA annual meeting.

May 1961: **Elias Snitzer** of American Optical Corporation published the first electromagnetic theory of optical fibers, titled "Cylindrical dielectric waveguide modes," in the *Journal of the Optical Society of America* (Snitzer 1961) with the plots of polarization distribution of different modes.

May 1961: **Snitzer** and **Harold Osterberg** published the results of the first experimental measurement of polarization directions of different fiber modes in the same journal.

January 1971: **L.G. Cohen** of Bell Labs published a paper in *The Bell System Technical Journal* titled "Measured attenuation and depolarization of light transmitted along glass fibers" describing his measurements of Stokes parameters and depolarization of a linearly polarized input light after passing through different multimode optical fibers.

February 1972: **Felix Kapron**, **Nicholas Borrelli**, and **Donald Keck** of Corning and **W.O. Schlosser** of Bell Labs first reported investigation of PMD effect on data transmission limit of fiber-optic communication systems and estimated an upper information rate of over **1000** Gb/s due to the PMD. **Felix Kapron, Nicholas Borrelli**, and **Donald Keck** first reported linear birefringence induced by transverse stress and circular birefringence induced by twist in single mode fibers and the polarization evolution along a single mode fiber with the induced linear and circular birefringence.

March 1973: **R.H. Stolen** and **A. Ashkin** of Bell Labs first reported light induced birefringence in a single mode optical fiber via optical Kerr effect.

October 1975: **A. Rapp** and **H. Harms** of Siemens AG first reported the polarization maintaining capability of an optical fiber by launching a linear polarization into the fast axis of a graded-index fiber having a fairly large linear birefringence that resulted from mechanical stress created during fabrication.

March 1976: **H. Harms**, **A. Rapp**, and **K. Kemper** of Siemens AG first reported the magneto-optic properties in an optical fiber and presented the theory to describe the polarization evolution of light under the influence of both Faraday Effect and linear birefringence inside fiber.

October 1977: **A. Simon** and **R. Ulrich** first reported a study of the polarization evolution along a single mode fiber with both electro-optic effect induced linear birefringence and Faraday Effect induced circular birefringence.

January 1978: **A.M. Smith** of Central Electricity Research Laboratories in England reported the first use of a single mode fiber for current sensing using the Faraday Effect.

March 1978: **V. Ramaswamy** and **W.G. French** first concluded that the form birefringence resulting from the non-circular shape of the core was not sufficient to maintain the state of polarization in single mode fibers.

August 1978: S.C. Rashleigh and R. Ulrich of Max-Plant Institute first coined the term polarization mode dispersion (PMD) to describe the birefringence induced group delay between two orthogonal polarizations in a single mode fiber in a paper. They also reported a **first PMD compensator** using the ±68° double twist midway of a single mode fiber in the same paper.

October 1978: **Stolen, Ramaswamy, Kaiser, and Pleibel** reported a first PM fiber using non-circular cladding to cause the anisotropic strain and hence increase the birefringence.

December 1978: **M. Johnson** submitted the report on the first polarization controller using electro-magnet based fiber squeezers to Applied Optics and the paper was published in May 1979.

February 1979: **Kaminow** and **Ramaswamy** first theoretically predicted that non-circular cladding fibers could be fabricated from the usual low-loss high-silica glasses to have a strain birefringence as large as 5×10^{-4} resulting from the large differential thermal expansion induced strain.

July 1979: **R. Ulrich** and **A. Simon** of Max-Plant Institute first reported twist induced birefringence in single mode optical fibers and described the ±68° double twist for interchanging the fast and slow polarization modes.

October 1979: **Kaminow, Simpson, and Presby** successfully achieved a modal birefringence of 3×10^{-4} using the differential thermal expansion in a germanosilicate fiber based on earlier prediction.

October 1979: **R. Stolen** of Bell Labs first reported polarization effects in fiber Raman and Brillouin Lasers.

December 1979: **R. Ulrich** of Max-Plant Institute reported the first polarization stabilizer using the polarization controller of M. Johnson.

June 1980: **R. Ulrich, S.C. Rashleigh, and W. Eickhoff** of Max-Plant Institute reported first bend induced birefringence in single mode optical fibers.

August 1980: **A.M. Smith** of Central Electricity Research Laboratories in England first reported polarization evolution in a fiber with both bend and twist.

September 1980: **A.M. Smith** reported detailed study on pressure induced birefringence.

November 1980: **Luc B. Jeunhomme** and **M. Monerie** of CNET Lab, France, first proposed circular polarization maintaining fiber.

July 1981: **T. Hosaka, K. Okamoto, T. Miya, Y. Sasaki**, and **T. Edahiro** of NTT, Japan, reported the first PANDA fiber which used differential thermal expansion of strain rods to introducing birefringence in the fiber core based on the earlier work of Bell Labs.

December 1981: **Kumar** and **Ulrich** found that for an optical fiber pressed into a V-groove, the induced birefringence could be zero for a V-groove angle of 60° in the absence of friction.

References

Cohen, L.G. (1971). Measured attenuation and depolarization of light transmitted along glass fibers. *Bell Syst. Tech. J.* 50: 23–42.

Dyott, R.B. and Stern, J.R. (1971). Group delay in glass-fibre wave-guide. *Electron. Lett.* 7: 82–84.

Harms, H., Rapp, A., and Kemper, K. (1976). Magnetooptical properties of index-gradient optical fibers. *Appl. Opt.* 15 (3): 799–801.

van Heel, A.C.S. (1954). A new method of transporting optical images without aberrations. *Nature* 173: 39.

Hicks, J. and Potter, R. (1959). Fiber optics IV Electromagnetic theory, FB39. *Program of the 1959 Annual Meeting of the Optical Society of America*, Ottawa, Canada (8–10 October 1959). p. 507.

Hocker, G.B. (1979). Fiber-optic sensing of pressure and temperature. *Appl. Opt.* 18 (9): 1445–1448.

Hopkins, H.H. and Kapany, N.S. (1954). A flexible fiber scope using static scanning. *Nature* 173: 39.

Hopkins, H.H. and Kapany, N.S. (1955). Transparent fibres for the transmission of optical images. *Optica Acta: Int. J. Opt.* 1 (4): 164–170. https://doi.org/10.1080/713818685.

Hosaka, T., Okamoto, K., Miya, T. et al. (1981). Low-loss single polarization fibers with asymmetrical strain birefringence. *Electron. Lett.* 17 (15): 530–531.

Jeunhomme, L. and Monerie, M. (1980). Polarization-maintaining singlemode fiber cable design. *Electron. Lett.* 16 (24): 921–922.

Johnson, M. (1979). In-line fiber-optical polarization transformer. *Appl. Opt.* 18 (9): 1288–1289.

Kaminow, I.P. and Ramaswamy, V. (1979). Single-polarization optical fibers: slab model. *Appl. Phys. Lett.* 34 (4): 268–270.

Kaminow, I.P., Simpson, J.R., and Presby, H.M. (1979). Strain birefringence in single polarization germanosilicate optical fibers. *Electron. Lett.* 15 (21): 677–679.

Kao, K.C. and Hockham, G.A. (1966). Dielectric-fibre surface waveguides for optical frequencies. *IEE Proc.* 133, Pt. J (3): 191–198.

Kapron, F.P. and Keck, D.E. (1971). Pulse transmission through a dielectric optical waveguide. *Appl. Opt.* 10: 1519–1523.

Kapron, F., Borrelli, N., and Donald Keck, D. (1972). Birefringence in dielectric optical waveguides. *IEEE J. Quantum Electron.* QE-8 (2): 222–225.

Kumar, A. and Ulrich, R. (1981). Birefringence of optical fiber pressed into a V-groove. *Opt. Lett.* 6 (12): 644–646.

Lindberg, D.C. (1976). *Theories of Vision from al-Kindi to Kepler*, 94–99. Chicago: University of Chicago Press https://en.wikipedia.org/wiki/History_of_optics.

Morton, O. (2009). *Eating the Sun: How Plants Power the Planet*. Harper Collins Publishers.

Norman, S., Payne, D., and Adams, M. (1979). Fabrication of single-mode fibres exhibiting extremely low polarisation birefringence. *Electron. Lett.* 15 (11): 309–311.

Potter, R. and Hicks, J. (1959). Fiber optics III Electromagnetic theory, FB38. *Program of the 1959 Annual Meeting of the Optical Society of America*, Ottawa, Canada (8–10 October 1959). p. 507.

Ramaswamy, V. and French, W.G. (1978). Influence of noncircular core on the polarization performance of single node fibers. *Electron. Lett.* 14 (5): 143–144.

Rapp, A. and Harms, H. (1975). Polarization optics of index-gradient optical waveguide fibers. *Appl. Opt.* 14 (10): 2406–2410.

Rashleigh, S.C. and Ulrich, R. (1978). Polarization mode dispersion in single-mode fibers. *Opt. Lett.* 3 (2): 60–63.

Schlosser, W.O. (1972). Delay distortion in weakly guiding optical fibers due to elliptic deformation of the boundary. *The Bell System Technical Journals* 51 (2): 487–492.

Schneider, H., Harms, H., Papp, A., and Aulich, H. (1978). Low-birefringence single-mode optical fibers: preparation and polarization characteristics. *Appl. Opt.* 17 (19): 3035–3037.
Smith, A.M. (1978). Polarization and magnetooptic properties of single-mode optical fiber. *Appl. Opt.* 17 (1): 52–56.
Smith, A.M. (1980a). Single-mode fiber pressure sensitivity. *Electron. Lett.* 16 (20): 773–774.
Smith, A.M. (1980b). Birefringence induced by bends and twists in single-mode optical fiber. *Appl. Opt.* 19 (15): 2606–2610.
Snitzer, E. (1959a). Optical waveguide modes in small glass fibers I, Theoretical, TB36. *Program of the 1959 Annual Meeting of the Optical Society of America*, Ottawa, Canada (8–10 October 1959). p. 1128.
Snitzer, E. (1959b). Optical coupling between two parallel dielectric waveguide TB38. *Program of the 1959 Annual Meeting of the Optical Society of America*, Ottawa, Canada (8–10 October 1959). p. 1128.
Snitzer, E. (1961). Cylindrical dielectric waveguide modes. *JOSA* 51 (5): 491–498.
Snitzer, E. and Osterberg, H. (1961). Observed dielectric waveguide modes in the visible spectrum. *JOSA* 51 (5): 499–505.
Snitzer, E., Osterberg, H., and Polanyi M., et al. (1959). Optical waveguide modes in small glass fibers II, Experimental, TB37. *Program of the 1959 Annual Meeting of the Optical Society of America*, Ottawa, Canada (8–10 October 1959). p. 1128.
Steinberg, R. and Giallorenzi, T. (1976). Performance limitations imposed on optical waveguide switches and modulators by polarization. *Appl. Opt.* 15 (10): 2440–2453.
Stichler, C. (2002). Grass growth and development. Texas Cooperative Extension, Texas A&M University.
Stolen, R. and Ashkin, A. (1973). Optical Kerr effect in glass waveguide. *Appl. Phys. Lett.* 22 (6): 294–296.
Stolen, R.H., Ramaswamy, V., Kaiser, P., and Pleibel, W. (1978). Linear polarization in birefringent single-mode fibers. *Appl. Phys. Lett.* 33 (8): 699–701.
Ulrich, R. (1979). Polarization stabilization on single-mode fiber. *Appl. Phys. Lett.* 35 (11): 840–842.
Vali, V. and Shorthill, R. (1976). Fiber ring interferometer. *Appl. Opt.* 15 (5): 1099–1100.

Further Reading

Al-Amri, M.D., El-Gomati, M.M., and Zubairy, S. (2016). *Optics in Our Time*. https://doi.org/10.1007/978-3-319-31903-2. Springer Open.
Bayvel, P. (2018). Kuen Charles Kao (1933–2018). *Nature* 563: 326.
Davidson, M.W. (2015). Pioneers in optics: Jacques Babinet. *Micros. Today* 23 (5) (September 2015): 45.
Goldstein, D. (2003). *Polarized Light*, 2e. Marcel Dekker, Inc.
Hecht, J. (1999). *City of Light: The Story of Fiber Optics*. Oxford University Press.
Hecht, E. and Zajac, A. (1974). *Optics*, 1–10. Addison-Wesley Publishing Co., 4th printing.
Memorial Tributes (2016). *Elias Snitzer*, vol. 20. The National Academies Press.
Wikipedia. (n.d.): http://en.wikipedia.org.

2

Polarization Basics

2.1 Introduction to Polarization

As described in Chapter 1, in his wave theory of light published in 1820, **Augustin-Jean Fresnel** explained the nature of polarization by supposing that light waves were purely transverse. He also coined the terms linear polarization (LP), circular polarization, and elliptical polarization. Now let's use a rope to get simple pictures of different polarization states, as illustrated in Figure 2.1.

Figure 2.1a shows a vertically vibrating rope wave propagating down the rope generated by wiggling an end of rope up and down. When such a rope wave is projected onto a screen perpendicular to the direction of wave propagation, one expects to see a vertical line moving up and down. Such a rope wave is said to be vertically polarized, or have a linear vertical polarization (LVP). Similarly, by holding an end of a rope and wiggling left and right, a horizontally vibrating rope wave can be generated, as shown in Figure 2.1b. Such a rope wave is said to be horizontally polarized, or have a linear horizontal polarization (LHP). By analogy, by wiggling a rope end in a clockwise circular motion, a rope wave with a circular motion can be generated, with the projection of the wave on a screen perpendicular to the direction of wave propagation tracing out a clockwise rotating circle, as shown in Figure 2.1c. With the thumb pointing against wave's propagation direction with one's left hand, one finds that the rest of four fingers curve in the direction of the rotation. Therefore, we call such a wave left-hand circularly polarized waves or a wave with left-hand circular polarization (LCP). Similarly, by wiggling a rope end in a counter-clockwise circular motion, a rope wave with a circular motion can be generated, with its projection on the screen tracing out a counter-clockwise rotating circle, as shown in Figure 2.1d. With the thumb pointing against wave's propagation direction with one's right hand, one finds that the rest of four fingers curve in the direction of the rotation. Therefore, we call such a wave right-hand circularly polarized waves or a wave with right-hand circular polarization (RCP). In general, by wiggling a rope end in an elliptical motion either clockwise or counter-clockwise, left-hand and right-hand elliptically polarized waves, or left-hand and right-hand elliptical polarizations, can be generated, respectively, as shown in Figure 2.1e,f. Note that the linear and circular polarizations described earlier are the special cases of elliptical polarizations. In short, a polarization describes the motion of a wave in a plane perpendicular to its propagation direction.

Another example is the water wave generated by throwing a small rock in a quiet pond, which propagates away from the center where the rock hits the water to form expanding circles. One can see that the water wave vibrates up and down locally and therefore is said to have a LVP.

Polarization Measurement and Control in Optical Fiber Communication and Sensor Systems, First Edition.
X. Steve Yao and Xiaojun (James) Chen.

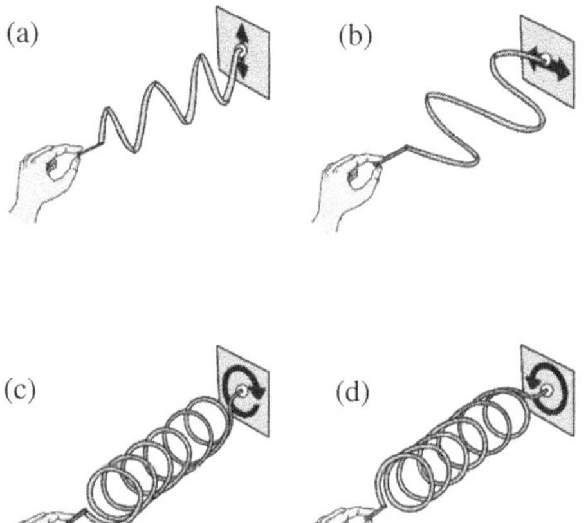

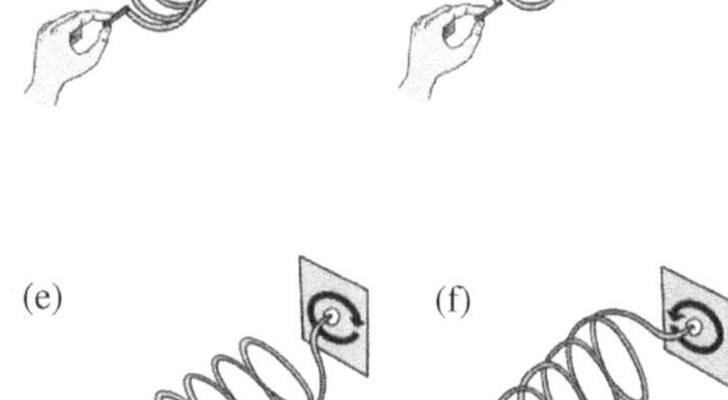

Figure 2.1 Illustration of waves generated by wiggling a rope with different polarizations. (a) Vertical; (b) Horizontal; (c) Left-hand circular; (d) Right-hand circular; (e) Left-hand elliptical; and (f) Right-hand elliptical.

2.2 The Degenerate Polarization States of Light

As the transverse wave, light also behaves similar to the rope waves, having linear, circular, and elliptical polarizations. The major difference is that a light wave oscillates trillion times faster than the rope waves and its direction of oscillation (polarization) cannot be directly observed with naked eyes. Special tools or devices must be used to facilitate the indirect observation or measurement of different polarization states, as will be discussed in Chapters 8 and 9.

To have a better understanding of light wave's polarization, let's proceed to describe the polarization properties of a light wave mathematically. As described in Chapter 1, Fresnel's wave theory of light was unified with electricity and magnetism by Maxwell's classical theory of electromagnetic radiation. According to Maxwell's electromagnetic theory, light is an oscillating electromagnetic wave with both the electrical field $\mathbf{E}(r, t)$ and magnetic field $\mathbf{H}(r, t)$ obeying the wave equation, where r is the space coordinate of the observer and t is time. Because $\mathbf{E}(r, t)$ and $\mathbf{H}(r, t)$ are always orthogonal to each other in space, knowing the polarization behavior of one of them is sufficient for both. In addition, because the light under study mostly propagates in free space or in dielectric media with no free charges, it is more convenient to treat the electrical field than the magnetic field. Therefore, we will only use electric field to represent light in this book unless specifically stated otherwise. For a transverse plane wave propagating in the z direction in a homogeneous and isotropic medium, each of the x and y components of the electrical field E_x and E_y in a plane perpendicular to z must follow the wave equation:

$$\nabla^2 E_i(z,t) = \frac{1}{v}\frac{\partial^2 E_i(z,t)}{\partial t^2} \qquad i = x, y \tag{2.1}$$

where v is the velocity of the wave propagating in the z direction, and the space coordinate r in the argument of E_x and E_y is replaced with z, recognizing the assumption of a plane wave propagating in the z direction.

The solutions to Eq. (2.1) can be expressed as

$$E_x(z,t) = E_{x0}\cos(\omega t - kz + \varphi_{x0}) \tag{2.2a}$$

$$E_y(z,t) = E_{y0}\cos(\omega t - kz + \varphi_{y0}) \tag{2.2b}$$

where E_{i0} and φ_{i0} are amplitudes and phases of the oscillating electrical fields $E_i(z, t)$ $(i = x, y)$, and k is the corresponding propagation constants which can be expressed as

$$k = \frac{\omega}{v} = \frac{\omega}{c}n \tag{2.3}$$

In Eq. (2.3), c is the speed of light in vacuum and n is the index of refraction of the medium for the electrical field. One may verify the solutions by plugging Eq. (2.2a) or (2.2b) into Eq. (2.1) to see that Eq. (2.1) indeed holds. The two orthogonally oscillating waves are illustrated in Figure 2.2.

In general, the electrical field vector in the XY plane can be expressed as the vectorial summation of the field components oscillating in x and y directions as

$$\mathbf{E}(z,t) = E_x(z,t)\hat{\boldsymbol{x}} + E_y(z,t)\hat{\boldsymbol{y}} \tag{2.4}$$

where $\hat{\boldsymbol{x}}$ and $\hat{\boldsymbol{y}}$ are unit vectors in x and y directions, respectively. Equation (2.4) basically reflects Fresnel's understanding of the light wave: "*the act of polarization consists not in creating transverse motions, but in decomposing them in two fixed, mutually perpendicular directions, and in separating the two components.*"

Corresponding to the rope waves described earlier, let's examine the following special cases

(1) $E_{y0} = 0$. In this case, only the horizontally oscillating wave exists. Projecting the oscillation on a screen perpendicular to the beam direction at any location z, one can observe line oscillating back and forth, as shown in Figure 2.3a. This is the horizontally polarized light, similar to the horizontally polarized rope wave earlier, and is denoted as linear horizontal polarization (LHP).
(2) $E_{x0} = 0$. In this case, only the vertically oscillating wave exists. Projecting the oscillation on a screen perpendicular to the beam direction at any location z, one can observe line oscillating up and down, as shown in Figure 2.3b. This is the vertically polarized light, similar to the vertically polarized rope wave earlier, and is denoted as linear vertical polarization (LVP).

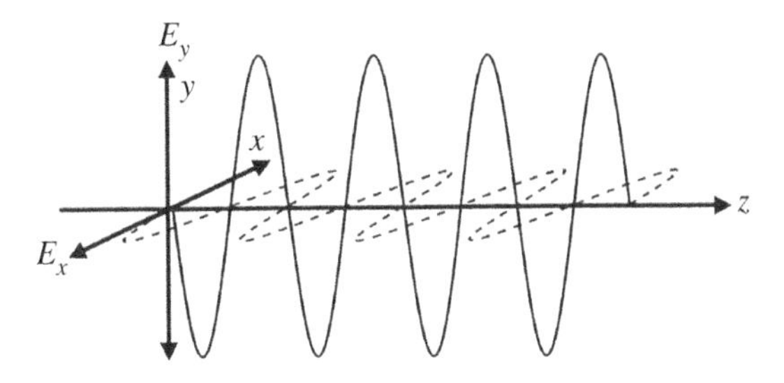

Figure 2.2 Illustration of two transverse waves propagating in the z direction.

(3) $E_{x0} = E_{y0} \neq 0$ and $\varphi_y = \varphi_x$. The two orthogonally oscillating waves are in phase. Projecting the oscillations on a screen perpendicular to the beam direction at any location z, one can observe the combined wave tracing a line oscillating at +45° from the x axis, as shown in Figure 2.3c. This is the 45° polarized light, similar to the 45° polarized rope wave earlier, and is denoted as the linear +45 polarization (L + 45P). Conversely, one may realize that a light wave of L + 45P polarization can be decomposed into two waves of orthogonal polarizations of equal amplitude oscillating in phase.

(4) $E_{x0} = E_{y0} \neq 0$ and $\delta = \varphi_y - \varphi_x = \pi$. The two orthogonally oscillating waves are out of phase. Projecting the oscillations on a screen perpendicular to the beam direction at any location z, one can observe the combined wave tracing a line oscillating at −45° from the x axis, as shown in Figure 2.3d. This is the −45° polarized light, similar to the −45° polarized rope wave earlier, and is denoted as the linear −45 polarization (L − 45P). Conversely, one may realize that a light wave of L − 45P polarization can be decomposed into two waves of orthogonal polarizations of equal amplitude oscillating out of phase.

(5) $E_{x0} = E_{y0} \neq 0$ and $\delta = \varphi_y - \varphi_x = \pi/2$. Looking at the beam in the $-z$ direction and projecting the oscillations on a screen perpendicular to the beam direction at any location z, one can observe the combined wave tracing a clockwise rotating circle, as shown in Figure 2.3e. With the thumb pointing against the beam (in the $-z$ direction toward the paper) with one's right hand, one finds that the rest of four fingers curve in the direction of the rotation. Therefore, this clockwise circularly polarized light, similar to the counter-clockwise circularly polarized rope wave in Figure 2.1d (recognizing that the optical beam here has the opposite propagation direction of the rope wave), is denoted as right-hand circular polarization (RCP). Conversely, one may realize that a light wave of RCP polarization can be decomposed into two waves of orthogonal polarizations of equal amplitude oscillating with a $\pi/2$ phase difference.

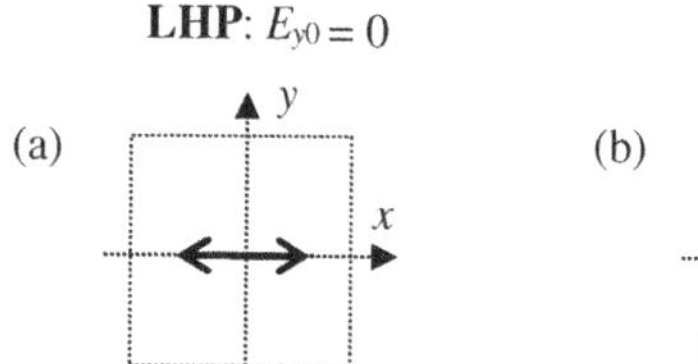

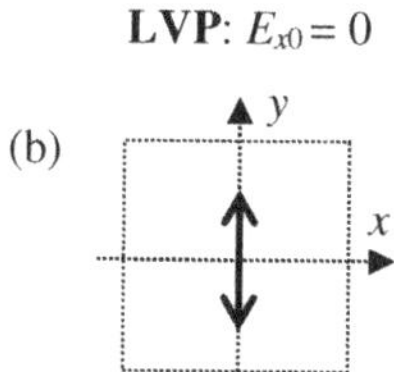

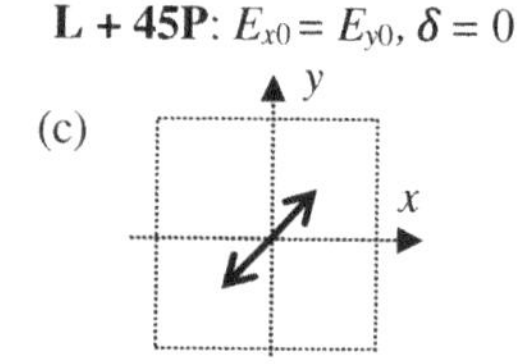

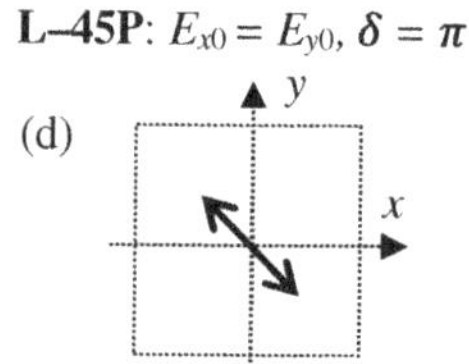

RCP: $E_{x0} = E_{y0}$, $\delta = \pi/2$

(e)

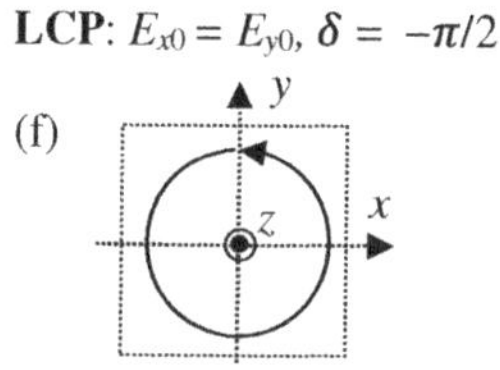

Figure 2.3 Illustration of six degenerate polarization states. The optical beam is propagating in the z direction, pointing to the observer. (a) Linear horizontal polarization; (b) Linear vertical polarization; (c) Linear 45° polarization; (d) Linear −45° polarization; (e) Right-hand circular polarization; (f) Left-hand circular polarization.

(6) $E_{x0} = E_{y0} \neq 0$ and $\delta = \varphi_y - \varphi_x = -\pi/2$. Looking at the beam in the $-z$ direction and projecting the oscillations on a screen perpendicular to the beam direction at any location z, one can observe the combined wave tracing a counter-clockwise rotating circle, as shown in Figure 2.3f. With the thumb pointing against the beam with one's left hand, one finds that the rest of four fingers curve in the direction of the rotation. Therefore, this counter-clockwise circularly polarized light, similar to the clockwise circularly polarized rope wave of Figure 2.1c (recognizing that the optical beam here has the opposite propagation direction of the rope wave), is denoted as left-hand circular polarization (LCP). Conversely, one may realize that a light wave of LCP can be decomposed into two waves of orthogonal polarizations of equal amplitude oscillating with a $-\pi/2$ phase difference.

These six special cases are generally referred to as six degenerate polarization states or six distinctive states of polarization (SOPs). As will be discussed in Chapter 9, the generation of these six SOPs is important for the analysis of polarization related parameters of the light sources as well as the media affecting the SOP.

2.3 The Polarization Ellipse of Light

With the pictures of the polarization states of light in mind, now let's mathematically look at the electrical fields in a plane at a particular location z_0 perpendicular to the light beam in a more general way. Equations (2.2a) and (2.2b) can be rewritten as

$$\frac{E_x(t)}{E_{x0}} = cos(\omega t + \varphi_x) \tag{2.5a}$$

$$\frac{E_y(t)}{E_{y0}} = cos(\omega t + \varphi_y) \tag{2.5b}$$

where $\varphi_i = \varphi_{i0} - kz_0$ $(i = x, y)$ are the phases of the electrical fields oscillating in x and y directions at location z_0.

Expanding the right-hand side of Eq. (2.5a) and (2.5b) with the identity $cos(\omega t + \varphi_j) = cos\omega t cos\varphi_j - sin\omega t sin\varphi_j$, one obtains the following equation with some mathematical manipulations:

$$\left(\frac{E_x}{E_{x0}}\right)^2 + \left(\frac{E_y}{E_{y0}}\right)^2 - 2\frac{E_x}{E_{x0}}\frac{E_y}{E_{y0}}cos\delta = sin^2\delta \tag{2.6}$$

where

$$\delta = \varphi_y - \varphi_x \tag{2.7}$$

Equation (2.6) is recognized as **the equation of ellipse**, indicating that the optical field traces out an ellipse in the observation plane perpendicular to the optical beam, as shown in Figure 2.4. It can be seen that the ellipse has a major axis of $2a$ and a minor axis of $2b$, which are rotated from the x axis with an orientation angle of ψ $(0 \leq \psi \leq \pi)$. In general, an ellipse can be fully characterized by three parameters, the orientation angle ψ, the sense of rotation, and the ellipticity angle α defined as

$$\alpha = tan^{-1}\left(\frac{\pm b}{a}\right) \quad \left(-\frac{\pi}{4} \leq \alpha \leq \frac{\pi}{4}\right) \tag{2.8}$$

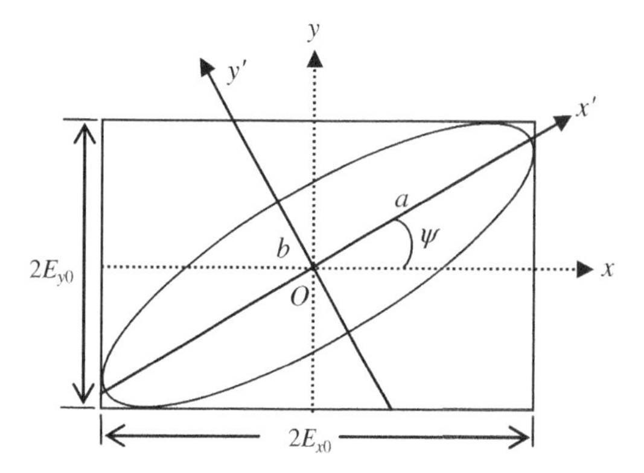

Figure 2.4 The polarization ellipse of a light wave showing a major axis of $2a$, a minor axis of $2b$, and an orientation angle of ψ.

For a linear polarization, $\alpha = 0$, and for a circular polarization ($a = b$), $\alpha = \pm\pi/4$. As will be seen next, the sign of the ellipticity angle defines the sense of rotation.

As one can see, all the six degenerate polarization states described in Section 2.2 are special cases of the polarization ellipse: the LHP and LVP correspond to the cases of ($\alpha = 0$, $\psi = 0$) and ($\alpha = 0$, $\psi = \pi/2$) respectively; the L + 45P and L − 45P correspond to the cases of ($\alpha = 0$, $\psi = \pm\pi/4$); and the LCP and RCP correspond to the cases of ($\alpha = \pm\pi/4$, $\psi = 0$).

Now let's proceed to find out the relations between the ellipse parameters (α, ψ) and the electrical field parameters (δ, E_{x0}, E_{y0}) in Eq. (2.6). One may observe that when the phase difference $\delta = \pm\pi/2$, the cross term in Eq. (2.6) is zero and the expression for the polarization ellipse has the standard form:

$$\left(\frac{E_x}{E_{x0}}\right)^2 + \left(\frac{E_y}{E_{y0}}\right)^2 = 1 \tag{2.9}$$

Therefore, in the rotated coordinates (x', y') shown in Figure 2.4, the polarization ellipse should also have the standard form:

$$\left(\frac{E_{x'}}{a}\right)^2 + \left(\frac{E_{y'}}{b}\right)^2 = 1 \tag{2.10}$$

which means the phase difference $\delta' = \delta_{x'} - \delta_{y'}$ of the electrical fields in the (x', y') coordinates must also be $\pm\pi/2$:

$$E_{x'}(t) = a\cos(\omega t + \varphi_0) \tag{2.11a}$$

$$E_{y'}(t) = b\cos(\omega t + \varphi_0 \mp \pi/2) = \pm b\sin(\omega t + \varphi_0) \tag{2.11b}$$

It can be seen from Eq. (2.11b) that the "+" sign in $\delta' = \pm\pi/2$ means that $E_{y'}(t)$ is ahead of $E_{x'}(t)$, responding to a LCP, and the "−" sign in $\delta' = \pm\pi/2$ means that $E_{y'}(t)$is lagging $E_{x'}(t)$, corresponding to a RCP. Consequently, the "+" sign in front of b corresponds to a RCP and the "−" sign corresponds to a LCP.

As shown in Figure 2.4, the coordinate system (x', y') is rotated by an angle of ψ with respect to the coordinate system (x, y); therefore the field components $E_{x'}(t)$ and $E_{y'}(t)$ can be expressed as

$$E_{x'}(t) = E_x cos\psi + E_y sin\psi \tag{2.12a}$$

$$E_{y'}(t) = -E_x sin\psi + E_y cos\psi \tag{2.12b}$$

Substituting Eqs. ((2.5a) and (2.5b)) and ((2.11a) and (2.11b)) in Eqs. (2.12a) and (2.12b), with some mathematical manipulation as shown in Appendix 2.A (Born and Wolf 1980, p. 26), we obtain the expressions relating the orientation angle ψ and the ellipticity angle α with the electrical field parameters (δ, E_{x0}, E_{y0}) as

$$tan2\psi = \frac{2E_{x0}E_{y0}}{E_{x0}^2 - E_{y0}^2} cos\delta = tan2\beta \, cos\delta, \quad 0 \leq \psi \leq \pi \tag{2.13a}$$

$$sin2\alpha = \frac{2E_{x0}E_{y0}}{E_{x0}^2 + E_{y0}^2} sin\delta = sin2\beta \, sin\delta, \quad \frac{-\pi}{4} \leq \alpha \leq \frac{\pi}{4} \tag{2.13b}$$

where β is an auxiliary angle defined as

$$tan\beta = \frac{E_{yo}}{E_{x0}} \quad 0 \leq \beta \leq \frac{\pi}{2} \tag{2.13c}$$

The ellipticity angle defined in Eq. (2.8) can be rewritten here as

$$tan\alpha = \frac{\pm b}{a} \quad \frac{-\pi}{4} \leq \alpha \leq \frac{\pi}{4} \tag{2.13d}$$

Other useful relations obtained in Appendix 2.A are

$$a^2 + b^2 = E_{x0}^2 + E_{y0}^2 \tag{2.13e}$$

$$A = \pi ab = \pi E_{x0} E_{yo} sin\delta \tag{2.13f}$$

where $a^2 + b^2$ and $E_{x0}^2 + E_{y0}^2$ stand for the optical power and A is the area of the polarization ellipse. Equation (2.13e) simply implies the optical power is conserved regardless of the coordinate system chosen, which is expected.

Now let's discuss the following scenarios:

(1) $\delta = 0$ or π
One gets $\alpha = 0$, $A = 0$, and $\psi = \pm\beta = \pm tan^{-1}(E_{yo}/E_{x0})$, indicating a linear polarization (LP) with both the ellipticity α and area A equal to zero, as shown in Figure 2.5a,b.

(2) $\delta = \frac{+\pi}{2}$ or $\frac{-\pi}{2}$ $\left(\frac{3\pi}{2}\right)$,
One gets $\psi = 0$ and $\alpha = \pm\beta$, indicating a right-hand elliptical polarization (REP) and left-hand elliptical polarization (LEP), respectively, represented in a coordinate system (x, y) coinciding with the coordinate systems (x', y'), as shown in Figure 2.5c,d. In general, the "+" sign of α $(\alpha \geq 0)$ indicates the right hand rotation, while the "−" sign $(\alpha \leq 0)$ indicates the left-hand rotation, as mentioned in the discussion in Eq. (2.8). When $E_{x0} = E_{y0}$, $\alpha = \pm 45°$, the REP and LEP reduce to RCP and LCP.

(1) $0 \leq \delta \leq \frac{\pi}{2}$ or $\pi \leq \delta \leq \frac{3\pi}{2}$
One gets $(\psi \geq 0, \alpha \geq 0)$ or $(\psi \leq 0, \alpha \leq 0)$. $(\psi \geq 0, \alpha \geq 0)$ indicates a REP with a positive orientation angle as shown in Figure 2.5e, while $(\psi \leq 0, \alpha \leq 0)$ indicates a LEP with a negative orientation angle as shown in Figure 2.5f.

(2) $\frac{\pi}{2} \leq \delta \leq \pi$ or $\frac{3\pi}{2} \leq \delta \leq 2\pi$
One gets $(\psi \leq 0, \alpha \geq 0)$ or $(\psi \geq 0, \alpha \leq 0)$. $(\psi \leq 0, \alpha \geq 0)$ indicates a REP with a negative orientation angle as shown in Figure 2.5g, while $(\psi \geq 0, \alpha \leq 0)$ indicates a LEP with a positive orientation angle as shown in Figure 2.5h.

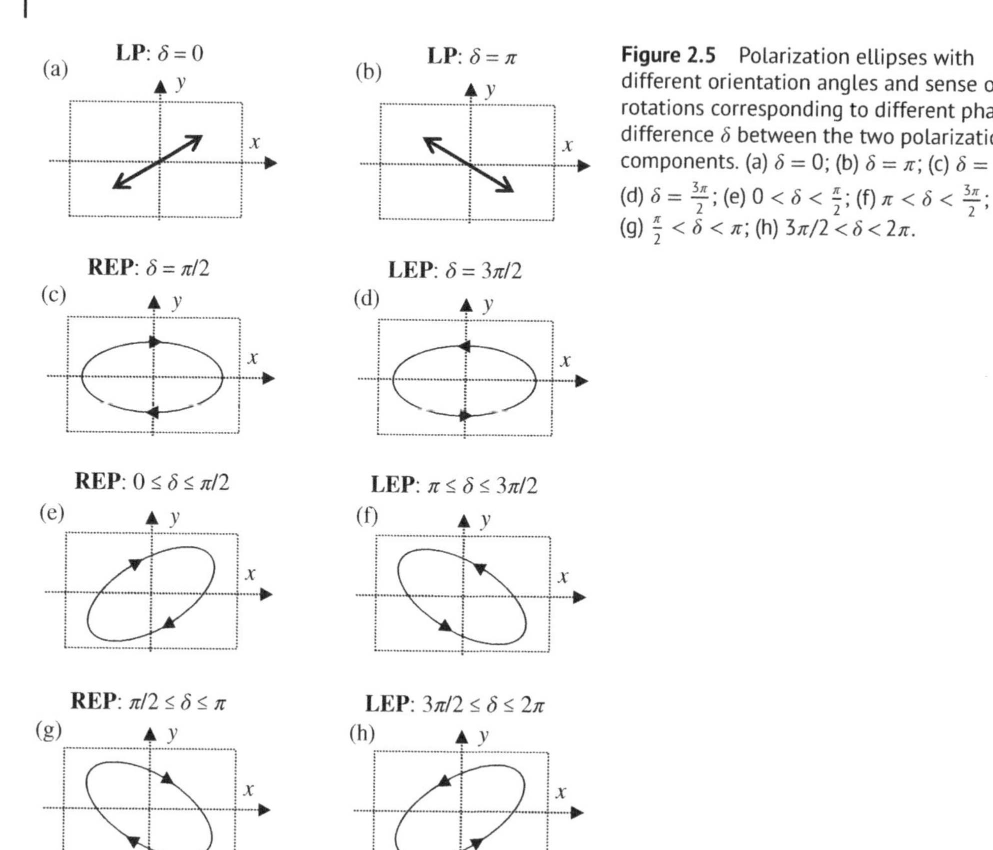

Figure 2.5 Polarization ellipses with different orientation angles and sense of rotations corresponding to different phase difference δ between the two polarization components. (a) $\delta = 0$; (b) $\delta = \pi$; (c) $\delta = \frac{\pi}{2}$; (d) $\delta = \frac{3\pi}{2}$; (e) $0 < \delta < \frac{\pi}{2}$; (f) $\pi < \delta < \frac{3\pi}{2}$; (g) $\frac{\pi}{2} < \delta < \pi$; (h) $3\pi/2 < \delta < 2\pi$.

2.4 Poincaré Sphere Presentation of Polarization

The two-dimensional (2D) polarization ellipse is an excellent way to visualize an SOP. However, since each SOP must be represented with a unique ellipse, it is tedious to present the polarization evolution of a light beam when it passes through multiple polarization-altering optical components. In order to overcome this difficulty, a single three-dimensional (3D) sphere was suggested to represent all possible SOPs, as shown in Figure 2.6a. Such a sphere is called **Poincaré sphere**, named after the French mathematician **Jules Henri Poincaré**, who first introduced it to describe the polarized light in 1892.

Equations (2.13a) and (2.13b) indicate that a SOP can be fully described with two parameters: the orientation angle ψ and the ellipticity angle α, which can be represented in the Cartesian coordinate system (x, y, z), as shown in Figure 2.6a. The Cartesian coordinates are related to the spherical coordinates by

$$x = cos2\alpha\, cos2\psi \quad 0 \le \psi \le \pi \tag{2.14a}$$

$$y = cos2\alpha\, sin2\psi \quad -\frac{\pi}{4} \le \alpha \le \frac{\pi}{4} \tag{2.14b}$$

$$z = sin2\alpha \tag{2.14c}$$

$$x^2 + y^2 + z^2 = 1 \tag{2.14d}$$

Equation (2.14d) indicates that the sphere has a unit radius.

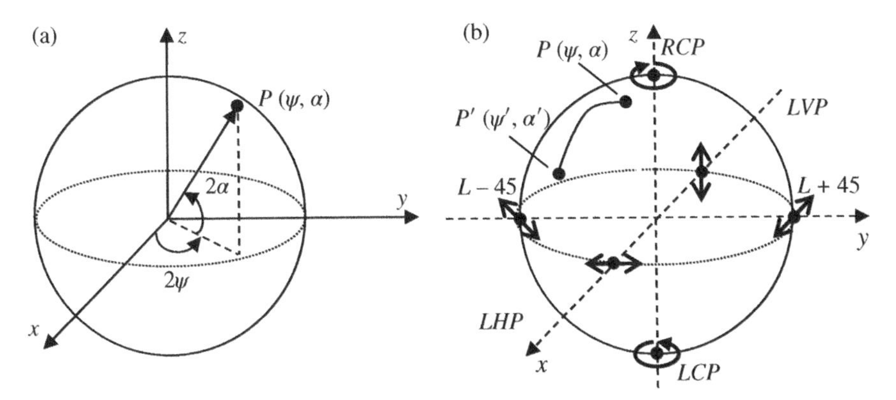

Figure 2.6 (a) Illustration of a Poincaré sphere showing a point P on the sphere with 2ψ as the azimuthal angle and 2α as the latitude angle. (b) Presentation of six degenerate SOPs on a Poincaré sphere, as well as the polarization evolution between two SOPs.

The six degenerate SOPs can now be presented on a same graph, as shown in Figure 2.6b, instead of with six different 2D graphs shown in Figure 2.3. These six special points are (0, 0) for LHP, ($\pi/2$, 0) for LVP, ($+\pi/4$, 0) for L + 45P, ($-\pi/4$, 0) for L – 45P, (0, $+\pi/4$) for RCP, and (0, $-\pi/4$) for LCP. In fact, all SOPs can be represented as a point on the Poincaré sphere, which is much easier to see in polarization evolution from $P(\psi, \alpha)$ to point $P'(\psi', \alpha')$ with a trace connecting the two points, as shown in Figure 2.6b.

Now let's look at three special traces for polarization evolution. The first corresponds to the case that the SOPs on the trace have the same ellipticity $\alpha = \pi/8$, but different orientation angles ψ. This trace should be a circle of latitude with an angle of 45° from the (x, y) plane, as shown in Figure 2.7a. Table 2.1 shows the polarization ellipse parameters of four representative SOP points on the trace. In Figure 2.7a, the shape and sense of rotation of each corresponding polarization ellipse are also shown on each SOP point.

The second corresponds to the case that the SOPs on the trace have the same orientation angle of $\pi/8$, but different ellipticity angles. Such a trace should be half of the circle of longitude 45° from the (x, z) plane, as shown in Figure 2.7b. Table 2.2 shows the polarization ellipse parameters of five

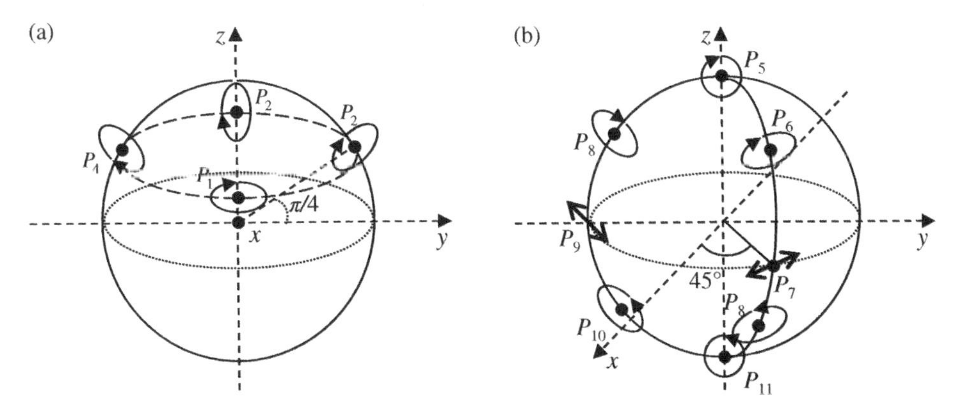

Figure 2.7 (a) The SOP evolution trace on a Poincaré sphere with an ellipticity angle of $\pi/4$ and varying orientation angle. (b) The SOP evolution traces with fixed orientation angles at 45° and –90°, but varying ellipticity angles.

Table 2.1

	P_1	P_2	P_3	P_4
2ψ	0	$\pi/2$	π	$-\pi/2$
2α	$\pi/4$	$\pi/4$	$\pi/4$	$\pi/4$

Table 2.2

	P_5	P_6	P_7	P_8	P_{11}
2ψ		$\pi/4$	$\pi/4$	$\pi/4$	
2α	$\pi/2$	$\pi/3$	0	$-\pi/3$	$-\pi/2$

Table 2.3

	P_5	P_8	P_9	P_{10}	P_{11}
2ψ		$-\pi/2$	$-\pi/2$	$-\pi/2$	
2α	$\pi/2$	$\pi/4$	0	$-\pi/4$	$-\pi/2$

representative SOPs on the trace. To have a better visualization, the shape and sense of rotation of each corresponding polarization ellipse are also shown on each SOP point in Figure 2.7b.

The third corresponds to the case that the SOPs on the trace have the same orientation angle of $-\pi/4$, but different ellipticity angles. Such a trace should be half of the circle of longitude $-90°$ from the (x, z) plane, as shown in Figure 2.7b. Table 2.3 shows the polarization ellipse parameters of five representative SOPs on the trace. The corresponding polarization ellipse is also shown on each SOP point in Figure 2.7b.

2.5 Degree of Polarization (DOP)

In the discussions earlier, light is assumed to be fully polarized. However, natural light or sun light is completely unpolarized, and can be viewed as the mix of a large amount of light waves with different SOPs randomly distributed on the Poincaré sphere. This can be understood microscopically by considering that the Sun has zillions of light emitting elements, such as atoms, dipoles, and ions at very high temperatures to emit lights of all possible SOPs with random phases, as well as different wavelengths.

As mentioned in Chapter 1, humans did not know much about polarized light until Étienne-Louis Malus discovered the polarization of light by reflection and by double refraction in crystals, in 1809 and 1810, respectively. In fact, Thomas Bartholin first discovered in 1669 that when a single incident beam of natural light propagated through a rhombohedral calcite crystal, two beams emerged, as shown in Figure 2.8. Such a phenomenon could be explained by assuming that the crystal has two indices of refraction n_o and n_e for lights polarized perpendicular to the plane of optical axis and in the plane of optical axis, respectively, and therefore is termed double refraction. The plane of optical axis here is defined as the plane containing the optical axis

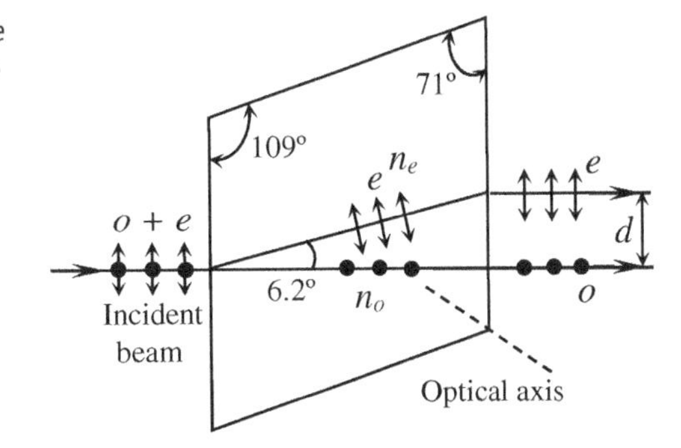

Figure 2.8 Illustration of a calcite crystal to separate an incident beam of unpolarized light into two beams of orthogonally polarized light, the o-ray and e-ray. When rotating the crystal about the beam, the o-ray stays stationary while the e-ray circles around the o-ray. Source: Hecht and Zajac 1974/Pearson.

and the incident beam, and is the same as the plane of incidence in Figure 2.8. When rotating the crystal about the optical beam, one beam stays stationary while the other circles around the stationary beam following the rotation of the crystal. The stationary beam appeared to be "normal" and therefore is called ordinary ray or *o*-ray, while the other beam looks unusual and therefore is called extra ordinary ray or *e*-ray, per the suggestion of Bartholin. As shown in Figure 2.8, the walk-off angle between the two rays is 6.2° in a calcite crystal with $n_o = 1.6853$ and $n_e = 1.4864$ at wavelength of 589.3 nm so that the beam separation d at the exit face of the crystal with a length of L is

$$d = L \cdot tan(6.2°) = 0.109L \tag{2.15}$$

For a crystal length of 10 mm, the beam separation is 1.09 mm. See Chapter 5 for more details about double refraction.

At the beginning of the nineteenth century, the only way to analyze the polarization behavior of light was with a calcite crystal. By using it as a polarizer, Malus was able to observe and analyze the polarization properties of sun light reflected from the church's glass windows, as shown in Figure 2.9a, by rotating the calcite polarizer around the optical beam while observing the brightness or intensity of light passing through the calcite. As he noticed that the brightness of the two beams exiting from the crystal went opposite ways as the crystal is rotated, one went stronger while the

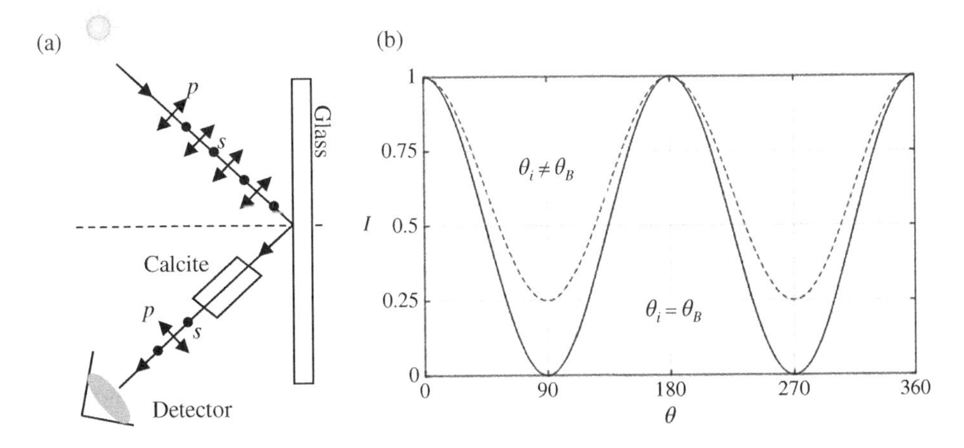

Figure 2.9 (a) Illustration of a sun ray is reflected by a glass plate to be analyzed after passing through a calcite crystal (polarizer); (b) Intensity of light as a function of polarizer rotation angle. Solid line: Beam incident at the Brewster angle $\theta_i = \theta_B$. Dashed line: beam incident at an arbitrary angle θ_i.

other went to weaker. As will be discussed in detail in Section 2.9.1 and in Chapter 4, when a light beam is reflected from a glass or water surface, the polarization perpendicular to the plane of incidence (the "*s*" polarization) reflects more than the polarization in the plane of incidence (the "*p*" polarization). At the Brewster angle, only the "*s*" polarization is reflected to obtain a fully polarized light. Malus discovered that the intensity I of one of the beams passing through the calcite polarizer was a sinusoidal function of the rotation angle θ, as shown in Figure 2.9b.

$$I(\theta) = I_0 \cos^2\theta \tag{2.16}$$

This is the well-known Malus' law, which can be explained by Fresnel's formula for reflection and refraction in Chapter 4. When the incident beam is at the Brewster angle, the reflected light is fully polarized, with the intensity swinging from zero to the maximum. However, at other incident angles, the polarization is partially polarized, with the intensity swings from a non-zero minimum to the maximum. This non-zero minimum can be considered as the unpolarized light. If one holds the calcite polarizer to directly analyze the sun light, no intensity variation will be observed and therefore the sun light is considered totally unpolarized. Note that nowadays, high quality sheet polarizers, such as the early polymer based Polaroid™ from Polaroid Corporation or the recent glass based Polarcor™ from Corning, are commercially available. They are of low cost, can be cut into different sizes, and are much easier to use than the calcite polarizers.

In order to quantitatively describe the partially polarized light, degree of polarization (DOP) is introduced and is defined as the fraction of the total intensity I_{total} that is carried by the polarized component of the light:

$$DOP = \frac{I_{pol}}{I_{total}} = \frac{I_{pol}}{I_{pol} + I_{unpol}} \tag{2.17}$$

where I_{pol} and I_{unpol} are intensities of polarized and unpolarized portions of a light beam, respectively, and $I_{total} = I_{pol} + I_{unpol}$. A perfectly polarized light has a *DOP* of 100%, whereas an unpolarized light, such as the sun light, has a *DOP* of 0%. A partially polarized light beam therefore can be represented by a superposition of a polarized and unpolarized component, having a DOP somewhere in between 0% and 100%.

For the case of Figure 2.9 in which the *s* and *p* polarization components have no fixed phase relationship (instead their phase difference rapidly varying with time), they cannot be recombined into an elliptical polarization as described in Section 2.3. Therefore the maximum and minimum intensities I_{max} and I_{min} passing through the calcite polarizer in Figure 2.9b are $I_{max} = I_{pol} + I_{unpol}/2$ and $I_{min} = I_{unpol}/2$, which lead to

$$I_{pol} = I_{max} - I_{min} \tag{2.18a}$$

$$I_{total} = I_{max} + I_{min} \tag{2.18b}$$

$$DOP = \frac{I_{max} - I_{min}}{I_{max} + I_{min}} \tag{2.18c}$$

For the solid line in Figure 2.9b, $I_{max} = 1$ and $I_{min} = 0$, resulting in a $DOP = 1$ or 100%. For the dashed line in Figure 2.9b, $I_{max} = 1$ and $I_{min} = 0.25$, resulting in a $DOP = 0.6$ or 60%.

Note that Eqs. (2.18a), (2.18b), and (2.18c) are obtained for completely incoherent sunlight. As will be discussed in Chapter 7, for a partially incoherent light, Eq. (2.18a) and (2.18b) is valid only when a polarization scrambler or a polarization controller is placed in front of the polarizer to generate the I_{max} and I_{min} intensities detected after the polarizer. In general, for a highly coherent laser light, the *DOP* is always 100%, while for amplified spontaneous emission (ASE) noise from an optical amplifier, the *DOP* is close to 0%. For an optical signal generated from a laser in an optical fiber communication system, which has been amplified by one or more optical amplifiers, the *DOP* is generally between 100% and 0%. The more optical amplifiers the signal goes through, the more the ASE noises it gets, and the poorer its signal-to-noise ratio (SNR) and DOP. Therefore, DOP can be used to measure the SNR in an optical fiber communication link, as will be discussed in Section 8.4.4.

The act of degrading the DOP of a light signal is called depolarization. In addition to the process of adding incoherent noise to the light signal described earlier, a light signal can also be depolarized by other processes, such as by scattering from a biological tissue as in optical coherent tomography (OCT) or by changing the relative delay between the orthogonal polarization components beyond the coherence length of the light source.

Conversely, the act of increasing the DOP of a light signal is often called re-polarization. For example, a low DOP light signal can be re-polarized by reflecting from a glass surface, passing through a polarizer, or by reducing the relative delay between the orthogonal polarization components within the coherence length of the light source. These topics will be further explored in Chapters 7 and 8.

2.6 Birefringence

The index of refraction or refractive index n of a medium or material is defined as the speed of light c in vacuum over the speed v light in the medium:

$$n = \frac{c}{v} \tag{2.19}$$

The refractive index of an optically dense medium is generally higher than that of a less dense medium. In isotropic materials, light travels at the same speed regardless its polarization and therefore has the same refractive index for different polarizations. However, in anisotropic materials, light may travel at different speeds depending on its polarization with respect to a symmetry axis of the material (the optical axis) and therefore the material may have two or more refractive indices.

As discussed in Section 2.5, the calcite crystal has two refractive indices n_o and n_e for two orthogonal linear polarizations, and therefore is said to be double refractive, bi-refractive, or birefringent. In general, birefringence is the optical property of a material having a refractive index that depends on the polarization and propagation direction of light and is often quantified as the maximum difference between refractive indices exhibited by the material. Crystals with non-cubic structures are often birefringent, as are plastics and glasses under mechanical stress.

In general, the transverse direction corresponding to the polarization of larger refractive index is recognized as the slow axis because it has a slower speed according to Eq. (2.19). Correspondingly, the direction corresponding to the smaller refractive index is recognized as the fast axis. Similar to Eqs. (2.2) and (2.3), one may decompose an optical field propagating in the birefringence material into two orthogonal components $E_s(z, t)$ and $E_f(z, t)$ along the slow and fast axes as

$$E_s(z, t) = E_{s0}\cos\left(\omega t - \frac{\omega}{c} n_s z + \varphi_0\right) \tag{2.20a}$$

$$E_f(z, t) = E_{f0}\cos\left(\omega t - \frac{\omega}{c} n_f z + \varphi_0\right) \tag{2.20b}$$

where E_{s0} and E_{f0} are the corresponding field amplitudes, and φ_0 is the initial phase of the two polarization components at $z = 0$, and n_s and n_f are the refractive indices associated with the slow and fast axes. At a distance L into the linear birefringence material, the phase difference between the slow and fast components δ_{sf} is

$$\delta_{sf} = \frac{\omega}{c} L(n_s - n_f) = \frac{2\pi}{\lambda} \Delta n L \tag{2.20c}$$

where $\Delta n = n_s - n_f$ is the birefringence. Note that this phase difference δ_{sf} is often referred to as **retardation**. Therefore, the act of linear birefringence is to introduce an additional phase difference $\frac{2\pi}{\lambda}\Delta n L$ between two orthogonal linear polarization components of a light beam to alter its initial SOP according to Eqs. (2.13a)–(2.13d), in which (E_{x0}, E_{y0}, δ) are replaced by $(E_{s0}, E_{f0}, \delta_{sf})$. Note that the increase of distance L may cause both the orientation angle ψ and ellipticity angle α to change, resulting in a circular trace on the Poincaré sphere. In particular, when $E_{s0} = E_{f0}$, the SOP traces out a big circle enclosing the north and south poles on the Poincaré sphere. Detailed discussions on how a linear birefringence affect the polarization of an optical signal will be presented in Chapter 4.

Most birefringent materials exhibit linear birefringence, as the case of calcite crystal in which the two orthogonal linear polarizations have two different refractive indexes. There also exist materials with circular birefringence in which two orthogonal circular polarizations have two different refractive indexes. For example, optical activity in quartz crystals is a form of circular birefringence, which causes a linearly polarized light beam to rotate its polarization as it propagates along the optical axis of the crystal. Another well known example of optical activity is the liquid crystal commonly used in displays in our smart phones, computer screens, and television sets. The Faraday effect describes magnetic field induced circular birefringence, which rotates the major axis of an elliptical polarization with an applied magnetic field. How do these circular birefringence affect the polarization will be discussed in detail in Chapters 4 and 6.

As will be discussed in Chapter 3, when an optical fiber is under transverse mechanical stress or when its core or cladding is not perfectly circular, birefringence will appear. Similar to the optical signal in a birefringence crystal, the birefringence in the optical fiber will cause the SOP of the optical signal to vary as it propagates in the fiber. In addition, for a long optical fiber with a length of L, the two orthogonal polarization components of a pulsed optical signal projected on the slow and fast axes will experience a time delay τ as the pulse propagates in the fiber:

$$\tau = \frac{(n_s - n_f)L}{c} = \frac{\Delta n L}{c} \tag{2.21}$$

which eventually will lead to pulse spreading and signal distortion (Kapron et al. 1972; Schlosser 1972). This phenomenon is called polarization mode dispersion (PMD) first investigated by F. Kapron, N. Borrelli, and D. Keck, and W. Schlosser in 1972, and the term PMD was first coined by S.C. Rashleigh and R. Ulrich in 1978. The relative delay between the two polarization components is called the differential group delay (DGD).

2.7 Photoelasticity or Photoelastic Effect

Photoelasticity was first discovered by David Brewster (1815), which describes the birefringence induced in certain materials, particularly initially isotropic materials, by an applied stress. Specifically, when a material, such as a piece of glass or a piece of plastic, is under an external or internal mechanical stress, the refractive index in the direction x of the stress will be different from that in the direction y orthogonal to the stress and the birefringence obeys Brewster's law (or the stress-optic law):

$$\Delta n = n_x - n_y = C_B(T_x - T_y) \tag{2.22a}$$

where C_B is the stress-optical coefficient, n_x and n_y are the refractive indices, and T_x and T_y are the principal stresses in x and y directions, respectively. Some materials have a larger refractive index if the direction of the stress and other materials is opposite.

In general, the birefringence induced by the photoelastic effect can be described by a photoelastic tensor. For an isotropic material, such as a fused silica, the stress-optic coefficient can be expressed as

$$C_B = \frac{n^3}{2}\frac{(1+\sigma)(p_{12} - p_{11})}{Y} \tag{2.22b}$$

where n, Y, σ, and p_{ij} $(i = 1, j = 1, 2)$ are the refractive index, Young's modulus, Poisson's ratio, and photoelastic tensor elements of the material, respectively. For a transverse line force f applied to a length L of a single mode fiber with a diameter of d, the induced birefringence can be expressed as (Smith 1980):

$$\Delta n = \frac{8C_B}{\pi}\left(\frac{f}{d}\right) \tag{2.23a}$$

where the line force f is defined as the applied force F divided by the length L of the fiber

$$f = \frac{F}{L} \tag{2.23b}$$

As will be discussed throughout this book, the photoelastic effect is extremely important for understanding the polarization variations in optical fibers, for sensing the transverse stress, as well as for making devices for control the SOP in optical fibers. More detailed discussions on the topic can be found in Section 5.4.

2.8 Dichroism, Diattenuation, and Polarization Dependent Loss

Diattenuation is the property of an optical element or system whereby the power transmittance of the exiting beam depends on the polarization state of the incident beam. The transmitted optical power is a maximum P_{max} for one incident polarization state, and a minimum P_{min} for the orthogonal state. The diattenuation D is defined as

$$D = \frac{P_{max} - P_{min}}{P_{max} + P_{min}} \tag{2.24}$$

An example of diattenuation is the transmission of the refracted beam when a light beam is incident on a glass slab from air for the case shown in Figure 2.9a. Because the reflected beam contains more "s" polarization component, the refracted beam must contain more "p" polarization component. The diattenuation reaches maximum at Brewster's angle because "s" component reflection is maximum.

Diattenuator is any homogeneous polarization element which displays significant diattenuation and minimal retardance. Polarizers have a diattenuation close to one, but nearly all optical interfaces are weak diattenuators. Examples of diattenuators include polarizers, dichroic materials, metal and dielectric interfaces with reflection and transmission differences described by Fresnel equations, thin films (homogeneous and isotropic), and diffraction gratings.

Dichroism is the material property of displaying differential attenuation between two orthogonal polarization states during propagation in a medium. For each direction of propagation, dichroic media have two modes of propagation with different absorption coefficients. Examples of dichroic materials include sheet polarizers and dichroic crystals such as tourmaline. Dichroism generally includes the contribution from diattenuation or diattenuators.

Polarization dependent loss (PDL) is a term used in fiber optics to describe the differential insertion loss between two orthogonal polarization states during propagation in a fiber, expressed in dB, which is essentially the same as dichroism.

$$PDL = 10log\frac{P_{max}}{P_0} - 10log\frac{P_{min}}{P_0} = 10log\frac{P_{max}}{P_{min}} \tag{2.25}$$

It is important to point out that, unlike birefringence, the presence of diattenuation, dichroism, or PDL destroys orthogonality of polarization multiplexed signals, causing the loss of certain information in the optical signal, which makes it impossible to fully recover the signal through digital signal processing (DSP). Mathematically, their presence destroys the unitary properties of Jones and Muller matrices, making the signal processing very difficult. In contrast, the presence of birefringence and PMD does not affect the orthogonality of polarization multiplexed signals (the unitary properties of Jones and Muller matrices). These points will be further discussed in detail in Chapter 4.

2.9 Polarization Properties of Reflected and Refracted Light

2.9.1 Reflection

As mentioned previously, when a beam is incident from a medium with a refractive index of n_i to another medium with a refractive index of n_t, as shown in Figure 2.10, the reflection coefficients r_p and r_s of the p-ray and the s-ray are different, which are given by the Fresnel equations:

$$r_p \equiv \frac{E_{pr}}{E_{p0}} = \frac{n_t cos\,\theta_i - n_i cos\,\theta_t}{n_t cos\,\theta_i + n_i cos\,\theta_t} \tag{2.26a}$$

$$r_s \equiv \frac{E_{sr}}{E_{s0}} = \frac{n_i cos\,\theta_i - n_t cos\,\theta_t}{n_i cos\,\theta_i + n_t cos\,\theta_t} \tag{2.26b}$$

where E_{pr} and E_{p0} are the reflected and incident amplitudes of the p-ray, respectively, E_{sr} and E_{s0} are the reflected and incident amplitudes of the s-ray, respectively, and θ_i and θ_t are the incident and refraction angles shown in Figure 2.10, which follow Snell's law of refraction:

$$n_i sin\theta_i = n_t sin\theta_t \tag{2.27}$$

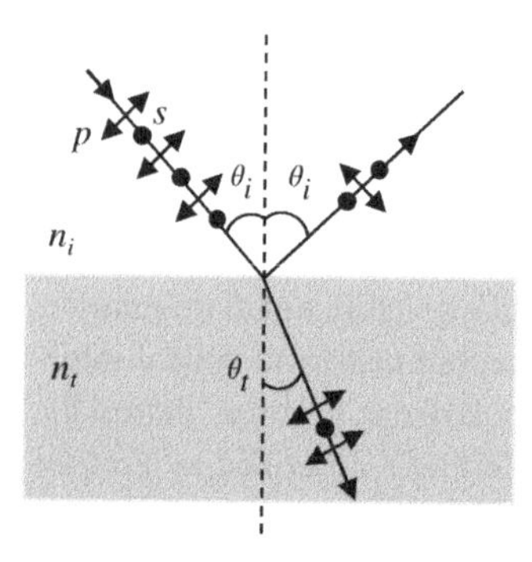

Figure 2.10 Reflection and refraction of an optical beam at an interface of two media with different refractive indexes of n_i and n_t. s-ray is defined as the light beam polarized perpendicular to the plane of incident, while the p-ray is defined as the beam polarized in the plane of incidence (the plane containing both the incident beam and the surface normal at the point of incidence).

Because of the differences in the reflection coefficients, as a light beam reflected from a flat medium, such as the surface of a glass or a body of water, its polarization will be changed. In other words, the reflection from a surface at an angle can induce PDL in the beam. More detailed discussions on this topic will be given in Chapter 4.

2.9.2 Refraction

Similarly, for the diffracted beam, the p- and s-rays also have different transmission coefficients t_p and t_s given by the Fresnel equations.

$$t_p \equiv \frac{E_{pt}}{E_{p0}} = \frac{2n_i cos\theta_i}{n_t cos\theta_i + n_i cos\theta_t} = \frac{\sqrt{sin2\theta_i sin2\theta_t}}{sin(\theta_i + \theta_t)cos(\theta_i - \theta_t)} \tag{2.28a}$$

$$t_s \equiv \frac{E_{st}}{E_{s0}} = \frac{2n_i cos\theta_i}{n_i cos\theta_i + n_t cos\theta_t} = \frac{\sqrt{sin2\theta_i sin2\theta_t}}{sin(\theta_i + \theta_t)} \tag{2.28b}$$

where E_{pt} and E_{st} are amplitudes of the refracted p-ray and s-ray, respectively. Therefore, when an optical beam enters from a first medium to a second medium of different refractive index at an angle, its polarization state will change because the relative amplitudes between the p- and s-polarizations are changed. When a beam enters and passes through a parallel glass plate, its polarization state after exiting the plate will also be changed due to the different transmissions between the two polarization components, which will also induce a PDL, as will be discussed in more detail in Chapter 6.

2.A Appendix

Substituting Eqs. (2.5a), (2.5b), (2.11a), and (2.11b) in Eqs. (2.12a) and (2.12b),

$$acos(\omega t + \varphi_0) = E_{x0}cos(\omega t + \varphi_x)cos\psi + E_{y0}cos(\omega t + \varphi_y)sin\psi \tag{2.A.1a}$$

$$\pm bsin(\omega t + \varphi_0) = -E_{x0}cos(\omega t + \varphi_x)sin\psi + E_{y0}cos(\omega t + \varphi_y)cos\psi \tag{2.A.1b}$$

Substituting $cos(\omega t + \varphi_j) = cos\omega t cos\varphi_j - sin\omega t sin\varphi_j$ and $sin(\omega t + \varphi_j) = sin(\omega t)\cos\varphi_j + cos(\omega t)\sin\varphi_j$ ($j = 0, x, y$) in Eqs. (2.A.1a) and (2.A.1b), and comparing the coefficients of $cos(\omega t)$ and $sin(\omega t)$ in Eq. (2.A.1a), one obtains

$$acos\delta' = E_{x0}cos\delta_x cos\psi + E_{y0}cos\delta_y sin\psi \tag{2.A.2a}$$

$$asin\delta' = E_{x0}sin\delta_x cos\psi + E_{y0}sin\delta_y sin\psi \tag{2.A.2b}$$

Similarly, comparing the coefficients of $cos(\omega t)$ and $sin(\omega t)$ in Eq. (2.A.1b), one obtains

$$\pm bcos\delta' = E_{x0}sin\delta_x sin\psi - E_{y0}sin\delta_y cos\psi \tag{2.A.3a}$$

$$\pm bsin\delta' = -E_{x0}cos\delta_x sin\psi + E_{y0}cos\delta_y cos\psi \tag{2.A.3b}$$

Squaring and adding Eqs. (2.A.2a) and (2.A.2b) yields

$$a^2 = E_{x0}^2 cos^2\psi + E_{y0}^2 sin^2\psi + E_{x0}E_{y0}sin2\psi cos\delta \tag{2.A.4a}$$

Similarly, squaring and adding Eqs. (2.A.3a) and (2.A.3b) yields

$$b^2 = E_{x0}^2 sin^2\psi + E_{y0}^2 cos^2\psi - E_{x0}E_{y0}sin2\psi cos\delta \tag{2.A.4b}$$

Adding the two equations earlier yields:

$$a^2 + b^2 = E_{x0}^2 + E_{y0}^2 \tag{2.A.5}$$

where $a^2 + b^2$ and $E_{x0}^2 + E_{y0}^2$ stand for the optical power. Equation (2.A.5) implies the optical power is conserved regardless of the coordinate system chosen, which is expected. Dividing Eq. (2.A.3a) by Eqs. (2.A.2a) and (2.A.3b) by Eq. (2.A.2b) yields

$$\pm\frac{b}{a} = \frac{E_{x0}sin\delta_x sin\psi - E_{y0}sin\delta_y cos\psi}{E_{x0}cos\delta_x cos\psi + E_{y0}cos\delta_y sin\psi} = \frac{-E_{x0}cos\delta_x sin\psi + E_{y0}cos\delta_y cos\psi}{E_{x0}sin\delta_x cos\psi + E_{y0}sin\delta_y sin\psi} \tag{2.A.6}$$

Using Eq. (2.A.6), one may obtain the orientation angle ψ of the ellipse as a function of the electrical field parameters (δ, E_{x0}, E_{y0}) in Eq. (2.6)

$$tan2\psi = \frac{2E_{x0}E_{y0}}{E_{x0}^2 - E_{y0}^2}cos\delta = tan2\beta\, cos\delta \tag{2.A.7a}$$

where β is an auxiliary angle defined as

$$tan\beta = \frac{E_{yo}}{E_{x0}} \tag{2.A.7b}$$

References

Born, M. and Wolf, E. (1980). *Principles of Optics*, 6e. Oxford: Pergamon Press.

Brewster, D. (1815). Experiments on the depolarization of light as exhibited by various mineral, animal and vegetable bodies with a reference of the phenomena to the general principle of polarization. *Philos. Trans.*, Photoelestivity 105: 29–53.

Hecht, E. and Zajac, A. (1974). *Optics*. 4th printing. Addison-Wesley Publishing Co.

Kapron, F., Borrelli, N., and Donald Keck, D. (1972). Birefringence in dielectric optical waveguides. *IEEE J. Quantum Electron.* QE-8 (2): 222–225.

Schlosser, W.O. (1972). Delay distortion in weakly guiding optical fibers due to elliptic deformation of the boundary. *Bell Syst. Tech. J.* 51 (2): 487–492.

Smith, A.M. (1980). Single-mode fiber pressure sensitivity. *Electron. Lett.* 16 (20): 773–774.

Further Reading

Bertholds, A. and Dandliker, R. (1988). Determination of the individual strain-optic coefficients in single-mode optical fibres. *J. Lightwave Technol.* 6 (1): 17–20.

Brewster, D. (1816). On the communication of the structure of doubly-refracting crystals to glass, murite of soda, flour spar, and other substances by mechanical compression and dilation. *Philos. Trans.*, Photoelestivity 106: 156–178.

Goldstein, D. (2003). *Polarized Light*, 2e. New York: Marcel Dekker, Inc.

Kumar, A. and Ghatak, A. (2011). *Polarization of Light with Applications in Optical Fibers*. Tutorial texts in Optical Engineering, Volume TT90. SPIE Press.

3

Polarization Effects Unique to Optical Fiber Systems

3.1 Polarization Variation in Optical Fibers

Unlike in free-space, the state of polarization (SOP) of the light propagating in a single mode (SM) optical fiber is generally not stable (Leo et al. 2003; Peterson et al. 2004; Boroditsky et al. 2005; Krummrich et al. 2005). In fact, it is quite sensitive to almost any disturbances on the fiber, such as pressing, bending, twisting, and even touching the fiber. It is also sensitive to temperature and pressure variations. As shown in Figure 3.1, the SOP of the light changes rapidly as it propagates in the fiber due to different perturbations along the fiber. When the perturbations change with time, so does the SOP. Many non-contacting perturbations, such as acoustic vibration, temperature, magnetic field, and electrical field, can also cause SOP variations of the light propagating inside the fiber. In general, SOP variations induced by the temperature fluctuations are slow, depending on the rate of temperature change. On the other hand, SOP variations induced by vibration have relatively high changing rates, such as those caused by wind blowing on aerial fiber cables installed on utility poles or by train vibration on fiber cables buried along the railway tracks. Mechanical shock, such as a screw driver dropping on a spool of dispersion compensating fiber, can cause even faster SOP variations, at rates up to 280 krad/s (280×10^3 rad/s) (Krummrich and Kotten 2004). The fastest SOP variations are caused by direct lightning strikes on aerial fiber cables, with rates estimated to be more than 2 Mrad/s (2×10^6 rad/s) (Kuschnerov and Herrmann 2016). Here the SOP changing rate is defined as the change of the solid angle of an SOP point on the Poincaré sphere per second.

The root cause of the SOP variation in a single mode fiber can be attributed to birefringence induced by various perturbations randomly distributed at different locations along the fiber. Therefore, SOP variations of light transmitted through a long length optical fiber are generally random, covering the whole Poincaré sphere, as shown in Figure 3.2a. The rates of the SOP variations in long haul optical fiber communication systems have been found to follow the Rayleigh distribution (Leo et al. 2003; Boroditsky et al. 2005; Krummrich et al. 2005), as shown in Figure 3.2b. Of cause, lightning strikes on aerial fiber cables during a thunder storm may induce sudden surges of electrical current flowing in the metal armor coiled around the fiber cables or huge electrical field across the fiber, causing extremely fast SOP variations, which in turn may cause disruption of internet services.

In the absence of magnetic and electrical fields as in majority situations, the birefringence is almost all induced by the photoelastic effect discussed in Section 2.7, which is extremely sensitive to transverse stresses on the fiber, whether it from bending, twisting, or pressure on the fiber. Although the temperature change itself does not directly contribute to birefringence, it affects the stress level inside the fiber due to any differential thermal expansion not symmetric about the

Polarization Measurement and Control in Optical Fiber Communication and Sensor Systems, First Edition.
X. Steve Yao and Xiaojun (James) Chen.

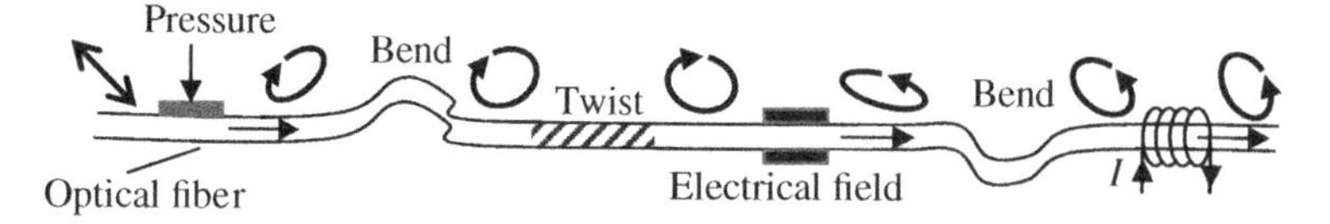

Figure 3.1 Illustration of SOP variations along a single mode fiber under different perturbations at different locations. The SOPs at different locations will also vary with time.

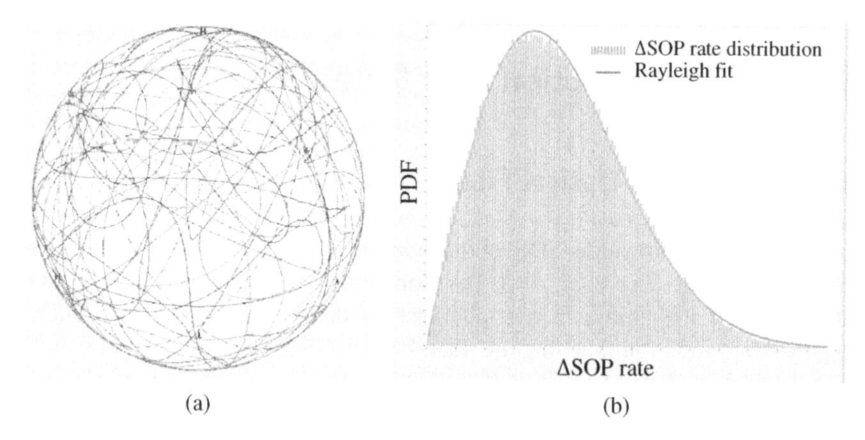

Figure 3.2 Typical SOP variations in a long haul optical fiber link: (a) Poincaré sphere presentation of the SOP variation, (b) the corresponding SOP changing rate ($\Delta SOP/\Delta t$) distribution largely following a Rayleigh function.

fiber axis, which in turn influence the birefringence induced by such internal or external stresses via the photoelastic effect.

The rapid SOP variations caused by lightning strikes are attributed to the optical Kerr effect (Stolen and Ashkin 1973) and the Faraday effect (Simon and Ulrich 1977). The optical Kerr effect describes the linear birefringence induced by an electrical field, while the Faraday Effect describes circular birefringence induced by magnetic field. More details will be discussed in Section 3.3.

3.2 Polarization Eigenmodes in a Single Mode Optical Fiber

Excellent discussions on the origin of polarization variations in optical fibers were provided by S.C. Rashleigh (1983), some of which will be adopted in this section and in Sections 3.3 and 3.4. A single mode fiber only supports the lowest order mode HE_{11} (or LP_{01}), which can be arbitrarily chosen to have its transverse electric field polarized predominantly along the along the x-direction, as shown in Figure 3.3a, for the case that no birefringence is present in the fiber. Note that although the electrical field vectors away from the center of the fiber core are curved with certain y-direction components, the average field vector is linear in the x-direction, because y-components are canceled out due to the symmetrical distribution about the x-axis. The orthogonal polarization along the y-direction shown in Figure 3.3b is an independent mode, with the average field vector aligned in the y-direction. For convenience, these two orthogonal modes constitute the two polarization eigenmodes of the single mode fiber. Any other pair of orthogonal polarization modes, such as right-hand circular and left-hand circular polarizations (RCP and LCP), can also be chosen as the polarization modes.

Figure 3.3 Electrical field distributions of the x-polarized (a) and y-polarized (b) light of the fundamental HE_{11} mode in a single mode fiber.

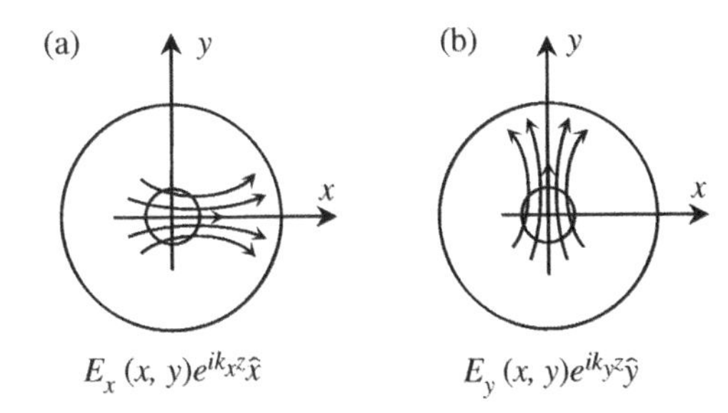

For the cases with birefringence present in the fiber, the polarization eigenmodes can be chosen for convenience to be linear along the birefringence directions x and y if the birefringence is of linear, or to be RCP and LCP if the birefringence is of circular. In a lossless single mode fiber with linear birefringence, the electrical field vector of a monochromatic light wave of an arbitrary polarization propagating along the fiber (the z-direction) can be represented as the linear superposition of two linear polarization eigenmodes as (Ulrich 1977; Ulrich and Simon 1979)

$$\mathbf{E}(x,y,z) = c_x E_x(x,y)e^{ik_xz}\hat{\mathbf{x}} + c_y E_y(x,y)e^{ik_yz}\hat{\mathbf{y}} \tag{3.1a}$$

where c_x and c_y, $E_x(x, y)$ and $E_y(x, y)$, $k_x = \left(\frac{2\pi}{\lambda_0}\right) n_x$ and $k_y = \left(\frac{2\pi}{\lambda_0}\right) n_y$, $\hat{\mathbf{x}}$ and $\hat{\mathbf{y}}$, n_x and n_y are the complex amplitude coefficients, spatial variations, propagation constants, unit vectors, and effective indexes of the two linear polarization eigenmodes, respectively.

Similarly, in a lossless single mode fiber with circular birefringence, the electrical field vector of a monochromatic light wave of an arbitrary polarization propagating along the fiber can be represented as the linear superposition of two circular polarization eigenmodes as

$$\mathbf{E}(x,y,z) = c_r E_r(x,y)e^{ik_rz}\hat{\mathbf{r}} + c_r E_l(x,y)e^{ik_lz}\hat{\mathbf{l}} \tag{3.1b}$$

where c_r and c_l, $E_r(x, y)$ and $E_l(x, y)$, $k_r = \left(\frac{2\pi}{\lambda_0}\right) n_r$ and $k_l = \left(\frac{2\pi}{\lambda_0}\right) n_l$, $\hat{\mathbf{r}}$ and $\hat{\mathbf{l}}$, and n_r and n_l are the complex amplitude coefficients, spatial variations, propagation constants, unit vectors, and effective indexes of the RCP and LCP polarization eigenmodes, respectively.

The SOP of any of the polarization modes is described by the average direction of its electric field vector and is linear in either x- or y-direction, and the SOP of the propagating light wave is determined by the complex ratio $(c_x/c_y)e^{-i(k_y-k_x)z}$ or $(c_r/c_l)e^{-i(k_l-k_r)z}$. In ideal fibers with perfect rotational symmetry, the two modes are degenerate with $k_x = k_y$ (or $k_r = k_l$) and any SOP the fiber would stay unchanged as it propagates along the fiber. In practice, imperfections, such as asymmetrical lateral stress, and noncircular core, break the circular symmetry of the ideal fiber and lift the degeneracy of the two modes. They propagate with different phase velocities and the difference between their effective refractive indexes is the fiber birefringence

$$\Delta n_{BL} = n_y - n_x \text{ (for linear birefringence)} \tag{3.2a}$$

$$\Delta n_{BC} = n_r - n_l \text{ (for circular birefringence)} \tag{3.2b}$$

$$\Delta \beta_{Bi} = \left(\frac{2\pi}{\lambda_0}\right) \Delta n_{Bi} \tag{3.2c}$$

where the index i is either L or C for linear and circular birefringence, respectively, and the x-mode or RCP mode is taken to be the fast mode. If light is injected into the fiber so that both modes are excited, one will slip in phase relative to the other as they propagate. When this phase difference is an integral number of 2π, the two modes will beat and, at this point, the input polarization state will be reproduced. Therefore the effect of a (uniform) birefringence is to cause a general polarization state to evolve through a periodic sequence of states as it propagates. The length over which this beating occurs is the fiber beat length L_{Bi}

$$L_{Bi} = \frac{2\pi}{\Delta\beta_{Bi}} = \frac{\lambda_0}{\Delta n_{Bi}} \tag{3.2d}$$

Early single-mode fibers have beat lengths around 10 cm, while more recent single mode fibers have beat lengths around 2 m, although they may not necessarily be uniform along the fiber. Such beat lengths correspond to a linear birefringence in the range of $10^{-7} < \Delta n_{BL} < 10^{-5}$. This birefringence is a factor of 10^{-4} to 10^{-2} smaller than the index difference between the core and cladding regions of typical single-mode fibers.

3.3 Birefringence Contributions in Optical Fibers

As mentioned previously, in an ideal optical fiber with perfect circular symmetry, there is no birefringence. Birefringence is introduced whenever the circular symmetry of the ideal fiber is broken, producing an anisotropic refractive index distribution in the core region. This asymmetry may either come from the geometrical deformation of the fiber core, or from the material anisotropy induced by a mechanical stress via the photo-elastic effect, by a lateral displacement of the light-field induced dipole moments in the fiber material (SiO_2) due to the Lorentz force via the magneto-optic effect (the Faraday Effect), or by a lateral displacement of the light-field induced dipole moments in the fiber material due to the external electric field via the electro-optic effect (the optical Kerr effect).

Figure 3.4a–l shows the basic mechanisms for introducing different birefringences in the fiber. The first four are internal to the fiber, being introduced during fiber fabrication, while the remainder results from some external actions on the fiber. Among the first four, (a) is due to the non-symmetric fiber core geometry, while the other three are due to internal lateral stress induced by the non-circular cladding or by stress rods with different thermal coefficients as that of the fiber cladding and the core. The majority of the mechanisms introduce a linear birefringence, while only two (the twist and magnetic field) introduce a circular birefringence. In the following discussions, the linear birefringence Δn_j is defined as $\Delta n_j = n_{jx} - n_{jy}$, where $j = a, b, \ldots l$, x is the horizontal axis and y is the vertical axis, as shown in the first row of Figure 3.4. $\Delta n_j > 0$ indicates the slow axis is in the x direction or the fast axis is in the y direction, while $\Delta n_j < 0$ indicates the opposite.

3.3.1 Noncircular Core

The geometrical anisotropy of a noncircular core represented by a major and a minor core dimensions $2a$ and $2b$ shown in Figure 3.4a introduces a linear birefringence in the fiber. The light travels fastest when polarized along the direction of the smallest transverse dimension $2b$ of the core and often referred to as the fast axis of the birefringence. Generally, this birefringence cannot be represented by a simple expression as it depends strongly on the wavelength at which the fiber is being operated. However, if the operating wavelengths are near the higher mode cutoff of the fiber, the induced linear birefringence Δn_{a1} for single-mode fibers with small-core ellipticity can be approximated by

$$\Delta n_{a1} \approx 0.2\left(\frac{a}{b} - 1\right)(\Delta n_{cc})^2 \tag{3.3a}$$

where Δn_{cc} is the refractive index difference between the core and the cladding of the fiber and $(a/b - 1) \ll 1$ was assumed. It is evident that the induced birefringence is very sensitive to the fiber

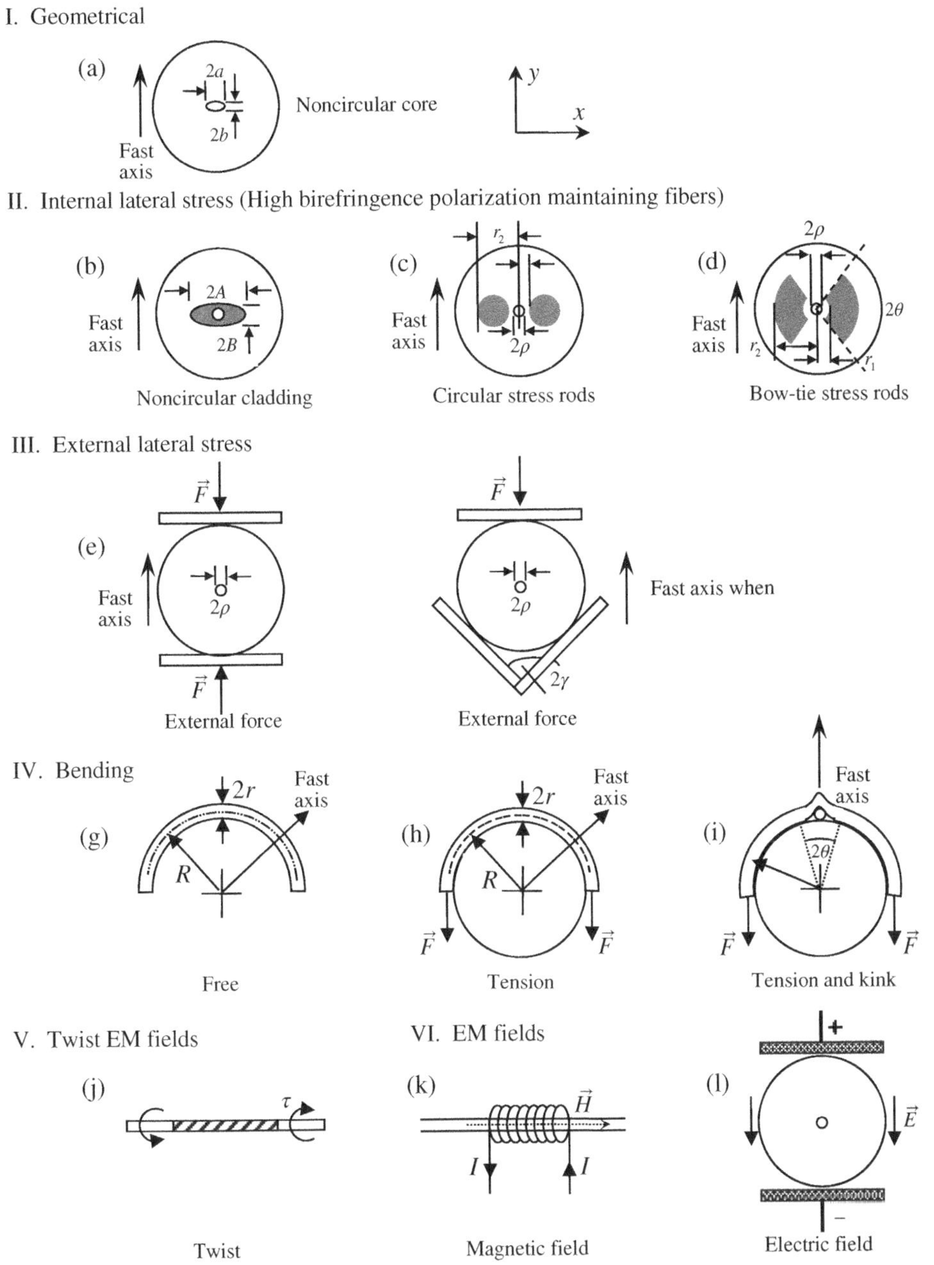

Figure 3.4 Mechanisms for introducing birefringence inside a single mode fiber.

core ellipticity $(a/b-1)$. It can be seen from Eq. (3.2d), in order to fabricate a low birefringence single-mode fiber with a beat length L_B of 50 m at 1550 nm, the birefringence Δn_{a1} must be less than 0.31×10^{-7}, corresponding to an ellipticity of the fiber core of 0.62% from Eq. (3.3a), assuming a typical core-cladding refractive index difference of $\Delta n_{cc}=5\times10^{-3}$.

For fiber cores with large ellipticity for making a high birefringence (HiBi) fiber, $(a/b-1)>1$, the induced birefringence is no longer sensitive to the ellipticity. For the ellipticity in the range of

$2 < (a/b - 1) < 6$, the maximum birefringence which can be introduced by the elliptical core at the wavelengths below the higher order mode cutoff can be approximated as

$$\Delta n_{a2} \approx 0.25(\Delta n_{cc})^2 \tag{3.3b}$$

To make an HiBi fiber with a beat-length L_B of 2 mm at 1550 nm, the birefringence Δn_{a2} should be at 7.8×10^{-4} from Eq. (3.2d), requiring a large core–cladding refractive index difference of 5.6%. With such a large core-cladding index difference, the core dimensions must be extremely small for maintaining single-mode transmission in the fiber, on the order of 2 μm for $2a$ and 0.5 μm or less for $2b$, which is not practical.

3.3.2 Internal Lateral Stress

Any asymmetrical transverse stress introduces a linear birefringence via photo-elastic effect. The stress may be frozen internally in the fiber during fabrication as the result of different thermal expansions between differently doped noncircularly symmetric regions of the fiber, as shown in Figure 3.4b–d. The typical thermal expansion coefficient of the fused silica (SiO_2) for making the fiber is $\alpha_{si} = 5.4 \times 10^{-7}$/°C and those of the doping materials are $\alpha_G = 7 \times 10^{-6}$/°C for GeO_3, $\alpha_B = 10 \times 10^{-6}$/°C for B_2O_3, and $\alpha_P = 14 \times 10^{-6}$/°C for P_2O_5 (Guan et al. 2005, pp. 240–254). The HiBi optical fibers made with internal lateral stresses described next are generally called polarization maintaining (PM) fiber because when an input light with a linear polarization is aligned with one of the birefringence axes of the fiber, it can propagate through the fiber without changing its SOP. Such fibers are extremely useful for sensor systems, such as fiber optic gyroscopes (FOGs), and for pigtailing polarization sensitive devices, such as electro-optic modulators, polarizers, and polarization beam splitters.

3.3.2.1 Elliptical Cladding

In the case of Figure 3.4b, there are two claddings, one is a more heavily doped elliptical inner cladding constrained by an outer (usually silica) cladding to produce the stress asymmetry. If the elliptical cladding region has major and minor diameters of $2A$ and $2B$, respectively, and the core is perfectly circular, the induced birefringence can be expressed as (Chu and Sammut 1984; Guan et al. 2005; Tsai et al. 1991)

$$\Delta n_b = C_\sigma \Delta\alpha_b \Delta T \frac{(A - B)}{(A + B)} \left[1 - \frac{3}{2}\frac{AB(A + B)}{b^3}\right] \tag{3.4}$$

where b is fiber cladding radius, $\Delta\alpha_b = \alpha_{1b} - \alpha_{2b}$ is the difference between the thermal expansion coefficient of the inner elliptical cladding α_{1b} and that of the outer cladding α_{2b}, $\Delta T = T_o - T_s$ is the difference of the fiber operating temperature T_o and the softening (or drawing) temperature T_s of the heavily doped inner cladding material, which is always a negative number, and C_σ is a proportional constant relating to the photo-elastic effect:

$$C_\sigma = \frac{1}{2} n_0^3 (p_{11} - p_{12}) \frac{(1 + \nu_p)}{(1 - \nu_p)} \tag{3.5}$$

In Eq. (3.5), n_0 is the average refractive index of the fiber, p_{11} and p_{12} are the components of the strain-optical tensor, and ν_p is Poisson's ratio of the fiber material. For fused silica, $n_0 = 1.46$, $p_{11} = 0.12$, $p_{12} = 0.27$, and $\nu_p = 0.17$, resulting in $C_\sigma \approx -0.23$, a negative number. It can be seen from Eq. (3.4), the larger the ellipticity A/B, the stronger the birefringence. The last term of Eq. (3.4) is approximately 0.54 when taking $A = 50$ μm, $B = 10$ μm, and $2b = 125$ μm.

If the minor radius B of the inner elliptical cladding is much larger than that of the fiber core ρ, i.e. $B \gg \rho$, the stress induced birefringence Δn_b is constant everywhere within the elliptical cladding and the mode field is totally confined within the elliptical cladding.

When $\Delta\alpha > 0$, indicating that the thermal expansion coefficient α_{1c} of the elliptical cladding is larger than that of the outer circular cladding α_{2c}, the fast axis is along the minor ellipse diameter (y direction in Figure 3.4b) as in the case when the elliptical cladding is doped with phosphorus, boron, or germanium. When $\Delta\alpha < 0$, as in the case for fluorine doping in the elliptical cladding, the fast axis is along the major ellipse diameter (x direction). The typical birefringence produced with the elliptical cladding is around 1×10^{-4} and the PM fiber with the elliptical cladding is often referred to as Tiger-eye fiber or Tiger fiber, probably due to the similarity of its shape to the eyes of a tiger. iXblue in France is among very few companies in the world making such PM fibers.

3.3.2.2 Circular Stress Rods

As shown in Figure 3.4c, there are two circular rods with different thermal expansion coefficient from that of the cladding and core are embedded in the fiber cladding. The differential thermal expansion produces a stress in the fiber across the fiber core after the fiber is cooled from the drawing temperature T_s around 1000 °C to the operation temperature T_0, which is frozen inside the fiber as long as the fiber is below the softening temperature of the fiber material. The stress in turn produces a birefringence Δn_c in the fiber core via the photoelastic effect which can be described by the following expression (Guan et al. 2005; Chu and Sammut 1984; Tsai et al. 1991):

$$\Delta n_c = C_\sigma \Delta\alpha_c \Delta T \left[4\left(\frac{r_2 - r_1}{r_2 + r_1}\right)^2 - \frac{3}{4}\left(\frac{r_2^2 - r_1^2}{b^2}\right)^2 \right] = C_\sigma \Delta\alpha_c \Delta T \left(\frac{d_1}{d_2}\right)^2 \left[1 - 3\left(\frac{d_2}{b}\right)^4\right] \tag{3.6}$$

where $\Delta\alpha_c$ is the difference between the thermal expansion coefficient of the stress rods and that of the fiber cladding, r_1 and r_2 are the distances of the inner and outer faces of the stress rods from the fiber core, as marked on Figure 3.4c, b is the diameter of the fiber cladding, d_1=$(r_2 - r_1)/2$ is the radius of the stress rods, and d_2=$(r_2 + r_1)/2$ is the distance of a stress rod from the fiber core center. It can be seen from Eq. (3.6) that birefringence increases as the stress rods get larger in size or get closer to the fiber core. However, in practice the rods cannot be too close to the fiber core because this will induce large attenuation of light propagating in the fiber, and therefore a balance must be reached when designing the fiber. Typical fiber parameters are $2b = 125$ μm, $r_1 = 10$ μm, and $r_2 = 50$ μm and the typical birefringence can be realized with the stress rods approach is on the order of 5×10^{-4}, with the slow axis along the direction connecting the center of the two stress rods (meaning that the refractive index along this direction is larger than that along the orthogonal direction) (Guan et al. 2005). Probably because the stress rods look like Panda eyes, the resulting PM fibers are often nicknamed PANDA fibers. Nowadays most PM fibers in the world are produced with the circular stress rods approach and are therefore are of PANDA type.

3.3.2.3 Bow-Tie Shaped Stress Rods

It is not difficult to imagine that birefringence in a fiber can also be produced by stress rods of other shapes, as long as the rods have differential thermal expansion from that of the fiber cladding. Figure 3.4d illustrates the cross section of a bow-tie PM fiber in which the birefringence is induced by two bow-tie shaped stress rods. At the center of the fiber core, the birefringence Δn_d can be expressed as (Guan et al. 2005; Chu and Sammut 1984; Tsai et al. 1991)

$$\Delta n_d = C_\sigma \Delta\alpha_d \Delta T \sin 2\theta \left[2 ln\left(\frac{r_2}{r_1}\right) - \frac{3}{2b^4}\left(r_2^4 - r_1^4\right) \right] \tag{3.7}$$

where $\Delta\alpha_d$ is the difference between the thermal expansion coefficient of the bow-tie rods and that of the fiber cladding, r_1 and r_2 are the distances of the inner and outer faces of the bow-tie rods from the fiber core center, respectively, as marked on Figure 3.4d, 2θ is the sector angle of the bow-tie rods, and b is the radius of the fiber cladding. As can be seen from Eq. (3.7), the closer the bow-tie rods to the fiber core, the larger the induced birefringence; the larger the rods, the larger the induced birefringence is. Again, the rods cannot be too close to the fiber core because they will introduce high transmission loss in the fiber core. Typical bow-tie fibers have the following parameters: $2b = 125\ \mu m$, $r_1 = 7.5\ \mu m$, $r_2 = 47.5\ \mu m$, $2\theta = 90°$, with a typical birefringence on the order of 5×10^{-4} and having the slow axis along the direction connecting the center of the two stress rods (Guan et al. 2005). Fibercore in England is one of the few companies in the world making bow-tie PM fibers.

3.3.3 External Lateral Stress

3.3.3.1 Fiber Between Parallel Plates

As illustrated in Figure 3.4e, pressing or squeezing a fiber between a pair of parallel plates will produce a birefringence via the photo-elastic effect, which can be described by the following expression (Smith 1980a):

$$\Delta n_e = 4C_\sigma(1 - \nu_p)\frac{f}{\pi Ed} \tag{3.8}$$

where E is Young's modulus of the fiber material ($E = 7.75\times10^9\ \text{kg/m}^2$ for fused silica), d is the total fiber diameter (including coating), and f is the line force (defined as the force applied to the fiber divided by the length of the fiber subject to the force) applied in the y direction. The positive value of Δn_e indicates that the fast axis of the birefringence is in the direction of applied compression force, or vertical in the case of Figure 3.4e. The magnitude of such squeezing induced birefringence is on the order of 10^{-7} to 10^{-4}, depending on the size of the applied line force, as will be described in Chapter 10.

3.3.3.2 Fiber in an Angled V-Groove

As illustrated in Figure 3.4f, pressing a fiber in an angled V-groove in the y direction with a flat plate can also produce a birefringence via the photoelastic effect, however, with a much reduced value depending on the V-groove angle 2γ. This birefringence Δn_f can be described by the following expression (Kumar and Ulrich 1981; Maystre and Bertholds 1987):

$$\Delta n_f = -2C_\sigma(1 - \nu_p)\left(1 - \frac{\cos 2\gamma}{\sin\gamma + \mu\cos\gamma}\right)\frac{f}{\pi Ed} \tag{3.9a}$$

where μ is the friction coefficient between the fiber and the walls of the V-groove. In Eq. (3.9a), the force is in the y direction and a minus sign indicates that the fast axis is in the x direction. In the absence of friction, the induced birefringence can be simplified to

$$\Delta n_f = -2C_\sigma(1 - \nu_p)\left(1 - \frac{\cos 2\gamma}{\sin\gamma}\right)\frac{f}{\pi Ed} \tag{3.9b}$$

It can be seen from Eq. (3.9b) that the induced birefringence is zero when the V-groove angle 2γ is 60°. When the V-groove angle 2γ is less than 60°, $\cos 2\gamma < \sin\gamma$, resulting in a minus value of Δn_f or the fast axis is in the x direction. On the other hand, when the V-groove angle 2γ is larger than 60°, $\cos 2\gamma > \sin\gamma$, resulting in a positive value of Δn_f or the fast axis is in the y direction.

In the other limit of complete friction, $\mu cos\gamma \gg sin\gamma$ in Eq. (3.9a), the friction coefficient μ can be written as (Kumar and Ulrich 1981)

$$\mu = \frac{1}{tan\gamma} \tag{3.9c}$$

Substituting Eq. (3.9c) in Eq. (3.9a) yields

$$\Delta n_f = -2C_\sigma(1 - v_p)(1 - cos2\gamma \cdot sin\gamma)\frac{f}{\pi Ed} \tag{3.9d}$$

It can be seen from Eq. (3.9d) that when a large friction is present, such as the case when fiber coating is not removed, it is impossible to avoid birefringence when pressing the fiber into a V-groove. More detailed theory and experiments on the subject will be discussed in Chapter 10. In particular, if a fiber is clamped in a pair of identical V-grooves, the optimal groove-angle for a zero clamp-force induced birefringence is 90°.

3.3.4 Fiber Bending

3.3.4.1 Pure-Bend

When a fiber is freely bent, as shown in Figure 3.4g, the outer portion of the fiber is in tension, while the inner portion of the fiber is in compression. Such a differential tension produces a second-order transverse stress (Ulrich et al. 1980; Smith 1980b) in the fiber, which in turn induces a linear birefringence:

$$\Delta n_g = \frac{1}{2}C_\sigma(1 - v_p)\left(\frac{r}{R}\right)^2 \tag{3.10a}$$

where r is the radius of the fiber, R is the radius of the bending, and the fast axis of this induced birefringence is in the radial direction of the bending. For a standard telecommunication fiber (SMF-28) with a diameter $2r$ of 250 μm, this bending induced birefringence at a wavelength of 1550 nm has been accurately measured to be (Feng et al. 2018)

$$\Delta n_g = 6.61 \times 10^{-10}\left(\frac{1}{R}\right)^2 \tag{3.10b}$$

where R has a unit of meter. The details of the measurement method will be discussed in Chapter 10. For a large coil with a diameter R of 10 m, the birefringence induced in the standard single mode fiber is 6.61×10^{-8}, corresponding to a beat length of 235 m. For a small coil with a radius R of 1 cm, the induced birefringence is 6.61×10^{-6}, corresponding to a beat length of 23.5 cm.

3.3.4.2 Bending with Tension

Winding a fiber with an axial tension F around a drum introduces a linear birefringence, in addition to the pure bending birefringence of Eq. (3.10a), as shown in Figure 3.4h. This situation corresponds to the cases of winding fiber coils for applications such as FOGs, hydrophones, or electric current sensors. Such a birefringence Δn_h induced by bending with tension results from the lateral force exerted by the drum on the fiber in reaction to the tensile force F and can be expressed as (Rashleigh and Ulrich 1980)

$$\Delta n_h = C_\sigma(2 - 3v_p)\frac{r}{R}\sigma_F \tag{3.11a}$$

$$\sigma_F = \frac{F}{\pi r^2 E} \tag{3.11b}$$

where σ_F is the mean axial strain in the fiber induced by the tension F. For a moderately large axial tension $\sigma_F = 0.5\%$ applied to a large diameter coil with a diameter $2R$ of 1 m, the birefringence Δn_h from Eq. (3.11a) far exceeds the pure bending induced birefringence Δn_g of Eq. (3.10a). For a small diameter coil with a diameter $2R$ of 1 cm and an axial strain σ_F of 0.5%, the values of Δn_g and Δn_h are comparable.

3.3.4.3 Bending with a Kink

When a fiber is wound with tension F on a drum having a sharp ridge with a height of H, as shown in Figure 3.4i, the transverse stresses are re-distributed over the short fiber length in the vicinity of the ridge to introduce additional birefringence adding to both Δn_g and Δn_h described earlier (Ulrich and Rashleigh 1982). This corresponds to the situation of winding a fiber coil on a drum, during which the fiber on the upper layer is crossed on the fiber in the lower layer. As will be discussed in Chapter 10, not only such a kink causes a sharp local birefringence that induces polarization crosstalk in a PM fiber coil, but also temperature instability of the phase of the lightwave propagating inside the fiber as the temperature is changed, eventually degrading the performance of the system deploying the fiber coil.

It can be shown (Ulrich and Rashleigh 1982) that this kink birefringence Δn_i results from both the increased curvature of the fiber in the vicinity of the kink and the localized lateral force F_0 exerted by the ridge on the fiber. Δn_i varies along the fiber length $2\theta R$ corresponding to the arc angle 2θ in Figure 3.4i and vanishes outside this region. The actual dependence of Δn_i along this fiber length is complicated and its average value $\overline{\Delta n_i}$ over the length $2\theta R$ can be used to quantitatively describe its value, although the detailed spatial information of Δn_l is lost.

$$\overline{\Delta n_i} = C_\sigma \sqrt{\frac{HF}{2\pi RE}} = \frac{1}{2} C_\sigma \sqrt{\frac{H\sigma_F}{R}\frac{r}{R}} \tag{3.12}$$

where $HRF \gg E\pi r^4/8$ is assumed. Although this $\overline{\Delta n_i}$ exists over only a very short fiber length of $2\theta R$, it can be quite large. For example, for $H = 2r = 100\ \mu\text{m}$, $\overline{\Delta n_i} \approx 1.35 \times 10^{-8}$ when $2R = 1$ m, and $\overline{\Delta n_i} \approx 1.35 \times 10^{-5}$ when $2R = 1$ cm. In compact fiber coils, the number of kinks can be very large when the coils are not properly wound and each such kink can introduce a polarization crosstalk peak, as will be discussed in Chapter 10.

3.3.5 Fiber Twist

Twist is one of the imperfections present on any real optical fiber, which can be produced several ways either intentionally or unintentionally, such as fiber rotation during the fiber drawing process, winding the fiber on a spool, rotation motion when connecting two fibers, fiber arrangement on an optical table, etc. The influence of twist on the polarization is intimately linked with the linear birefringence already existing in the fiber due to deviations from a circular shape of the fiber core and internal or external stresses described earlier. Three typical situations may be distinguished, depending on the relative magnitudes of the twist rate τ (rad/m) measured in number of turns per unit length and the linear birefringence Δn_{jL} or the linear retardation per unit length $\Delta\beta_{jL} = 2\pi\Delta n_{jL}/\lambda$ (Rashleigh 1983).

In the case of weak twist, $|\tau| \ll |\Delta\beta_L|$, the linear birefringence is dominant and the birefringence axes are simply rotated by the twist. With respect to the birefringence axes, the polarization evolves essentially identically as it would do in the non-twisted fiber, or the polarization is simply twisted with the fiber. For example, for a PM fiber with a large linear birefringence, the polarization aligned with the slow (or fast) axis simple rotates with the slow (fast) axis when the fiber is twisted (Simon and Ulrich 1977; Ulrich and Simon 1979).

At medium twist rate, $|\tau| \approx |\beta_{jL}|$, the share strain in the twisted fiber gives rise to an optical activity with a circular birefringence Δn_{jC} or a circular retardation per unit length $\Delta\beta_{jC} = 2\pi\Delta n_{jC}/\lambda$ proportional to the twist,

$$\Delta\beta_{jC} = g\tau \tag{3.13a}$$

$$\Delta n_{jC} = n_{rc} - n_{lc} = \frac{\lambda}{2\pi} g\tau \tag{3.13b}$$

where n_{rc} and n_{lc} are the refractive indexes of the RCP and LCP eigen polarizations, respectively, and g is the gyration coefficient expressed as

$$g = -n_0^2 p_{44} = -\frac{1}{2} n_0^2 (p_{11} - p_{12}) \tag{3.13c}$$

For a fused silica fiber, $\frac{1}{2}(p_{11} - p_{12}) = -0.075$ and $n_0 \approx 1.46$, $g \approx 0.16$. Therefore, the major axis of an elliptically polarized light will be rotated at a rate of α_j (rad/m) along a twisted optical fiber with a twist rate of τ (rad/m) given by

$$\alpha_j = \frac{1}{2} g\tau \tag{3.14}$$

As an example, from Eq. (3.14) one can see that when a fiber is rotated 10°, the polarization rotates 0.8°. From Eq. (3.13b), a twist of 10 turns/m in a fiber operating at a wavelength of 1.5 μm, the circular birefringence induced by the twist is 2.4×10^{-6}. A positive value of g indicates that the left circular component propagates faster. If the fiber twist is counterclockwise (with respect to an observer looking against the propagation direction of light), the polarization rotation is also counterclockwise, and vice versa. In practice, in combination with the linear birefringence, elliptical birefringence results, causing rather complicated polarization evolution when light is propagating in the fiber, which will be discussed in detail in Chapter 4.

At strong twist rate, $|\tau| \gg \Delta\beta_L$, the situation becomes simple again because the polarization evolution is dominated by the induced circular birefringence Δn_C. The large twist may be introduced by spin twisting the fiber preform at a high rate while drawing the fiber. The preform used can be either that for making a SM fiber or that for making a PM fiber; however, the latter is more widely adopted in the industry because the resulting spun fiber is much less sensitive to bending induced birefringence than that made with the former (Laming and Payne 1989). The spun fibers made with the SM fiber preform and PM fiber preform are often referred to as the low birefringence (LoBi) and HiBi spun fibers, respectively. When it is not spin twisted during the drawing process, the PM fiber perform can produce a PM fiber with a large linear birefringence (defined as the unspun linear birefringence Δn_0). However, with the preform spinning during the drawing process, the average linear birefringence of the resulting spun fiber over a length much larger than the spin pitch is expected to approach zero, although the local linear birefringence still remains at the level of the unspun linear birefringence (Laming and Payne 1989). In fact, the large local linear birefringence renders the HiBi spun fiber less susceptible to bend and stress perturbations, while the near-zero average linear birefringence helps to preserve the linear relationship between the Faraday rotation angle and the longitudinal magnetic field, which is critical for electrical current and magnetic field sensing applications. For a HiBi spun fiber, the residual linear birefringence $\delta_j(L)$, and the polarization rotation angle $\varphi_j(L)$ after a fiber length of L can be expressed as (Barlow et al. 1981; Laming and Payne 1989)

$$\delta_j(L) = 2\sin^{-1}\left\{\frac{1}{\sqrt{1+(2\tau/\Delta\beta_0)^2}}\sin\frac{\sqrt{\Delta\beta_0^2+4\tau^2}}{2}L\right\} \tag{3.15a}$$

$$\theta_j(L) = \tau L - \tan^{-1}\left\{\frac{2\tau/\Delta\beta_0}{\sqrt{1+(2\tau/\Delta\beta_0)^2}}\tan\frac{\sqrt{\Delta\beta_0^2+4\tau^2}}{2}L\right\} + n\pi \tag{3.15b}$$

where $\Delta\beta_0 = 2\pi\Delta n_0/\lambda$ is the original retardation per unit length of the unspun PM fiber. Detailed analysis and measurement of spun optical fibers can be found in Chapter 4 and more in Chapter 9, respectively.

3.3.6 Electrical and Magnetic Fields

3.3.6.1 Axial Magnetic Field (The Faraday Effect)

As shown in Figure 3.4k, in a magneto-optical material, such as a silica fiber, a magnetic field H can produce a circular birefringence $\Delta n_{kC} = n_{rc} - n_{lc}$ via the Faraday effect, which produces a retardation $\Delta\beta_{kC}L$ (Simon and Ulrich 1977):

$$\Delta\beta_{kC}L = \frac{2\pi\Delta n_{kC}}{\lambda} = 2V_F HL \tag{3.16a}$$

where V_F is the Verdet constant of the material and L is the length of the material subject to the magnetic field H. Such a circular retardation will cause a linear polarization to rotate by an angle of $\varphi_K(L)$ in case there is no linear birefringence in the fiber:

$$\theta_K(L) = V_F HL \tag{3.16b}$$

It is important to note that Faraday rotation is non-reciprocal with respect to the magnetic field, that is, the direction of polarization rotation depends on the direction of the magnetic field H observed facing the light propagation direction. If the polarization rotates θ_K degrees counter-clockwise when the light propagates in the direction of H, it will rotate $-\theta_K$ degrees clockwise if light reverses its propagation direction. The net rotation angle with respect to a fixed reference is, therefore, $2\theta_K$. This nonreciprocal property is frequently used in making optical isolators and circulators, as well as optical magnetic and current sensors, as will be discussed further in Chapters 6, 7, and 9.

Optical cables for telecommunications generally have a metal armor coiled around multiple optical fibers for protection, as well as for transporting electrical power. As mentioned in Section 3.1, lightning strikes during thunder storms can induce large current surges flowing in the metal armor and thus produce surges of magnetic field along the fiber, resulting in extremely fast polarization rotations of the optical signals propagating in the fiber via the Faraday effect (Kuschnerov and Herrmann 2016). Such fast polarization jumps may disrupt the operation of polarization locking algorithm in a coherent communication system and eventually cause data service interruptions.

3.3.6.2 Transversal Electrical Field

As illustrated in Figure 3.4l, when a length of optical fiber passes through a pair of parallel plates with an applied electrical field E_l, a linear birefringence Δn_l will be introduced in the fiber via the electro-optic Kerr effect (Stolen and Ashkin 1973). The corresponding retardation is given by (Simon and Ulrich 1977)

$$\Delta\beta_l L = \frac{2\pi\Delta n_l}{\lambda} L = 2\pi B_l E_l^2 L \tag{3.17}$$

where B_l is the Kerr electrooptic constant of the fiber material. For fused silica fibers, B_l is too small for practical applications, except for birefringence measurements via polarization modulation. However, the huge electrical fields produced by lightning strikes across the optical fiber can induce fast birefringence changes, which in turn cause rapid polarization variations of light signals transmitting in the fiber, resulting in internet or telephone service interruptions, as mentioned at the beginning of the chapter.

3.4 Polarization Impairments in Optical Fiber Systems

The externally induced birefringences in the optical fiber discussed earlier may cause uncontrollable polarization variations in an optical fiber communication system, directly or together with other degrading polarization related effects to impair the proper operation of the system (Lichtman 1995). In fact, the higher the data transmission speed in the fiber communication system, the worse the polarization related impairments are. Such impairments include the polarization mode dispersion (PMD) in optical fibers, polarization dependent loss (PDL) in passive optical components, polarization dependent wavelength (PDW) responses of filters, polarization dependent modulation (PDM) in electro-optic modulators, polarization dependent gain (PDG) in optical amplifiers, and polarization dependent responsivity (PDR) of photodetectors. The cause for majority of these polarization impairments is the imperfection of the optical fibers. If the fibers were perfect, the SOP of the light signal transmitting in the fiber would remain constant and the effects of PMD, PDL, PDW, PDM, PDG, and PDR could easily be eliminated. Unfortunately, the SOP of light propagating in a length of standard communication fiber varies along the fiber due to the random birefringence induced by the thermal stress, mechanical stress, irregularities of the fiber core, and electrical and magnetic fields. Generally, along the fiber as well as at the output end of the fiber the light is elliptically polarized, with varying degrees of ellipticity, and with the major elliptical axis at an arbitrary angle relative to some reference orientation. Worst of all, the induced birefringence changes with temperature, pressure, stress, lightning strike, fiber motions, and other environmental variations, making polarization related impairments time dependent.

3.4.1 Polarization Mode Dispersion (PMD)

Birefringence is the origin of PMD. Before the deployment of coherent detection schemes, PMD was often cited as the critical hurdle for the high bit rate transmission systems (10 Gb/s and higher) after chromatic dispersion and fiber nonlinearity impairments are successfully managed (Kogelinik et al. 2002). As will be discussed in Chapter 8, coherent detection enables the cost-effective compensation of PMD in the digital domain; however, if the PMD in the system is too large or changes too rapidly, the PMD compensation algorithm may temporarily lose track and cause service interruptions.

As illustrated in Figure 3.5, a fiber link with birefringence distributed along its length can be considered as a concatenation of many, randomly oriented retardation plates. In the absence of PDL or PDG in the fiber link to be discussed next, these retardation plates are optically equivalent to a single retardation plate with an effective birefringence and a pair of effective orthogonal

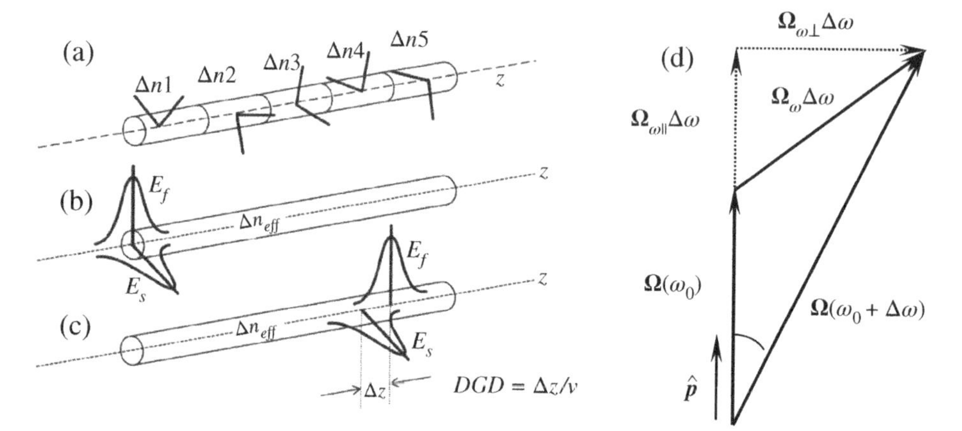

Figure 3.5 (a) A fiber can be considered as a concatenation of many randomly oriented retardation plates. (b) These plates are optically equivalent to a single retardation plate with an effective birefringence and a pair of effective orthogonal principal axes. (c) An optical pulse is decomposed into two polarization components along the two axes and delayed from each other at the exit by the amount equal to DGD. (d) Schematic diagram of PMD vector and the first- and second-order PMD.

principal axes (together as an effective birefringence vector) for a given optical wavelength (see Jones Theorem I in Chapter 4). Upon entering the retardation plate, an optical pulse is decomposed into two polarization components along the two axes. Because the two components travel with different speeds in the retardation plate, they exit the plate with a relative time delay called differential group delay (DGD). When the DGD is comparable with the bit separation of a data stream, bit error rate (BER) may significantly increase. Such a signal pulse spreading due to the effective refractive index difference is called polarization mode dispersion (PMD), as introduced in Section 2.6, with Δn in Eq. (2.21) being replaced by Δn_{eff}.

The simple picture of Figure 3.5a is based on Jones Theorem I for a light wave of single wavelength; however, signals pulses (such as return-to-zero [RZ] or non-return-to zero [NRZ] optical pulses) in an optical fiber communication system generally have a relatively large bandwidth containing many wavelength components, which have different effective birefringence vectors when transmitting through the optical fiber. Therefore, Figure 3.5b,c may not be accurate in describing the effect of PMD on signal pulses transmitting in a long optical fiber.

Referring to Figure 3.5a, a signal pulse with a particular input SOP is first split into a local fast- and slow-component by the first birefringence segment. After passing through the second segment, each of the fast- and slow-component of the previous segment will be further split into a second set of fast- and slow-component, with the process repeated in the following birefringence segments till the pulse exits from the fiber. Therefore, PMD in the fiber is the combined effect of local birefringence and random polarization mode coupling along the fiber, which causes random spreading of optical pulses (Xie 2016, p. 206). Because of such random polarization mode coupling and pulse splitting, the behavior of the pulse propagating in a long optical fiber is quite complex. Despite the complication, the propagation of the optical pulse can still be described by the principal states model originally developed by Poole and Wagner (1986), which states that even for a signal pulse with a finite bandwidth, there still exist a pair of orthogonal polarization states at the input with which a signal pulse is undistorted to the first order after transmitting through the fiber. That is, the fiber can still be considered for a pulsed signal as a single retardation plate with an effective birefringence and a pair of effective orthogonal principal axes, similar to that of Figure 3.5b,c following Jones Theorem I. This pair of orthogonal polarization states is called the principal state of polarization (PSP).

For a signal of finite bandwidth, PMD can be characterized in Stokes space by the PMD vector $\mathbf{\Omega}$ (Kogelinik et al. 2002)

$$\mathbf{\Omega} = \Delta\tau\hat{\mathbf{p}} \tag{3.18a}$$

where $\Delta\tau$ is DGD of a fiber segment and is the unit vector pointing to the slow axis of the PSP. In the time domain, PMD causes pulse splitting and broadening, which induces intersymbol interference (ISI) in optical communication systems (Xie 2016). As will be described in detail in Chapter 4, in the frequency domain, PMD causes the changes of the output polarization vector $\mathbf{S}$ of a signal with frequency, which can be described as

$$\frac{d\mathbf{S}}{d\omega} = \mathbf{\Omega} \times \mathbf{S} \tag{3.18b}$$

Equation (3.18b) indicates that the output SOPs are different for different frequency components even though they have the same SOP at the input. In other words, depolarization occurs due to the PMD. Taking the Taylor expansion of $\mathbf{\Omega}$, one obtains (Foschini and Poole 1991)

$$\mathbf{\Omega}(\omega_0 + \Delta\omega) = \mathbf{\Omega}(\omega_0) + \mathbf{\Omega}_{\omega}(\omega_0)\Delta\omega + \frac{1}{2}\mathbf{\Omega}_{\omega\omega}(\omega_0)\Delta\omega^2 + \cdots \tag{3.19a}$$

where $\mathbf{\Omega}_{\omega}$ and $\mathbf{\Omega}_{\omega\omega}$ stand for the first and second order derivatives, respectively, and $\mathbf{\Omega}(\omega_0)$, $\mathbf{\Omega}_{\omega}$, and $\mathbf{\Omega}_{\omega\omega}$ are termed as the first-, second-, and third-order PMD, respectively. From Eq. (3.18a), the second-order PMD can be expressed as

$$\mathbf{\Omega}_{\omega} = \frac{d\mathbf{\Omega}}{d\omega} = \Delta\tau_{\omega}\hat{\mathbf{p}} + \Delta\tau\hat{\mathbf{p}}_{\omega} \tag{3.19b}$$

where $\Delta\tau_{\omega} = d(\Delta\tau)/d\omega$ and $\hat{\mathbf{p}}_{\omega} = d\hat{\mathbf{p}}/d\omega$. Equation (3.19b) indicates that the second-order PMD has two terms, with the first term parallel to the direction of the first-order PMD, $\mathbf{\Omega}(\omega_0)$, denoted as $\mathbf{\Omega}_{\omega\|}$, while the second perpendicular to $\mathbf{\Omega}(\omega_0)$, denoted as $\mathbf{\Omega}_{\omega\perp}$, as shown in Figure 3.5d.

$$\mathbf{\Omega}_{\omega\|} = \Delta\tau_{\omega}\hat{\mathbf{p}} \qquad \mathbf{\Omega}_{\omega\perp} = \Delta\tau\hat{\mathbf{p}}_{\omega} \tag{3.19c}$$

The parallel term $\mathbf{\Omega}_{\omega\|} = \Delta\tau_{\omega}\hat{\mathbf{p}}$ is also called the polarization dependent chromatic dispersion (PDCD) because it describes the delay $\Delta\tau$ change as a function of frequency, while the perpendicular term $\mathbf{\Omega}_{\omega\perp} = \Delta\tau\hat{\mathbf{p}}_{\omega}$ is called depolarization because it describes the PSP direction $\hat{\mathbf{p}}$ change as a function of frequency. The characteristics of the first-order and second-order PMD, including their statistical properties, have been extensively studied and well understood. In particular, the value of the first-order PMD or DGD $\Delta\tau$ of a fiber generally varies with time because the magnitude and orientation of the equivalent retardation plates changes with time. It has been shown that the DGD in installed fiber communication links follows a Maxwell probability distribution, with the probability density function $PDF(\Delta\tau)$ expressed as (Kogelinik et al. 2002)

$$PDF(\Delta\tau) = 2\left(\frac{4}{\pi}\right)^2 \frac{(\Delta\tau)^2}{(\overline{\Delta\tau})^3} exp\left[-\frac{4(\Delta\tau)^2}{\pi(\overline{\Delta\tau})^2}\right] \tag{3.20a}$$

where $\overline{\Delta\tau}$ is the mean value of $\Delta\tau$. Figure 3.6 shows the plot of Eq. (3.20a) for the case of $\overline{\Delta\tau} = 50$ ps, which indicates that the probability density function (PDF) of the optical fiber link having a DGD of 50 ps is 0.1874. If a system has a DGD tolerance of 150 ps, three times of the mean DGD, the probability of the system service interruption is

$$PDF(\Delta\tau > 3\overline{\Delta\tau}) = 2\left(\frac{4}{\pi}\right)^2 \int_{150}^{\infty} \frac{(\Delta\tau)^2}{(\overline{\Delta\tau})^3} exp\left[-\frac{4(\Delta\tau)^2}{\pi(\overline{\Delta\tau})^2}\right] d(\Delta\tau) = 4.2 \times 10^{-5} \tag{3.20b}$$

which corresponds to 22 minutes per year.

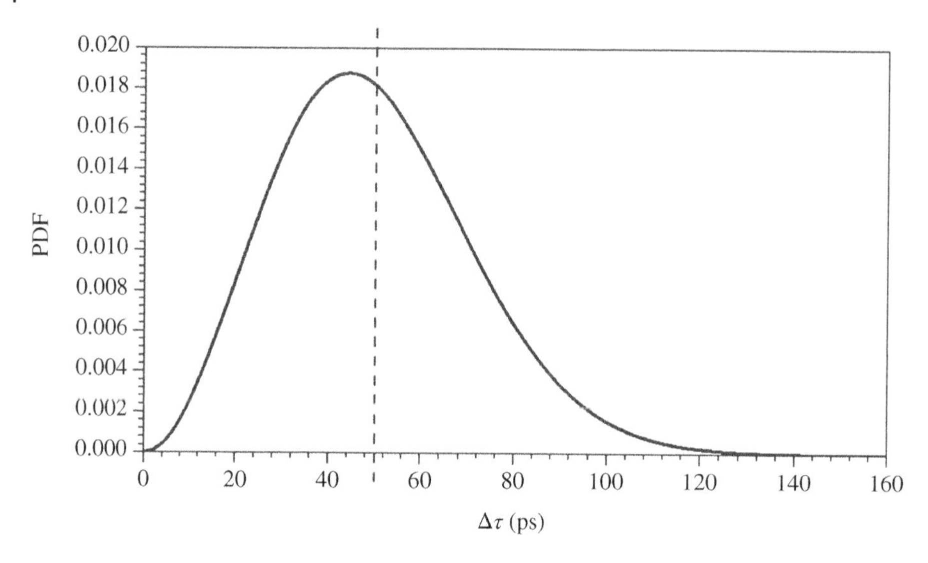

Figure 3.6 Probability density function (PDF) of DGD for the case that the mean DGD equals 50 ps, which follows a Maxwellian distribution.

The root-mean-square (rms) value of DGD, $\sqrt{\langle[\Delta\tau(z)]^2\rangle}$, is often referred to as first order PMD, where the bracket $\langle[\Delta\tau(z)]^2\rangle$ stands for ensemble average of $[\Delta\tau(z)]^2$ of all fiber segments. Similar to the famous random walk problem in which the rms distance of a drunken man to the origin is proportional to the square-root of the number of walking steps, in a long fiber span, the DGD accumulates as a three-dimensional random-walk, and on average is proportional to the square-root of the number of cascaded retardation plates, or equivalently, to the square-root of the fiber length for the case of strong polarization coupling (Poole 1988; Poole and Nagel 1997):

$$PMD = \sqrt{\langle[\Delta\tau(z)]^2\rangle} = \beta\sqrt{z} \tag{3.20c}$$

where β is a proportional constant relating to the birefringence per unit length and the random polarization coupling factor, with a unit of ps/$\sqrt{z}$. More detailed discussions on PMD will be included in Chapter 4.

Contrary to the case of a pure retardation plate, the DGD and the principal axes of the fiber link depend on the wavelength and fluctuate in time as a result of temperature variations and external constraints. Consequently, the corresponding pulse broadening is random, both as a function of wavelength at a given time and as a function of time at a given wavelength. As a rule of thumb, the maximum tolerable DGD value is 14% of the bit duration to ensure an outage probability of less than 5 minutes per year at a 3-dB power penalty. This translates to 14 ps for a 10 Gb/s system and 3.5 ps for a 40 Gb/s system. Unfortunately, for link distance greater than 300 km, 20% of the installed fiber plant around year 2000 was not suitable for 10 Gb/s transmission and 75% was not suitable for 40 Gb/s transmission at this outage tolerance level with the direct detection system or the on–off-key (OOK) system. Therefore, PMD compensation was required for these old fiber links at that time.

Unlike the effects of chromatic dispersion and fiber nonlinearity, which are deterministic and stable in time, the PMD-induced penalty can be totally absent at any given moment and adversely large enough several days later to cause an unacceptable BER for no apparent reason (Chbat 2000). To ensure an acceptable outage probability for the fiber optic system, PMD compensation was required to be dynamic in nature and adaptive to the random time variations (Ono and Yano 2000).

Figure 3.7 A PMD compensation configuration in direct detection systems. DPC: Dynamic polarization controller, vDGD: variable differential group delay line, DSP: digital signal processing.

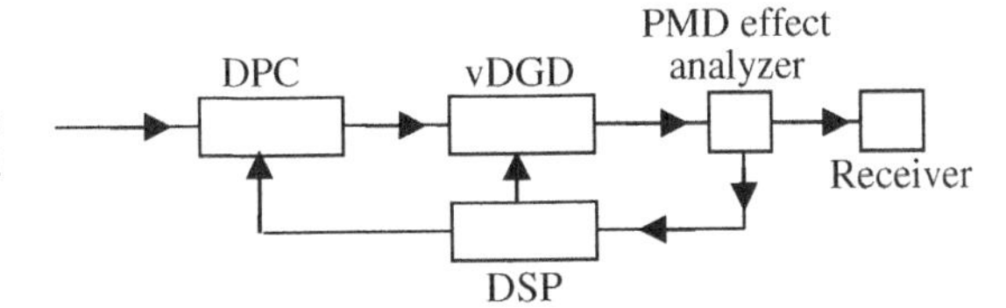

Figure 3.7 shows a generic PMD compensation scheme containing three key components (Buelow 2000; Watley et al. 2000): (i) a dynamic polarization controller (DPC), (ii) a variable differential group delay (vDGD) element, (iii) a PMD effect analyzer, and (iv) a digital feedback circuit. Specifically, the PMD analyzer detects the PMD effect on the signal and sends the information to the digital signal processing (DSP) circuit on the feedback circuit, which then sends instructions to the DPC and vDGD to adjust polarization and DGD to minimize the PMD effect. Such a configuration is often referred to as PMD nulling technique (Kogelinik et al. 2002) and particular implementation examples will be discussed in Chapter 7.

Note that with the deployment of coherent detection and polarization division multiplexing (PDMux), PMD can be completely compensated in the electrical domain using DSP in principle. However, it does not mean that PMD does not cause any performance degradations (Xie 2016).

3.4.2 Polarization Dependent Loss (PDL)

The PDL of an optical component is defined as the difference between the maximum and the minimum insertion losses for all possible input SOPs, as shown in Figure 3.8, and can be obtained with Eq. (2.25): $PDL = 10\,log\,(P_{max}/P_{min})$ after measuring the maximum and minimum optical power after the device under test (DUT) as the input polarization is scrambled for all possible SOPs. Optical components with PDL act as partial polarizers with two orthogonal axes (either linear or circular). A light signal experiences a maximum loss if its SOP is aligned with one axis and a minimum loss if aligned with the orthogonal axis. Almost all fiber optic components have PDL and the causes may be different for different components. First of all, as discussed in Chapter 2, when light passes from an optical medium with an index of n_1 to another medium with an index n_2, reflection occurs. The reflection coefficients for the polarization states perpendicular and parallel to the plane of incidence are different if the angle of incidence is not normal (many fiber components have angled input and output surfaces for increased return loss.) Such a difference in reflection results in a difference in transmission loss or PDL. For example, an 8° angle polished connector (FC/APC or SC/APC type connectors) has a PDL of 0.022 dB. Fiber grating based devices may also exhibit PDL if the grating is not normal to the fiber longitudinal axis. Second, for many optical components, such as isolators and circulators, birefringent crystals are often used. Because a birefringent

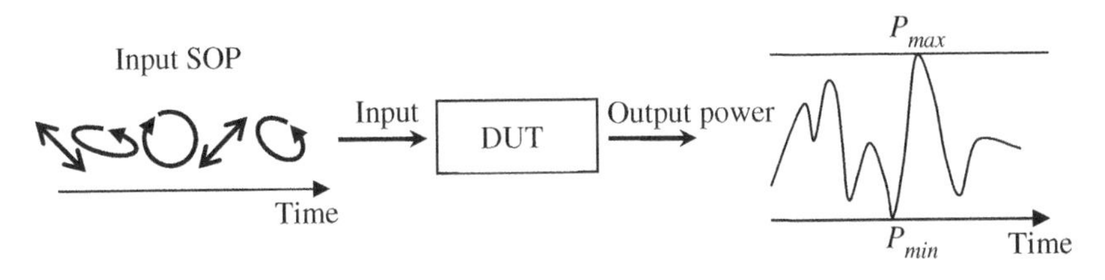

Figure 3.8 Illustration of different polarizations experience different transmission losses passing through a device under test (DUT).

crystal has two principal axes with different indexes of refraction, n_0 and n_e, the Fresnel reflection coefficients of two polarization states perpendicular and parallel to the principal axes are different even at normal incidence, resulting in different transmission losses. The corresponding PDL can be obtained from the Fresnel refraction formulae as

$$PDL = 10log\left|\frac{n_e(n_0+1)^2}{n_0(n_e+1)^2}\right| \tag{3.21}$$

Anti-reflection (AR) coatings can greatly reduce the reflection; however, they may not totally eliminate PDL because the optimal coating layer thickness is determined by the refractive index of the coated material: it is either optimized for n_0 or n_e. Diffraction grating based optical components or instruments generally have high PDL because the diffraction efficiencies for the two polarization states perpendicular and parallel to the plane of incidence are different. Finally, any fiber component containing a dichroic material also has PDL. A dichroic material has two principal axes with different absorption or attenuation coefficients. The principal axes can either be linear (linear dichroism) or circular (circular dichroism). For example, the $LiNbO_3$ waveguide made with the proton exchange method exhibits strong linear dichroism and acts just like a polarizer.

Table 3.1 lists the typical PDL values of fiber optical components in telecommunication bandwidths around 1310 and 1550 nm.

In a fiber link which contains many optical components with different PDL values oriented in different directions, as shown in Figure 3.9a, the total PDL value depends on the SOP of the light signal before each component C_i (i = 1–N) with respect to the PDL direction of the component, where the PDL direction of each component is defined to be aligned with the maximum power transmission direction (Damask 2005). As can be seen in Figure 3.9a, due to the birefringence (or PMD) in the fiber between any two components, the SOP vector before each component may vary randomly, causing power transmission variations after each component. As shown in Figure 3.9b, the maximum PDL value is equal to the summation of the PDL values of all the components, which can be considered as the PDLs of all the components are aligned. In particular, the maximum PDL is the difference between two insertion loss measurements of the link: the first one is when SOP of the light before each component is aligned with the minimum loss

Table 3.1 Typical PDL values of common optical components.

Component	Typical PDL value (dB)
1 m single mode fiber	<0.02
10 km single mode fiber	<0.05
PC type connector	0.005–0.02
APC type connector	0.02–0.06
50% fused coupler, 1310 or 1550 nm	0.1–0.2
50% fused coupler, dual 1310 and 1550 nm	0.15–0.3
90/10 fused coupler, 90% through path	0.02
90/10 fused coupler, 10% cross path	0.1
Isolator	0.05–0.3
3-Port circulator	0.1–0.2
Dense wavelength division multiplexer	0.05–0.15
Polarizer	30–50

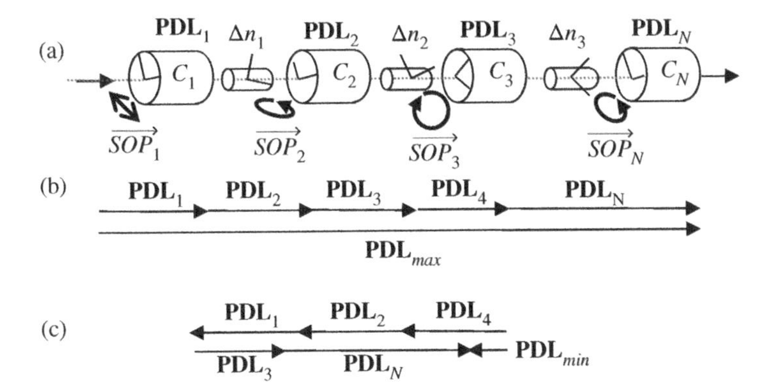

Figure 3.9 Illustration of concatenation of PDL vectors in an optical fiber link. (a) An optical fiber link with multiple components with PDL linked by optical fibers with different birefringence (or PMD vectors) to cause random SOP change; (b) The maximum PDL obtained when the SOP before each component is such that all PDLs are aligned up; (c) The minimum PDL obtained when the SOP before each component is such that the summation yields the least PDL value.

axis of the component and the second one is when SOP of the light before each component is aligned with the maximum loss axis. On the other hand, the minimum PDL value corresponds to the case that the SOPs of a light signal before all the PDL components are arranged such that the PDLs cancel one another out, as shown in Figure 3.9c. The net residual PDL is the minimum PDL of the link. In all other situations, the PDL value of the whole optical fiber system varies between the maximum and minimum PDL values, depending on the local SOP before each component.

In principle, according to Jones Theorem III to be discussed in Chapter 4, corresponding to a given wavelength and birefringence condition of the fibers between any two components with PDL, the total PDL of the fiber link shown in Figure 3.9a is equivalent to an effective PDL. In fact, even for optical signals with finite bandwidths, it is demonstrated that any linear optics system with interleaved PMD and PDL is equivalent to another linear optic system where all the PMD comes first, followed by some PDL; however, these "global PMD" and "global PDL" are not simply to sum of the ones appearing in the first system (Gisin 2005).

3.4.3 Polarization Dependent Gain (PDG)

The gain of an optical amplifier for the stronger polarization component is less than that for the weaker component (because the stronger component saturates the gain more) and the gain difference is called polarization dependent gain (PDG). One cause for the PDG is that the cross sections of the stimulated emission for different polarization states are different. This polarization hole burning always gives more gain to the weaker polarization component and thus tends to cause the polarization state to change with time. In addition, when the input SOP changes, the signal gain may increase temporarily, and then come down in a short period of time. Consequently, the polarization hole burning always encourages polarization fluctuations in a fiber laser system and thus causes mode-hopping and increases super-mode noise in a mode-locked laser. An optical amplifier may at the same time exhibit PDL effect. For example, couplers and isolators are generally contained in an Er^{+} doped fiber amplifier (EDFA) and the presence of PDL in these components gives rise to the apparent PDL of the amplifier. Even in semiconductor optical amplifiers (SOAs), the facets of the semiconductor chip are generally angle cleaved to prevent optical feedback into the amplifier. As discussed previously, these angled interfaces exhibit large PDL, which directly contributes to PDG of the SOA.

In a fiber optic link with many optical amplifiers and many components with PDL, the effect of PDG can be significant at some moments and negligible at other times. When a large number of optical amplifiers are cascaded in a long haul fiber link, the performance degradation caused by PDG is significant even though each amplifier may have a very small PDG (~0.1 dB). The performance degradation becomes even worse when PDG is combined with the PMD and PDL of the fiber and other components in the link. Polarization scrambling at a frequency above amplifier's response rate (inverse of amplifier's upper energy level life time, ~500 Hz for EDFA) has proven to be effective in mitigating PDG impairment in long haul systems. A factor of 2 increase in system *Q* factor was demonstrated in an 8100-km link containing 181 EDFAs with such a scheme (Bruyere et al. 1994).

3.4.4 Polarization Dependent Wavelength (PDW)

In a wavelength division multiple (WDM) access communications systems, optical filters are frequently used, such as the WDM multiplexers, WDM demultiplexers, gain flattening filters, wavelength interleavers, wavelength routers, and add-drop filters. These spectral management components are based on the following possible technologies: thin film filters, diffraction gratings, fiber gratings (long period and Bragg grating), planer waveguides on silica (SiO_2) or silicon (Si), and birefringence crystals. The spectral responses of these devices all have slight polarization dependencies in terms of center wavelength, spectral shape, spectral ripple, and spectral phase. Cascading many of these components in tandem may cause a large polarization dependent effect on the spectral and spectral phase, or polarization dependent spectral distortion of the signals.

3.4.5 Polarization Dependent Modulation (PDM)

In addition to PDL, external modulators, such as $LiNbO_3$ based electro-optical modulators and semiconductor electro-absorption modulators, also exhibit PDM in that the modulation depth of signals with different polarization states is different. As a result, the amplitude of the received data bits varies when the SOP of light before entering the modulator varies due to the fluctuation of temperature or other external constraints on the fiber, resulting in BER fluctuation. To assist polarization alignment, most Ti-indiffused $LiNbO_3$ modulators embed a polarizer at the input and output of the waveguide and thus convert the PDM problem into a more easily identified PDL problem. The $LiNbO_3$ modulators made with proton exchange process act like a polarizer themselves without the embedded polarizer. In general, a PM fiber is attached at the input end of the modulator to make sure the polarization entering the modulator is properly aligned with the intended axis of the modulator.

3.4.6 Polarization Dependent Responsivity (PDR)

Photodetectors generally have different efficiencies for the photon to electron conversion or optical power to photocurrent conversion. Such a conversion efficiency is called responsivity with the unit of ampere per watt (A/W) and the maximum difference in efficiency for all possible polarizations is called PDR, which is on the order of 0.01–0.1 dB in most cases. In optical fiber communication systems, such small PDR values are trivial and can be ignored. However, for measurement and sensing applications, such a small PDR is critical and must be carefully managed. For example, for a PDL measurement instrument, the presence of PDR directly contributes to the PDL measurement inaccuracy.

Many factors may contribute to the PDR, including the polarization dependence of the quantum conversion efficiency, the differential Fresnel reflections when light incident on the detector chip at an angle, and strain or stress induced when dicing or bonding the detector wafer into small pieces. Generally, small area detectors have larger PDR while larger area detectors have smaller PDR probably due to less dicing or bonding induced stress. Therefore test equipment generally adopt large area detectors despite their relatively large dark current and higher cost. Anti-reflection coatings may be used to reduce the differential reflection induced PDR for cases with angled incident light, which is not effective for cases with light of normal incidence. It is possible to identify the axes of the maximum and minimum responsivities of a detector (the PDR axis) and place an angled glass window in front of the detector chip to produce a reflection induced PDR to balance out the PDR of the chip itself. It is also possible to tilt photodetector (PD) chip at an angle to purposely induce a PDR with an axis orthogonal to the original PDR axis to balance out the original PDR.

3.5 Polarization Multiplexing

In addition to the degrading effects to the optical fiber systems, polarization can be beneficial for certain application if utilized properly. One of such applications is polarization multiplexing in which two orthogonal polarization channels of the same wavelength carrying different information are combined by a polarization beam combiner into a single optical fiber to be transmitted to a remote location before being demultiplexed into two separate channels, as shown in Figure 3.10. This way, the information transmission capacity is doubled.

At the receiving end of Figure 3.10, a polarization controller is used to adjust the SOPs of the two signal channels to be linear and aligned with the axes of the polarization beam splitter (PBS) so that they can be separated by the PBS with minimum crosstalk. This process is called polarization demultiplexing. The error signal for controlling the polarization controller can be derived several ways, including the low frequency correlation noise between two channels, which will be minimized when the two polarization channels are properly separated with a low crosstalk. Another approach is to add a low frequency modulation, or pilot tone, in TX2. A detection circuit in RX1 extracts this pilot tone and feeds back to the polarization controller to minimize the pilot tone and therefore ensuring the proper separation of the two polarization channels. The BER of RX1 before the forward error correction (FEC) in the receiver can also be used as the error signal for the

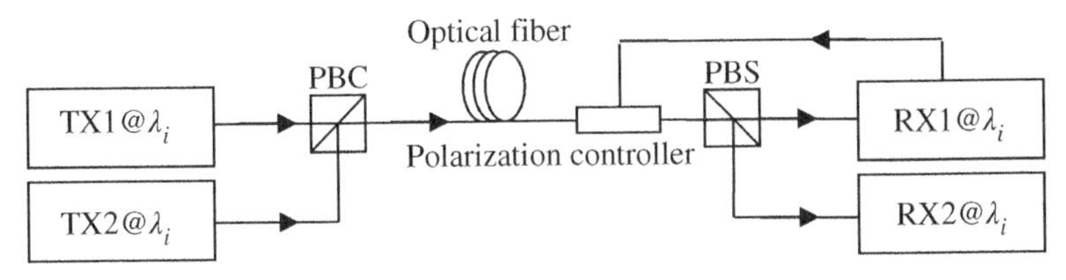

Figure 3.10 Illustration of a simplified polarization division multiplexing system. TX and RX stand for transmitter and receiver, respectively.

feedback control of the polarization controller for minimizing the BER. Other feedback schemes will be discussed in detail in Section 7.1.6. Note that in coherent detection systems, the retardation information of the two polarization channel can be extracted digitally so that the polarization demultiplexing can be achieved in the digital domain via DSP with a virtual polarization controller, without the need of a real physical polarization controller.

As pointed out in Chapter 2, unlike birefringence or PMD, the presence of PDL in the transmission link destroys orthogonality of polarization multiplexed signals, resulting in crosstalk between the two polarization channels even with optimized polarization adjustment. In the coherent detection system (Xie 2016), PDL causes the loss of certain information in the optical signal, which makes it impossible to fully demultiplex the two polarization channels through DSP. Mathematically, the presence of PDL destroys the unitary properties of Jones and Muller matrices, making the signal processing very difficult. In addition, PDL causes the power and signal-to-noise ratio (SNR) to fluctuate and repolarize amplified spontaneous emission (ASE) noise, which cannot be compensated digitally even in the coherent detection systems (Xie 2016). In contrast, the presence of birefringence or PMD does not affect the orthogonality of polarization multiplexed signals, nor the unitary properties of Jones and Muller matrices, which can be fully compensated digitally in polarization multiplexed coherent detection systems in principle. Therefore, PDL poses fundamental limitation on polarization multiplexed coherent detection systems.

3.6 Polarization Issues Unique to Optic Fiber Sensing System

All the polarization impairments in the optical fiber communication systems discussed in Section 3.4 may affect signals in optic fiber sensing system. In particular, fiber Bragg gratings (FBGs) are often used in optical fiber sensing systems to sense temperature, strain, and stress. PDW and PDL may degrade system performance due to polarization variations, causing measurement uncertainties. These degrading effects can be mitigated passively with depolarizers or actively with polarization scramblers, as will be discussed in Chapters 6 and 7.

Another polarization impairment unique to the fiber sensor system is the polarization fading in an optical fiber interferometer based sensor system, as shown in Figure 3.11. In general, the SOPs of the light signals in the two interferometer arms are different and fluctuate, causing signal instability. In the worst case that the light signals in the two interference arms have orthogonal polarizations, no interference signal will be observed, resulting in total signal loss. As can be seen in Chapters 6 and 7, the polarization fading problem can be resolved either passively with passive components, such as Faraday rotators and polarization diversity detectors, or actively with polarization track schemes.

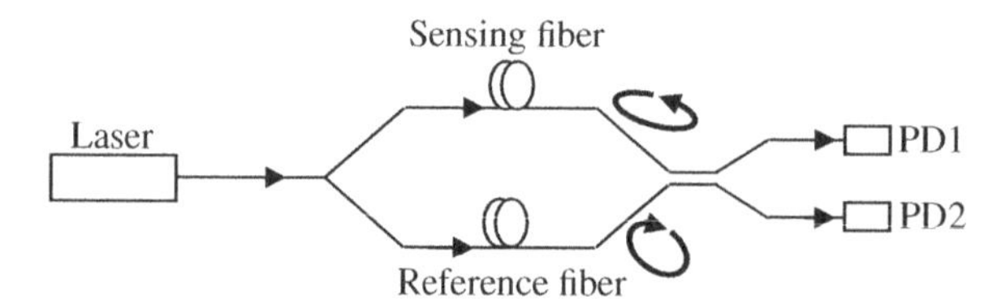

Figure 3.11 Illustration of polarization fading in an optical fiber interferometer.

3.7 Polarization Issues Unique to Microwave Photonics Systems

PMD described in Section 3.4 may induce modulation fading and phase instability in microwave photonics systems (Poole and Darcie 1993; Adamczyk et al. 2001; Hofstetter et al. 1995). When an optical carrier at circular frequency ω_0 is modulated by a Mach-Zehnder (MZ) modulator with a radio-frequency (RF) ω_m shown in Figure 3.12a, two or more modulation side bands $\omega_0 \pm \omega_m$ will be generated around ω_0, as shown in Figure 3.12b. The beating of the carrier and the modulation sidebands in the photodetector at the receiving end converts the modulation into the amplitude modulated RF signal. In an optic fiber link with PMD, the carrier and the modulation sidebands may experience different retardation as they propagate inside fiber due to the frequency difference. At the input end, the SOPs of both the carrier ω_0 and the modulation sideband $\omega_0 \pm \omega_m$ are the same, as shown in Figure 3.12c. However, as they propagate inside the fiber with PMD, the different retardations they experience will cause their SOP to differ and therefore the beat between the carrier and a sideband will be reduced in most cases, as shown in Figure 3.12c, and can be totally diminished if their SOPs differ by 90°.

To simplify the discussion, let's just look at the case of a single modulation sideband, such as the case using a single sideband modulator. Let τ_s and τ_f be the group delays of the signal polarized along the slow and fast axes, respectively, the electrical field at the end of the fiber can be expressed as

$$\mathbf{E}_{out}(t) = cos\theta \left[E_0 + E_m e^{i\omega_m(t-\tau_s)}\right] e^{-i\omega_0(t-\tau_s)}\hat{\mathbf{s}} + sin\theta \left[E_0 + E_m e^{i\omega_m(t-\tau_f)}\right] e^{-i\omega_0(t-\tau_f)}\hat{\mathbf{f}} \tag{3.22a}$$

In the photodetector at the receiving end, the detected photocurrent I_{ph} can be expressed as

$$I_{ph} = \eta|\mathbf{E}_{out}(t)|^2 = \eta \left[cos^2\theta\left|E_0 + E_m e^{i\omega_m(t-\tau_s)}\right|^2 + sin^2\theta\left|E_0 + E_m e^{i\omega_m(t-\tau_f)}\right|^2\right] \tag{3.22b}$$

where η is the responsivity of the photodetector. Equation (3.22b) can be simplified for the case $\theta = 45°$ as

$$I_{ph} = \eta \left[E_0^2 + E_m^2 + 2E_0E_m cos\left(\frac{\omega_m \Delta\tau}{2}\right) cos\omega_m(t-\tau)\right] \tag{3.22c}$$

where $\Delta\tau = \tau_s - \tau_f$ is the DGD of the fiber and $\tau = (\tau_s + \tau_f)/2$ is the average group delay. Clearly, the last term vanishes when $\omega_m \Delta\tau = \pi$, corresponding to the total modulation fading described earlier.

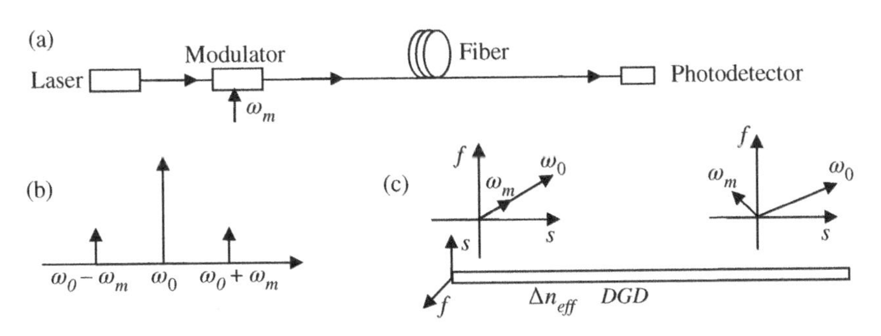

Figure 3.12 (a) An externally modulated microwave photonics link with a modulation frequency of ω_m. (b) Illustration of the modulation sidebands. (c) The fiber with PMD can be viewed as a birefringence material with an effective birefringence.

PMD can also induce phase instability of the received RF signal V_{RF}. When the light is polarized along the slow axis ($\theta = 0$), it experiences a phase difference of $\omega_m \tau_s$ between the carrier and the modulation sideband, resulting in a phase of $\omega_m \tau_s$ of the received RF signal. Similarly, when the light is polarized along the slow axis ($\theta = 90°$), it experiences a phase difference of $\omega_m \tau_f$ between the carrier and the modulation sideband, resulting in a phase of $\omega_m \tau_f$ of the received RF signal. For an arbitrary SOP, the phase of the received RF signal is in between the two extreme values. Because in a real fiber system with PMD, the orientation and the magnitude of the effective birefringence or DGD vary with time, resulting in a fluctuating phase of the received RF signal. Mathematically, the received RF signal V_{RF} can be obtained from Eq. (3.22b) as

$$V_{RF} \propto I_{ph} = 2\eta E_0 E_m [cos^2\theta\ cos(\omega_m t - \omega_m \tau_s) + sin^2\theta cos(\omega_m t - \omega_m \tau_f)] \tag{3.22d}$$

In obtaining (3.22d), only the time varying terms are kept. From Eq. (3.22d) one can see that when θ, τ_s, or τ_f varies, the phase of the RF signal fluctuates. Similar to the digital fiber communication system, the PMD effect in a RF system can also be compensated (Yu et al. 2004).

References

Adamczyk, O.H., Sahin, A.B., Yu, Q. et al. (2001). Statistics of PMD-induced power fading for intensity-modulated doublesideband and single-sideband microwave and millimeter-wave signals. *IEEE Trans. Microwave Theory Tech.* 49: 1962–1967.

Barlow, A.J., Ramskov-Hansen, J.J., and Payne, D.N. (1981). Birefringence and polarisation mode-dispersion in spun single-mode fibers. *Appl. Opt.* 20: 2962–2968.

Boroditsky, M., Brodsky, M., Frigo, N.J., et al. (2005). Polarization dynamics in installed fiberoptic systems. *2005 IEEE LEOS Annual Meeting Conference Proceedings, Sydney, Australia, 22-28 Oct. 2005.*

Bruyere, F., Audouin, O., Letellier, V. et al. (1994). Demonstration of an optimal polarization scrambler for long-haul optical amplifier systems. *IEEE Photonics Technol. Lett.* 6: 1153–1155.

Buelow, H. (2000). PMD mitigation techniques and their effectiveness in installed fiber. *Optical Fiber Communication Conference 2000, Baltimore, Maryland United States, 7 March 2000.*

Chbat, M. (2000). Managing polarization mode dispersion. *Photonics Spectra* (June), Pittsfield, MA, USA, pp. 100–104.

Chu, P. and Sammut, R. (1984). Analytical method for calculation of stresses and material birefringence in polarization maintaining optical fiber. *J. Lightwave Technol.* 2 (5): 650–662.

Damask, J.N. (2005). Properties of polarization-dependent loss and polarization-mode dispersion, Chapter 8. In: *Polarization Optics in Telecommunications*, 297. Springer Science + Business Media, Inc.

Feng, T., Shang, Y., Wang, X. et al. (2018). Distributed polarization analysis with binary polarization rotators for the accurate measurement of distance-resolved birefringence along a single-mode fiber. *Opt. Express* 26 (20): 25989–26002.

Foschini, G.J. and Poole, C.D. (1991). Statistical theory of polarization dispersion in single mode fibers. *J. Lightwave Technol.* 9: 1439–1456.

Gisin, N. (2005). PMD & PDL. In: *Polarization Mode Dispersion* (ed. A. Galtarossa and C. Menyuk), 113–125. Springer Science + Business Media, Inc.

Guan, R., Zhu, F., Gan, Z. et al. (2005). Stress birefringence analysis of PM optical fibers. *Opt. Fiber Technol.* 11: 240–254.

Hofstetter, R., Schmuck, H., and Heidemann, R. (1995). Dispersion effects in optical millimeter-wave systems using self-heterodyne method for transport and generation. *IEEE Trans. Microwave Theory Tech.* 43: 2263–2269.

Kogelinik, H., Jopson, R., and Nelson, L. (2002). Polarization-mode dispersion, Chapter 15. In: *Optical Fiber Communications, IVB* (ed. I. Kaminov and T. Li), 725–861. San Diego: Academic Press.

Krummrich, P.M. and Kotten, K. (2004). Extremely fast (microsecond scale) polarization changes in high speed long hail WDM transmission systems. *Proceedings of Optical Fiber Communication Conference 2004, Los Angeles, California, United States, 22 February 2004.*

Krummrich, P.M., Schmidt, E.D., Weiershausen, W., and Mattheus, A. (2005). Field trial on statistics of fast polarization changes in long haul WDM transmission systems. *Proceedings of Optical Fiber Communication Conference 2005, Anaheim, California United States, 6 March 2005.*

Kumar, A. and Ulrich, R. (1981). Birefringence of optical fiber pressed into a V-groove. *Opt. Lett.* 6: 644–646.

Kuschnerov, M. and Herrmann, M. (2016). Lightning affects coherent optical transmission in aerial fiber. *Lightwave* (2 March), pp. 1–7. https://www.lightwaveonline.com/print/content/16654079.

Laming, R.I. and Payne, D.N. (1989). Electric current sensors employing spun highly birefringent optical fibers. *J. Lightwave Technol.* 7 (12): 2084–2094.

Leo, P.J., Gray, G.R., Simer, G.J., and Rochford, K.B. (2003). State of polarization changes: classification and measurement. *J. Lightwave Technol.* 21: 2189–2193.

Lichtman, E. (1995). Limitations imposed by polarization-dependent gain and loss on all-optical ultra-long communication systems. *J. Lightwave Technol.* 13: 906–913.

Maystre, F. and Bertholds, A. (1987). Zero-birefringence optical fiber holder. *Opt. Lett.* 12: 126–128.

Ono, T. and Yano, Y. (2000). 10 Gb/s PMD compensation field experiment over 452 km using principal state transmission method. *Optics & Photonics News* (May). p. 61.

Peterson, D.L., Leo, P.J., and Rochford, K.B. (2004). Field measurements of state of polarization and PMD from a tier-1 carrier. *Proceedings of Optical Fiber Communication Conference 2004, Los Angeles, California United States, 22 February 2004.*

Poole, C.D. (1988). Statistical treatment of polarization dispersion in single mode fiber. *Opt. Lett.* 13: 687–689.

Poole, C.D. and Darcie, T.E. (1993). Distortion related to polarization-mode dispersion in analog lightwave systems. *J. Lightwave Technol.* 11: 1749–1759.

Poole, C.D. and Nagel, J.A. (1997). Polarization effects in lightwave systems. In: *Optical Fiber Communications IIIA* (ed. I.P. Kaminow and T.L. Koch), 114–161. Academic Press.

Poole, C.D. and Wagner, R.E. (1986). Phenomenological approach to polarization dispersion in long single-mode fibers. *Electron. Lett.* 22: 1029–1030.

Rashleigh, S.C. (1983). Origins and control of polarization effects in single-mode fibers. *J. Lightwave Technol.* LT-1 (2): 312–331.

Rashleigh, S.C. and Ulrich, R. (1980). High-birefringence in tension coiled single-mode fibers. *Opt. Lett.* 5: 354–356.

Simon, A. and Ulrich, R. (1977). Evolution of polarization along a single-mode fiber. *Appl. Phys. Lett.* 31 (8): 517–520.

Smith, A.M. (1980a). Single-mode fiber pressure sensitivity. *Electron. Lett.* 16 (20): 773–774.

Smith, A.M. (1980b). Birefringence induced by bends and twists in single-mode optical fiber. *Appl. Opt.* 19 (15): 2606–2611.

Stolen, R. and Ashkin, A. (1973). Optical Kerr effect in glass waveguide. *Appl. Phys. Lett.* 22 (6): 294–296.

Tsai, K., Kim, K., Morse, T. et al. (1991). General solutions for stress-induced polarization in optical fibers. *J. Lightwave Technol.* 9 (1): 7–17.

Ulrich, R. (1977). Representation of codirectional coupled waves. *Opt. Lett.* 1 (3): 109–111.

Ulrich, R. and Rashleigh, S.C. (1982). Polarization coupling in kinked single-mode fiber. *IEEE J. Quantum Electron.* 18: 2032–2039.

Ulrich, R. and Simon, A. (1979). Polarization optics of twisted single-mode fibers. *Appl. Opt.* 18 (13): 2241–2251.

Ulrich, R., Rashleieh, S.C., and Eickhoff, W. (1980). Bending-induced birefringence in single mode fiber. *Opt. Lett.* 5: 273–275.

Watley, D.A., Farley, K., Lee, W., Bordogna, G., Shaw, B. and Hadjifotiou, A. et al. (2000). Field evaluation of an optical PMD compensator using an installed 10 Gb/s system. *Optical Fiber Communication Conference 2000, Baltimore, Maryland United States, 7 March 2000.*

Xie, C. (2016). Polarization and nonlinear impairments in fiber communication systems. In: *Enabling Technologies for High Spectral-Efficiency Coherent Optical Communication Networks* (ed. X. Zhou and C. Xie), 201–246. Hoboken, NJ: Wiley.

Yu, C., Yu, Q., Pan, Z. et al. (2004). Optically compensating the PMD-induced RF power fading for single-sideband subcarrier-multiplexed systems. *IEEE Photonics Technol. Lett.* 16 (1): 341–343.

4

Mathematics for Polarization Analysis

A vast amount of knowledge on the mathematical treatments of polarization issues have been accumulated since Malus' early investigation on the topic, especially since optical fiber communications became a reality in the early 1970s. In this chapter, we describe mathematical methods for polarization analysis, including detailed derivations of almost all major polarization related formulae or conclusions for interested readers to have a "one-stop shop" to find out not only what they are, but also how they are arrived at, without having to search for the answers scattered in dozens of publications. In addition, we stick to using the most basic mathematics in these derivations with which most engineering students or professionals with an engineering degree are familiar with, avoiding involving more advanced mathematics, such as those used in Spin-vector calculus (Damask 2005) for people familiar with quantum mechanics, although the Spin-vector calculus may be more elegant mathematically in treating polarization problems.

As discussed in Chapter 2, the polarization of a monochromatic light wave in a homogeneous medium can be visually represented by a polarization ellipse. However, it is difficult to use a polarization ellipse to quantitatively describe how the polarization evolves along an optical path or with time. This chapter will introduce two quantitative methods to describe the polarization state of light and its evolution: the Jones calculus and the Mueller calculus. They are both matrix methods, in which the polarization of a light wave is represented by a two-dimensional (2D) complex vector (the Jones vector) or a four-dimensional real vector (the Stokes vector). An optical device, for example, a retardation plate, a partial polarizer, or a rotator, can be represented by a matrix mathematically. The polarization vectors of emerging light will be the multiplication of the matrix representing the optical element with the polarization vector of the incident light. Any optical system containing a serial of optical devices arranged along the direction of light propagation can be described by a single matrix, which is the multiplication of all the matrices of the optical devices.

The Stokes vector is a four-dimensional real vector defined by four or six light intensity measurements after the light beam passes through a 0°, 90°, 45°, or 135° linear polarizer, or right- and left-handed circular polarizer, sequentially, as will be discussed later. The corresponding Mueller matrix of an optical system is a 4×4 real matrix. The Jones vector is a 2D complex vector defined by the amplitudes and relative phase shift of the x and y components of the electrical field of a fully polarized light, and a 2×2 complex matrix (the Jones matrix) describes the corresponding optical system.

As mentioned earlier, the Stokes vector comes from the experimental measurements of optical power, regardless of whether the light is polarized or not; therefore, the corresponding Mueller calculus can be used to describe both the polarization and depolarization properties of an optical system. On the other hand, the Jones calculus is only applicable to the characterization of fully polarized monochromatic light in an optical system.

Polarization Measurement and Control in Optical Fiber Communication and Sensor Systems, First Edition.
X. Steve Yao and Xiaojun (James) Chen.

The Jones calculus is more powerful for problems related to the coherent superposition of two or more beams because the phase information is preserved in the Jones calculus. On the contrary, the Stokes vector loses the phase information and cannot be used to describe the coherent phenomenon.

4.1 Jones Vector Representation of Monochromatic Light

4.1.1 Jones Vector

As we discussed in Chapter 2, the monochromatic plane wave is a transverse wave. Suppose it propagates along the z-axis, with its electric field at position (0, 0, z) oscillating in the x–y plane and passing point (0, 0, z), as shown in Figure 4.1. The electric field vector in the x–y plane can be represented as the superposition of two mutually orthogonal linearly polarized electric fields, $\mathbf{E}_x$ and $\mathbf{E}_y$, which oscillate along x- and y-axes, respectively:

$$\mathbf{E}(z,t) = \mathbf{E}_x + \mathbf{E}_y = A_x cos(\omega t - k_0 z + \varphi_x)\hat{\mathbf{e}}_x + A_y cos(\omega t - k_0 z + \varphi_y)\hat{\mathbf{e}}_y \tag{4.1}$$

where $\hat{\mathbf{e}}_x$ and $\hat{\mathbf{e}}_y$ are the unit vectors along the positive x and y axes, respectively, A_x and A_y are the amplitudes of the corresponding linearly polarized electric fields, ω is the angular frequency of light, k_0 is the free-space wave number equal to $2\pi/\lambda_0$ for wavelength λ_0, and φ_x and φ_y are the absolute phases of two components of the electric field.

According to Euler's formula, Eq. (4.1) can be rewritten as

$$\mathbf{E}(z,t) = Re\left(A_x e^{i(\omega t - k_0 z + \varphi_x)}\right)\hat{e}_x + Re\left(A_y e^{i(\omega t - k_0 z + \varphi_y)}\right)\hat{e}_y \tag{4.2}$$

It is often preferable to express Eq. (4.1) in its equivalent complex form for easy mathematical manipulation:

$$\mathbf{E} = A_x e^{i(\omega t - k_0 z + \varphi_x)}\hat{e}_x + A_y e^{i(\omega t - k_0 z + \varphi_y)}\hat{e}_y \tag{4.3}$$

Both the real and complex forms of the electric field represented by Eqs. (4.2) and (4.3) are mathematically valid solutions of Maxwell's equations in a homogeneous medium; therefore, as long as one deals with linear equations, all the algebraic manipulations can be carried out using the complex electrical field presentation, and the physical quantities can be obtained by taking the real parts of the complex results.

R. Clark Jones (1941) proposed to use a 2×1 column complex vector to represent the complex electric field, with the first entry being the x component of the electric field and the second entry being the y component of the electric field:

$$\begin{pmatrix} E_x \\ E_y \end{pmatrix} = \begin{pmatrix} A_x e^{i(-k_0 z + \varphi_x)} e^{i\omega t} \\ A_y e^{i(-k_0 z + \varphi_y)} e^{i\omega t} \end{pmatrix} = e^{i\omega t}\begin{pmatrix} A_x e^{i\delta_x(z)} \\ A_x e^{i\delta_y(z)} \end{pmatrix} \tag{4.4}$$

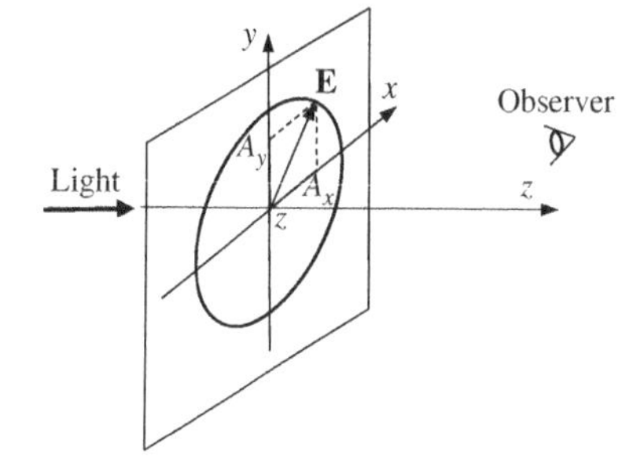

Figure 4.1 Transverse electric field of a monochromatic plane wave in the right-handed coordinates system.

where $\delta_x(z) = -k_0 z + \varphi_x$ and $\delta_y(z) = -k_0 z + \varphi_y$ are the absolute phases of the x and y components of the electric field at position z and time $t = 0$. For a monochromatic light wave, its polarization state (polarization ellipse) is determined only by the amplitudes A_x, A_y and the phase difference $\delta = \delta_y - \delta_x$. The common phase factor $e^{i\omega t}$ in both components does not change the relative values of the two vector components in Eq. (4.4); therefore, it can be neglected so that the vector in Eq. (4.4) can be reduced to

$$\mathbf{J} = \begin{pmatrix} j_x \\ j_y \end{pmatrix} = \begin{pmatrix} A_x e^{i\delta_x(z)} \\ A_y e^{i\delta_y(z)} \end{pmatrix} \tag{4.5}$$

where the vector $\left(A_x e^{i\delta_x(z)},\ A_x e^{i\delta_y(z)}\right)^T$ is called the Jones vector (Jones 1941; Born and Wolf 1999). It is often useful to denote the Jones vector in its "standard normalized" form, where the vector is reduced to the simplest form with a magnitude of unity. This is easily accomplished by dividing the Jones vector with its magnitude $|\mathbf{J}|$

$$|\mathbf{J}| = \sqrt{A_x^2 + A_y^2} \tag{4.6}$$

If we denote the conjugate transpose of Jones vector $\mathbf{J}$ as $\mathbf{J}^\dagger$ $[\mathbf{J}^\dagger = \begin{pmatrix} j_x^* & j_y^* \end{pmatrix}]$, and define the inner product "$\cdot$" of two-column complex vectors $\mathbf{J}_1$ and $\mathbf{J}_2$ as

$$\mathbf{J}_1 \cdot \mathbf{J}_2 = \mathbf{J}_1^\dagger \mathbf{J}_2 = \begin{pmatrix} j_{1x}^* & j_{1y}^* \end{pmatrix} \begin{pmatrix} j_{2x} \\ j_{2y} \end{pmatrix} = j_{1x}^* j_{2x} + j_{1y}^* j_{2y} \tag{4.7}$$

the magnitude of a Jones vector of Eq. (4.6) can be written as the inner product of $\mathbf{J}$ and itself.

$$\sqrt{\mathbf{J} \cdot \mathbf{J}} = \sqrt{\mathbf{J}^\dagger \mathbf{J}} = \sqrt{j_x^* j_x + j_y^* j_y} = \sqrt{A_x^2 + A_y^2} = |\mathbf{J}| \tag{4.8}$$

Therefore the normalized form $\mathbf{J}_n$ of a Jones vector in Eq. (4.5) can be written as

$$\mathbf{J}_n = \frac{\mathbf{J}}{\sqrt{\mathbf{J}^\dagger \mathbf{J}}} = \begin{pmatrix} \frac{A_x}{\sqrt{A_x^2 + A_y^2}} e^{i\delta_x(z)} \\ \frac{A_y}{\sqrt{A_x^2 + A_y^2}} e^{i\delta_y(z)} \end{pmatrix} = \begin{pmatrix} a_x e^{i\delta_x(z)} \\ a_y e^{i\delta_y(z)} \end{pmatrix} \tag{4.9}$$

where $a_x^2 + a_y^2 = 1$ and $a_x \geq 0$, $a_y \geq 0$. Define an angle $\beta = arctan(a_x/a_y)$ $(0 \leq \beta \leq \pi/2)$, the normalized Jones vector $\mathbf{J}_n$ can also be written as

$$\mathbf{J}_n = \begin{pmatrix} cos\beta e^{i\delta_x(z)} \\ sin\beta e^{i\delta_y(z)} \end{pmatrix} \tag{4.10}$$

By extracting the common factor $e^{i\delta_x(z)}$, Eqs. (4.9) and (4.10) can be rewritten as

$$\mathbf{J}_n = e^{i\delta_x(z)} \begin{pmatrix} a_x \\ a_y e^{i\delta(z)} \end{pmatrix} \tag{4.11}$$

Or

$$\mathbf{J}_n = e^{i\delta_x(z)} \begin{pmatrix} cos\beta \\ sin\beta e^{i\delta(z)} \end{pmatrix} \tag{4.12}$$

where $\delta(z) = \delta_y(z) - \delta_x(z)$. Note that the two expressions are equivalent, and either of them can be used in practice. The common factor $e^{i\delta_x(z)}$ can be often omitted because it does not affect the description of the polarization. Table 4.1 lists the normalized Jones vectors of frequently used polarization states.

Table 4.1 Normalized Jones vectors of selected polarization states.

Polarization states	Normalized Jones vector
1. LHP: linearly polarized light along x-axis ($a_x = 1, a_y = 0$, any δ) or ($\beta = \pi/2$, any δ)	$\mathbf{J}_{LHP} = \begin{pmatrix} 1 \\ 0 \end{pmatrix}$
2. LVP: linearly polarized light along y-axis ($a_x = 0, a_y = 1$, any δ) or ($\beta = \pi/2$, any δ)	$\mathbf{J}_{LVP} = \begin{pmatrix} 0 \\ 1 \end{pmatrix}$
3. L + 45P: linearly polarized light at 45° from x-axis ($a_x = a_y = \sqrt{2}/2, \delta = 2n\pi$) or ($\beta = \pi/4, \delta = 2n\pi$)	$\mathbf{J}_{+45} = \frac{1}{\sqrt{2}} \begin{pmatrix} 1 \\ 1 \end{pmatrix}$
4. L − 45P: linearly polarized at −45° from x-axis ($a_x = a_y = \sqrt{2}/2, \delta = (2n+1)\pi$) or ($\beta = \pi/4, \delta = (2n+1)\pi$)	$\mathbf{J}_{-45} - \frac{1}{\sqrt{2}} \begin{pmatrix} 1 \\ -1 \end{pmatrix}$
5. Arbitrarily orientated linear polarized light ($a_x^2 + a_y^2 = 1, \delta = n\pi$) or ($0 \leq \beta \leq \pi/2, \delta = n\pi$)	$\mathbf{J}_{linear} = \begin{pmatrix} cos\beta \\ \pm sin\beta \end{pmatrix}$
6. RCP: right-handed circularly polarized light ($a_x = a_y = \sqrt{2}/2, \delta = \pi/2 + 2n\pi$) or ($\beta = \pi/4, \delta = \pi/2 + 2n\pi$)	$\mathbf{J}_{RCP} = \frac{1}{\sqrt{2}} \begin{pmatrix} 1 \\ i \end{pmatrix}$
7. LCP: left-handed circularly polarized light ($a_x = a_y = \sqrt{2}/2, \delta = -\pi/2 + 2n\pi$) or ($\beta = \pi/4, \delta = -\pi/2 + 2n\pi$)	$\mathbf{J}_{LCP} = \frac{1}{\sqrt{2}} \begin{pmatrix} 1 \\ -i \end{pmatrix}$
8. Arbitrary elliptically polarized light	$\mathbf{J} = \begin{pmatrix} cos\beta \\ sin\beta e^{i\delta(z)} \end{pmatrix}$

4.1.2 Orthogonality of Jones Vectors

Two Jones vectors are orthogonal if their inner product (dot product) equals zero. From Table 4.1, one finds that the inner products of $\mathbf{J}_{LHP} \cdot \mathbf{J}_{LVP}$, $\mathbf{J}_{+45} \cdot \mathbf{J}_{-45}$, and $\mathbf{J}_{RCP} \cdot \mathbf{J}_{LCP}$ are all zero, i.e. LHP and LVP, L + 45P and L − 45P, RCP and LCP are mutually orthogonal pairs of polarization states. For an arbitrary elliptical polarization $\mathbf{J} = \begin{pmatrix} j_x & j_y \end{pmatrix}^T$, it is easy to find its corresponding orthogonal Jones vector as

$$\mathbf{J}_{\perp} = \begin{pmatrix} j_y^* \\ j_x^* e^{i\pi} \end{pmatrix} \text{ or } \begin{pmatrix} j_y^* \\ -j_x^* \end{pmatrix} \tag{4.13}$$

because

$$\mathbf{J}_{\perp} \cdot \mathbf{J} = \mathbf{J}_{\perp}^{\dagger} \mathbf{J} = \begin{pmatrix} j_y & -j_x \end{pmatrix} \begin{pmatrix} j_x \\ j_y \end{pmatrix} = j_y j_x - j_x j_y = 0 \tag{4.14}$$

For example, $\begin{pmatrix} sin\beta e^{-i\delta(z)} & -cos\beta \end{pmatrix}^T$ is an orthogonal vector of the elliptical polarization $\begin{pmatrix} cos\beta & sin\beta e^{i\delta(z)} \end{pmatrix}^T$.

Any two orthogonal Jones vectors can construct an orthogonal basis of 2D complex space. Any other Jones vector can be represented as the linear superposition of these two orthogonal vectors. Let $\mathbf{J}_1$ and $\mathbf{J}_2$ be the two normalized mutually orthogonal Jones vectors; then any Jones vector $\mathbf{J}$ can be represented as

$$\mathbf{J} = c_1 \mathbf{J}_1 + c_2 \mathbf{J}_2 \tag{4.15}$$

where the complex coefficients c_1 and c_2 can be calculated from the following inner products

$$\mathbf{J}_1^{\dagger} \mathbf{J} = c_1 \mathbf{J}_1^{\dagger} \mathbf{J}_1 + c_2 \mathbf{J}_1^{\dagger} \mathbf{J}_2 = c_1 \times 1 + c_2 \times 0 = c_1 \tag{4.16}$$

$$\mathbf{J}_2^{\dagger} \mathbf{J} = c_1 \mathbf{J}_2^{\dagger} \mathbf{J}_1 + c_2 \mathbf{J}_2^{\dagger} \mathbf{J}_2 = c_1 \times 0 + c_2 \times 1 = c_2 \tag{4.17}$$

One generally selects {LHP, LVP} as a pair of orthogonal basis to represent a Jones vector. However, sometimes it is more convenient to use a different orthogonal basis to simplify the deviation and the final expressions. For example, when studying the polarization properties of an optically active medium, one often uses {RCP, LCP} as the orthogonal basis. For example, a linear polarization can be represented as the superposition of a right- and a left-handed circular polarization:

$$\mathbf{J}_{linear} = c_1\mathbf{J}_{RCP} + c_2\mathbf{J}_{LCP} \tag{4.18}$$

Substituting the normalized Jones vector of linear polarization (see Table 4.1) in Eqs. (4.16) and (4.17), one finds the coefficients c_1 and c_2 of an arbitrary linear polarization in the orthogonal basis {RCP, LCP} by

$$c_1 = \mathbf{J}^{\dagger}_{RCP}\mathbf{J}_{linear} = \frac{\sqrt{2}}{2}(1, -i)\begin{pmatrix} cos\beta \\ \pm sin\beta \end{pmatrix} = \frac{\sqrt{2}}{2}(cos\beta \mp isin\beta) = \frac{\sqrt{2}}{2}e^{\mp i\beta} \tag{4.19}$$

$$c_2 = \mathbf{J}^{\dagger}_{LCP}\mathbf{J}_{linear} = \frac{\sqrt{2}}{2}(1, i)\begin{pmatrix} cos\beta \\ \pm sin\beta \end{pmatrix} = \frac{\sqrt{2}}{2}(cos\beta \pm isin\beta) = \frac{\sqrt{2}}{2}e^{\pm i\beta} \tag{4.20}$$

As an example, the horizontal and vertical linear polarization $\mathbf{J}_{LHP}$ $(\beta = 0)$ and $\mathbf{J}_{LVP}$ $(\beta = \pi/2)$ can be represented by

$$\mathbf{J}_{LHP} = \begin{pmatrix} 1 \\ 0 \end{pmatrix} = \frac{\sqrt{2}}{2}(\mathbf{J}_{LCP} + \mathbf{J}_{RCP}) \tag{4.21}$$

$$\mathbf{J}_{LVP} = \begin{pmatrix} 0 \\ 1 \end{pmatrix} = \frac{\sqrt{2}}{2}i(\mathbf{J}_{LCP} - \mathbf{J}_{RCP}) \tag{4.22}$$

4.1.3 Linear Independence of Jones Vectors

In a real three-dimensional (3D) space, if a vector equals another vector multiplied by a real constant, we call these two vectors as parallel. Similarly, in a complex 2D space, two vectors are called linearly dependent if one can find a complex number to make one of them equal to the multiplication of the other vector with the complex number. Mathematically, one can judge whether two vectors $\mathbf{J}_1 = \begin{pmatrix} j_{1x} & j_{1y} \end{pmatrix}^T$ and $\mathbf{J}_2 = \begin{pmatrix} j_{2x} & j_{2y} \end{pmatrix}^T$ are linearly dependent by calculating the determinant of the $\mathbf{V}$ matrix constructed from $\mathbf{J}_1$ and $\mathbf{J}_2$ by

$$\mathbf{V} = \begin{pmatrix} \mathbf{J}_1 & \mathbf{J}_2 \end{pmatrix} = \begin{pmatrix} j_{1x} & j_{2x} \\ j_{1y} & j_{2y} \end{pmatrix} \tag{4.23}$$

That is, the first column of $\mathbf{V}$ is vector $\mathbf{J}_1$, and the second column is the vector $\mathbf{J}_2$. If the determinant of the $\mathbf{V}$ matrix is zero, $(det(\mathbf{V}) = j_{1x}j_{2y} - j_{2x}j_{1y} = 0)$, then $\mathbf{J}_1$ and $\mathbf{J}_2$ are linearly dependent; if $det(\mathbf{V}) \neq 0$, $\mathbf{J}_1$ and $\mathbf{J}_2$ are linearly independent. For example, the matrix $\mathbf{V}$ of $\mathbf{J}_{RCP}$ and $\mathbf{J}_{LCP}$ is

$$\mathbf{V} = \begin{pmatrix} \frac{\sqrt{2}}{2} & \frac{\sqrt{2}}{2} \\ \frac{\sqrt{2}}{2}i & -\frac{\sqrt{2}}{2}i \end{pmatrix} \tag{4.24}$$

Because its determinant $det(\mathbf{V}) = -i \neq 0$, $\mathbf{J}_{RCP}$ and $\mathbf{J}_{LCP}$ are linearly independent.

As another example, the $\mathbf{V}$ matrix of $\mathbf{J}_{LHP}$ and $\mathbf{J}_{RCP}$ is

$$\mathbf{V} = \begin{pmatrix} 1 & \frac{\sqrt{2}}{2} \\ 0 & \frac{\sqrt{2}}{2}i \end{pmatrix} \tag{4.25}$$

Again, because its determinant $det(\mathbf{V}) = i\sqrt{2}/2 \neq 0$, $\mathbf{J}_{LHP}$ and $\mathbf{J}_{RCP}$ are also linearly independent.

4.2 Jones Matrix of Optical Devices

In Section 4.1, we have defined a 2×1 complex Jones vector to describe the state of polarization (SOP) of a monochromatic light beam. In this section, we will introduce how to use a matrix to represent the transformation of an incident SOP to an exit SOP after the light beam passes through an optical system. R. Clark Jones (1941) has already proved that this transformation can be represented as the following matrix multiplication:

$$\begin{pmatrix} E_{out,x} \\ E_{out,y} \end{pmatrix} = \begin{pmatrix} m_{11} & m_{12} \\ m_{21} & m_{22} \end{pmatrix} \begin{pmatrix} E_{in,x} \\ E_{in,y} \end{pmatrix} \tag{4.26}$$

or

$$\mathbf{J}_{out} = \mathbf{M}\mathbf{J}_{in} \tag{4.27}$$

where the 2×2 matrix $\mathbf{M}$ is called the Jones Matrix of the optical system, and the entries m_{11}, m_{12}, m_{21}, and m_{22} are complex numbers determined by the optical properties of the medium where the light propagates. In the following discussions, we will derive the Jones matrix $\mathbf{M}$ of some basic optical elements, including the retarder, the partial polarizer, the rotator, and the ideal mirror.

When the monochromatic light beam propagates in an optical medium, it interacts with the medium, and its polarization changes along the beam propagation path. Its SOP can be altered by (i) the changes of the relative amplitude of the two orthogonal components due to the anisotropy of optical absorption in the medium; (ii) the changes of the relative phase between the two orthogonal components because of the birefringence; (iii) the changes of the orientation of the two orthogonal components because of the optical activity or Faraday effect; and (iv) the changes of both relative phase and amplitude caused by the reflection at the boundary of two different media.

If an optical device causes different attenuations of two orthogonal polarization components, it is called a partial polarizer or diattenuator, as mentioned in Chapter 2. If an optical element introduces a differential phase shift between two orthogonal polarization components, it is called a retarder, a wave plate, a compensator, or a phase shifter. An optical component that changes the orientation of two orthogonal polarization components in space is called a rotator.

4.2.1 Jones Matrix of Optical Elements

4.2.1.1 Linear Retarder and Linear Partial Polarizer in Principal Coordinates

A linear retardation plate, sometimes also called a retarder, a wave plate, or a phase shifter, is a transparent plate made of an anisotropic optical crystal with its optical axis parallel to the plate surfaces, as shown in Figure 4.2. Consider a monochromatic plane wave propagates along $+z$ direction to pass through a retardation plate with its transmission surfaces perpendicular to the light propagating direction, the electric field of light can be decomposed into two linear orthogonal components, one experiencing a lower refractive index (fast-axis), and the other experiencing a higher refractive index (slow axis). In the principal coordinates, the fast axis is aligned to the x-axis, while the slow axis is parallel to the y-axis. After light passes through the plate, the electric field of the emerging light can be written as

$$\mathbf{E} = e^{-i\frac{2\pi(n_x - i\kappa_x)d}{\lambda_0}} E_{in,x}\hat{\mathbf{e}}_x + e^{-i\frac{2\pi(n_y - i\kappa_y)d}{\lambda_0}} E_{in,y}\hat{\mathbf{e}}_y \tag{4.28}$$

Or in matrix form:

$$\mathbf{E} = \begin{pmatrix} E_{out,x} \\ E_{out,y} \end{pmatrix} = \begin{pmatrix} e^{-i\frac{2\pi(n_x - i\kappa_x)d}{\lambda_0}} & 0 \\ 0 & e^{-i\frac{2\pi(n_y - i\kappa_y)d}{\lambda_0}} \end{pmatrix} \begin{pmatrix} E_{in,x} \\ E_{in,y} \end{pmatrix} \tag{4.29}$$

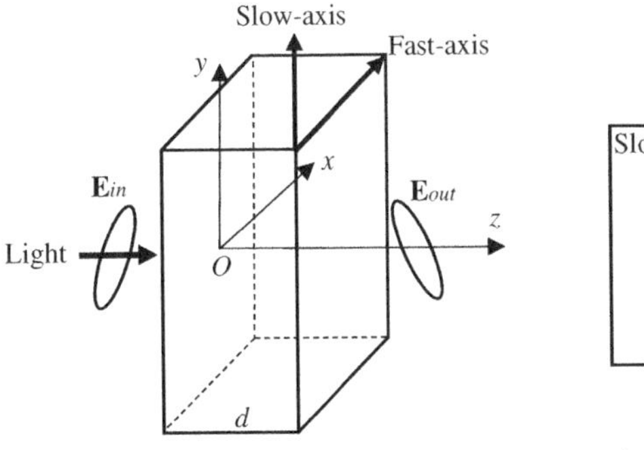

(a) Retarder or partial polarizer

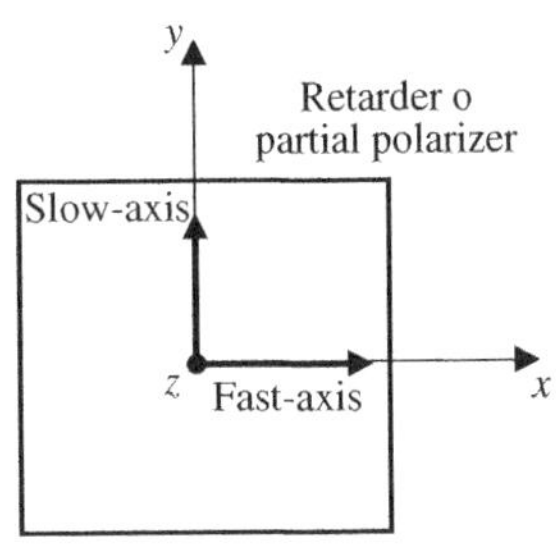

(b) Principal coordinate system

Figure 4.2 A retarder or partial polarizer in principal coordinates.

where n_x and n_y are the refractive indices when the electric field oscillates along retarder x and y principal axes, respectively; κ_x and κ_y represent the principal attenuation coefficients of the plate, assuming the wave plate has diattenuation or differential loss.

4.2.1.2 Ideal Retarder in Principal Coordinates

An ideal retarder has no differential loss between the two principal polarization states, i.e. $\kappa_x = \kappa_y = \kappa_0$ so that Eq. (4.29) can be reduced to

$$\begin{pmatrix} E_{out,x} \\ E_{out,y} \end{pmatrix} = e^{-i\frac{2\pi}{\lambda_0}\frac{n_x+n_y}{2}d - \frac{2\pi}{\lambda_0}\kappa_0 d} \begin{pmatrix} e^{i\frac{\delta}{2}} & 0 \\ 0 & e^{-i\frac{\delta}{2}} \end{pmatrix} \begin{pmatrix} E_{in,x} \\ E_{in,y} \end{pmatrix} \tag{4.30}$$

where the phase retardation $\delta = \frac{2\pi}{\lambda_0}\frac{n_y - n_x}{2}d$ is caused by the linear birefringence of retarder. The exponential $e^{-i\frac{2\pi}{\lambda_0}\frac{n_x+n_y}{2}d - \frac{2\pi}{\lambda_0}\kappa_0 d}$ contains the absolute phase and loss of both electric field components, which can be omitted if one is only interested in the polarization states of light, as discussed in Section 4.1. Therefore, Eq. (4.30) is simplified to

$$\mathbf{E}_{out} = \begin{pmatrix} e^{i\frac{\delta}{2}} & 0 \\ 0 & e^{-i\frac{\delta}{2}} \end{pmatrix} \mathbf{E}_{in} \tag{4.31}$$

Consequently, the retardation plate can be represented by the following diagonal matrix

$$\mathbf{M}_{RP} = \begin{pmatrix} e^{i\frac{\delta}{2}} & 0 \\ 0 & e^{-i\frac{\delta}{2}} \end{pmatrix} \tag{4.32}$$

4.2.1.3 The Ideal Partial Polarizer in Principal Coordinates

For an ideal partial polarizer, there is no phase difference between the two orthogonal polarization states, i.e. $n_x = n_y = n_0$. Equation (4.29) can be reduced to

$$\begin{pmatrix} E_{out,x} \\ E_{out,y} \end{pmatrix} = e^{-i\frac{2\pi n_0}{\lambda_0}d} \begin{pmatrix} p_1 & 0 \\ 0 & p_2 \end{pmatrix} \begin{pmatrix} E_{in,x} \\ E_{in,y} \end{pmatrix} \tag{4.33}$$

where $0 \le p_1 = e^{-\frac{2\pi}{\lambda_0}\kappa_x d} \le 1$ and $0 \le p_2 = e^{-\frac{2\pi}{\lambda_0}\kappa_y d} \le 1$, p_1^2 and p_2^2 are the transmittances when the electric field oscillates along the x or y-axis, respectively. Omitting the common phase factor $e^{-i\frac{2\pi n_0}{\lambda_0}d}$, the partial polarizer can be represented by the following real diagonal matrix

$$\mathbf{M}_P = \begin{pmatrix} p_1 & 0 \\ 0 & p_2 \end{pmatrix} \tag{4.34}$$

4.2.1.4 Jones Matrix of a Rotator

When light propagates in an optically active media or a Faraday rotator along the +z axis, the electric field will be rotated around the z-axis without changing the shape of the polarization ellipse (see Figure 4.3a). Assuming that the incident light field $\mathbf{E}_{in}$ is the superposition of two mutually orthogonal linearly polarized electric fields $\mathbf{E}_{in,x} = \varepsilon_{in,x}\hat{\mathbf{e}}_x$ and $\mathbf{E}_{in,y} = \varepsilon_{in,y}\hat{\mathbf{e}}_y$ (see Figure 4.3b), where $\hat{\mathbf{e}}_x$ and $\hat{\mathbf{e}}_y$ are the unit vectors along the positive x and y-axes, respectively, one obtains

$$\mathbf{E}_{in} = \mathbf{E}_{in,x} + \mathbf{E}_{in,y} = \varepsilon_{in,x}\hat{\mathbf{e}}_x + \varepsilon_{in,y}\hat{\mathbf{e}}_y \tag{4.35}$$

After light passes through the rotator plate, $\hat{\mathbf{e}}_x$ and $\hat{\mathbf{e}}_y$ will be rotated to $\hat{\mathbf{e}}'_x$ and $\hat{\mathbf{e}}'_y$, respectively. Let the rotation angle be β, which is positive for counterclockwise rotation and negative for clockwise rotation, $\hat{e}'_x$ and $\hat{e}'_y$ can be expressed as (see Figure 4.3c)

$$\hat{e}'_x = cos\beta\hat{e}_x + sin\beta\hat{e}_y \tag{4.36}$$

$$\hat{e}'_y = -sin\beta\hat{e}_x + cos\beta\hat{e}_y \tag{4.37}$$

Thus we have

$$\mathbf{E}_{out} = \varepsilon_{in,x}\hat{\mathbf{e}}'_x + \varepsilon_{in,y}\hat{\mathbf{e}}'_y = (\varepsilon_{in,x}cos\beta - \varepsilon_{in,y}sin\beta)\hat{\mathbf{e}}_x + (\varepsilon_{in,x}sin\beta + \varepsilon_{in,y}cos\beta)\hat{\mathbf{e}}_y \tag{4.38}$$

which is equivalent to the following matrix transformation:

$$\begin{pmatrix} \varepsilon_{out,x} \\ \varepsilon_{out,y} \end{pmatrix} = \begin{pmatrix} cos\beta & -sin\beta \\ sin\beta & cos\beta \end{pmatrix} \begin{pmatrix} \varepsilon_{in,x} \\ \varepsilon_{in,y} \end{pmatrix} = \mathbf{M}_R(\beta)\mathbf{J}_{in} \tag{4.39}$$

where $(\varepsilon_{out,x}, \varepsilon_{out,y})^T$ and $(\varepsilon_{in,x}, \varepsilon_{in,y})^T$ are the Jones vectors of the emerging light and incident light, respectively, and the transformation between them is represented by the rotation matrix $\mathbf{M}_R(\beta)$:

$$\mathbf{M}_R(\beta) = \begin{pmatrix} cos\beta & -sin\beta \\ sin\beta & cos\beta \end{pmatrix} \tag{4.40}$$

At the fundamental physics level, polarization rotation in an optically active medium is caused by circular birefringence. In an optically active medium, right- and left-handed circular polarizations have different refractive indices n_R and n_L, respectively, which will cause a phase difference δ between LCP and RCP polarization states after light passes through it. The incident electric field $\mathbf{E}_{in}$ can be represented by the linear superposition of RCP and LCP polarization vectors according to Eqs. (4.21) and (4.22):

$$\mathbf{E}_{in} = \varepsilon_{in,x}\mathbf{J}_{LHP} + \varepsilon_{in,y}\mathbf{J}_{LVP} = \frac{\sqrt{2}}{2}(\varepsilon_{in,x} - i\varepsilon_{in,y})\mathbf{J}_{RCP} + \frac{\sqrt{2}}{2}(\varepsilon_{in,x} + i\varepsilon_{in,y})\mathbf{J}_{LCP} \tag{4.41}$$

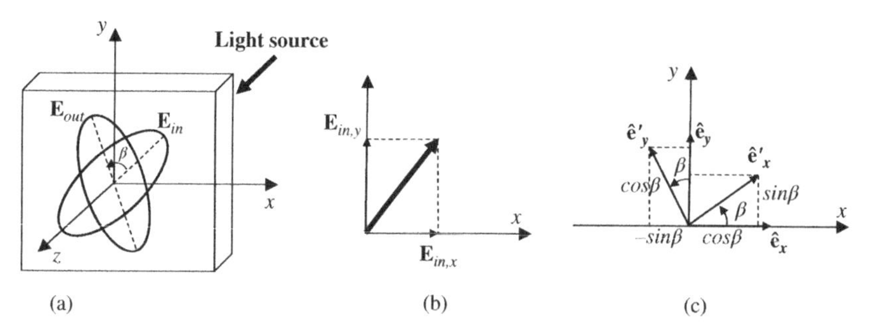

Figure 4.3 Polarization rotator. (a) Polarization ellipse is rotated without shape change. (b) Two orthogonal components of an electric field. (c) Rotation of unit vectors of two orthogonal linear polarization states.

After light passes through the rotator, the emerging electric field becomes

$$\begin{aligned}
\mathbf{E}_{out} &= \frac{\sqrt{2}}{2}(\varepsilon_{in,x} - i\varepsilon_{in,y})e^{-i\frac{2\pi}{\lambda_0}n_R \mathrm{d}}\mathbf{J}_{RCP} + \frac{\sqrt{2}}{2}(\varepsilon_{in,x} + i\varepsilon_{in,y})e^{-i\frac{2\pi}{\lambda_0}n_L \mathrm{d}}\mathbf{J}_{LCP} \\
&= \frac{1}{2}(\varepsilon_{in,x} - i\varepsilon_{in,y})e^{-i\frac{2\pi}{\lambda_0}n_R \mathrm{d}}\begin{pmatrix}1\\ i\end{pmatrix} + \frac{1}{2}(\varepsilon_{in,x} + i\varepsilon_{in,y})e^{-i\frac{2\pi}{\lambda_0}n_L \mathrm{d}}\begin{pmatrix}1\\ -i\end{pmatrix} \\
&= \frac{1}{2}e^{-i\frac{2\pi}{\lambda_0}n_R \mathrm{d}}\begin{pmatrix}1 & -i\\ i & 1\end{pmatrix}\begin{pmatrix}\varepsilon_{in,x}\\ \varepsilon_{in,y}\end{pmatrix} + \frac{1}{2}e^{-i\frac{2\pi}{\lambda_0}n_L \mathrm{d}}\begin{pmatrix}1 & i\\ -i & 1\end{pmatrix}\begin{pmatrix}\varepsilon_{in,x}\\ \varepsilon_{in,y}\end{pmatrix}
\end{aligned} \tag{4.42}$$

where λ_0 is the wavelength of light in vacuum and d is the thickness of the rotator plate. Let $\delta = \delta_{LCP} - \delta_{RCP} = -\frac{2\pi}{\lambda_0}n_L \mathrm{d} + \frac{2\pi}{\lambda_0}n_R \mathrm{d} = \frac{2\pi}{\lambda_0}(n_R - n_L)\mathrm{d}$, then the aforementioned equation can be reduced to

$$\begin{aligned}
\mathbf{E}_{out} &= e^{-i\frac{2\pi}{\lambda_0}\frac{(n_R+n_L)\mathrm{d}}{2}}\left[\frac{1}{2}e^{-i\frac{\delta}{2}}\begin{pmatrix}1 & -i\\ i & 1\end{pmatrix} + \frac{1}{2}e^{i\frac{\delta}{2}}\begin{pmatrix}1 & i\\ -i & 1\end{pmatrix}\right]\begin{pmatrix}\varepsilon_{in,x}\\ \varepsilon_{in,y}\end{pmatrix} \\
&= e^{-i\frac{2\pi}{\lambda_0}\frac{(n_R+n_L)\mathrm{d}}{2}}\begin{pmatrix}\cos(\delta/2) & -\sin(\delta/2)\\ \sin(\delta/2) & \cos(\delta/2)\end{pmatrix}\begin{pmatrix}\varepsilon_{in,x}\\ \varepsilon_{in,y}\end{pmatrix}
\end{aligned} \tag{4.43}$$

Neglecting the common phase factor $e^{-i\frac{2\pi}{\lambda_0}\frac{(n_R+n_L)\mathrm{d}}{2}}$, Eq. (4.43) can be reduced to

$$\begin{pmatrix}\varepsilon_{out,x}\\ \varepsilon_{out,y}\end{pmatrix} = \begin{pmatrix}\cos\left(\frac{\delta}{2}\right) & -\sin\left(\frac{\delta}{2}\right)\\ \sin\left(\frac{\delta}{2}\right) & \cos\left(\frac{\delta}{2}\right)\end{pmatrix}\begin{pmatrix}\varepsilon_{in,x}\\ \varepsilon_{in,x}\end{pmatrix} \quad \text{or } \mathbf{J}_{out} = \mathbf{M}_R\left(\frac{\delta}{2}\right)\mathbf{J}_{in} \tag{4.44}$$

Comparing Eqs. (4.40) and (4.44), one concludes that a circular retardation plate with a circular phase difference $\delta = \delta_{LCP} - \delta_{RCP} = \frac{2\pi}{\lambda_0}(n_R - n_L)\mathrm{d}$ is equivalent to a rotator with a counterclockwise rotation angle of $\delta/2$.

4.2.1.5 Jones Matrix Transformation Between Two Reference Frames

In general, we represent the Jones vector or Jones matrix in the laboratory (x, y, z) Cartesian coordinates, where $+z$ is the propagation direction of light, the x-axis is horizontal, and the y-axis is upward vertical. When deriving the Jones matrix of the retarder and the partial polarizer in the principal coordinate system earlier, we assume the principal axes are parallel to the x- and y-axes of laboratory coordinates. In the following discussions, we will derive the general Jones matrix formulism when the principal axes are rotated at an angle about origin O. As shown in Figure 4.4, the principal coordinate (x', y', z) and laboratory (x, y, z) coordinate system have the same z-axis and origin O, but x' and y' axes are rotated counterclockwise in the x–y plane by an angle θ. Assume an electric field vector $\mathbf{E}$ in the (x', y') coordinate system is expressed as

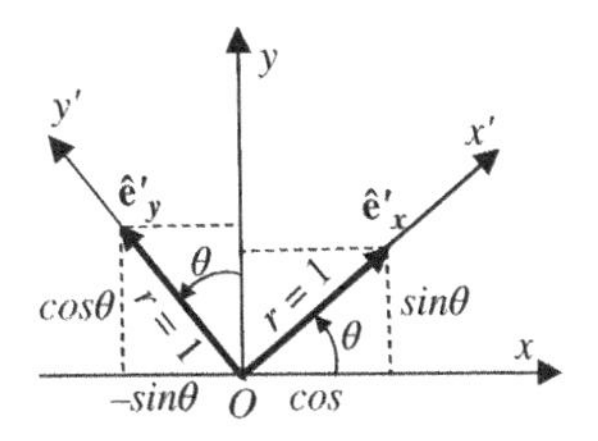

Figure 4.4 Rotation between principal coordinates and laboratory Cartesian coordinate.

$$\mathbf{E} = \varepsilon'_x\hat{\mathbf{e}}'_x + \varepsilon'_y\hat{\mathbf{e}}'_y \tag{4.45}$$

The same electric field in the (x, y) coordinates system is

$$\mathbf{E} = \varepsilon_x\hat{\mathbf{e}}_x + \varepsilon_y\hat{\mathbf{e}}_y \tag{4.46}$$

where $\hat{\mathbf{e}}'_x$ and $\hat{\mathbf{e}}'_y$ are the unit vectors along the positive x' and y'-axes, respectively, while $\hat{\mathbf{e}}_x$ and $\hat{\mathbf{e}}_y$ are the unit vectors along the positive x and y-axes. Inspecting Figure 4.4, one finds that $(\hat{\mathbf{e}}'_x, \hat{\mathbf{e}}'_y)$

and ($\hat{\mathbf{e}}_x$, $\hat{\mathbf{e}}_y$) are connected via the following relations

$$\hat{\mathbf{e}}'_x = cos\theta\hat{\mathbf{e}}_x + sin\theta\hat{\mathbf{e}}_y \quad (4.47)$$

$$\hat{\mathbf{e}}'_y = -sin\theta\hat{\mathbf{e}}_x + cos\theta\hat{\mathbf{e}}_y \quad (4.48)$$

Substituting Eqs. (4.47) and (4.48) in (4.45), one obtains

$$\mathbf{E} = \varepsilon'_x\hat{\mathbf{e}}'_x + \varepsilon'_x\hat{\mathbf{e}}'_y = \left(\varepsilon'_x cos\theta - \varepsilon'_y sin\theta\right)\hat{\mathbf{e}}_x + \left(\varepsilon'_x sin\theta + \varepsilon'_y cos\theta\right)\hat{\mathbf{e}}_y \quad (4.49)$$

The matrix form of the transformation between two vectors can be written as

$$\begin{pmatrix}\varepsilon_x\\ \varepsilon_y\end{pmatrix} = \begin{pmatrix}cos\theta & -sin\theta\\ sin\theta & cos\theta\end{pmatrix}\begin{pmatrix}\varepsilon'_x\\ \varepsilon'_y\end{pmatrix} = \mathbf{R}(\theta)\begin{pmatrix}\varepsilon'_x\\ \varepsilon'_y\end{pmatrix} \quad (4.50a)$$

$$\begin{pmatrix}\varepsilon'_x\\ \varepsilon'_y\end{pmatrix} = \begin{pmatrix}cos\theta & sin\theta\\ -sin\theta & cos\theta\end{pmatrix}\begin{pmatrix}\varepsilon_x\\ \varepsilon_y\end{pmatrix} = \mathbf{R}(-\theta)\begin{pmatrix}\varepsilon_x\\ \varepsilon_y\end{pmatrix} \quad (4.50b)$$

where $\mathbf{R}$ is a 2×2 rotation matrix

$$\mathbf{R}(\theta) = \begin{pmatrix}cos\theta & -sin\theta\\ sin\theta & cos\theta\end{pmatrix} \quad (4.51)$$

in 2D space with the following properties:

$$\mathbf{R}(-\theta) = \mathbf{R}^{-1}(\theta) \quad \text{or} \quad \mathbf{R}(\theta)\mathbf{R}(-\theta) = \mathbf{R}(-\theta)\mathbf{R}(\theta) = \mathbf{I} \quad (4.52)$$

$$\mathbf{R}(\theta_1)\mathbf{R}(\theta_2) = \mathbf{R}(\theta_1 + \theta_2) \quad (4.53)$$

In Eq. (4.52), $\mathbf{I}$ represents a 2×2 identity matrix. Assuming the optical system can be represented by a matrix $\mathbf{M}'$ in the (x', y', z) reference frame, the transformation between the input and the output Jones vectors $\mathbf{J}'_{in}$ and $\mathbf{J}'_{out}$ in the same reference frame can be written as

$$\mathbf{J}'_{out} = \mathbf{M}'\mathbf{J}'_{in} \quad (4.54)$$

Substitution of Eq. (4.50b) in (4.54) yields

$$\mathbf{R}(-\theta)\mathbf{J}_{out} = \mathbf{M}'\mathbf{R}(-\theta)\mathbf{J}_{in} \rightarrow \mathbf{R}(\theta)\mathbf{R}(-\theta)\mathbf{J}_{out} = \mathbf{R}(\theta)\mathbf{M}'\mathbf{R}(-\theta)\,\mathbf{J}_{in}$$
$$\rightarrow \mathbf{J}_{out} = \mathbf{R}(\theta)\,\mathbf{M}'\mathbf{R}(-\theta)\mathbf{J}_{in} = \mathbf{M}\mathbf{J}_{in} \quad (4.55)$$

where $\mathbf{J}_{out}$ and $\mathbf{J}_{in}$ are the incident and emerging Jones vectors in the laboratory reference frame, respectively. Therefore the transformation between the Jones matrix $\mathbf{M}$ in the laboratory reference frame and the $\mathbf{M}'$ in principal reference frame can be obtained as

$$\mathbf{M} = \mathbf{R}(\theta)\mathbf{M}'\mathbf{R}(-\theta) \quad (4.56)$$

For example, the Jones matrix of a linear retardation plate with a fast-axis at an angle θ from the x-axis can be obtained from

$$\mathbf{M}_{RP}(\theta,\delta) = \mathbf{R}(\theta)\begin{pmatrix}e^{\frac{i\delta}{2}} & 0\\ 0 & e^{-\frac{i\delta}{2}}\end{pmatrix}\mathbf{R}(-\theta) \quad (4.57)$$

Or

$$\mathbf{M}_{RP}(\theta,\delta) = \begin{pmatrix}cos^2\theta e^{i\frac{\delta}{2}} + sin^2\theta e^{-i\frac{\delta}{2}} & i(sin2\theta)sin\frac{\delta}{2}\\ i(sin2\theta)sin\frac{\delta}{2} & cos^2\theta e^{-i\frac{\delta}{2}} + sin^2\theta e^{+i\frac{\delta}{2}}\end{pmatrix} \quad (4.58)$$

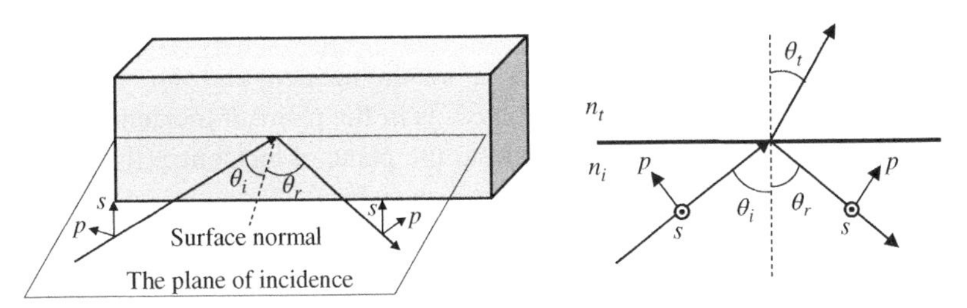

Figure 4.5 Illustration of reflection and refraction. θ_i, θ_r, and θ_t are the angles of incidence, reflection, and refraction, respectively. *s* and *p* denote the polarizations of the *s* and *p* rays, respectively. The *s* ray is polarized perpendicular to the plane of incidence and the *p* ray is polarized in the plane of incidence.

Similarly, by substituting Eq. (4.34) in Eq. (4.56), one obtains the general Jones matrix of a partial polarizer

$$\mathbf{M}_P(\theta) = \mathbf{R}(\theta) \begin{pmatrix} p_1 & 0 \\ 0 & p_2 \end{pmatrix} \mathbf{R}(-\theta) \tag{4.59}$$

Or

$$\mathbf{M}_P(\theta) = \begin{pmatrix} p_1 cos^2\theta + p_2 sin^2\theta & (p_1 - p_2) sin\theta\, cos\theta \\ (p_1 - p_2) sin\theta\, cos\theta & p_1 sin^2\theta + p_2 cos^2\theta \end{pmatrix} \tag{4.60}$$

For an ideal polarizer, $p_1 = 1$ and $p_2 = 0$, Eq. (4.60) can be reduced to

$$\mathbf{M}_P = \begin{pmatrix} cos^2\theta & sin\theta\, cos\theta \\ sin\theta\, cos\theta & sin^2\theta \end{pmatrix} \tag{4.61}$$

4.2.2 Jones Matrix of Reflection

When a ray of light strikes onto a smooth interface between two homogeneous media of different refractive indexes, it will be split into reflected and refracted beams, as shown in Figure 4.5. The geometrical relationship of these three beams is determined by the law of reflection and Snell's Law.

4.2.2.1 Law of Reflection

The reflected ray is in the plane of incidence and the angle of reflection θ_r equals to the angle of incidence θ_i.

$$\theta_i = \theta_r \tag{4.62}$$

4.2.2.2 Snell's Law

The refracted ray is in the plane of incidence and obeys

$$n_i sin\theta_i = n_t sin\theta_t \tag{4.63}$$

Here the plane of incidence is defined as the plane containing the incident ray and the surface normal at the point of incidence; the angle of incidence θ_i is the angle from the surface normal to incident ray; the angle of reflection θ_r is the angle from the surface normal to the reflected beam; the angle of refraction θ_t is the angle from the surface normal to the refracted beam; n_i is the refractive index of optical medium in which the incident ray propagates; n_t is the refractive index of optical medium in which the refracted ray propagates.

4.2.2.3 Fresnel's Equations

Let's define the Cartesian coordinates as shown in Figure 4.5 for the incident and the reflected beams. $+z$ is the direction of light propagation, while the x-axis is in the plane of incidence (the p ray polarization direction) and the y-axis is perpendicular to the plane of incidence (the s ray polarization direction). The electric field of the reflected beam is given by

$$\mathbf{E}_r = \begin{pmatrix} \varepsilon_{r,p} \\ \varepsilon_{r,s} \end{pmatrix} = \begin{pmatrix} r_p & 0 \\ 0 & r_s \end{pmatrix} \begin{pmatrix} \varepsilon_{i,p} \\ \varepsilon_{i,s} \end{pmatrix} \tag{4.64}$$

where $\varepsilon_{r,s}$ and $\varepsilon_{i,s}$ are the amplitudes of the reflection and incident electric fields of the s-polarization (or the TE-wave), respectively, and $\varepsilon_{r,p}$ and $\varepsilon_{i,p}$ are the amplitudes of the reflection and incident electric fields of the p-polarization (or the TM-wave), respectively. Finally, r_p and r_s are the reflection coefficients of the p- and s-rays, respectively, and can be determined by the Fresnel formulae (Born and Wolf 1999):

$$r_p = \frac{\varepsilon_{r,p}}{\varepsilon_{i,p}} = \frac{n_t \cos\theta_i - n_i \cos\theta_t}{n_t \cos\theta_i + n_i \cos\theta_t}$$
$$r_s = \frac{\varepsilon_{r,s}}{\varepsilon_{i,s}} = \frac{n_i \cos\theta_i - n_t \cos\theta_t}{n_i \cos\theta_i + n_t \cos\theta_t} \tag{4.65}$$

Except at normal incidence ($\theta_i = 0°$), the absolute values of r_p and r_s are not equal in general such that the reflection can be considered a partial polarizer because its Jones matrix has the same form as partial polarizer (see Eq. (4.34)). This relates to how Malus discovered polarization in 1808 by observing the reflections from the windows of a church in front of his residence by rotating a polarizer (made with a natural calcite crystal), as described in Chapters 1 and 2, because different polarizations have different reflection coefficients.

Since θ_t is related to θ_i through Snell's law (4.63), it can be eliminated in Eq. (4.65) using

$$\cos\theta_t = \sqrt{1 - \frac{n_i^2 \sin^2\theta_i}{n_t^2}} \tag{4.66}$$

Consequently, Eq. (4.65) can also be written in a form only containing incidence angle θ_i:

$$r_p = \frac{n^2 \cos\theta_i - \sqrt{n^2 - \sin^2\theta_i}}{n^2 \cos\theta_i + \sqrt{n^2 - \sin^2\theta_i}}$$
$$r_s = \frac{\cos\theta_i - \sqrt{n^2 - \sin^2\theta_i}}{\cos\theta_i + \sqrt{n^2 - \sin^2\theta_i}}, \quad \text{where } n = n_t/n_i \tag{4.67}$$

Figure 4.6 shows the plots of r_p and r_s as a function of the incident angle θ_i for reflection at the interface between glass and air with refractive indexes of 1.5 and 1, respectively. It can be seen that before Brewster's angle (to be discussed shortly), the p- and s-rays have the opposite signs, indicating a phase difference of π for the two reflected rays.

The reflectivity of the s- and p-rays can be calculated with r_s^2 and r_p^2 in Eq. (4.120), respectively. Figure 4.7 shows the reflectivity of the two polarizations as a function of the incidence angle θ_i at the glass to air interface.

Now, we will consider the Jones matrix of the reflection with some particular incident angles:

Case 1: Normal incidence ($\theta_i = 0$)

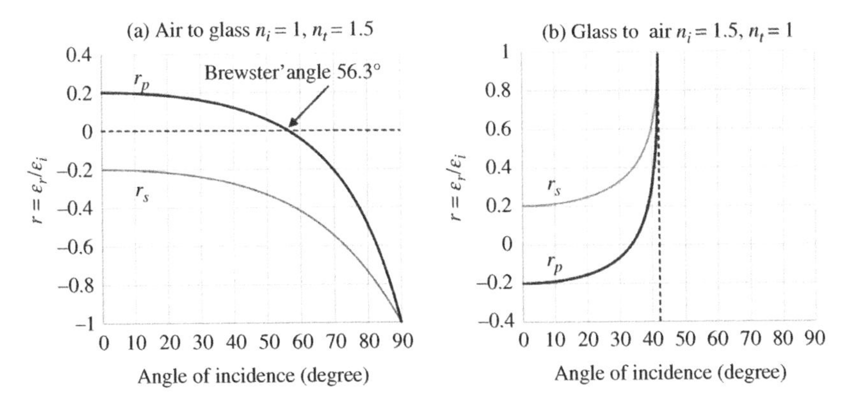

Figure 4.6 Electric-field amplitude ratio of incident to emerging light as a function of the incident angle. (a) External reflection: air ($n_i = 1$) to a glass ($n_t = 1.5$). (b) Internal reflection: from glass ($n_i = 1.5$) to air ($n_t = 1$).

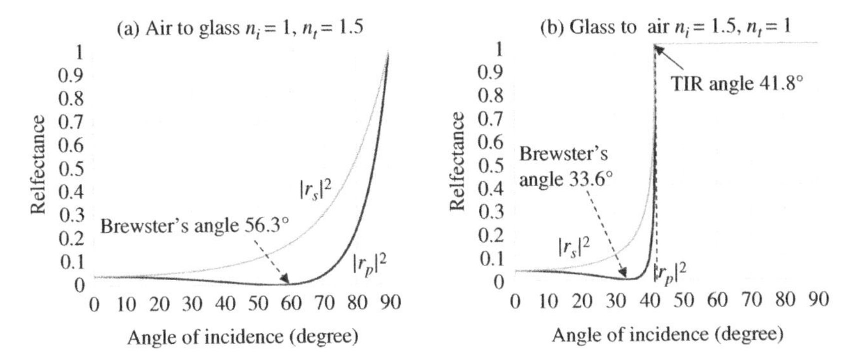

Figure 4.7 Reflectivity as a function of the angle of incidence. (a) External reflection: air ($n_i = 1$) to glass ($n_t = 1.5$). (b) Internal reflection: glass ($n_i = 1.5$) to air ($n_t = 1$).

When the incident ray is normal to the interface of the two media, the law of reflection and Snell's law lead to $\theta_i = \theta_r = \theta_t = 0$. The Jones matrix of reflection at normal incidence between two media can be obtained from Eq. (4.64) or (4.67) as

$$\text{External reflection } (n_t > n_i): \begin{pmatrix} |r| & 0 \\ 0 & -|r| \end{pmatrix}, \quad \text{where } r = r_p = -r_s = \frac{n_t - n_i}{n_t + n_i} \tag{4.68}$$

$$\text{Internal reflection } (n_t < n_i): e^{i\pi} \begin{pmatrix} |r| & 0 \\ 0 & -|r| \end{pmatrix}, \quad \text{where } r = r_p = -r_s = \frac{n_t - n_i}{n_t + n_i} \tag{4.69}$$

Here $|r|^2$ is the reflectivity of the interface. For example, reflectivity at the interface between a glass or the end of an optical fiber ($n_g \approx 1.5$) and air ($n_{air} = 1$) is about 4%, i.e. 4% of the incident optical power will be reflected, corresponding to a return loss of −14 dB if expressed in dB ($10\, log\, |r|^2$), which is commonly used in fiber optics for estimating the fiber end reflection. The comparison of Eqs. (4.68) and (4.69) shows that the internal reflection generates a common phase shift of π, which should be counted for multilayer reflection film design.

In practical applications, metals are often used for mirror coating because of their wide spectral range and high reflectivity. Since metals are electrically conductive, their refractive indices are complex and can be written as $n_m - i\kappa_m$ where κ_m is the attenuation coefficient. Therefore, the reflection coefficient for metals becomes

$$r = \frac{n_m - i\kappa_m - n_i}{n_m - i\kappa_m + n_i} \tag{4.70}$$

For example, the complex refractive index of Gold is $n_m - i\kappa_m = 0.58 - i9.8$ at the wavelength of 1.55 μm. From Eq. (4.70), the reflectivity of a Gold coated mirror at 1.55 μm is $|r|^2 \approx 98\%$. Therefore, Gold coated mirrors can be approximated to an ideal mirror with a Jones matrix in the following text:

$$\mathbf{M}_{mirror} = \begin{pmatrix} 1 & 0 \\ 0 & -1 \end{pmatrix} \tag{4.71}$$

Case 2: Incident at Brewster's angle

From the Fresnel Eq. (4.67), there exists an incident angle θ_B to make the reflectivity of the p-ray be zero ($r_p = 0$), which is called Brewster's angle, as shown in Figures 4.6 and 4.7. The s- and p-rays are out of phase when the incident angle is less than Brewster's angle, and become in phase when the incident angle is larger than Brewster's angle. Substituting Snell's Eq. (4.63) into (4.65) and letting $r_p = 0$, one obtains

$$\theta_t + \theta_B = 90^\circ \quad \text{and} \quad \theta_B = atan\left(\frac{n_t}{n_i}\right) \tag{4.72}$$

The corresponding Jones matrix of reflection is

$$\begin{pmatrix} 0 & 0 \\ 0 & r_s \end{pmatrix} \tag{4.73}$$

At the Brewster angle θ_B, the reflection of p-polarized light vanishes for incident light of any polarization, with the reflection being totally s-polarized light. A stack of thin glass plates can be used as a polarizer if the light incident at Brewster's angle. Another important application of Brewster's angle is that it enables one to get the precise refractive index of an optical material by measuring its Brewster's angle without having to polish it to a prism.

Case 3: Total internal reflection

When light propagates from a high-refractive-index medium to a low-refractive index medium ($n_i > n_t$), there exists an angle called the critical angle θ_c defined by

$$sin\theta_c = \frac{n_t}{n_i} \tag{4.74}$$

When $\theta \geq \theta_c$, r_p and r_s in Fresnel's formulae (4.67) become a complex exponential:

$$r_p = \frac{n^2 \cos\theta_i - i\sqrt{sin^2\theta_i - n^2}}{n^2 \cos\theta_i + i\sqrt{sin^2\theta_i - n^2}} = e^{i\delta_p}$$

$$r_s = \frac{\cos\theta_i - i\sqrt{sin^2\theta_i - n^2}}{\cos\theta_i + i\sqrt{sin^2\theta_i - n^2}} = e^{i\delta_s} \tag{4.75}$$

where $n = n_t/n_i$. It can be seen from Eq. (4.75) that $|r_p|^2 = |r_s|^2 = 1$, indicating that the energy of light is totally reflected. The total internal reflection (TIR) also introduces a retardation $\delta_p - \delta_s$

between the s and p polarizations, with the corresponding Jones matrix expressed as

$$\mathbf{M}_{total} = \begin{pmatrix} e^{i\delta_p} & 0 \\ 0 & e^{i\delta_s} \end{pmatrix}$$
$$tan\left(\frac{\delta_p}{2}\right) = \frac{-\sqrt{sin^2\theta_i - n^2}}{n^2 cos\,\theta_i}$$
$$tan\left(\frac{\delta_s}{2}\right) = \frac{-\sqrt{sin^2\theta_i - n^2}}{cos\,\theta_i} \tag{4.76}$$

It is clear that the total reflection is equivalent to a retardation plate with a phase difference $\delta = \delta_s - \delta_p$ given by

$$tan\left(\frac{\delta}{2}\right) = \frac{tan\left(\frac{\delta_s}{2}\right) - tan\left(\frac{\delta_p}{2}\right)}{1 + tan\left(\frac{\delta_s}{2}\right) tan\left(\frac{\delta_p}{2}\right)} = \frac{cos\theta_i\sqrt{sin^2\theta_i - n^2}}{sin^2\theta_i} \tag{4.77}$$

The phase shift between s- and p-polarizations as a function of the angle of incidence is plotted in Figure 4.8 for the case of TIR at the glass–air interface inside the glass with a refractive index of 1.5. The maximum phase shift δ_{max} can be obtained by letting the derivative of Eq. (4.77) be zero:

$$tan\left(\frac{\delta_{max}}{2}\right) = \frac{n_{glass}^2 - 1}{2n_{glass}} \quad \text{at} \quad sin^2\theta_i = \frac{2}{1 + n_{glass}^2} \tag{4.78}$$

Utilizing the phase retardation produced by the TIR, Fresnel invented a quart-wave retarder with a glass rhomb to generate circularly polarized light from a linearly polarized light, as shown in Figure 4.9a. In the rhomb, the lower and upper TIRs each generates a 45° phase shift for a total of 90° phase retardation. By putting two Fresnel's rhombs together, a half-wave retarder can be made, as shown in Figure 4.9b.

Compared with the retarders made from birefringence crystals, Fresnel rhomb retarders have the advantages of no wavelength dependence and extremely low-temperature dependence, which is attractive for applications sensitive to the retardation variations and ultra-wide bandwidth applications.

In summary, the SOP of a light beam changes upon reflection at a specular interface between two media with different refractive indexes because the s- and the p-rays experience different reflectivities and phases. In general, the Jones matrix representing the reflection in Eq. (4.64) can be written

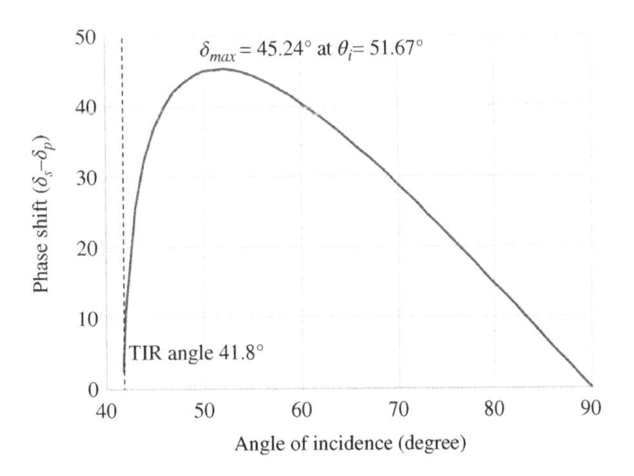

Figure 4.8 Phase shift between s- and p-polarizations as a function of the angle of incidence when light is reflected on the glass–air interface inside glass with refractive index of 1.5.

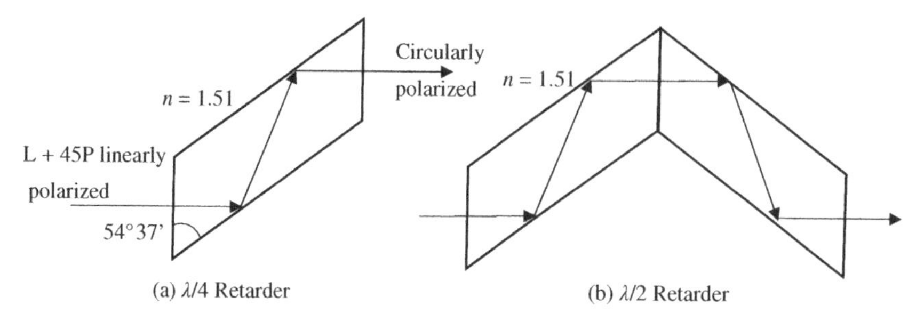

Figure 4.9 Fresnel's rhomb retarder.

as the multiplication of two matrices; one is of a partial polarizer and the other a retardation plate

$$\begin{pmatrix} r_p & 0 \\ 0 & r_s \end{pmatrix} = \begin{pmatrix} |r_p| & 0 \\ 0 & |r_s| \end{pmatrix} \begin{pmatrix} e^{i\delta_p} & 0 \\ 0 & e^{i\delta_s} \end{pmatrix}$$

In other words, the reflection is equivalent to a system containing a partial polarizer and a retardation plate with their principal axes parallel to each other, the combination of two basic optical elements for describing polarization evolution.

4.2.3 Polarization Compensation of Reflection

When a light beam is reflected by a reflector, its SOP changes because of the differences in amplitude and in phase of the s- and p-polarization components of the light beam. By adding a second reflector identical to the first reflector in the optical path, as shown in Figure 4.10, the amplitude and phase differences can be compensated if the following conditions can be met:

1. The incident plane of the 2nd reflection is perpendicular to that of the first reflection.
2. The incident angle of the 2nd reflection is equal to that of the first reflection.

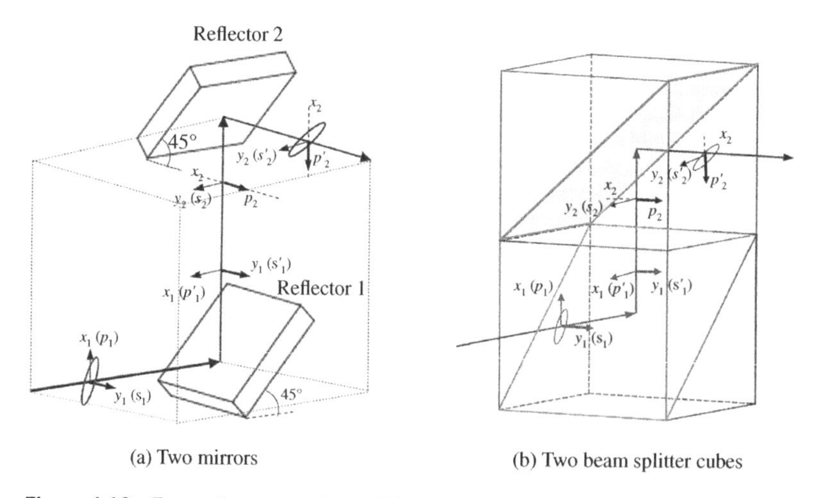

Figure 4.10 Two reflectors work as a 90-degree rotator.

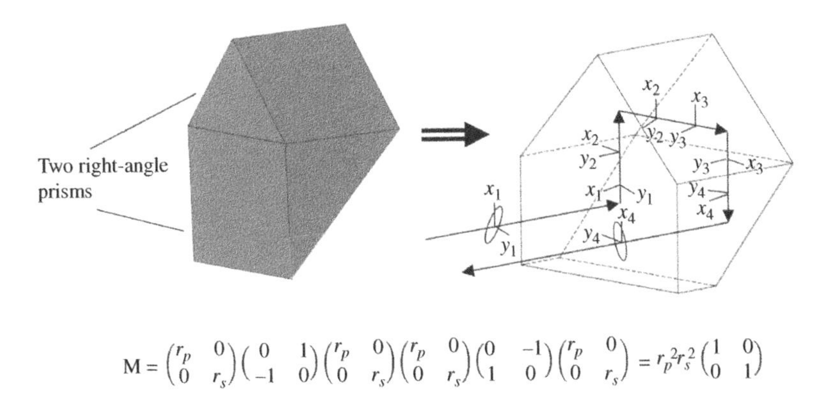

Figure 4.11 Polarization-maintaining retroreflector consisting of two right-angle prisms.

In other words, the s-polarization of the 1st reflection becomes the p-polarization of the 2nd reflection, while the p-polarization of the 1st reflection becomes the s-polarization of the 2nd reflection. Therefore, the matrix representing these two sequential reflections can be written as

$$\mathbf{M} = \begin{pmatrix} r_p & 0 \\ 0 & r_s \end{pmatrix} \begin{pmatrix} 0 & -1 \\ 1 & 0 \end{pmatrix} \begin{pmatrix} r_p & 0 \\ 0 & r_s \end{pmatrix} = r_p r_s \begin{pmatrix} 0 & -1 \\ 1 & 0 \end{pmatrix} \tag{4.79}$$

One may recognize that it is a 90° rotation matrix, i.e. this optical system consisting of these two reflectors is equivalent to a 90° polarization rotator. After a light beam passes through this system of two reflectors, its SOP is rotated 90° without changing the shape of polarization ellipse.

It is possible to use two pairs of reflectors to build a polarization-maintaining (PM) retroreflector, as shown in Figure 4.11, which consists of two right-angle prisms. Interested readers are encouraged to read the US patent 7,723,670 awarded to the authors of this book (Yan et al. 2008). Such a PM reflector was successfully used in a large range polarization mode dispersion (PMD) emulator to be discussed in Chapter 7 for minimizing the device size.

4.2.4 Polarization Properties of Corner-Cube Retroreflector

The corner reflector is a kind of retroreflector with three mutually perpendicular and intersected flat surfaces AOB, BOC, and COA, as shown in Figure 4.12a. A light beam incident on surface ABC is reflected internally three times, once by each surface, before exiting from the same surface. The reflected light is at the same angle as the incident beam, even if the incident light is not normal to surface ABC. Therefore, a retroreflector is often used as the mirror in a Michelson interferometer because it does not require the perfect alignment in order for the beam to return in the same direction. Another attractive feature of a retroreflector is that it may be placed on a moving platform to allow a light beam returns toward its origin despite the changes in the retroreflector's orientation.

When viewed from the ABC plane in Figure 4.12a, the incident and exiting points on ABC are center-reversed about O, as shown in Figure 4.12b. Surface ABC can be divided into six pie-slice segments, and the internal reflection sequences of the light beam are listed in Table 4.2. With an actual corner reflector, the angles of the returned six beams corresponding to the six incident beams

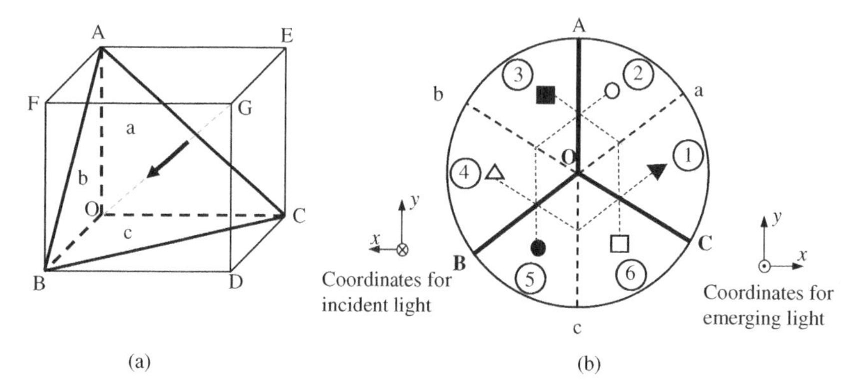

Figure 4.12 (a) Geometry of the corner-cube reflector, with AOB, BOC, and COA being the three reflecting faces mutually perpendicular to each other, and ABC the incident and exiting face. (b) Six segments on the front face ABC of a retroreflector. The incident and exiting points on the front face ABC are center-reversed. A light beam incident on segment 1, 3, or 5 undergoes clockwise internal reflections, while on segment 2,4, or 6 undergoes counterclockwise internal reflections.

Table 4.2 Internal reflection sequence in a corner cube reflector.

Incident segment	Reflection surface sequence	Exiting segment	Rotation direction
1	a → c → b	4	Clockwise
2	a → b → c	5	Counterclockwise
3	b → a → c	6	Clockwise
4	b → c → a	1	Counterclockwise
5	c → b → a	2	Clockwise
6	c → a → b	3	Counterclockwise

a, surface COA; b, surface AOB; c, surface BOC.

may be slightly different because of the manufacturing angle errors of surfaces COA, AOB, and BOC, as well as their surface finish variations.

There are two types of corner reflectors, distinguished by whether the space before the three reflection surfaces is hollow or solid. A hollow reflector is constructed with three mutually perpendicular mirrors while a solid corner reflector is made by cutting a solid glass cube diagonally, as shown in Figure 4.12a. Therefore, solid corner reflectors are often called corner cube retroreflectors. Inside a corner cube, the incident light beam is reflected on three glass–air surfaces by TIR. Since TIR causes a phase shift between *s*- and *p*-components of the electric field, the polarization states of the reflected beams from the corner cube cannot be maintained with respect to the incident beam in general. The polarization properties of the corner cube were reported previously (Liu and Azzam 1997; Zhu et al. 2014). Liu et al. found that there exist two linear polarization eigenvectors when a light beam enters the front surface in segment 1 or 4 of a corner cube, with their orientations shown in Figure 4.13.

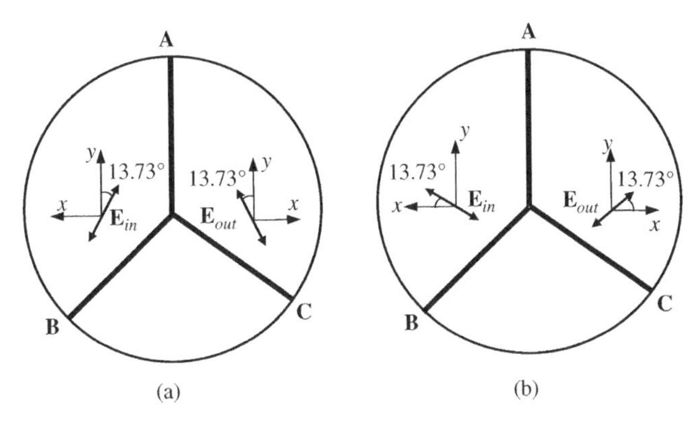

Figure 4.13 Illustration of the pair of linear polarization eigenvectors when an incident light beam enters the corner cube in segment 1 or 4 shown in Figure 4.12b. (a) Linear eigenvector tilted 13.73° from the vertical direction for a corner cube made of BK7 glass with $n = 1.5$ at 1550 nm. (b) Linear eigenvector tilted 13.73° from the horizontal direction.

4.3 Jones Matrix of Multi-element Optical Systems

4.3.1 Jones Equivalent Theorems

In Section 4.2, we introduce how to describe the polarization properties of an individual optical element using a 2×2 Jones matrix. In general, there are only three basic optical elements: retardation plate, rotator, and partial polarizer. Assume a light beam passes through an optical system cascaded with a series of n optical elements, represented by Jones matrices $\mathbf{M}_1$, $\mathbf{M}_2$, ..., $\mathbf{M}_n$; its electric field after passing through each optical element can be expressed as

$$\begin{aligned} \mathbf{E}_1 &= \mathbf{M}_1 \mathbf{E}_{in} \\ \mathbf{E}_2 &= \mathbf{M}_2 \mathbf{E}_1 \\ &\vdots \\ \mathbf{E}_{out} &= \mathbf{M}_n \mathbf{E}_{n-1} \end{aligned} \tag{4.80}$$

By substituting each of the equations in (4.80) in the one following it, one obtains

$$\begin{aligned} \mathbf{E}_{out} &= \mathbf{M}_n \mathbf{M}_{n-1} \cdots \mathbf{M}_1 \mathbf{E}_{in} = \mathbf{M} \mathbf{E}_{in} \\ \mathbf{M} &= \mathbf{M}_n \mathbf{M}_{n-1} \cdots \mathbf{M}_1 \end{aligned} \tag{4.81}$$

The derivation earlier proves that an optical system consisting of a series of n optical elements can be represented by a 2×2 complex matrix that is the multiplication of the matrices of all the optical elements successively from right to left in the order along the optical beam. Jones further established three equivalent theorems for using a minimum number of equivalent optical devices to fully describe an optical system of multiple optical elements at a specific wavelength (Hurwitz and Jones 1941):

(I) For a light beam of a given wavelength, an optical system containing any number of retardation plates and rotators is optically equivalent to a system containing only two elements: a retardation plate and a rotator.

$$\mathbf{M} = \mathbf{R}(\alpha) \mathbf{M}_{RP}(\theta, \delta) = \mathbf{M}_{RP}(\theta + \alpha, \delta) \mathbf{R}(\alpha) \tag{4.82}$$

(II) For a light beam of a given wavelength, an optical system containing any number of partial polarizers and rotators is optically equivalent to a system containing only two elements: a partial polarizer and a rotator.

$$\mathbf{M} = \mathbf{R}(\alpha)\mathbf{M}_P(\theta, p_1, p_2) = \mathbf{M}_P(\theta + \alpha, p_1, p_2)\mathbf{R}(\alpha) \tag{4.83}$$

(III) For the light beam of a given wavelength, an optical system containing any number of retardation plates, partial polarizers, and rotators is optically equivalent to a system containing at most four elements: two retardation plates, one partial polarizer, and one rotator. The partial polarizer must be between the two retardation plates and the rotator can be at any position in the system. In a large and finite class of cases, the rotator is not necessary. Mathematically, the optical system can be represented by

$$\mathbf{M} = \mathbf{U}\mathbf{M}_p(p_1, p_2)\mathbf{V} \tag{4.84}$$

where **U** and **V** are two unitary matrices, and $\mathbf{M}_P = \begin{pmatrix} p_1 & 0 \\ 0 & p_2 \end{pmatrix}$ is the Jones matrix of the partial polarizer with its principal axes being parallel to the x and y-axes.

4.3.2 Properties of Optical System Containing Only Retarders and Rotators

In linear algebra, a complex square matrix **U** is unitary if its conjugation transpose $\mathbf{U}^\dagger$ is its inverse, that is,

$$\mathbf{U}\mathbf{U}^\dagger = \mathbf{U}^\dagger\mathbf{U} = \mathbf{I} \tag{4.85}$$

For rotation matrix $\mathbf{R}(\theta)$, it is easy to find that $\mathbf{R}^\dagger(\theta) = \mathbf{R}(-\theta)$, $\mathbf{R}(\theta)\mathbf{R}^\dagger(\theta) = \mathbf{R}(\theta)\mathbf{R}(-\theta) = \mathbf{R}(0) = \mathbf{I}$, and $\mathbf{R}^\dagger(\theta)\mathbf{R}(\theta) = \mathbf{R}(-\theta)\mathbf{R}(\theta) = \mathbf{R}(0) = \mathbf{I}$, so that the rotation matrix is unitary. Let $\Gamma(\delta) = \begin{pmatrix} e^{i\delta/2} & 0 \\ 0 & e^{-i\delta/2} \end{pmatrix}$ be the Jones matrix of a retardation plate in the principal coordinates; it is easy to prove that $\Gamma^\dagger(\delta) = \Gamma(-\delta)$, $\Gamma(\delta)\Gamma^\dagger(\delta) = \Gamma(\delta)\Gamma(-\delta) = \mathbf{I}$, and $\Gamma^\dagger(\delta)\Gamma(\delta) = \mathbf{I}$. So that $\Gamma(\delta)$ is unitary. The generalized Jones matrix of a retardation plate with its fast axis at angle θ, Eq. (4.57) can be rewritten as $\mathbf{M}_{RP} = \mathbf{R}(\theta)\Gamma(\delta)\mathbf{R}(-\theta)$. Its conjugation transpose is $\mathbf{M}_{RP}^\dagger = \mathbf{R}^\dagger(-\theta)\Gamma^\dagger(\delta)\mathbf{R}^\dagger(\theta) = \mathbf{R}(\theta)\Gamma(-\delta)\mathbf{R}(-\theta)$. Finally, one arrives at

$$\begin{aligned} \mathbf{M}_{RP}\mathbf{M}_{RP}^\dagger &= \mathbf{R}(\theta)\boldsymbol{\Gamma}(\delta)\,\mathbf{R}(-\theta)\mathbf{R}(\theta)\boldsymbol{\Gamma}(-\delta)\mathbf{R}(-\theta) = \mathbf{I} \\ \mathbf{M}_{RP}^\dagger\mathbf{M}_{RP} &= \mathbf{R}(\theta)\boldsymbol{\Gamma}(-\delta)\mathbf{R}(-\theta)\mathbf{R}(\theta)\boldsymbol{\Gamma}(\theta)\,\mathbf{R}(-\theta) = \mathbf{I} \end{aligned} \tag{4.86}$$

Therefore, the Jones matrix of any retardation plate is also unitary. Now let us examine an optical system consisting of only rotators and retarders, which can be represented by the multiplication of a series of unitary matrices $\mathbf{U}_1, \mathbf{U}_2, \ldots, \mathbf{U}_n$. Let $\mathbf{U} = \mathbf{U}_n \ldots \mathbf{U}_2\mathbf{U}_1$, then we have

$$\begin{aligned} \mathbf{U}_N\mathbf{U}_N^\dagger &= (\mathbf{U}_n \ldots \mathbf{U}_2\mathbf{U}_1)(\mathbf{U}_n \ldots \mathbf{U}_2\mathbf{U}_1)^\dagger = \mathbf{U}_n \ldots \mathbf{U}_2\mathbf{U}_1\mathbf{U}_1^\dagger\mathbf{U}_2^\dagger \;\ldots\; \mathbf{U}_n^\dagger = \mathbf{I} \\ \mathbf{U}_N^\dagger\mathbf{U}_N &= (\mathbf{U}_n \ldots \mathbf{U}_2\mathbf{U}_1)^\dagger(\mathbf{U}_n \ldots \mathbf{U}_2\mathbf{U}_1) = \mathbf{U}_1^\dagger\,\mathbf{U}_2^\dagger \ldots \mathbf{U}_n^\dagger\,\mathbf{U}_n \ldots \mathbf{U}_2\mathbf{U}_1 = \mathbf{I} \end{aligned} \tag{4.87}$$

This means that the Jones matrix of an optical system consisting of only retarders and rotators is also unitary, which has the following properties according to linear algebra:

- The power of light is invariable after passing through a unitary optical system because

$$P \propto \mathbf{E}_{out}^\dagger\,\mathbf{E}_{out} = (\mathbf{U}\mathbf{E}_{in})^\dagger(\mathbf{U}\mathbf{E}_{in}) = \mathbf{E}_{in}^\dagger\mathbf{U}^\dagger\mathbf{U}\mathbf{E}_{in} = \mathbf{E}_{in}^\dagger\mathbf{I}\mathbf{E}_{in} = \mathbf{E}_{in}^\dagger\mathbf{E}_{in} \tag{4.88}$$

- The inner product of two Jones vectors (polarization states) is preserved after light passing through a unitary optical system because

$$\mathbf{E}_{out1} \cdot \mathbf{E}_{out2} = \mathbf{E}_{out1}^{\dagger}\mathbf{E}_{out2} = (\mathbf{U}\mathbf{E}_{in1})^{\dagger}(\mathbf{U}\mathbf{E}_{in2}) = \mathbf{E}_{in1}^{\dagger}\mathbf{U}^{\dagger}\mathbf{U}\mathbf{E}_{in2} = \mathbf{E}_{in1}^{\dagger}\mathbf{E}_{in2} = \mathbf{E}_{in1} \cdot \mathbf{E}_{in2} \quad (4.89)$$

It means that the relative polarization angle between two polarizations remains the same after they transmit in the same optical medium, especially the polarization orthogonality of the two polarizations.

- The determinant of matrix $\mathbf{U}$ equals to unity, i.e. $|det(\mathbf{U})| = 1$.
- Any 2×2 unitary matrix can be expressed by

$$\mathbf{U} = \begin{pmatrix} u & v \\ -e^{i\varphi}u & e^{i\varphi}u^* \end{pmatrix} \quad \text{where } |u|^2 + |v|^2 = 1 \quad (4.90)$$

The matrix depends on four real parameters: the phase of u and v, the relative magnitude between u and v, and the angle φ. Let $a = e^{-i\varphi/2}u$ and $b = e^{-i\varphi/2}v$, and omit the constant phase factor $e^{i\varphi/2}$, the unitary matrix in (4.90) is equivalent to

$$\mathbf{U} = \begin{pmatrix} a & b \\ -b^* & a^* \end{pmatrix} \quad \text{where } |a|^2 + |b|^2 = 1 \text{ and } det(\mathbf{U}) = 1 \quad (4.91)$$

The matrix can also be written in the following alternative forms:

$$\mathbf{U} = \begin{pmatrix} cos\alpha & -sin\alpha \\ sin\alpha & cos\alpha \end{pmatrix} \begin{pmatrix} e^{i\delta_1} & 0 \\ 0 & e^{i\delta_2} \end{pmatrix} \begin{pmatrix} cos\beta & sin\beta \\ -sin\beta & cos\beta \end{pmatrix} \quad (4.92)$$

There are many other equivalent forms for unitary matrices, which are not listed here, for we will not use them in this book.

We note that the equivalent form of Eq. (4.92) is expressed as the multiplication of a rotator $\mathbf{R}(\alpha)$, a retarder, and another rotator $\mathbf{R}(-\beta)$. Suppose there is a retardation plate

$$\mathbf{M}_{RP}(\beta, \delta) = \mathbf{R}(\beta) \begin{pmatrix} e^{\frac{i\delta}{2}} & 0 \\ 0 & e^{-\frac{i\delta}{2}} \end{pmatrix} \mathbf{R}(-\beta) \quad (4.93)$$

Let $\delta = \delta_1 - \delta_2$, Eq. (4.92) can be rewritten as

$$\mathbf{U} = e^{i\frac{\delta_1+\delta_2}{2}}\mathbf{R}(\alpha)\mathbf{M}_{RP}(0,\delta)\mathbf{R}(-\beta) = e^{i\frac{\delta_1+\delta_2}{2}}\mathbf{R}(\alpha)\mathbf{R}(-\beta)\mathbf{M}_{RP}(\beta,\delta) = e^{i\frac{\delta_1+\delta_2}{2}}\mathbf{R}(\alpha-\beta)\mathbf{M}_{RP}(\beta,\delta) \quad (4.94)$$

The phase factor in front of the equation can be omitted. This equation means that one can always express an optical system containing retarders and rotators as the multiplication of one retardation plate and one rotator, which proves the first Jones theorem.

The following are some useful devices constructed with basic polarization elements discussed earlier:

4.3.2.1 A Variable Rotator Constructed with Three Retarders

As shown in Figure 4.14, one can construct a rotator by inserting a variable retarder between two crossed quarter-wave plates (QWPs), with its fast-axis orientated to 45° from the fast-axis of QWPs. The rotation angle equals to one half of the phase retardation of the inserted retardation plate.

Let $\Gamma(\delta)$ represent the Jones matrix of retardation plate with the fast-axis parallel to the x-axis, and $\mathbf{M}_{RP}(\theta, \delta)$ denote the matrix of the retardation plate with phase retardation δ and fast axis at an

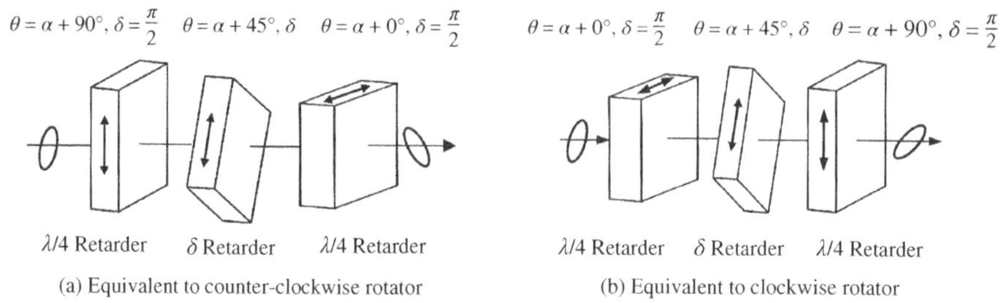

Figure 4.14 Construction of a rotator using three wave plates. δ is the retardation of wave plate, θ is the orientation angle of fast-axis of wave plate; α is an arbitrary angle.

angle θ from the x-axis. For the configuration of Figure 4.12a, one gets the following transformation matrix between the incident and emerging light beams

$$\begin{aligned}\mathbf{M}(90°,45°,0°) &= \mathbf{M}_{RP}\left(\alpha,\frac{\pi}{2}\right)\mathbf{M}_{RP}\left(\alpha+\frac{\pi}{4},\delta\right)\mathbf{M}_{RP}\left(\alpha+\frac{\pi}{2},\frac{\pi}{2}\right)\\ &= \underline{\mathbf{R}(\alpha)\boldsymbol{\Gamma}\left(\frac{\pi}{2}\right)\mathbf{R}(-\alpha)}\ \underline{\mathbf{R}\left(\alpha+\frac{\pi}{4}\right)\boldsymbol{\Gamma}(\delta)\mathbf{R}\left(-\alpha-\frac{\pi}{4}\right)}\\ &\quad\times\underline{\mathbf{R}\left(\alpha+\frac{\pi}{2}\right)\boldsymbol{\Gamma}\left(\frac{\pi}{2}\right)\mathbf{R}\left(-\alpha-\frac{\pi}{2}\right)}\end{aligned} \tag{4.95}$$

Since $\mathbf{R}(\beta_1)\mathbf{R}(\beta_2) = \mathbf{R}(\beta_1+\beta_2)$, $\mathbf{M}(90°, 45°, 0°)$ can be reduced to

$$\begin{aligned}\mathbf{M}(90°,45°,0°) &= \mathbf{R}(\alpha)\underline{\boldsymbol{\Gamma}\left(\frac{\pi}{2}\right)}\ \underline{\mathbf{R}\left(\frac{\pi}{4}\right)\boldsymbol{\Gamma}(\delta)\mathbf{R}\left(-\frac{\pi}{4}\right)}\ \underline{\mathbf{R}\left(\frac{\pi}{2}\right)\boldsymbol{\Gamma}\left(\frac{\pi}{2}\right)\mathbf{R}\left(-\frac{\pi}{2}\right)}\mathbf{R}(-\alpha)\\ &= \mathbf{R}(\alpha)\underline{\boldsymbol{\Gamma}\left(\frac{\pi}{2}\right)\mathbf{M}_{RP}\left(\frac{\pi}{4},\delta\right)\mathbf{M}_{RP}\left(\frac{\pi}{2},\frac{\pi}{2}\right)}\mathbf{R}(-\alpha)\end{aligned} \tag{4.96}$$

Substituting (4.58) in (4.96), $\mathbf{M}(90°, 45°, 0°)$ can be reduced to

$$\mathbf{M}(90°,45°,0°) = \mathbf{R}(\alpha)\begin{pmatrix} cos\frac{\delta}{2} & -sin\frac{\delta}{2}\\ sin\frac{\delta}{2} & cos\frac{\delta}{2}\end{pmatrix}\mathbf{R}(-\alpha) = \begin{pmatrix} cos\frac{\delta}{2} & -sin\frac{\delta}{2}\\ sin\frac{\delta}{2} & cos\frac{\delta}{2}\end{pmatrix} = \begin{pmatrix} cos\beta & -sin\beta\\ sin\beta & cos\beta\end{pmatrix} \tag{4.97}$$

It is clear that the matrix represents a counter-clockwise rotation matrix with a rotation angle β equal to one half of the retardation of the middle wave plate.

Similarly, for the case that the orientation angles of the 1st QWP, the variable retardation plate, and the 2nd QWP are 0°, 45°, and 90°, respectively, as shown in Figure 4.12b, the transformation matrix $\mathbf{M}(0°, 45°, 90°)$ can be obtained as

$$\mathbf{M}(0°,45°,90°) = \begin{pmatrix} cos\frac{\delta}{2} & sin\frac{\delta}{2}\\ -sin\frac{\delta}{2} & cos\frac{\delta}{2}\end{pmatrix} = \begin{pmatrix} cos(-\beta) & -sin(-\beta)\\ sin(-\beta) & cos(-\beta)\end{pmatrix} \tag{4.98}$$

It is equivalent to a clockwise rotator with a rotation angle $\beta = \frac{\delta}{2}$.

4.3.2.2 A Variable Wave Plate Constructed with a Rotator Between Two Quarter-Wave Plates

Let's consider the configuration of Figure 4.15, in which a rotator is inserted between two crossed QWPs. It can be shown next that this optical system is equivalent to a variable retardation plate with its fast-axis oriented at 45° from the axis of the wave plates.

Again, let $\mathbf{M}_{RP}(\theta, \delta)$ represent the matrix of the retardation plate with phase retardation δ and fast axis at an angle θ from the x-axis, $\mathbf{R}(\beta)$ be the rotation matrix with counterclockwise rotation

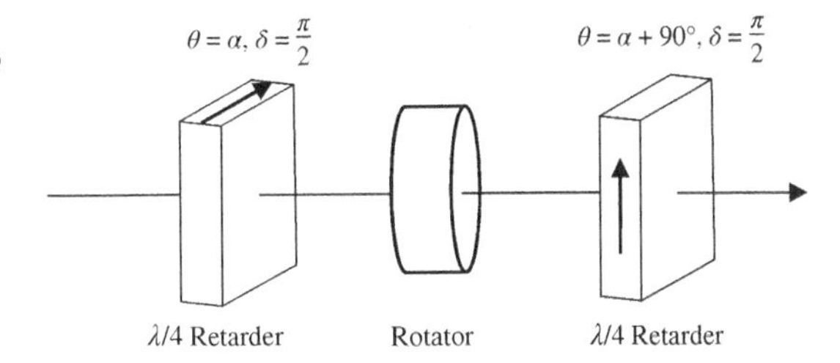

Figure 4.15 A variable wave plate constructed by inserting a rotator between two quarter-wave plates.

angle β, the Jones matrix **M** of the optical system can be written as

$$\mathbf{M} = \mathbf{M}_{RP}\left(\alpha + \frac{\pi}{2}, \frac{\pi}{2}\right)\mathbf{R}(\beta)\mathbf{M}_{RP}\left(\alpha, \frac{\pi}{2}\right)$$
$$= \mathbf{R}\left(\alpha + \frac{\pi}{2}\right)\mathbf{M}_{RP}\left(0, \frac{\pi}{2}\right)\underline{\mathbf{R}(-\alpha - \pi/2)\mathbf{R}(\beta)}\,\underline{\mathbf{R}(\alpha)\mathbf{M}_{RP}\left(0, \frac{\pi}{2}\right)\mathbf{R}(-\alpha)} \tag{4.99}$$

By using the relation $\mathbf{R}(\beta_1)\mathbf{R}(\beta_2) = \mathbf{R}(\beta_1 + \beta_2)$, Eq. (4.99) can be simplified to

$$\mathbf{M} = \mathbf{R}\left(\alpha + \frac{\pi}{4}\right)\underline{\mathbf{M}_{RP}\left(\frac{\pi}{4}, \frac{\pi}{2}\right)\mathbf{R}(\beta)\mathbf{M}_{RP}\left(-\frac{\pi}{4}, \frac{\pi}{2}\right)}\mathbf{R}\left(-\alpha - \frac{\pi}{4}\right) \tag{4.100}$$

Substituting Eqs. (4.51) and (4.58) into Eq. (4.100), one obtains

$$\mathbf{M} = \mathbf{R}\left(\alpha + \frac{\pi}{4}\right)\begin{pmatrix} e^{i\beta} & 0 \\ 0 & e^{-i\beta} \end{pmatrix}\mathbf{R}\left(-\alpha - \frac{\pi}{4}\right) \tag{4.101}$$

Comparing (4.57) and (4.101), one can see that **M** is a variable retardation plate with the retardation equal to the double rotation angle of the rotator inserted between the two QWPs. Such a variable wave plate can be used to construct Faraday rotator based polarization controllers, as will be discussed in Chapters 6 and 7.

4.3.3 Eigenvector and Eigenvalue of an Optical System

A retardation plate is prepared from a birefringence optical crystal with its optic axis parallel to the incident and emerging surfaces. If the polarization of incident light is linearly polarized along the fast or slow-axis of the retarder, the polarization of emerging light will remain the same as that of the incident light (as shown in Figure 4.16a). For a diattenuation plate, the emerging light has the same SOP when the polarization plane is parallel to one of the diattenuation principal axes, although the intensity is attenuated. Similarly, a right- or left-handed circularly polarized light suffers no change in its SOP after passing through a polarization rotator or an optically active crystal. Such a non-changing polarization state is called the eigen polarization state of the corresponding optical device. In general, if there exists an SOP for the incident light, which remains unchanged

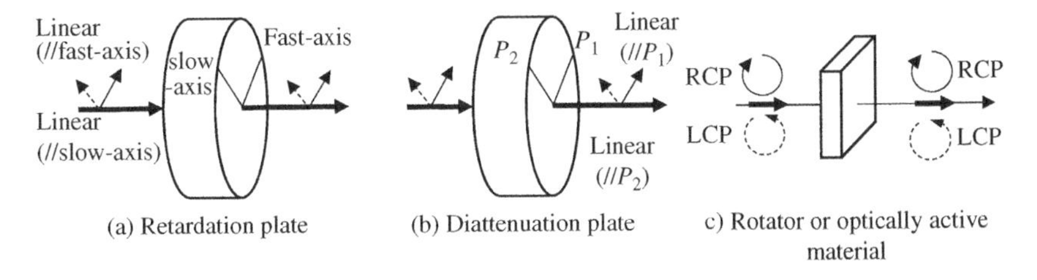

Figure 4.16 Eigen polarization states of optical devices. (a) Retardation plate: liner polarizations parallel to slow- or fast-axis. (b) Diattenuation plate: linear polarizations parallel to diattenuation principal axes. (c) Rotator: right and left-handed circular polarizations.

after transmitting through an optical device, such an SOP is called the eigenstate of polarization. The corresponding Jones vector is called the Jones eigenvector of the optical system. Mathematically, the output Jones vector equals the input Jones vector multiplied by a complex constant. Let $\mathbf{M}$ be the matrix of an optical system, $\mathbf{J}_i$ is one of its eigenvectors, from the definition of eigenvector, one has

$$\mathbf{M}\mathbf{J}_i = \lambda_i \mathbf{J}_i \tag{4.102}$$

Or

$$(\mathbf{M} - \lambda_i \mathbf{I})\mathbf{J}_i = 0 \tag{4.103}$$

where λ_i is a constant called the eigenvalue corresponding to the eigenvector $\mathbf{J}_i$. The condition for Eq. (4.103) to possess a non vanishing solution is

$$det(\mathbf{M} - \lambda_i \mathbf{I}) = det\begin{pmatrix} m_{11} - \lambda_i & m_{12} \\ m_{21} & m_{22} - \lambda_i \end{pmatrix} = 0 \tag{4.104}$$

Or

$$(m_{11} - \lambda_i)(m_{22} - \lambda_i) - m_{12}m_{21} = 0 \tag{4.105a}$$

Therefore one may find the eigenvalues λ_1 and λ_2 by solving Eq. (4.105a). The corresponding normalized eigenvectors can be solved by using the solutions of λ_1 and λ_2 in Eq. (4.102)

$$\mathbf{J}_i = \frac{1}{\sqrt{(m_{11} - \lambda_i)^2 + m_{12}^2}} \begin{pmatrix} -m_{12} \\ m_{11} - \lambda_i \end{pmatrix}, \quad i = 1, 2 \tag{4.105b}$$

It should be emphasized that the eigenvector corresponding to the eigenvalue λ_i is not unique. For example, the multiplication of an eigenvector by a non-zero complex constant is also the eigenvector of the matrix. In practice, normalized vectors are often used to represent eigenvectors.

4.3.3.1 The Eigenvalues and Eigenvectors of Retardation Plate

The Jones matrix of a retardation plate is

$$\mathbf{M}_{RP}(\theta, \delta) = \mathbf{R}(\theta)\begin{pmatrix} e^{i\frac{\delta}{2}} & 0 \\ 0 & e^{-i\frac{\delta}{2}} \end{pmatrix}\mathbf{R}(-\theta) = \mathbf{R}(\theta)\mathbf{M}_{RP}(0, \delta)\mathbf{R}(-\theta) \tag{4.106}$$

It is difficult to directly calculate its eigenvalues from the determinant equation $det(\mathbf{M}_{RP}(\theta, \delta) - \lambda I) = 0$ for an arbitrary orientation angle θ. In the following derivation, we will show that the eigenvalues are not related to its optic axis orientation angle. Assuming that the eigenvalues and eigenvectors of $\mathbf{M}_{RP}(\theta, \delta)$ are $\lambda_{\theta,i}$ $(i = 1, 2)$ and $\mathbf{J}_{\theta,i}$ $(i = 1, 2)$, respectively, from the definition of eigenvectors, we have:

$$\mathbf{M}_{RP}(\theta, \delta)\mathbf{J}_{\theta,i} = \lambda_{\theta,i}\mathbf{J}_{\theta,i} \Rightarrow \mathbf{R}(\theta)\mathbf{M}_{RP}(0, \delta)\mathbf{R}(-\theta)\mathbf{J}_{\theta,i} = \lambda_{\theta,i}\mathbf{J}_{\theta,i} \tag{4.107}$$

Multiply $\mathbf{R}(-\theta)$ on both sides of the aforementioned equation, we get

$$\mathbf{M}_{RP}(0, \delta)[\mathbf{R}(-\theta)\mathbf{J}_{\theta,i}] = \lambda_i[\mathbf{R}(-\theta)\mathbf{J}_{\theta,i}] \tag{4.108}$$

Comparing Eqs. (4.107) and (4.108), one finds that $\mathbf{M}_{RP}(\theta, \delta)$ and $\mathbf{M}_{RP}(0, \delta)$ have the same eigenvalues. Therefore, finding the eigenvalues of $\mathbf{M}_{RP}(\theta, \delta)$ is reduced to solve the eigen equation of $\mathbf{M}_{RP}(0, \delta)$:

$$det[\mathbf{M}_{RP}(0, \delta) - \lambda \mathbf{I}] = det\begin{pmatrix} e^{i\frac{\delta}{2}} - \lambda & 0 \\ 0 & e^{-i\frac{\delta}{2}} - \lambda \end{pmatrix} = 0 \tag{4.109}$$

The two eigenvalues λ_1 and λ_2 are solved to be $e^{i\delta/2}$ and $e^{-i\delta/2}$, respectively, and the corresponding normalized eigenvectors of $\mathbf{M}_{RP}(0, \delta)$ are linear horizontal polarization $\mathbf{J}_{0,1} = \begin{pmatrix} 1 & 0 \end{pmatrix}^T$ and linear vertical polarization $\mathbf{J}_{0,2} = \begin{pmatrix} 0 & 1 \end{pmatrix}^T$. Multiple $\mathbf{R}(\theta)$ on both sides of the eigenvalue equation of $\mathbf{M}_{RP}(0, \delta)$, we have

$$\mathbf{M}_{RP}(0, \delta)\,\mathbf{J}_{0,i} = \lambda_{0,i}\mathbf{J}_{0,i} \Rightarrow \mathbf{R}(\theta)\mathbf{M}_{RP}(0, \delta)\mathbf{R}(-\theta)\mathbf{R}(\theta)\,\mathbf{J}_{0,i} = \lambda_{0,i}\mathbf{R}(\theta)\mathbf{J}_{0,i} \Rightarrow$$
$$\mathbf{M}_{RP}(\theta, \delta)\left[\mathbf{R}(\theta)\,\mathbf{J}_{0,i}\right] = \lambda_{0,i}\left[\mathbf{R}(\theta)\mathbf{J}_{0,i}\right] \tag{4.110}$$

It is clear from the aforementioned equation that one can get the eigenvectors of $\mathbf{M}_{RP}(\theta, \delta)$ by rotating the eigenvectors of $\mathbf{M}_{RP}(0, \delta)$ by an angle of θ:

$$\mathbf{J}_{\theta,1} = \mathbf{R}(\theta)\begin{pmatrix} 1 \\ 0 \end{pmatrix} = \begin{pmatrix} cos\theta \\ sin\theta \end{pmatrix} \quad \text{for eigenvalue } \lambda_1 = e^{i\frac{\delta}{2}}$$
$$\mathbf{J}_{\theta,2} = \mathbf{R}(\theta)\begin{pmatrix} 0 \\ 1 \end{pmatrix} = \begin{pmatrix} -sin\theta \\ cos\theta \end{pmatrix} \quad \text{for eigenvalue } \lambda_2 = e^{-i\frac{\delta}{2}} \tag{4.111}$$

4.3.3.2 The Eigenvalues and Eigenvectors of a Unitary Matrix

In (4.87), we have shown the optical system containing only retarders and rotators can be represented by a 2×2 unitary matrix. In the following, we will show that the eigenvalues of this 2×2 unitary matrix have an exponential form $e^{i\alpha}$, and its two eigenvectors are mutually orthogonal.

Let λ_i $(i = 1, 2)$ be the eigenvalues of a 2×2 unitary matrix $\mathbf{U}$, and the corresponding 2D column eigenvectors are $\mathbf{J}_1$ and $\mathbf{J}_2$, respectively. From the definition of eigenvalues and eigenvectors, we have

$$\mathbf{U}\mathbf{J}_i = \lambda_i\mathbf{J}_i \tag{4.112}$$

and

$$(\mathbf{U}\mathbf{J}_i)^\dagger = (\lambda_i\mathbf{J}_i)^\dagger \Rightarrow \mathbf{J}_i^\dagger\mathbf{U}^\dagger = \lambda_i^*\mathbf{J}_i^\dagger \tag{4.113}$$

Multiplication of the two equations earlier leads to

$$\mathbf{J}_i^\dagger\mathbf{U}^\dagger\mathbf{U}\mathbf{J}_i = \lambda_i^*\lambda_i\mathbf{J}_i^\dagger\mathbf{J}_i \Rightarrow \mathbf{J}_i^\dagger\mathbf{J}_i = \lambda_i^*\lambda_i\mathbf{J}_i^\dagger\mathbf{J}_i \tag{4.114}$$

For any non-zero Jones vector $\mathbf{J}_i$, we must have $\lambda_i^*\lambda_i = 1$, i.e. the modulus of λ_i $(i = 1, 2)$ is unity and can be written as $e^{i\delta_i}$. From Eq. (4.112), we have $\mathbf{J}_i = e^{i\delta_i}\mathbf{U}^\dagger\mathbf{J}_i$, and the inner product of these two eigenvectors is

$$\mathbf{J}_1^\dagger\mathbf{J}_2 = \left(e^{i\delta_1}\mathbf{U}^\dagger\mathbf{J}_1\right)^\dagger\left(e^{i\delta_2}\mathbf{U}^\dagger\mathbf{J}_2\right) = e^{i(\delta_2-\delta_1)}\mathbf{J}_1^\dagger\,\mathbf{U}\mathbf{U}^\dagger\mathbf{J}_2 = e^{i(\delta_2-\delta_1)}\mathbf{J}_1^\dagger\mathbf{J}_2 \tag{4.115}$$

Because $\delta_1 \neq \delta_2$, $\mathbf{J}_1^\dagger\mathbf{J}_2$ must be zero, i.e. the eigenvectors $\mathbf{J}_1$ and $\mathbf{J}_2$ of the unitary matrix must be mutually orthogonal.

4.3.3.3 Obtaining Jones Matrix from Eigenvectors and Eigenvalues

Assuming a given optical system with a Jones Matrix $\mathbf{M}$ is found by experiment to have eigenvectors

$$\mathbf{J}_1 = \begin{pmatrix} a_1 \\ b_1 \end{pmatrix} \quad \mathbf{J}_2 = \begin{pmatrix} a_2 \\ b_2 \end{pmatrix} \tag{4.116}$$

with the corresponding eigenvalues λ_1 and λ_2, we have

$$\mathbf{M}\mathbf{J}_1 = \lambda_1\mathbf{J}_1$$
$$\mathbf{M}\mathbf{J}_2 = \lambda_2\mathbf{J}_2 \tag{4.117}$$

Let's construct a matrix **V** from the two eigenvectors

$$\mathbf{V} = \begin{pmatrix} a_1 & a_2 \\ b_1 & b_2 \end{pmatrix} \quad \text{and} \quad \mathbf{V}^{-1} = \frac{1}{a_1 b_2 - a_2 b_1} \begin{pmatrix} b_2 & -a_2 \\ -b_1 & a_1 \end{pmatrix} \tag{4.118}$$

With matrix **V**, Eq. (4.117) can be written as

$$\mathbf{MV} = \mathbf{V} \begin{pmatrix} \lambda_1 & 0 \\ 0 & \lambda_2 \end{pmatrix} \tag{4.119}$$

Multiplying $\mathbf{V}^{-1}$ on both sides of Eq. (4.119), we obtain

$$\mathbf{M} = \mathbf{V} \begin{pmatrix} \lambda_1 & 0 \\ 0 & \lambda_2 \end{pmatrix} \mathbf{V}^{-1} = \frac{1}{a_1 b_2 - a_2 b_1} \begin{pmatrix} \lambda_1 a_1 b_2 - \lambda_2 a_2 b_1 & -(\lambda_1 - \lambda_2) a_1 a_2 \\ (\lambda_1 - \lambda_2) b_1 b_2 & \lambda_2 a_1 b_2 - \lambda_1 a_2 b_1 \end{pmatrix} \tag{4.120}$$

It shows that the Jones matrix of an optical system can be obtained by measuring its eigenvectors and eigenvalues, although the measurements are difficult in practice.

4.3.4 Transmission Properties of an Optical System Including Partial Polarizers

As discussed in Chapter 3, an optical fiber communication or sensing system contains many components with polarization dependent loss (PDL), which are called partial polarizers in Jones' calculus. The random cascading effect of the multiple PDL will cause signal fluctuations and degrade system performance. In addition, the optical fiber system may also have PMDs distributed along the fiber, which are called retarders in the Jones calculus. The presence of PMD causes signal distortions and polarization fluctuations. PDL and PMD are two major polarization impairments degrading the performance of the optical system. In this section, we will introduce a mathematical method to treat PDL.

From the Jones equivalent theorem III, for the light beam of a given wavelength, an optical system containing any number of retardation plates, partial polarizers, and rotators can be represented by the multiplication of a unitary matrix, a real diagonal matrix representing the partial polarizer, and another unitary matrix:

$$\mathbf{M} = \mathbf{U}\mathbf{M}_p(p_1, p_2)\mathbf{V} \tag{4.121}$$

The power P of emerging light can be obtained as

$$\begin{aligned} P &= \mathbf{J}_{out}^{\dagger}\mathbf{J}_{out} = (\mathbf{MJ}_{in})^{\dagger}\mathbf{MJ}_{in} = \mathbf{J}_{in}^{\dagger}\mathbf{M}^{\dagger}\mathbf{MJ}_{in} = \mathbf{J}_{in}^{\dagger}[\mathbf{UM}_p\mathbf{V}]^{\dagger}[\mathbf{UM}_p\mathbf{V}]\mathbf{J}_{in} \\ &= \mathbf{J}_{in}^{\dagger}\mathbf{V}^{\dagger}\mathbf{M}_p^2\mathbf{VJ}_{in} = \mathbf{J}_{in}^{\dagger}\mathbf{HJ}_{in} \quad \text{where } \mathbf{M}_p^2 = \begin{pmatrix} p_1^2 & 0 \\ 0 & p_2^2 \end{pmatrix} \end{aligned} \tag{4.122}$$

In Eq. (4.122), $\mathbf{H} = \mathbf{M}^{\dagger}\mathbf{M} = \mathbf{V}^{\dagger}\mathbf{M}_p^2\mathbf{V}$ is Hermitian since $[\mathbf{V}^{\dagger}\mathbf{M}_p^2\mathbf{V}]^{\dagger} = \mathbf{V}^{\dagger}\mathbf{M}_p^2\mathbf{V}$. Assume the matrix **H** has two real eigenvalues λ_1, λ_2, and the corresponding normalized eigenvectors are $\mathbf{J}_1$ and $\mathbf{J}_2$, according to the definition of eigenvalues and eigenvectors, we have

$$\mathbf{HJ}_i = \lambda_i\mathbf{J}_i \;\; \text{or} \;\; \mathbf{V}^{\dagger}\mathbf{M}_p^2\mathbf{VJ}_i = \lambda_i\mathbf{J}_i \tag{4.123}$$

Multiplying **V** on both sides of the second equation in (4.123), one finds

$$\mathbf{M}_p^2(\mathbf{VJ}_i) = \lambda_i(\mathbf{VJ}_i) \tag{4.124}$$

It is clear that λ_i $(i = 1, 2)$ are also the eigenvalues of $\mathbf{M}_p^2$. It is easy to find the eigenvalues of $\mathbf{M}_p^2$ are $\lambda_1 = p_1^2$ and $\lambda_2 = p_2^2$, which are also the eigenvalues of matrix **H**: $\lambda_1 = p_1^2$ and $\lambda_2 = p_2^2$.

In the following, we will prove that the eigenvectors $\mathbf{J}_1$ and $\mathbf{J}_2$ of $\mathbf{H}$ are mutually orthogonal. According to the eigenvalue equation of $\mathbf{H}$, we have

$$\mathbf{H}\mathbf{J}_1 = \lambda_1 \mathbf{J}_1 \Rightarrow \mathbf{J}_1^\dagger = \frac{1}{\lambda_1^*}(\mathbf{H}\mathbf{J}_1)^\dagger = \frac{1}{\lambda_1^*}\mathbf{J}_1^\dagger \mathbf{H}^\dagger \tag{4.125}$$

Then, the inner product of $\mathbf{J}_1$ and $\mathbf{J}_2$ is

$$\mathbf{J}_1^\dagger \mathbf{J}_2 = \frac{1}{\lambda_1^*}\mathbf{J}_1^\dagger \mathbf{H}^\dagger \mathbf{J}_2 \tag{4.126}$$

Since the matrix $\mathbf{H}$ is Hermitian, i.e. $\mathbf{H} = \mathbf{H}^\dagger$, Eq. (4.126) becomes

$$\mathbf{J}_1^\dagger \mathbf{J}_2 = \frac{1}{\lambda_1^*}\mathbf{J}_1^\dagger \mathbf{H}\mathbf{J}_2 = \frac{\lambda_2}{\lambda_1^*}\mathbf{J}_1^\dagger \mathbf{J}_2 \tag{4.127}$$

Since $\lambda_2/\lambda_1^* = p_2^2/p_1^2 \neq 1$, we have $\mathbf{J}_1^\dagger \mathbf{J}_2 = 0$, i.e. the two eigenvectors of $\mathbf{H}$ are orthogonal to each other and can be considered as an orthogonal basis in a 2D complex space such that their linear supposition can be used to represent the polarization state of the incident light:

$$\mathbf{J}_{in} = c_1 \mathbf{J}_1 + c_2 \mathbf{J}_2 \tag{4.128}$$

According to (4.122), the transmittance of the optical system can be obtained as

$$T = \frac{\mathbf{J}_{out}^\dagger \mathbf{J}_{out}}{\mathbf{J}_{in}^\dagger \mathbf{J}_{in}} = \frac{\mathbf{J}_{in}^\dagger \mathbf{H}\mathbf{J}_{in}}{\mathbf{J}_{in}^\dagger \mathbf{J}_{in}} = \frac{c_1 c_1^* p_1^2 + c_2 c_2^* p_2^2}{c_1 c_1^* + c_2 c_2^*} = p_1^2 (cos\theta)^2 + p_2^2 (sin\theta)^2 \tag{4.129}$$

where $cos\theta = \sqrt{c_1 c_1^* / \left(c_1 c_1^* + c_2 c_2^*\right)}$ and $sin\theta = \sqrt{c_2 c_2^* / \left(c_1 c_1^* + c_2 c_2^*\right)}$.

If $p_1 > p_2$, $T_{max} = p_1^2$ when $cos\theta = 1$ ($\mathbf{J}_{in}$ is parallel to eigenvector $\mathbf{J}_1$), and $T_{min} = p_2^2$ when $sin\theta = 1$ ($\mathbf{J}_{in}$ is parallel to eigenvector $\mathbf{J}_2$).

If $p_2 > p_1$, $T_{max} = p_2^2$ when $sin\theta = 1$ ($\mathbf{J}_{in}$ is parallel to eigenvector $\mathbf{J}_2$), and $T_{min} = p_1^2$ when $cos\theta = 0$ ($\mathbf{J}_{in}$ is parallel to eigenvector $\mathbf{J}_1$).

In summary, the power of emerging light of an optical system containing a partial polarizer varies between $T_{max}P_0$ and $T_{min}P_0$ for different polarization states of incident light. When the input polarization states are parallel to the eigenstates of matrix $\mathbf{H}(\mathbf{M}^\dagger\mathbf{M})$, the emerging light reaches the maximum or minimum power. The ratio of maximum and minimum light power of emerging light is called polarization dependent loss (PDL), which can be calculated from

$$PDL = 10log\left(\frac{P_{max}}{P_{min}}\right) = 10log\left(\frac{T_{max}}{T_{min}}\right) = 10\left|log\left(\frac{p_2^2}{p_1^2}\right)\right| \tag{4.130}$$

The T_{max} and T_{min} can also be determined by calculating the trace and determinant of matrix $\mathbf{H}$:

$$tr(\mathbf{H}) = tr\left(\mathbf{V}^\dagger \mathbf{M}_p^2 \mathbf{V}\right) = p_1^2\left(v_{11}v_{11}^* + v_{12}v_{12}^*\right) + p_2^2\left(v_{21}v_{21}^* + v_{22}v_{22}^*\right) \tag{4.131}$$

where v_{ij} ($i = 1, 2; j = 1, 2$) are the elements of matrix $\mathbf{V}$. Because $\mathbf{V}$ is unitary, we have

$$\mathbf{V}^\dagger \mathbf{V} = \begin{pmatrix} 1 & 0 \\ 0 & 1 \end{pmatrix} \Rightarrow \begin{cases} v_{11}v_{11}^* + v_{12}v_{12}^* = 1 \\ v_{21}v_{21}^* + v_{22}v_{22}^* = 1 \end{cases} \tag{4.132}$$

And the trace of $\mathbf{H}$ can be reduced to

$$tr(\mathbf{H}) = p_1^2 + p_2^2 \tag{4.133}$$

The determinant of **H** is

$$det(\mathbf{H}) = det\left(\mathbf{V}^{\dagger}\mathbf{M}_p^2\mathbf{V}\right) = det(\mathbf{V}^{\dagger})det\left(\mathbf{M}_p^2\right) det(\mathbf{V}) = det(\mathbf{V}^{\dagger}\mathbf{V})p_1^2p_2^2 = p_1^2p_2^2 \tag{4.134}$$

Because now T_{max} is the maximum of p_1^2 and p_2^2, and T_{min} is the minimum of p_1^2 and p_2^2 from Eq. (4.129), one can get the following equations for T_{max} and T_{min} after solving Eqs. (4.133) and (4.134)

$$\begin{aligned} T_{max} &= \frac{tr(\mathbf{H}) + \sqrt{[tr(\mathbf{H})]^2 - 4det(\mathbf{H})}}{2} \\ T_{min} &= \frac{tr(\mathbf{H}) - \sqrt{[tr(\mathbf{H})]^2 - 4det(\mathbf{H})}}{2} \end{aligned} \tag{4.135}$$

Unlike unitary optical systems, when light passes through an optical system containing a partial polarizer, the emerging optical power will vary when the polarization state of incident light changes. Assume two incident orthogonal Jones vectors are $\mathbf{J}_{in1} = \left(cos\theta \quad sin\theta e^{i\delta}\right)^T$ and $\mathbf{J}_{in2} = \left(-sin\theta \quad cos\theta e^{i\delta}\right)^T$. After passing through a partial polarizer with its principal axes parallel to the *x*- and *y*-axes, the emerging Jones vectors become $\mathbf{J}_{out1}$ and $\mathbf{J}_{out2}$, and can be represented by

$$\begin{aligned} \mathbf{J}_{out1} &= \mathbf{M}\mathbf{J}_{in1} = \begin{pmatrix} p_1 & 0 \\ 0 & p_2 \end{pmatrix}\begin{pmatrix} cos\theta \\ sin\theta e^{i\delta} \end{pmatrix} = \begin{pmatrix} p_1 cos\theta \\ p_2 sin\theta e^{i\delta} \end{pmatrix} \\ \mathbf{J}_{out2} &= \mathbf{M}\mathbf{J}_{in2} = \begin{pmatrix} p_1 & 0 \\ 0 & p_2 \end{pmatrix}\begin{pmatrix} -sin\theta \\ cos\theta e^{i\delta} \end{pmatrix} = \begin{pmatrix} -p_1 sin\theta \\ p_2 cos\theta e^{i\delta} \end{pmatrix} \end{aligned} \tag{4.136}$$

The inner product of these two output Jones vectors is

$$\mathbf{J}_{out1} \cdot \mathbf{J}_{out2} = \mathbf{J}_{out1}^{\dagger}\mathbf{J}_{out2} = \frac{1}{2}\left(p_2^2 - p_1^2\right) sin2\theta \tag{4.137}$$

Since $p_1 \neq p_2$, the inner product between two emerging polarization states is not equal to zero except $\theta = n\pi/2$, and reaches maximum $\left(p_2^2 - p_1^2\right)$ when $sin2\theta = 1$ or $\theta = \pi/4$. That is, in general, the two orthogonal polarization states are no longer orthogonal after they pass through a partial polarizer if they are not aligned to the principal axes of the partial polarizer. Let's define the angle β between these two Jones vectors as

$$cos\beta = \frac{|\mathbf{J}_{out1} \cdot \mathbf{J}_{out2}|}{|\mathbf{J}_{out1}||\mathbf{J}_{out2}|} \tag{4.138}$$

Substitution of $\theta = \pi/4$ and (4.136) in (4.138) yields

$$cos\beta_{min} = \left|\frac{p_2^2 - p_1^2}{p_2^2 + p_1^2}\right| \tag{4.139}$$

From the relationship (4.130) between PDL, p_1, and p_2, Eq. (4.139) can be rewritten as

$$cos\beta_{min} = \left|\frac{p_2^2/p_1^2 - 1}{p_2^2/p_1^2 + 1}\right| = \frac{10^{PDL/10} - 1}{10^{PDL/10} + 1} \tag{4.140}$$

Figure 4.17 shows the plot of the minimum angle β_{min} as a function of PDL, which indicates that the presence of partial polarizer or PDL destroys the orthogonality of two originally orthogonal polarizations. Therefore, as mentioned in Chapter 3, in a polarization multiplexed optical fiber communication system, the two input orthogonally polarized channels will no longer be

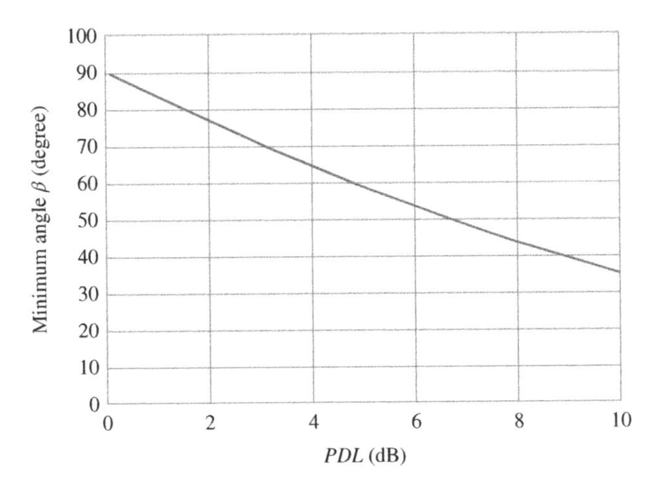

Figure 4.17 The degradation of polarization orthogonality of lights after passing through a partial polarizer, showing the minimum angle deviation from 90° as a function of PDL.

orthogonal after transmitting through an optical fiber with PDL. At the output of the fiber, the two signal channels can no longer be demultiplexed without crosstalk. Unlike the degrading effects of PMD, which can be compensated digitally in a polarization multiplexed coherent detection system because PMD does not destroy the orthogonality of the multiplexed signal channels, the degrading effect of PDL cannot be compensated digitally, although it may be mitigated to an extent through digital signal processing.

4.3.5 Experimental Measurement of Jones Matrix

Jones Matrix is a 2×2 complex matrix; eight real parameters are needed to determine the matrix, and seven real parameters are required if the absolute phase is omitted. Assuming the Jones matrix of optical device under test (DUT) is

$$\mathbf{M} = \begin{pmatrix} m_{11} & m_{12} \\ m_{21} & m_{22} \end{pmatrix} = A \begin{pmatrix} x_1 & x_2 \\ x_3 & 1 \end{pmatrix} \tag{4.141}$$

A known input Jones vector can be expressed as

$$\mathbf{J}_{in,i} = a_i \begin{pmatrix} cos\theta_i \\ sin\theta_i e^{i\delta_i} \end{pmatrix} \tag{4.142}$$

and the corresponding output Jones vector can be written as

$$\mathbf{J}_{out,i} = \begin{pmatrix} E_{i,x} \\ E_{i,y} \end{pmatrix} \quad \text{and} \quad k_i = \frac{E_{i,x}}{E_{i,y}} \tag{4.143}$$

where i stands for ith SOP. From $\mathbf{J}_{out} = \mathbf{M}\mathbf{J}_{in}$ and $E_{ix} = k_i E_{iy}$ $(i = 1, 2, 3)$, one gets (see Figure 4.18)

$$x_i cos\theta_i + x_i sin\theta_i e^{i\delta_i} - x_3 cos\theta_i k_i = k_i sin\theta_i e^{i\delta_i} \tag{4.144}$$

Three mutually linearly independent input Jones vectors are required to find the solutions of x_1, x_2, and x_3:

$$\begin{cases} x_1 cos\theta_1 + x_2 sin\theta_1 e^{i\delta_1} - x_3 cos\theta_1 k_1 = k_1 sin\theta_1 e^{i\delta_1} \\ x_1 cos\theta_2 + x_2 sin\theta_2 e^{i\delta_2} - x_3 cos\theta_2 k_2 = k_2 sin\theta_2 e^{i\delta_2} \\ x_1 cos\theta_3 + x_2 sin\theta_3 e^{i\delta_3} - x_3 cos\theta_3 k_3 = k_3 sin\theta_3 e^{i\delta_3} \end{cases} \tag{4.145}$$

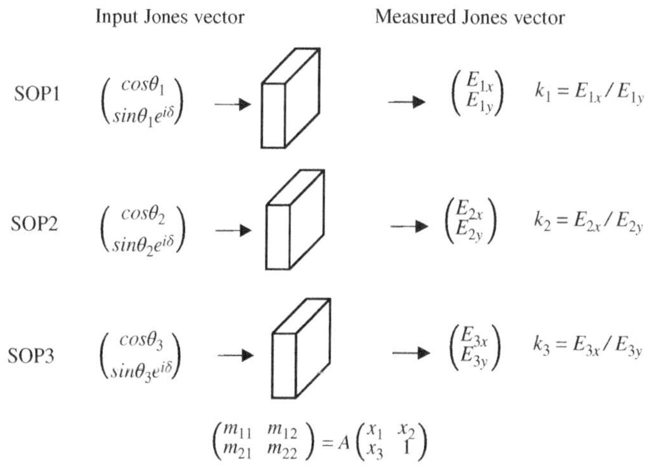

Input Jones vector — Measured Jones vector

SOP1 $\begin{pmatrix} cos\theta_1 \\ sin\theta_1 e^{i\delta} \end{pmatrix} \rightarrow \rightarrow \begin{pmatrix} E_{1x} \\ E_{1y} \end{pmatrix}$ $k_1 = E_{1x}/E_{1y}$

SOP2 $\begin{pmatrix} cos\theta_2 \\ sin\theta_2 e^{i\delta} \end{pmatrix} \rightarrow \rightarrow \begin{pmatrix} E_{2x} \\ E_{2y} \end{pmatrix}$ $k_2 = E_{2x}/E_{2y}$

SOP3 $\begin{pmatrix} cos\theta_3 \\ sin\theta_3 e^{i\delta} \end{pmatrix} \rightarrow \rightarrow \begin{pmatrix} E_{3x} \\ E_{3y} \end{pmatrix}$ $k_3 = E_{3x}/E_{3y}$

$$\begin{pmatrix} m_{11} & m_{12} \\ m_{21} & m_{22} \end{pmatrix} = A \begin{pmatrix} x_1 & x_2 \\ x_3 & 1 \end{pmatrix}$$

Figure 4.18 Experimentally obtaining Jones Matrix. Three input SOPs are required.

From Cramer's rule, the solutions of the aforementioned complex linear equations can be found to be

$$x_1 = \frac{\begin{vmatrix} k_1 sin\,\theta_1 e^{i\delta_1} & sin\,\theta_1 e^{i\delta_1} & -cos\,\theta_1 k_1 \\ k_2 sin\,\theta_2 e^{i\delta_2} & sin\,\theta_2 e^{i\delta_2} & -cos\,\theta_2 k_2 \\ k_3 sin\,\theta_3 e^{i\delta_3} & sin\,\theta_3 e^{i\delta_3} & -cos\,\theta_3 k_3 \end{vmatrix}}{\begin{vmatrix} cos\,\theta_1 & sin\,\theta_1 e^{i\delta_1} & -cos\,\theta_1 k_1 \\ cos\,\theta_2 & sin\,\theta_2 e^{i\delta_2} & -cos\,\theta_2 k_2 \\ cos\,\theta_3 & sin\,\theta_3 e^{i\delta_3} & -cos\,\theta_3 k_3 \end{vmatrix}} \quad x_2 = \frac{\begin{vmatrix} cos\,\theta_1 & k_1 sin\,\theta_1 e^{i\delta_1} & -cos\,\theta_1 k_1 \\ cos\,\theta_2 & k_2 sin\,\theta_2 e^{i\delta_2} & -cos\,\theta_2 k_2 \\ cos\,\theta_3 & k_3 sin\,\theta_3 e^{i\delta_3} & cos\,\theta_3 k_3 \end{vmatrix}}{\begin{vmatrix} cos\,\theta_1 & sin\,\theta_1 e^{i\delta_1} & -cos\,\theta_1 k_1 \\ cos\,\theta_2 & sin\,\theta_2 e^{i\delta_2} & -cos\,\theta_2 k_2 \\ cos\,\theta_3 & sin\,\theta_3 e^{i\delta_3} & -cos\,\theta_3 k_3 \end{vmatrix}}$$

$$x_3 = \frac{\begin{vmatrix} cos\,\theta_1 & sin\,\theta_1 e^{i\delta_1} & k_1 sin\,\theta_1 e^{i\delta_1} \\ cos\,\theta_2 & sin\,\theta_2 e^{i\delta_2} & k_2 sin\,\theta_2 e^{i\delta_2} \\ cos\,\theta_3 & sin\,\theta_3 e^{i\delta_3} & k_3 sin\,\theta_3 e^{i\delta_3} \end{vmatrix}}{\begin{vmatrix} cos\,\theta_1 & sin\,\theta_1 e^{i\delta_1} & -cos\,\theta_1 k_1 \\ cos\,\theta_2 & sin\,\theta_2 e^{i\delta_2} & -cos\,\theta_2 k_2 \\ cos\,\theta_3 & sin\,\theta_3 e^{i\delta_3} & -cos\,\theta_3 k_3 \end{vmatrix}} \tag{4.146}$$

If we select the three input polarization states as linear polarizations with azimuth angles of 0°, 90°, and 45°, we have

$$\begin{aligned} \mathbf{J}_{in,1} &= a_1 \begin{pmatrix} 1 \\ 0 \end{pmatrix} : \quad \theta_1 = 0, \ \delta_1 = 0 \\ \mathbf{J}_{in,2} &= a_2 \begin{pmatrix} 0 \\ 1 \end{pmatrix} : \quad \theta_2 = \frac{\pi}{2}, \ \delta_2 = 0 \\ \mathbf{J}_{in,3} &= a_3 \begin{pmatrix} \frac{\sqrt{2}}{2} \\ \frac{\sqrt{2}}{2} \end{pmatrix} : \quad \theta_3 = \frac{\pi}{4}, \ \delta_3 = 0 \end{aligned} \tag{4.147}$$

The solutions in Eq. (4.146) can be reduced to

$$x_1 = \frac{k_1(k_3 - k_2)}{(k_1 - k_3)} = k_1 k_4 \qquad x_2 = k_2 \qquad x_3 = \frac{(k_3 - k_2)}{(k_1 - k_3)} = k_4 \tag{4.148}$$

The Jones matrix of the optical DUT is then

$$\mathbf{M} = A \begin{pmatrix} k_1 k_4 & k_2 \\ k_4 & 1 \end{pmatrix} \tag{4.149}$$

where A is a complex constant undetermined so far, whose phase relates to the common phase retardation which cannot be determined through polarization and light intensity measurements. The absolute value of A determines the transmission factor of the optical DUT and can be obtained by any single measure of the transmission factor. For example, the transmission factor for the horizontally linearly polarized incident light $\mathbf{E}_{in,1} = (1, 0)^T$ is

$$T_1 = \frac{\text{emerging intensity}}{\text{incident intensity}} = \frac{|\mathbf{E}_{out,1}|^2}{|\mathbf{E}_{in,1}|^2} = |A|^2 |k_4|^2 \left(1 + |k_1|^2\right) \tag{4.150}$$

Finally, one arrives at

$$|A|^2 = \frac{T_1}{|k_4|^2 \left(1 + |k_1|^2\right)} \tag{4.151}$$

4.3.6 Jones Calculus in Retracing Optical Path

The Jones matrices discussed earlier are for the forward (FW) propagation; the transformation from the FW Jones matrix to the backward (BW) case is different for reciprocal and nonreciprocal media, which will be discussed in detail in this section (Pistoni 1995). In reciprocal media, the transformation with the defined coordinates in Figure 4.19 is

$$\mathbf{M}_{FW} = \begin{pmatrix} m_1 & m_4 \\ m_3 & m_2 \end{pmatrix} \Rightarrow \mathbf{M}_{BW} = \begin{pmatrix} m_1 & -m_3 \\ -m_4 & m_2 \end{pmatrix} \tag{4.152}$$

In nonreciprocal media, such as those with the Faraday Effect, the transformation is

$$\mathbf{M}_{FW} = \begin{pmatrix} m_1 & m_4 \\ m_3 & m_2 \end{pmatrix} \Rightarrow \mathbf{M}_{BW} = \begin{pmatrix} m_1 & -m_4 \\ -m_3 & m_2 \end{pmatrix} \tag{4.153}$$

When $\mathbf{M}$ is composed of a series of matrices cascaded together that include reciprocal and nonreciprocal polarization elements, each matrix must be transformed individually in order to obtain a new combined matrix. The BW Jones matrices of basic optical devices are listed as follows:

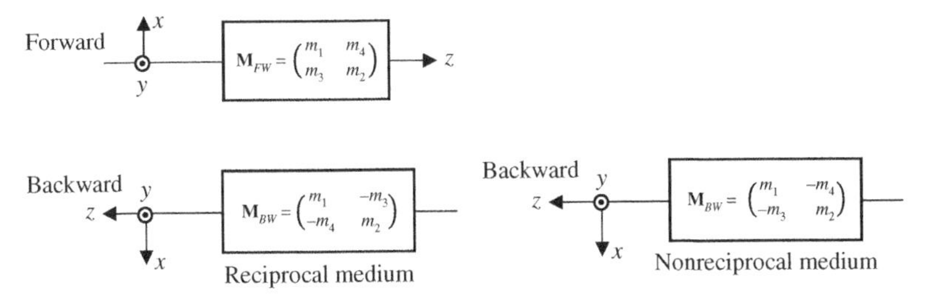

Figure 4.19 Jones matrix in retracing optical path.

Forward (FW)	Backward (BW)	
Rotator $\mathbf{R}_{FW}(\theta) = \begin{pmatrix} cos\theta & -sin\theta \\ sin\theta & cos\theta \end{pmatrix}$	In reciprocal media (e.g. optically active media) $\mathbf{R}_{BW}(\theta) = \mathbf{R}_{FW}(\theta) = \begin{pmatrix} cos\theta & -sin\theta \\ sin\theta & cos\theta \end{pmatrix}$	(4.154)
	In nonreciprocal media (e.g. Faraday rotators) $\mathbf{R}_{BW}(\theta) = \mathbf{R}_{FW}(-\theta) = \begin{pmatrix} cos\theta & sin\theta \\ -sin\theta & cos\theta \end{pmatrix}$	(4.155)
Linear retardation plate $\mathbf{M}_{RP,FW}(\theta, \delta) = \mathbf{R}_{FW}(\theta) \begin{pmatrix} e^{\frac{i\delta}{2}} & 0 \\ 0 & e^{-\frac{i\delta}{2}} \end{pmatrix} \mathbf{R}_{FW}(-\theta)$	$\mathbf{M}_{RP,BW}(\theta, \delta) = \mathbf{R}_{FW}(-\theta) \begin{pmatrix} e^{\frac{i\delta}{2}} & 0 \\ 0 & e^{-\frac{i\delta}{2}} \end{pmatrix} \mathbf{R}_{FW}(\theta) = \mathbf{M}_{RP,FW}(-\theta, \delta)$	(4.156)
Partial polarizer $\mathbf{M}_{P,FW}(\theta, \delta) = \mathbf{R}_{FW}(\theta) \begin{pmatrix} p_1 & 0 \\ 0 & p_1 \end{pmatrix} \mathbf{R}_{FW}(-\theta)$	Partial polarizer $\mathbf{M}_{P,BW} = \mathbf{R}_{FW}(-\theta) \begin{pmatrix} p_1 & 0 \\ 0 & p_1 \end{pmatrix} \mathbf{R}_{FW}(\theta) = \mathbf{M}_{P,FW}(-\theta, \delta)$	(4.157)

4.3.6.1 Jones Matrix of a Double-Pass Optical System with a Mirror

The Jones matrix of a double pass optical system with a mirror, with the FW and BW propagation coordinates as shown in Figure 4.20, can be represented by the Jones matrix in the following text:

$$\mathbf{M} = \mathbf{M}_{BW}\mathbf{M}_{mirror}\mathbf{M}_{FW} \tag{4.158}$$

Case 1: $\mathbf{M}_{FW}$ is a reciprocal rotator (circular retardation plate)

If the device before the mirror is a reciprocal rotator (see Figure 4.21), from Eq. (4.154), Eq. (4.158) becomes

$$\mathbf{M}_{rotator+mirror} = \mathbf{R}(\theta) \begin{pmatrix} 1 & 0 \\ 0 & -1 \end{pmatrix} \mathbf{R}(\theta) = \begin{pmatrix} 1 & 0 \\ 0 & -1 \end{pmatrix} \tag{4.159}$$

where the relationship $\mathbf{M}_{mirror}\mathbf{R}(\theta) = \mathbf{R}(-\theta)\mathbf{M}_{mirror}$ is used. Equation (4.159) shows that the rotation caused by the reciprocal rotator is completely canceled out when the light double passes the rotator due to the mirror.

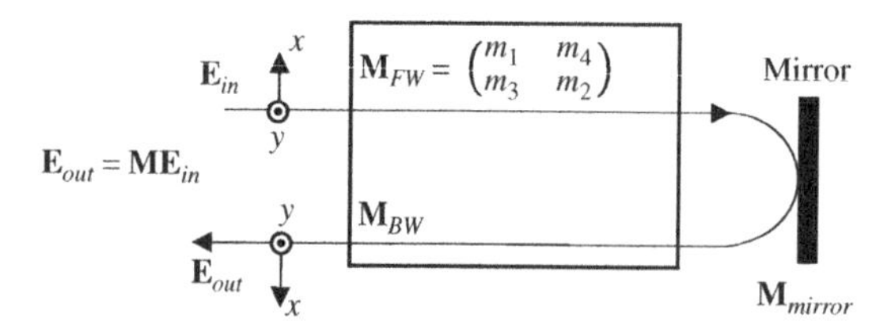

Figure 4.20 Jones matrix of a double pass optical system with a mirror.

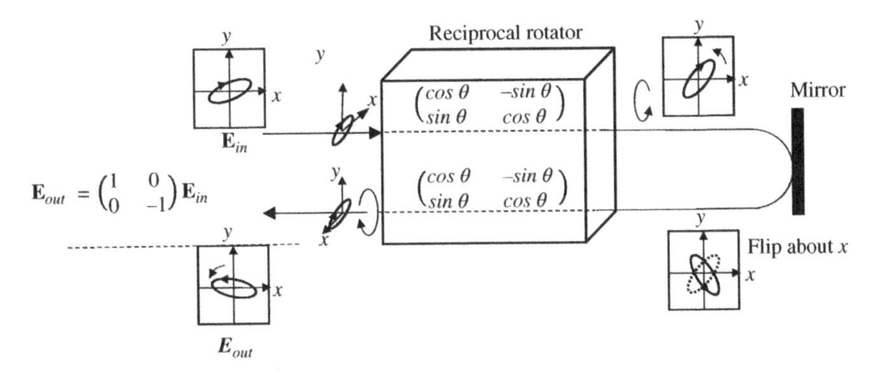

Figure 4.21 Retroreflection by rotator + mirror.

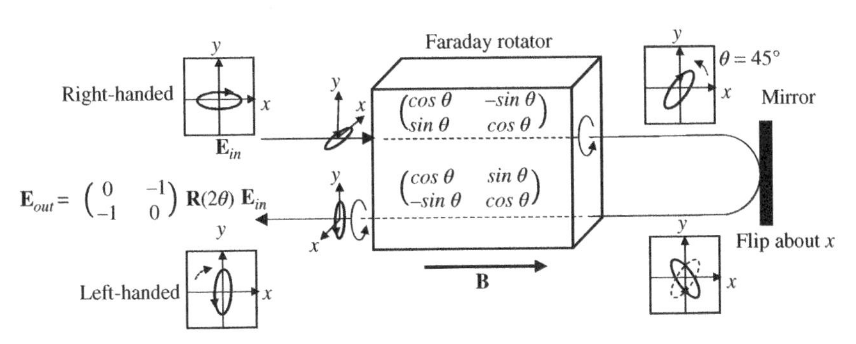

Figure 4.22 Retroreflection by a Faraday rotator + mirror. **B** is the external magnetic field.

Case 2: $\mathbf{M}_{FW}$ *is a Faraday rotator*

When we consider the device before the mirror to be a Faraday rotator (see Figure 4.22), from Eq. (4.155), the corresponding Jones matrix of the assembly of Faraday rotator and mirror is

$$\mathbf{M}_{Faraday\ rotator+mirror} = \mathbf{R}(-\theta)\begin{pmatrix}1 & 0\\ 0 & -1\end{pmatrix}\mathbf{R}(\theta) = \begin{pmatrix}1 & 0\\ 0 & -1\end{pmatrix}\mathbf{R}(2\theta)$$
$$= \begin{pmatrix}\cos 2\theta & -\sin 2\theta\\ -\sin 2\theta & -\cos 2\theta\end{pmatrix} \tag{4.160}$$

where the relationship $\mathbf{M}_{mirror}\mathbf{R}(\theta) = \mathbf{R}(-\theta)\mathbf{M}_{mirror}$ is again used. Equation (4.160) indicates that the SOP rotation angle caused by a Faraday rotator is doubled upon the reflection from a mirror, however, with a handedness reversal. When $\theta = \pi/4$, the corresponding retroreflector is called Faraday mirror and can be represented by

$$\mathbf{M}_{Faraday\ mirror} = \mathbf{M}_{\pi/4\ Faraday\ rotator+mirror} = \begin{pmatrix}0 & -1\\ -1 & 0\end{pmatrix} \tag{4.161}$$

Jones vector's transformations from input to output by the Faraday mirror are $E_{out,x} = -E_{in,y}$ and $E_{out,y} = -E_{in,x}$, which is equivalent to flipping polarization about $y = -x$ (135° line), i.e. if the incident light is LVP, the emerging light will be LHP, and vice versa.

Case 3: $\mathbf{M}_{FW}$ *is a linear retardation plate (waveplate)*
According to Eq. (4.158), the Jones matrix of retroreflector composed of linear retardation plate and mirror is

$$\begin{aligned}
\mathbf{M}_{RP+mirror} &= \mathbf{M}_{RP,BW}(\theta,\delta)\mathbf{M}_{mirror}\mathbf{M}_{RP,FW}(\theta,\delta) \\
&= \mathbf{M}_{RP,FW}(-\theta,\delta)\mathbf{M}_{mirror}\mathbf{M}_{RP,FW}(\theta,\delta) \\
&= \underline{\mathbf{R}_{FW}(-\theta)\begin{pmatrix} e^{i\delta/2} & 0 \\ 0 & e^{-i\delta/2}\end{pmatrix}\mathbf{R}_{FW}(\theta)}\,\mathbf{M}_{mirror}\underline{\mathbf{R}_{FW}(\theta)\begin{pmatrix} e^{i\delta/2} & 0 \\ 0 & e^{-i\delta/2}\end{pmatrix}\mathbf{R}_{FW}(-\theta)} \\
&= \mathbf{M}_{mirror}\mathbf{R}_{FW}(\theta)\begin{pmatrix} e^{i\delta/2} & 0 \\ 0 & e^{-i\delta/2}\end{pmatrix}\mathbf{R}_{FW}(-\theta)\,\mathbf{R}_{FW}(\theta)\begin{pmatrix} e^{i\delta/2} & 0 \\ 0 & e^{-i\delta/2}\end{pmatrix}\mathbf{R}_{FW}(-\theta) \\
&= \mathbf{M}_{mirror}\mathbf{R}_{FW}(\theta)\begin{pmatrix} e^{i\delta} & 0 \\ 0 & e^{-i\delta}\end{pmatrix}\mathbf{R}_{FW}(-\theta) \\
&= \mathbf{M}_{mirror}\mathbf{M}_{RP,FW}(\theta,2\delta)
\end{aligned} \tag{4.162}$$

where the relationships $\mathbf{M}_{mirror}\mathbf{R}(\theta) = \mathbf{R}(-\theta)\mathbf{M}_{mirror}$ and $\mathbf{M}_{mirror}\mathbf{M}_{RP,\,FW}(0,\ \delta) = \mathbf{M}_{RP,\,FW}(0,\ \delta)\mathbf{M}_{mirror}$ are used to commute $\mathbf{M}_{mirror}$ to the left side in (4.162). We note that the $\mathbf{M}_{mirror}$ is equivalent to a half-wave plated with its fast-axis is parallel to the x-axis, the optical system composed of a retardation plate, and an ideal mirror is equivalent to an optic system containing a half-wave plate and a retardation plate with the doubled retardation of the FW propagating light.

When $\theta = \pi/4$ and $\delta = \pi/2$ (see Figure 4.23), the corresponding retroreflector is called quarter-wave plate mirror (QWM), and its Jones matrix becomes

$$\mathbf{M}_{QWM} = \mathbf{M}_{mirror}\mathbf{M}_{RP,FW}\left(\frac{\pi}{4},\pi\right) = \begin{pmatrix} 0 & i \\ -i & 0\end{pmatrix} \tag{4.163}$$

This QWM will swap x and y components of the incident light and introduce a π phase shift between these two components. For vertically or horizontally polarized incident light, a QWM is equivalent to a Faraday mirror, i.e. it converts a LVP to LHP, and vice versa. For elliptical polarization, the output polarization is

$$\mathbf{E}_{out} = \mathbf{M}_{QWM}\begin{pmatrix} cos\alpha \\ sin\alpha e^{i\delta}\end{pmatrix} = e^{i\left(\delta+\frac{\pi}{2}\right)}\begin{pmatrix} sin\alpha \\ cos\alpha e^{i(\pi-\delta)}\end{pmatrix} \tag{4.164}$$

i.e. the QWM flips the polarization ellipse about $y = x$ line without changing the sense of rotation.

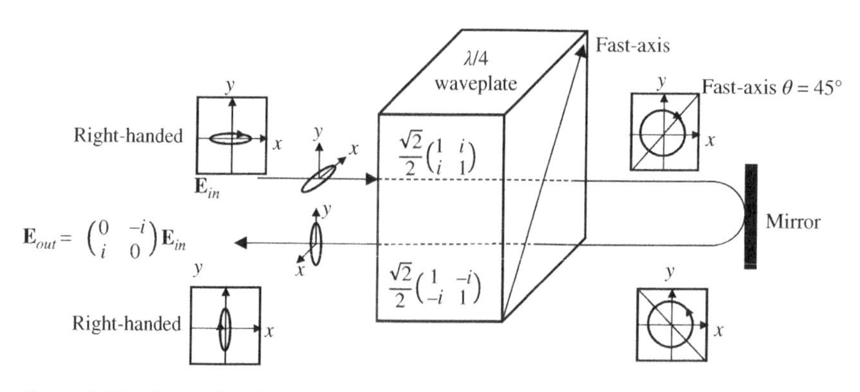

Figure 4.23 Retroreflecting by quarter-wave plate + mirror.

Case 4: $\mathbf{M}_{FW}$ *containing linear retardation plates and reciprocal rotators*
The optical system composed of linear retardation plates and reciprocal rotators can be represented by a unitary matrix. From Eq. (4.94), one obtains the FW Jones matrix of the systems as

$$\mathbf{M}_{FW} = \mathbf{R}(\alpha)\mathbf{M}_{RP,FW}(0,\delta)\mathbf{R}(\beta) \tag{4.165}$$

The corresponding BW matrix is

$$\mathbf{M}_{BW} = \mathbf{R}(\beta)\mathbf{M}_{RP,FW}(0,\delta)\mathbf{R}(\alpha) \tag{4.166}$$

Therefore, the Jones matrix of the returning beam, including the mirror, can be written as

$$\mathbf{M} = \mathbf{M}_{BW}\mathbf{M}_{mirror}\mathbf{M}_{FW} = \mathbf{R}(\beta)\mathbf{M}_{RP,FW}(0,\delta)\mathbf{R}(\alpha)\mathbf{M}_{mirror}\mathbf{R}(\alpha)\mathbf{M}_{RP,FW}(0,\delta)\mathbf{R}(\beta) \tag{4.167}$$

Since $\mathbf{R}(\theta)\mathbf{M}_{mirror} = \mathbf{M}_{mirror}\mathbf{R}(-\theta)$ and $\mathbf{M}_{RP,\,FW}(0,\delta)\mathbf{M}_{mirror} = \mathbf{M}_{RP,\,FW}(0,\delta)\mathbf{M}_{mirror}$, one obtains

$$\begin{aligned}\mathbf{M} &= \mathbf{R}(\beta)\mathbf{M}_{RP,FW}(0,\delta)\mathbf{R}(\alpha)\mathbf{R}(-\alpha)\mathbf{M}_{mirror}\mathbf{M}_{RP,FW}(0,\delta)\mathbf{R}(\beta)\\ &= \mathbf{R}(\beta)\mathbf{M}_{RP,FW}(0,\delta)\mathbf{M}_{RP,FW}(0,\delta)\mathbf{R}(-\beta)\mathbf{M}_{mirror}\\ &= \mathbf{M}_{RP,FW}(\beta,2\delta)\,\mathbf{M}_{mirror}\end{aligned}$$

or

$$\mathbf{M} = \mathbf{M}_{mirror}\mathbf{M}_{RP,FW}(-\beta,2\delta) \tag{4.168}$$

Equation (4.168) shows that a general retroreflector composed of linear retardation plates, reciprocal rotators (circular retardation plate), and a mirror is equivalent to the superposition of a linear retardation plate and a mirror, with the reciprocal rotation (circular retardation) being completely canceled out by the mirror while the total retardation being doubled.

Case 5: Retracing Optical System with a Reciprocal Medium and a Faraday Mirror
The Jones matrix of retracing optical system with a reciprocal medium and a Faraday mirror as shown in Figure 4.24 can be written as

$$\begin{aligned}\mathbf{M} &= \mathbf{M}_{BW}\mathbf{M}_{Faraday\ mirror}\mathbf{M}_{FW}\\ &= \begin{pmatrix} m_1 & -m_3\\ -m_4 & m_2\end{pmatrix}\begin{pmatrix} 0 & -1\\ -1 & 0\end{pmatrix}\begin{pmatrix} m_1 & m_4\\ m_3 & m_2\end{pmatrix} = (m_1m_2 - m_3m_4)\begin{pmatrix} 0 & -1\\ -1 & 0\end{pmatrix}\end{aligned} \tag{4.169}$$

It is clear that the optical system is equivalent to a Faraday mirror and a neutral attenuator, with any differential attenuation or PDL being neutralized. For the optical system only containing linear retarders and rotators, the FW matrix is unitary and can be represented by

$$\mathbf{M}_{FW} = \begin{pmatrix} a & b\\ -b^* & a^*\end{pmatrix} \quad \text{with } aa^* + bb^* = 1 \tag{4.170}$$

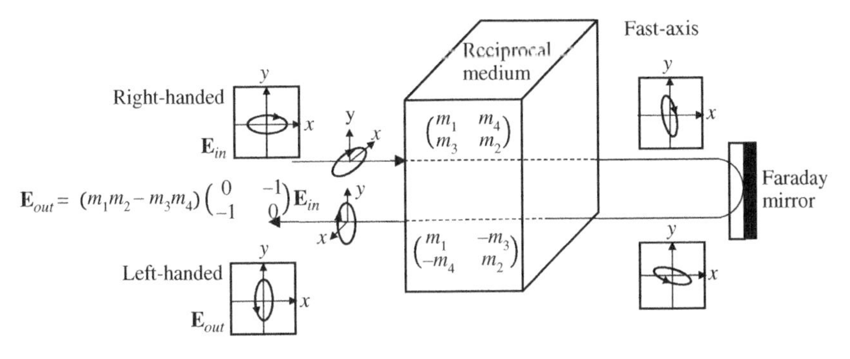

Figure 4.24 Retracing optical system with a reciprocal medium and a Faraday mirror.

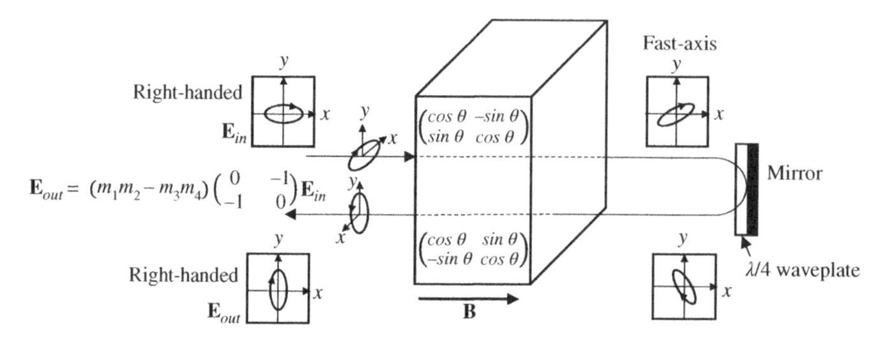

Figure 4.25 Retracting optical system with quarter-wave plate mirror.

Substitution (4.170) in (4.169) yields

$$\mathbf{M} = (aa^* + bb^*)\begin{pmatrix} 0 & -1 \\ -1 & 0 \end{pmatrix} = \begin{pmatrix} 0 & -1 \\ -1 & 0 \end{pmatrix} = \mathbf{M}_{Faraday\ mirror} \tag{4.171}$$

Equation (4.171) states that the linear and circular birefringences are totally canceled out by a Faraday mirror in a retracing optical system.

Case 6: Cancelation of Faraday rotation with a quarter waveplate mirror

The Jones matrix of retracing optical system consisting of Faraday rotator and QWM as shown in Figure 4.25 can be expressed as

$$\mathbf{M} = \mathbf{M}_{BW\ Faraday\ rotator}\mathbf{M}_{QWM}\mathbf{M}_{FW\ Faraday\ rotator} \tag{4.172}$$

Substitution of Eqs. (4.155) and (4.163) in (4.172) yields

$$\mathbf{M} = \begin{pmatrix} \cos\theta & \sin\theta \\ -\sin\theta & \cos\theta \end{pmatrix}\begin{pmatrix} 0 & i \\ -i & 0 \end{pmatrix}\begin{pmatrix} \cos\theta & -\sin\theta \\ \sin\theta & \cos\theta \end{pmatrix} = \begin{pmatrix} 0 & i \\ -i & 0 \end{pmatrix} \tag{4.173}$$

Equation (4.173) implies that the effect of a Faraday rotator can be completely canceled out by a QWM.

4.3.7 N-Matrix and Polarization Evolution

While light propagates in a continuous medium, such as an optical fiber, the linear and circular birefringences at different positions generally vary along the light propagation path. Therefore, the corresponding Jones matrix is not a constant matrix for it is position-dependent. In the following discussions, we will introduce the **N**-matrix method invented by R. Clark Jones (1948) to build and solve the Jones matrix differential equation to describe the SOP evolution when light propagates in a continuous medium.

Suppose a monochromatic plane wave propagates along the z-axis in the Cartesian coordinate, and the Jones vector of light at $z = 0$ is a constant vector $\mathbf{E}_{in}$, then the Jones vector $\mathbf{E}(z)$ at position z can be written as

$$\mathbf{E}(z) = \mathbf{M}(z)\mathbf{E}_{in} \tag{4.174}$$

where $\mathbf{M}(z)$ is the Jones matrix of the medium between the origin to z. Let $\Delta\mathbf{M}(z)$ represent the Jones matrix of a thin slice of the medium whose surfaces have the coordinates z and $z + \Delta z$ (see Figure 4.26), then the derivative of matrix $\mathbf{M}(z)$ can be defined by

$$\frac{d\mathbf{M}(z)}{dz} = \lim_{\Delta z \to 0}\frac{\mathbf{M}(z+\Delta z) - \mathbf{M}(z)}{\Delta z} = \lim_{\Delta z \to 0}\frac{\Delta\mathbf{M}(z)\mathbf{M}(z) - \mathbf{M}(z)}{\Delta z} = \lim_{\Delta z \to 0}\frac{\Delta\mathbf{M}(z) - \mathbf{I}}{\Delta z}\mathbf{M}(z) \tag{4.175}$$

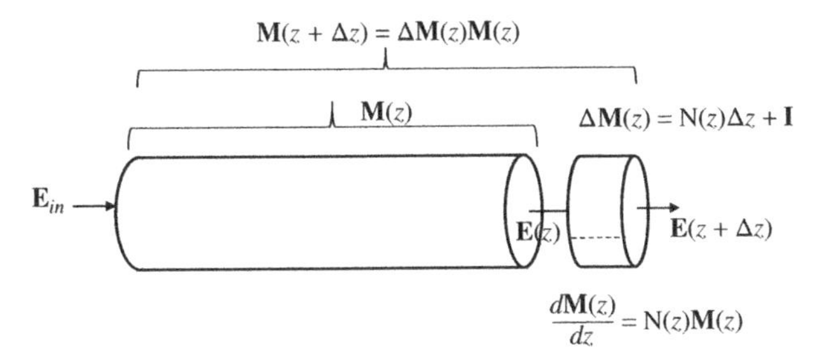

Figure 4.26 The relationship between distributed Jones matrix **M**(*z*) and **N**-matrix at position *z*.

Define **N**-matrix is defined as

$$\mathbf{N}(z) = \lim_{\Delta z \to 0} \frac{\Delta\mathbf{M}(z) - \mathbf{I}}{\Delta z} \tag{4.176}$$

Substitution of the definition of **N**(z) into (4.175) yields the differential equation of **M**(z)

$$\frac{d\mathbf{M}(z)}{dz} = \mathbf{N}(z)\mathbf{M}(z) \tag{4.177}$$

or

$$\mathbf{N}(z) = \frac{d\mathbf{M}(z)}{dz}\mathbf{M}^{-1}(z) \tag{4.178}$$

If one already knows the distribution of **N**(z) along z, the **M**(z) can be calculated from the following integration

$$\mathbf{M}(z) = \left[exp\left(\int_{z_0}^{z} \mathbf{N}(z)dz\right)\right]\mathbf{M}(z_0) \tag{4.179}$$

where the matrix exponential can be expressed with the Taylor series:

$$exp(\boldsymbol{X}) = \mathbf{I} + \sum_{n=1}^{\infty} \frac{\boldsymbol{X}^n}{n!} \tag{4.180}$$

From (4.174) and (4.177), the evolution of polarization can be represented by the **N**-matrix:

$$\frac{d\mathbf{E}(z)}{dz} = \frac{d\mathbf{M}(z)}{dz}\mathbf{E}_{in} = \frac{d\mathbf{M}(z)}{dz}[\mathbf{M}^{-1}(z)\mathbf{E}(z)] = \mathbf{N}(z)\mathbf{E}(z) \tag{4.181}$$

Generally speaking, **M**(z) and **N**(z) are position dependent, and no analytical solution is available. However, if **N**(z) is independent of z, for example, in a linear or circular birefringence crystal, a spun fiber, a PM fiber, the matrix **M**(z) of optical system can be analytically expressed in terms of z and the matrix elements of the **N**-matrix. Before further derivation, we first introduce two properties of the **N**-matrix useful in the derivation.

(1) If the **N**-matrix is independent of z, matrices **M** and **N** have the same eigenvectors
(2) If the **N**-matrix is independent of z, the eigenvalue λ_M of the **M** and eigenvalue λ_N of the **N**-matrix are related by

$$\lambda_M = e^{\lambda_N z} \tag{4.182}$$

Proof: Assuming ε_M is an eigenvector of the matrix **M**, which does not change with z by definition, then we have

$$\mathbf{M}\varepsilon_M = \lambda_M \varepsilon_M \tag{4.183}$$

Differentiation of Eq. (4.183) yields

$$\frac{d\mathbf{M}}{dz}\varepsilon_M = \frac{d\lambda_M}{dz}\varepsilon_M \tag{4.184}$$

Reformatting Eq. (4.183) by multiplying the inverse of matrix $\mathbf{M}$ yields

$$\varepsilon_M = \lambda_M \mathbf{M}^{-1}\varepsilon_M \tag{4.185}$$

Substituting Eq. (4.185) on the left-hand side of Eq. (4.184), one finds

$$\frac{d\mathbf{M}}{dz}\lambda_M \mathbf{M}^{-1}\varepsilon_M = \frac{d\lambda_M}{dz}\varepsilon_M \Longrightarrow \frac{d\mathbf{M}}{dz}\mathbf{M}^{-1}\varepsilon_M = \left(\frac{1}{\lambda_M}\frac{d\lambda_M}{dz}\right)\varepsilon_M \tag{4.186}$$

According to the relationship Eq. (4.178), i.e. $\mathbf{N} = \frac{d\mathbf{M}}{dz}\mathbf{M}^{-1}$, Eq. (4.186) can be reformed to

$$\mathbf{N}\varepsilon_M = \left(\frac{1}{\lambda_M}\frac{d\lambda_M}{dz}\right)\varepsilon_M = \lambda_N\varepsilon_M \tag{4.187}$$

Equation (4.187) shows that ε_M is also the eigenvector of the $\mathbf{N}$-matrix, with an eigenvalue λ_N relating to λ_M, the eigenvalue of matrix $\mathbf{M}$, by

$$\left(\frac{1}{\lambda_M}\frac{d\lambda_M}{dz}\right) = \lambda_N \tag{4.188}$$

If the $\mathbf{N}$-matrix is independent of z, its eigenvalues λ_N must also be a constant, which leads to the solution of the differential equation of (4.188) as

$$\lambda_M = e^{\lambda_N z} \tag{4.189}$$

In deriving Eq. (4.189), the boundary condition λ_M $(z = 0) = 1$ has been used.

4.3.7.1 Expression of M in Terms of N

Assuming matrix $\mathbf{M}(z)$ has eigenvalues λ_{M1} and λ_{M2}, with the two corresponding eigenvectors of

$$\varepsilon_{M1} = \begin{pmatrix} a_{M1} \\ b_{M1} \end{pmatrix} \text{ and } \varepsilon_{M2} = \begin{pmatrix} a_{M2} \\ b_{M2} \end{pmatrix} \tag{4.190}$$

According to Eq. (4.120), $\mathbf{M}$ can be expressed uniquely by

$$\mathbf{M} = \frac{1}{a_{M1}b_{M2} - a_{M2}b_{M1}}\begin{pmatrix} \lambda_{M1}a_{M1}b_{M2} - \lambda_{M2}a_{M2}b_{M1} & -(\lambda_{M1} - \lambda_{M2})a_{M1}a_{M2} \\ (\lambda_{M1} - d\lambda_{M2})b_{M1}b_{M2} & \lambda_{M2}a_{M1}b_{M2} - \lambda_{M1}a_{M2}b_{M1} \end{pmatrix} \tag{4.191}$$

Assuming the corresponding $\mathbf{N}$-matrix is

$$\mathbf{N} = \begin{pmatrix} n_{N1} & n_{N4} \\ n_{N3} & n_{N2} \end{pmatrix} \tag{4.192}$$

one obtains its determinant equation as

$$(n_{N1} - \lambda_N)(n_{N2} - \lambda_N) - n_{N3}n_{N4} = 0 \tag{4.193}$$

Solving Eq. (4.193) yields the eigenvalues

$$\lambda_N = \frac{(n_{N1} + n_{N2}) \pm \sqrt{(n_{N1} - n_{N2})^2 + (4n_{N3}n_{N4})}}{2} \tag{4.194}$$

Let

$$T_N = \frac{(n_{N1} + n_{N2})}{2} \text{ and } Q_N = \sqrt{\left(\frac{n_{N1} - n_{N2}}{2}\right)^2 + n_{N3}n_{N4}} \tag{4.195}$$

The eigenvalues of $\mathbf{N}(z)$ can be reformatted to

$$\lambda_N = T_N \pm Q_N \tag{4.196}$$

The corresponding eigenvectors are

$$\varepsilon_N = \begin{pmatrix} \frac{1}{2}(n_{N1} - n_{N2}) \pm Q_N \\ n_3 \end{pmatrix} \tag{4.197}$$

which are also eigenvectors of matrix $\mathbf{M}$ (see Eq. (4.187))

$$\varepsilon_M = \varepsilon_N = \begin{pmatrix} \frac{1}{2}(n_{N1} - n_{N2}) \pm Q_N \\ n_{N3} \end{pmatrix} \tag{4.198}$$

From Eq. (4.189), the eigenvalues of $\mathbf{M}$ can be obtained as

$$\lambda_M = e^{(T_N \pm Q_N)\, z} \tag{4.199}$$

Substitution of Eqs. (4.197) and (4.199) in Eq. (4.191) yields

$$\mathbf{M}(z) = e^{T_N z} \begin{pmatrix} coshQ_N z + \frac{1}{2}(n_{N1} - n_{N2})\frac{sinhQ_N z}{Q_N} & n_{N4}\frac{sinhQ_N z}{Q_N} \\ n_{N3}\frac{sinhQ_N z}{Q_N} & coshQ_n z - \frac{1}{2}(n_{N1} - n_{N2})\frac{sinhQ_N z}{Q_N} \end{pmatrix} \tag{4.200}$$

It should be noted that the expression of $\mathbf{M}(z)$ defined earlier is symmetrical with the parameter Q_N, i.e. $\mathbf{M}_{-Q_N} = \mathbf{M}_{Q_N}$. The function $e^{T_N z}$ is the common phase change and can be omitted.

4.3.7.2 Circular Retardation Plate

As a specific example of Eq. (4.200), consider a circular birefringence retarder with circular birefringence $2\Delta\rho$ and thickness z; its Jones matrix [see Eq. (4.44)] can be written as

$$\mathbf{M}(z) = \mathbf{R}(\Delta\rho z) \tag{4.201}$$

where $\mathbf{R}(\Delta\rho z)$ represents the rotation matrix with rotation angle equal to $\Delta\rho z$. Through substitution of Eq. (4.201) in Eq. (4.178), we can find the $\mathbf{N}$-matrix of the circular retarder is

$$\mathbf{N} = \left[\frac{d}{dz}\mathbf{R}(\Delta\rho z)\right]\mathbf{R}^{-1}(\Delta\rho z) = \Delta\rho \begin{pmatrix} -sin(\Delta\rho z) & -cos(\Delta\rho z) \\ cos(\Delta\rho z) & -sin(\Delta\rho z) \end{pmatrix} \begin{pmatrix} cos(\Delta\rho z) & sin(\Delta\rho z) \\ -sin(\Delta\rho z) & cos(\Delta\rho z) \end{pmatrix} \tag{4.202}$$

which can be further reduced to

$$\mathbf{N} = \begin{pmatrix} 0 & -\Delta\rho \\ \Delta\rho & 0 \end{pmatrix} = \Delta\rho\mathbf{R}\left(\frac{\pi}{2}\right) \tag{4.203}$$

Substitutions of (4.203) and λ_N in (4.196) yield

$$\mathbf{M}(z) = \begin{pmatrix} cos(\Delta\rho z) & -sin(\Delta\rho z) \\ sin(\Delta\rho z) & cos(\Delta\rho z) \end{pmatrix} \tag{4.204}$$

The $\mathbf{M}(z)$ is periodic in z with a periodicity of $2\pi/\Delta\rho$. If this circular retarder is a piece of fiber, the phase difference between two orthogonal states is $2\Delta\rho z$, as can be seen from the discussions of Eqs. (4.43) and (4.44), and the corresponding period is called circular beat length L_c and

$$L_c = \frac{\pi}{\Delta\rho} \tag{4.205}$$

4.3.7.3 Linear Retardation Plate

For the linear birefringence retarder with a thickness of z and a linear birefringence of $\Delta\beta$, with its fast-axis oriented at θ from the x-axis, the corresponding Jones matrix can be written as

$$\mathbf{M}(z) = \mathbf{R}(\theta)\mathbf{G}(\Delta\beta z)\mathbf{R}(-\theta) \tag{4.206}$$

where $\mathbf{G}(\Delta\beta z)$ is the linear retarder with the fast-axis at zero degrees. The corresponding $\mathbf{N}$-matrix can be expressed as from Eq. (4.178)

$$\begin{aligned}\mathbf{N} &= \frac{d\mathbf{M}(z)}{dz}\mathbf{M}^{-1}(z) = \left[\mathbf{R}(\theta)\frac{d\mathbf{G}(\Delta\beta z)}{dz}\mathbf{R}(-\theta)\right][\mathbf{R}(\theta)\mathbf{G}(-\Delta\beta z)\mathbf{R}(-\theta)] \\ &= \mathbf{R}(\theta)\frac{d\mathbf{G}(\Delta\beta z)}{dz}\mathbf{G}(-\Delta\beta z)\ \mathbf{R}(-\theta)\end{aligned} \tag{4.207}$$

where

$$\mathbf{G}(\Delta\beta z) = \begin{pmatrix} e^{i\frac{\Delta\beta}{2}z} & 0 \\ 0 & e^{-i\frac{\Delta\beta}{2}z}\end{pmatrix},\ \frac{d\mathbf{G}(\Delta\beta z)}{dz} = \begin{pmatrix} i\frac{\Delta\beta}{2}e^{i\frac{\Delta\beta}{2}z} & 0 \\ 0 & -i\frac{\Delta\beta}{2}e^{-i\frac{\Delta\beta}{2}z}\end{pmatrix} \text{ and } \mathbf{R}(\theta) = \begin{pmatrix} cos\theta & -sin\theta \\ sin\theta & cos\theta\end{pmatrix} \tag{4.208}$$

Substitution of (4.208) in (4.207) yields

$$\mathbf{N} = \begin{pmatrix} i\frac{\Delta\beta}{2}cos2\theta & i\frac{\Delta\beta}{2}sin2\theta \\ i\frac{\Delta\beta}{2}sin2\theta & -i\frac{\Delta\beta}{2}cos2\theta\end{pmatrix} = \frac{\Delta\beta}{2}cos2\theta\begin{pmatrix} i & 0 \\ 0 & -i\end{pmatrix} + \frac{\Delta\beta}{2}sin2\theta\begin{pmatrix} 0 & i \\ i & 0\end{pmatrix} \tag{4.209}$$

where the first term $\frac{\Delta\beta}{2}cos2\theta\begin{pmatrix} i & 0 \\ 0 & -i\end{pmatrix}$ describes a linear birefringence coefficient at zero degrees and we name $\Delta\beta_0/2 = (\Delta\beta/2)\,cos\,2\theta$ as the linear birefringence along zero-degree. The second term describes a linear birefringence coefficient at 45°, and we define $\Delta\beta_{45}/2 = (\Delta\beta/2)\,sin\,2\theta$ as the linear birefringence along 45°. From (4.196), (4.197), and (4.209), we can obtain the eigenvalues of $\mathbf{N}(z)$ as

$$\lambda_N = \pm i\sqrt{\left(\frac{\Delta\beta_0}{2}\right)^2 + \left(\frac{\Delta\beta_{45}}{2}\right)^2} = \pm\frac{\Delta\beta}{2}i, \text{ and } T_N = 0, \quad Q_N = \frac{\Delta\beta}{2}i \tag{4.210}$$

and two orthogonal linear eigenvectors: $(cos\theta\ \ sin\theta)^T$ and $(-sin\theta\ \ cos\theta)^T$

By substituting (4.209) and (4.210) in (4.200), one can get the matrix of linear retarder at z:

$$\mathbf{M}(z) = \begin{pmatrix} cos\frac{\Delta\beta}{2}z + i\,cos\,2\theta\,sin\frac{\Delta\beta}{2}z & isin2\theta\,sin\frac{\Delta\beta}{2}z \\ isin2\theta\,sin\frac{\Delta\beta}{2}z & cos\frac{\Delta\beta}{2}z - icos2\theta\,sin\frac{\Delta\beta}{2}z\end{pmatrix} \tag{4.211}$$

The $\mathbf{M}(z)$ is periodic in z with a periodicity of $2\pi/(\Delta\beta/2)$. If this linear retarder is a piece of fiber, the phase difference between two orthogonal linear states is $\Delta\beta z$, the corresponding period is called linear beat length L_p, and

$$L_p = \frac{2\pi}{\Delta\beta} \tag{4.212}$$

4.3.7.4 Elliptical Retardation Plate

For the elliptical birefringence retarder plate with thickness z, there are two orthogonal elliptical eigenstates ε_1 and ε_2, and the differential retardation $\Delta\eta z$ will be generated between these two eigenstates after light passes through the plate. Its matrix can be written as the multiplication of a circular retarder with circular birefringence $2\Delta\rho$, and a linear retarder with linear birefringence $\Delta\beta$ with its fast-axis oriented at θ:

$$\mathbf{M}(z) = \mathbf{R}(\Delta\rho z)\mathbf{R}(\theta)\mathbf{G}(\Delta\beta z)\mathbf{R}(\theta) \tag{4.213}$$

where $\mathbf{R}$ is the rotation matrix and $\mathbf{G}$ is the linear retarder with the fast-axis at zero degree. The corresponding $\mathbf{N}$-matrix can be written as

$$\mathbf{N}(z) = \frac{d\mathbf{M}(z)}{dz}\mathbf{M}^{-1}(z) = \left[\frac{d\mathbf{R}(\Delta\rho z)}{dz}\mathbf{R}(\theta)\mathbf{G}(\Delta\beta z)\mathbf{R}(\theta) + \mathbf{R}(\Delta\rho z)\mathbf{R}(\theta)\frac{d\mathbf{G}(\Delta\beta z)}{dz}\mathbf{R}(\theta)\right]\mathbf{M}^{-1}(z) \tag{4.214}$$

where $\mathbf{M}^{-1}(z)$ can be obtained from (4.213) as

$$\mathbf{M}^{-1}(z) = \mathbf{R}(-\theta)\mathbf{G}(-\Delta\beta z)\mathbf{R}(-\theta)\mathbf{R}(-\Delta\rho z) \tag{4.215}$$

Substitution of (4.215) in (4.214) yields

$$\mathbf{N}(z) = \frac{d\mathbf{R}(\Delta\rho z)}{dz}\mathbf{R}(-\Delta\rho z) + \mathbf{R}(\Delta\rho z + \theta)\frac{d\mathbf{G}(\Delta\beta z)}{dz}\mathbf{G}(-\Delta\beta z)\mathbf{R}(-\Delta\rho z - \theta) \tag{4.216}$$

Note that the first term on the right-hand side of Eq. (4.216) is the $\mathbf{N}$-matrix of a circular retarder, and the second item represents the $\mathbf{N}$-matrix of a linear retarder. According to (4.203) and (4.209), we have

$$\begin{aligned}\mathbf{N}(z) &= \Delta\rho\begin{pmatrix}0 & -1\\ 1 & 0\end{pmatrix} + \frac{\Delta\beta}{2}cos2(\Delta\rho z + \theta)\begin{pmatrix}i & 0\\ 0 & -i\end{pmatrix} + \frac{\Delta\beta}{2}sin2(\Delta\rho z + \theta)\begin{pmatrix}0 & i\\ i & 0\end{pmatrix}\\ &= \frac{\Delta\gamma}{2}\begin{pmatrix}0 & -1\\ 1 & 0\end{pmatrix} + \frac{\Delta\beta_{0^\circ}}{2}\begin{pmatrix}i & 0\\ 0 & -i\end{pmatrix} + \frac{\Delta\beta_{45^\circ}}{2}\begin{pmatrix}0 & i\\ i & 0\end{pmatrix}\\ &= \frac{1}{2}\begin{pmatrix}i\Delta\beta_{0^\circ} & -\Delta\gamma + i\Delta\beta_{45^\circ}\\ \Delta\gamma + i\Delta\beta_{45^\circ} & -i\Delta\beta_{0^\circ}\end{pmatrix}\end{aligned} \tag{4.217}$$

where the first term describes a circular birefringence defined as $\Delta\gamma = 2\Delta\rho$; the second term is a linear birefringence coefficient at zero degrees, which is defined as $\Delta\beta_{0^\circ} = \Delta\beta\ cos2(\Delta\rho z + \theta)$. The third term describes a linear birefringence coefficient at 45°, which is defined as $\Delta\beta_{45^\circ} = \Delta\beta cos\ 2(\Delta\rho z + \theta)$ at 45°. According to Eqs. (4.196), (4.197), and (4.209), one can find the eigenvalues λ_N and normalized eigenvectors ε_N of the $\mathbf{N}$-matrix as following:

$$\lambda_N = \frac{\pm i\sqrt{\Delta\gamma^2 + \Delta\beta_{0^\circ}^2 + \Delta\beta_{45^\circ}^2}}{2} = \frac{\pm i\sqrt{\Delta\gamma^2 + \Delta\beta^2}}{2} = \pm\frac{\Delta\Gamma}{2}i \tag{4.218a}$$

$$\text{where } \Delta\Gamma = \sqrt{\Delta\gamma^2 + \Delta\beta_{0^\circ}^2 + \Delta\beta_{45^\circ}^2} = \sqrt{4\Delta\rho^2 + \Delta\beta^2} \tag{4.218b}$$

$$\varepsilon_N = \frac{1}{\Delta\Gamma\sqrt{2 + 2\Delta\beta_{0^\circ}/\Delta\Gamma}}\begin{pmatrix}(\Delta\beta_{0^\circ} \pm \Delta\Gamma)i\\ \Delta\gamma + i\Delta\beta_{45^\circ}\end{pmatrix} \tag{4.218c}$$

Substituting eigenvalues λ_N and (4.217) in (4.200), we can obtain the matrix $\mathbf{M}(z)$ of elliptical retarder as

$$\mathbf{M}(z) = \begin{pmatrix}cos\frac{\Delta\Gamma z}{2} + i\frac{\Delta\beta_{0^\circ}}{\Delta\Gamma}sin\frac{\Delta\Gamma z}{2} & \frac{(-\Delta\gamma + i\Delta\beta_{45^\circ})}{\Delta\Gamma}sin\frac{\Delta\Gamma z}{2}\\ \frac{(\Delta\gamma + i\Delta\beta_{45^\circ})}{\Delta\Gamma}sin\frac{\Delta\Gamma z}{2} & cos\frac{\Delta\Gamma z}{2} - i\frac{\Delta\beta_{0^\circ}}{\Delta\Gamma}sin\frac{\Delta\Gamma z}{2}\end{pmatrix} \tag{4.219}$$

The matrix with a given z will cause a phase shift $\Delta\Gamma z$ between two orthogonal polarization eigenstates. Therefore, $\Delta\Gamma$ can be named as the elliptical birefringence; its magnitude can be obtained from (4.218). If this elliptical retarder is a piece of fiber, the phase difference between two orthogonal elliptical states will periodically change along light propagating direction, and the corresponding period L_e is called elliptical beat length, which is defined as:

$$L_e = \frac{2\pi}{\Delta\Gamma} \tag{4.220}$$

The orthogonal elliptical states will change periodically along light propagating direction. Substituting Eqs. (4.205), (4.212), and (4.218b) in Eq. (4.220), one obtains the following relationship:

$$\frac{1}{L_e^2} = \frac{1}{L_p^2} + \frac{1}{L_c^2} \tag{4.221}$$

or

$$L_e = \frac{L_p L_c}{\sqrt{L_c^2 + L_p^2}} \tag{4.222}$$

4.3.8 Jones Matrix of Twisted Fiber

Next, as an example, we derivate the Jones matrix of the high birefringence (Hi-Bi) spun optical fiber, which is mainly used in electric current sensors, as will be discussed in more detail in Section 9.4. The Hi-Bi spun optical fiber is made by spinning a high linear-birefringence fiber preform during the drawing process so as to impart a build-in rapid rotation of the birefringence axis. A model of spun fiber is shown in Figure 4.27; it is equivalent to a series of thin elliptical birefringence retardation plates with the linear birefringence $\Delta\beta = 2\pi/L_B$ and circular birefringence $\Delta\alpha = \pi/L_c$ induced rotation cascaded together sequentially. The birefringence axis of each thin plate is rotated with an angle $\rho\Delta z$, where ρ is the spin twist rate with a unit of radian/meter, Δz is the thickness of the thin plate at z. We can always select a proper x- and y-axes to make the orientation angle of the first thin wave plate be zero so that the orientation angle of the thin wave plate at position z will be ρz. The following describes step-by-step procedures for obtaining the analytical solution of twisted fiber as a function fiber manufacturing parameters (L_B, L_c, ρ), which is difficult to find in open literatures.

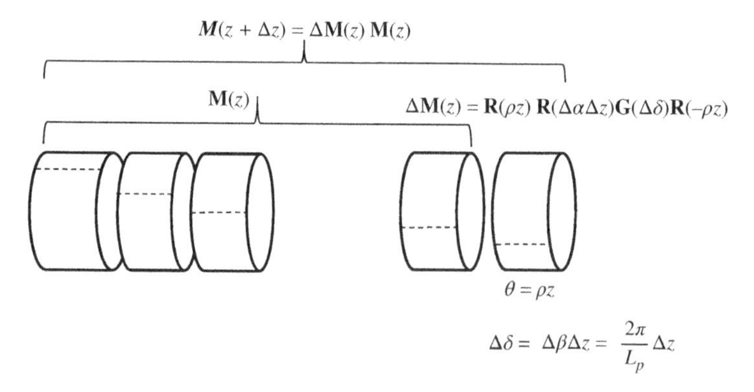

Figure 4.27 The model of a spun highly birefringence fiber. The birefringence axis of each thin plate is rotated with an angle $\rho\Delta z$. Assuming the birefringence axis angle is zero from the x-axis at $z = 0$, the axis will be ρz at position z.

Step 1: Find the **N**-matrix $\mathbf{N}(z)$ of the spun fiber.

$$\mathbf{N}(z) = \mathbf{R}(\rho z)\mathbf{N}_0\mathbf{R}(-\rho z), \quad \text{where } \mathbf{N}_0 = \begin{pmatrix} i\Delta\beta/2 & -\Delta\alpha \\ \Delta\alpha & -i\Delta\beta/2 \end{pmatrix} \tag{4.223}$$

Proof:

$$\mathbf{N}(z) = \lim_{\Delta z\to 0}\frac{\Delta\mathbf{M}(z) - \mathbf{I}}{\Delta z} = \lim_{\Delta z\to 0}\frac{\mathbf{R}(\rho z)\mathbf{R}(\Delta\alpha\Delta z)\mathbf{G}(\Delta\beta\Delta z)\mathbf{R}(-\rho z) - \mathbf{I}}{\Delta z}$$

$$= \lim_{\Delta z\to 0}\frac{\mathbf{R}(\rho z)\begin{pmatrix} e^{i(\Delta\beta\Delta z/2)} - 1 & -\Delta\alpha\Delta z e^{-i(\Delta\beta\Delta z/2)} \\ \Delta\alpha\Delta z e^{i(\Delta\beta\Delta z/2)} & e^{-i(\Delta\beta\Delta z/2)} - 1 \end{pmatrix}\mathbf{R}(-\rho z)}{\Delta z}$$

$$= \lim_{\Delta z\to 0}\frac{\mathbf{R}(\rho z)\begin{pmatrix} i\Delta\beta\Delta z/2 & -\Delta\alpha\Delta z \\ \Delta\alpha\Delta z & -i\Delta\beta\Delta z/2 \end{pmatrix}\mathbf{R}(-\rho z)}{\Delta z} = \mathbf{R}(\rho z)\begin{pmatrix} i\Delta\beta/2 & -\Delta\alpha \\ \Delta\alpha & -i\Delta\beta/2 \end{pmatrix}\mathbf{R}(-\rho z)$$

where **R** is a rotation matrix and **G** is the matrix of linear retardation plate with the fast axis parallel to the x-axis.

Step 2: Define a Jones vector $\varepsilon(z) = \mathbf{R}(-\rho z)\mathbf{E}(z)$ and obtain its derivative as

$$\frac{d\varepsilon(z)}{dz} = \frac{d\mathbf{R}(-\rho z)}{dz}\mathbf{E}(z) + \mathbf{R}(-\rho z)\frac{d\mathbf{E}(z)}{dz} \tag{4.224}$$

Step 3: The differential Eq. (4.224) can be reduced to

$$\frac{d\varepsilon(z)}{dz} = \boldsymbol{\eta}\varepsilon(z) \text{ where } \boldsymbol{\eta} = \begin{pmatrix} i\Delta\beta/2 & \rho - \Delta\alpha \\ -\rho + \Delta\alpha & -i\Delta\beta/2 \end{pmatrix} \tag{4.225}$$

Proof:
From Eq. (4.181), we get $\frac{d\mathbf{E}(z)}{dz} = \mathbf{N}(z)\mathbf{E}(z)$, Substituting it and Eq. (4.223) in (4.224), we obtain

$$\begin{aligned}\left(\frac{d\varepsilon(z)}{dz}\right) &= \frac{d\mathbf{R}(-\rho z)}{dz}\mathbf{E}(z) + \mathbf{R}(-\rho z)\frac{d\mathbf{E}(z)}{dz} = \frac{d\mathbf{R}(-\rho z)}{dz}\mathbf{E}(z) + \mathbf{R}(-\rho z)\mathbf{N}(z)\mathbf{E}(z) \\ &= \frac{d\mathbf{R}(-\rho z)}{dz}\mathbf{E}(z) + \mathbf{R}(-\rho z)\mathbf{R}(\rho z)\mathbf{N}_0\mathbf{R}(-\rho z)\mathbf{E}(z) \\ &= \left[\frac{d\mathbf{R}(-\rho z)}{dz}\mathbf{R}(\rho z) + \mathbf{N}_0\right]\varepsilon(z) = \left[-\rho\,\mathbf{R}\left(\frac{\pi}{2}\right) + \mathbf{N}_0\right]\varepsilon(z) \\ &= \begin{pmatrix} \frac{i\Delta\beta}{2} & \rho - \Delta\alpha \\ -(\rho - \Delta\alpha) & -\frac{i\Delta\beta}{2} \end{pmatrix}\varepsilon(z)\end{aligned} \tag{4.226}$$

Step 4: The Jones matrix $\mathbf{M}(z)$ of the Hi-Bi spun fiber system is

$$\mathbf{M}(z) = \mathbf{R}(\rho z)\mathbf{e}^{\boldsymbol{\eta} z} \tag{4.227}$$

Proof:
Since $\boldsymbol{\eta}$ is a constant, we have the solution of the differential Eq. (4.225),

$$\varepsilon(z) = \mathbf{e}^{\boldsymbol{\eta} z}\varepsilon(z = 0) \quad\Rightarrow\quad \mathbf{R}(-\rho z)\mathbf{E}(z) = \mathbf{e}^{\boldsymbol{\eta} z}\mathbf{E}(0)$$

$$\Rightarrow \mathbf{E}(z) = \mathbf{R}(\rho z)\,\mathbf{e}^{\boldsymbol{\eta} z}\mathbf{E}(0) \quad\Rightarrow\quad \mathbf{M}(z) = \mathbf{R}(\rho z)\,\mathbf{e}^{\boldsymbol{\eta} z} \quad \text{because } \mathbf{E}(z) = \mathbf{M}(z)\mathbf{E}(0)$$

Step 5: Calculate the exponential matrix $\mathbf{e}^{\boldsymbol{\eta}z}$ by finding the eigenvalues and eigenvectors of the matrix $\boldsymbol{\eta}$:

$$\mathbf{e}^{\boldsymbol{\eta}z} = \begin{pmatrix} cos\gamma z + i\frac{\Delta\beta/2}{\gamma}sin\gamma z & \frac{\rho-\Delta\alpha}{\gamma}sin(\gamma z) \\ -\frac{\rho-\Delta\alpha}{\gamma}sin(\gamma z) & cos\gamma z - i\frac{\Delta\beta/2}{\gamma}sin\gamma z \end{pmatrix} \tag{4.228}$$

Proof:

The eigenvalues λ_i $(i = 1, 2)$ and eigenvectors $\boldsymbol{V}_i$ $(i = 1, 2)$ of $\boldsymbol{\eta} = \begin{pmatrix} i\Delta\beta/2 & \rho-\Delta\alpha \\ -(\rho-\Delta\alpha) & -i\Delta\beta/2 \end{pmatrix}$ are

$$\lambda_1 = \gamma i = \frac{1}{2}\sqrt{\Delta\beta^2 + 4(\rho-\Delta\alpha)^2}i \qquad \text{eigenvector:}$$
$$\mathbf{v}_1 = \frac{1}{\sqrt{(\Delta\beta/2+\gamma)^2+\rho^2}}[(\Delta\beta/2+\gamma)i, -(\rho-\Delta\alpha)]^T \tag{4.229}$$

$$\lambda_2 = -\gamma i = -\frac{1}{2}\sqrt{\Delta\beta^2 + 4(\rho-\Delta\alpha)^2}\,i \qquad \text{eigenvector:}$$
$$\mathbf{v}_2 = \frac{1}{\sqrt{(\Delta\beta/2+\gamma)^2+\rho^2}}[(\rho-\Delta\alpha), -(\Delta\beta/2+\gamma)i]^T \tag{4.230}$$

Let's construct a new matrix $\mathbf{V} = \begin{pmatrix}\mathbf{v}_1 & \mathbf{v}_2\end{pmatrix}$, which is unitary because $\mathbf{V}^\dagger\mathbf{V} = \mathbf{V}\mathbf{V}^\dagger = \mathrm{I}$. Therefore, we can transform the matrix $\boldsymbol{\eta}$ to a diagonal matrix by left-multiplicating $\mathbf{V}^\dagger$ and right-multiplicating $\mathbf{V}$ because

$$\boldsymbol{\eta}\mathbf{V} = \mathbf{V}\begin{pmatrix}\lambda_1 & 0 \\ 0 & \lambda_2\end{pmatrix} \Rightarrow \mathbf{V}^\dagger\boldsymbol{\eta}\mathbf{V} = \begin{pmatrix}\lambda_1 & 0 \\ 0 & \lambda_2\end{pmatrix}$$

Since $[\mathbf{V}^\dagger(\boldsymbol{\eta}z)\mathbf{V}]^n = \mathbf{V}^\dagger(\boldsymbol{\eta}z)\mathbf{V}\,\mathbf{V}^\dagger(\boldsymbol{\eta}z)\mathbf{V}\ldots\mathbf{V}^\dagger(\boldsymbol{\eta}z)\mathbf{V} = \mathbf{V}^\dagger(\boldsymbol{\eta}z)^n\mathbf{V}$ then the exponential matrix $\mathbf{e}^{\boldsymbol{\eta}z}$ can be obtained as

$$\mathbf{e}^{\boldsymbol{\eta}z} = \mathbf{I} + \sum_{n=1}^{\infty}\mathbf{V}\mathbf{V}^\dagger\frac{(\boldsymbol{\eta}z)^n}{n!}\mathbf{V}\mathbf{V}^\dagger = \mathbf{V}\left[I + \sum_{n=1}^{\infty}\frac{(\mathbf{V}^\dagger\boldsymbol{\eta}z\mathbf{V})^n}{n!}\right]\mathbf{V}^\dagger = \mathbf{V}\left[I + \sum_{n=1}^{\infty}\frac{\begin{pmatrix}(\lambda_1 z)^n & 0 \\ 0 & (\lambda_2 z)^n\end{pmatrix}}{n!}\right]\mathbf{V}^\dagger$$
$$= \mathbf{V}\begin{pmatrix}e^{i\gamma z} & 0 \\ 0 & e^{-i\gamma z}\end{pmatrix}\mathbf{V}^\dagger \tag{4.231}$$

By substitution of Eqs. (4.229) and (4.230) in Eq. (4.231), one obtains

$$\mathbf{e}^{\boldsymbol{\eta}z} = \mathbf{V}\begin{pmatrix}e^{i\gamma z} & 0 \\ 0 & e^{-i\gamma z}\end{pmatrix}\mathbf{V}^\dagger$$
$$= \frac{1}{(\Delta\beta/2+\gamma)^2+\rho^2}\begin{bmatrix}(\Delta\beta/2+\gamma)i & \rho-\Delta\alpha \\ -(\rho-\Delta\alpha) & -(\Delta\beta/2+\gamma)i\end{bmatrix}\begin{bmatrix}e^{i\gamma z} & 0 \\ 0 & e^{-i\gamma z}\end{bmatrix}\begin{bmatrix}-(\Delta\beta/2+\gamma)i & -(\rho-\Delta\alpha) \\ \rho-\Delta\alpha & (\Delta\beta/2+\gamma)i\end{bmatrix}$$
$$= \begin{pmatrix} cos\gamma z + i\frac{\Delta\beta/2}{\gamma}sin\gamma z & \frac{\rho-\Delta\alpha}{\gamma}sin(\gamma z) \\ -\frac{\rho-\Delta\alpha}{\gamma}sin(\gamma z) & cos\gamma z - i\frac{\Delta\beta/2}{\gamma}sin\gamma z \end{pmatrix} \tag{4.232}$$

Step 6: Reformat the matrix $\mathbf{e}^{\boldsymbol{\eta}z}$ to the multiplication of a rotation matrix and a retardation plate.

Since $\mathbf{e}^{\eta z}$ is a unitary matrix, it is equivalent to cascading a rotator and a retardation plate. Assuming that the rotator has a rotation angle of Ψ, and the retardation plate has a retardation of δ and an orientation angle of Φ, the matrix $\mathbf{e}^{\eta z}$ can be expressed as

$$\mathbf{e}^{\eta z} = \mathbf{R}(\Psi)\mathbf{R}(\Phi)\mathbf{G}(\delta)\mathbf{R}(-\Phi) = \mathbf{R}(\Psi+\Phi)\mathbf{G}(\delta)\mathbf{R}(-\Phi) \tag{4.233}$$

where $\mathbf{G}(\delta)$ is the Jones matrix of the wave plate with a retardation δ and its fast-axis aligned to the x-axis. Substituting matrices $\mathbf{R}(\Psi+\Phi)$, $\mathbf{R}(\Phi)$, and $\mathbf{G}(\delta)$ in Eqs. (4.32) and (4.40), we obtain

$$\mathbf{e}^{\eta z} = \begin{pmatrix} cos(\Psi)cos\delta/2 + icos(\Psi+2\Phi)sin\delta/2 & -sin(\Psi)cos\delta/2 + isin(\Psi+2\Phi)sin\delta/2 \\ sin(\Psi)cos\delta/2 + isin(\Psi+2\Phi)sin\delta/2 & cos(\Psi)cos\delta/2 - icos(\Psi+2\Phi)sin\delta/2 \end{pmatrix}$$
$$= \begin{pmatrix} m_{11} & m_{12} \\ m_{21} & m_{22} \end{pmatrix} \tag{4.234}$$

Equating the corresponding matrix elements on both sides of the equation earlier yields

$$tan(\Psi+2\Phi) = \frac{m_{21}+m_{12}}{m_{11}-m_{22}}, \quad tan(\Psi) = \frac{m_{21}-m_{12}}{m_{11}+m_{22}} \tag{4.235}$$

Substitution of the matrix elements m_{11}, m_{12}, m_{21}, and m_{22} obtained from (4.232) in Eq. (4.235), we arrive at

$$tan(\Psi) = \frac{m_{21}-m_{12}}{m_{11}+m_{22}} = \frac{-(\rho-\Delta\alpha)sin(\gamma z)}{\gamma\ cos(\gamma z)} = \frac{-\frac{2(\rho-\Delta\alpha)}{\Delta\beta}}{\sqrt{1+\left(\frac{2(\rho-\Delta\alpha)}{\Delta\beta}\right)^2}}tan(\gamma z) \tag{4.236}$$

$$tan(\Psi+2\Phi) = \frac{m_{21}+m_{12}}{m_{11}-m_{22}} = 0 \quad \text{i.e. } \Psi = -2\Phi + n\pi \tag{4.237}$$

Since $\Psi = 2\Phi$, Eq. (4.234) can be reduced to

$$\mathbf{e}^{\eta z} = \begin{pmatrix} cos(\Psi)cos\frac{\delta}{2} + isin\frac{\delta}{2} & -sin(\Psi)cos\frac{\delta}{2} \\ sin(\Psi)cos\frac{\delta}{2}sin\frac{\delta}{2} & cos(\Psi)cos\frac{\delta}{2} - isin\frac{\delta}{2} \end{pmatrix} \tag{4.238}$$

Therefore we can find the retardation of $\mathbf{G}(\delta)$ by equating the corresponding elements in Eqs. (4.228) and (4.238):

$$sin\frac{\delta}{2} = Im(m_{11}) = \frac{1}{\sqrt{1+[2(\rho-\Delta\alpha)/\Delta\beta]^2}}sin\gamma z \tag{4.239}$$

Now, the angles Ψ, Φ, and retardation δ in (4.233) are all obtained.

Step 7: Get $\mathbf{M}(z)$ with the form of $\mathbf{R}(\Omega)\mathbf{M}_{wp}$

$$\mathbf{M}(z) = \mathbf{R}(\rho z)\mathbf{e}^{\eta z} = \mathbf{R}(\rho z)\mathbf{R}(\Psi)\mathbf{R}(\Phi)\mathbf{G}(\delta)\mathbf{R}(-\Phi) = \mathbf{R}(\rho z+\Psi)\mathbf{R}(\Phi)\mathbf{G}(\delta)\mathbf{R}(-\Phi) \tag{4.240}$$

Let $\Omega = \rho z + \Psi$ and substitute the expressions of $\mathbf{R}(\Phi)$ and $\mathbf{G}(\delta)$ in Eq. (4.240), we obtain

$$\mathbf{M}(z) = \begin{pmatrix} cos\Omega & -sin\Omega \\ sin\Omega & cos\Omega \end{pmatrix} \begin{pmatrix} cos(\delta/2) + isin(\delta/2)cos(2\Phi) & isin(\delta/2)sin(2\Phi) \\ isin(\delta/2)sin(2\Phi) & cos\delta/2 - isin(\delta/2)cos(2\Phi) \end{pmatrix} \tag{4.241}$$

where

$$\delta = 2sin^{-1}\left[\frac{sin(\gamma z)}{\sqrt{1+[2(\rho-\Delta\alpha)/\Delta\beta]^2}}\right] \tag{4.242}$$

$$\Omega = \rho z + tan^{-1}\left[\frac{-2(\rho - \Delta\alpha)/\Delta\beta}{\sqrt{1 + [2(\rho - \Delta\alpha)/\Delta\beta]^2}} tan(\gamma z)\right] + n\pi \tag{4.243}$$

$$\Phi = \frac{\rho z - \Omega}{2} + \frac{m}{2}\pi \tag{4.244}$$

$$\gamma = \frac{1}{2}\sqrt{\Delta\beta^2 + 4(\rho - \Delta\alpha)^2} \tag{4.245}$$

These equations are equivalent to Eqs. (3.15a) and (3.15b). In summary, the Hi-Bi spun optical fiber can be represented by two lumped birefringent elements, a retarder δ with its fast-axis orientated at angle Φ, and a rotator $\Omega(z)$, which can be described by Eq. (4.241) or $\mathbf{R}(\rho z)\mathbf{e}^{\boldsymbol{\eta} z}$ of Eq. (4.227). In the following discussions, we will describe the polarization evolution under different conditions:

Case 1: $\Delta\beta \gg \rho - \Delta\alpha$ (PM fiber)

Under the conditions that the local linear birefringence $\Delta\beta$ is much larger than the spin twist rate and the local circular birefringence, the fiber is equivalent to a PM fiber with its slow axis being twisted, which was qualitatively discussed in Section 3.3.5. Since $\Delta\beta \gg \rho - \Delta\alpha$, the local elliptical birefringence 2γ is approximately equal to $\Delta\beta$. Substituting (4.228) and $\gamma = \Delta\beta/2$ in (4.241), one obtains the Jones matrix of the fiber:

$$\mathbf{M}(z) = \mathbf{R}(\rho z)\mathbf{e}^{\boldsymbol{\eta} z} = \mathbf{R}(\rho z)\begin{pmatrix} e^{i\frac{\Delta\beta}{2}z} & \frac{2(\rho - \Delta\alpha)}{\Delta\beta} sin\left(\frac{\Delta\beta}{2}z\right) \\ -\frac{2(\rho - \Delta\alpha)}{\Delta\beta} sin\left(\frac{\Delta\beta}{2}z\right) & e^{-i\frac{\Delta\beta}{2}z} \end{pmatrix} \tag{4.246}$$

Suppose a linearly polarized light is launched into the fiber at position $z = 0$, with its polarization aligned parallel to the fast-axis (x-axis) of the local retarder at $z = 0$. Then the Jones vector at z is

$$\mathbf{E} = \mathbf{M}(z)\begin{pmatrix} 1 \\ 0 \end{pmatrix} = \mathbf{R}(\rho z)\begin{pmatrix} e^{i\frac{\Delta\beta}{2}z} \\ -\frac{2(\rho - \Delta\alpha)}{\Delta\beta} sin\left(\frac{\Delta\beta}{2}z\right) \end{pmatrix} \tag{4.247}$$

Let's define a new local coordinate by rotating the x-axis to an angle ρz and assume the electric field in the new local coordinates is $\mathbf{E}'$, i.e. $\mathbf{E}' = \mathbf{R}(-\rho z)\mathbf{E}$. Substituting it in Eq. (4.247), we have

$$\mathbf{E}' = \mathbf{R}(-\rho z)\mathbf{E} = \begin{pmatrix} e^{i\frac{\Delta\beta}{2}z} \\ -\frac{2(\rho - \Delta\alpha)}{\Delta\beta} sin\left(\frac{\Delta\beta}{2}z\right) \end{pmatrix} = \begin{pmatrix} E'_x \\ E'_y \end{pmatrix} \tag{4.248}$$

In practice, the beat length L_p of PM fiber is on the order of 5 mm, i.e. the local birefringence $\Delta\beta = 2\pi/L_p$ is about 400π/meter. Assume the combination of spin twist and circular birefringence $\rho - \Delta\alpha$ is 10π/meter (5 turns/meter), and then the magnitude factor $\frac{2(\rho - \Delta\alpha)}{\Delta\beta}$ of $\mathbf{E}'_y$ in (4.248) is about 1/20. Thus the power ratio between slow-axis and fast-axis will be $[\mathbf{E}'_y(\mathbf{E}'_y)^*] / [\mathbf{E}'_x(\mathbf{E}'_x)^*] = (1/400)sin^2\left(\frac{\Delta\beta}{2}z\right)$; therefore, most of energy maintains in the state linearly polarized along the fast-axis, and the input linear polarization along the fast-axis is maintained.

Case 2: Large twist rate and low local linear birefringence

In the limit of a high twist rate $\rho - \Delta\alpha \gg \Delta\beta$, from Eq. (4.242), the linear retardation becomes

$$\delta = 2sin^{-1}\left[\frac{sin(\gamma z)}{\sqrt{1 + [2(\rho - \Delta\alpha)/\Delta\beta]^2}}\right] \approx \frac{\Delta\beta}{\rho - \Delta\alpha} sin(\gamma z) \tag{4.249}$$

and the rotation

$$\Omega = (\rho - \gamma)z + n\pi \approx (\Delta\alpha)z + n\pi \tag{4.250}$$

In this case, the linear retardation is considerably reduced by the twist so that launching a linearly polarized light into the fiber results in a near-linear output. The corresponding polarization is continuously rotated along the fiber at the rotating rate α (rotation angle per unit length). That is, the fiber is equivalent a rotator (circular retarder).

4.4 Mueller Matrix Representation of Optical Devices

4.4.1 Definition of Mueller Matrix

In addition to the Jones vectors, one may also represent the polarization states using the Stokes parameters S_0, S_1, S_2, and S_3. For the monochromatic light, Stokes parameters are related to the Jones vector elements by

$$\begin{aligned}
&\mathbf{E} = \begin{pmatrix} E_x \\ E_y \end{pmatrix} = e^{i\phi}\begin{pmatrix} A_x \\ A_y e^{i\delta} \end{pmatrix} \\
&S_0 = A_x^2 + A_y^2 = E_x E_x^* + E_y E_y^* \\
&S_1 = A_x^2 - A_y^2 = E_x E_x^* - E_y E_y^* \\
&S_2 = 2A_x A_y cos\delta = E_x E_y^* + E_y E_x^* \\
&S_3 = 2A_x A_y sin\delta = i\left(E_x E_y^* - E_y E_x^*\right)
\end{aligned} \tag{4.251}$$

This equation can also be expressed in the matrix form as

$$\begin{pmatrix} S_1 \\ S_2 \\ S_3 \\ S_4 \end{pmatrix} = \mathbf{A}(\mathbf{E} \otimes \mathbf{E}^*) \tag{4.252}$$

where $\otimes$ denotes the Kronecker tensor product, and $\mathbf{A}$ is a 4×4 matrix:

$$\mathbf{A} = \begin{pmatrix} 1 & 0 & 0 & 1 \\ 1 & 0 & 0 & -1 \\ 0 & 1 & 1 & 0 \\ 0 & i & -i & 0 \end{pmatrix} \qquad \mathbf{E} \otimes \mathbf{E}^* = \begin{pmatrix} E_x E_x^* \\ E_x E_y^* \\ E_y E_x^* \\ E_y E_y^* \end{pmatrix} \tag{4.253}$$

Recall that in Jones Calculus, the transformation of the polarization state in an optical system from the incident field $\mathbf{E}$ to the emerging field $\mathbf{E}'$ can be described by a 2×2 complex matrix $\mathbf{M}_J$:

$$\mathbf{E}' = \mathbf{M}_J \mathbf{E} \tag{4.254}$$

Similar to the Jones vector, the polarization state can also be represented by a 4×1 real column vector called Stokes vector with entries equal to the Stokes parameters. Assuming that a polarization state can be expressed as the linear combination of the four parameters of input polarization state, the input SOP and output SOP can be related by a 4×4 matrix as

$$\mathbf{S}_{out} = \mathbf{M}\mathbf{S}_{in} \tag{4.255}$$

$$\begin{pmatrix} S_{out,0} \\ S_{out,1} \\ S_{out,2} \\ S_{out,3} \end{pmatrix} = \begin{pmatrix} m_{00} & m_{01} & m_{02} & m_{03} \\ m_{10} & m_{11} & m_{12} & m_{13} \\ m_{20} & m_{21} & m_{22} & m_{23} \\ m_{30} & m_{31} & m_{32} & m_{33} \end{pmatrix} \begin{pmatrix} \mathbf{S}_{in,0} \\ \mathbf{S}_{in,1} \\ \mathbf{S}_{in,2} \\ \mathbf{S}_{in,3} \end{pmatrix} \tag{4.256}$$

where $\mathbf{S}_{in}$ and $\mathbf{S}_{out}$ are the input and output Stokes vectors, and $\mathbf{M}$ is the 4×4 matrix known as the Mueller matrix of the optical system under investigation. In the following discussions, we will derive the Mueller matrices of the basic optical elements (retarder, rotator, partial polarizer, and mirror) when the incident light is monochromatic.

From Eq. (4.252), we can represent the output Stokes vector by the corresponding Jones vector as

$$\mathbf{S}_{out} = \mathbf{A}\left(\mathbf{E}_{out} \otimes \mathbf{E}^*_{out}\right) \tag{4.257}$$

Substitution of $\mathbf{E}_{out} = \mathbf{M}_J\mathbf{E}_{in}$ in (4.257) yields

$$\mathbf{S}_{out} = \mathbf{A}(\mathbf{M}_J\mathbf{E}_{in}) \otimes \left(\mathbf{M}^*_J\mathbf{E}^*_{in}\right) = \mathbf{M}\mathbf{S}_{in} \tag{4.258}$$

According to the property of Kronecker product $(\mathbf{AC})\otimes(\mathbf{BD}) = (\mathbf{A}\otimes\mathbf{B})(\mathbf{C}\otimes\mathbf{D})$, Eq. (4.258) can be re-written as

$$\mathbf{S}_{out} = \mathbf{A}\left(\mathbf{M}_J \otimes \mathbf{M}^*_J\right)\left(\mathbf{E}_{in} \otimes \mathbf{E}^*_{in}\right) = \mathbf{A}\left(\mathbf{M}_J \otimes \mathbf{M}^*_J\right)\mathbf{A}^{-1}\mathbf{A}\left(\mathbf{E}_{in} \otimes \mathbf{E}^*_{in}\right) = \mathbf{A}\left(\mathbf{M}_J \otimes \mathbf{M}^*_J\right)\mathbf{A}^{-1}\mathbf{S}_{in} \tag{4.259}$$

where $\mathbf{S}_{in} = \mathbf{A}\left(\mathbf{E}_{in} \otimes \mathbf{E}^*_{in}\right)$ has been used during derivation. Comparing the left side of (4.255) and (4.259), we get the transformation from the Jones matrix $\mathbf{M}_J$ to Mueller matrix $\mathbf{M}$

$$\mathbf{M} = \mathbf{A}\left(\mathbf{M}_J \otimes \mathbf{M}^*_J\right)\mathbf{A}^{-1} \tag{4.260}$$

where the inverse of matrix $\mathbf{A}$ is

$$\mathbf{A}^{-1} = \frac{1}{2}\begin{pmatrix} 1 & 1 & 0 & 0 \\ 0 & 0 & 1+i & 1-i \\ 0 & 0 & 1-i & -1+i \\ 1 & -1 & 0 & 0 \end{pmatrix} \tag{4.261}$$

Finding the Mueller matrix through the transformation (4.260) is complicated, especially for a system containing many optical devices. Fortunately, this processing can be simplified. For example, if an optical system is represented by the multiplication of a serial of n Jones matrices: $\mathbf{M}_J = \mathbf{M}_{J,n}\mathbf{M}_{J,n-1}\ldots\mathbf{M}_{J,1}$, and the corresponding Mueller matrix $\mathbf{M}_i$ for each $\mathbf{M}_{J,i}$ is known, we can directly express the Mueller matrix $\mathbf{M}$ as $\mathbf{M} = \mathbf{M}_n\mathbf{M}_{n-1}\ldots\mathbf{M}_2\mathbf{M}_1$. The proof is given as follows.

Suppose an optical system contains two optical devices with Jones matrices representation $\mathbf{M}_{J1}$ and $\mathbf{M}_{J2}$, and the corresponding Mueller matrices are $\mathbf{M}_1$ and $\mathbf{M}_2$, respectively. According to the relationship (4.260), the Mueller matrix corresponding to the Jones matrix $\mathbf{M}_J = \mathbf{M}_{J2}\mathbf{M}_{J1}$ can be expressed as

$$\mathbf{M} = \mathbf{A}(\mathbf{M}_{J2}\mathbf{M}_{J1}) \otimes \left(\mathbf{M}^*_{J2}\mathbf{M}^*_{J1}\right)\mathbf{A}^{-1} \tag{4.262}$$

According to the property of Kronecker product $(\mathbf{A}\otimes\mathbf{B})(\mathbf{C}\otimes\mathbf{D}) = (\mathbf{AC})\otimes(\mathbf{BD})$, the Mueller matrix can be reduced to

$$\begin{aligned}\mathbf{M} &= \mathbf{A}(\mathbf{M}_{J2}\mathbf{M}_{J1}) \otimes \left(\mathbf{M}_{J2}^{*}\mathbf{M}_{J1}^{*}\right)\mathbf{A}^{-1}\\ &= \mathbf{A}\left(\mathbf{M}_{J2}\otimes\mathbf{M}_{J2}^{*}\right)\left(\mathbf{M}_{J1}\otimes\mathbf{M}_{J1}^{*}\right)\mathbf{A}^{-1}\\ &= \mathbf{A}\left(\mathbf{M}_{J2}\otimes\mathbf{M}_{J2}^{*}\right)\mathbf{A}^{-1}\mathbf{A}\left(\mathbf{M}_{J1}\otimes\mathbf{M}_{J1}^{*}\right)\mathbf{A}^{-1}\end{aligned} \tag{4.263}$$

Noting that

$$\begin{aligned}\mathbf{M}_1 &= \mathbf{A}\left(\mathbf{M}_{J1}\otimes\mathbf{M}_{J1}^{*}\right)\mathbf{A}^{-1}\\ \mathbf{M}_2 &= \mathbf{A}\left(\mathbf{M}_{J2}\otimes\mathbf{M}_{J2}^{*}\right)\mathbf{A}^{-1}\end{aligned} \tag{4.264}$$

Equation (4.263) can be represented as

$$\mathbf{M} = \mathbf{M}_2\mathbf{M}_1 \tag{4.265}$$

Similarly, it can be readily shown next that if an optical system contains a serial of n optical devices represented by the Jones matrix multiplication $\mathbf{M}_{Jn}$, $\mathbf{M}_{Jn-1}$, …, $\mathbf{M}_{J1}$, the corresponding Mueller matrix $\mathbf{M}$ can also be represented by the multiplication of all Mueller matrices $\mathbf{M}_n\mathbf{M}_{n-1}\ldots\mathbf{M}_2\mathbf{M}_1$, where $\mathbf{M}_n$, $\mathbf{M}_{n-1}$, …, $\mathbf{M}_1$ are the individual Mueller matrices:

$$\begin{array}{ll} \text{Jones matrix representation} & \text{Mueller matrix representation}\\ \mathbf{M}_{J2\leftarrow J1} = \mathbf{M}_{J2}\mathbf{M}_{J1} & \Rightarrow \mathbf{M}_{2\leftarrow 1} = \mathbf{M}_2\mathbf{M}_1\\ \mathbf{M}_{J3\leftarrow J1} = \mathbf{M}_{J3}\mathbf{M}_{J2\leftarrow J1} & \Rightarrow \mathbf{M}_{3\leftarrow 1} = \mathbf{M}_3\mathbf{M}_{2\leftarrow 1} = \prod_{i=3}^{1}\mathbf{M}_i\\ \mathbf{M}_{J4\leftarrow J1} = \mathbf{M}_{J4}\mathbf{M}_{J3\leftarrow J1} & \Rightarrow \mathbf{M}_{4\leftarrow 1} = \mathbf{M}_4\mathbf{M}_{3\leftarrow 1} = \prod_{i=4}^{1}\mathbf{M}_i\\ \vdots & \vdots\\ \mathbf{M}_{J(n-1)\leftarrow J1} = \mathbf{M}_{J(n-1)}\mathbf{M}_{J(n-2)\leftarrow J1} & \Rightarrow \mathbf{M}_{(n-1)\leftarrow 1} = \mathbf{M}_{n-1}\mathbf{M}_{(n-2)\leftarrow 1} = \prod_{i=n-1}^{1}\mathbf{M}_i\\ \mathbf{M}_{Jn\leftarrow J1} = \mathbf{M}_{Jn}\mathbf{M}_{J(n-1)\leftarrow J1} & \Rightarrow \mathbf{M} = \mathbf{M}_{n\leftarrow 1} = \mathbf{M}_n\mathbf{M}_{(n-1)\leftarrow 1} = \prod_{i=n}^{1}\mathbf{M}_i \end{array} \tag{4.266}$$

4.4.2 Mueller Matrix of Optical Elements

4.4.2.1 Mueller Matrix of a Retarder with a Horizontal Fast-Axis

For a retardation plate with its fast-axis parallel to the x-axis, the Jones vector $\mathbf{J}_{out}$ of the emerging light can be written as

$$\begin{pmatrix} E_{out,x}\\ E_{out,y}\end{pmatrix} = \begin{pmatrix} e^{i\frac{\delta}{2}} & 0\\ 0 & e^{-i\frac{\delta}{2}}\end{pmatrix}\begin{pmatrix} E_{in,x}\\ E_{in,y}\end{pmatrix} \quad \text{where } (\delta = \varphi_x - \varphi_y) \tag{4.267}$$

Substitution of Eq. (4.267) in Eq. (4.251) yields

$$\begin{aligned} S_{out,0} &= E_{out,x}E_{out,x}^{*} + E_{out,y}E_{out,y}^{*} = E_{in,x}E_{in,x}^{*} + E_{in,y}E_{in,y}^{*} = S_{in,0}\\ S_{out,1} &= E_{out,x}E_{out,x}^{*} - E_{out,y}E_{out,y}^{*} = E_{in,x}E_{in,x}^{*} - E_{in,y}E_{in,y}^{*} = S_{in,1}\\ S_{out,2} &= E_{out,x}E_{out,y}^{*} + E_{out,y}E_{out,x}^{*} = e^{i\delta}E_{in,x}E_{in,y}^{*} + e^{-i\delta}E_{in,y}E_{in,x}^{*} = cos(\delta)S_{in,2} + sin(\delta)S_{in,3}\\ S_{out,3} &= i\left(E_{out,y}E_{out,x}^{*} - E_{out,x}E_{out,y}^{*}\right) = i\left(e^{-i\delta}E_{in,y}E_{in,x}^{*} - e^{i\delta}E_{in,x}E_{in,y}^{*}\right) = -sin(\delta)S_{in,2} + cos(\delta)S_{in,3}\end{aligned} \tag{4.268}$$

Expressing Eq. (4.268) in a matrix form, one obtains

$$\begin{pmatrix} S_{out,0} \\ S_{out,1} \\ S_{out,2} \\ S_{out,3} \end{pmatrix} = \begin{pmatrix} 1 & 0 & 0 & 0 \\ 0 & 1 & 0 & 0 \\ 0 & 0 & cos\delta & sin\delta \\ 0 & 0 & -sin\delta & cos\delta \end{pmatrix} \begin{pmatrix} S_{in,0} \\ S_{in,1} \\ S_{in,2} \\ S_{in,3} \end{pmatrix} \tag{4.269}$$

By comparing with (4.255), one finds the Mueller matrix of a retarder with a horizontal fast-axis as

$$\mathbf{M}_{RP}(\theta = 0, \delta) = \begin{pmatrix} 1 & 0 & 0 & 0 \\ 0 & 1 & 0 & 0 \\ 0 & 0 & cos\delta & sin\delta \\ 0 & 0 & -sin\delta & cos\delta \end{pmatrix} \tag{4.270}$$

where θ is the angle from the x-axis to the fast-axis. One may note that the expression (4.270) can be reformatted as

$$\begin{pmatrix} 1 & \mathbf{0}^T \\ \mathbf{0} & \mathbf{R}_1(-\delta) \end{pmatrix} \quad \text{with} \quad \mathbf{R}_1(-\delta) = \begin{pmatrix} 1 & 0 & 0 \\ 0 & cos\delta & sin\delta \\ 0 & -sin\delta & cos\delta \end{pmatrix} \tag{4.271}$$

where $\mathbf{0} = \begin{pmatrix} 0 & 0 & 0 \end{pmatrix}^T$ and $\mathbf{R}_1(-\delta)$ is the rotation matrix with a rotation angle of $-\delta$ about s_1-axis in the 3D space defined by the Poincaré sphere.

4.4.2.2 Mueller Matrix of Rotator

Now we consider a rotator with rotation angle θ from the x-axis, we can find the corresponding Jones matrix from expression (4.40). Thus, we can get the emergent Jones vector of the rotator by

$$\begin{pmatrix} E_{out,x} \\ E_{out,y} \end{pmatrix} = \begin{pmatrix} cos\theta & -sin\theta \\ sin\theta & cos\theta \end{pmatrix} \begin{pmatrix} E_{in,x} \\ E_{in,y} \end{pmatrix} \tag{4.272}$$

Substituting Eq. (4.272) into Eq. (4.251), we have

$$\begin{cases} S_{out,0} = S_{in,0} \\ S_{out,1} = cos(2\theta)\, S_{in,1} - sin(2\theta)\, S_{in,2} \\ S_{out,2} = sin(2\theta) S_{in,1} + cos(2\theta) S_{in,2} \\ S_{out,3} = S_{in,3} \end{cases} \tag{4.273}$$

It can be written in the following matrix format

$$\begin{pmatrix} S_{out,0} \\ S_{out,1} \\ S_{out,2} \\ S_{out,3} \end{pmatrix} = \begin{pmatrix} 1 & 0 & 0 & 0 \\ 0 & cos2\theta & -sin2\theta & 0 \\ 0 & sin2\theta & cos2\theta & 0 \\ 0 & 0 & 0 & 1 \end{pmatrix} \begin{pmatrix} S_{in,0} \\ S_{in,1} \\ S_{in,2} \\ S_{in,3} \end{pmatrix} = \mathbf{M}_R(\theta) \begin{pmatrix} S_{in,0} \\ S_{in,1} \\ S_{in,2} \\ S_{in,3} \end{pmatrix} \tag{4.274}$$

Therefore, the Mueller matrix of a rotator can be obtained as

$$\mathbf{M}_R(\theta) = \begin{pmatrix} 1 & 0 & 0 & 0 \\ 0 & cos2\theta & -sin2\theta & 0 \\ 0 & sin2\theta & cos2\theta & 0 \\ 0 & 0 & 0 & 1 \end{pmatrix} \tag{4.275}$$

One may note Eq. (4.275) can be reformed to

$$\begin{pmatrix} 1 & \mathbf{0}^T \\ \mathbf{0} & \mathbf{R}_3(2\theta) \end{pmatrix} \text{ with } \mathbf{R}_3(2\theta) = \begin{pmatrix} cos2\theta & -sin2\theta & 0 \\ sin2\theta & cos2\theta & 0 \\ 0 & 0 & 1 \end{pmatrix} \tag{4.276}$$

where $\mathbf{0} = \begin{pmatrix} 0 & 0 & 0 \end{pmatrix}^T$ and $\mathbf{R}_3(2\theta)$ is the rotation matrix about the s_3-axis with a rotation angle of 2θ in the 3-D space defined by the Poincaré sphere.

4.4.2.3 Mueller Matrix of a Partial Polarizer

In the Jones calculus, a partial polarizer with attenuation coefficients p_1 and p_2 along two orthogonal axes can be mathematically expressed by a diagonal 2×2 real matrix

$$\begin{pmatrix} E_{out,x} \\ E_{out,y} \end{pmatrix} = \begin{pmatrix} p_1 & 0 \\ 0 & p_2 \end{pmatrix} \begin{pmatrix} E_{in,x} \\ E_{in,y} \end{pmatrix} \tag{4.277}$$

Substitution of Eq. (4.277) in Eq. (4.295) yields

$$\begin{aligned} S_{out,0} &= p_1^2 E_{in,x} E_{in,x}^* + p_2^2 E_{in,y} E_{in,y}^* \\ &= \frac{(p_1^2 + p_2^2)}{2} \left(E_{in,x} E_{in,x}^* + E_{in,y} E_{in,y}^* \right) + \frac{(p_1^2 - p_2^2)}{2} \left(E_{in,x} E_{in,x}^* - E_{in,y} E_{in,y}^* \right) \\ &= \frac{(p_1^2 + p_2^2)}{2} S_{in,0} + \frac{(p_1^2 - p_2^2)}{2} S_{in,1} \\ S_{out,1} &= p_1^2 E_{in,x} E_{in,x}^* - p_2^2 E_{in,y} E_{in,y}^* = \frac{(p_1^2 - p_2^2)}{2} S_{in,0} + \frac{(p_1^2 + p_2^2)}{2} S_{in,1} \\ S_{out,2} &= p_1 p_2 E_{in,x} E_{in,y}^* + p_1 p_2 E_{in,y} E_{in,x}^* = p_1 p_2 S_{in,2} \\ S_{out,3} &= i p_1 p_2 E_{in,x} E_{in,y}^* - i p_1 p_2 E_{in,y} E_{in,x}^* = p_1 p_2 S_{in,3} \end{aligned} \tag{4.278}$$

It has the following matrix representation:

$$\begin{pmatrix} S_{out,0} \\ S_{out,1} \\ S_{out,2} \\ S_{out,3} \end{pmatrix} = \frac{1}{2} \begin{pmatrix} p_1^2 + p_2^2 & p_1^2 - p_2^2 & 0 & 0 \\ p_1^2 - p_2^2 & p_1^2 + p_2^2 & 0 & 0 \\ 0 & 0 & 2p_1 p_2 & 0 \\ 0 & 0 & 0 & 2p_1 p_2 \end{pmatrix} \begin{pmatrix} S_{in,0} \\ S_{in,1} \\ S_{in,2} \\ S_{in,3} \end{pmatrix} \tag{4.279}$$

Therefore the Mueller matrix of the partial polarizer can be obtained as

$$\mathbf{M}_P(\theta = 0) = \frac{1}{2}\begin{pmatrix} p_1^2 + p_2^2 & p_1^2 - p_2^2 & 0 & 0 \\ p_1^2 - p_2^2 & p_1^2 + p_2^2 & 0 & 0 \\ 0 & 0 & 2p_1p_2 & 0 \\ 0 & 0 & 0 & 2p_1p_2 \end{pmatrix} \tag{4.280}$$

where θ is the angle from the x-axis to the principal axis with transmission $= p_1^2$.

4.4.2.4 Mueller Matrix of Retardation Plate with Fast Axis at θ from the x-Axis

From Eq. (4.57), the Jones matrix of a retardation plate with its fast-axis at θ from the x-axis can be written as

$$\mathbf{M}_{J,RP} = \mathbf{M}_{J,R}(\theta)\mathbf{M}_{J,RP}(\theta = 0)\mathbf{M}_{J,R}(-\theta) \tag{4.281}$$

where the subscript J represents the Jones matrix. According to (4.266), the corresponding Mueller matrix can be written as

$$\mathbf{M}_{RP} = \mathbf{M}_R(\theta)\mathbf{M}_{RP}(\theta = 0)\mathbf{M}_R(-\theta) \tag{4.282}$$

Substitution of Eqs. (4.270) and (4.275) in Eq. (4.282) yields

$$\mathbf{M}_{RP} = \begin{pmatrix} 1 & 0 & 0 & 0 \\ 0 & cos2\theta & -sin2\theta & 0 \\ 0 & sin2\theta & cos2\theta & 0 \\ 0 & 0 & 0 & 1 \end{pmatrix}\begin{pmatrix} 1 & 0 & 0 & 0 \\ 0 & 1 & 0 & 0 \\ 0 & 0 & cos\delta & sin\delta \\ 0 & 0 & -sin\delta & cos\delta \end{pmatrix}\begin{pmatrix} 1 & 0 & 0 & 0 \\ 0 & cos2\theta & sin2\theta & 0 \\ 0 & -sin2\theta & cos2\theta & 0 \\ 0 & 0 & 0 & 1 \end{pmatrix} \tag{4.283}$$

Further deviation gives

$$\mathbf{M}_{RP} = \begin{pmatrix} 1 & 0 & 0 & 0 \\ 0 & cos4\theta sin^2\frac{\delta}{2} + cos^2\frac{\delta}{2} & sin4\theta sin^2\frac{\delta}{2} & -sin2\theta\ sin\delta \\ 0 & sin4\theta sin^2\frac{\delta}{2} & -cos4\theta sin^2\frac{\delta}{2} + cos^2\frac{\delta}{2} & cos2\theta\ sin\delta \\ 0 & sin2\theta\ sin\delta & -cos2\theta\ sin\delta & cos\delta \end{pmatrix} \tag{4.284}$$

It can be found from (4.284):

$$tr(\mathbf{M}_{RP}) = 2\,cos^2\left(\frac{\delta}{2}\right) + cos\delta + 1 = 2(cos\delta + 1) \tag{4.285}$$

Therefore, if the Mueller matrix of a linear retarder is experimentally measured, its retardation can be obtained from (4.285) as

$$\delta = arccos\left(\frac{tr(\mathbf{M}_{RP})}{2} - 1\right) \tag{4.286}$$

4.4.2.5 Mueller Matrix of the Partial Polarizer with Fast Axis at θ from the x-Axis

From Eq. (4.59), the Jones matrix of a partial polarizer with its attenuation axis p_1 oriented at θ from the x-axis is

$$\mathbf{M}_{J,P} = \mathbf{M}_{J,R}(\theta)\mathbf{M}_{J,P}(\theta = 0)\mathbf{M}_{J,R}(-\theta) \tag{4.287}$$

where the subscript J represents the Jones matrix. According to (4.266), the corresponding Mueller matrix can be written as

$$\mathbf{M}_P = \mathbf{M}_R(\theta)\mathbf{M}_P(\theta = 0)\mathbf{M}_R(-\theta) \tag{4.288}$$

Substitution of Eqs. (4.275) and (4.280) in Eq. (4.288) yields

$$\begin{pmatrix} 1 & 0 & 0 & 0 \\ 0 & cos2\theta & -sin2\theta & 0 \\ 0 & sin2\theta & cos2\theta & 0 \\ 0 & 0 & 0 & 1 \end{pmatrix} \frac{1}{2} \begin{pmatrix} p_1^2+p_2^2 & p_1^2-p_2^2 & 0 & 0 \\ p_1^2-p_2^2 & p_1^2+p_2^2 & 0 & 0 \\ 0 & 0 & 2p_1p_2 & 0 \\ 0 & 0 & 0 & 2p_1p_2 \end{pmatrix} \begin{pmatrix} 1 & 0 & 0 & 0 \\ 0 & cos2\theta & sin2\theta & 0 \\ 0 & -sin2\theta & cos2\theta & 0 \\ 0 & 0 & 0 & 1 \end{pmatrix} \tag{4.289}$$

and

$$\mathbf{M}_p = \frac{1}{2} \begin{pmatrix} p_1^2+p_2^2 & cos2\theta\,(p_1^2-p_2^2) & sin2\theta\,(p_1^2-p_2^2) & 0 \\ cos2\theta\,(p_1^2-p_2^2) & cos^2(2\theta)\,(p_1^2+p_2^2)+2sin^2(2\theta)p_1p_2 & cos2\theta\,sin\,2\theta(p_1-p_2)^2 & 0 \\ sin2\theta\,(p_1^2-p_2^2) & cos2\theta\,sin\,2\theta(p_1-p_2)^2 & sin^2(2\theta)\,(p_1^2+p_2^2)+2cos^2(2\theta)p_1p_2 & 0 \\ 0 & 0 & 0 & 2p_1p_2 \end{pmatrix} \tag{4.290}$$

For an ideal polarizer with $p_1 = 1$ and $p_2 = 0$, the matrix earlier can be simplified to

$$\mathbf{M}_{polarizer}(\theta) = \frac{1}{2} \begin{pmatrix} 1 & cos2\theta & sin2\theta & 0 \\ cos2\theta & cos^2(2\theta) & cos2\theta sin2\theta & 0 \\ sin2\theta & cos2\theta sin2\theta & sin^2(2\theta) & 0 \\ 0 & 0 & 0 & 0 \end{pmatrix} \tag{4.291}$$

If the incident light is a linearly polarized, its Stokes vector will be

$$\mathbf{E} = I_0 \begin{pmatrix} cos\alpha \\ sin\alpha \end{pmatrix} \to \begin{pmatrix} I_0 \\ I_0 cos2\alpha \\ I_0 sin2\alpha \\ 0 \end{pmatrix} \tag{4.292}$$

where α is the orientation angle of the polarization plane from the x-axis. By placing a polarizer at an angle of θ from the x-axis, the emerging Stokes vector is

$$\begin{pmatrix} S_0 \\ S_1 \\ S_2 \\ S_3 \end{pmatrix} = \mathbf{M}_{polarizer}(\theta) \begin{pmatrix} I_0 \\ I_0 cos2\alpha \\ I_0 sin2\alpha \\ 0 \end{pmatrix} = I_0 \begin{pmatrix} cos^2(\alpha-\theta) \\ cos2\theta cos^2(\alpha-\theta) \\ sin2\theta cos^2(\alpha-\theta) \\ 0 \end{pmatrix} = I_0 cos^2(\alpha-\theta) \begin{pmatrix} 1 \\ cos2\theta \\ sin2\theta \\ 0 \end{pmatrix} \tag{4.293}$$

Since $S_3 = 0$ in (4.293), the emergent light is a linearly polarized light with polarization plane at θ from x-axis, and the corresponding intensity of emerging light is equal to S_0 in (4.293), i.e.

$$I = S_0 = I_0 cos^2(\alpha - \theta) \tag{4.294}$$

This equation is the well-known Malus's law.

In summary, the Mueller matrices of all the basic optical elements are listed in Table 4.3, together with the corresponding Jones matrix in the table for convenience.

Table 4.3 Summary of the Jones and Mueller matrices.

	Jones matrix	Mueller matrix
Partial polarizer with principal parallel to x-axis	$\begin{pmatrix} p_1 & 0 \\ 0 & p_2 \end{pmatrix}$	$\frac{1}{2}\begin{pmatrix} p_1^2+p_2^2 & p_1^2-p_2^2 & 0 & 0 \\ p_1^2-p_2^2 & p_1^2+p_2^2 & 0 & 0 \\ 0 & 0 & 2p_1p_2 & 0 \\ 0 & 0 & 0 & 2p_1p_2 \end{pmatrix}$
Linear polarizer at angle = 0	$\begin{pmatrix} 1 & 0 \\ 0 & 0 \end{pmatrix}$	$\frac{1}{2}\begin{pmatrix} 1 & 1 & 0 & 0 \\ 1 & 1 & 0 & 0 \\ 0 & 0 & 0 & 0 \\ 0 & 0 & 0 & 0 \end{pmatrix}$
Linear polarizer at angle θ	$\begin{pmatrix} cos^2\theta & sin\theta\ cos\theta \\ sin\theta\ cos\theta & sin^2\theta \end{pmatrix}$	$\frac{1}{2}\begin{pmatrix} 1 & cos\ 2\theta & sin\ 2\theta & 0 \\ cos\ 2\theta & cos^2(2\theta) & cos2\theta\ sin\ 2\theta & 0 \\ sin\ 2\theta & cos\ 2\theta\ sin\ 2\theta & sin^2(2\theta) & 0 \\ 0 & 0 & 0 & 0 \end{pmatrix}$
Circular polarizers (RCP and LCP)	$RCP: \frac{1}{2}\begin{pmatrix} 1 & -i \\ 1 & -i \end{pmatrix}$ $LCP: \frac{1}{2}\begin{pmatrix} 1 & i \\ 1 & i \end{pmatrix}$	$RCP: \frac{1}{2}\begin{pmatrix} 1 & 0 & 0 & 1 \\ 0 & 0 & 0 & 0 \\ 1 & 0 & 0 & 1 \\ 0 & 0 & 0 & 0 \end{pmatrix}$ $LCP: \frac{1}{2}\begin{pmatrix} 1 & 0 & 0 & -1 \\ 0 & 0 & 0 & 0 \\ 1 & 0 & 0 & -1 \\ 0 & 0 & 0 & 0 \end{pmatrix}$
Half-wave linear retarder with the fast-axis at 0	$\begin{pmatrix} 1 & 0 \\ 0 & -1 \end{pmatrix}$	$\begin{pmatrix} 1 & 0 & 0 & 0 \\ 0 & 1 & 0 & 0 \\ 0 & 0 & -1 & 0 \\ 0 & 0 & 0 & -1 \end{pmatrix}$ rotate π about S_1

Half-wave linear retarder with the fast-axis at 45°	$\begin{pmatrix} 0 & i \\ i & 0 \end{pmatrix}$	$\begin{pmatrix} 1 & 0 & 0 & 0 \\ 0 & -1 & 0 & 0 \\ 0 & 0 & 1 & 0 \\ 0 & 0 & 0 & -1 \end{pmatrix}$ rotate π about S_2
Quarter-wave linear retarder with fast-axis at 0	$\begin{pmatrix} e^{i\pi/4} & 0 \\ 0 & e^{-i\pi/4} \end{pmatrix}$	$\begin{pmatrix} 1 & 0 & 0 & 0 \\ 0 & 1 & 0 & 0 \\ 0 & 0 & 0 & 1 \\ 0 & 0 & -1 & 0 \end{pmatrix}$ rotate $\pi/2$ about S_1
Rotator with rotation angle θ, or left circular with retardance $\theta = \delta/2$	$\mathbf{R}(\theta) = \begin{pmatrix} \cos\theta & -\sin\theta \\ \sin\theta & \cos\theta \end{pmatrix}$	$\begin{pmatrix} 1 & 0 & 0 & 0 \\ 0 & \cos 2\theta & -\sin 2\theta & 0 \\ 0 & \sin 2\theta & \cos 2\theta & 0 \\ 0 & 0 & 0 & 1 \end{pmatrix}$ rotate 2θ about S_3
General linear retarder; retardance δ, fast axis at angle θ from the x-axis	$\mathbf{R}(\theta)\begin{pmatrix} e^{i\delta/2} & 0 \\ 0 & e^{-i\delta/2} \end{pmatrix}\mathbf{R}(-\theta)$ $\begin{pmatrix} \cos^2\theta e^{i\frac{\delta}{2}} + \sin^2\theta e^{-i\frac{\delta}{2}} & i(\sin 2\theta)\sin\frac{\delta}{2} \\ i(\sin 2\theta)\sin\frac{\delta}{2} & \cos^2\theta e^{-i\frac{\delta}{2}} + \sin^2\theta e^{+i\frac{\delta}{2}} \end{pmatrix}$	$\begin{pmatrix} 1 & 0 & 0 & 0 \\ 0 & \cos 4\theta \sin^2\delta/2 + \cos^2\delta/2 & \sin 4\theta \sin^2\delta/2 & -\sin 2\theta \sin\delta \\ 0 & \sin 4\theta \sin^2\delta/2 & -\cos 4\theta \sin^2\delta/2 + \cos^2\delta/2 & \cos 2\theta \sin\delta \\ 0 & \sin 2\theta \sin\delta & -\cos 2\theta \sin\delta & \cos\delta \end{pmatrix}$ Rotate δ about $(1, \cos 2\theta, \sin 2\theta, 0)^T$
Elliptical retarder	$\begin{pmatrix} a & b \\ -b^* & a^* \end{pmatrix}$ where $aa^* + bb^* = 1$	

(Continued)

Table 4.3 (Continued)

	Jones matrix	Mueller matrix
Mirror	$\begin{pmatrix} 1 & 0 \\ 0 & -1 \end{pmatrix}$	$\begin{pmatrix} 1 & 0 & 0 & 0 \\ 0 & 1 & 0 & 0 \\ 0 & 0 & -1 & 0 \\ 0 & 0 & 0 & -1 \end{pmatrix}$ rotate π about S_1
Faraday mirror	$\begin{pmatrix} 0 & -1 \\ -1 & 0 \end{pmatrix}$	$\begin{pmatrix} 1 & 0 & 0 & 0 \\ 0 & -1 & 0 & 0 \\ 0 & 0 & 1 & 0 \\ 0 & 0 & 0 & -1 \end{pmatrix}$ rotate π about S_2
Quarter-wave plate mirror (45° linear quarter-wave plate mirror)	$\begin{pmatrix} 0 & i \\ -i & 0 \end{pmatrix}$	$\begin{pmatrix} 1 & 0 & 0 & 0 \\ 0 & -1 & 0 & 0 \\ 0 & 0 & -1 & 0 \\ 0 & 0 & 0 & 1 \end{pmatrix}$ rotate π about S_3
Backward matrix of reciprocal medium	$\begin{pmatrix} m_1 & m_4 \\ m_3 & m_2 \end{pmatrix}_{forward}$ $\begin{pmatrix} m_1 & -m_3 \\ -m_4 & m_2 \end{pmatrix}_{backward}$	Not available
Backward matrix of nonreciprocal medium	$\begin{pmatrix} m_1 & m_4 \\ m_3 & m_2 \end{pmatrix}_{forward}$ $\begin{pmatrix} m_1 & -m_4 \\ -m_3 & m_2 \end{pmatrix}_{backward}$	Not available
Backward of rotator	$\mathbf{R}(\theta)_{reciprocal}$ or $\mathbf{R}(-\theta)_{nonreciprocal}$	$\mathbf{R}(2\theta)_{reciprocal}$ or $\mathbf{R}(-2\theta)_{nonreciprocal}$

4.5 Polarization Evolution in Optical Fiber

4.5.1 Rotation Matrix Representation of the Unitary Optical System

In this section, we introduce the matrix representation of an optical fiber. In most practical applications, the optical fiber can be considered as an optical system without PDL so that we can represent it with a unitary matrix in Jones Calculus. According to the properties of the unitary matrix, the optical fiber is equivalent to an optical system containing only two elements: one is a retardation plate and the other a rotator, and the corresponding Jones matrix is unitary and can be written as

$$\mathbf{M}_J = \mathbf{M}_{J,R}(\beta)\mathbf{M}_{J,RP}(\alpha,\delta) = \mathbf{M}_{J,R}(\beta)\mathbf{M}_{J,R}(\alpha)\mathbf{M}_{J,RP}(\alpha=0,\delta)\mathbf{M}_{J,R}(-\alpha) \tag{4.295}$$

where the subscript J represents a 2×2 Jones matrix, β is the rotation angle of the rotator, δ is the retardation of the retardation plate, and α is the angle of its fast axis from the x-axis. According to Eq. (4.266), the 4×4 Mueller matrix of the corresponding optical system can be represented by the multiplications of four 4×4 real matrices:

$$\mathbf{M} = \mathbf{M}_R(\beta)\mathbf{M}_R(\alpha)\mathbf{M}_{RP}(\alpha=0,\delta)\mathbf{M}_R(-\alpha) \tag{4.296}$$

Substituting 4.271 and (4.276) in (4.296), we have

$$\mathbf{M} = \begin{pmatrix} 1 & \mathbf{0}^T \\ \mathbf{0} & \mathbf{R}_3(2\beta) \end{pmatrix}\begin{pmatrix} 1 & \mathbf{0}^T \\ \mathbf{0} & \mathbf{R}_3(2\alpha) \end{pmatrix}\begin{pmatrix} 1 & \mathbf{0}^T \\ \mathbf{0} & \mathbf{R}_1(-\delta) \end{pmatrix}\begin{pmatrix} 1 & \mathbf{0}^T \\ \mathbf{0} & \mathbf{R}_3(-2\alpha) \end{pmatrix} \tag{4.297}$$

where vector $\mathbf{0} = (0,0,0)^T$. It is not difficult to find that $\mathbf{M}$ can be written as

$$\mathbf{M} = \begin{pmatrix} 1 & \mathbf{0}^T \\ \mathbf{0} & \mathbf{R}_{3\times3} \end{pmatrix} \quad \text{where} \quad \mathbf{R}_{3\times3} = \mathbf{R}_3(2\beta)\mathbf{R}_3(2\alpha)\mathbf{R}_1(-\delta)\mathbf{R}_3(-2\alpha) \tag{4.298}$$

where $\mathbf{R}_{3\times3}$ is the multiplication of four 3×3 rotation matrices in the 3D space defined by the Poincaré sphere, and it is still a rotation matrix in 3D space. The Stokes parameters $\mathbf{S}$ of incident and Stokes parameters $\mathbf{S}'$ of emergent light are related by

$$\begin{pmatrix} S_0' \\ S_1' \\ S_2' \\ S_3' \end{pmatrix} = \mathbf{M}\begin{pmatrix} S_0 \\ S_1 \\ S_2 \\ S_3 \end{pmatrix} = \begin{pmatrix} 1 & \mathbf{0}^T \\ \mathbf{0} & \mathbf{R}_{3\times3} \end{pmatrix}\begin{pmatrix} S_0 \\ S_1 \\ S_2 \\ S_3 \end{pmatrix} \tag{4.299}$$

Therefore, we have

$$S_0' = S_0 \text{ and } \begin{pmatrix} S_1' \\ S_2' \\ S_3' \end{pmatrix} = \mathbf{R}_{3\times3}\begin{pmatrix} S_1 \\ S_2 \\ S_3 \end{pmatrix} \tag{4.300}$$

Note that the first element S_0 of the Stokes parameters in Eq. (4.299) is invariant; therefore, it can be separated out in Eq. (4.300) when investigating the SOP evolution of a unitary optical system, which can be fully described by a 1×3 vector $\begin{pmatrix} S_1 & S_2 & S_3 \end{pmatrix}^T$, with the transformation between the incident and emerging SOPs represented by a 3×3 rotation matrix. Figure 4.28 illustrates the relative relationship of rotation axes and angles on the Poincaré sphere for different typical optical elements. Three elementary rotation matrices about s_1-, s_2-, and s_3-axes in the Stokes space are also listed in Table 4.4 to be re-used in the future. For convenience, the corresponding Jones matrices are also given in the table.

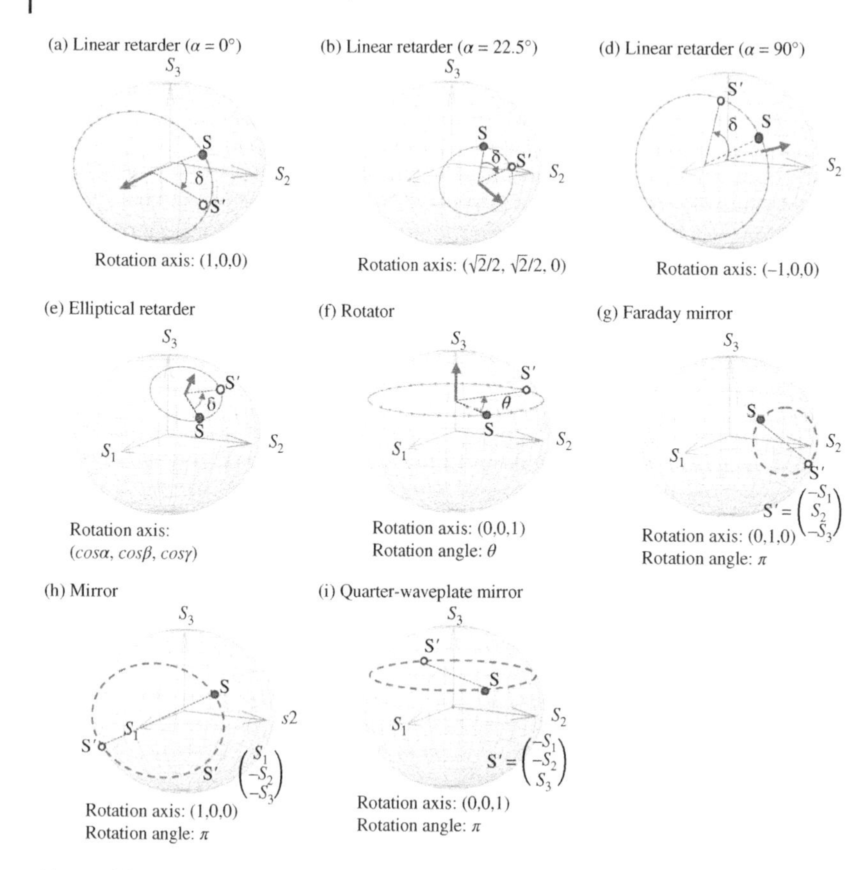

Figure 4.28 The rotation axis and rotation angle of unitary optical devices on Poincaré sphere.

Table 4.4 Elementary rotation matrix and the corresponding unitary matrix.

2 × 2 Unitary matrix	**3 × 3 Rotation matrix**
$\mathbf{M}_{J,RP}\left(\alpha=\frac{\pi}{2},\delta\right)=\begin{pmatrix} e^{-\frac{\delta}{2}} & 0 \\ 0 & e^{\frac{i\delta}{2}} \end{pmatrix}$ (slow axis is parallel to x-axis)	$\mathbf{R}_1(\delta)=\begin{pmatrix} 1 & 0 & 0 \\ 0 & \cos\delta & -\sin\delta \\ 0 & \sin\delta & \cos\delta \end{pmatrix}$
$\mathbf{M}_{J,RP}\left(\alpha=-\frac{\pi}{4},\delta\right)=\begin{pmatrix} \cos\frac{\delta}{2} & -i\sin\frac{\delta}{2} \\ -i\sin\frac{\delta}{2} & \cos\frac{\delta}{2} \end{pmatrix}$ (slow-axis is at $\frac{\pi}{4}$)	$\mathbf{R}_2(\delta)=\begin{pmatrix} \cos\delta & 0 & \sin\delta \\ 0 & 1 & 0 \\ -\sin\delta & 0 & \cos\delta \end{pmatrix}$
$\mathbf{M}_{J,R}=\begin{pmatrix} \cos\frac{\theta}{2} & -\sin\frac{\theta}{2} \\ \sin\frac{\theta}{2} & \cos\frac{\theta}{2} \end{pmatrix}$ (rotator with rotation angle $\frac{\theta}{2}$)	$\mathbf{R}_3(\theta)=\begin{pmatrix} \cos\theta & -\sin\theta & 0 \\ \sin\theta & \cos\theta & 0 \\ 0 & 0 & 1 \end{pmatrix}$

The main properties of 3D rotation matrix are listed in the following without proving:

(1) A square matrix **R** is a rotation matrix if and only if

$$\mathbf{R}^T = \mathbf{R}^{-1} \tag{4.301}$$

and its determinant $det\mathbf{R} = 1$

(2) $\mathbf{R}(\hat{n}, \theta)$ denotes for the 3D rotation, where $\hat{n}$ is the unit vector of the rotation axis and the counterclockwise rotation is defined to be positive, while the direction of the axis is determined by the right-hand rule

(3) The rotation matrix is not communitive: in general,

$$\mathbf{R}_1\mathbf{R}_2 \neq \mathbf{R}_2\mathbf{R}_1 \tag{4.302}$$

(4) Multiplication of two rotation matrices with the same rotation axis obeying the following rule:

$$\mathbf{R}(\hat{n}, \theta_1)\mathbf{R}(\hat{n}, \theta_2) = \mathbf{R}(\hat{n}, \theta_1 + \theta_2) \tag{4.303}$$

(5) The rotation matrix is an orthogonal matrix with its inverse as

$$[\mathbf{R}(\hat{n}, \theta)]^{-1} = [\mathbf{R}(\hat{n}, \theta)]^T = \mathbf{R}(\hat{n}, -\theta) = \mathbf{R}(-\hat{n}, \theta) \tag{4.304}$$

(6) The rotation angle of a rotation matrix can be expressed as

$$Tr\mathbf{R}(\hat{n}, \theta) = 1 + 2cos\theta \tag{4.305}$$

(7) Eigenvalues of rotation matrix with rotation angle θ are

$$\lambda_k = 1, e^{i\theta}, \text{or } e^{-i\theta} \ (0 \leq \theta < \pi) \tag{4.306}$$

(8) The rotation axis vector of a rotation matrix is

$$\hat{\boldsymbol{n}} = (R_{32} - R_{23}, R_{13} - R_{31}, R_{21} - R_{12})^T \tag{4.307}$$

where R_{ij} are matrix elements

(9) For a given unit rotation axis $\hat{\boldsymbol{n}} = \begin{pmatrix} n_1 & n_2 & n_3 \end{pmatrix}^T$ and a rotation angle θ, one can obtain the corresponding rotation matrix $\mathbf{R}(\hat{n}, \theta)$, as:

$$\mathbf{R} = (\hat{\boldsymbol{n}}\,\hat{\boldsymbol{n}} + sin\theta\hat{\boldsymbol{n}} \times -cos\theta\hat{\boldsymbol{n}} \times \hat{\boldsymbol{n}} \times) \tag{4.308}$$

where $\hat{\boldsymbol{n}}\,\hat{\boldsymbol{n}} = \begin{pmatrix} n_1^2 & 0 & 0 \\ 0 & n_3^2 & 0 \\ 0 & 0 & n_3^2 \end{pmatrix} \quad \hat{\boldsymbol{n}} \times = \begin{pmatrix} 0 & -n_3 & n_2 \\ n_3 & 0 & -n_1 \\ -n_2 & n_1 & 0 \end{pmatrix}$

4.5.2 Infinitesimal Rotation and Rotation Vector in Fiber

4.5.2.1 Infinitesimal Rotation and Rotation Vector

In Section 4.5.1, we understand that the matrix for representing a unitary optical system is a rotation matrix in Stokes space. However, the number of matrix elements has always been more than the number of independent variables. Therefore, various subsidiary conditions have to be invoked to determine the matrix. To simplify the problem, one may ask whether the rotation can be associated with a vector. For example, we can define a vector with the direction along the rotation axis and the magnitude being the rotation angle about the rotation axis. The answer is "no" for finite rotation, and "yes" for infinitesimal rotation (Goldstein 2001).

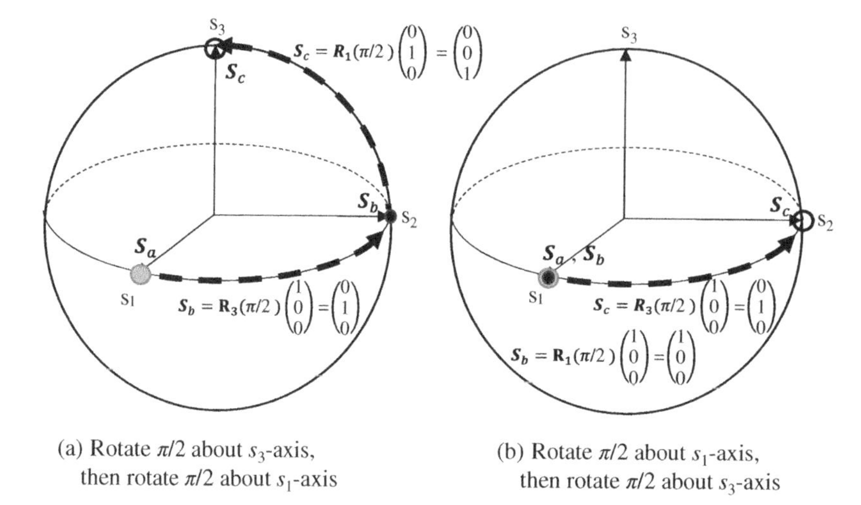

Figure 4.29 Finite rotation dependent on the order of the rotations. $\mathbf{R}_1(\pi/2)\mathbf{R}_3(\pi/2)$ transforms $(1,0,0)^T$ to $(0,0,1)^T$, however, $\mathbf{R}_3(\pi/2)\mathbf{R}_1(\pi/2)$ transforms $(1,0,0)^T$ to $(0,1,0)^T$.

Assume **A** and **B** are two such "vectors" associated with transformations $\mathbf{R}_\mathrm{A}$ and $\mathbf{R}_\mathrm{B}$ with finite rotations. In order to qualify them as vectors, they must be commutative in addition:

$$\mathbf{A}+\mathbf{B}=\mathbf{B}+\mathbf{A}$$

It has been shown previously that a rotation $\mathbf{R}_\mathrm{A}$ performed after another $\mathbf{R}_\mathrm{B}$ corresponds to the product $\mathbf{R}_\mathrm{B}\,\mathbf{R}_\mathrm{A}$ of the two matrices. Unfortunately, the matrix multiplication is not commutative, i.e. $\mathbf{R}_\mathrm{A}\cdot\mathbf{R}_\mathrm{B}\neq\mathbf{R}_\mathrm{B}\cdot\mathbf{R}_\mathrm{A}$ such that the two operations **A** and **B** are not commutative. Consequently, they cannot be considered as vectors when performing the two operations. Figure 4.29 gives a simple example to show that the finite rotation depends on the order of the rotation and is not commutative.

When considering an infinitesimal rotation, for example, with slight changes in fiber length or optical frequency, the Stokes vector **S** will be rotated to **S′** (see Figure 4.30a) with an extremely small difference between **S** and **S′**. The **S′** can be written as

$$\mathbf{S}'=\mathbf{S}+\begin{pmatrix} d\epsilon_{11} & d\epsilon_{12} & d\epsilon_{13}\\ d\epsilon_{21} & d\epsilon_{22} & d\epsilon_{23}\\ d\epsilon_{31} & d\epsilon_{32} & d\epsilon_{33}\end{pmatrix}\mathbf{S}=(\mathbf{I}+d\epsilon)\mathbf{S} \tag{4.309}$$

Equation (4.309) states that the typical form for the matrix of an infinitesimal transformation is $\mathbf{I}+\epsilon$; i.e. it is almost the identity transformation, differing at most by an infinitesimal operator. If $\mathbf{I}+\epsilon_1$ and $\mathbf{I}+\epsilon_2$ are two infinitesimal transformations, then we have

$$\begin{aligned}(\mathbf{I}+d\epsilon_1)(\mathbf{I}+d\epsilon_2)&=\mathbf{I}+\mathbf{I}(d\epsilon_1)+\mathbf{I}(d\epsilon_2)+(d\epsilon_1)(d\epsilon_2)=\mathbf{I}+\epsilon_1+\epsilon_2\\ (\mathbf{I}+d\epsilon_2)(\mathbf{I}+d\epsilon_1)&=\mathbf{I}+\mathbf{I}(d\epsilon_2)+\mathbf{I}(d\epsilon_1)+(d\epsilon_2)(d\epsilon_1)=\mathbf{I}+\epsilon_1+\epsilon_2\end{aligned} \tag{4.310}$$

Neglecting the high-order infinitesimals $(d\epsilon_1)(d\epsilon_2)$ and $(d\epsilon_2)(d\epsilon_1)$ in Eq. (4.310), $(\mathbf{I}+d\epsilon_1)(\mathbf{I}+d\epsilon_2)$ is equal to $\mathbf{I}+d\epsilon_2+d\epsilon_1$, the sequence of these two operations can be reversed for infinitesimal rotations; in other words, they commute.

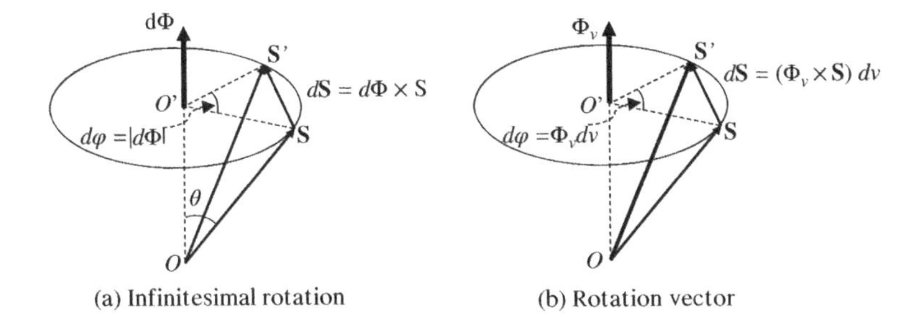

Figure 4.30 Infinitesimal rotation $d\mathbf{\Phi}$ and rotation rate vector $\mathbf{\Phi}_\nu = d\mathbf{\Phi}/d\nu$ defined in (4.325).

Let $\mathbf{R}_\Delta = (\mathbf{I} + d\epsilon)$ be the transformation of the infinitesimal rotation; its inverse is then

$$\mathbf{R}_\Delta^{-1} = (\mathbf{I} - d\epsilon) \tag{4.311}$$

As a proof, one notes that $\mathbf{R}_\Delta \mathbf{R}_\Delta^{-1} = (\mathbf{I} + d\epsilon)(\mathbf{I} - d\epsilon) = \mathbf{I} - (d\epsilon)(d\epsilon) = \mathbf{I}$, which is in agreement with the definition of a matrix inverse. On the other hand, $\mathbf{R}$ is a rotation matrix, from Eq. (4.301), its transpose $\mathbf{R}^T$ is equal to its inverse so that we have

$$\mathbf{R}_\Delta^{-1} = \mathbf{R}_\Delta^T = (\mathbf{I} + d\epsilon)^T = \mathbf{I} + (d\epsilon)^T \tag{4.312}$$

Substituting (4.311) in (4.312), one finds

$$d\epsilon = -(d\epsilon)^T \quad \text{i.e.} \quad \begin{pmatrix} d\epsilon_{11} & d\epsilon_{12} & d\epsilon_{13} \\ d\epsilon_{21} & d\epsilon_{22} & d\epsilon_{23} \\ d\epsilon_{31} & d\epsilon_{32} & d\epsilon_{33} \end{pmatrix} = \begin{pmatrix} -d\epsilon_{11} & -d\epsilon_{21} & -d\epsilon_{31} \\ -d\epsilon_{12} & -d\epsilon_{22} & -d\epsilon_{32} \\ -d\epsilon_{13} & -d\epsilon_{23} & -d\epsilon_{33} \end{pmatrix} \tag{4.313}$$

Since the elements of $d\epsilon$ are real number, to hold Eq. (4.313) requires the matrix $d\epsilon$ to be a skew-symmetric matrix, i.e. all of the diagonal elements of $d\epsilon$ are zero, and off-diagonal elements has the relationship $d\epsilon_{ij} = -d\epsilon_{ji}$. Therefore, the infinitesimal matrix ϵ can be reduced to the following general form

$$d\epsilon = \begin{pmatrix} 0 & -d\epsilon_3 & d\epsilon_2 \\ -d\epsilon_3 & 0 & -d\epsilon_1 \\ -d\epsilon_2 & d\epsilon_1 & 0 \end{pmatrix} \tag{4.314}$$

which is equivalent to the cross product of an infinitesimal rotation vector $d\mathbf{\Phi}$, i.e. $d\epsilon = d\mathbf{\Phi}\,\times$, where $d\mathbf{\Phi}$ is

$$d\mathbf{\Phi} = \begin{pmatrix} d\epsilon_1 & d\epsilon_2 & d\epsilon_3 \end{pmatrix}^T \tag{4.315}$$

The proof of $d\epsilon = d\mathbf{\Phi}\,\times$ is shown next:

$$(d\epsilon)\mathbf{S} = \begin{pmatrix} 0 & -d\epsilon_3 & d\epsilon_2 \\ -d\epsilon_3 & 0 & -d\epsilon_1 \\ -d\epsilon_2 & d\epsilon_1 & 0 \end{pmatrix} \mathbf{S} = \begin{pmatrix} -d\epsilon_3\, S_2 + d\epsilon_2 S_3 \\ -d\epsilon_3\, S_1 - d\epsilon_1\, S_3 \\ -d\epsilon_2\, S_1 + d\epsilon_1\, S_3 \end{pmatrix} \tag{4.316}$$

$$d\mathbf{\Phi} \times \mathbf{S} = \begin{vmatrix} \mathbf{i} & \mathbf{j} & \mathbf{k} \\ d\epsilon_1 & d\epsilon_2 & d\epsilon_3 \\ S_1 & S_2 & S_3 \end{vmatrix} = \begin{pmatrix} -d\epsilon_3 S_2 + d\epsilon_2 S_3 \\ -d\epsilon_3 S_1 - d\epsilon_1 S_3 \\ -d\epsilon_2 S_1 + d\epsilon_1 S_3 \end{pmatrix} \tag{4.317}$$

where **i**, **j**, and **k** are the unit vectors along the s_1-, s_2-, and s_3-axes in the Stokes space. One may note that the right side of (4.316) and (4.317) is equal and therefore

$$(d\epsilon)\mathbf{S} = d\mathbf{\Phi} \times \mathbf{S} \tag{4.318}$$

From Eq. (4.309), we have $d\mathbf{S} = \mathbf{S}' - \mathbf{S} = (\mathbf{I} + d\epsilon)\mathbf{S} - \mathbf{S} = (d\epsilon)\mathbf{S}$. Substituting it in (4.318), we arrive at the two equivalent expressions for an infinitesimal change $d\mathbf{S}$:

$$d\mathbf{S} = (d\epsilon)\mathbf{S} \tag{4.319}$$

$$d\mathbf{S} = d\mathbf{\Phi} \times \mathbf{S} \tag{4.320}$$

In the aforementioned equations, one may obtain $d\mathbf{S}$ either from matrix multiplication or by vector cross product in the Stokes space. Figure 4.30a shows the relative geometric relationship of the Stokes vector $\mathbf{S}$, infinitesimal rotation vector $d\Phi$, and infinitesimal Stokes vector variation $d\mathbf{S}$. The cross product follows the right-hand rule, i.e. $d\mathbf{S}$, is in the direction of a right-hand screw that advances as $d\Phi$ is bent into $\mathbf{S}$. After some geometric manipulations, the magnitude of $d\mathbf{\Phi}$ is found to be equal to the rotation angle of $\mathbf{S}$ around $d\mathbf{\Phi}$,

$$d\varphi = \frac{|d\mathbf{S}|}{|O'S|} = \frac{|d\mathbf{\Phi} \times \mathbf{S}|}{|\mathbf{S}|sin\theta} = \frac{|d\mathbf{\Phi}||S|sin\theta}{|\mathbf{S}|sin\theta} = |d\mathbf{\Phi}| \tag{4.321}$$

In summary, the infinitesimal derivation $d\mathbf{S}$ of a Stokes vector on the Poincaré sphere can be represented by two equivalent methods. One is by the multiplication of infinitesimal matrix $d\epsilon$ and the Stokes vector $\mathbf{S}$ (see Eq. (4.319)). The other is the cross product of an infinitesimal rotation vector $d\mathbf{\Phi}$ and the Stokes vector $\mathbf{S}$. If one has already obtained the vector $d\mathbf{\Phi}$ by measuring the SOP trajectories, one can find the matrix $d\epsilon$ according to Eqs. (4.314) and (4.315). On the other hand, one may also find the $d\mathbf{\Phi}$ and $d\epsilon$ by measuring the distributed rotation matrix $\mathbf{R}$ along the fiber by using the following relationships between $\mathbf{R}$ and $d\mathbf{\Phi}$ and $d\epsilon$.

An infinitesimal rotation may occur in an optical fiber when the light propagates a small step dz in the fiber or when its frequency changes by a small amount $d\omega$. In general, let ν represent z or ω, the infinitesimal rotation $d\mathbf{S}$ represented by Eqs. (4.319) and (4.320) can be rewritten as

$$\begin{aligned} d\mathbf{S}(\nu) &= d\epsilon(\nu)\,\mathbf{S} \\ d\mathbf{S}(\nu) &= d\mathbf{\Phi}(\nu) \times \mathbf{S} \end{aligned} \tag{4.322}$$

The corresponding derivative of $\mathbf{S}(\nu)$ with respect to the variable ν is

$$\begin{aligned} \frac{d\mathbf{S}(\nu)}{d\nu} &= \frac{d\epsilon(\nu)}{d\nu}\mathbf{S} \\ \frac{d\mathbf{S}(\nu)}{d\nu} &= \mathbf{\Phi}_\nu \times \mathbf{S} \end{aligned} \tag{4.323}$$

$\frac{d\epsilon(\nu)}{d\nu}$ and $\mathbf{\Phi}_\nu\times$ can be obtained using (4.314) and (4.315):

$$\frac{d\epsilon(\nu)}{d\nu} = \begin{pmatrix} 0 & -\frac{d\epsilon_3}{d\nu} & \frac{d\epsilon_2}{d\nu} \\ \frac{d\epsilon_3}{d\nu} & 0 & -\frac{d\epsilon_1}{d\nu} \\ -\frac{d\epsilon_2}{d\nu} & \frac{d\epsilon_1}{d\nu} & 0 \end{pmatrix} \tag{4.324}$$

$$\mathbf{\Phi}_\nu \equiv \frac{d\mathbf{\Phi}}{d\nu} = \left(\frac{d\epsilon_1}{d\nu} \quad \frac{d\epsilon_2}{d\nu} \quad \frac{d\epsilon_3}{d\nu}\right)^T \tag{4.325}$$

If the transformation matrix $\mathbf{R}(\nu)$ between the $\mathbf{S}(\nu)$ and input polarization $\mathbf{S}_{in}$ is already obtained

$$\mathbf{S}(\nu) = \mathbf{R}(\nu)\mathbf{S}_{in} \tag{4.326}$$

Taking the derivative of both sides of (4.326), one gets

$$\frac{d\mathbf{S}(v)}{dv} = \frac{d\mathbf{R}(v)}{dv}\mathbf{S}_{in} \tag{4.327}$$

From Eq. (4.326), $\mathbf{S}_{in} = \mathbf{R}^{-1}(v)\mathbf{S}(v)$. Substituting it in Eq. (4.327) yields

$$\frac{d\mathbf{S}(v)}{dv} = \frac{d\mathbf{R}(v)}{dv}\mathbf{R}^{-1}(v)\mathbf{S}(v) \tag{4.328}$$

Comparing (4.279) and (4.328), we get the following equivalent expression

$$\mathbf{\Phi}_v \times = \frac{d\mathbf{R}(v)}{dv}\mathbf{R}^{-1}(v) = \frac{d\epsilon(v)}{dv} \tag{4.329}$$

Let's consider the evolution of $\mathbf{S}(v)$. Recognize the existence of two eigenstates $\pm\mathbf{\Phi}_v/|\mathbf{\Phi}_v|$, which do not change when v changes by a small amount $d\omega$. All other states will move about the axis $\pm\mathbf{\Phi}_v/|\mathbf{\Phi}_v|$ with a rotation rate of $|\mathbf{\Phi}_v|$ (see Figure 4.30b). As will be discussed in Section 4.5.3, $\mathbf{\Phi}_v$ either represents the local birefringence vector when v is the position z or the PMD vector when v is the optical angular frequency.

4.5.3 Birefringence Vector and Polarization Evolution Along with Fiber

Let's consider a short fiber section between z and $z + dz$, which can be regarded as the concatenation of a linear retarder with its slow-axis at θ from the x-axis and a rotator (or circular retarder). From Eq. (4.298) and the rotation matrix in Table 4.4, the corresponding transportation $\mathbf{R}_\Delta$ of this fiber section can be written as

$$\begin{aligned}\mathbf{R}_\Delta(z) &= \mathbf{R}_3[\Delta\alpha(z)]\,\mathbf{R}_3[2\theta(z)]\mathbf{R}_1[-\Delta\delta(z)]\mathbf{R}_3[-2\theta(z)] \\ &= \begin{pmatrix} cos\Delta\alpha(z) & -sin\Delta\alpha(z) & 0 \\ sin\Delta\alpha(z) & cos\Delta\alpha(z) & 0 \\ 0 & 0 & 1 \end{pmatrix}\mathbf{R}_3[2\theta(z)]\begin{pmatrix} 1 & 0 & 0 \\ 0 & cos\Delta\delta(z) & -sin\Delta\delta(z) \\ 0 & sin\Delta\delta(z) & cos\Delta\delta(z) \end{pmatrix}\mathbf{R}_3[-2\theta(z)]\end{aligned} \tag{4.330}$$

where

$$\Delta\alpha(z) = \frac{2\pi}{\lambda_0}[n_R(z) - n_L(z)]\Delta z = \alpha(z)\Delta z \tag{4.331}$$

$$\Delta\delta(z) = \frac{2\pi}{\lambda_0}[n_s(z) - n_f(z)]\Delta z = \beta_l(z)\Delta z \tag{4.332}$$

where n_R and n_L are the eigen refractive indices of the circular retarder, n_s and n_f are the eigen refractive indices of the linear retarder, α is the local differential circular propagation constant between the right- and left-handed circular polarizations representing the circular birefringence at position z, and $\beta_l(z)$ is the local differential linear propagation constant between the slow- and fast-axis representing the linear birefringence between z and $z + \Delta z$. One may always select Δz sufficiently short in order to make $\Delta\alpha$ and $\Delta\delta$ sufficiently small so that the following approximations hold: $cos(\Delta\alpha) = 1$, $sin(2\Delta\alpha) = 2\Delta\alpha$, $cos(\Delta\delta) = 1$ and $sin(\Delta\delta) = \Delta\delta$. Equation (4.330) then becomes

$$\mathbf{R}_\Delta(z) = \begin{pmatrix} 1 & -\alpha(z)\Delta z & \beta_l(z)\Delta z sin\ 2\theta(z) \\ \alpha(z)\Delta z & 1 & -\beta_l(z)\Delta z cos\ 2\theta(z) \\ -\beta_l(z)\Delta z sin\ 2\theta(z) & \beta_l(z)\Delta z cos\ 2\theta(z) & 1 \end{pmatrix} \tag{4.333}$$

Recognizing $\mathbf{R}_\Delta$ is the transformation matrix between polarization $\mathbf{S}(z)$ at z and $\mathbf{S}(z + dz)$ at $z + dz$, one gets

$$\mathbf{S}(z + dz) = \mathbf{R}_\Delta(z) \cdot \mathbf{S}(z) \tag{4.334}$$

According to the definition of the derivative, one can write the change rate of $\mathbf{S}(z)$ with respect to z by

$$\frac{d\mathbf{S}(z)}{dz} = \lim_{\Delta z \to 0} \frac{\mathbf{S}(z+dz) - \mathbf{S}(z)}{\Delta z} = \lim_{\Delta z \to 0} \frac{(\mathbf{R}_{\Delta}(z) - \mathbf{I})}{\Delta z}\mathbf{S}(z) \tag{4.335}$$

Substituting (4.333) in (4.335), one obtains

$$\frac{d\mathbf{S}(z)}{dz} = \begin{pmatrix} 0 & -\alpha(z) & \beta_l sin2\theta(z) \\ \alpha(z) & 0 & -\beta_l cos2\theta(z) \\ -\beta_l sin2\theta(z) & \beta_l cos2\theta(z) & 0 \end{pmatrix} \mathbf{S}(z) \tag{4.336}$$

Define the birefringence vector as

$$\boldsymbol{\beta}(z) = \left[\beta_l(z)cos2\theta(z) \quad \beta_l sin2\theta(z) \quad \alpha(z)\right]^T \tag{4.337}$$

According to Eqs. (4.323), (4.324), and (4.325), the derivative of Stokes vector in Eq. (4.336) can be rewritten as the following vector's cross product:

$$\frac{d\mathbf{S}(z)}{dz} = \boldsymbol{\beta}(z) \times \mathbf{S}(z) \tag{4.338}$$

If the distribution of birefringence vector $\boldsymbol{\beta}(z)$ along the fiber is already known, it is possible to construct the trajectory of $\mathbf{S}(z)$ for any given input SOP. This geometrical construction of $\mathbf{S}(z)$ as a succession of rotations give us a clear insight into the various polarization effects, which is particularly convenient if $\boldsymbol{\beta}(z)$ is a constant or varies with certain signature patterns along the fiber. In the following, we will discuss the construction for linear birefringence, circular birefringence, and twist, considering them first individually and then in combination, the detailed discussion can also be found in the paper Ulrich and Simon (1979).

4.5.3.1 Linear Birefringence

Consider a straight fiber without circular birefringent and twist, and assume its linear birefringence is uniformly distributed, for example, a PM fiber. Following Eq. (4.337), the birefringence vector $\boldsymbol{\beta}_l$ is a constant vector

$$\boldsymbol{\beta}_l = \left(\beta_l cos2\theta \quad \beta_l sin2\theta \quad 0\right)^T \tag{4.339}$$

It lies in the equatorial plane with an amplitude of $|\boldsymbol{\beta}_l| = \beta_l$ and with its slow axis oriented at 2θ from s_1-axis. When the input polarization $\mathbf{S}(0)$ is parallel to $\boldsymbol{\beta}_l$ and $\mathbf{S}(0)$ is linear with its polarization plane parallel to the fiber's slow or fast axis, $d\mathbf{S}(z)/dz$ in Eq. (4.338) is equal to zero and the trajectory of $\mathbf{S}(z)$ becomes a point locating at $\mathbf{S}(0)$, i.e. the polarization is maintained along with the fiber. If the initial polarization $\mathbf{S}(0)$ is not parallel to $\boldsymbol{\beta}_l$, $\mathbf{S}(z)$ will form a cone around $\boldsymbol{\beta}_l$ with rotation rate β_l, and the trajectory of $\mathbf{S}(z)$ is a circular ace on the Poincaré sphere, as shown in Figure 4.31a. With some simple geometrical derivation, we can get the radius of the circle is

$$r = \frac{|\boldsymbol{\beta}_l \times \mathbf{S}(0)|}{|\boldsymbol{\beta}_l|} \tag{4.340}$$

or the half angle ξ of the cone

$$\xi = arccos\left|\frac{\boldsymbol{\beta}_l \cdot \mathbf{S}(0)}{|\boldsymbol{\beta}_l||\mathbf{S}(0)|}\right| \tag{4.341}$$

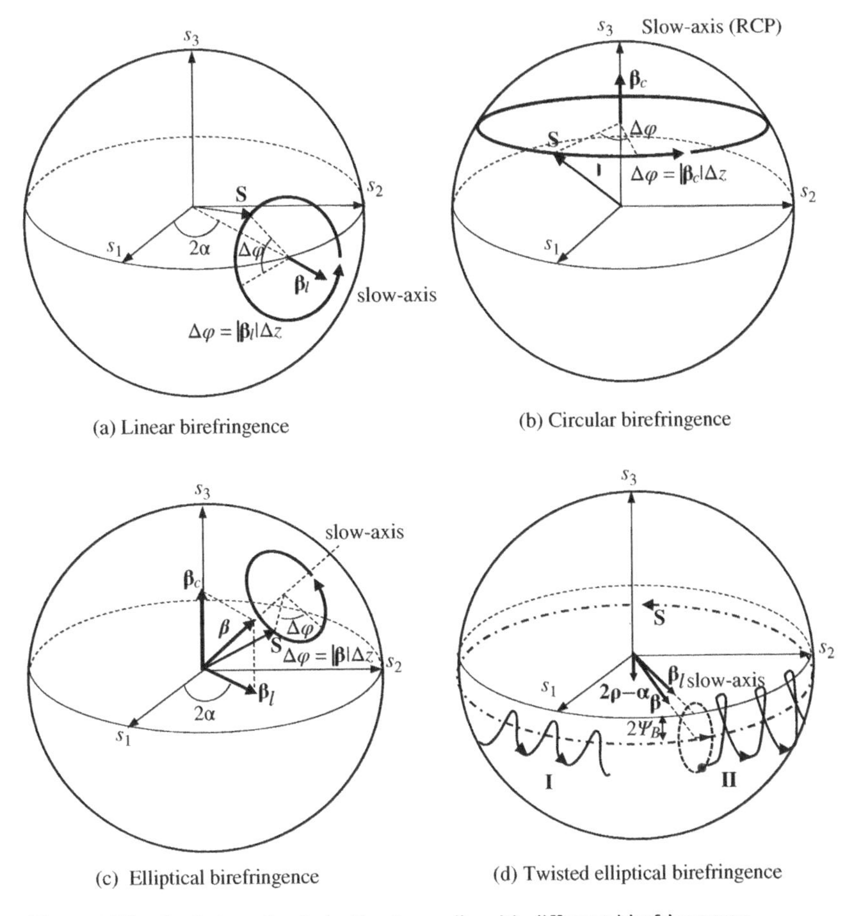

Figure 4.31 Evolution of polarization in media with different birefringences.

In practice, stretching or heating a PM fiber can cause the change of fiber length, and the corresponding polarization at the output end of fiber will draw an arc on the unit Poincaré sphere. The radius of the circle can be used to estimate the polarization extinct ratio (PER) by

$$PER = 10\left|log\frac{(sin\,\xi/2)^2}{(cos\,\xi/2)^2}\right| = 10\left|log\frac{1-cos\,\xi}{1+cos\,\xi}\right| = 10\left|log\frac{1-\sqrt{1-r^2}}{1+\sqrt{1-r^2}}\right| \tag{4.342}$$

As the angle $\Delta\varphi$ increases beyond 2π, the SOP circle repeats itself on the Poincaré sphere, as shown in Figure 4.31a, which means that SOP varies periodically with z. We also noticed that the SOP varies periodically along z with a periodicity of $2\pi/|\boldsymbol{\beta}_l|$, which is called linear beat length of the fiber L_p

$$L_p = \frac{2\pi}{|\boldsymbol{\beta}_l|} \tag{4.343}$$

4.5.3.2 Circular Birefringence (Optical Activity)

When considering a fiber with only local circular birefringence without linear birefringence, the local circular birefringence vector can be obtained from Eq. (4.337):

$$\boldsymbol{\beta}_c = \begin{pmatrix}0 & 0 & \alpha(z)\end{pmatrix}^T \tag{4.344}$$

which means that the circular birefringence vector $\boldsymbol{\beta}_c$ coincides with the s_3-axis, and that the trajectory of SOP is an arc or circle parallel to the equator, as shown in Figure 4.31b. The rotation rate of SOP along the s_3-axis is $|\boldsymbol{\beta}_c|$. For the linear SOP input, the plane of polarization rotates at the rate $|\boldsymbol{\beta}_c|/2$. The sign of $\boldsymbol{\beta}_c$ is positive when the polarization rotates counterclockwise.

In an optical fiber, the circular birefringence is mainly caused by twisting induced optical activity or the Faraday rotation induced by an external magnetic field, as discussed in Chapter 3.

One can note that the SOP in fiber varies periodically along z with a period L_c, which is called the circular beat length of the fiber and is related to $\boldsymbol{\beta}_c$ by

$$|\boldsymbol{\beta}_c|L_c = 2\pi \tag{4.345}$$

4.5.3.3 Elliptical Birefringence (Optical Activity)

Elliptical birefringence results from the superposition of linear and circular birefringence. In a fiber having both linear birefringence (due to asymmetrical lateral stress and noncircular core) and circular birefringence (due to twist or Faraday rotation), the corresponding elliptical birefringence vector (see (4.337)) is

$$\boldsymbol{\beta}(z) = \boldsymbol{\beta}_l(z) + \boldsymbol{\beta}_c(z) \tag{4.346}$$

For the case where $\boldsymbol{\beta}$ does not depend on z, we noted that there are existing two eigenvectors with elliptical polarization $\pm\boldsymbol{\beta}/|\boldsymbol{\beta}|$, which propagate unchanged, and the $\boldsymbol{\beta}$ has a larger propagation constant than that of $-\boldsymbol{\beta}$. When the input SOP is not aligned to $\boldsymbol{\beta}$ or $-\boldsymbol{\beta}$, the trajectory of $\mathbf{S}(z)$ is an arc on a circle, as shown in Figure 4.31c. The $\mathbf{S}(z)$ is a periodic function with the period L_e, called elliptical beat length. The L_e can be determined by

$$|\boldsymbol{\beta}|L_e = 2\pi \tag{4.347}$$

From Eqs. (4.339) and (4.344), one can find that $\boldsymbol{\beta}_l$ and $\boldsymbol{\beta}_c$ are mutually orthonormal such that the elliptical beat length defined in Eq. (4.347) can be expressed as:

$$L_e = \frac{2\pi}{|\boldsymbol{\beta}|} = \frac{2\pi}{\sqrt{|\boldsymbol{\beta}_l|^2 + |\boldsymbol{\beta}_c|^2}} \tag{4.348}$$

4.5.3.4 Elliptical Birefringence with Twist

When an optical fiber with birefringence is twisted, the evaluation $\mathbf{S}(z)$ is complicated due to the fact that the birefringence vector is not a constant with z. Here we focus on the spun fiber, a practically very important case, in which a uniform counterclockwise twist ρ is applied to a fiber that has initially uniform linear birefringence $\beta_l(z)$. For such a fiber, the rotation matrix $\mathbf{R}_\Delta$ between z and $z+dz$ can be written as from Eq. (4.298):

$$\mathbf{R}_\Delta(z) = \mathbf{R}_3(\Delta\alpha)\mathbf{R}_3(2\rho z)\mathbf{R}_1(-\Delta\delta)\mathbf{R}_3(-2\rho z) \tag{4.349}$$

where ρ is the spin twist rate defined as the rotation angle per unit length, $\Delta\alpha$ and $\Delta\delta$ are the circular and linear birefringence defined in Eqs. (4.331) and (4.332), respectively. Following the similar procedure for the deviation of Eqs. (4.330)–(4.336), we can obtain

$$\frac{d\mathbf{S}(z)}{dz} = \mathbf{R}_\in \mathbf{S}(z) \tag{4.350}$$

$$\mathbf{R}_\in = \begin{pmatrix} 0 & -\alpha(z) & \beta_l sin 2\rho z \\ \alpha(z) & 0 & -\beta_l cos 2\rho z \\ -\beta_l sin 2\rho z & \beta_l cos 2\rho z & 0 \end{pmatrix}$$

and

$$\frac{d\mathbf{S}(z)}{dz} = \boldsymbol{\beta}(z) \times \mathbf{S}(z) \tag{4.351}$$

$$\boldsymbol{\beta}(z) = \left[\beta_l cos2\rho z \quad \beta_l sin2\rho z \quad \alpha(z)\right]^T$$

where $\alpha(z) = \Delta\alpha(z)/\Delta z$ and $\beta_l = \Delta\delta(z)/\Delta z$ from Eqs. (4.331) and (4.332). It is difficult to get the visual picture of the trajectory of SOP in the fixed coordinate frame because $\mathbf{R}_{\in}$ and $\boldsymbol{\beta}(z)$ varies with position. Fortunately, we find that $\mathbf{R}_{\in}$ and $\boldsymbol{\beta}(z)$ can be transformed to be a constant in the local reference frame whose x-axis is rotated with the fiber twist. The Stokes vector in the new reference frame can be written as

$$\mathbf{Y}(z) = \mathbf{R}_3(-2\rho z)\mathbf{S}(z) \tag{4.352}$$

Taking the derivative with respect to z, we have

$$\frac{d\mathbf{Y}(z)}{dz} = \frac{d\mathbf{R}_3(-2\rho z)}{dz}\mathbf{S}(z) + \mathbf{R}_3(-2\rho z)\frac{d\mathbf{S}(z)}{dz} \tag{4.353}$$

Substituting $\mathbf{S}(z) = \mathbf{R}_3(2\rho z)\mathbf{Y}(z)$ and (4.350) in (4.353), one obtains

$$\frac{d\mathbf{Y}(z)}{dz} = \left[\frac{d\mathbf{R}_3(-2\rho z)}{dz}\mathbf{R}_3(2\rho z) + \mathbf{R}_3(-2\rho z)\mathbf{R}_{\in}\mathbf{R}_3(2\rho z)\right]\mathbf{Y}(z) \tag{4.354}$$

By substituting $\mathbf{R}_{\in}$ in (4.350) and $\mathbf{R}_3$ in Table 4.4, $d\mathbf{Y}(z)/dz$ can be further simplified to

$$\frac{d\mathbf{Y}(z)}{dz} = \begin{pmatrix} 0 & -(\alpha - 2\rho) & 0 \\ -2\rho + \alpha & 0 & -\beta_l \\ 0 & \beta_l & 0 \end{pmatrix}\mathbf{Y}(z) \tag{4.355}$$

Define the corresponding birefringence vector $\boldsymbol{\beta}'(z)$ in local reference frame as

$$\boldsymbol{\beta}' = \left[\beta_l \quad 0 \quad \alpha - 2\rho\right]^T$$

$$|\boldsymbol{\beta}'| = \sqrt{\beta_l^2 + (\alpha - 2\rho)^2} \tag{4.356}$$

According to Eq. (4.279), the derivative of Stokes vector $\mathbf{Y}(z)$ is equivalent to the following vector cross product:

$$\frac{d\mathbf{Y}(z)}{dz} = \boldsymbol{\beta}'(z) \times \mathbf{Y}(z) \tag{4.357}$$

It is clear that $\boldsymbol{\beta}'(z)$ is a constant in the local rotating reference frame. It can be recognized that there exists two linear polarization states $\mathbf{Y}(0) = \pm\boldsymbol{\beta}'/|\boldsymbol{\beta}'|$ at $z = 0$, which are unchanged when propagating in the fiber while viewed in the rotating frame. The corresponding polarization state in the laboratory frame is $\mathbf{S}(z) = \mathbf{R}_3(2\rho z)\mathbf{Y}(0) = (\pm\mathbf{R}_3(2\rho z)(\boldsymbol{\beta}'/|\boldsymbol{\beta}'|)$, the trajectory of $\mathbf{S}(z)$ is a circle parallel to the equatorial plane (the chain line in Figure 4.25d) with a rotation rate of 2ρ. For all other states deviated from $\pm\boldsymbol{\beta}'$, $\mathbf{Y}(z)$ rotates about $\boldsymbol{\beta}'$ and forms a cone with constant a cone half angle $\xi = arcos(\mathbf{Y}(0)\cdot\boldsymbol{\beta}')$. To find the trajectory of $\mathbf{S}(z)$, let's switch back from the local rotating sphere to the fixed sphere, on which a general $\mathbf{S}(z)$ moves on the end circle of the cone with the rotation rate $|\boldsymbol{\beta}'| = \sqrt{\beta_l^2 + (\alpha - 2\rho)^2}$, while the cone axis moves on a larger circle (chain line in Figure 4.29d) parallel to the equator with a latitude of

$$2\psi_B = arctan\left[\frac{(\alpha - 2\rho)}{\beta_l}\right] \tag{4.358}$$

Therefore, $\mathbf{S}(z)$ is generally a cycloid. It is interesting to consider the evolution of $\mathbf{S}(z)$ in the limits of weak and strong twists. In a PM fiber, the twist is weak because $\beta_l \gg 2\rho$ and α, then $|\boldsymbol{\beta}'| \approx \beta_l$ and $\psi_B = 0$, so that the evolution of $\mathbf{S}(z)$ in the rotating frame remains practically unaffected by the twist. Therefore, the polarization plate is rotated as if it were rigidly attached to the fiber. In spun low birefringence (LoBi) fiber, $\alpha \gg \beta_l$, then $|\boldsymbol{\beta}'| \approx (\alpha - 2\rho)$, and the ψ_B in Eq. (4.358) is close to $\pi/2$. In the rotating frame, polarization evolution approaches a uniform rotation about s_3-axis at a rate of $(\alpha - 2\rho)$. Correspondingly, on the laboratory fixed sphere, the SOPs rotate at a rate of α.

4.5.4 PMD Vector and Polarization Evolution with Optical Frequency

As introduced in Section 3.4, the PMD causes an optical pulse to spread in the time domain and may impair the performance of the optical fiber telecommunication systems. In deployed optical fibers, there always exist intrinsic and external stress-induced birefringences along the fiber, which results in different group velocities and the corresponding arrival times for different polarization components of the signal pulses. In this section, we will introduce the relationship between birefringence and PMD, and the methods to measure PMD in optical frequency domain.

4.5.4.1 Principal States of Polarization and PMD Vector

In 1989, Poole et al. proposed a phenomenological model of polarization dispersion for long fiber lengths (Poole et al. 1989). For any linear optical transmission medium with no PDL, there are existing orthogonal input states of polarization, and the corresponding output states of polarization are also orthogonal. These two pairs of polarization states show no dependence on optical frequency to the first-order approximation. These two states are called the Principal States of Polarization (PSP). A differential group delay (DGD) exists between signals launched along with one PSP and its orthogonal complement.

According to Eq. (4.323), the evolution of polarization with frequency in Stokes space can be represented by

$$\frac{d\mathbf{S}(\omega)}{d\omega} = \boldsymbol{\Omega} \times \mathbf{S}(\omega) \tag{4.359}$$

where $\boldsymbol{\Omega}$ is a vector in Stokes space. In the following discussions, we will show that $\boldsymbol{\Omega}$ is parallel to the PSP of the fiber, and its magnitude is equal to the DGD between two principal states.

Let's consider a monochromatic optical field of frequency ω at the input of a single-mode fiber without PDL. Assume there exist two normalized input PSPs $\hat{\varepsilon}_1$ and $\hat{\varepsilon}_2$, respectively, then the corresponding Jones vectors of the two polarization states at fiber output are

$$\begin{aligned} \mathbf{E}_{out1} &= e^{i\varphi_1(\omega)}\mathbf{U}\hat{\varepsilon}_1 \\ \mathbf{E}_{out2} &= e^{i\varphi_2(\omega)}\mathbf{U}\hat{\varepsilon}_2 \end{aligned} \tag{4.360}$$

where the matrix $\mathbf{U}$ represents the Jones matrix of the fiber. It should be emphasized that $\mathbf{E}_{out1}$ and $\mathbf{E}_{out2}$ are not necessary to be parallel to $\hat{\varepsilon}_1$ and $\hat{\varepsilon}_2$ in general. We can express Eq. (4.360) in matrix form:

$$\mathbf{V}_{out} = \mathbf{U}\mathbf{V}_{in}\mathbf{M} \tag{4.361}$$

where

$$\mathbf{V}_{out} \equiv \begin{pmatrix} \mathbf{E}_{out1} & \mathbf{E}_{out2} \end{pmatrix} = \begin{pmatrix} E_{out1x} & E_{out2x} \\ E_{out1y} & E_{out2y} \end{pmatrix}$$

$$\mathbf{V}_{in} \equiv (\hat{\varepsilon}_1 \ \ \hat{\varepsilon}_2) = \begin{pmatrix} \varepsilon_{1x} & \varepsilon_{2x} \\ \varepsilon_{1y} & \varepsilon_{2y} \end{pmatrix}$$

$$\mathbf{M} = \begin{pmatrix} e^{i\varphi_1(\omega)} & 0 \\ 0 & e^{i\varphi_2(\omega)} \end{pmatrix}$$

Because $\mathbf{E}_{out1}$ and $\mathbf{E}_{out2}$, $\hat{\varepsilon}_1$ and $\hat{\varepsilon}_2$ are two pairs of normalized orthogonal vectors, the matrices $\mathbf{V}_{out}$ and $\mathbf{V}_{in}$ are unitary matrices since

$$\mathbf{V}^{\dagger}_{out}\mathbf{V}_{out} = \begin{pmatrix} E^*_{out1x} & E^*_{out1y} \\ E^*_{out2x} & E^*_{out2y} \end{pmatrix} \begin{pmatrix} E_{out1x} & E_{out2x} \\ E_{out1y} & E_{out2y} \end{pmatrix} = \begin{pmatrix} \mathbf{E}^{\dagger}_{out1}\mathbf{E}_{out1} & \mathbf{E}^{\dagger}_{out1}\mathbf{E}_{out2} \\ \mathbf{E}^{\dagger}_{out2}\mathbf{E}_{out1} & \mathbf{E}^{\dagger}_{out2}\mathbf{E}_{out2} \end{pmatrix} = \begin{pmatrix} 1 & 0 \\ 0 & 1 \end{pmatrix}$$

$$\mathbf{V}^{\dagger}_{in}\mathbf{V}_{in} = \begin{pmatrix} \hat{\varepsilon}^{\dagger}_1 \\ \hat{\varepsilon}^{\dagger}_2 \end{pmatrix} (\hat{\varepsilon}_1 \ \ \hat{\varepsilon}_2) = \begin{pmatrix} \hat{\varepsilon}^{\dagger}_1\hat{\varepsilon}_1 & \hat{\varepsilon}^{\dagger}_1\hat{\varepsilon}_2 \\ \hat{\varepsilon}^{\dagger}_2\hat{\varepsilon}_1 & \hat{\varepsilon}^{\dagger}_2\hat{\varepsilon}_2 \end{pmatrix} = \begin{pmatrix} 1 & 0 \\ 0 & 1 \end{pmatrix} \tag{4.362}$$

If we already know $\mathbf{V}_{out}$ and $\mathbf{V}_{in}$, we can find the matrix $\mathbf{U}$ by multiplying $\mathbf{M}^{-1}\mathbf{V}^{\dagger}_{in}$ on both sides of Eq. (4.361)

$$\mathbf{U} = \mathbf{V}_{out}\mathbf{M}^{-1}\mathbf{V}^{\dagger}_{in} = \mathbf{V}_{out} \begin{pmatrix} e^{-i\varphi_1(\omega)} & 0 \\ 0 & e^{-\varphi_2(\omega)} \end{pmatrix} \mathbf{V}^{\dagger}_{in} = e^{-i\frac{\varphi_1(\omega)+\varphi_2(\omega)}{2}} \mathbf{V}_{out} \begin{pmatrix} e^{-\frac{i\delta}{2}} & 0 \\ 0 & e^{\frac{i\delta}{2}} \end{pmatrix} \mathbf{V}^{\dagger}_{in} \tag{4.363}$$

where $\delta = \frac{\varphi_1(\omega)-\varphi_2(\omega)}{2}$. Omitting the common phase factor $e^{-i\frac{\varphi_1(\omega)+\varphi_2(\omega)}{2}}$ the matrix $\mathbf{U}$ of fiber is then

$$\mathbf{U} = \mathbf{V}_{out}\mathbf{M}_{\delta}(-\delta)\mathbf{V}^{\dagger}_{in} \tag{4.364}$$

where $\mathbf{M}_{\delta}(-\delta)$ is the retarder with retardation of δ and its slow-axis parallel to the x-axis. Since $\mathbf{V}_{out}$ and $\mathbf{V}^{\dagger}_{in}$ are unitary matrices, as we discussed in Section 4.5.1, their corresponding Mueller matrices are rotation matrices in Stokes space, which are assumed to be $\mathbf{R}_{out}$ and $\mathbf{R}_{in}$. The rotation matrix representation in Stokes space of the $\mathbf{U}$ matrix can be written as

$$\mathbf{R} = \mathbf{R}_{out}\mathbf{R}_1(\delta)\mathbf{R}^{-1}_{in} \tag{4.365}$$

Substituting it in (4.328), the derivative of Stokes vector at the output of fiber is

$$\frac{d\mathbf{S}(\omega)}{d\omega} = \frac{d\mathbf{R}(\omega)}{d\omega}\mathbf{R}^{-1}(\omega)\mathbf{S}(\omega) = \left\{ \mathbf{R}_{out}\frac{d\mathbf{R}_1[\delta(\omega)]}{d\omega}\mathbf{R}^{-1}_1[\delta(\omega)]\mathbf{R}^{-1}_{out} \right\} \mathbf{S}(\omega) \tag{4.366}$$

where the relation $\frac{d\mathbf{R}_{out}}{d\omega} = \frac{d(\mathbf{R}^{-1}_{in})}{d\omega} = 0$ has been used because $\mathbf{R}_{out}$ and $\mathbf{R}^{-1}_{in}$ are related to the input and output PSPs (see Eq. (4.361)) and are optically frequency independent. Substitution of $\mathbf{R}_1[\delta(\omega)]$ (see Table 4.4) in (4.366) yields

$$\frac{d\mathbf{S}(\omega)}{d\omega} = \mathbf{R}_{out} \begin{pmatrix} 0 & 0 & 0 \\ 0 & 0 & -\Delta\tau \\ 0 & \Delta\tau & 0 \end{pmatrix} \mathbf{R}^{-1}_{out}\mathbf{S}(\omega) = \mathbf{R}_{out}\left[\Delta\boldsymbol{\tau} \times \left(\mathbf{R}^{-1}_{out}\mathbf{S}\right)\right] \tag{4.367}$$

$$\Delta\boldsymbol{\tau} = (\Delta\tau, 0, 0), \ \text{and} \ \Delta\tau = \frac{d\delta(\omega)}{d\omega} = \frac{d[\varphi_2(\omega)] - d[\varphi_1(\omega)]}{d\omega} \tag{4.368}$$

Since $\mathbf{R}(\mathbf{a}\times\mathbf{b}) = (\mathbf{Ra})\times(\mathbf{Rb})$, Eq. (4.367) can be simplified to

$$\frac{d\mathbf{S}(\omega)}{d\omega} = \mathbf{R}_{out}\Delta\boldsymbol{\tau} \times \left(\mathbf{R}_{out}\mathbf{R}^{-1}_{out}\mathbf{S}\right) = \mathbf{R}_{out}\Delta\boldsymbol{\tau} \times \mathbf{S} \tag{4.369}$$

The phase change $\varphi_i(\omega)$ of the principal state is equal to the multiplication of the corresponding propagation constant k_i and fiber length L, i.e. $\varphi_i(\omega) = k_iL$, $i = 1, 2$. Note that $d\omega/dk_i = v_{gi}$ is the group velocity of the ith PSP, so that $d[\varphi_i(\omega)]/d\omega = L/(d\omega/dk_i) = L/v_{gi} = \tau_i$ is the time of light propagation from the input to output. In other words, $d[\varphi_i(\omega)]/d\omega$ is the group delay of the ith PSP after light passes through the fiber with a length of L. Thus, $\Delta\tau$ in Eq. (4.368) can be simplified to

$$\Delta\tau = \tau_1 - \tau_2 \tag{4.370}$$

where $\Delta\tau$ is called the differential group delay (DGD) between two PSPs. The rotation matrix $\mathbf{R}_{out}$ in Stokes space is related to the 2×2 complex matrix $\mathbf{V}_{\text{out}}$ defined by Eq. (4.361). Note that $\mathbf{V}_{\text{out}}(1\ \ 0)^T = \mathbf{E}_{out1}$, $\mathbf{V}_{\text{out}}(0\ \ 1)^T = \mathbf{E}_{out2}$, i.e. $\mathbf{V}_{\text{out}}$ transforms the horizontal linear polarization to the principal state $\mathbf{E}_{out1}$ with group delay τ_1, and vertical linear polarization to principal state $\mathbf{E}_{out2}$ with group delay τ_2. In Stokes space, the same transformation should be obtained, i.e. $\mathbf{R}_{out}(1, 0, 0)^T = \mathbf{S}_{out1}$ and $\mathbf{R}_{out}(-1, 0, 0)^T = \mathbf{S}_{out2}$ where $\mathbf{S}_{out1}$ and $\mathbf{S}_{out2}$ are the Stokes vector corresponding to the principal states $\mathbf{E}_{out1}$ and $\mathbf{E}_{out2}$ in Jones vector space, respectively. It is clear $\mathbf{S}_{out1}$ and $\mathbf{S}_{out1}$ are in opposite direction. Therefore, $\mathbf{R}_{out}\Delta\boldsymbol{\tau}$ in Eq. (4.369) is equal to $\Delta\tau\mathbf{R}_{out}(1, 0, 0)^T = \Delta\tau\mathbf{S}_{out1}$. Define the vector $\mathbf{\Omega}(\omega)$ as follows

$$\mathbf{\Omega}(\omega) = \begin{cases} |\Delta\tau|\mathbf{S}_{out1} & \text{when } \tau_1 \geq \tau_2 \\ |\Delta\tau|\mathbf{S}_{out2} & \text{when } \tau_1 < \tau_2 \end{cases} \tag{4.371}$$

the polarization evolution Eq. (4.369) can be reformatted as

$$\frac{d\mathbf{S}(\omega)}{d\omega} = \mathbf{\Omega}(\omega) \times \mathbf{S}(\omega) \tag{4.372}$$

where the Stokes vector $\mathbf{\Omega}(\omega)$ is called PMD vector, which has the magnitude $|\mathbf{\Omega}(\omega)| = |\tau_1 - \tau_2| = DGD$, and is parallel to the unit vector $\hat{\mathbf{p}}$ along the PSP with a slower group velocity. $\mathbf{\Omega}(\omega)$ may also be represented by $\Delta\tau\hat{\mathbf{p}}$. The geometrical relationship between PMD vector and the change of $\mathbf{S}(\omega)$ with frequency is indicated in Figure 4.32. One may recognize that there are two principal states $\pm\hat{\mathbf{p}}$, which are unchanged when the optical frequency is varied. The state $\hat{\mathbf{p}}$ propagates at a slower group velocity than the state $-\hat{\mathbf{p}}$. If the $\mathbf{\Omega}(\omega)$ is a constant with respect to ω, for example, a retarder, a PM fiber, or a short length of SM fiber, $\mathbf{S}(\omega)$ moves along a cone around $\mathbf{\Omega}$ while ω changes, the half-angle β of the cone in Stokes space equals $\beta = arccos[\hat{\mathbf{p}} \cdot \mathbf{S}(\omega)]$. The tip of $\mathbf{S}(\omega)$ draws an arc on a circle on the Poincaré sphere. The circle becomes a great circle when the polarization of light is equally split into two principal states. From Eq. (4.372), the change $d\mathbf{S}(\omega)$ is proportional to $d\omega$; therefore, the output SOPs are different for different frequency components even though they have the same SOP at the input, which results in depolarization of pulsed signals in the fiber.

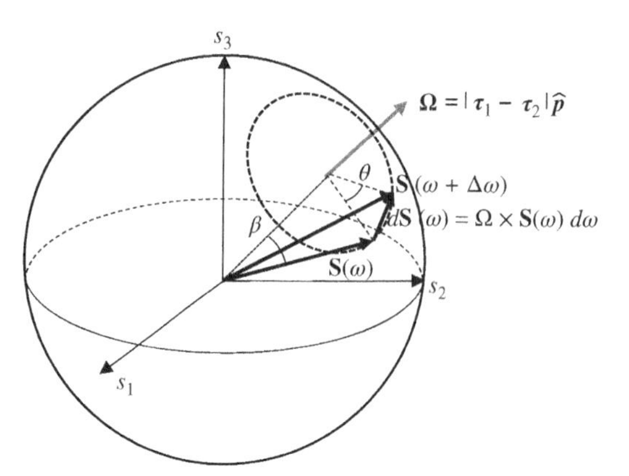

Figure 4.32 SOP evolution vs frequency on Poincaré sphere. The output polarizations $\mathbf{S}(\omega)$ is rotated to $\mathbf{S}(\omega + \Delta\omega)$ about PMD vector Ω, and the corresponding.

For a long single mode fiber, PMD vector $\boldsymbol{\Omega}(\omega)$ is a function of ω, and both its magnitude and direction vary with ω. When higher-order PMD effects are considered, $\boldsymbol{\Omega}(\omega)$ is usually called the "first-order" PMD vector, its frequency derivative $\boldsymbol{\Omega}'(\omega)$ is called the "second-order" PMD vector. If we limit ourselves to the second-order only, the PMD characteristics of a fiber can be described by the following expression for $\boldsymbol{\Omega}(\omega)$

$$\boldsymbol{\Omega}(\omega) = \Delta\tau\hat{\mathbf{p}} + \boldsymbol{\Omega}_{\omega}(\omega)d\omega \tag{4.373}$$

where

$$\boldsymbol{\Omega}_{\omega}(\omega) = \frac{d\boldsymbol{\Omega}_{\omega}(\omega)}{d\omega} = \left[\frac{d\Delta\tau}{d\omega}\hat{\mathbf{p}} + \Delta\tau\frac{d\hat{\mathbf{p}}}{d\omega}\right] \tag{4.374}$$

The derivative $\boldsymbol{\Omega}_{\omega}(\omega)$ is called the second order PMD vector. The first item $(d\Delta\tau/d\omega)\hat{\mathbf{p}}$ in the right side of Expression (4.374) is related to the derivatives of group velocities of principal states and called the polarization dependent chromatic dispersion (PDCD). The second item $\Delta\tau(d\hat{\mathbf{p}}/d\omega)$ represents the change rate of the PMD vector direction. It is named depolarization because it causes different polarization output for different frequency components even though the input polarization at ω is parallel to one of the principal states.

4.5.4.2 PMD Vector Concatenation Rules

For a long single-mode fiber, one may consider the fiber as an assembly of concatenated fiber sections. Let's first consider the case of two concatenated fiber sections, each having a known PMD vector, as shown in Figure 4.33. Assume the rotation matrices for the sections are $\mathbf{R}_a$ and $\mathbf{R}_a$, the corresponding PMD vectors are $\boldsymbol{\Omega}_a$ and $\boldsymbol{\Omega}_b$, and the total PMD vector of the system is $\boldsymbol{\Omega}$. From Eqs. (4.328) and (4.372), the following relationships can be obtained:

$$\lambda_0 = 2\pi\frac{c}{\omega} \tag{4.375}$$

$$\frac{d\mathbf{R}_a}{d\omega}\mathbf{R}_a^{-1} = \boldsymbol{\Omega}_a(\omega)\times \tag{4.376}$$

$$\frac{d\mathbf{R}_b}{d\omega}\mathbf{R}_b^{-1} = \boldsymbol{\Omega}_b(\omega)\times \tag{4.377}$$

$$\boldsymbol{\Omega}\times = \frac{d\mathbf{R}}{d\omega}\mathbf{R}^{-1} = \frac{d\mathbf{R}_b\mathbf{R}_a}{d\omega}(\mathbf{R}_b\mathbf{R}_a)^{-1} = \frac{d\mathbf{R}_b}{d\omega}\mathbf{R}_b^{-1} + \mathbf{R}_b\frac{d\mathbf{R}_a}{d\omega}\mathbf{R}_a^{-1}\mathbf{R}_a^{-1} \tag{4.378}$$

Similar to the derivation from Eqs. (4.367)–(4.369), $\mathbf{R}_b[\boldsymbol{\Omega}_a(\omega)\times]\mathbf{R}_b^{-1} = \mathbf{R}_b\boldsymbol{\Omega}_a(\omega)\times$; thus, we can simplify Eq. (4.378) into the form

$$\boldsymbol{\Omega}\times = \boldsymbol{\Omega}_b(\omega)\times + \mathbf{R}_b[\boldsymbol{\Omega}_a(\omega)\times]\mathbf{R}_b^{-1} \tag{4.379}$$

Similar to the derivation from Eqs. (4.367)–(4.369), $\mathbf{R}_b\ [\boldsymbol{\Omega}_a(\omega)\times]\mathbf{R}_b^{-1} = \mathbf{R}_b\boldsymbol{\Omega}_a(\omega)\times$; thus, we can simplify Eq. (4.378) into the form

$$\boldsymbol{\Omega} = \boldsymbol{\Omega}_b(\omega) + \mathbf{R}_b\boldsymbol{\Omega}_a(\omega) \tag{4.380}$$

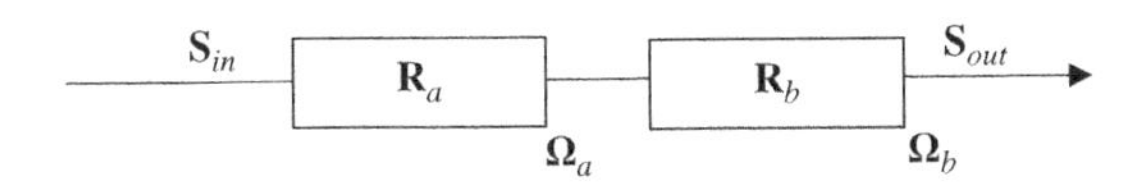

Figure 4.33 PMD of two concatenated sections. $\mathbf{R}_a$ and $\mathbf{R}_b$ are the rotation matrix of section a and section b, and the corresponding PMD vector are $\boldsymbol{\Omega}_a$ and $\boldsymbol{\Omega}_b$.

This is the basic concatenation rule. As an example, we consider two PM fibers. The first PM fiber's DGD is τ_a, with its slow axis at θ from the x-axis. The second PM fiber's DGD is τ_b with its slow axis parallel to the x-axis. Their corresponding rotation matrices are

$$\begin{aligned}\mathbf{R}_a &= \mathbf{R}_3(2\theta)\mathbf{R}_1(\delta_a)\mathbf{R}_3(-2\theta)\\ \mathbf{R}_b &= \mathbf{R}_1(\delta_b)\end{aligned} \tag{4.381}$$

Substituting (4.381) in Eqs. (4.376)–(4.378), one finds the PMD vector for each PM section as

$$\frac{d\mathbf{R}_a}{d\omega}\mathbf{R}_a^{-1} = \begin{pmatrix} 0 & 0 & \Delta\tau_a sin2\theta \\ 0 & 0 & -\Delta\tau_a cos2\theta \\ -\Delta\tau_a sin2\theta & \Delta\tau_a cos2\theta & 0 \end{pmatrix} \Rightarrow \mathbf{\Omega}_a(\omega) = \begin{pmatrix} \Delta\tau_a cos2\theta \\ \Delta\tau_a sin2\theta \\ 0 \end{pmatrix} \tag{4.382}$$

$$\frac{d\mathbf{R}_b}{d\omega}\mathbf{R}_b^{-1} = \begin{pmatrix} 0 & 0 & 0 \\ 0 & 0 & -\Delta\tau_b \\ 0 & \Delta\tau_b & 0 \end{pmatrix} \Rightarrow \mathbf{\Omega}_b(\omega) = \begin{pmatrix} \Delta\tau_b \\ 0 \\ 0 \end{pmatrix} \tag{4.383}$$

Substitution of Eqs. (4.382) and (4.383) in (4.379) yields the total PMD vector

$$\begin{aligned}\mathbf{\Omega} &= \mathbf{\Omega}_b(\omega) + \mathbf{R}_b\mathbf{\Omega}_a(\omega) = \begin{pmatrix} \Delta\tau_a cos\,2\theta + \Delta\tau_b \\ \Delta\tau_a cos\,\Delta_b\,sin2\theta \\ \Delta\tau_a sin\,\Delta_b\,sin2\theta \end{pmatrix}\\ \Delta\tau &= |\mathbf{\Omega}| = \sqrt{(\Delta\tau_a)^2 + (\Delta\tau_b)^2 + 2\Delta\tau_a\Delta\tau_b cos\,2\theta}\end{aligned} \tag{4.384}$$

And the second-order PMD vector is

$$\begin{aligned}\mathbf{\Omega}_\omega(\omega) &= \frac{d\mathbf{\Omega}_\omega(\omega)}{d\omega} = \frac{d\Delta_b}{d\omega}\begin{pmatrix} 0 \\ -\Delta\tau_a sin\,\Delta_b sin2\theta \\ \Delta\tau_a cos\Delta_b sin2\theta \end{pmatrix} = \begin{pmatrix} 0 \\ -\Delta\tau_a\Delta\tau_b sin\Delta_b sin2\theta \\ \Delta\tau_a\Delta\tau_b cos\Delta_b sin2\theta \end{pmatrix}\\ |\mathbf{\Omega}_\omega(\omega)| &= \Delta\tau_a\Delta\tau_b|sin2\theta|\end{aligned} \tag{4.385}$$

In summary, the total DGD of two concatenated PM fibers is the function of angle θ between their slow-axes, which reaches a maximum value of $\Delta\tau_a + \Delta\tau_b$ when their slow-axes are parallel ($\theta = 0$), and a minimum value of $|\Delta\tau_a - \Delta\tau_b|$ when their slow-axes are mutually orthogonal ($\theta = 90°$). The magnitude of DGD is independent of frequency such that the PDCD component of its second-order PMD $\mathbf{\Omega}_\omega$ is zero. The depolarization of component of the second-order PMD is $\Delta\tau_a\Delta\tau_b\,sin\,2\theta$, which is also a function of the angle θ between the slow axes of the two PM fiber sections. The second-order PMD is equal to zero when $\theta = 0$ and reaches a maximum value of $\Delta\tau_a\Delta\tau_b$ when $\theta = 45°$.

4.6 Polarimetric Measurement of PMD

As discussed in Section 4.5, when the optical frequency sweeps, the output polarization of a fiber link rotates about the PMD vector $\mathbf{\Omega}$. In general, both the magnitude and direction of the PMD vector $\mathbf{\Omega}$ are frequency dependent such that the trace of the output polarization on the Poincaré sphere is not a perfect circle in general, as shown in Figure 4.34. However, if the frequency step size is sufficiently small, the SOP can be considered rotating about PMD vector $\mathbf{\Omega}(\omega_i)$, with the adjacent SOPs on the arc of a circle, as shown in Figure 4.34c. Careful examination of Figure 4.32 reveals

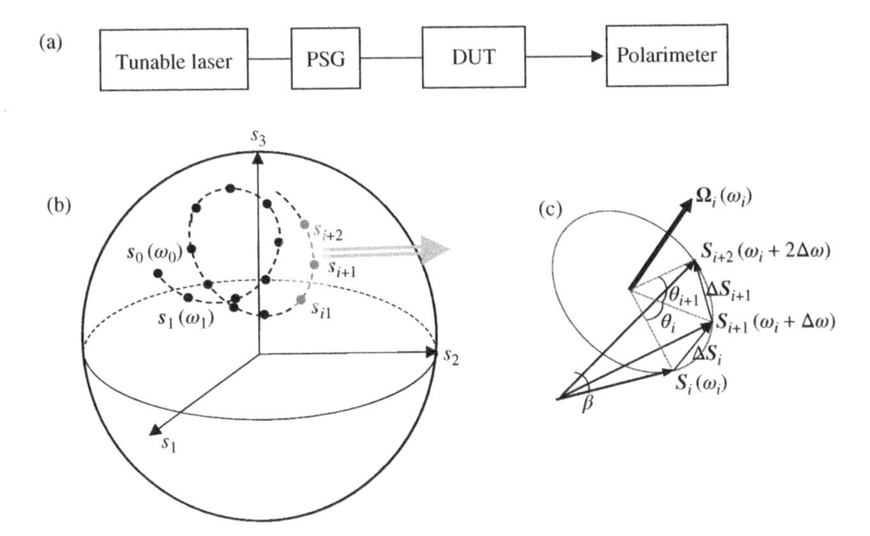

Figure 4.34 PMD measurement by scanning optical frequency. (a) Measurement setup. Narrow linewidth tunable laser sweeps with a frequency step $\Delta\omega$. PSG generates desired SOPs which are ensured to be the same for different optical frequencies at the input of DUT. (b) The SOP evolution trace when the optical frequency sweeps with a step $\Delta\omega$, with each solid dot representing a measured SOP at $\omega_i = \omega_0 + i\Delta\omega$. (c) The geometric relationship between PMD vector $\mathbf{\Omega}_i$ and measured Stokes parameter $\mathbf{S}_i$. **PSG**: polarization state generator.

that the PMD vector at the frequency ω_i in (4.372) can be related to the SOP change $d\mathbf{S}(\omega_i)$ caused by a frequency change $d\omega$ as

$$\frac{d\mathbf{S}(\omega_i)}{d\omega} = \mathbf{\Omega}(\omega_i) \times \mathbf{S}(\omega_i) \tag{4.386}$$

where ω_i is the angular optical frequency of the light. The magnitude of PMD vector (DGD) can be obtained by

$$\left|\frac{\Delta\theta_i}{\Delta\omega}\right| = |\mathbf{\Omega}(\omega_i)| = \Delta\tau \tag{4.387}$$

where $\Delta\theta_i$ is the angle of rotation of the output SOP about the PMD vector $\mathbf{\Omega}(\omega_i)$. The DGD $\Delta\tau$ can be found by measuring $\Delta\theta_i/\Delta\omega$. Since it is impossible to identify the difference between $|\Delta\theta_i|$ and $|2\pi - \Delta\theta_i|$, one has to ensure $\Delta\omega$ to be sufficiently small such that $|\Delta\theta_i| < \pi$. Therefore the DGD measurement range is limited by the frequency sweeping step size $\Delta\omega$ to

$$DGD < \frac{\pi}{|\Delta\omega|} \tag{4.388}$$

Because $\omega = 2\pi f = 2\pi c/\lambda$, we have $|\Delta\omega| = 2\pi c\Delta\lambda/\lambda^2$, where c is the light speed in vacuum, and λ_0 is the wavelength of light in vacuum, the DGD measurement range limited by the wavelength step $\Delta\lambda$ can be obtained from (4.388) as

$$DGD < \frac{\lambda^2}{2c\Delta\lambda} \sim \begin{cases} 4.0 \text{ ps/mm} & \text{when } \lambda = 1550 \text{ nm} \\ 2.9 \text{ ps/mm} & \text{when } \lambda = 1310 \text{ nm} \end{cases} \tag{4.389}$$

In general, the PMD vector $\mathbf{\Omega}(\omega_i)$ is wavelength dependent. After $\mathbf{\Omega}(\omega_i)$ is obtained, the second-order PMD can be calculated using

$$\mathbf{SOPMD}(\omega_i) = \frac{\mathbf{\Omega}(\omega_{i+1}) - \mathbf{\Omega}(\omega_i)}{\Delta\omega} \tag{4.390}$$

From Eq. (4.374), the chromatic dispersion PDCD and depolarization components can be obtained using

$$\begin{aligned}\mathbf{PDCD}(\omega_i) &= \frac{|\boldsymbol{\Omega}(\omega_{i+1})| - |\boldsymbol{\Omega}(\omega_i)|}{\Delta\omega}\\ \text{Depolarization} &= |\boldsymbol{\Omega}(\omega_i)|\frac{\hat{\mathbf{p}}(\omega_{i+1}) - \hat{\mathbf{p}}(\omega_i)}{\Delta\omega}\end{aligned} \tag{4.391}$$

where $\hat{\mathbf{p}}\left(\omega_i\right)$ is DUT's PSP at ω_i, which is the unit vector parallel to the positive direction of $\boldsymbol{\Omega}(\omega_i)$.

In the following, we will introduce several similar techniques of polarimetric measurement of DGD. They are the Poincaré arc (PA) method, Poincaré sphere analysis (PSA) method, Mueller Matrix Method (MMM), and Jones Matrix Eigenanalysis (JME). The distinction between these polarimetric techniques is how they measure $\Delta\theta/\Delta\omega$; all of these techniques use the same setup as shown in Figure 4.34a. The measurement begins with launching a desired polarized light into the DUT and measuring the output polarization states $\mathbf{S}(\omega)$ as a function of optical frequency. Different techniques use different ways to find the angle change $\Delta\theta$ from the variations of $\mathbf{S}(\omega)$.

4.6.1 Poincaré Sphere Arc Method

Poincaré arc method measures $\Delta\theta_i/\Delta\omega$ without needing to know the polarization state at input of the DUT. Assume $\mathbf{S}_a(\omega_i)$ and $\mathbf{S}_b(\omega_i)$ are two output Stokes vectors corresponding to two different input polarization states and already measured under different optical frequencies. Then, the relationship (4.386) can be approximated by

$$\frac{\Delta\mathbf{S}_a(\omega_i)}{\Delta\omega} = \boldsymbol{\Omega}(\omega_i) \times \mathbf{S}_a(\omega_i) \tag{4.392}$$

$$\frac{\Delta\mathbf{S}_b(\omega_i)}{\Delta\omega} = \boldsymbol{\Omega}(\omega_i) \times \mathbf{S}_b(\omega_i) \tag{4.393}$$

where $\Delta\mathbf{S}_a(\omega_i) = \mathbf{S}_a(\omega_i + \Delta\omega) - \mathbf{S}_a(\omega_i)$, $\Delta\mathbf{S}_b(\omega_i) = \mathbf{S}_b(\omega_i + \Delta\omega) - \mathbf{S}_b(\omega_i)$. Taking the cross product of (4.392) and (4.393), and using $\mathbf{a} \times (\mathbf{b} \times \mathbf{c}) = \mathbf{b}(\mathbf{c}\cdot\mathbf{a}) - \mathbf{c}(\mathbf{a}\cdot\mathbf{b})$, one obtains

$$\begin{aligned}\frac{\Delta\mathbf{S}_a(\omega_i)}{\Delta\omega} \times \frac{\Delta\mathbf{S}_b(\omega_i)}{\Delta\omega} &= [\boldsymbol{\Omega}(\omega_i) \times \mathbf{S}_a(\omega_i)] \times [\boldsymbol{\Omega}(\omega_i) \times \mathbf{S}_b(\omega_i)]\\ &= \boldsymbol{\Omega}(\omega_i)\,\{[\boldsymbol{\Omega}(\omega_i) \times \mathbf{S}_a(\omega_i)] \cdot \mathbf{S}_b(\omega_i)\} - \mathbf{S}_b(\omega_i)\{[\boldsymbol{\Omega}(\omega_i) \times \mathbf{S}_a(\omega_i)] \cdot \boldsymbol{\Omega}(\omega_i)\}\\ &= \boldsymbol{\Omega}(\omega_i)\left[\frac{\Delta\mathbf{S}_a(\omega_i)}{\Delta\omega} \cdot \mathbf{S}_b(\omega_i)\right]\end{aligned} \tag{4.394}$$

Divide both sides of (4.394) with $\frac{\Delta\mathbf{S}_a(\omega_i)}{\Delta\omega} \cdot \mathbf{S}_b(\omega_i)$, the PMD vector $\boldsymbol{\Omega}(\omega_i)$ is obtained

$$\boldsymbol{\Omega}(\omega_i) = \frac{1}{\Delta\omega}\frac{\Delta\mathbf{S}_a(\omega_i) \times \Delta\mathbf{S}_b(\omega_i)}{\Delta\mathbf{S}_a(\omega_i) \cdot \mathbf{S}_b(\omega_i)} \tag{4.395}$$

It should be noted Eqs. (4.392) and (4.393) require the assumption that

$$\left|\frac{\Delta\mathbf{S}(\omega_i)}{\Delta\omega}\right| = \left|\frac{d\mathbf{S}(\omega_i)}{d\omega}\right| \tag{4.396}$$

This is only true when the frequency step $\Delta\omega$ is sufficiently small such that $\Delta\theta$ is close to zero. This condition is difficult to maintain in practice; therefore, the Poincaré arc method is only used for PMD estimation, not accurate measurement.

4.6.2 Poincaré Sphere Analysis

The PSA technique uses three orthogonal polarizations in the Stokes space as inputs, for example, horizontal linear (LHP), +45° linear (L + 45P), and right-handed circular (RCP) polarization states. Assume there is no PDL in the DUT, their corresponding three output normalized polarization states $\mathbf{h}(\omega_i)$, $\mathbf{q}(\omega_i)$, $\mathbf{c}(\omega_i)$ must also be orthogonal to each other in the Stokes space such that $\mathbf{h}(\omega_i)\times\mathbf{q}(\omega_i) = \mathbf{c}(\omega_i)$. From the geometric relationships (see Figure 4.35), the rotation angle change for $\Delta\theta_i$ can be obtained from

$$sin\left(\frac{\Delta\theta_i}{2}\right) = \frac{\Delta\mathbf{h}(\omega_i)/2}{|\mathbf{h}(\omega_i) - [\mathbf{h}(\omega_i)\cdot\hat{\mathbf{p}}_i]\hat{\mathbf{p}}_i|} \tag{4.397}$$

where $\hat{\mathbf{p}}_i$ is the unit vector along $\boldsymbol{\Omega}(\omega_i)$. Squaring Eq. (4.397) yields

$$\begin{aligned}
|\Delta\mathbf{h}(\omega_i)|^2 &= 4sin^2\left(\frac{\Delta\theta_i}{2}\right)\{\mathbf{h}(\omega_i) - [\mathbf{h}(\omega_i)\cdot\hat{\mathbf{p}}_i]\hat{\mathbf{p}}_i\}\cdot\{\mathbf{h}(\omega_i) - [\mathbf{h}(\omega_i)\cdot\hat{\mathbf{p}}_i]\hat{\mathbf{p}}_i\} \\
&= 4sin^2\left(\frac{\Delta\theta_i}{2}\right)\{1 - [\mathbf{h}(\omega_i)\cdot\hat{\mathbf{p}}_i]^2\} \\
&= 4sin^2\left(\frac{\Delta\theta_i}{2}\right)(1 - cos^2\beta_h)
\end{aligned} \tag{4.398}$$

where $cos\beta_h = \mathbf{h}(\omega_i)\cdot\hat{\mathbf{p}}_i$ and β_h is the angle between unit vectors $\mathbf{h}(\omega_i)$ and $\hat{\mathbf{p}}_i$. Repeating the derivation of (4.397) and (4.398), we can get the similar equations for Stokes vectors $\mathbf{q}(\omega_i)$, $\mathbf{c}(\omega_i)$:

$$|\Delta\mathbf{q}(\omega_i)|^2 = 4sin^2\left(\frac{\Delta\theta_i}{2}\right)(1 - cos^2\beta_q) \tag{4.399}$$

$$|\Delta\mathbf{c}(\omega_i)|^2 = 4sin^2\left(\frac{\Delta\theta_i}{2}\right)(1 - cos^2\beta_c) \tag{4.400}$$

where β_q and β_c are the angles between $\hat{\mathbf{p}}_i$ and $\mathbf{q}(\omega_i)$, $\hat{\mathbf{p}}_i$ and $\mathbf{c}(\omega_i)$, respectively. Adding Eqs. (4.398)–(4.400), one finds the following relation:

$$|\Delta\mathbf{h}(\omega_i)|^2 + |\Delta\mathbf{q}(\omega_i)|^2 + |\Delta\mathbf{c}(\omega_i)|^2 = 4sin^2\left(\frac{\Delta\theta_i}{2}\right)\{3 - [cos^2\beta_h + cos^2\beta_q + cos^2\beta_c]\} \tag{4.401}$$

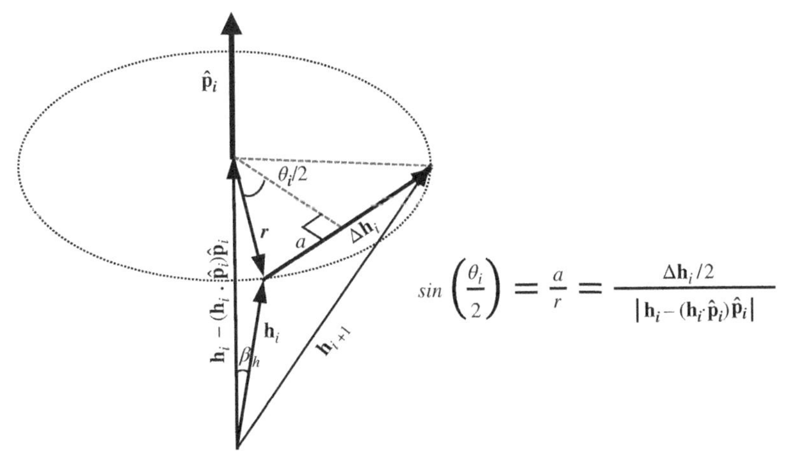

Figure 4.35 Geometric relationships of PSP $\hat{\mathbf{p}}_i$, output polarizations $\boldsymbol{h}_i(\omega_i)$ and $\boldsymbol{h}_i(\omega_i + \Delta\omega)$, where $r = \boldsymbol{h}_i - (\mathbf{h}_i\cdot\hat{\mathbf{p}}_i)\hat{\mathbf{p}}_i$, a equals the half of the length $|\Delta\boldsymbol{h}_i| = |\boldsymbol{h}_i(\omega_i) - \boldsymbol{h}_i(\omega_i + \Delta\omega)|$.

Since the vectors $\mathbf{h}(\omega_i)$, $\mathbf{q}(\omega_i)$, and $\mathbf{c}(\omega_i)$ are orthogonal to one another, the angles β_h, β_q, and β_c can be considered as the direction cosines of PMD vector in the coordinates constructed $\mathbf{h}(\omega_i)$, $\mathbf{q}(\omega_i)$, and $\mathbf{c}(\omega_i)$. Therefore, $cos^2\beta_h + cos^2\beta_q + cos^2\beta_c = 1$. Substituting this relationship in (4.401), we obtain $\Delta\tau$ as

$$\Delta\tau = \left|\frac{\Delta\theta_i}{\Delta\omega}\right| = \frac{2}{\Delta\omega} arcsin \frac{1}{2}\sqrt{\frac{1}{2}\left[|\Delta\mathbf{h}(\omega_i)|^2 + |\Delta\mathbf{q}(\omega_i)|^2 + |\Delta\mathbf{c}(\omega_i)|^2\right]} \tag{4.402}$$

The PSP is also found geometrically as the axis about which $\mathbf{h}(\omega_i)$, $\mathbf{q}(\omega_i)$, and $\mathbf{c}(\omega_i)$ rotate:

$$PSP = \hat{\mathbf{p}}_i = \frac{\mathbf{\Omega}(\omega_i)}{|\mathbf{\Omega}(\omega_i)|} = \frac{\mathbf{u}(\omega_i)}{|\mathbf{u}(\omega_i)|} \tag{4.403}$$

with

$$\mathbf{u} = (\mathbf{c}\cdot\Delta\mathbf{q})\mathbf{h} + (\mathbf{h}\cdot\Delta\mathbf{c})\mathbf{q} + (\mathbf{q}\cdot\Delta\mathbf{h})\mathbf{c} \tag{4.404}$$

where the angular frequency argument ω_i in $\mathbf{h}(\omega_i)$, $\mathbf{q}(\omega_i)$, and $\mathbf{c}(\omega_i)$ is hidden in order to make the equation easy to read. It should be noted that only two distinctive input SOPs are required for taking the measurement, although three mutually orthogonal SOP vectors are used in Eqs. (4.402) and (4.404), because the three mutually orthogonal SOP vectors can be constructed using two arbitrary SOP vectors, as described in the following text. Assume $\mathbf{a}(\omega_i)$ and $\mathbf{b}(\omega_i)$ are two output Stokes vectors corresponding to two distinctive input SOPs, one may construct $\mathbf{h}(\omega_i)$, $\mathbf{q}(\omega_i)$, and $\mathbf{c}(\omega_i)$ mathematically from $\mathbf{a}(\omega_i)$ and $\mathbf{b}(\omega_i)$ as follows

$$\mathbf{h} = \mathbf{a} \tag{4.405}$$

$$\mathbf{q} = (\mathbf{a}\times\mathbf{b})\times\frac{\mathbf{a}}{|\mathbf{a}\times\mathbf{b}|} \tag{4.406}$$

$$\mathbf{c} = \mathbf{h}\times\mathbf{q} \tag{4.407}$$

The advantage of the operation earlier is that the two input SOPs are not required to be orthogonal, so long as they are sufficiently different from each other, which make the accuracy of the SOP generated by the PSG in Figure 4.34a less demanding. For example, they have a 3D angle larger than 45° from each other on the Poincaré sphere.

4.6.3 Mueller Matrix Method

The MMM measures PMD through measuring the rotation matrix related to the change of optical frequency. In the absence of PDL, the Mueller matrix representation of the DUT can be reduced to a 3×3 rotation matrix. The details of measuring the Mueller matrix are discussed in Section 9.3.2.2. Assuming the Mueller matrix is already measured at optical frequency ω_i and $\omega_i + \Delta\omega$, the corresponding output and input polarization states are related by

$$\mathbf{S}(\omega_i) = \mathbf{M}_{R,3\times3}(\omega_i)\mathbf{S}_{in} \Rightarrow \mathbf{M}^{-1}_{R,3\times3}(\omega_i)\mathbf{S}(\omega_i) = \mathbf{S}_{in} \tag{4.408}$$

$$\mathbf{S}(\omega_i + \Delta\omega) = \mathbf{M}_{R,3\times3}(\omega_i + \Delta\omega)\mathbf{S}_{in} \Rightarrow \mathbf{M}^{-1}_{R,3\times3}(\omega_i + \Delta\omega)\mathbf{S}(\omega_i + \Delta\omega) = \mathbf{S}_{in} \tag{4.409}$$

The subscript "$R, 3\times3$" represents the matrix is a 3×3 rotation matrix in the Stokes space. Equating (4.408) and (4.409), one obtains

$$\mathbf{S}(\omega_i + \Delta\omega) = \mathbf{M}_{R,3\times3}(\omega_i + \Delta\omega)\mathbf{M}^{-1}_{R,3\times3}(\omega_i)\mathbf{S}(\omega_i) = \mathbf{M}_\Delta(\omega_i)\mathbf{S}(\omega_i) \tag{4.410}$$

where $\mathbf{M}_\Delta(\omega_i) = \mathbf{M}_{R,3\times3}(\omega_i + \Delta\omega)\mathbf{M}^{-1}_{R,3\times3}(\omega_i)$. Since the matrices $\mathbf{M}_{R,3\times3}(\omega_i + \Delta\omega)$ and $\mathbf{M}^{-1}_{R,3\times3}$ are both 3×3 rotation matrices, the matrix $\mathbf{M}_\Delta$ is also a rotation matrix in Stokes space. Therefore, from Eq. (4.286), the precession angle of $\Delta\theta_i$ of the output polarization state about PMD vector $\mathbf{\Omega}(\omega_i)$ is given by

$$cos(\Delta\theta_i) = \frac{1}{2}\left(Tr\mathbf{M}_\Delta(\omega_i) - 1\right) \tag{4.411}$$

and

$$\Delta\tau = \left|\frac{acos\left(\frac{1}{2}(Tr\mathbf{M}_\Delta(\omega_i) - 1)\right)}{\Delta\omega}\right| \tag{4.412}$$

The PSP is the rotation axis of SOP change, which is the eigenvector of $\mathbf{M}_\Delta$ with an eigenvalue of unity. Both the MMM and the PSA method described in Section 4.6.2 assume the absence of PDL in DUT and they should give the same result in theory. However, the PSA method does not require the knowledge of the input SOP, while the MMM needs to know the input SOP in order to measure the Mueller matrix of DUT. Therefore, the PSA method is more convenient and faster than the MMM. On the other hand, when PDL cannot be ignored, the measurement accuracy of PSA gets worse. For the MMM, the matrix $\mathbf{M}_\Delta(\omega_i)$ cannot be represented by a 3×3 rotation matrix when PDL cannot be ignored. It becomes a 4×4 matrix, and the PMD vector becomes a complex vector, as discussed in detail in Section 9.3.2.2.

4.6.4 Jones Matrix Eigenanalysis

JME (Heffner 1992) measures the $\Delta\theta_i/\Delta\omega$ and principal states by calculating the eigenvalues and eigenvector of the 2×2 complex transformation matrix between two output Jones vectors with different frequencies. Assume the Jones matrix of the DUT is already measured at optical frequencies ω_i and $\omega_i + \Delta\omega$, the corresponding output and input Jones vectors of SOPs are related by

$$\mathbf{J}_{out}(\omega_i) = \mathbf{M}_J(\omega_i)\mathbf{J}_{in} \Rightarrow \mathbf{M}_J^{-1}(\omega_i)\mathbf{J}_{out}(\omega_i) = \mathbf{J}_{in} \tag{4.413}$$

$$\mathbf{J}_{out}(\omega_{i+1}) = \mathbf{M}_J(\omega_i + \Delta\omega)\mathbf{J}_{in} \Rightarrow \mathbf{M}_J^{-1}(\omega_{i+1})\mathbf{J}_{out}(\omega_{i+1}) = \mathbf{J}_{in} \tag{4.414}$$

The $\mathbf{M}_J(\omega_i)$ is the 2×2 complex Jones matrix of DUT, $\mathbf{J}_{in}$ is the Jones vector of the incident light which is frequency independent. $\mathbf{J}_{out}(\omega_i)$ and $\mathbf{J}_{out}(\omega_{i+1})$ are the Jones vectors at frequencies ω_i and ω_{i+1} at the output of DUT. Equating Eqs. (4.413) and (4.414), one obtains

$$\mathbf{J}_{out}(\omega_{i+1}) = \mathbf{M}_J(\omega_{i+1})\mathbf{M}_J^{-1}(\omega_i)\mathbf{J}_{out}(\omega_i) = \mathbf{\Gamma}(\overline{\omega}_i)\mathbf{J}_{out}(\omega_i) \tag{4.415}$$

where $\mathbf{\Gamma}(\overline{\omega}_i) = \mathbf{M}_J(\omega_{i+1})\mathbf{M}_J^{-1}(\omega_i)$. The two eigenvectors of the transformation matrix $\mathbf{\Gamma}(\omega_i)$ are the Jones vectors which remain unchanged when optical frequency varies about $\overline{\omega}_i$. In other words, the eigenvectors of $\mathbf{\Gamma}(\overline{\omega}_i)$ are the PSP of the DUT at the frequency $\overline{\omega}_i$. The corresponding eigenvalues are $\rho_1 = e^{i\tau_1\Delta\omega}$ and $\rho_2 = e^{i\tau_2\Delta\omega}$, where $\tau_1 - \tau_2$ the group delay between two PSP; therefore

$$\Delta\tau(\overline{\omega}_i) = |\tau_1(\overline{\omega}_i) - \tau_2(\overline{\omega}_i)| = \left|\frac{Arg[\rho_1(\overline{\omega}_i)/\rho_2(\overline{\omega}_i)]}{\Delta\omega}\right| \tag{4.416}$$

where *Arg* stands for the argument of a complex number. Therefore, the JME method allows the full $\mathbf{\Omega}(\overline{\omega}_i)$ to be calculated after the transformation matrix $\mathbf{\Gamma}(\overline{\omega}_i)$ is obtained from the Jones matrix measurements at two optical frequencies.

4.7 Polarization Properties of Quasi-monochromatic Light

In Sections 4.1–4.6 of this chapter, the light has been assumed to be monochromatic. As described in Chapter 2, such a strictly monochromatic light is always polarized, i.e. the endpoint of its electric field vector at a given position in space moves periodically around an ellipse with time. However, in reality, one often encounters unpolarized light, such as the sunlight. In this case, the endpoint of light's electric field vector may be viewed to move quite randomly, with no periodic trajectory. In general, the variation of the electric field vectors of light is neither completely regular nor totally random, or the light is partially polarized. For example, an unpolarized light becomes partially polarized after being reflected from a glass window because different polarization components have different reflection coefficients, as described by the Fresnel formulae (4.65). Conversely, a polarized broadband light becomes partially polarized after passing through a birefringence crystal or a depolarizer, as will be discussed in Chapter 6.

As mentioned in Chapter 1, the wave theory developed by Fresnel was completely successful in describing polarized light (as seen in Sections 4.1–4.6), however, was unable to satisfactorily describe interference experiments with unpolarized or partially polarized light. In fact, Fresnel himself was not able to mathematically describe the unpolarized light until his early death in 1827. The problem remained unsolved until George Stokes abandoned Fresnel's fruitless attempts to describe unpolarized light with amplitudes; instead, he used six measured intensities, or observables, known as Stokes parameters to define the polarization properties of light 35 years later.

In this section, we will introduce two mathematic methods to describe partial polarization. One is the coherency matrix introduced by Emil Wolf (Born and Wolf 1999), and another is the Stokes parameters introduced by Stokes. They are equivalent mathematically and can be determined by light intensity measurements.

4.7.1 Analytic Signal Representation of Polychromatic Light

In discussing monochromatic light wave, it is often convenient to represent a real electric field $E^{(r)}{}_i = A_x \cos(\omega t - k_0 z + \varphi_x)$ as the real part of an associated complex wave function $E_i = A_x[cos(\omega t - k_0 z + \varphi_x) + icos(\omega t - k_0 z + \varphi_x)] = A_x e^{i(\omega t - k_0 z + \varphi_x)}$. In this section, it will be helpful again to deploy such a complex presentation to discuss polychromatic light waves.

Let's consider a polychromatic plane wave propagating in a homogeneous medium in the direction $+z$. At a fixed point in space, the real transverse field that obeys Maxwell's equations can be expressed as

$$\mathbf{E}^{(r)}(t) = E_x^{(r)}(t)\hat{\mathbf{e}}_x + E_y^{(r)}(t)\hat{\mathbf{e}}_y \tag{4.417}$$

where $E_j^{(r)}$ $(j = x, y)$ represents that the field is a real function. Assume $E_j^{(r)}$ is square-integrable, it may be expressed in the form of a Fourier integral

$$E_j^{(r)}(t) = \int_{-\infty}^{\infty} \varepsilon(\upsilon)e^{i2\pi\upsilon t}d\upsilon \tag{4.418}$$

where $\varepsilon(\upsilon)$ can be expressed as the inverse Fourier integral

$$\varepsilon_j(\upsilon) = \int_{-\infty}^{\infty} E_j^{(r)}(t)e^{-i2\pi\upsilon t}dt \tag{4.419}$$

Since $E_j^{(r)}(t)$ is real, one obtains from Eq. (4.419) that the complex conjugate of $\varepsilon_j(\upsilon)$ is equal to $\varepsilon_j(-\upsilon)$, i.e.

$$\varepsilon_j^*(\upsilon) = \varepsilon_j(-\upsilon) \tag{4.420}$$

Therefore integral (4.418) can be rewritten as

$$\begin{aligned} E_j^{(r)}(t) &= \int_0^{\infty} \varepsilon_j(\upsilon)e^{i2\pi\upsilon t}d\upsilon + \int_0^{\infty} \varepsilon_j^*(\upsilon)e^{-i2\pi\upsilon t}d\upsilon \\ &= \int_0^{\infty} \left[\varepsilon_j(\upsilon)e^{i2\pi\upsilon t} + \varepsilon_j^*(\upsilon)e^{-i2\pi\upsilon t}\right] d\upsilon \\ &= 2Re\int_0^{\infty} \varepsilon_j(\upsilon)e^{i2\pi\upsilon t}d\upsilon \end{aligned} \tag{4.421}$$

Assume $\varepsilon_j(\upsilon) = (a_j(\upsilon)/2)e^{i\varphi_j(\upsilon)}$ where $\upsilon \geq 0$, and $a_j(\upsilon)$ is a non-negative real number, and substitute it in Eq. (4.421); the real electrical field $E_j^{(r)}(t)$ can be expressed by the following Fourier cosine integral:

$$E_j^{(r)}(t) = \int_0^{\infty} a_j(\upsilon)cos[2\pi\upsilon t + \varphi_j(\upsilon)]d\upsilon \tag{4.422}$$

Similarly, one may introduce Fourier sine integral function as

$$E_j^{(i)}(t) = 2Im\int_0^{\infty} \varepsilon_j(\upsilon)e^{i2\pi\upsilon t}d\upsilon = \int_0^{\infty} a_j(\upsilon)sin[2\pi\upsilon t + \varphi_j(\upsilon)]d\upsilon \tag{4.423}$$

Consequently one may construct a complex analytic signal $E_j(t)$ as

$$E_j(t) \equiv E_j^{(r)}(t) + iE_j^{(i)}(t) \tag{4.424}$$

From (4.422) and (4.423), the complex field can be expressed as

$$E_j(t) \equiv 2\int_0^{\infty} \varepsilon_j(\upsilon)e^{i2\pi\upsilon t}d\upsilon \tag{4.425}$$

Since from (4.424) υ only takes non-negative values for $\varepsilon_j(\upsilon)$, (4.419) can be re-written next for clarity:

$$\varepsilon_j(\upsilon) = \begin{cases} \int_{-\infty}^{\infty} E_j^{(r)}(t)e^{-i2\pi\upsilon t}dt & \text{when } \upsilon \geq 0 \\ 0 & \text{when } \upsilon < 0 \end{cases} \tag{4.426}$$

4.7.1.1 Quasi-monochromatic Light

In most applications, such as in optical fiber communications, the light signal is considered as quasi-monochromatic waves whose amplitude is zero outside a narrow frequency range of width $\Delta\upsilon$ that is much smaller compared with that of the mean frequency $\overline{\upsilon}$. Therefore (4.424) the complex analytic electric field can be written as

$$E_j(t) = e^{i2\pi\overline{\upsilon}t}\int_0^{\infty} 2\varepsilon_j(\upsilon)e^{i2\pi(\upsilon-\overline{\upsilon})t}d\upsilon = a_j(t)e^{i[2\pi\overline{\upsilon}t-\varphi_j(t)]} \tag{4.427}$$

where $a_j(t)e^{-\varphi_j(t)} = \int_0^{\infty} 2\varepsilon_j(\upsilon)e^{i2\pi(\upsilon-\overline{\upsilon})t}d\upsilon$, with $a_j(t) \geq 0$ and $\varphi_j(t)$ is real. Let $\mu = \upsilon - \overline{\upsilon}$, $a_j(t)e^{-i\varphi_j(t)}$ can be re-written as

$$a_j(t)e^{-i\varphi_j(t)} = \int_{-\overline{\upsilon}}^{\infty} g_j(\mu)e^{i2\pi\mu t}d\mu \tag{4.428}$$

where

$$g_j(\mu) = 2\varepsilon_j(\mu + \overline{\upsilon})$$

Since the spectral amplitude is assumed to differ appreciably from zero only in the neighborhood of $\upsilon = \bar{\upsilon}$, $|g(\mu)|$ will be appreciable only near $\mu = 0$. Therefore, $a_j(t)$ and $\varphi_j(t)$ vary slowly in comparison with $cos(2\pi\bar{\upsilon}t)$ and $sin(2\pi\bar{\upsilon}t)$. Consequently, the real and imaginary parts of $E_j(t)$ can be written as

$$E_j^{(r)}(t) = a_j(t)cos[2\pi\bar{\upsilon}t - \varphi_j(t)]$$
$$E_j^{(i)}(t) = a_j(t)sin[2\pi\bar{\upsilon}t - \varphi_j(t)] \tag{4.429}$$

These formulae indicate that $E_j^{(r)}(t)$ and $E_j^{(i)}(t)$ are modulated signals with a carrier frequency $\bar{\upsilon}$, an envelope $a_j(t)$, and a phaser $\varphi_j(t)$.

4.7.2 Coherency Matrix

Let's consider a quasi-monochromatic light wave propagating in the positive z-direction, with its electric field vector at $z = 0$ expressed as

$$\mathbf{E} = E_x\hat{\mathbf{e}}_x + E_y\hat{\mathbf{e}}_y \tag{4.430}$$

$$E_x = a_1(t)e^{i[2\pi\bar{\nu}t-\phi_1(t)]} \text{ and } E_y = a_2(t)e^{i[2\pi\bar{\nu}t-\phi_2(t)]} \tag{4.431}$$

where $\bar{\nu}$ is the mean frequency of quasi-monochromatic light, E_x and E_y are the "analytic signals" associated with the real electrical field components $Re(E_x) = a_1(t)cos(2\pi\bar{\nu}t - \phi_1)$ and $Re(E_y) = a_2(t)cos(2\pi\bar{\nu}t - \phi_2)$. If the light is strictly monochromatic, a_1, a_2, ϕ_1, and ϕ_2 are constants, and the light is completely polarized. For a quasi-monochromatic light wave, these quantities are time dependent; however, they change insignificantly when the observation time is shorter than the coherence time of light.

Let's consider the light intensity measured by a photodetector after the light passes through a retarder and a polarizer, as shown in Figure 4.36. The retarder's slow-axis is vertical, and its retardation δ is adjustable. A polarizer is placed after the retarder with the transmission axis at θ from the x-axis. The polarizer is rotatable so that one can measure the light intensity with different polarizer orientations. After the light passes through the polarizer, the electric field in the transmission direction can be written as

$$E_p = cos\theta E_x + sin\theta e^{-i\delta}E_y \tag{4.432}$$

and the intensity detected by the photodetector can be written as

$$\begin{aligned} I(\theta, \Delta) &= \langle E_p(\theta,\delta)E_p^*(\theta,\delta)\rangle \\ &= \langle E_xE_x^*\rangle\, cos^2(\theta) + \langle E_yE_y^*\rangle\, sin^2(\theta) + e^{i\delta}\,\langle E_xE_y^*\rangle\, cos\theta\, sin\theta \\ &\quad + e^{-i\delta}\,\langle E_x^*E_y\rangle\, cos\theta\, sin\theta \end{aligned} \tag{4.433}$$

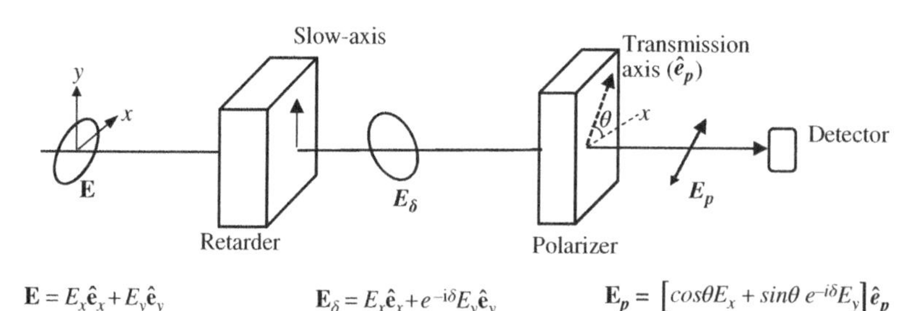

Figure 4.36 The electric fields of quasi-monochromatic light while passing through a retarder and polarizer.

where the angle brackets represent the average over the detector's integration time.

$$\left\langle E_i E_j^* \right\rangle = \lim_{T\to\infty} \frac{1}{T} \int_{-\frac{T}{2}}^{\frac{T}{2}} E_i(t)E_j^*(t)dt \quad \text{where } i = x \text{ or } y, \text{ and } j = x \text{ or } y \tag{4.434}$$

It is clear that the light intensity $I(\theta, \delta)$ at the given angle θ and retardation δ is determined by four parameters $\langle E_x E_x^* \rangle$, $\langle E_y E_y^* \rangle$, $\langle E_x E_y^* \rangle$, and $\langle E_x^* E_y \rangle$. Based on these four parameters, Wolf constructed the following matrix $\mathbf{J}_c$

$$\mathbf{J}_c = \begin{pmatrix} J_{xx} & J_{xy} \\ J_{yx} & J_{yy} \end{pmatrix} = \begin{pmatrix} \langle E_x E_x^* \rangle & \langle E_x E_y^* \rangle \\ \langle E_y E_x^* \rangle & \langle E_y E_y^* \rangle \end{pmatrix} = \begin{pmatrix} \langle a_1^2 \rangle & \langle a_1 a_2 e^{i(\phi_2-\phi_1)} \rangle \\ \langle a_1 a_2 e^{-i(\phi_2-\phi_1)} \rangle & \langle a_2^2 \rangle \end{pmatrix} \tag{4.435}$$

The matrix $\mathbf{J}_c$ is called the coherency matrix, where the subscript c is used to distinguish the coherency matrix from the Jones vector and matrix. It is clear that the diagonal elements of $\mathbf{J}_c$ are real, and their sum (the trace of the matrix) is equal to the total intensity of the light, i.e.

$$Tr(\mathbf{J}_c) = J_{xx} + J_{yy} = \langle E_x E_x^* \rangle + \langle E_y E_y^* \rangle = I \tag{4.436}$$

The nondiagonal elements of $\mathbf{J}_c$ are in general complex, and they are conjugates of each other, i.e. $J_{xy} = J_{yx}^*$; therefore, the matrix $\mathbf{J}_c$ is a Hermitian matrix (because of $\mathbf{J}_c = \mathbf{J}_c^{\dagger}$). Thus Eq. (4.433) can be rewritten as

$$I(\theta, \delta) = J_{xx}\cos^2(\theta) + J_{yy}\sin^2(\theta) + e^{i\delta}J_{xy}\cos\theta\,\sin\theta + e^{-i\delta}J_{xy}^*\cos\theta\,\sin\theta \tag{4.437}$$

The elements of $\mathbf{J}_c$ are measurable by measuring the output light intensities $I(\theta, \delta)$ of the system with different retardation δ and polarizer angle θ, which may be selected in many different ways. The $I(\theta, \delta)$ measurements suggested by Wolf are

$$\begin{aligned}
J_{xx} &= I(0,0) \\
J_{yy} &= I\left(\frac{\pi}{2}, 0\right) \\
J_{xy} &= \frac{1}{2}\left[I\left(\frac{\pi}{4}, 0\right) - I\left(\frac{3\pi}{4}, 0\right)\right] - \frac{1}{2}i\left[I\left(\frac{\pi}{4}, \frac{\pi}{2}\right) - I\left(\frac{3\pi}{4}, \frac{\pi}{2}\right)\right] \\
J_{yx} &= \frac{1}{2}\left[I\left(\frac{\pi}{4}, 0\right) - I\left(\frac{3\pi}{4}, 0\right)\right] + \frac{1}{2}i\left[I\left(\frac{\pi}{4}, \frac{\pi}{2}\right) - I\left(\frac{3\pi}{4}, \frac{\pi}{2}\right)\right]
\end{aligned} \tag{4.438}$$

where $J_{ij} = \left\langle E_i E_j^* \right\rangle$. $I(0,0), I\left(\frac{\pi}{4}, 0\right), I\left(\frac{\pi}{2}, 0\right)$, and $I\left(\frac{3\pi}{4}, 0\right)$ are the intensities of light after passing through a zero-retardation plate and a polarizer with an orientation angle θ of 0, $\pi/4$, $\pi/2$, and $3\pi/4$, respectively. One may notice that only a polarizer is needed to obtain J_{xx}, J_{yy}, and the real part of J_{xy} (or J_{yx}). $I\left(\frac{\pi}{4}, \frac{\pi}{2}\right)$ and $I\left(\frac{3\pi}{4}, \frac{\pi}{2}\right)$ are the measured intensities of light after passing through a QWP ($\Delta = \frac{\pi}{2}$) followed by a polarizer at ±45°, corresponding to a right-handed circular polarizer and left-handed polarizer. These two measurements determine the imaginary part of J_{xy}. Although six measurements are suggested earlier, in principle, only four intensity measurements are necessary to determine the four parameters of the coherency matrix, for example, the four measurements using three linear polarizers orientated at 0, $\pi/4$, $\pi/2$, and one right-handed circular polarizer, respectively. From Eq. (4.437), one may obtain four equations after doing these four measurements and find the corresponding solutions to be

$$\begin{aligned}
J_{xx} &= I(0,0) \\
J_{yy} &= I\left(\frac{\pi}{2}, 0\right) \\
Re(J_{xy}) &= -\frac{1}{2}I(0,0) - \frac{1}{2}I\left(\frac{\pi}{2}, 0\right) + I\left(\frac{\pi}{4}, 0\right) \\
Im(J_{xy}) &= \frac{1}{2}I(0,0) + \frac{1}{2}I\left(\frac{\pi}{2}, 0\right) - I\left(\frac{\pi}{4}, \frac{\pi}{2}\right)
\end{aligned} \tag{4.439}$$

4.7.2.1 Completely Unpolarized Light

Completely unpolarized light, sometimes also called natural light, contains waves polarized in all directions in the plane perpendicular to the beam direction with the same probability such that any retardation plate in the optical path doesn't change this symmetry and a polarizer simply cuts the intensity in half regardless of its azimuth angle. This means that the measured light intensity (4.437) is always constant for retardation plate and any polarizer orientation, i.e.

$$I(\theta, \delta) = const \tag{4.440}$$

From Eq. (4.438), one can find the non-diagonal elements of its coherency matrix are zero ($J_{xy} = J_{yx} = 0$), and the diagonal elements are equal, i.e. $J_{xx} = J_{yy}$. Substituting these relationships and $I_0 = J_{xx} + J_{yy}$ in (4.433), one obtains the coherency matrix of a completely unpolarized light as

$$\mathbf{J}_{c,unpol} = \frac{1}{2} I_0 \begin{pmatrix} 1 & 0 \\ 0 & 1 \end{pmatrix} \tag{4.441}$$

4.7.2.2 Completely Polarized Light

As discussed in (4.431), the x and y components of the electric field of a quasi-monochromatic light are

$$E_x = a_1(t)e^{i[2\pi\bar{\nu}t - \phi_1(t)]} \text{ and } E_y = a_2(t)e^{i[2\pi\bar{\nu}t - \phi_2(t)]} \tag{4.442}$$

In general, a_1, a_2, ϕ_1, and ϕ_2 are time dependent. If the light is completely polarized, its polarization state should be time-independent; this means the ratio of amplitudes $|E_x|/|E_y|$ and the phase difference should be constant with time. Consequently, we have

$$\frac{a_2(t)}{a_1(t)} = q, \quad \Delta\phi = \phi_1(t) - \phi_2(t) = \chi \tag{4.443}$$

where q and χ are constants, then the coherency matrix $\mathbf{J}_c$ of Eq. (4.437) becomes

$$\mathbf{J}_{c,pol} = \begin{pmatrix} J_{xx} = \langle a_1^2(t) \rangle & J_{xy} = q \langle a_1^2(t) \rangle e^{i\chi} \\ J_{yx} = q \langle a_1^2(t) \rangle e^{-i\chi} & J_{yy} = q^2 \langle a_1^2(t) \rangle \end{pmatrix} \tag{4.444}$$

It can be found from Eq. (4.430) that

$$det(\mathbf{J}_c) = J_{xx}J_{yy} - J_{xy}J_{yx} = 0 \tag{4.445}$$

i.e. the determinant of the coherency matrix is zero when light is completely polarized. In comparison, from Eq. (4.441), one finds that the determinant of natural light is 1. Therefore, one may judge the degree of polarization (DOP) by checking the determinant of the coherency matrix. When $det(\mathbf{J}_c) = 1$, light is completely unpolarized; when $det(\mathbf{J}_c) = 0$, light is completely polarized; when $0 < det(\mathbf{J}_c) < 1$, light is partially polarized.

4.7.2.3 Partially Polarized Light and Degree of Polarization

A partially polarized light can be regarded as the sum of a completely unpolarized light and a completely polarized light, which are independent of each other and the decomposition is unique. Mathematically a partially polarized light can be represented as

$$\mathbf{J}_c = \mathbf{J}_{c,unpol} + \mathbf{J}_{c,pol} \tag{4.446}$$

$$\mathbf{J}_{c,unpol} = \begin{pmatrix} A & 0 \\ 0 & A \end{pmatrix} \text{ and } \mathbf{J}_{c,pol} = \begin{pmatrix} B & D \\ D^* & C \end{pmatrix}$$

With $A \geq 0, B \geq 0, C \geq 0$, and $BC - DD^* = 0$

where $\mathbf{J}_c$ is the coherency matrix of quasi-monochromatic light under investigation; $\mathbf{J}_{c,\,unpol}$ represents the completely unpolarized portion with the form obtained in (4.441); $\mathbf{J}_{c,\,pol}$ represents the completely polarized portion with the form of (4.444). In order to quantitatively express the percentage of polarized light in total light, the term DOP is defined, which is ratio of the intensity I_{pol} of the polarized portion to the total intensity I_{total}:

$$DOP = \frac{I_{pol}}{I_{total}} \tag{4.447}$$

In the following, we derive the general formula to calculate DOP without the need for decomposing $\mathbf{J}_c$ to A, B, C, and D. From Eq. (4.436) and the definition (4.446), we can obtain the following relationships

$$\begin{aligned} I_{total} &= Tr(\mathbf{J}_c) = J_{xx} + J_{yy} = (A+B) + (A+C) = 2A + B + C \\ det(\mathbf{J}_c) &= J_{xx}J_{yy} - J_{xy}J_{xy}^* = (A+B)(A+C) - DD^* = A^2 + A(B+C) \\ I_{pol} &= Tr(\mathbf{J}_{pol}) = B + C \end{aligned} \tag{4.448}$$

Taking the square of both sides of the first equation in (4.448), one gets

$$I_{total}^2 = 4det(\mathbf{J}_c) + {I_{pol}}^2 \tag{4.449}$$

Therefore, the DOP defined in (4.447) can be written as

$$DOP = \frac{I_{pol}}{I_{total}} = \frac{\sqrt{I_{total}^2 - 4det(\mathbf{J}_c)}}{I_{total}} = \sqrt{1 - \frac{4det(\mathbf{J}_c)}{Tr^2(\mathbf{J}_c)}} \tag{4.450}$$

For natural light, according to (4.441), $4det(\mathbf{J}_c) = I_0^2$ and $Tr^2(\mathbf{J}_c) = I_0^2$, so that its DOP is equal to zero. For a completely polarized light, the determinant of corresponding coherency matrix equals zero (see (4.445)), resulting in a DOP of 1.

4.7.2.4 Coherency Matrix of the Superposition of Individual Waves

Let's consider the case that several light waves propagating in the same direction are superposed. Assuming $E_x^{(i)}$ and $E_y^{(i)}$ $(i = 1, 2, \ldots, n)$ are the components of the electric field of the individual waves, the components of the electric vector of the resulting wave are

$$E_x = \sum_{i=1}^{n} E_x^{(i)}, \quad E_y = \sum_{i=1}^{n} E_y^{(i)} \tag{4.451}$$

From the definition of coherency matrix in (4.435), the elements of its coherency matrix are given by

$$\begin{aligned} J_{kl} &= \left\langle \sum_{i=1}^{n} E_k^{(i)} \sum_{j=1}^{n} \left[E_l^{(j)}\right]^* \right\rangle \\ &= \sum_{i=1}^{n} \left\langle E_k^{(i)} \left[E_l^{(i)}\right]^* \right\rangle + \sum_{i=1}^{n} \sum_{j \neq i, j=1}^{n} \left\langle E_k^{(i)} \left[E_l^{(j)}\right]^* \right\rangle \end{aligned} \tag{4.452}$$

where $k = x, y$ and $l = x, y$. If the waves are independent of each other, each term $\left\langle E_k^{(i)} \left[E_l^{(j)}\right]^* \right\rangle$ with $i \neq j$ is zero, Eq. (4.452) can be simplified to

$$J_{kl} = \left\langle \sum_{i=1}^{n} E_k^{(i)} \left[E_l^{(i)}\right]^* \right\rangle = \sum_{i=1}^{n} J_{kl}^{(i)} \tag{4.453}$$

where $J_{kl}^{(i)}$ are the elements of the coherency matrix of the ith wave. Equation (4.453) shows that the coherency matrix of the combined wave is equal to the sum of the coherency matrices of all the individual waves.

4.7.2.5 Superposition of Two Individual Waves with Mutually Orthogonal Polarizations

Let's consider the light with intensity I_0, which is superposed by two individual light waves with equal intensity and mutually orthogonal polarizations. Assume the Jones vector of one wave is $\mathbf{E}_1 = \frac{I_0}{2}(cos\alpha, sin\alpha e^{i\delta})^T$, then the polarization of another wave will be $\mathbf{E}_2 = \frac{I_0}{2}(sin\alpha e^{-i\delta}, -cos\alpha)^T$ (see formula (4.13)). According to the definition of the coherency matrix (4.435), the coherency matrices $\mathbf{J}_1$ and $\mathbf{J}_2$ of these two individual waves are

$$\mathbf{J}_1 = \frac{I_0}{2}\begin{pmatrix} cos^2\alpha & cos\alpha\, sin\alpha e^{-i\delta} \\ cos\alpha\, sin\alpha e^{i\delta} & sin^2\alpha \end{pmatrix} \tag{4.454}$$

$$\mathbf{J}_2 = \frac{I_0}{2}\begin{pmatrix} sin^2\alpha & -cos\alpha\, sin\alpha e^{-i\delta} \\ -cos\alpha\, sin\alpha e^{i\delta} & cos^2\alpha \end{pmatrix} \tag{4.455}$$

From Eq. (4.455), we get the coherency matrix of the superposition of these two waves as

$$\mathbf{J} = \mathbf{J}_1 + \mathbf{J}_2 = \frac{I_0}{2}\begin{pmatrix} 1 & 0 \\ 0 & 1 \end{pmatrix} \tag{4.456}$$

By comparing (4.456) and (4.441), $\mathbf{J}$ has the same format with the coherency matrix of natural light, i.e. a wave of natural light, of intensity I_0, is equivalent to two independent orthogonally polarized waves, each of intensity $I_0/2$. For example, one may choose a pair of horizontally and vertically polarized light beams or right-handed and left-handed circularly polarized light beams to make a non-polarized light beam.

4.7.3 The Stokes Parameters of Quasi-monochromatic Plane Wave

As discussed in Section 4.6.1, a quasi-monochromatic plane wave may be characterized by four real parameters, for example, $J_{xx}, J_{yy}, Re(J_{xy})$, and $Im(J_{xy})$. In his investigation of partially polarized light, Stokes introduced four Stokes parameters to represent the polarization states. For monochromatic light, the definition of the Stokes parameters is given in Chapter 2. For quasi-monochromatic wave, the Stokes parameters become

$$\begin{aligned} S_0 &= \langle a_1^2 \rangle + \langle a_2^2 \rangle \\ S_1 &= \langle a_1^2 \rangle - \langle a_2^2 \rangle \\ S_2 &= 2\langle a_1 a_2 cos(\delta)\rangle \\ S_3 &= 2\langle a_1 a_2 sin(\delta)\rangle \end{aligned} \tag{4.457}$$

where a_1, a_2, and $\delta = \phi_1 - \phi_2$ are the instantaneous parameters of quasi-monochromatic wave, which are defined in (4.431). Comparing the definitions (4.435) and (4.457), one obtains the following transformation between coherency matrix and the Stokes parameters:

$$\begin{aligned} S_0 &= J_{xx} + J_{yy} \\ S_1 &= J_{xx} - J_{yy} \\ S_2 &= J_{xy} + J_{yx} \\ S_3 &= i(J_{xy} - J_{yx}) \end{aligned} \tag{4.458}$$

and

$$\begin{aligned} J_{xx} &= \frac{1}{2}(S_0 + S_1) \\ J_{yy} &= \frac{1}{2}(S_0 - S_1) \\ J_{xy} &= \frac{1}{2}(S_2 - iS_3) \\ J_{yx} &= \frac{1}{2}(S_2 + iS_3) \end{aligned} \tag{4.459}$$

Like the coherency matrix can be determined from optical power measurements (see relationship (4.438)), the Stokes parameters can also be represented by optical power measurements. Substitution of Eq. (4.438) in Eq. (4.458) yields

$$\begin{aligned} S_0 &= I(0,0) + I\left(\frac{\pi}{2},0\right) \\ S_1 &= I(0,0) - I\left(\frac{\pi}{2},0\right) \\ S_2 &= I\left(\frac{\pi}{4},0\right) - I\left(\frac{3\pi}{4},0\right) \\ S_3 &= I\left(\frac{\pi}{4},\frac{\pi}{2}\right) - I\left(\frac{3\pi}{4},\frac{\pi}{2}\right) \end{aligned} \tag{4.460}$$

where $I(\theta, \delta)$ are defined in Figure 4.24 and Eq. (4.437). $I(0, 0)$, $\left(\frac{\pi}{4},0\right)$, $I\left(\frac{\pi}{2},0\right)$, and $I\left(\frac{3\pi}{4},0\right)$ are the measured intensities when light passes a linear polarizer with transmission direction at 0, $\pi/4$, $\pi/2$, and $\frac{3\pi}{4}$, respectively. $I\left(\frac{\pi}{4},\frac{\pi}{2}\right)$ and $I\left(\frac{3\pi}{4},\frac{\pi}{2}\right)$ are the measured intensities when the light passes right-hand circular and left-hand circular polarizers, respectively. Equation (4.439) indicates that four measurements are required to determine the coherency matrix after light passing through three linear polarizers orientated to 0, $\pi/4$, $\pi/2$, and one right-hand circular polarizer, respectively. As with the coherency matrix, the Stokes can be determined by these four measurements. Substituting (4.439) in (4.458), one can find

$$\begin{aligned} S_0 &= I(0,0) + I\left(\frac{\pi}{2},0\right) \\ S_1 &= I(0,0) - I\left(\frac{\pi}{2},0\right) \\ S_2 &= -I(0,0) - I\left(\frac{\pi}{2},0\right) + 2I\left(\frac{\pi}{4},0\right) \\ S_3 &= -I(0,0) - I\left(\frac{\pi}{2},0\right) + I\left(\frac{\pi}{4},\frac{\pi}{2}\right) \end{aligned} \tag{4.461}$$

It can also be rearranged to a matrix relation:

$$\begin{pmatrix} S_0 \\ S_1 \\ S_2 \\ S_3 \end{pmatrix} = \begin{pmatrix} 1 & 1 & 0 & 0 \\ 1 & -1 & 0 & 0 \\ -1 & -1 & 2 & 0 \\ -1 & -1 & 0 & 2 \end{pmatrix} \begin{pmatrix} I(0,0) \\ I\left(\frac{\pi}{2},0\right) \\ I\left(\frac{\pi}{4},0\right) \\ I\left(\frac{\pi}{4},\frac{\pi}{2}\right) \end{pmatrix} \tag{4.462}$$

Let us now consider the properties of the Stokes parameters with different degrees of polarization.

4.7.3.1 Completely Unpolarized Light

When light is completely unpolarized, all of six measurements in (4.460) are the same, similar to that in (4.440), so that $S_0 = I(0,0) + I\left(\frac{\pi}{2},0\right) = J_{xx} + J_{yy}$, which is the light intensity, and $S_1 = S_2 = S_3 = 0$.

4.7.3.2 Completely Polarized Light

When light is completely polarized, the following relationship can be obtained from Eq. (4.458)

$$S_0^2 = J_{xx}^2 + J_{yy}^2 + 2J_{xx}J_{yy} \tag{4.463}$$

$$S_1^2 + S_2^2 + S_3^2 = J_{xx}^2 + J_{yy}^2 - 2J_{xx}J_{yy} + 4J_{xy}J_{yx} \tag{4.464}$$

From (4.445), $J_{xx}J_{yy} = J_{xy}J_{yx}$ for a completely polarized light; hence Eq. (4.464) can be rewritten as

$$S_1^2 + S_2^2 + S_3^2 = J_{xx}^2 + J_{yy}^2 + 2J_{xx}J_{yy} \tag{4.465}$$

Comparing (4.463) and (4.465), we obtain the following relationship for completely polarized light:

$$S_0^2 = S_1^2 + S_2^2 + S_3^2 \text{ or } S_0 = \sqrt{S_1^2 + S_2^2 + S_3^2} \tag{4.466}$$

4.7.3.3 Partially Polarized Light

For partially polarized light, one can decompose Stokes parameters to

$$\begin{pmatrix} S_0 \\ S_1 \\ S_2 \\ S_3 \end{pmatrix} = \begin{pmatrix} S_0 - \sqrt{S_1^2 + S_2^2 + S_3^2} \\ 0 \\ 0 \\ 0 \end{pmatrix} + \begin{pmatrix} \sqrt{S_1^2 + S_2^2 + S_3^2} \\ S_1 \\ S_2 \\ S_3 \end{pmatrix} \tag{4.467}$$

The first term on the left hand side represents the Stokes parameters of completely unpolarized portion of the light, and the second term describes the Strokes parameters of completely polarized portion; therefore the DOP of the light can be obtained as

$$DOP = \frac{I_{polarized}}{I_{total}} = \frac{\sqrt{S_1^2 + S_2^2 + S_3^2}}{S_0} \tag{4.468}$$

This relationship can also be derived from the coherency matrix representation (4.450) of DOP. Substituting (4.459) in $det(\mathbf{J}_c) = J_{xx}J_{yy} - J_{xy}J_{yx}$, we can obtain

$$det(\mathbf{J}_c) = \frac{1}{4}\left(S_0^2 - S_1^2 - S_2^2 - S_3^2\right) \tag{4.469}$$

Finally, substituting (4.469) in the DOP formula (4.450), one arrives at

$$DOP = \frac{\sqrt{I_{total}^2 - 4det(\mathbf{J}_c)}}{I_{total}} = \frac{\sqrt{S_1^2 + S_2^2 + S_3^2}}{S_0} \tag{4.470}$$

This is the same relationship given in (4.468). For completely unpolarized light, DOP is 0 because $S_1 = S_2 = S_3 = 0$; for a completely polarized light, DOP is 1 because $S_0 = \sqrt{S_1^2 + S_2^2 + S_3^2}$; for a partial polarized light, $0 < \sqrt{S_1^2 + S_2^2 + S_3^2} < S_0$, the corresponding DOP has a value between zero and one.

4.7.4 Depolarization of Polychromatic Plane Wave by Birefringence Media

In this section we consider the polarization change after a polychromatic plane wave passing through a birefringence device. In fiber optical applications, the light at the output of a light source is generally completely polarized, i.e. all of the frequency components of light have the same polarization state. After passing through a birefringence media, the phase difference $\Delta\phi(\upsilon)$ between the x and y components shifts $2\pi\upsilon\frac{\Delta nl}{c}$ due to the birefringence. This phase shift $\Delta\phi(\upsilon)$ variations are frequency dependent, resulting in the polarization states of different frequency components not being the same anymore, i.e. the light becomes partially polarized.

In measurement, we are unable to follow the instantaneous amplitude variation of a light wave, and only its power is measurable. Therefore, one has to describe polarization states of polychromatic light using measurable optical power, like coherency matrix and Stokes parameters. In the following, we will first introduce the mathematical relationship between complex electric field and power spectral density (PSD). As an example of PSD's application, we then discuss the SOP and the DOP change when a polychromatic light propagates through a birefringence media.

4.7.4.1 Power Spectral Density

The instantaneous intensity of a light wave is proportional to the square of its electric field $|E|^2$. Omitting the factor of proportionality, the intensity of light can be written as $I = |E|^2$. In practice, the integration time T of a photodetector to measure the power light is much longer than the oscillation period of the light wave (~fs). The observed power intensity of light is the time average of instantaneous optical intensity. Let's consider a polychromatic plane wave propagating a homogeneous medium in the direction $+z$. Assuming the real transverse electric field at a given position is

$$\mathbf{E}^{(r)}(t) = E_x^{(r)}(t)\hat{\mathbf{e}}_x + E_y^{(r)}(t)\hat{\mathbf{e}}_y \tag{4.471}$$

Then, the light intensity of the x and y components can be expressed as the following time average

$$I_h = \left\langle E_h^{(r)}(t)E_h^{(r)}(t)\right\rangle = \lim_{T\to\infty}\frac{1}{T}\int_{-\frac{T}{2}}^{\frac{T}{2}} E_h^{(r)2}(t)dt \quad \text{where } h = x, y \tag{4.472}$$

In order to apply continuous Fourier transform on the electric field, we define a truncated field $E_{T,h}^{(r)}(t)$ as

$$E_{T,h}^{(r)}(t) = \begin{cases} E_h^{(r)}(t) & |t| \le \frac{T}{2} \\ 0 & |t| > \frac{T}{2} \end{cases} \tag{4.473}$$

The truncated field $E_{T,h}^{(r)}(t)$ is square-integrable; therefore, it can be expressed in the form of a Fourier integral

$$E_{T,h}^{(r)}(t) = \int_{-\infty}^{\infty} \varepsilon_h(\upsilon)e^{i2\pi\upsilon t}d\upsilon \tag{4.474}$$

and the magnitude $\varepsilon_h(\upsilon)$ of a frequency component can be obtained from the inverse Fourier transform of $E_{T,h}^{(r)}(t)$

$$\varepsilon_h(\upsilon) = \int_{-\infty}^{\infty} E_{T,h}^{(r)}(t)e^{-i2\pi\upsilon t}dt \tag{4.475}$$

According to (4.422)–(4.424), and letting $E_{T,h}^{(i)}(t)$ be the associated function of $E_{T,h}^{(r)}(t)$, one obtains the corresponding complex analytic signal

$$E_{T,h}(t) = E_{T,h}^{(r)}(t) + iE_{T,h}^{(i)}(t) = 2\int_{0}^{\infty} \varepsilon_h(\upsilon)e^{i2\pi\upsilon t}d\upsilon \tag{4.476}$$

Substituting (4.473)–(4.475) in (4.472), we obtain the intensity of light

$$I_h = 2\lim_{T\to\infty}\frac{1}{T}\int_{-\infty}^{\infty} E_{T,h}^{(r)}(t)\left[E_{T,h}^{(r)}(t)\right]^* dt = 2\lim_{T\to\infty}\frac{1}{T}\int_{-\infty}^{\infty} E_{T,h}^{(r)}(t)\left[\int_{-\infty}^{\infty}\varepsilon_h^*(\upsilon)e^{-i2\pi\upsilon t}d\upsilon\right]dt$$

$$= 2\lim_{T\to\infty}\frac{1}{T}\int_{-\infty}^{\infty}\varepsilon_h^*(\upsilon)\left[\int_{-\infty}^{\infty} E_{T,h}^{(r)}(t)e^{-i2\pi\upsilon t}dt\right]d\upsilon$$

$$= 2\lim_{T\to\infty}\frac{1}{T}\int_{-\infty}^{\infty}\varepsilon_h(\upsilon)\varepsilon_h^*(\upsilon)d\upsilon = 4\int_0^{\infty} S_h(\upsilon)d\upsilon \tag{4.477}$$

where

$$S_h(\upsilon) = \lim_{T\to\infty}\frac{|\varepsilon_h(\upsilon)|^2}{T} \tag{4.478}$$

represents the PSD, which is defined as the optical power per unit optical frequency interval and can be measured by an optical spectrometer. Two typical spectral profiles of lasers are

$$\text{Gaussian spectrum}\qquad S_h(\upsilon) \sim \frac{2\sqrt{ln2}}{\sqrt{\pi}\Delta\upsilon}e^{-\left[\frac{2\sqrt{ln2}}{\Delta\upsilon}(\upsilon-\upsilon_0)\right]^2} \tag{4.479}$$

$$\text{Lorentzian spectrum}\qquad S_h(\upsilon) \sim \frac{1}{\pi}\frac{\Delta\upsilon/2}{(\upsilon-\upsilon_0)^2+(\Delta\upsilon/2)^2} \tag{4.480}$$

A basic property of spectral density follows Parseval's theorem and has the following relationships

$$\int_{-\infty}^{\infty}\left|E_{T,h}^{(r)}(t)\right|^2 dt = \int_{-\infty}^{\infty}\left|E_{T,h}^{(i)}(t)\right|^2 dt = \frac{1}{2}\int_0^{\infty} E_{T,h}(t)E_{T,h}^*(t)dt = 2\int_{-\infty}^{\infty}|\varepsilon_h(\upsilon)|^2 d\upsilon \tag{4.481}$$

Substituting (4.481) into (4.477), we get the following relationship between light intensity and $E_{T,h}^{(r)}$, $E_{T,h}^{(i)}$, and $E_{T,h}(t)$:

$$I_h = 2\left\langle\left|E_{T,h}^{(r)}(t)\right|^2\right\rangle = 2\left\langle\left|E_{T,h}^{(i)}(t)\right|^2\right\rangle = \left\langle|E_{T,h}(t)|^2\right\rangle = 4\int_0^{\infty} S_h(\upsilon)d\upsilon \tag{4.482}$$

4.7.4.2 Polarization State of a Polychromatic Light After Passing Through a Birefringence Medium

Let's consider a completely polarized polychromatic light passing through a birefringence medium with a length of l, a pair of principal axes along x and y-axes, and the corresponding refractive indexes of n_x and n_y.

As shown in Figure 4.37, assuming the complex electrical vector of incident light at $z = 0$ is

$$\mathbf{E}(z=0) = E_{T,x}(t,0)\hat{\mathbf{e}}_x + E_{T,y}(t,0)\hat{\mathbf{e}}_y \tag{4.483}$$

where $\hat{\mathbf{e}}_x$ and $\hat{\mathbf{e}}_y$ are the unit vector along $+x$ and $+y$ axes, and $E_{T,x}(t,0)$ and $E_{T,y}(t,0)$ are the electrical field's complex components along x and y directions at position $z = 0$. According to (4.476), they can be expressed by the following Fourier integral

$$E_{T,h}(t,0) = 2\int_0^{\infty}\varepsilon_h(\upsilon)e^{i2\pi\upsilon t}d\upsilon,\quad h = x, y \tag{4.484}$$

Since the incident light is completely polarized, the ratio between the x and y components of the electric field, $\varepsilon_y(\upsilon)/\varepsilon_x(\upsilon)$, is a complex constant and is independent of the optical frequency υ, i.e.

$$\frac{\varepsilon_y(\upsilon)}{\varepsilon_x(\upsilon)} = \frac{sin\alpha}{cos\alpha}e^{i\varphi_0}\quad\text{where } 0 \le \alpha \le \frac{\pi}{2} \tag{4.485}$$

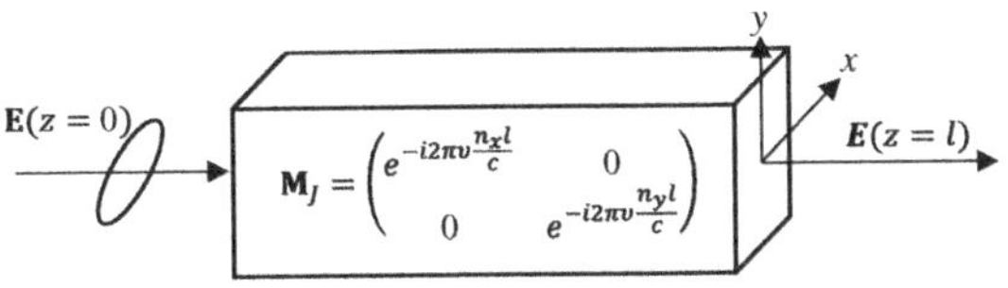

$\mathbf{E}(z=0)$ $\quad \mathbf{M}_J = \begin{pmatrix} e^{-i2\pi\upsilon\frac{n_x l}{c}} & 0 \\ 0 & e^{-i2\pi\upsilon\frac{n_y l}{c}} \end{pmatrix}$ $\quad y \quad x \quad \mathbf{E}(z=l)$

$\mathbf{E}(z=0) = E_{Tx}(z=0)\hat{\mathbf{e}}_x + E_{Ty}(z=0)\hat{\mathbf{e}}_y$

$E_{Tx}(z=0) = \cos\alpha\left(2\int_0^\infty \varepsilon(\upsilon)\, e^{i2\pi\upsilon t} d\upsilon\right)$

$E_{Ty}(z=0) = 2\sin\alpha\, e^{i\varphi_0}\left(2\int_0^\infty \varepsilon(\upsilon)\, e^{i2\pi\upsilon t} d\upsilon\right)$

$\mathbf{E}(z=l) = E_{Tx}(z=l)\hat{\mathbf{e}}_x + E_{Ty}(z=l)\hat{\mathbf{e}}_y$

$E_{Tx}(z=l) = 2\cos\alpha\int_0^\infty \varepsilon(\upsilon)\, e^{-i2\pi\upsilon\frac{n_x l}{c}} e^{i2\pi\upsilon t} d\upsilon$

$E_{Ty}(z=l) = 2\sin\alpha\, e^{i\varphi_0}\int_0^\infty \varepsilon(\upsilon)\, e^{-i2\pi\upsilon\frac{n_x l}{c}} e^{i2\pi\upsilon t} d\upsilon$

Figure 4.37 A polychromatic light passes through a linear birefringence medium with a length of l. At $z = 0$, the incident light is completely polarized; at $z = l$, the birefringence causes the differential phase shifts of different frequency components to vary with length, i.e. the polarization states of different frequency components are different and the light is depolarized by the birefringence medium.

Let $\varepsilon_x(\upsilon) = \cos\alpha\varepsilon(\upsilon)$, the x- and y-components of electrical field $\mathbf{E}(z=0)$ can be written as

$$E_{Tx}(t,0) = \cos\alpha E_T(t) \tag{4.486}$$

$$E_{Ty}(t,0) = \sin\alpha e^{i\varphi_0} E_T(t) \tag{4.487}$$

where

$$E_T(t) = 2\int_0^\infty \varepsilon(\upsilon)(\upsilon)e^{i2\pi\upsilon t} d\upsilon \tag{4.488}$$

From (4.482), one gets the relationship between the total power P_0 and complex electric field $E_T(t)$ as

$$P_0 = P_x + P_y = \langle |E_{Tx}(t,0)|^2\rangle + \langle |E_{Ty}(t,0)|^2\rangle = \langle |E_T(t)|^2\rangle = 4\int_0^\infty S(\upsilon)d\upsilon \tag{4.489}$$

After light passes through the birefringence medium, phase shifts $\varphi_x = -i2\pi\upsilon\tau_x$ and $\varphi_y = -i2\pi\upsilon\tau_y$ are introduced to all of the frequency components of the x and y components of the electrical field, which can be expressed as

$$E_{Tx}(z=l) = 2\cos\alpha\int_0^\infty \varepsilon(\upsilon)e^{i2\pi\upsilon(t-\tau_x)} d\upsilon = \cos\alpha E_T(t-\tau_x) \tag{4.490}$$

$$E_{Ty}(z=l) = 2\sin\alpha e^{i\varphi_0}\int_0^\infty \varepsilon(\upsilon)e^{i2\pi\upsilon(t-\tau_y)} d\upsilon = \sin\alpha e^{i\varphi_0} E_T(t-\tau_y) \tag{4.491}$$

where $\tau_x = n_x l/c$ and $\tau_y = n_y l/c$ are the times required for the x and y polarization components transmitting through the medium with a length of l, respectively.

According to (4.435), the coherency matrix at $z = l$ can be written as

$$\mathbf{J}_c = \begin{pmatrix} J_{xx} & J_{xy} \\ J_{yx} & J_{yy} \end{pmatrix} = \begin{pmatrix} \langle E_x E_x^*\rangle & \langle E_x E_y^*\rangle \\ \langle E_y E_x^*\rangle & \langle E_x E_y^*\rangle \end{pmatrix} \tag{4.492}$$

Substituting (4.486)–(4.492) and using (4.491), we can obtain the elements of the coherency matrix as following.

$$J_{xx} = \langle E_x E_x^*\rangle = \cos^2\alpha P_0 \tag{4.493}$$

$$J_{yy} = \langle E_y E_y^*\rangle = \sin^2\alpha P_0 \tag{4.494}$$

$$J_{xy} = \langle E_x E_y^* \rangle = cos\alpha\ sin\alpha e^{-i\varphi_0} \langle E_T(t-\tau_x)\, E_T^*(t-\tau_y) \rangle$$
$$= cos\alpha\ sin\alpha e^{-i\varphi_0} \langle E_T(t+\Delta\tau) E_T^*(t) \rangle \tag{4.495}$$

$$J_{yx} = E_y E_x^* = cos\alpha\ sin\alpha e^{i\varphi_0} \langle E_T(t-\tau_y) E_T^*(t-\tau_x) \rangle$$
$$= cos\alpha\ sin\alpha e^{i\varphi_0} \langle E_T(t)\, E_T^*(t+\Delta\tau) \rangle \tag{4.496}$$

where $\Delta\tau = \tau_y - \tau_x$ and the correlation function $\langle E_T(t)\, E_T^*(t+\Delta\tau) \rangle$ is

$$\begin{aligned} \langle E_T(t+\Delta\tau)E_T^*(t) \rangle &= \lim_{T\to\infty} \frac{1}{T} \int_{-\infty}^{\infty} \left[2\int_0^{\infty} \varepsilon(\upsilon) e^{i2\pi\upsilon(t+\Delta\tau)} d\upsilon \right] \left[2\int_0^{\infty} \varepsilon^*(\upsilon') e^{-i2\pi\upsilon' t} d\upsilon' \right] dt \\ &= \lim_{T\to\infty} \frac{4}{T} \int_0^{\infty} \int_0^{\infty} \varepsilon(\upsilon)\varepsilon^*(\upsilon') e^{i2\pi\upsilon\Delta\tau} \left[\int_{-\infty}^{\infty} e^{i2\pi(\upsilon-\upsilon')t} dt \right] d\upsilon d\upsilon' \\ &= \lim_{T\to\infty} \frac{4}{T} \int_0^{\infty} \varepsilon(\upsilon)\varepsilon^*(\upsilon') e^{i2\pi\upsilon\Delta\tau} \delta(\upsilon-\upsilon') d\upsilon d\upsilon' = 4\int_0^{\infty} S(\upsilon) e^{i2\pi\upsilon\Delta\tau} d\upsilon \end{aligned} \tag{4.497}$$

where $\delta(\upsilon - \upsilon')$ is the Dirac delta function. Define the self-correlation function $\gamma(\Delta\tau)$

$$\gamma(\Delta\tau) = \frac{\langle E_T(t+\Delta\tau)E_T^*(t) \rangle}{\langle E_T(t)E_T^*(t) \rangle} = \frac{\langle E_T(t+\Delta\tau)E_T^*(t) \rangle}{P_0} \tag{4.498}$$

Substituting (4.493)–(4.495), and (4.499) in the coherency matrix in (4.492), one obtains

$$\mathbf{J}_c = P_0 \begin{pmatrix} cos^2\alpha & cos\alpha\ sin\alpha e^{-i\varphi_0}\gamma(\Delta\tau) \\ cos\alpha\ sin\alpha e^{i\varphi_0}\gamma^*(\Delta\tau) & sin^2\alpha \end{pmatrix} \tag{4.499}$$

From Eq. (4.450), the *DOP* of the emerging light is

$$DOP = \sqrt{1 - \frac{4det(\mathbf{J}_c)}{Tr^2(\mathbf{J}_c)}} = \sqrt{(cos2\alpha)^2 + (sin2\alpha)^2\gamma(\Delta\tau)\gamma^*(\Delta\tau)} \tag{4.500}$$

Substituting the coherency matrix (4.499) into Eq. (4.458), one finds the Stokes parameters of emerging light as

$$\mathbf{SOP} = P_0 \begin{pmatrix} 1 \\ cos2\alpha \\ sin2\alpha Re\left(e^{-i\varphi_0}\gamma(\Delta\tau)\right) \\ -sin2\alpha Im\left(e^{-i\varphi_0}\gamma(\Delta\tau)\right) \end{pmatrix} \tag{4.501}$$

When electric field of incident light has equal x and y components, for example, the incident light is +45° linearly polarized, Eqs. (4.499)–(4.501) can be reduced to

$$\mathbf{J}_c = \frac{P_0}{2} \begin{pmatrix} 1 & e^{-i\varphi_0}\gamma(\Delta\tau) \\ e^{i\varphi_0}\gamma^*(\Delta\tau) & 1 \end{pmatrix} \tag{4.502}$$

$$DOP = |\gamma(\Delta\tau)| \tag{4.503}$$

$$\mathbf{SOP} = P_0 \begin{pmatrix} 1 \\ 0 \\ Re\left(e^{-i\varphi_0}\gamma(\Delta\tau)\right) \\ -Im\left(e^{-i\varphi_0}\gamma(\Delta\tau)\right) \end{pmatrix} \tag{4.504}$$

4.7.4.3 Polychromatic Light with Rectangular Spectrum

As an example of the self-correlation function $\gamma(\Delta\tau)$, we firstly consider a polychromatic light source with a rectangular frequency spectrum, which has a center frequency of υ_0 and bandwidth of $\Delta\upsilon_R$. The optical power outside of the bandwidth is zero, and each frequency component has equal power inside the bandwidth; thus, the PSD can be expressed as

$$S_R(\upsilon) = \begin{cases} 1 & |\upsilon - \upsilon_0| \leq \dfrac{\Delta\upsilon_R}{2} \\ 0 & |\upsilon - \upsilon_0| > \dfrac{\Delta\upsilon_R}{2} \end{cases} \tag{4.505}$$

where R denotes the rectangular spectrum. By substituting (4.505) in (4.498), one obtains the following self-correlation function $\gamma_R(\Delta\tau)$

$$\gamma_R(\Delta\tau) = e^{i2\pi\upsilon_0\tau}\frac{sin(\pi\Delta\upsilon\Delta\tau)}{\pi\Delta\upsilon\Delta\tau} \tag{4.506}$$

The absolute value of $\gamma(\tau)$ is plotted in Figure 4.38a. As can be seen, it becomes zero whenever $\Delta\upsilon\Delta\tau$ is an integer, and the corresponding DOP reaches a minimum, as shown in Figure 4.38b. At the first zero, the corresponding delay $\Delta\tau_c$ equals to $1/\Delta\upsilon$ and is called the coherence time of the light source. The product of coherence time $\Delta\tau_c$ and light speed c is called coherence length, which can also be represented as $l_c = c/\Delta\upsilon$. In practice, one often uses wavelength linewidth $\Delta\lambda$ to describe the spectral width. Taking the derivative of both sides of the relation $\upsilon = c/\lambda$, one finds $c/\Delta\upsilon = \lambda^2/\Delta\lambda$, and the coherence length l_c can also be expressed as $l_c = \lambda^2/\Delta\lambda$. In the following, we listed the expressions about coherence time and length for convenience

$$\Delta\tau_c = \frac{1}{\Delta\upsilon} = \frac{1}{c}\frac{\lambda^2}{\Delta\lambda} = \frac{l_c}{c} \tag{4.507}$$

$$l_c = c\Delta\tau = \frac{C}{\Delta\upsilon} = \frac{\lambda^2}{\Delta\lambda} \tag{4.508}$$

Case 1

When $\Delta\upsilon\Delta\tau = 1$, the value of $\gamma(\Delta\tau)$ goes to zero. The Stokes parameters become

$$SOP = P_0\begin{pmatrix} 1 \\ cos2\alpha \\ 0 \\ 0 \end{pmatrix} \tag{4.509}$$

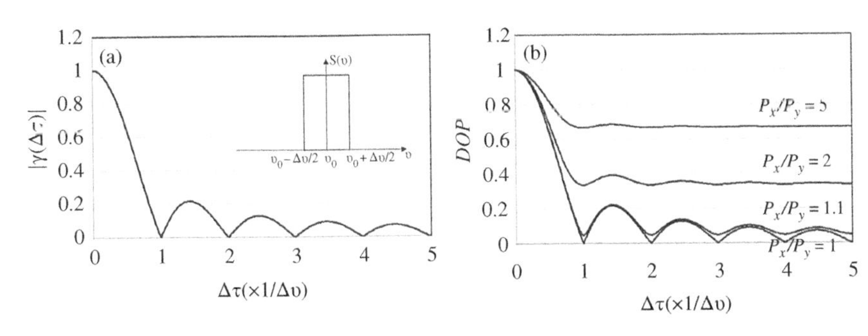

Figure 4.38 (a) Plot of the self-correlation function $|\gamma(\Delta\tau)|$ vs. $\Delta\tau\Delta\upsilon$, with the power spectral density shown in the inset. $|\gamma(\Delta\tau)|$ reaches zero whenever $\Delta\upsilon\Delta\tau$ equals to an integer. (b) DOP of the emerging light vs. $\Delta\tau$ with different power ratios $P_x/P_y = ctan(\alpha)$. $DOP = 0$ when $\gamma(\Delta\tau) = 0$ and $P_x/P_y = 1$ and $\Delta\upsilon\Delta\tau$ equals to an integer, indicating complete depolarization.

And the corresponding DOP becomes

$$DOP = P_0 cos^2 2\alpha \tag{4.510}$$

When $\alpha = \pi/4$, i.e. the x and y components of the electric field have the same power, DOP becomes zero, and the light is completely depolarized.

Case 2: $\Delta\upsilon\Delta\tau \ll 1$

When $\Delta\upsilon\Delta\tau \ll 1$, $\gamma(\Delta\tau)$ in Eq. (4.506) can be approximated to

$$\gamma(\tau) = e^{i2\pi\upsilon_0\Delta\tau} \tag{4.511}$$

Substituting (4.511) into (4.500) and (4.501), one finds the DOP of emerging light equals one, and the corresponding Stokes parameters are reduced to

$$SOP = P_0 \begin{pmatrix} 1 \\ cos2\alpha \\ sin2\alpha\ cos(\varphi_0 - 2\pi\upsilon_0\Delta\tau) \\ sin2\alpha\ sin(\varphi_0 - 2\pi\upsilon_0\Delta\tau) \end{pmatrix} = P_0 \begin{pmatrix} 1 \\ cos2\alpha \\ sin2\alpha\ cos\left(\varphi_0 - 2\pi\dfrac{\Delta nl}{\lambda_0}\right) \\ sin2\alpha\ sin\left(\varphi_0 - 2\pi\dfrac{\Delta nl}{\lambda_0}\right) \end{pmatrix} \tag{4.512}$$

By comparing (4.512) and (4.457), it is clear the Stokes parameters of polychromatic light are reduced to the format of a quasi-monochromatic light when $\Delta\upsilon\Delta\tau \ll 1$. From the relationships (4.507) and (4.508), the condition $\Delta\upsilon\Delta\tau \ll 1$ is equivalent to $\Delta\tau \ll \Delta\tau_c$ or $\Delta l \ll l_c$, i.e. when the optical path length difference between two eigen polarization states is much smaller than the coherence length of the polychromatic light source, the light can be considered quasi-monochromatic, with its frequency represented by its center frequency υ_0.

4.7.4.4 Polychromatic Light with Gaussian or Lorentzian Spectrum

When the polychromatic light source has a Gaussian or Lorentzian shape spectrum, its self-correlation functions can be obtained by substituting Eqs. (4.479) and (4.480) in (4.498)

$$\text{Gaussian spectrum:}\quad \gamma_G(\Delta\tau) = e^{i2\pi\upsilon_0 - \frac{\pi^2}{4ln2}\left(\frac{\Delta\tau}{\tau_c}\right)^2} \quad \text{where } \tau_G = \frac{1}{\Delta\upsilon_G} \tag{4.513}$$

$$\text{Lorentzian spectrum:}\quad \gamma_L(\Delta\tau) = e^{i2\pi\upsilon_0 - \pi\frac{|\Delta\tau|}{\tau_c}} \quad \text{where } \tau_L = \frac{1}{\Delta\upsilon_L} \tag{4.514}$$

where $\Delta\upsilon$ is the full width at half maximum (FWHM) of the power density spectrum (see (4.479)), the subscripts G and L denote for the spectrum with Gaussian or Lorentzian profile, respectively. During derivation, the following integrations are used:

$$\int_{-\infty}^{\infty} e^{-ax^2} e^{i2\pi xt} dx = \sqrt{\frac{\pi}{a}}\, e^{-\pi^2 t^2/a} \tag{4.515}$$

$$\int_{-\infty}^{\infty} \frac{1}{\pi}\frac{\Delta\upsilon/2}{(x - x_0)^2 + (\Delta\upsilon/2)^2} e^{i2\pi xt} dx = e^{i2\pi x_0 t}\, e^{-\pi\Delta\upsilon t} \tag{4.516}$$

Figure 4.39a plots the self-correlation functions $\gamma_G(\Delta\tau)$ and $\gamma_L(\Delta\tau)$ vs. differential time $\Delta\tau$. By substituting (4.513) and (4.514) in (4.500), the DOP curves after light pass through a birefringence crystal and are plotted in Figure 4.39b with different $\Delta\tau$. It is clear that the DOP of emerging light is practically zero when $\Delta\tau > 2/\Delta\upsilon$ and light is equally split to two eigen polarization states at entrance face.

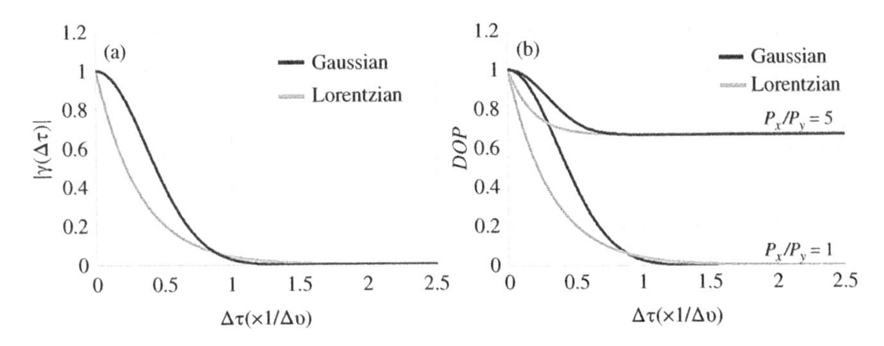

Figure 4.39 (a) Plots of the self-correlation function $|\gamma(\Delta\tau)|$ vs. $\Delta\tau$ of Gaussian and Lorentzian spectra. $|\gamma(\Delta\tau)|$ is practically zero when $\Delta\tau$ is larger than $2/\Delta\upsilon$. (b) DOP of the emerging light vs. $\Delta\tau$ with different power ratios $P_x/P_y = ctan(\alpha)$. $DOP = 0$ when $P_x/P_y = 1$ and $\Delta\tau$ is larger than $2/\Delta\upsilon$, indicating complete depolarization.

References

Born, M. and Wolf, E. (1999). *Principles of Optics: Electromagnetic Theory of Propagation, Interference and Diffraction of Light*, 7e, 42. Cambridge: Cambridge University Press.

Damask, J.N. (2005). *Polarization Optics in Telecommunications*. Springer Science + Business Media, Inc.

Goldstein, H. (2001). *Classical Mechanics*, 3e: Section 4.28. Pearson.

Heffner, B.L. (1992). Automated measurement of polarization mode dispersion using Jones matrix eigenanalysis. *IEEE Photonics Technol. Lett.* 4: 1066–1069.

Hurwitz, H. and Jones, R.C. (1941). A new calculus for the treatment of optical systems, II. Proof of three general equivalence theorems. *J. Opt. Soc. Am.* 31 (7): 493–499.

Jones, R.C. (1941). A new calculus for the treatment of optical systems, I. Description and discussion of the calculus. *J. Opt. Soc. Am.* 31: 488–493.

Jones, R.C. (1948). A new calculus for the treatment of optical systems, VII. Properties of the N-matrices. *J. Opt. Soc. Am.* 38 (8): 671–683.

Liu, J. and Azzam, R.M.A. (1997). Polarization properties of corner-cube retroreflectors: theory and experiment. *Appl. Opt.* 36 (7): 1553–1559.

Pistoni, N.C. (1995). Simplified approach to the Jones calculus in retracing optical circuits. *Appl. Opt.* 34: 7870–7876.

Poole, C.D., Bergano, N.S., Wagner, R.E., and Schulte, H.J. (1989). Polarization dispersion and principal states in a 147 km undersea lightwave cable. *J. Lightwave Technol.* LT-7: 1185–1190.

Ulrich, R. and Simon, A. (1979). Polarization optics of twisted single-mode fibers. *Appl. Opt.* 18: 2241–2251.

Yan, L., Chen, X., and Yao, X.S. (2008). Optical differential group delay module with folded optical path. US Patent 7,723,670, filed 26 March 2008 and issued 25 May 2010.

Zhu, M., Li, Y., and Ellis, J.D. (2014). Polarization model for total internal reflection-based retroreflectors. *Opt. Eng.* 53: 06410.1.

Further Reading

Cyr, A.G., and Schinn, G.W. (1999). Stokes parameter analysis method, the consolidated test method for PMD measurements. *Proceedings 15th National Fiber Optics Engineers Conference*, Chicago. p. 280.

Gordon, J.P. and Kogelnik, H. (2000). PMD fundamentals: polarization mode dispersion in optical fibers. *Proc. Natl. Acad. Sci. U.S.A.* 97 (9): 4541–4550.

Jones, R.C. (1941). A new calculus for the treatment of optical systems, III The Sohncke theory of optical activity. *J. Opt. Soc. Am.* 31 (7): 500–503.

Jones, R.C. (1942). A new calculus for the treatment of optical systems, IV. *J. Opt. Soc. Am.* 32 (8): 486–493.

Jones, R.C. (1947a). A new calculus for the treatment of optical systems, V. A more general formulation and description of another calculus. *J. Opt. Soc. Am.* 37 (2): 107–110.

Jones, R.C. (1947b). A new calculus for the treatment of optical systems, VI. Experimental determination of the matrix. *J. Opt. Soc. Am.* 37 (2): 110–112.

Jopson, R.M. (1999). Measurement of second-order polarization-mode dispersion vectors in optical fibers. *IEEE Photonics Technol. Lett.* 11: 1153–1155.

Williams, P.A. (2004). PMD measurement techniques and how to avoid pitfalls. *J. Opt. Fiber Commun. Rep.* 1: 84–105.

5

Polarization Properties of Common Anisotropic Media

A dielectric medium may be isotropic or anisotropic. If the atoms or molecules in the medium are randomly located in space and oriented in random directions, the medium must be isotropic. Gases, liquids, and glasses are these kinds of materials. The macroscopic physical properties of isotropic media are also isotropic. A light beam propagating in an isotropic medium has the same phase velocity in all directions. Therefore, the polarization state remains unchanged during propagation.

On the other hand, if the atoms, ions, or molecules in the medium are located in space according to a regular periodic pattern and are oriented in certain regular directions, the medium is, in general, anisotropic. Examples of anisotropic materials include optical crystals such as quartz, calcite, potassium dihydrogen phosphate (KDP), $LiNbO_3$, and YVO_4, as well as liquid crystals. Their optical properties depend on the direction of propagation, as well as the polarization of the light waves entering the medium. They generally exhibit some interesting optical phenomena, including double refraction, spatial walk-off, polarization rotation, and electro-optical and acoustic-optical effects. Based on these properties, many useful optical devices can be made with optical crystals, such as wave plates, polarization rotators, prism polarizers, optical displacers, optical isolators, optical circulators, phase modulators, etc. In this chapter, we review the basics of the electromagnetic theory of light wave interaction with anisotropic materials, including the concepts of index ellipsoid and the phenomena of optical activity, electro-optical effect, photoelastic effect, and their applications. The detailed derivations of formulae and conclusions in this chapter can be found in *Principle of Optics* by M. Born and E. Wolf (1999) and in *Optical Waves in Crystals* by A. Yariv and P. Yeh (1984).

5.1 Plane Waves in Anisotropic Media

The interaction between light and medium can be described by the well-known Maxwell's equations (in MKS units) as follows:

$$\begin{aligned}
&\nabla \cdot \mathbf{D} = \rho && \text{(Gauss's law)}\\
&\nabla \cdot \mathbf{B} = 0 && \text{(Gauss's law for magnetism)}\\
&\nabla \times \mathbf{E} = -\frac{\partial \mathbf{B}}{\partial t} && \text{(Faraday's law)}\\
&\nabla \times \mathbf{H} = \frac{\partial \mathbf{D}}{\partial t} + \mathbf{J} && \text{(Ampere's law)}
\end{aligned} \tag{5.1}$$

where $\mathbf{E}$ is the electric field, $\mathbf{D}$ is the electric displacement, $\mathbf{H}$ is the magnetic field, $\mathbf{B}$ is magnetic induction, ρ is the free-charge density, and $\mathbf{J}$ is the free-current density. Most optical crystals are

Polarization Measurement and Control in Optical Fiber Communication and Sensor Systems, First Edition.
X. Steve Yao and Xiaojun (James) Chen.

insulators and magnetically isotropic, i.e. there are no free charge and current in them ($\rho = 0$ and $\mathbf{J} = 0$), and the $\mathbf{B}$ and $\mathbf{H}$ are mutually parallel. Therefore, the equations in (5.1) can be reduced to

$$\nabla \cdot \mathbf{D} = 0 \tag{5.2}$$

$$\nabla \cdot \mathbf{B} = 0 \tag{5.3}$$

$$\nabla \times \mathbf{E} = -\frac{\partial \mathbf{B}}{\partial t} \tag{5.4}$$

$$\nabla \times \mathbf{H} = \frac{\partial \mathbf{D}}{\partial t} \tag{5.5}$$

5.1.1 Dielectric Tensor and Its Symmetry

In general, the electric displacement vector $\mathbf{D}$ in a dielectric medium is not in the direction of vector $\mathbf{E}$. The relationship between $\mathbf{D}$ and $\mathbf{E}$ can be approximately described by the following linear equation:

$$D_i = \sum_j \varepsilon_{ij} E_j \quad \text{where } i = x,\ y,\ \text{or } z \tag{5.6}$$

with the matrix format of

$$\mathbf{D} = \boldsymbol{\varepsilon}\mathbf{E} = \begin{pmatrix} \varepsilon_{xx} & \varepsilon_{xy} & \varepsilon_{xz} \\ \varepsilon_{yx} & \varepsilon_{yy} & \varepsilon_{yz} \\ \varepsilon_{zx} & \varepsilon_{zy} & \varepsilon_{zz} \end{pmatrix} \begin{pmatrix} E_x \\ E_y \\ E_z \end{pmatrix} \tag{5.7}$$

The nine quantities ε_{ij} are constant of the medium and constitute the 3×3 dielectric tensor $\boldsymbol{\varepsilon}$ (second-order tensor), and the vector $\mathbf{D}$ is the product of the tensor $\boldsymbol{\varepsilon}$ with $\mathbf{E}$. For a transparent and non-absorptive medium, $\boldsymbol{\varepsilon}$ is real, and the conservation of electromagnetic energy requires

$$\varepsilon_{ij} = \varepsilon_{ji} \tag{5.8}$$

This means that the dielectric tensor $\boldsymbol{\varepsilon}$ is symmetric and has only six independent elements. Because a symmetric 3×3 matrix has real mutually orthogonal eigenvectors in Cartesian coordinates, in principal coordinates, $\boldsymbol{\varepsilon}$ can be reduced to a diagonal matrix

$$\boldsymbol{\varepsilon} = \begin{pmatrix} \varepsilon_x & 0 & 0 \\ 0 & \varepsilon_y & 0 \\ 0 & 0 & \varepsilon_z \end{pmatrix} \tag{5.9}$$

The relationship between ε_x, ε_y, and ε_z is determined by the structure of the crystal lattice. For example, in cubic crystals, the three principal axes are physically equivalent; therefore, $\varepsilon_x = \varepsilon_y = \varepsilon_z$. Table 5.1 lists the crystal symmetries and the corresponding dielectric tensors of different types of crystals.

5.1.2 Plane Wave Propagation in Anisotropic Media

In an anisotropic medium such as a crystal, the phase velocity of light depends on its state of polarization as well as its propagation direction. Considering a monochromatic plane wave with angular frequency $\omega = 2\pi\upsilon$ and phase velocity c/n propagating in the direction of the unit wave-normal $\hat{\boldsymbol{k}}$, the vectors $\mathbf{E}$, $\mathbf{D}$, $\mathbf{H}$, and $\mathbf{B}$ are in complex notation proportional to $exp[i(\omega t - \mathbf{k} \cdot \boldsymbol{r})]$ where $\mathbf{k}$ is the wave vector $\mathbf{k} = \omega n / c\hat{\mathbf{k}}$. Thus, in Eqs. (5.2)–(5.5), the operation $\partial/\partial t$ is always equivalent

Table 5.1 Dielectric tensors of crystals with different crystal symmetries.

Optical symmetry	Crystal symmetry	Dielectric tensor[a]
Isotropic	Cubic	$\begin{pmatrix} \varepsilon & 0 & 0 \\ 0 & \varepsilon & 0 \\ 0 & 0 & \varepsilon \end{pmatrix} = \varepsilon_0 \begin{pmatrix} n^2 & 0 & 0 \\ 0 & n^2 & 0 \\ 0 & 0 & n^2 \end{pmatrix}$
Anisotropic (uniaxial)	Tetragonal Hexagonal Trigonal	$\begin{pmatrix} \varepsilon_x & 0 & 0 \\ 0 & \varepsilon_x & 0 \\ 0 & 0 & \varepsilon_z \end{pmatrix} = \varepsilon_0 \begin{pmatrix} n_o^2 & 0 & 0 \\ 0 & n_o^2 & 0 \\ 0 & 0 & n_e^2 \end{pmatrix}$
Anisotropic (biaxial)	Triclinic Monoclinic Orthorhombic	$\begin{pmatrix} \varepsilon_x & 0 & 0 \\ 0 & \varepsilon_y & 0 \\ 0 & 0 & \varepsilon_z \end{pmatrix} = \varepsilon_0 \begin{pmatrix} n_x^2 & 0 & 0 \\ 0 & n_y^2 & 0 \\ 0 & 0 & n_z^2 \end{pmatrix}$

a) Note: $n_i^2 = \varepsilon_i/\varepsilon_0$, where $i = x, y$, or z.

to the multiplication of $i\omega$, while the operation ∇ is equivalent to $-i\mathbf{k}$. Maxwell Eqs. (5.2)–(5.5) become

$$\mathbf{k} \cdot \mathbf{D} = 0 \Rightarrow \mathbf{k} \perp \mathbf{D} \tag{5.10}$$

$$\mathbf{k} \cdot \mathbf{B} = 0 \Rightarrow \mathbf{B} \perp \mathbf{k} \tag{5.11}$$

$$\mathbf{k} \times \mathbf{E} = \omega\mathbf{B} \Rightarrow \mathbf{B} \perp \mathbf{E} \tag{5.12}$$

$$\mathbf{k} \times \mathbf{H} = -\omega\mathbf{D} \Rightarrow \mathbf{H} \perp \mathbf{D} \Rightarrow \mathbf{B} \perp \mathbf{D} \tag{5.13}$$

where $\mathbf{B} \perp \mathbf{D}$ is due to the assumption that the anisotropic optical media are magnetic isotropic, which results in $\mathbf{B} = \mu\mathbf{H}$. The electromagnetic energy flows in the direction of Poynting vector **S**, which is

$$\mathbf{S} = \mathbf{E} \times \mathbf{H} \Rightarrow \mathbf{S} \perp \mathbf{E} \text{ and } \mathbf{S} \perp \mathbf{H} \tag{5.14}$$

The geometric relationships of **k**, **E**, **D**, **B**, **H**, and **S** are shown in Figure 5.1. One may easily find the following from Eqs. (5.10)–(5.14): (i) **D**, **E**, and **S** are coplanar and perpendicular to **B** and **H**. (ii) **D** is perpendicular to **k**. (iii) **E** is not parallel to **D** and is not perpendicular to **k** in general. (iv) Energy flow (ray direction) may not necessarily have the same propagation direction **k** of the phase front.

5.1.2.1 Fresnel's Equation of Wave Normals

By eliminating **B** in Eqs. (5.12) and (5.13), one obtains

$$\mathbf{k} \times (\mathbf{k} \times \mathbf{E}) + \omega^2\mu\mathbf{D} = \mathbf{k} \times (\mathbf{k} \times \mathbf{E}) + \omega^2\mu\varepsilon\mathbf{E} = 0 \tag{5.15}$$

where Eq. (5.7) and $\mathbf{B} = \mu\mathbf{H}$ are used in the derivation. Substituting $\mathbf{k} \times (\mathbf{k} \times \mathbf{E}) = \mathbf{k}(\mathbf{E} \cdot \mathbf{k}) - k^2\mathbf{E}$ and Eq. (5.9) in Eq. (5.15), one obtains

$$E_i = \frac{k_i(\mathbf{E} \cdot \mathbf{k})}{k^2 - \omega^2\mu\varepsilon_i} \quad i = x, y, z \tag{5.16}$$

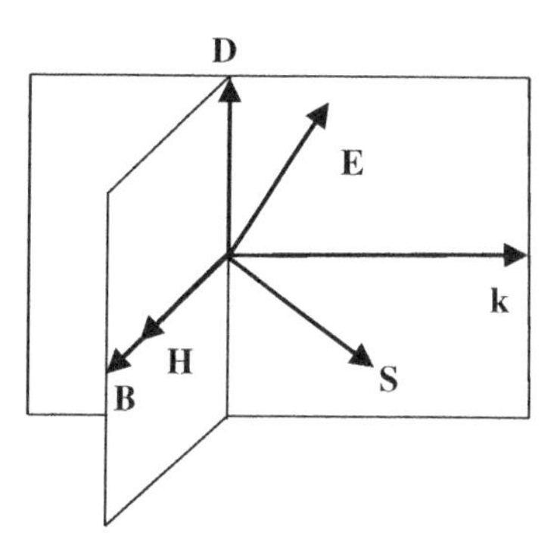

Figure 5.1 The geometric relationships of directions of wave front normal **k**, electric displacement **D**, electric field **E**, energy flow (Poynting vector **S**), magnetic field **H**, and magnetic induction **B**.

Assuming that the refractive index related to the wave vector **k** is n and the unit vector along **k** is $\hat{\boldsymbol{k}}$, one obtains

$$\mathbf{k} = \frac{\omega}{c} n\hat{\mathbf{k}} = \frac{\omega}{\upsilon_p}\hat{\boldsymbol{k}} \tag{5.17}$$

where $\upsilon_p = c/n$ is the phase velocity along $\hat{\boldsymbol{k}}$. Solve Eq. (5.16) with $\hat{\boldsymbol{k}} \cdot \hat{\boldsymbol{k}} = 1$, $k^2 = \mathbf{k} \cdot \mathbf{k} = \left(\frac{\omega}{c}n\right)^2$, and $k_i = \frac{\omega}{c} n\hat{k}_i$, one obtains

$$\frac{\hat{k}_x^2}{n^2 - c^2\mu\varepsilon_x} + \frac{\hat{k}_y^2}{n^2 - c^2\mu\varepsilon_y} + \frac{\hat{k}_z^2}{n^2 - c^2\mu\varepsilon_z} = \frac{1}{n^2} \tag{5.18}$$

Multiplying both sides of (5.18) with n^2 and applying $\hat{k}_x^2 + \hat{k}_y^2 + \hat{k}_z^2 = 1$, we obtain

$$\frac{\hat{k}_x^2}{1/n^2 - 1/(c^2\mu\varepsilon_x)} + \frac{\hat{k}_y^2}{1/n^2 - 1/(c^2\mu\varepsilon_y)} + \frac{\hat{k}_z^2}{1/n^2 - 1/(c^2\mu\varepsilon_z)} = 0 \tag{5.19}$$

We can define three principal velocities of propagation as follows

$$\upsilon_x = \frac{1}{\sqrt{\mu\varepsilon_x}}, \quad \upsilon_y = \frac{1}{\sqrt{\mu\varepsilon_y}}, \quad \upsilon_z = \frac{1}{\sqrt{\mu\varepsilon_z}} \tag{5.20}$$

Equations (5.17) and (5.19) can be rewritten in the form

$$E_i = \frac{\upsilon_i^2 \hat{\mathrm{k}}_i(\mathbf{E} \cdot \hat{\boldsymbol{k}})}{\upsilon_i^2 - \upsilon_p^2} \tag{5.21}$$

$$\frac{\hat{k}_x^2}{\upsilon_p^2 - \upsilon_x^2} + \frac{\hat{k}_y^2}{\upsilon_p^2 - \upsilon_y^2} + \frac{\hat{k}_z^2}{\upsilon_p^2 - \upsilon_z^2} = 0 \tag{5.22}$$

It should be emphasized that υ_x, υ_y, and υ_z are the constants defined by (5.20), which are not components of a vector. Equations (5.18), (5.19), and (5.22) are the equivalent forms of Fresnel's equation of wave normal. Equation (5.22) is a quadratic equation in υ_p^2 and there are two phase velocity solutions $|\upsilon_p|$. With each of the two values of $|\upsilon_p|$, one can find the ratio of $E_x : E_y : E_z$ (direction of **E**) from (5.21), then the corresponding ratio of **D** can be obtained from Eq. (5.7). Since these ratios are real, **E** and **D** are linearly polarized. Thus, we conclude that an anisotropic medium permits two monochromatic plane waves with two different linear polarizations and two different velocities to propagate in any given direction. It can be shown that the two directions of the electric displacement vector **D** corresponding to a given propagation direction $\hat{\boldsymbol{k}}$ are perpendicular to each other.

5.1.3 The Index Ellipsoid

The surface of the constant energy density U_e in **D** space is given by

$$U_e = \frac{1}{2}\mathbf{E}\cdot\mathbf{D} \tag{5.23}$$

Substituting Eqs. (5.7) and (5.9) in (5.23), one finds

$$\frac{D_x^2}{\varepsilon_x} + \frac{D_y^2}{\varepsilon_y} + \frac{D_z^2}{\varepsilon_y} = 2U_e \tag{5.24}$$

Let $x = D_x/\sqrt{2\varepsilon_0 U_e}$, $y = D_y/\sqrt{2\varepsilon_0 U_e}$, and $z = D_z/\sqrt{2\varepsilon_0 U_e}$, and consider them as Cartesian coordinates in space. Further, let $n_x = \sqrt{\varepsilon_x/\varepsilon_0}$, $n_y = \sqrt{\varepsilon_y/\varepsilon_0}$, and $n_z = \sqrt{\varepsilon_z/\varepsilon_0}$; Eq. (5.24) can be rewritten as

$$\frac{x^2}{n_x^2} + \frac{y^2}{n_y^2} + \frac{z^2}{n_z^2} = 1 \tag{5.25}$$

Equation (5.25) represents an ellipsoid, as shown in Figure 5.2, with the principal axes parallel to x, y, and z-directions, and the corresponding lengths of the principal axes are $2n_x$, $2n_y$, and $2n_z$, respectively. This ellipsoid is known as index ellipsoid, which is mainly used to find the polarization directions of **D** and corresponding refractive indices associated with the two independent plane waves for a given phase front propagation direction $\hat{\mathbf{k}}$. This can be done by following procedures in reference to Figure 5.2.

(1) Draw the propagation direction vector $\hat{\mathbf{k}}$ with the start point at the origin.
(2) Draw a plane through the origin that is normal to the direction of $\hat{\mathbf{k}}$.
(3) Find the intersection ellipse between the plane obtained in step (2) and the index ellipsoid. Find the lengths of semiaxes n_1 and n_2, respectively.
(4) One of the eigenmodes is the linearly polarized light wave whose electric displacement $\mathbf{D}_1$ is in the direction of the semiaxis with the length of n_1.
(5) The other eigenmode is the linearly polarized light whose electric displacement $\mathbf{D}_2$ is in the direction of the semiaxis with the length of n_2.

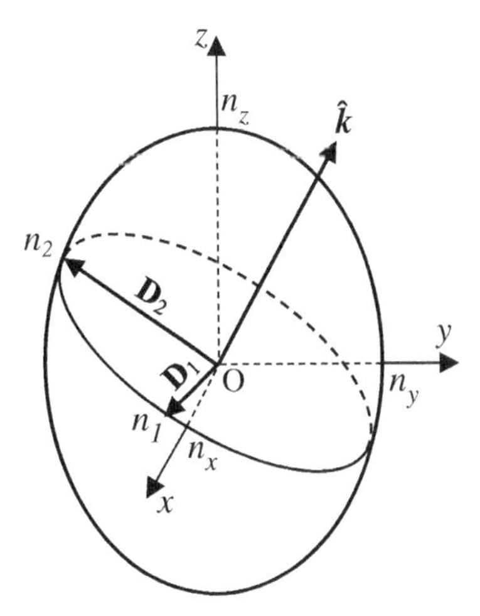

Figure 5.2 Index ellipsoid. The inner ellipse is the intersection between the ellipsoid and the plane which passes the origin and is normal to the propagation direction $\hat{k}$ of the light's phase front.

5.2 Optical Properties of Anisotropic Crystals

5.2.1 Light Propagation in Uniaxial Crystals

As listed in Table 5.1, the crystals with tetragonal (such as YVO_4 and MgF_2), hexagonal (such as high quartz), or trigonal (such as low quartz, $LiNbO_3$, and Ruby) symmetry have the same dielectric constant along the x and y-axes in principal coordinates, when the axis of symmetry (c-axis) is chosen as the z-axis. Assuming $n_0 = \sqrt{\varepsilon_x/\varepsilon_0} = \sqrt{\varepsilon_y/\varepsilon_0}$ and $n_e = \sqrt{\varepsilon_z/\varepsilon_0}$, one can rewrite the index ellipsoid defined by (5.25) as

$$\frac{x^2}{n_o^2} + \frac{y^2}{n_o^2} + \frac{z^2}{n_e^2} = 1 \tag{5.26}$$

Figure 5.3a shows the index ellipsoids for positive ($n_e > n_o$) and negative ($n_e < n_o$) uniaxial crystals, respectively. The direction of phase front propagation is $\hat{\boldsymbol{k}}$. Since $n_x = n_y = n_o$, the ellipsoid is invariant under rotation about the z-axis, and the optical properties are only determined by the intersection angle θ between the z-axis and $\hat{\boldsymbol{k}}$. Following the procedures introduced in Section 5.1.3, we first find the intersection of the plane through the origin that is normal to $\hat{\boldsymbol{k}}$ with the index ellipsoid. The intersection is an ellipse which is shown in Figure 5.3a. The two semi-axes of the ellipse determine the following two allowed light waves propagating in such crystals.

5.2.1.1 Ordinary Ray

Ordinary ray is the light wave that is linearly polarized in the direction of the semiaxis OA (Figure 5.3a) of the ellipse. Its electric displacement vector $\mathbf{D}_o$ is parallel to OA and perpendicular to the plane containing the z-axis and $\hat{\boldsymbol{k}}$. For any direction of $\hat{\boldsymbol{k}}$, the vector $\mathbf{D}_o$ is always on the equatorial plane, and the corresponding index of refraction is a constant n_o. Therefore, the wave normal surface of the ordinary ray is a sphere and can be expressed as

$$n_o(\theta) = n_o \tag{5.27}$$

5.2.1.2 Extraordinary Ray

Extraordinary ray is the light wave that is linearly polarized in the direction of the semiaxis OB (Figure 5.3a) of the intersection ellipse. Its electric displacement vector $\mathbf{D}_e$ coincides with the semi-axis OB of the ellipse and is parallel to the plane containing the z-axis and $\hat{\boldsymbol{k}}$. The corresponding refractive index $n_e(\theta)$ depends on the angle θ between the z-axis and $\hat{\boldsymbol{k}}$ (see Figure 5.3b), and is given by

$$\frac{1}{n_e^2(\theta)} = \frac{cos^2(\theta)}{n_o^2} + \frac{sin^2(\theta)}{n_e^2} \tag{5.28}$$

Assuming $\hat{\boldsymbol{k}}$ is on the yz plane, one can find that the direction of the $\mathbf{D}$ vector of the extraordinary wave as $(0, -cos\theta, sin\theta)^T$. Substituting it, together with the dielectric tensor of uniaxial crystal in Table 5.1, in Eq. (5.7), one finds the direction of the electric field $\mathbf{E}$ of the extraordinary wave as

$$\mathbf{E}_e = \begin{pmatrix} 0 \\ -cos\theta/n_o^2 \\ sin\theta/n_e^2 \end{pmatrix} \tag{5.29}$$

One can prove that the electric fields $\mathbf{E}_e$ are in the tangent line of the ellipse, as shown in Figure 5.3c. The direction of the Poynting vector $\mathbf{S}_e$ can be obtained from $\mathbf{S}_e = \mathbf{E} \times \mathbf{H}$,

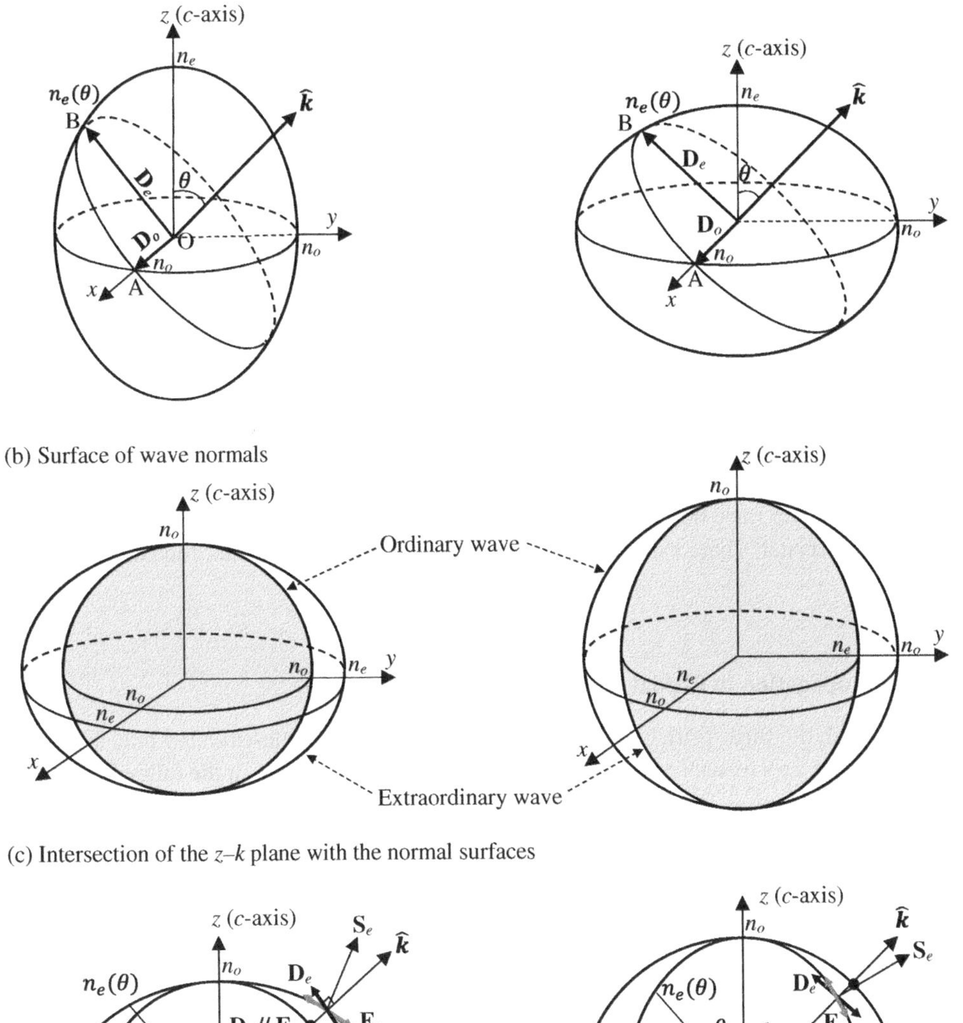

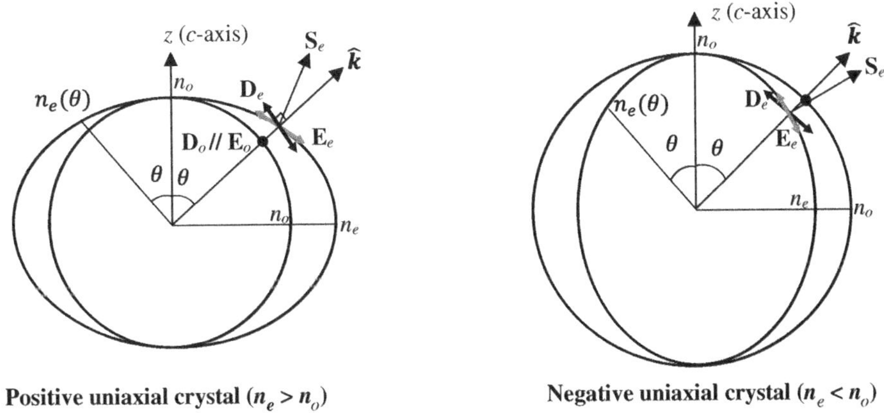

Figure 5.3 Index ellipsoid and the normal surfaces of uniaxial crystals, where *o* represents ordinary wave, *e* represents extraordinary wave, $\hat{k}$ is the propagation direction of the light's phase front, and $\mathbf{S}_e$ is the Poynting vector of the extraordinary wave.

where $\mathbf{H} = (1 \quad 0 \quad 0)^T$

$$\mathbf{S}_e = \begin{pmatrix} 0 \\ sin\theta/n_e^2 \\ cos\theta/n_o^2 \end{pmatrix} \tag{5.30}$$

Finally, one finds the angle α between $\hat{\boldsymbol{k}}$ and $\mathbf{S}_e$

$$tan(\alpha) = \pm\frac{|\mathbf{S}_e \times \hat{\boldsymbol{k}}|}{\mathbf{S}_e \cdot \hat{\boldsymbol{k}}} = \frac{(n_e^2 - n_o^2)\, tan(\theta)}{n_e^2 + n_o^2 tan^2(\theta)} \tag{5.31}$$

where θ is the angle between the c-axis and $\hat{\boldsymbol{k}}$. θ is positive for positive uniaxial crystals and negative for negative uniaxial crystals.

5.2.1.3 Optical Axis

From Eqs. (5.26) and (5.27), one finds that the refractive indices of ordinary and extraordinary rays are both equal to n_o when light propagates along z-axis (c-axis). In general, the direction in which two allowed light waves have the same refractive index (or the same phase velocity) is called the optical axis of the crystal. There exist only one optical axis in uniaxial crystals and two in biaxial crystals.

5.2.2 Light Propagation in Biaxial Crystals

In a biaxial crystal, the three principal indices n_x, n_y, and n_z are all different. One may label the principal coordinate axes in such a way that the three principal indices are in the following order

$$n_x < n_y < n_z \tag{5.32}$$

Let's consider the case where the light beam propagates in the x–z plane. From the index ellipsoid Eq. (5.25), one finds the refractive indices of two eigenmodes

$$n_1(\theta) = n_y \tag{5.33}$$

$$\frac{1}{n_2^2(\theta)} = \frac{cos^2(\theta)}{n_x^2} + \frac{sin^2(\theta)}{n_z^2} \tag{5.34}$$

The curves of $n_1(\theta)$ and $n_2(\theta)$ on the x–z plate are shown in Figure 5.4. It is clear there exist two optical axes which are located in the x–z plane. The tilt angle β of the optical axis from the z-axis can be obtained by substituting $n_2(\theta) = n_1(\theta) = n_y$ in (5.34):

$$cos^2\beta = \frac{\left(\frac{1}{n_y^2} - \frac{1}{n_z^2}\right)}{\left(\frac{1}{n_x^2} - \frac{1}{n_z^2}\right)} \tag{5.35a}$$

or

$$tan\beta = \frac{n_z}{n_x}\sqrt{\frac{n_y^2 - n_x^2}{n_z^2 - n_y^2}} \tag{5.35b}$$

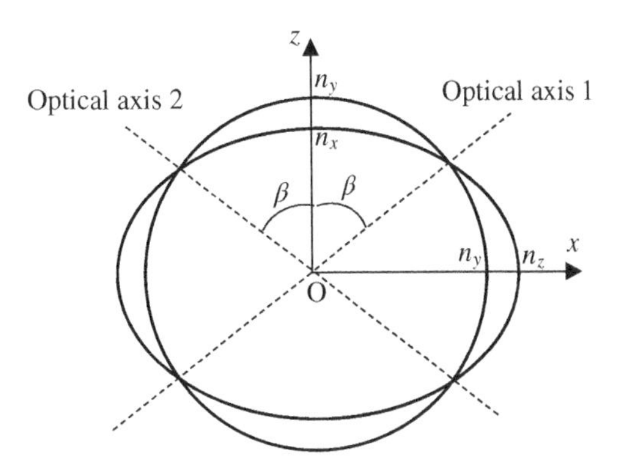

Figure 5.4 Intersection of the normal surface $(n(\theta)\hat{k})$ with the x–z plane for biaxial crystal with $n_x < n_y < n_z$. There exist two optical axes which are located in the x–z plane when $n_x < n_y < n_z$.

5.2.3 Double Refraction

When light enters an anisotropic crystal, the light ray will be split into two orthogonally polarized beams. Such a phenomenon is called double refraction. In a uniaxial crystal, the refracted waves will be a mixture of ordinary and extraordinary waves. The relationship between angles of incidence and refraction is determined by the boundary conditions, which requires that all the wave vectors lie in the plane of incidence and their tangential components along the boundary be the same. Let $\mathbf{k}_i$ be the propagation vector of the incident beam, $\mathbf{k}_1$ and $\mathbf{k}_2$ are the wave vectors of two eigenmodes of refraction; the boundary conditions requires the following relationship:

$$k_i sin\theta_i = k_1 sin\theta_1 = k_2 sin\theta_2 \tag{5.36}$$

where θ_i is the incident angle, and θ_1 and θ_2 are the angles of refraction of two eigenmodes of crystal. As an example, the double refraction at the boundary of a positive uniaxial crystal with its c-axis in the incidence plane is shown in Figure 5.5. To find the angle of refraction, one should first find the intersection of normal surfaces $(n_e\omega/c\hat{\mathbf{k}}$ and $n_o\omega/c\hat{\mathbf{k}})$ with the plane of incidence, where n_o

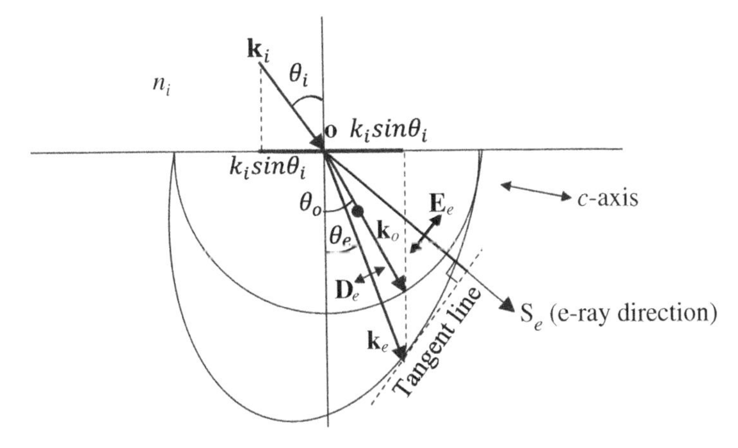

Figure 5.5 Double refraction at a boundary of a positive uniaxial crystal where the c-axis is in the incident plane. The intersection of the normal surface with the plane of incidence for the ordinary wave is a circle with a radius of $n_o k_o$. The intersection ellipse is for the extraordinary wave. The electric field **E** is parallel to the tangent line of the ellipse of the extraordinary wave. The Poynting vector $\mathbf{S}_e$ shows the ray direction of the extraordinary wave. The ray direction of the ordinary wave is parallel to $\mathbf{k}_o$.

and n_e are the refractive indices of ordinary and extraordinary, respectively. According to Eqs. (5.26) and (5.27), the intersection curves can be expressed as

$$\text{Ordinary wave}: \quad k_o = \frac{n_o \omega}{c} \tag{5.37}$$

$$\text{Extraordinary wave}: \quad \frac{1}{n_e^2(\theta)} = \frac{\cos^2(\beta - \theta)}{n_x^2} + \frac{\sin^2(\beta - \theta)}{n_z^2} \tag{5.38}$$

with the following the boundary conditions

$$k_i \sin\theta_i = k_o \sin\theta_o = k_e \sin\theta_e \tag{5.39}$$

where β is the angle from the vertical to the crystal's c-axis and θ is the angle of refraction. By solving Eqs. (5.36) and (5.37), one can find the refractive angles θ_o and θ_e. It should be emphasized that the ray direction (the Poynting vector $\mathbf{S}_e$) of the extraordinary wave is not parallel to the wave vector $\mathbf{k}_e$, in general. Therefore, one has to find the direction $\mathbf{S}_e$ from Eq. (5.9), $\mathbf{D}_e \cdot \mathbf{k}_e = 0$, $\mathbf{D}_e = \boldsymbol{\varepsilon}\mathbf{E}_e$, and $\mathbf{S}_e = \mathbf{E}_e \times \mathbf{H}$.

5.2.4 Spatial Walk-Off

Let's consider a light beam entering a uniaxial crystal with parallel entrance and exit faces, and the incident beam is normal to the boundary surface, as shown in Figure 5.6. According to the boundary condition (5.36), the wave vectors of ordinary and extraordinary rays in the crystal must be normal to the boundary surface. However, the electromagnetic energy propagation direction (ray

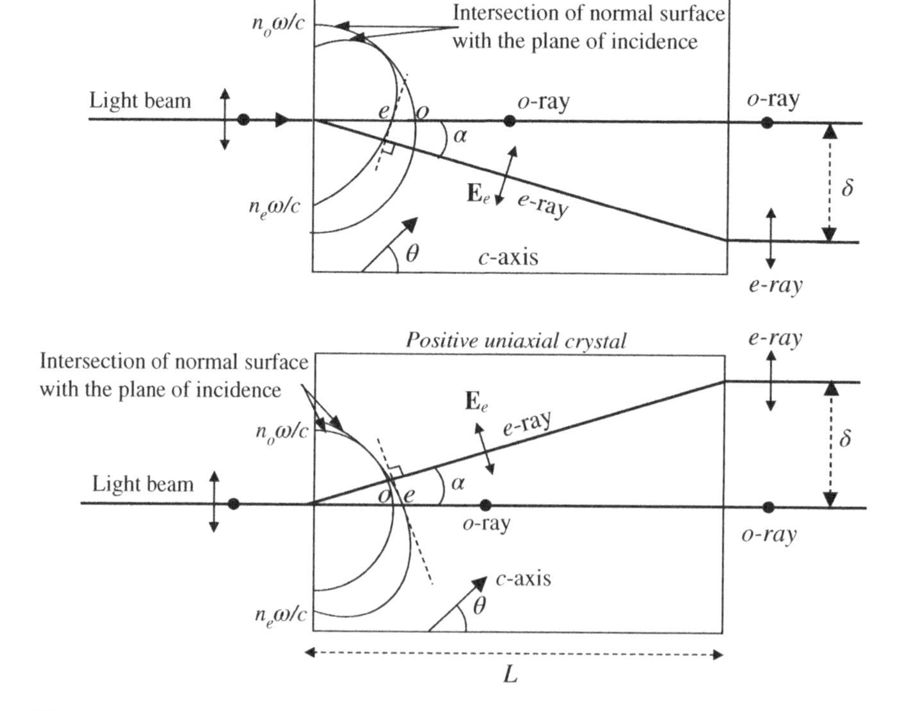

Figure 5.6 Spatial walk-off between the o-ray and the e-ray in a negative (top) and a positive (bottom) uniaxial crystal.

direction, or the Poynting vector direction) of extraordinary wave $\mathbf{S}_e$ is off the normal boundary, and the deviation angle α can be obtained from Eq. (5.31). Therefore, there will be two linearly polarized rays after the exit face. One coincides with the incident ray with polarization perpendicular to the plane containing the c-axis and incident light; this ray is called the ordinary ray (o-ray). Another ray is parallel with the o-ray; however, it is laterally shifted by a distance δ. This ray's polarization is parallel to the plane containing the c-axis and incident light, which is called the extraordinary ray (e-ray). The lateral position shift δ is called spatial walk-off distance, which can be calculated by

$$\delta = L\,|tan(\alpha)| = L\frac{|n_e^2 - n_o^2|\,tan(\theta)}{n_e^2 + n_o^2 tan^2(\theta)} \tag{5.40}$$

In practice, this piece of crystal is also called a polarization displacer, as will be discussed further in Section 6.3.1.

5.2.5 Optical Activity

Optical activity is the phenomenon where the plane of polarization (here is defined as the plane containing the polarization vector and the beam propagation direction) is rotated while a linear polarized light propagates in a certain optical medium being referred to as optically active. Optical activity was first observed in quartz crystals. Figure 5.7 illustrates the rotation of the plane of polarization by an optically active medium. The rotation angle θ is proportional to the path length L of light in the medium, which can be written as

$$\theta = \rho L \tag{5.41}$$

where ρ is the specific rotatory power that is defined as the amount of rotation per unit length. The sense of rotation is determined by an observer facing the approaching light beam. Therefore, the rotation will be canceled out in the case of reflection from the right end face in Figure 5.7. An optically active medium is called right-handed if the sense of rotation of the plane of polarization is counterclockwise or left-handed for the clockwise rotation.

Fresnel first recognized in 1825 that optical activity arose from circular birefringence. In an optically active medium, the eigenwaves of propagation are right and left-handed circularly polarized waves (RCP and LCP). As discussed in Section 4.1.2, any polarized light can be considered as the superposition of RCP and LCP. Assume the input light at $z = 0$ has horizontal linear polarization. From Eq. 4.21, its Jones vector can be written as

$$\mathbf{J}(z = 0) = \mathbf{J}_{LHP} = \begin{pmatrix} 1 \\ 0 \end{pmatrix} = \frac{\sqrt{2}}{2}(\mathbf{J}_{RCP} + \mathbf{J}_{LCP}) \tag{5.42}$$

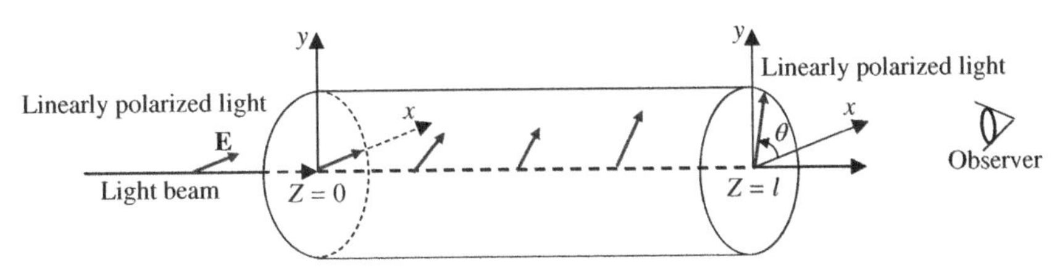

Figure 5.7 The plane of linear polarization rotates in right-handed optically active media.

where $\mathbf{J}_{RCP}$ and $\mathbf{J}_{LCP}$ are the Jones vectors of right and left-handed circular polarizations, respectively. At distance z in the medium, the Jones vector of light is

$$\mathbf{J}(z) = \frac{\sqrt{2}}{2}\left(e^{-i\frac{2\pi}{\lambda_0}zn_R}\mathbf{J}_{RCP} + e^{-i\frac{2\pi}{\lambda_0}zn_L}\mathbf{J}_{LCP}\right) \tag{5.43}$$

where n_R and n_L are the refractive indexes of RCP and LCP, respectively, λ_0 is the wavelength in vacuum. Substituting $\mathbf{J}_{RCP} = \frac{1}{\sqrt{2}}\begin{pmatrix}1\\i\end{pmatrix}$ and $\mathbf{J}_{LCP} = \frac{1}{\sqrt{2}}\begin{pmatrix}1\\-i\end{pmatrix}$ in Eq. (5.47), one obtains

$$\mathbf{J}(z) = e^{-i\frac{2\pi}{\lambda_0}z\frac{n_R+n_L}{2}}\begin{pmatrix}\cos\theta\\ \sin\theta\end{pmatrix}$$
$$\theta - \frac{\pi}{\lambda_0}z(n_R - n_L) \tag{5.44}$$

The Jones vector at z in Eq. (5.44) shows the polarization at any position z is still a linear polarization, with its plane of polarization rotated in the counterclockwise direction from x by an angle of θ. Therefore, the specific rotatory power is given by

$$\rho = \frac{\pi}{\lambda_0}(n_R - n_L) \tag{5.45}$$

5.3 Electro-optic Effect

5.3.1 General Description

As discussed in Section 5.1, the propagation of a light wave can be described completely by dielectric tensor $\boldsymbol{\varepsilon}$. The two directions of the eigenmodes of the light wave can be determined by using the index ellipsoid (see Section 5.1.3). In the principal coordinates, the index ellipsoid can be described by Eq. (5.25), i.e.

$$\frac{x^2}{n_x^2} + \frac{y^2}{n_y^2} + \frac{z^2}{n_z^2} = 1 \tag{5.46}$$

We list the equation of index ellipsoid repetitively here for the convenience of discussion. Let's define the dielectric impermeability tensor $\boldsymbol{\eta}$ as

$$\boldsymbol{\eta} = \varepsilon_0\boldsymbol{\varepsilon}^{-1} \tag{5.47}$$

Substituting (5.9) in (5.47), one finds the dielectric impermeability tensor $\boldsymbol{\eta}$ in the principal coordinates as

$$\boldsymbol{\eta} = \begin{pmatrix}1/n_x^2 & 0 & 0\\ 0 & 1/n_y^2 & 0\\ 0 & 0 & 1/n_z^2\end{pmatrix} \tag{5.48}$$

It is easier to use $\boldsymbol{\eta}$ to express the equation of index ellipsoid (5.46) as

$$\sum_{i=1}^{3}\eta_{ii}x_ix_i = 1 \tag{5.49}$$

where x, y, and z are represented by x_1, x_2, and x_3, respectively. According to the quantum theory of solids, the optical dielectric impermeability tensor depends on the distribution of charges in the crystal. The application of an external electric field will result in a redistribution of the bond charges and possibly a slight deformation of the ion lattice. The net result is the change in the optical impermeability tensor, which is known as the electro-optic effect. The dependence of $\boldsymbol{\eta}$ on applied electric field $\mathbf{E}$ can be represented by

$$\eta_{ij} = \eta_{ij}(0) + \sum_{k=1}^{3} r_{ijk}E_k + \sum_{k=1}^{3}\sum_{l=1}^{3} s_{ijkl}E_kE_l + \cdots, \quad i,j = 1,2,3 \tag{5.50}$$

where r_{ijk} is the linear (or Pockels) electro-optic coefficients, s_{ijkl} is the quadratic (or Kerr) electro-optic coefficients; the terms higher than the quadratic are neglected because these higher-order effects are too small for most applications.

The index ellipsoid of a crystal in the presence of an applied electric field is given by

$$\sum_{i=1}^{3}\sum_{j=1}^{3} \eta_{ij}x_ix_j = 1 \tag{5.51}$$

One should note that the dielectric impermeability tensor $\boldsymbol{\eta}$ has the same symmetry as the dielectric tensor $\boldsymbol{\varepsilon}$, which is a real 3×3 real symmetric matrix. Therefore, the $\boldsymbol{\eta}$ is symmetric, i.e. $\eta_{ij} = \eta_{ji}$; consequently, r_{ijk} and s_{ijkl} have the following symmetries:

$$r_{ijk} = r_{jik} \tag{5.52}$$

$$s_{ijkl} = s_{jikl} \tag{5.53}$$

$$s_{ijkl} = s_{ijlk} \tag{5.54}$$

It is often convenient to introduce the following contracted indices to abbreviate the notation

$$1 = (11), 2 = (22), 3 = (33), 4 = (32) = (23), 5 = (31) = (13), \text{and } 6 = (12) = (21) \tag{5.55}$$

Then Eqs. (5.46) and (5.50) can be simplified to

$$\eta_m = \eta_m(0) + \sum_{k=1}^{3} r_{mk}E_k + \sum_{k=1}^{3}\sum_{l=1}^{3} s_{mn}E_kE_l + \cdots \quad i,j = 1,2,3 \tag{5.56}$$

In Eq. (5.56), the second rank dielectric impermeability tensor $\boldsymbol{\eta}$ is represented by a 6×1 matrix; the third rank linear electro-optic tensor $\mathbf{r}$ is represented by a 6×3 matrix; the fourth rank quadratic electro-optic tensor $\mathbf{s}$ is represented by a 6×6 matrix. It should be noted that the 6×1, 6×3, and 6×6 matrices are just a matter of convenience, and they do not have the usual tensor transformation properties.

5.3.2 Linear Electro-optic Effect

In most applications of the electro-optic effect, the applied electric field is much smaller than the electric field inside the atom ($\sim 10^8$ V/cm), and the quadratic effect is expected to be small compared with the linear effect and is often neglected when the linear effect is present. It can be proven that the linear electro-optic effect vanishes in crystals with center symmetry, such that the quadratic effect becomes the dominant phenomenon.

By using the contracted indices (5.55) and substituting Eq. (5.56) in (5.51), the equation of index ellipsoid can be written as

$$\left(\frac{1}{n_x^2} + \sum_{k=1}^{3} r_{1k}E_k\right)x^2 + \left(\frac{1}{n_y^2} + \sum_{k=1}^{3} r_{2k}E_k\right)y^2 + \left(\frac{1}{n_z^2} + \sum_{k=1}^{3} r_{3k}E_k\right)z^2$$
$$+ 2yz\sum_{k=1}^{3} r_{4k}E_k + 2zx\sum_{k=1}^{3} r_{5k}E_k + 2xy\sum_{k=1}^{3} r_{6k}E_k = 1 \tag{5.57}$$

Equation (5.57) can also be expressed in the following format for convenience:

$$\eta_1 x^2 + \eta_2 y^2 + \eta_3 z^2 + 2\eta_4 yz + 2\eta_5 zx + 2\eta_6 xy = 1 \tag{5.58}$$

where η_m $(m = 1, \ldots, 6)$ can be obtained from

$$\begin{pmatrix} \eta_1 \\ \eta_2 \\ \eta_3 \\ \eta_4 \\ \eta_5 \\ \eta_6 \end{pmatrix} = \begin{pmatrix} 1/n_x^2 \\ 1/n_y^2 \\ 1/n_z^2 \\ 0 \\ 0 \\ 0 \end{pmatrix} + \begin{pmatrix} r_{11} & r_{12} & r_{13} \\ r_{21} & r_{22} & r_{23} \\ r_{31} & r_{32} & r_{33} \\ r_{41} & r_{42} & r_{43} \\ r_{51} & r_{52} & r_{53} \\ r_{61} & r_{62} & r_{63} \end{pmatrix} \begin{pmatrix} E_x \\ E_y \\ E_z \end{pmatrix} \tag{5.59}$$

In general, the terms with yz, zx, and xy are not zero when an external electric field is applied, i.e. the corresponding principal axes do not coincide with the original principal axes (x, y, z) without the external electric field.

5.3.2.1 Example: The Electro-optic Effect in $LiNbO_3$

Lithium niobate ($LiNbO_3$) is a frequently used electro-optic crystal for manufacturing phase modulators and high-speed polarization controllers for fiber optics applications. $LiNbO_3$ is a uniaxial crystal with a crystal symmetry of $3m$, which has the following nonzero electro-optic coefficients at 1550 nm: $r_{33} = 31$ pm/V, $r_{51} = 28$ pm/V, $r_{13} = 8.6$ pm/V, $r_{22} = 3.4$ pm/V. All other electro-optic coefficients are zero. Therefore, the corresponding electro-optic tensor is

$$\begin{pmatrix} 0 & -r_{22} & r_{13} \\ 0 & r_{22} & r_{13} \\ 0 & 0 & r_{33} \\ 0 & r_{51} & 0 \\ r_{51} & 0 & 0 \\ -r_{22} & 0 & 0 \end{pmatrix} \tag{5.60}$$

Substituting Eq. (5.60) in Eqs. (5.59) and (5.58), one obtains the equation of index ellipsoid of $LiNbO_3$ as

$$\left(\frac{1}{n_o^2} - r_{22}E_y + r_{13}E_z\right)x^2 + \left(\frac{1}{n_o^2} + r_{22}E_y + r_{13}E_z\right)y^2 + \left(\frac{1}{n_e^2} + r_{33}E_z\right)z^2 + 2r_{51}E_y yz + 2r_{51}E_x zx - 2r_{22}E_x xy = 1 \tag{5.61}$$

5.3.2.2 Case 1: Electric Field is Applied Along the z-Axis

When an electric field is applied along the z-axis, as shown in Figure 5.8, one obtains the following ellipsoid equation by substituting $\mathbf{E} = (0, 0, E_z)^T$ in (5.61):

$$\left(\frac{1}{n_o^2} + r_{13}E_z\right)x^2 + \left(\frac{1}{n_o^2} + r_{13}E_z\right)y^2 + \left(\frac{1}{n_e^2} + r_{33}E_z\right)z^2 \tag{5.62}$$

Since $r_{13}E_z \ll \frac{1}{n_o^2}$ and $r_{33}E_z \ll \frac{1}{n_e^2}$, one obtains the following approximations

$$\frac{1}{n_o^2} + r_{13}E_z \approx \frac{1}{(n_o')^2} \quad \text{where } n_o' = n_o - \frac{1}{2}n_o^3 r_{13}E_z \tag{5.63}$$

$$\frac{1}{n_e^2} + r_{33}E_z \approx \frac{1}{(n_e')^2} \quad \text{where } n_e' = n_e - \frac{1}{2}n_e^3 r_{33}E_z \tag{5.64}$$

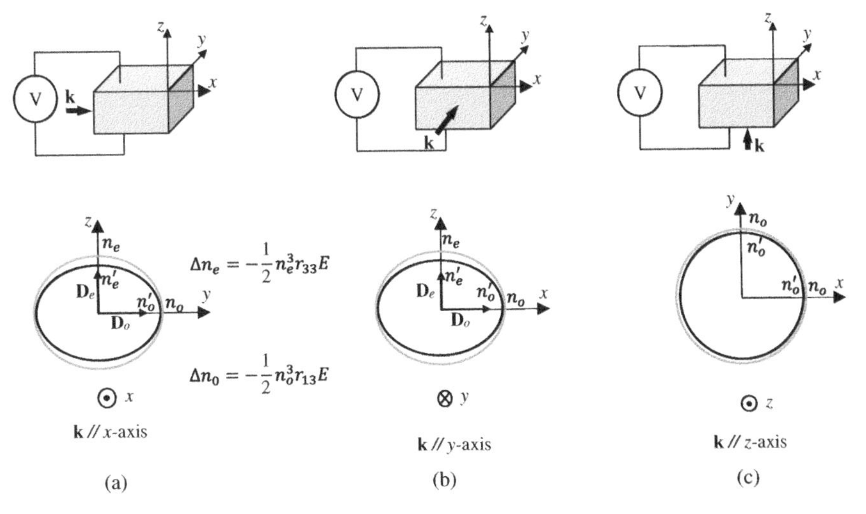

Figure 5.8 The index ellipses of $LiNbO_3$ when an electric field is applied along the x-axis. (a), (b), and (c) correspond to the cases of the light propagating along the x, y, and z-axes, respectively. The dark and gray curves represent the index ellipses with and without the external electric field applied, respectively.

Therefore, the ellipsoid Eq. (5.62) can be approximated by

$$\frac{x^2}{(n'_o)^2} + \frac{y^2}{(n'_o)^2} + \frac{z^2}{(n'_e)^2} = 1 \tag{5.65}$$

One may notice that the principal coordinates remain unchanged because no cross-terms appear in Eq. (5.65) so that the crystal remains uniaxial. The corresponding index ellipses with light propagating in the directions of the x, y, or z-axes are shown in Figure 5.8. When light propagates along the x or y-axis, the birefringence change caused by the external electric field along the z-axis is

$$\Delta n = \Delta n_e - \Delta n_o = (n'_e - n_e) - (n'_o - n_o) = -\frac{1}{2}\left(n_e^3 r_{33} - n_o^3 r_{13}\right) E_z \tag{5.66}$$

5.3.2.3 Case 2: Electric Field is Applied Along the *x*-Axis

When an electric field is applied along the x-axis, as shown in Figure 5.9, one obtains the following ellipsoid equation by substituting $\mathbf{E} = (E_x, 0, 0)^T$ in (5.61):

$$\frac{x^2}{n_o^2} + \frac{y^2}{n_o^2} + \frac{z^2}{n_e^2} + 2r_{51}E_x zx - 2r_{22}E_x xy = 1 \tag{5.67}$$

Let $x = 0$, $y = 0$, or $z = 0$ in Eq. (5.67), respectively, the index ellipse equations corresponding to light propagation directions of (100), (010), and (001) are as follows:

$$\mathbf{k}//x\text{-axis}: \quad \frac{y^2}{n_o^2} + \frac{z^2}{n_e^2} = 1 \tag{5.68}$$

$$\mathbf{k}//y\text{-axis}: \quad \frac{x^2}{n_o^2} + \frac{z^2}{n_e^2} + 2r_{51}E_x zx = 1 \tag{5.69}$$

$$\mathbf{k}//z\text{-axis}: \quad \frac{x^2}{n_o^2} + \frac{y^2}{n_o^2} - 2r_{22}E_x xy = 1 \tag{5.70}$$

From Eq. (5.68), one can see that no birefringence change is observed when light propagates along the x-axis, even when an external electric field in the x-axis direction is applied, as shown in Figure 5.9a. However, if the light propagates along the y or z-axis, the two eigenpolarizations are no

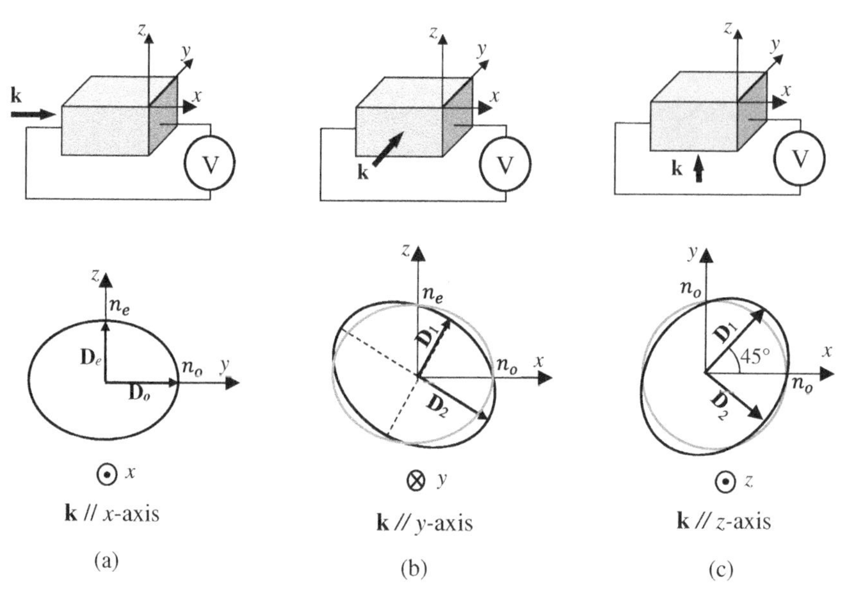

Figure 5.9 The index ellipses of $LiNbO_3$ when an electric field is applied along the x-axis. (a), (b), and (c) correspond to the cases where the light propagates along the x-, y-, and z-axes, respectively. The dark and gray curves are for situations with and without the external electric field applied, respectively.

longer parallel to the original principal axes of the crystals (see Figure 5.9b) because of the mixed terms in Eqs. (5.69) and (5.70). For the case where the light propagates along the z-axis, the new eigenpolarizations are at ±45° to the x-axis (see Figures 5.9c).

5.3.2.4 Case 3: Electric Field is Along the y-Axis

When an electric field is applied along the y-axis, as shown in Figure 5.10, one obtains the following ellipsoid equation by substituting $\mathbf{E} = (0, E_y, 0)^T$ in (5.61):

$$\left(\frac{1}{n_o^2} - r_{22}E_y\right)x^2 + \left(\frac{1}{n_o^2} + r_{22}E_y\right)y^2 + \left(\frac{1}{n_e^2}\right)z^2 + 2r_{51}E_y yz = 1 \tag{5.71}$$

Let $x = 0$, $y = 0$, or $z = 0$ in Eq. (5.70), the index ellipse equations corresponding to light propagation directions of (100), (010), and (001) are as follows:

$$\mathbf{k}//x\text{-axis}: \quad \left(\frac{1}{n_o^2} + r_{22}E_y\right)y^2 + \left(\frac{1}{n_e^2}\right)z^2 + 2r_{51}E_y yz = 1 \tag{5.72}$$

$$\mathbf{k}//y\text{-axis}: \quad \left(\frac{1}{n_o^2} - r_{22}E_y\right)x^2 + \left(\frac{1}{n_e^2}\right)z^2 = 1 \tag{5.73}$$

$$\mathbf{k}//z\text{-axis}: \quad \left(\frac{1}{n_o^2} - r_{22}E_y\right)x^2 + \left(\frac{1}{n_o^2} + r_{22}E_y\right)y^2 = 1 \tag{5.74}$$

Since $r_{22}E_y \ll \frac{1}{n_o^2}$, Eqs. (5.73) and (5.74) can be approximated as

$$\mathbf{k}//x\text{-axis}: \quad \frac{y^2}{(n'_y)^2} + \frac{z^2}{n_e^2} + 2r_{51}E_y yz = 1 \tag{5.75}$$

$$\mathbf{k}//y\text{-axis}: \quad \frac{x^2}{(n'_x)^2} + \frac{z^2}{n_e^2} = 1 \tag{5.76}$$

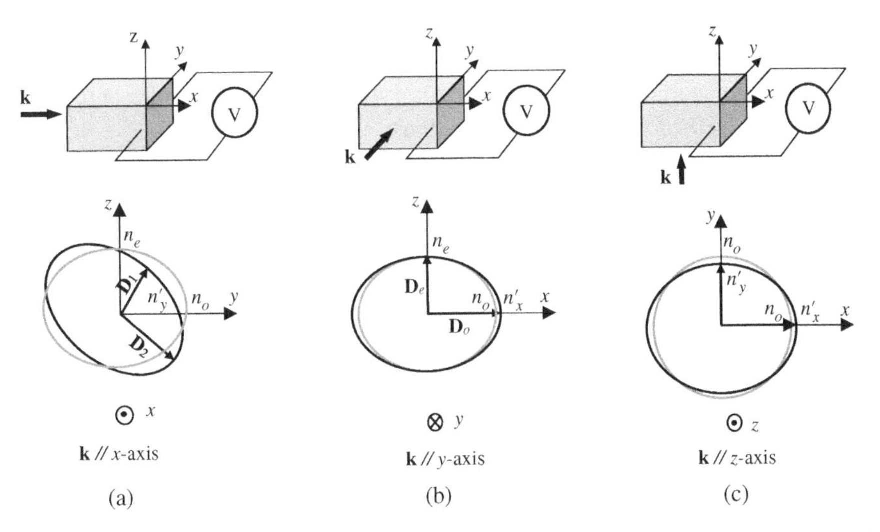

Figure 5.10 The index ellipses of $LiNbO_3$ when an electric field is applied along the y-axis. (a), (b), and (c) correspond to the cases of light propagating along the x-, y-, and z-axes. The dark and gray curves are for the situations with and without the external electric field applied, respectively.

$$\mathbf{k}//z\text{-axis}: \quad \frac{x^2}{(n'_x)^2} + \frac{y^2}{(n'_y)^2} = 1 \tag{5.77}$$

where

$$n'_x = n_o + \frac{1}{2} n_o^3 r_{22} E_y \tag{5.78}$$

$$n'_y = n_o - \frac{1}{2} n_o^3 r_{22} E_y \tag{5.79}$$

From Eq. (5.72), when the light propagates along the x-axis, the two eigenpolarizations will no longer be parallel with the crystal's original principal axes (see Figure 5.10a) because of the nonzero cross term. For the cases where light propagates in the y or z-axis, the eigenpolarizations are not changed by an external electric field along the y-axis because there are no mixed terms in (5.73) and (5.74). However, the magnitude of birefringence is changed due to an external electric field along the y-axis (see Figure 5.10b,c). From Eqs. (5.76) and (5.77), the birefringence changes can be calculated by

$$\mathbf{k}//y\text{-axis}: \quad \Delta n = \Delta n_z - \Delta n_x = 0 - (n'_x - n_o) = -\frac{1}{2} n_o^3 r_{22} E_y \tag{5.80}$$

$$\mathbf{k}//z\text{-axis}: \quad \Delta n = \Delta n_y - \Delta n_x = (n'_y - n_o) - (n'_x - n_o) = -n_o^3 r_{22} E_y \tag{5.81}$$

5.4 The Photoelastic Effect in Isotropic Media

The polarization state of a light wave is preserved while it propagates in an ideal isotropic media since its phase speed is independent of its polarization states. In general, there always exists mechanical strain in materials, which results in a change of the refractive index. Such a phenomenon is referred to as the photoelastic effect. In an optical fiber, lateral stress, such as bend and twist, may cause strains at the core area of the fiber, resulting in birefringence due to the photoelastic effect. In this section, we focus on the mathematical method to represent the relationship of birefringence, strain, and stress in isotropic media. Readers can find more detailed discussions of birefringences in optical fibers induced by different kinds of stresses in Section 3.3.

The photoelastic effect in a material couples the mechanical strain to the optical index of refraction, which can be described by the following linear equations:

$$\Delta\eta_{ij} = \left(\frac{1}{n^2}\right)_{ij} = p_{ijkl}S_{kl} \tag{5.82}$$

where i, j, k, and l can be one of x, y, and z. $\Delta\eta_{ij} = \eta_{ij} - \eta_{ij}(0)$ represents the change in optical impermeability tensor defined in (5.47), in which the 0 in $\eta_{ij}(0)$ stands for zero perturbation, S_{kl} is the second-rank strain tensor, and p_{ijkl} are strain-optic tensor. Both η_{ij} and p_{ijkl} are symmetric tensors, such that the indexes i and j as well as k and j can be permuted. Therefore, there are only 6 independent elements in η_{ij} tensor, and 36 independent elements in strain-optic tensor p_{ijkl}. Using the contracted indices defined in Eq. (5.55) for convenience, Eq. (5.82) can be written as

$$\begin{pmatrix}\eta_1\\ \eta_2\\ \eta_3\\ \eta_4\\ \eta_5\\ \eta_6\end{pmatrix} = \begin{pmatrix}1/n_1^2\\ 1/n_2^2\\ 1/n_3^2\\ 0\\ 0\\ 0\end{pmatrix} + \begin{pmatrix}p_{11} & p_{12} & p_{13} & p_{14} & p_{15} & p_{16}\\ p_{21} & p_{22} & p_{23} & p_{24} & p_{25} & p_{26}\\ p_{31} & p_{32} & p_{33} & p_{34} & p_{35} & p_{36}\\ p_{41} & p_{42} & p_{43} & p_{44} & p_{45} & p_{46}\\ p_{51} & p_{52} & p_{53} & p_{54} & p_{55} & p_{56}\\ p_{61} & p_{62} & p_{63} & p_{64} & p_{65} & p_{66}\end{pmatrix}\begin{pmatrix}S_1\\ S_2\\ S_3\\ S_4\\ S_5\\ S_6\end{pmatrix} \tag{5.83}$$

where n_1, n_2, and n_3 are the principal indices of refraction. In an isotropic material (for example, glass), there are only two independent elements in the corresponding strain-optic tensor, which has the following format:

$$\begin{pmatrix}p_{11} & p_{12} & p_{12} & 0 & 0 & 0\\ p_{12} & p_{11} & p_{12} & 0 & 0 & 0\\ p_{12} & p_{12} & p_{11} & 0 & 0 & 0\\ 0 & 0 & 0 & \frac{1}{2}(p_{11}-p_{12}) & 0 & 0\\ 0 & 0 & 0 & 0 & \frac{1}{2}(p_{11}-p_{12}) & 0\\ 0 & 0 & 0 & 0 & 0 & \frac{1}{2}(p_{11}-p_{12})\end{pmatrix} \tag{5.84}$$

If we choose the principal coordinates, the strain tensor becomes diagonal, i.e. $S_4 = S_5 = S_6 = 0$, and $\eta_4 = \eta_5 = \eta_6 = 0$, Eq. (5.83) becomes

$$\begin{pmatrix}\eta_1\\ \eta_2\\ \eta_3\\ \eta_4\\ \eta_5\\ \eta_6\end{pmatrix} = \begin{pmatrix}1/n_o^2 + p_{11}S_1 + p_{12}S_2 + p_{12}S_3\\ 1/n_o^2 + p_{12}S_1 + p_{11}S_2 + p_{12}S_3\\ 1/n_o^2 + p_{12}S_1 + p_{12}S_2 + p_{11}S_3\\ 0\\ 0\\ 0\end{pmatrix} \tag{5.85}$$

where n_o is the index of refraction of the isotropic substance.

Consider $(p_{11}S_1 + p_{12}S_2 + p_{12}S_3) \ll \frac{1}{n_o^2}$, $(p_{12}S_1 + p_{11}S_2 + p_{12}S_3) \ll \frac{1}{n_o^2}$, and $(p_{12}S_1 + p_{12}S_2 + p_{11}S_3) \ll \frac{1}{n_o^2}$, one gets the following approximations:

$$\left(\frac{1}{n_o^2} + p_{11}S_1 + p_{12}S_2 + p_{12}S_3\right) \approx \frac{1}{\left[n_o - \frac{1}{2}n_o^3(p_{11}S_1 + p_{12}S_2 + p_{12}S_3)\right]^2} = \frac{1}{n_x^2}$$

$$\left(\frac{1}{n_o^2} + p_{12}S_1 + p_{12}S_2 + p_{11}S_3\right) \approx \frac{1}{\left[n_o - \frac{1}{2}n_o^3(p_{12}S_1 + p_{11}S_2 + p_{12}S_3)\right]^2} = \frac{1}{n_y^2}$$
$$\left(\frac{1}{n_o^2} + p_{12}S_1 + p_{12}S_2 + p_{11}S_3\right) \approx \frac{1}{\left[n_o - \frac{1}{2}n_o^3(p_{12}S_1 + p_{12}S_2 + p_{11}S_3)\right]^2} = \frac{1}{n_z^2} \quad (5.86)$$

Therefore, the index ellipsoid (5.49) related to the strain tensor becomes

$$\frac{x^2}{n_x^2} + \frac{y^2}{n_y^2} + \frac{z^2}{n_z^2} = 1 \quad (5.87)$$

where n_x, n_y, and n_z are defined in Eq. (5.86).

5.4.1 Birefringence and Strain-Optical Tensor in Isotropic Material

Let's consider the case where light propagates in the direction of one of the principal axes of the strain tensor. When light propagates along the z-axis, from Eqs. (5.86) and (5.87), one finds the corresponding birefringence as

$$\Delta n_{xy} = n_y - n_x = \frac{1}{2}n_o^3(p_{11} - p_{12})(S_1 - S_2) \quad (5.88)$$

Similarly, when light propagates along the x-axis, the birefringence is

$$\Delta n_{yz} = n_z - n_y = \frac{1}{2}n_o^3(p_{11} - p_{12})(S_2 - S_3) \quad (5.89)$$

Finally, when light propagates along the y-axis, the birefringence is

$$\Delta n_{zx} = n_x - n_z = \frac{1}{2}n_o^3(p_{11} - p_{12})(S_3 - S_1) \quad (5.90)$$

5.4.2 Relationship Between Birefringence and Stress Tensor

For a linear isotropic material subjected only to compressive (normal) force, the stress and strain tensors are related by

$$S_1 = \frac{1}{E}[\sigma_1 - \upsilon(\sigma_2 + \sigma_3)]$$
$$S_2 = \frac{1}{E}[\sigma_2 - \upsilon(\sigma_1 + \sigma_3)]$$
$$S_3 = \frac{1}{E}[\sigma_3 - \upsilon(\sigma_1 + \sigma_2)] \quad (5.91)$$

where S_1, S_2, and S_3 are the strains in the directions of the x, y, and z-axes, respectively; σ_1, σ_2, and σ_3 are the stresses in the directions of x, y, and z-axes, respectively; E is Young's modulus, and υ is Poisson's ratio of the material, with both E and υ the same in all directions. Substitution of Eq. (5.91) in Eqs. (5.88)–(5.90) yields

$$\Delta n_{xy} = n_y - n_x = \frac{1}{2}\frac{n_o^3(p_{11} - p_{12})}{E}(1 + \upsilon)(\sigma_1 - \sigma_2) \quad (5.92)$$

$$\Delta n_{yz} = n_z - n_y = \frac{1}{2}n_o^3(p_{11} - p_{12})(1 + \upsilon)(\sigma_2 - \sigma_3) \quad (5.93)$$

$$\Delta n_{zx} = n_x - n_z = \frac{1}{2}n_o^3(p_{11} - p_{12})(1 + \upsilon)(\sigma_3 - \sigma_1) \quad (5.94)$$

Let $C = \frac{1}{2}\frac{n_o^3(p_{11}-p_{12})}{E}(1+\upsilon)$, Eqs. (5.92)–(5.94) can be simplified to

$$\Delta n_{xy} = C(\sigma_1 - \sigma_2) \tag{5.95}$$

$$\Delta n_{yz} = C(\sigma_2 - \sigma_3) \tag{5.96}$$

$$\Delta n_{zx} = C(\sigma_3 - \sigma_1) \tag{5.97}$$

These three relationships are called stress-optic law, with C being the stress-optic coefficient of the isotropic material (Dally and Riley 1991).

Reference

Dally, J.W. and Riley, W.F. (1991). *Experimental Stress Analysis*, 3e, Chapter 12. McGraw-Hill Education.

Further Reading

Born, M. and Wolf, E. (1999). Optics of crystals. In: *Principles of Optics*, 7e, Chapter 15, 790–852. Cambridge: Cambridge University Press.

Nye, J.F. (1985). *Physical Properties of Crystals*. Oxford, UK: Oxford University Press.

Yariv, M. and Yeh, P. (1984). *Optical Waves in Crystals: Propagation and Control of Laser Radiation*. New York: Wiley.

6

Polarization Management Components and Devices

As described in Chapter 3, the state of polarization (SOP) in a regular single mode (SM) is random due to the random birefringence distribution along the fiber. Any external perturbations on the fiber, such as pressing, bending, twisting, and even touching the fiber, would cause the local birefringence to vary and thus the SOP to fluctuate, creating problems for many applications requiring stable linear polarization, such as for fiber optic gyroscope (FOG) and optic current sensing applications (Jaecklin and Lietz 1972; Smith 1978). In addition, an optical fiber system may also suffer from polarization related impairments, such as polarization mode dispersion (PMD), polarization dependent loss (PDL), polarization dependent gain (PDG), etc., compromising the performance of the system. Therefore, it is important to be able to manage the polarization and the related impairments in real optical fiber systems, either passively or actively. In this chapter we will focus on components and devices for controlling or managing the SOP and polarization related impairments in optical fiber systems. We will also discuss devices enabled by polarization management for controlling the reflection and routing optical signals, such as isolators and circulators.

6.1 Polarization Management Fibers

6.1.1 Low Birefringence Fiber

The earliest attempts to tame the SOP instability issues were to make fibers with extremely low retardation (Schneider et al. 1978; Norman et al. 1979) on the order of 2.6°/m, corresponding to a beat length of 138 m. The term "beat length" is frequently used to describe the magnitude of birefringence or retardation in the fiber, corresponding to the fiber length with an accumulated retardation of 2π or 360°. Unfortunately, light propagating in such a low birefringence (LoBi) fiber is still sensitive to external stresses on the fiber because the stress induced birefringence may be on the same order of magnitude, making such LoBi fibers difficult to be used in practical applications for maintaining stable linear polarization.

6.1.2 Polarization Maintaining Fiber

An alternative approach was to make fibers with high linear birefringence (the high birefringence [HiBi] fiber), on the order of 10^{-5} or more, first fabricated by Stolen et al. (1978). In applications, the SOP of the input light is chosen to be linear and aligned with one of the birefringence axes, with the slow and fast axes having larger and smaller refractive indices, respectively. As long as the birefringence induced by external perturbations is much smaller than that of fiber's internal

Polarization Measurement and Control in Optical Fiber Communication and Sensor Systems, First Edition.
X. Steve Yao and Xiaojun (James) Chen.

birefringence, the SOP will remain linear as the light propagates along the fiber and continues to be aligned with the birefringence axis. As described in Section 3.3, such a HiBi fiber can be produced either with a highly elliptical cladding or with two stress rods on each side of the fiber core. Such a HiBi fiber is also frequently referred to as polarization maintaining fiber or PM fiber. In practice, PM fibers with a beat length on the orders of 1.5–5 mm at 1550 nm are commercially available.

6.1.3 Polarizing Fiber

Note that a PM fiber discussed earlier supports polarizations aligned in both the slow and fast axes (the two orthogonal polarization modes). External stresses may still cause light polarized in the slow axis to couple into the fast axis and vice versa, on the order of −60 dB (one millionth) to −10 dB (one tenth), depending on the severity and orientation angle of the stress with respect to one of the birefringence axes, as will be discussed in Chapter 10. For some applications requiring high polarization extinction ratio (PER), such as FOG, such a polarization cross-coupling cannot be tolerated. In order to eliminate optical power in the fast axis, an optical fiber can be made to only support the slow polarization mode with low transmission loss while leaving the fast polarization mode with high loss, especially when the fiber is bent with a certain radius. Such a fiber is called polarizing (PZ) fiber or single-polarization fiber, which can be made using the conventional technology with solid core or cladding (Okoshi 1981) or using photonic crystal technology with micro air-holes (Schreiber et al. 2005).

6.1.4 Spun Fiber

For some applications, such as fiber optic magnetic or electrical current sensing, it is important to have fibers with extremely low linear birefringence, because the linear birefringence in the fiber destroys the linear relationship between the Faraday rotation angle and the magnetic field, causing large sensing or measurement errors. To overcome the issues caused by the linear birefringence, the annealed fiber (Tang et al. 1991; Bohnert 2002), the low-stress fiber (Kurosawa, Yoshida, and Sakamoto 1995), and the spun fiber (Clarke 1993) were introduced. Eventually, the spun fiber seems to have gained acceptance (Laming and Payne 1989; Polynkin and Blake 2005) in the field for the applications of using Faraday effect for electrical current and magnetic field sensing, because it is effective in minimizing the linear birefringence by introducing a large amount of circular birefringence and can be consistently produced by multiple specialty fiber manufacturers (http://www.fibercore.com/product/spun-hibi-fiber; www.photonics.ixblue.com; http://www.YOFC.com/view/1650.html).

As discussed in Chapters 3 and 4, the spun fiber can be made by rotating a preform while drawing the fiber. This spin and draw process introduces a large amount of circular birefringence uniformly distributed along the fiber (Peng et al. 2013). The preform used can be either that for making a SM fiber or that for making a PM fiber; however, the latter is more widely adopted in the industry because the resulting spun fiber is much less sensitive to bending induced birefringence than that made with the former (Laming and Payne 1989). The spun fibers made with the SM fiber preform and PM fiber preform are often referred to as the LoBi and HiBi spun fibers, respectively. When it is not spun during the drawing process, the PM fiber preform can produce a PM fiber with a large linear birefringence (defined as the unspun linear birefringence Δn). However, with the preform spinning during the drawing process, the average linear birefringence of the resulting spun fiber over a length much larger than the spin pitch is expected to approach zero, although the local linear birefringence still remains at the level of the unspun linear birefringence (Barlow, Ramskov-Hansen, and Payne 1982). In fact, the large local linear birefringence renders the HiBi spun fiber less susceptible to bend and stress perturbations, while the near-zero average linear birefringence helps to preserve the linear relationship between the Faraday rotation angle and the longitudinal magnetic field, which is critical for electrical current and magnetic field sensing applications.

For a HiBi spun fiber, the residual linear birefringence and the polarization rotation angle after a fiber length of L can be expressed by Eqs. (3.15a) and (3.15b). It can be shown that when the spin twist rate ρ (rad/m) is much larger (four times or more) than the retardation per unit length $\Delta\beta$ (rad/m) of the unspun fiber, the linear relationship between the Faraday rotation angle $\Delta\varphi_F$ and the magnetic field H still holds (Yao et al. 2021):

$$\Delta\varphi_F \approx \frac{V_{H0}HL}{[1+(\Delta\beta/2\rho)^2]^{1/2}} = V_{eff}HL \tag{6.1}$$

where V_{H0} is the Verdet constant of an ideal fiber free of birefringence and $V_{eff}(\lambda) = \frac{V_{H0}}{[1+(\Delta\beta/2\rho)^2]^{1/2}}$ is the effective Verdet constant, which is always smaller than the Verdet constant V_{H0} of the ideal fiber and approach V_{H0} when the spin twist rate is infinitely large. The error of the approximation of Eq. (6.1) is about 3% for commercial spun fibers.

6.2 Polarizers

An ideal polarizer generally has two orthogonal axes that pass the light polarized along one of the axes with no attenuation and totally attenuates the light polarized along the orthogonal axis. PER is often used to characterize the polarizing ability of a polarizer, which is defined as the ratio of maximum power transmission T_{max} and minimum power transmission T_{min} of light polarized along the passing axis and the blocking axis, respectively.

$$PER = 10log\left(\frac{T_{max}}{T_{min}}\right) \tag{6.2}$$

A typically good polarizer has a PER on the order of 10^4 or 40 dB or better. As will be discussed in the following text, a low cost plastic sheet polarizer has a PER on the order of 20 dB while a high quality polarizer made with birefringence crystals has a PER up to 60 dB.

A polarizer can be linear, circular, or elliptical. A polarizer is called linear if the two orthogonal axes are linear, which can be represented by $\hat{x}$ and $\hat{y}$. On the other hand, a circular polarizer passes light of right-hand circular polarization (RCP) and blocks light of left-hand circular polarization (LCP) or vice versa, with its two orthogonal axes expressed as $(\hat{x}+i\hat{y})/\sqrt{2}$ and $(\hat{x}-i\hat{y})/\sqrt{2}$. In general, an elliptical polarizer passes light of right-hand elliptical polarization (REP) and blocks light of left-hand elliptical polarization (LEP) or vice versa, with its two orthogonal axes expressed as $(\hat{x}+i\alpha\hat{y})/\sqrt{2\alpha}$ and $(\alpha\hat{x}-i\hat{y})/\sqrt{2\alpha}$.

A polarizer is often called an analyzer when it is put in front of a photodetector to analyze the SOP. For example, as described in Chapters 4 and 8, by sequentially inserting a 0° linear polarizer, a 45° linear polarizer, a 90° linear polarizer, and an RCP polarizer in an optical beam and measuring its power, one can determine the SOP and degree of polarization (DOP) of the optical beam by obtaining the beam's Stokes parameters. As another example, by rotating a polarizer inserted in an optical beam before a power meter, one may determine the PER of a light beam, as will be discussed in more detail in Chapter 8.

6.2.1 Birefringence Crystal Polarizers

Commonly Used Birefringence Crystals The commonly used birefringence crystals for making fiber optic components for communication and sensing systems are Yttrium Vanadate (YVO_4), Rutile (TiO_2), Calcite ($CaCO_3$), and Lithium Niobate ($LiNbO_3$). Among them, YVO_4 is the most widely used crystal for making passive micro-optics components, followed by $LiNbO_3$, due to their excellent optical, mechanical, and chemical properties, as listed in Table 6.1. YVO_4 and TiO_2 are positive uniaxial crystals ($n_e > n_o$) while $CaCo_3$ and $LiNbO_3$ are negative ($n_e < n_o$). As mentioned in Chapter 1, the Calcite crystal is the first birefringence crystal discovered by man and is closely associated with the understanding of polarization, via the observation of double refraction, in the history.

Nicol Prism Polarizer Early polarizers were made with birefringence crystals, such as Nicol prism polarizer and Glan–Thompson prism polarizer, as shown in Figure 6.1. A **Nicol prism** was the first type of polarizing prism, invented in 1828 by William Nicol (1770–1851) of Edinburg. It consists of a rhombohedral crystal of Iceland Spar (a variety of calcite) that has been cut at an angle of 48° with respect to the optical axis (OA), cut again diagonally, and then rejoined using a layer of transparent Canada balsam (refractive index $n = 1.55$) as a cement, with its length three times of its width, as shown in Figure 6.1a. It is made in such a way that it eliminates one of the rays by total internal reflection, i.e. the ordinary ray (*o-ray*) is eliminated and only the extraordinary ray is transmitted through the prism.

In operation, unpolarized light ray enters through the left face of the crystal, as shown in the diagram, and is split into two orthogonally polarized, differently directed rays by the birefringence property of the calcite. The *o-ray* experiences a refractive index of $n_o = 1.6584$ in the calcite at

Table 6.1 Important properties of commonly used birefringence crystals in optical fiber systems.

	Yttrium vanadate (YVO_4)	Rutile (TiO_2)	Calcite ($CaCO_3$)	Lithium niobate ($LiNbO_3$)
n_o	1.9447@1550 nm	2.454@1530 nm	1.6346@1497 nm	2.2151@1440 nm
n_e	2.1486@1550 nm	2.710@1530 nm	1.4774@1497 nm	2.1413@1440 nm
$n_e - n_o$	0.2039@1550 nm	0.256@1530 nm	−0.1572@1497 nm	−0.0738@1440 nm
c-axis TEC (/°C)	11.4×10^{-6}	9.2×10^{-6}	26.3×10^{-6}	16.7×10^{-6}
a-axis TEC (/°C)	4.4×10^{-6}	7.1×10^{-6}	5.4×10^{-6}	7×10^{-6}
Mohs hardness	5	6.5	3	5
Transparency	0.4–5 μm	0.4–5 μm	0.35–2.3 μm	0.4–5 μm
Deliquescence	None	None	Weak	None

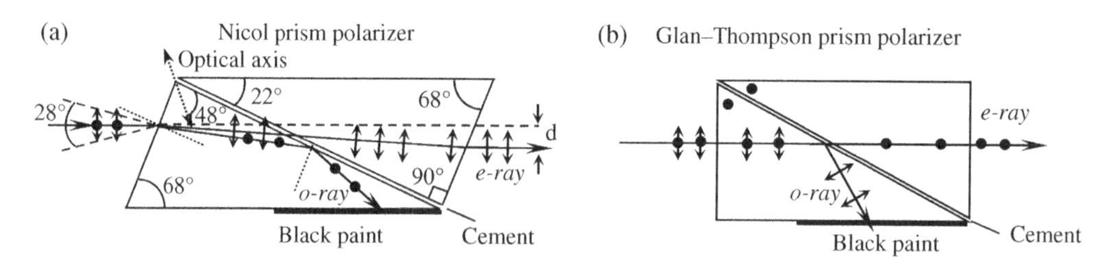

Figure 6.1 Illustrations of Nicol prism polarizer (a) and Glan–Thompson prism polarizer (b).

589.3 nm and undergoes total internal refraction at the calcite–cement interface because its angle of incidence at the glue layer ($n = 1.55$) exceeds the critical angle of 69° $\left(= asin\frac{1.55}{1.6584}\right)$ for the interface. It passes through the bottom side of the lower half of the prism before being absorbed by the black paint, as shown in Figure 6.1a. The *extraordinary* ray, or *e-ray*, experiences a lower refractive index ($n_e = 1.4864$ at 589.3 nm) in the calcite and is not totally reflected at the interface because it strikes the interface at a sub-critical angle. The *e-ray* merely undergoes a slight refraction, or bending, as it passes through the interface into the upper half of the prism. It finally leaves the prism as a ray of plane-polarized light, undergoing another refraction, as it exits the far right side of the prism, with a lateral displacement of *d* from the incident beam due to the refraction of the prism. The range of the incident beam angle for the Nicol prism is limited to 28° total or ±14° from the horizontal direction, as shown in Figure 6.1a, to avoid the transmission of the *o-ray* with the *e-ray* or to avoid the total internal reflection of the e-ray. Therefore, the Nicol prism polarizer cannot be used for a beam with a divergence angle larger than 28°, a drawback of such a polarizer.

Glan–Thompson Prism Polarizer A Glan–Thompson prism was devised to overcome the beam displacement issue of the Nicol prism. It consists of two right-angled calcite prisms that are cemented together by their long faces, as shown in Figure 6.1b. The optical axes of the calcite crystals are parallel and aligned perpendicular to the plane of reflection. Birefringence splits light entering the prism into two rays, experiencing different refractive indices; the *o-ray* is totally internally reflected from the calcite–cement interface, leaving the *e-ray* to be transmitted. Traditionally Canada balsam was used as the cement in assembling these prisms, but this has largely been replaced by synthetic polymers.

Glan–Foucault Prism Polarizer A Glan–Foucault prism (also called a Glan–air prism) is similar in construction to a Glan–Thompson prism in Figure 6.1b, except that two right-angled calcite prisms are spaced with an air gap instead of being cemented together. Total internal reflection of *o-ray* light at the air gap means that only *e-ray* light is transmitted straight through the prism.

Compared with the Glan–Thompson prism, the Glan–Foucault has a narrower acceptance angle over which it works, but because it uses an air gap rather than cement, much higher irradiances can be used without damage. The prism can thus be used with high power laser beams. The prism is also shorter (for a given usable aperture) than the Glan–Thompson design, and the deflection angle of the rejected beam can be made close to 90°, which is sometimes useful. Glan–Foucault prisms are not typically used as polarizing beamsplitters because while the transmitted beam is completely polarized, the reflected beam is not.

Glan–Taylor Prism Polarizer The Glan–Taylor prism is similar, except that the crystal axes and transmitted polarization direction are orthogonal to those of the Glan–Foucault prism. That is, the optical axis is in the plane of reflection parallel to the input side and the polarization of the transmitted light is an o-ray, in reference with Figure 6.1b. This yields higher transmission and better polarization of the reflected light. Calcite Glan–Foucault prisms are now rarely used, having been mostly replaced by Glan–Taylor polarizers and other more recent designs.

6.2.2 Sheet Polarizers

6.2.2.1 Film Polarizers

The crystal based polarizers are generally high cost and large thickness. For applications requiring large optical apertures, sheet polarizers of low cost and short optical path are preferred. According to Wikipedia (en.wikipedia.org/wiki/Polaroid_(polarizer) 2022), the sheet polarizer material was first patented in 1929 by Edwin H. Land and Joseph S. Friedman (Land 1951) and further developed by Land in 1932 at Polaroid Corporation. It consists of many microscopic crystals of iodoquinine sulfate (herapathite) embedded in a transparent nitrocellulose polymer film. The needle-like crystals are aligned during the manufacture of the film by stretching or by applying electric or magnetic fields. With the crystals aligned, the sheet is dichroic: it tends to absorb light, which is polarized parallel to the direction of crystal alignment but to transmit light which is polarized perpendicular to it, hence allowing the material to be used as a light polarizer. This material, known as *J-sheet*, was later replaced by the improved *H-sheet* Polaroid, invented in 1938 by Land. H-sheet is a polyvinyl alcohol (PVA) polymer impregnated with iodine. During manufacture, the PVA polymer chains are stretched such that they form an array of aligned, linear molecules in the material. The iodine dopant attaches to the PVA molecules and makes them conducting along the length of the chains. Light polarized parallel to the chains is absorbed, and light polarized perpendicular to the chains is transmitted. The latest type of film polarizers introduced by Polaroid is the *K-sheet* polarizer, which consists of aligned polyvinylene chains in a PVA polymer created by dehydrating PVA. This polarizer material is particularly resistant to humidity and heat.

Nowadays, many other companies, such as 3M™ Company, Dai Nippon Printing Co. Ltd., DuPont, BenQ Materials Corp., and American Polarizer, Inc., also make different linear and circular polarizer film sheets for various applications, including sunglasses, 3-dimensional (3D) movie glasses, polarizer filters for photography, and most importantly liquid crystal display (LCD) panels. Most of the film polarizers are made for visible wavelengths, with the *PER* in the range of 20–37 dB, and generally are not suited for optical fiber applications.

Note that the circular polarizer film is made by laminating a linear polarizer film with a quarter wave plate (QWP) film with its slow axis oriented ±45° from the passing axis of the polarizer, as discussed in Section 6.4.2.

6.2.2.2 Glass Polarizers

For applications requiring high PER, high dimensional stability, longevity, wide operation temperature range and the ability to withstand other severe environmental conditions for optical telecommunication, optical sensing, and optical instrumentation applications, these film type polarizers do not suffice. In the 1980s, Corning Inc. introduced a glass sheet polarizer trademarked Polarcor™ to overcome the shortcomings of the film polarizers, which could meet the stringent requirements of optical fiber communications. Polarcor is characterized by high extinction of up to 50 dB and low insertion loss (IL) throughout the 600–2300 nm wavelengths. This glass based polarizer has elongated silver crystals embedded in the material which strongly absorb light polarized along the long side of the silver crystal while has minimal absorption for light polarized along the short side of the crystal. This operation mechanism (resonant absorption by elongated silver crystals within the glass material) ensures the elimination of stray light, by absorbing the unwanted polarization. Since Polarcor is a solid glass product, it is extremely resistant to chemical, physical, and thermal damage, while exhibiting excellent optical properties. With extremely thin thicknesses, from 30 to 500 μm, the polarizing glass can be cut into a size down to 0.5 mm for integration into many fiber optic devices requiring small sizes. Since its introduction, Polarcor has been a keystone optical element in polarization-dependent isolators, optical modulators, polarimetry systems, ellipsometers, shutters, and many other polarization-based devices. A fiber optic in-line polarizer can be made by placing a small piece (~1 mm × 1 mm) polarizer chip between two fiber pigtailed collimators, as shown in Figure 6.7c. The fiber pigtails can be either SM fiber or PM fiber with the slow (or fast) axis aligned with the passing axis of the polarizer chip. The final *PER* and insertion loss of the device are 40 and 0.6 dB, respectively (Table 6.2).

Table 6.2 Typical specifications of Polarcor glass polarizers.

Bandwidth (nm)	600–1100	1275–1635	960–1160	1275–1345	1510–1590
PER (dB)	>40	>50	>23	>23	>23
Transmission without anti-reflection (AR)	>60%		>88.5%	>88.5%	>88.5%
Transmission with AR (2 sides)	>66%	>98.5%			
Refractive index	1.5218–1.5107	1.5083–1.5034			
Reflectance with AR per side	<1.0%	<0.2%			
Thickness (mm)	0.5 ± 0.05	0.5 ± 0.05	0.03 ± 0.01	0.03 ± 0.01	0.03 ± 0.01
TEC (/°C)	6.5×10^{-6}	6.5×10^{-6}	6.5×10^{-6}	6.5×10^{-6}	6.5×10^{-6}

TEC: Thermal expansion coefficient.

Other companies also produce similar glass polarizers. For example, the dichroic glass polarizers produced by Codixx AG are trademarked colorPol® and are made from a highly durable soda-lime glass containing silver nanoparticles. Like all dichroic polarizers they let the desired polarized light pass and absorb the unwanted polarization. Different types of polarizers are available to suit a wide field of applications operating at ultraviolet (UV) wavelength range (340–420 nm) and visible (VIS), near infrared (NIR), and MIR wavelength range (475–5.0 μm). They can be processed like glass or silicon wafers, while being thin like foil polarizers and withstand UV radiation and most chemicals without being damaged. They have a large acceptance angle of ±20° and a high accuracy of polarization axis.

6.2.3 Waveguide Polarizers

If an optical waveguide only supports a single polarization mode, such as transversal electric (TE) mode, and rejects or leaks the orthogonal mode, such as the transversal magnetic (TM) mode, it will function as a polarizer. A polarizing (PZ) fiber described previously in Section 6.1 is an example of such waveguide polarizers. Another example is the Lithium niobate ($LiNbO_3$) waveguide made with annealed proton exchange (APE) process (Rajneesh et al. 2009), which involves a replacement of Lithium ions (Li^+) by hydrogen ions, or protons (H^+). The replacement increases the extraordinary refractive index while slightly decreasing the ordinary refractive index. Therefore, an APE:$LiNbO_3$ waveguide supports only the extraordinary polarized light. If the waveguide is fabricated on the surface of an X-cut $LiNbO_3$ substrate and is aligned parallel to Y-axis, it will guide only the TE polarized light. Due to its long optical path length, an APE:$LiNbO_3$ waveguide polarizer can achieve extremely high *PER*, up to 90 dB and is commonly used in applications where extremely high PER is required, such as FOGs. An in-line fiber optic polarizer can be made by attaching fiber pigtails at the input and output ends of the waveguide. The insertion loss resulting from the waveguide attenuation and the coupling losses at the input and output interfaces is on the order of 3–4 dB, significantly higher than an in-line polarizer made with glass polarizer chip described previously.

Recently, many different waveguide polarizers are proposed and demonstrated with photonics integrated circuits (PICs) (Wu et al. 2020, 2021).

6.2.4 Other Types of Polarizers

There are many other types of polarizers, such as those made of wire-grid or made of a stack of glass plates at Brewster's angle. Because of the major performance issues, such as low PER or high transmission loss, these types of polarizers generally not have been adopted in the fiber optics industry and will not be discussed further in this book.

6.3 Polarization Beam Splitters/Combiners

As the name implies, a polarization beam splitter (PBS) separates an input beam into two beams with orthogonal polarizations, while a polarization beam combiner (PBC) combines two beams of orthogonal polarizations into a single beam. PBS and PBC are essentially the same device, but are used reversely. Therefore, we will just use PBS to refer to both PBS and PBC in this book. There are three types PBS in general, grouped by the way these devices are made, including those made with birefringence crystals, thin film coatings, and PIC.

6.3.1 Birefringence Crystal PBS

6.3.1.1 PBS with Angular Separation

Some of the crystal polarizers described in Section 6.2 can be used as a PBS if the totally internally reflected beam is not absorbed by the black paint, but used as the orthogonally polarized beam, although such a beam generally points in an awkward direction to be conveniently utilized. The PBSs described in this section are intended for use in both beams of orthogonal polarizations.

Figure 6.2a shows the construction and operation of the Rochon prism PBS. The device is constructed with two prisms made with birefringence crystals (here there are two negative birefringence crystals with $n_e < n_0$, where n_0 and n_e are the refractive indexes of the ordinary and extra-ordinary rays) of orthogonal optical axis orientations. The optical axis of the first prism is in the horizontal direction in the plane of the paper such that both of the orthogonal input polarizations experience n_0, while the optical axis in the second prism is perpendicular to the

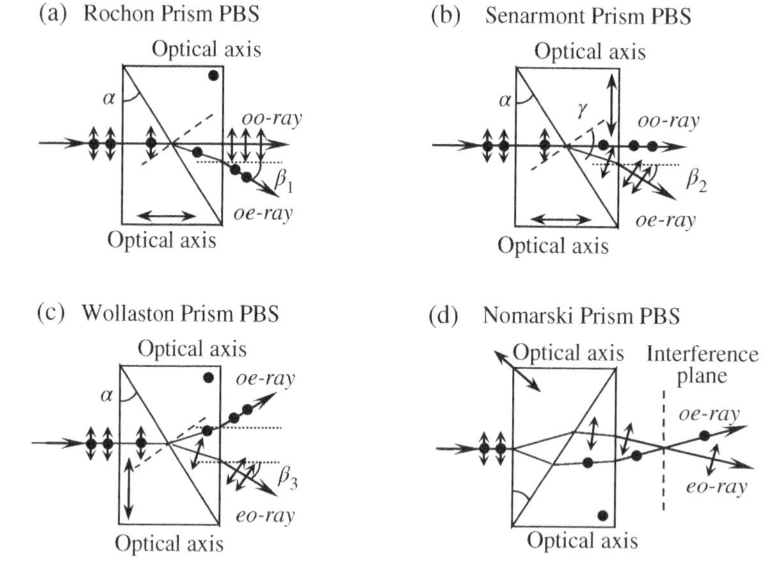

Figure 6.2 Illustrations of four different prism PBS' made with birefringence crystals, including Rochon prism PBS (a), Senarmont prism PBS (b), Wollaston prism PBS (c), and Nomarski prism PBS (d).

paper such that one of the polarizations continues to experience n_o (the oo-ray) and the other polarization experiences n_e (the oe-ray). At the interface between the two prisms, the oe-ray will be refracted according to Snell's law and enter the second prism at an angle and finally exits the second prism at a larger angle β_1, again following Snell's law. In fact, the first prism can be replaced with an isotropic material of similar refractive index, such as FK5 glass ($n = 1.4875$ at 0.589 nm). It can be easily shown that the angular deviation of the oe-ray from the oo-ray can be expressed as

$$\beta_1 = sin^{-1}\left\{\frac{n_e}{n_a}sin\left[sin^{-1}\left(\frac{n_o}{n_e}sin\alpha\right) - \alpha\right]\right\} \tag{6.3a}$$

where α is the base angle of the prisms, n_a is the refractive index of air, while n_o and n_e are the ordinary and extra-ordinary refractive indexes of the prisms, respectively. For a negative uniaxial crystal (such as calcite), $n_e < n_o$, the oe-ray deviates from the oo-ray downwards, as shown in Figure 6.2a. On the other hand, for a positive uniaxial crystal (such as YVO_4), $n_e > n_o$, the oe-ray bends upward from the oo-ray. As an example, for prisms made with calcite with a n_o of 1.6584 and a n_e of 1.4846 at 0.589 nm, $\beta_1 \approx 4°$.

The Senarmont prism PBS shown in Figure 6.2b is constructed similar to the Rochon prism, except that optical axis of the second prism is vertical in the plane of the paper. As a result, the light polarized perpendicular to the paper experiences n_o in the second prism (the oo-ray) and thus propagates without deviation, while the light polarized in the plane of the paper experiences n_e (the oe-ray) and enters the second prism with a refraction angle following Snell's law. Finally, this oe-ray exits the second prism with a second refraction. Similar to the Rochon prism, the oe-ray deviation angle β_2 can be expressed as

$$\beta_2 = sin^{-1}\left\{\frac{n_e(\gamma - \alpha)}{n_a}sin\left[sin^{-1}\left(\frac{n_o}{n_e}sin\alpha\right) - \alpha\right]\right\} \tag{6.3b}$$

where $n_e(\gamma - \alpha)$ is the refractive index of the oe-ray in the second prism and γ is oe-ray's refraction angle at the interface between the two prisms defined by Snell's law $n_o sin\alpha = n_e sin\gamma$. Similar to the Rochon prism, if the second prism is made with a negative uniaxial crystal ($n_e < n_o$), the oe-ray will bend downward and vice versa.

The most commonly used prism PBS in fiber optics is the Wollaston prism first developed by William Hyde Wollaston, as shown in Figure 6.2c. It also comprised two prisms, with the first prism having its optical axis vertical in the plane of the paper and the second prism having its optical axis perpendicular to the plane of the paper. The light polarized in the plane of the paper parallel to the optical axis of the first prism experiences n_e in the first prism and n_o in the second prism (the eo-ray). At the interface of the two prisms, it will be refracted downward if $n_o < n_e$ (positive uniaxial) and upward in $n_o > n_e$ (negative uniaxial), following Snell's law. On the other hand, the light polarized perpendicular to the plane of the paper first experiences n_e in the first prism and then n_o in the second prism (the oe-ray). At the interface of the two prisms, it will be refracted oppositely from that of the eo-ray according to Snell's law. Both the rays will undergo another refraction, and further deviate apart from each other. The angle separation $\Delta\beta$ between the two rays can be expressed as

$$\Delta\beta = sin^{-1}\left\{\frac{n_o}{n_a}sin\left[sin^{-1}\left(\frac{n_e}{n_o}sin\alpha\right) - \alpha\right]\right\} - sin^{-1}\left\{\frac{n_e}{n_a}sin\left[sin^{-1}\left(\frac{n_o}{n_e}sin\alpha\right) - \alpha\right]\right\} \tag{6.3c}$$

Figure 6.3a shows the beam crossing angle $\Delta\beta$ increases with the prism base angle α for a Wollaston prism made with YVO_4 crystals at a wavelength of 1550 nm. At the maximum base angle of 45°, the crossing angle can reach 23.76°. At a base angle of 9°, the beam crossing angle is 3.7°. Such

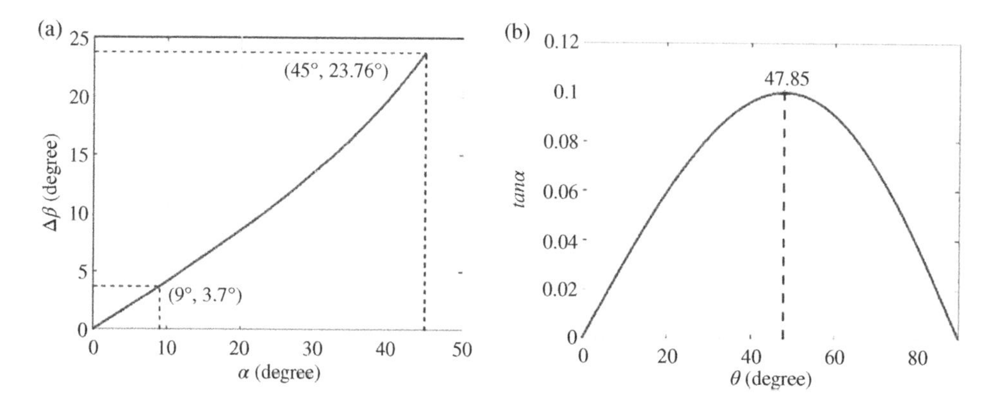

Figure 6.3 (a) The beam crossing angle $\Delta\beta$ of a Wollaston prism made with YVO_4 as a function of the base angle α at a wavelength of 1550 nm. (b) Beam displacement per unit length of crystal as a function of optical axis orientation angle for a polarization displacer made with YVO_4.

a Wollaston prism is widely used in the fiber optic industry to make fiber pigtailed PBS, as will be discussed next.

Another interesting prism PBS is the Nomarski prism named after George Nomarski, again comprising two prisms cemented together as shown in Figure 6.2d. The optical axis of the first prism is in the plane of the paper but is skewed, while the optical axis is perpendicular to the plane of the paper. The double refraction in the first prism causes light polarized perpendicular to the paper (the o-ray) to go downward and the light in the plane of the paper (the e-ray) to go upward if the prisms are made with a negative uniaxial crystal. After entering the second prism, o-ray becomes the e-ray (the oe-ray) and refracts downward while the e-ray becomes o-ray (the eo-ray) and refracts upward. These two beams will intersect at the outside of the crystal at a distance and a crossing angle determined by the thickness of the prisms. Nomarski prisms are often used in different interference contrast (DIC) microscopes.

6.3.1.2 PBS with Lateral Separation

This type of PBS is also called polarization beam displacer (PBD) and is widely used in making different fiber optic components, such as polarization insensitive optical circulators, as will be described in Section 6.6.5. Unlike the PBS with angular beam separation, here the two beams of orthogonal polarizations are parallel with a separation of d, as illustrated in Figure 6.4. Figure 6.4a shows a PBD made with a positive uniaxial crystal, such as YVO_4. The optical axis of the crystal

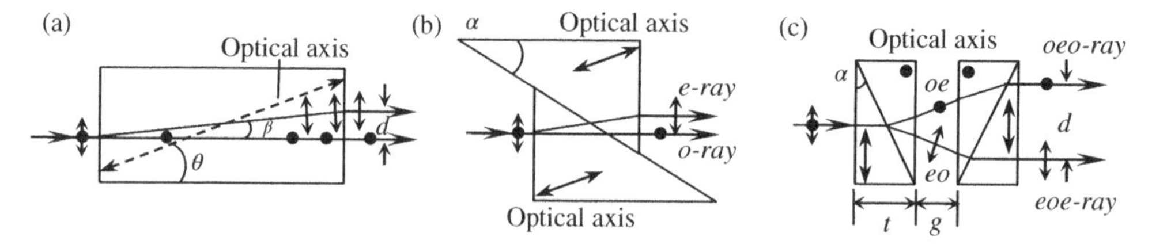

Figure 6.4 Illustrations of different polarization beam displacers (PBD). (a) A single crystal PBD made with a positive uniaxial crystal ($n_e > n_o$). (b) A variable beam separation PBD made with two identical prisms of positive uniaxial crystal. (c) A variable beam separation PBD made with two identical Wollaston prisms with one flipped 180° with respect to the other one. t and g are the thickness of each Wollaston prism and g is the gap between the prisms.

is in the plane of the paper at a skewed angle such that the o-ray propagates through the crystal without deviation, while the e-ray bends at an angle from the o-ray due to the double refraction. For a beam displacer made with a positive uniaxial crystal, the e-ray bends upwards, while for a beam displacer made with a negative uniaxial crystal, the e-ray bends downwards. As shown in Section 5.2, the beam separation d can be expressed as

$$d = L \cdot tan\beta \tag{6.4a}$$

$$tan\beta = \left(1 - \frac{n_o^2}{n_e^2}\right) \frac{tan\theta}{1 + \left(n_o^2/n_e^2\right) tan^2\theta} \tag{6.4b}$$

where β is the e-ray deviation angle, L is the crystal length, and θ is the optical axis orientation angle from the o-ray. It can be seen that when $n_e > n_o$, $\beta > 0$, the e-ray bends upwards; when $n_e < n_o$, $\beta < 0$, the e-ray bends downwards. Figure 6.4b shows that for a PBD made with YVO_4, the largest beam displacement occurs at an optical axis orientation angle of 47.85°. A 10 mm crystal length is required to achieve a beam displacement of 1 mm.

Because the displacement is linearly proportional to the path length L inside the crystal, the beam displacement d can be made variable by using two birefringence prisms with the same optical axis orientation, as shown in Figure 6.4b. The total path length can be controlled by sliding the two prisms with each other.

Two identical Wollaston prisms can also be used to make a variable PBD (Li, Li, and Zhu 2014), as shown in Figure 6.4c. The second Wollaston is flipped 180° from the first one and is separated from the first one with a gap g. It can be seen from the figure that the larger the gap, the larger the beam separation. In general, the beam separation can be expressed as

$$d = \frac{t \cdot tan\theta_{2oe} + g \cdot tan\varphi_{2oe}}{1 - tan\alpha \cdot tan\theta_{2oe}} + \frac{t \cdot tan\theta_{2eo} + g \cdot tan\varphi_{2eo}}{1 + tan\alpha \cdot tan\theta_{2eo}} \tag{6.5a}$$

where α is the base angle of the prism, t is the thickness of the Wollaston prism, and

$$\theta_{2oe} = sin^{-1}\left(\frac{n_o}{n_e} sin\alpha\right) - \alpha \tag{6.5b}$$

$$\theta_{2eo} = \alpha - sin^{-1}\left(\frac{n_e}{n_o} sin\alpha\right) \tag{6.5c}$$

$$\varphi_{2oe} = sin^{-1}\left\{n_e sin\left[sin^{-1}\left(\frac{n_o}{n_e} sin\alpha\right) - \alpha\right]\right\} \tag{6.5d}$$

$$\varphi_{2eo} = sin^{-1}\left\{n_o sin\left[\alpha - sin^{-1}\left(\frac{n_e}{n_o} sin\alpha\right)\right]\right\} \tag{6.5e}$$

It can be seen that the displacement d is linearly proportional to the gap g. Figure 6.5a shows the 3-dimensional (3D) plot of d as functions of g and α, while Figure 6.5b quantitatively shows the linear relation between d and g. Understandably, d increases with both g and α. In obtaining the plots, the prism thickness t is taken to be 10 mm and the parameters of YVO_4 at 1550 nm listed in Table 6.1 are used. A beam displacement of more than 10 mm can be achieved when the gap is set to be 20 mm using a pair of Wollaston prisms with a α of 45°.

6.3.2 Thin Film Coating PBS

Another method to make a PBS is by making use of Fresnel refractions in alternating high- and low-index film layers invented by Stephen MacNeille of Eastman Kodak Company in 1943

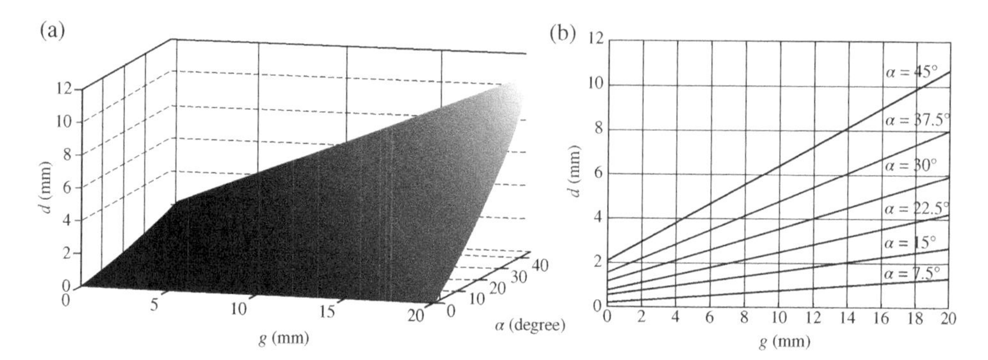

Figure 6.5 (a) 3D plot of Eq. (6.5a) showing the beam displacement d with g and α as variables. (b) Plots showing the displacement d is linearly proportional to the gap g for fixed prism base angles α.

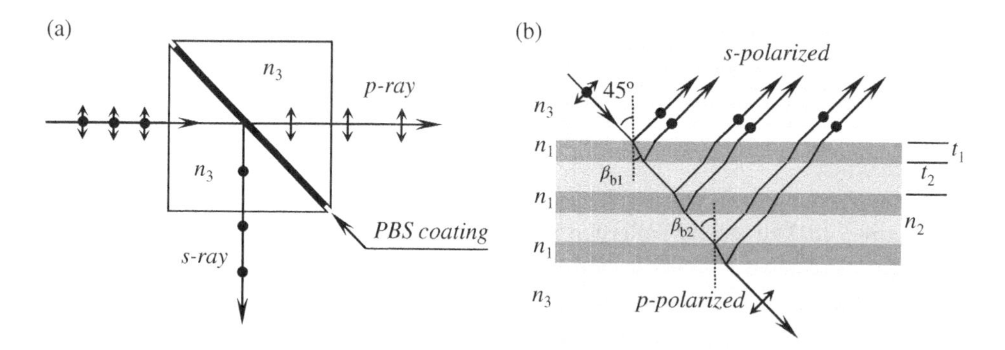

Figure 6.6 (a) Illustration of a polarization beam splitter cube made with a thin film of multiple layers of alternating high- and low-refractive index materials sandwiched between two prisms. (b) Detailed structure of the thin-film stack is designed to reflect s-polarized light at the interfaces of the layers with Brewster's angles and transmit p-polarized light.

(MacNeille 1943). As shown in Figure 6.6a, the thin-film stack is deposited on the hypotenuse face of an isosceles right angle prism, which is brought in contact with the hypotenuse face of a second isosceles right angle prism so that the reflected beam from the thin-film stack is at a right angle from the transmitted beam. Figure 6.6b illustrates a five-layer transparent thin-film stack having refractive indices n_1 and n_2, deposited between the prism pair having a refractive index n_3, where $n_1 \gg n_2$ and $n_1 > n_3$. All internal rays within the layers hit the next layer at Brewster's angle to assure the reflected light at each layer interface is completely s-polarized. If we choose $n_1 = 2.3$ (zinc sulfide) and $n_2 = 1.38$ (magnesium fluoride), then Brewster's angles β_{b1} and β_{b2} are given by

$$\beta_{b1} = atan\left(\frac{n_2}{n_1}\right) = 30.9° \tag{6.6a}$$

$$\beta_{b2} = atan\left(\frac{n_1}{n_2}\right) = 59.1° \tag{6.6b}$$

where $\beta_{b1} + \beta_{b2} = 90°$ and Snell's law is satisfied at each layer interface. To ensure a 45° incident angle at the thin-film stack of the light beam from the prism while satisfying a β_{b1} of 30.9°, the refractive index n_3 of the prism material must satisfy:

$$n_3 = n_1 \frac{sin\beta_{b1}}{sin45°} \approx 1.67 \tag{6.6c}$$

Since the 45° incident angle is not Brewster's angle, the incident ray is not completely *s*-polarized upon reflection at the first high index layer, which will compromise the PER of the resulting PBS. A suitable material for the prism is SF5 glass having a refractive index of 1.673.

To maximize the intensity of the reflected *s*-polarized ray at each layer, the layer thickness is controlled such that the rays reflected from all layers are all in phase. To achieve this, the physical thicknesses t_1 and t_2 of the layers are controlled to be

$$t_1 = \frac{\lambda}{4\sqrt{(n_1^2 + n_2^2)/n_1^2}} \tag{6.7a}$$

$$t_2 = \frac{\lambda}{4\sqrt{(n_1^2 + n_2^2)/n_2^2}} \tag{6.7b}$$

where λ is the wavelength of the incident light. It can be shown that for a stack of seven layers, approximately 50% of the incident unpolarized light is reflected as *s*-polarized, while the other half is transmitted as *p*-polarized light. Commercial PBS cubes of a modified MacNeille design can achieve a transmission extinction ratio of 30 dB.

6.3.3 Fiber Pigtailed Polarizers and PBS

For a polarization component to be used in an optical fiber communication or a sensing system, an input fiber and an output fiber must be attached to the component in the factory to make it easier to be used. Otherwise, it will be extremely difficult for an ordinary user to couple the light out from the input fiber, passing it through the component, and then couple the light into the output fiber. Such a device is often called "pigtailed" device and the input and out fibers are called "pigtails." Because the light exiting from a fiber generally diverges, it is difficult to directly couple the light back into another fiber without experiencing a high loss. The divergence angle of the exiting light is generally characterized by a quantity call numerical aperture (NA), which is defined as the sine of the half of the divergence angle. For a common SM fiber used in telecommunications industry (SMF-28) with a core diameter of 9 μm and cladding diameter of 125 μm, the *NA* is about 0.12 ± 0.02, corresponding to a full divergence angle of 13.75°. A fiber collimator is generally used to form a collimated beam at the fiber exit or focus a collimated beam into a fiber with a minimal loss. It is a critical component for making fiber optic devices and has been discussed in detail by Jay Damask (2005a).

Figure 6.7a shows a typical construction of a single fiber collimator consisting of a ferrule and a micro lens. The ferrule is generally made with glass or ceramic for great thermal stability, with a hole around 126 μm in diameter at the center for the fiber to go through. One end of the ferrule has an angle of 8°. After the fiber is affixed in the hole with adhesive, it is polished with the 8° cut surface to prevent light reflecting back into the fiber at the fiber interface. Finally, the ferrule is butt onto the end of a lens also having an 8° cut surface and is cemented together with an adhesive, with the fiber end on the focal point of the lens to assure that the light beam exiting from the lens is collimated. A second collimator can be used to couple the collimated light beam from a first collimator into the fiber pigtail of the second collimator. The beam diameter in general is about 0.4–0.5 mm and a working distance (defined as the separation distance between a pair of collimators with an acceptable insertion loss on the order of 0.5 dB) is from 5 to 20 mm.

Figure 6.7b shows a typical dual fiber collimator in which a ferrule holds two fibers separated by a distance *d* (typically 125 μm). Like in Figure 6.7a, the ferrule end has an 8° angled surface and is butt onto a lens with a complementary 8° angled surface after being polished. The two fiber ends

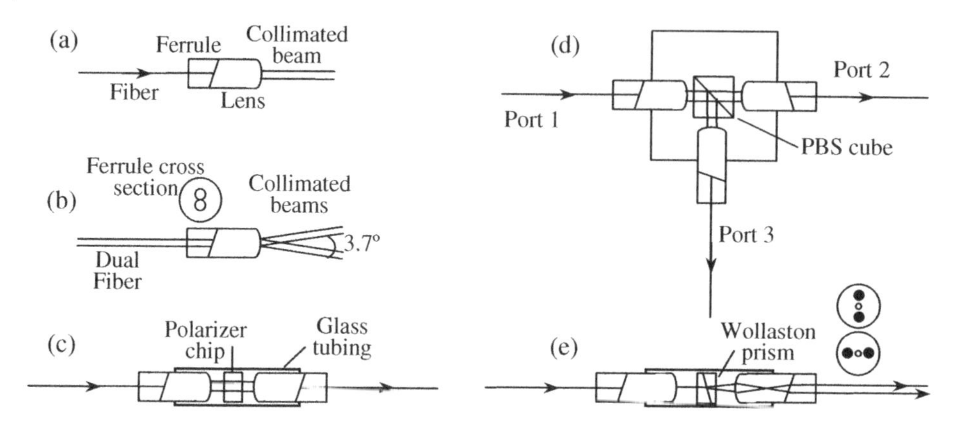

Figure 6.7 Illustration of pigtailing polarization devices. (a) The construction of a single fiber collimator; (b) the construction of a dual fiber collimator; (c) a typical fiber pigtailed polarizer; (d) a fiber pigtailed PBS made with a thin film PBS cube; and (e) a fiber pigtailed PBS made with a Wollaston prism and a dual fiber collimator. The axes of the two fibers in the dual fiber collimator are oriented orthogonally for the case of PM fiber pigtails.

are on the focal plane of the lens. Because the two fiber ends are symmetrical on each side of the optical axis of the lens, the exiting beams will be collimated and bend toward each other with a crossing angle, as shown in Figure 6.7b, which depends on the focal length of the lens and the fiber separation, but is typically 3.7°.

A pigtailed polarizer (sometimes is called in-line polarizer) is shown in Figure 6.7c in which an input collimator provides a collimated beam for passing through a polarizer chip (for example a Corning Polarcor glass polarizer chip) and is focused back into an output fiber via an output collimator. The two collimators and the polarizer chip are hosted in a glass tube and affixed together with a special adhesive with good temperature stability (such as Epotek™ 353ND epoxy). The input and output fibers can be either PM fiber or SM fiber. If a PM fiber is used, the slow axis of the PM fiber is generally aligned to the polarizer passing axis unless specified otherwise.

A pigtailed PBS based on a thin film PBS cube is shown in Figure 6.7d, in which an input fiber collimator provides a collimated beam. The *p*-polarization goes through the PBS cube and is focused by the first output collimator into a first output fiber. The *s*-polarization is reflected by the PBS cube before being focused by a second output collimator into a second output fiber. The input and output fibers can be either PM fiber or SM fiber. If the input fiber is PM fiber, its slow axis can be aligned with the *p*-polarization direction to direct the light all into Port 2, the *s*-polarization direction to direct the light all into Port 3, or 45° from the *s*-polarization direction to split half the light into Port 2 and the other half into Port 3.

The PBS described in Figure 6.7d is bulky and the PER is relatively poor, on the order of 20 dB, due to the limitations of thin film coating. Figure 6.7e shows a compact pigtailed PBS made with a micro Wollaston prism and a dual fiber collimator. To match the 3.7° crossing angle of the dual fiber collimator, the Wollaston prism is made with a proper base angle α, which is 9° if made with YVO_4, as shown in Figure 6.3a. Again, the input fiber and output fiber pigtails can be of SM or PM, depending on the application. For the case of using PM fiber pigtails, the slow axes of the two fibers in the dual fiber collimator must be oriented orthogonally. For this type of PBS, the PER is generally limited by the fiber axis alignment, on the order of 23–25 dB, although the PER of the Wollaston prism itself is more than 40 dB. For more advanced PBS design, please be referred to Yonglin Huang's patent (Huang and Xie 1999).

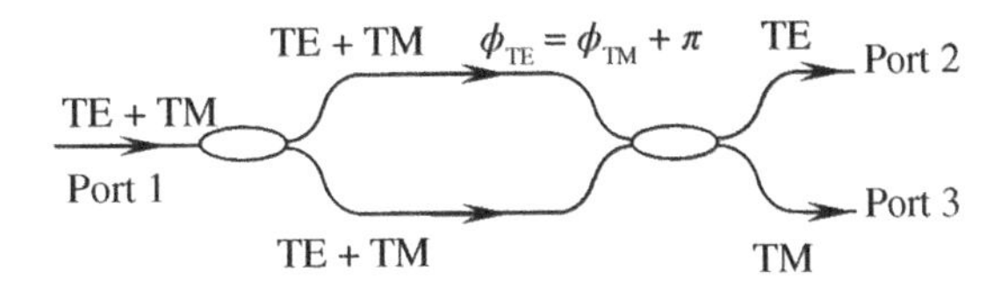

Figure 6.8 Illustration of an interferometer based PBS.

6.3.4 Waveguide PBS

Figure 6.8 shows a PBS made with a waveguide interferometer in which the phase difference between the TE and TM modes is π and the optical path length difference between the two arms for the TE or TM mode is $2n\pi$ (n is an integer) such that at Port 2 the TE mode has constructive interference while TM has destructive interference. However, at Port 3, TE mode has destructive interference while TM has constructive interference. Consequently, all optical power of TE polarization exits from Port 2 while all optical power of TM polarization exits from Port 3. Such a PBS requires precise phase control between the interference arms, as well as between the two polarizations and therefore is sensitive to temperature variations. In addition, such a PBS is also wavelength sensitive, resulting in limited operation bandwidth.

PBS with great performance can also be made in integrated photonics circuits, which will be discussed in a future edition of the book.

6.4 Linear Birefringence Based Polarization Management Components

6.4.1 Wave Plates

Wave plates are another key component for polarization control and management. As introduced in Chapter 4 and further illustrated in Figure 6.9, a wave plate is made of a birefringence material having a pair of orthogonal axes of different refractive indices. A light beam polarized along the axis of larger refractive index travels slower and accumulates more phase than that polarized along the axis of smaller refractive index. Therefore, the axis having a larger refractive index is referred to slow axis while the axis having a smaller refractive index is called fast axis. The phase difference δ between the two orthogonal polarizations is called retardation and can be expressed as

$$\delta = \frac{2\pi\Delta L(n_s - n_f)}{\lambda} = \frac{2\pi\Delta L\Delta n_B}{\lambda} \tag{6.8}$$

where ΔL is the thickness of the birefringence material, n_s and n_f are the refractive indices of the slow and fast axes, respectively, and $\Delta n_B = (n_s - n_f)$ is the birefringence. A wave plate is

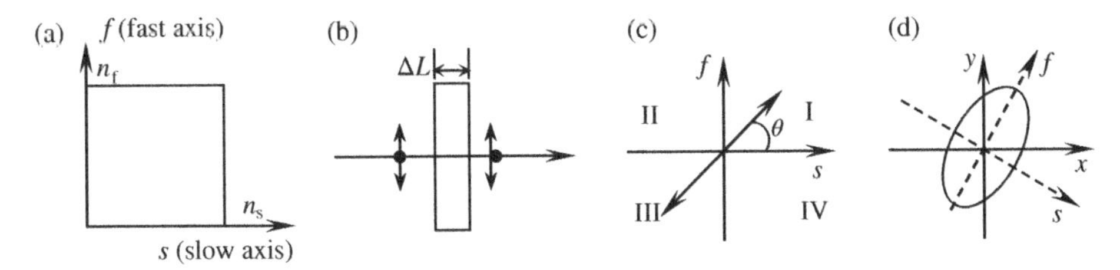

Figure 6.9 (a) Illustration of a wave plate having a slow and a fast axis; (b) Two orthogonal polarization components experience a slight delay after passing through a wave plate of thickness ΔL; (c) aligning a linear SOP θ degrees from the slow axis; (d) aligning the polarization ellipse having a major and minor axis of $2a$ and $2b$ with the slow and fast axes of a wave plate.

called quarter wave plate if the retardation $\delta = (2j+1)\pi/2$ because the optical path difference $\Delta L \Delta n_B = (2j+1)\lambda/4$, where λ is the wavelength of the light beam and j is an integer representing the wave plate order. For the case of $j = 0$, the corresponding wave plate is called zero order quarter wave plate (QWP). Similarly, a wave plate is called half wave plate if the retardation $\delta = (2j+1)\pi$ because the optical path difference is $(2j+1)\lambda/2$. For the case of $j = 0$, the corresponding wave plate is called zero order half-wave plate (HWP).

6.4.2 Polarization Manipulation with a Quarter-Wave Plate

A QWP can perform two important functions to manipulate the SOP, one is to convert a linear SOP into an elliptical SOP with its major axis aligned with the slow or fast axis of the wave plate, particularly a circular SOP, and the other one is to convert an elliptical SOP into a linear SOP, as discussed in the following text.

When a linear SOP is oriented θ degrees from the slow axis of a QWP, as shown in Figure 6.9c, its output SOP $\hat{e}_{out}$ from the wave plate can be expressed as

$$\varepsilon_{out} = sin\theta\hat{\mathbf{f}} + icos\theta\hat{\mathbf{s}} \tag{6.9}$$

where $\hat{\mathbf{f}}$ and $\hat{\mathbf{s}}$ are unit vectors representing the fast and slow axes of the wave plate. It can be easily seen from Eq. (6.9) that when θ is $\pm 45°$, $\varepsilon_{out} = (\pm\hat{\mathbf{f}} + i\hat{\mathbf{s}})/\sqrt{2}$, representing a left-hand circular SOP (LCP) and right-hand circular SOP (RCP), respectively, assuming light beam propagating toward the reader. For $\theta \neq \pm 45°$, ε_{out} is an elliptical SOP having different ellipticities and handedness depending on the value of θ, however, with the major axis aligned with the slow or fast axis of the QWP. When θ is in quadrants I or III, the output SOP is LEP, while θ is in quadrants II or IV, the output SOP is REP. In other words, when the slow (or fast) axis of the QWP is fixed, varying the angle of the input linear SOP with respect to the slow axis of the QWP from 0° to 180° can generate an elliptical SOP with any desired ellipticity and handedness, however, with a fixed orientation aligned with the slow axis of the QWP. If one has the freedom to rotate the slow axis of the QWP from 0° to 180°, then arbitrary elliptical SOP with any desired ellipticity, handedness, and orientation can be generated by changing the relative angle between the linear SOP and the slow axis of the QWP from 0° to 180°.

On the other hand, for an arbitrary elliptical SOP, a QWP can always be rotated around the optical beam to align its slow axis with the major or minor axis of the polarization ellipse of an arbitrary SOP, as shown in Figure 6.9d. In this case, the SOP of the input and output light can be expressed as

$$\varepsilon_{in} = a\hat{\mathbf{f}} + ib\hat{\mathbf{s}} \tag{6.10a}$$

$$\varepsilon_{out} = a\hat{\mathbf{f}} + ibe^{i\pi/2}\hat{\mathbf{s}} = a\hat{\mathbf{f}} - b\hat{\mathbf{s}} \tag{6.10b}$$

where a and b are half of the major and minor axes of the polarization ellipse, respectively. Clearly, the QWP has converted the elliptical SOP of $(a\hat{\mathbf{f}} + ib\hat{\mathbf{s}})$ into a linear SOP of $(a\hat{\mathbf{f}} - b\hat{\mathbf{s}})$. For a circular input SOP, $a = b$, the output SOP is therefore $(\hat{\mathbf{f}} - \hat{\mathbf{s}})$, which is a L − 45P SOP.

6.4.2.1 Circular and Elliptical Polarizer

If one puts a linear polarizer $\hat{\mathbf{p}} = (\hat{\mathbf{f}} + \hat{\mathbf{s}})$ behind the QWP, it will totally block the light ($\varepsilon_{out} \cdot \hat{\mathbf{p}} = 0$). That is, a QWP followed by a polarizer oriented polarizer with its passing axis oriented +45° from the slow axis of the QWP acts as a circular polarizer. In practice, people generally cement a QWP with a polarizer having its passing axis oriented ±45° from the slow axis of the QWP to form a circular polarizer. The +45° polarizer is to block an input circular SOP of $(\hat{\mathbf{f}} + i\hat{\mathbf{s}})$, while the −45° is to block an input SOP of $(\hat{\mathbf{f}} - i\hat{\mathbf{s}})$.

Similarly, one may make an elliptical polarizer to totally block an elliptical SOP $(a\hat{\mathbf{f}} + ib\hat{\mathbf{s}})$ using a QWP followed by a polarizer with its passing axis oriented in the direction of $\hat{\mathbf{p}} = (b\hat{\mathbf{f}} + a\hat{\mathbf{s}})$.

6.4.2.2 Anti-reflection or Anti-glare Film

One may use the circular polarizer in reverse direction to block the reflection from a reflecting surface, such as a window or a TV screen. When in use, the QWP is pressed against the reflecting surface. In practice, the light first passes through the polarizer oriented +45° from the slow axis of the QWP behind it to become linearly polarized at 45° before it is converted into a circular SOP by the QWP. Upon reflection, it is converted by the QWP into a linear SOP oriented −45° from the slow axis and finally is blocked by the +45° polarizer. Both the polarizer and the QWP can be made into thin films and be cemented together to make a composite circular polarizer film, as mentioned in Section 6.2.2. Such an anti-reflection film or circular polarizer film is widely used for car windows and display screens to cut down unwanted reflection.

6.4.3 Polarization Manipulation with a Half-Wave Plate

The main functions for an HWP are (i) to rotate the orientation of a linear SOP, (ii) to reverse the sense of rotation of an elliptical SOP aligned with the fast (slow) axis of the HWP, and (iii) to rotate the major axis of a polarization ellipse while reversing the sense of rotation.

The input SOP shown in Figure 6.10a can be expressed as

$$\varepsilon_{in} = cos\varphi\hat{\mathbf{f}} - sin\varphi\hat{\mathbf{s}} \tag{6.11a}$$

After the light beam passes through the HWP, the slow axis component will accumulate a relative phase of π such that the output SOP can be expressed as

$$\varepsilon_{out} = cos\varphi\hat{\mathbf{f}} - sin\varphi e^{\pm i\pi}\hat{\mathbf{s}} = cos\varphi\hat{\mathbf{f}} + sin\varphi\hat{\mathbf{s}} \tag{6.11b}$$

It can be seen by comparing Eq. (6.11b) with Eq. (6.11a) that the HWP has rotated the input SOP by an angle of 2φ with respect to the fast axis of the HWP, as shown in Figure 6.10a.

An input elliptical SOP with its major axis aligned with the slow and fast axes of the HWP, as shown in Figure 6.10b, can be expressed as

$$\varepsilon_{in} = a\hat{\mathbf{f}} \pm ib\hat{\mathbf{s}} \tag{6.11c}$$

where a and b are the major and minor axes of the polarization ellipse, respectively. After passing through the HWP with its major or minor axis aligned with the slow or fast axis of the HWP, the output SOP can be written as

$$\varepsilon_{out} = a\hat{\mathbf{f}} \pm ibe^{\pm i\pi}\hat{\mathbf{s}} = a\hat{\mathbf{f}} \mp ib\hat{\mathbf{s}} \tag{6.11d}$$

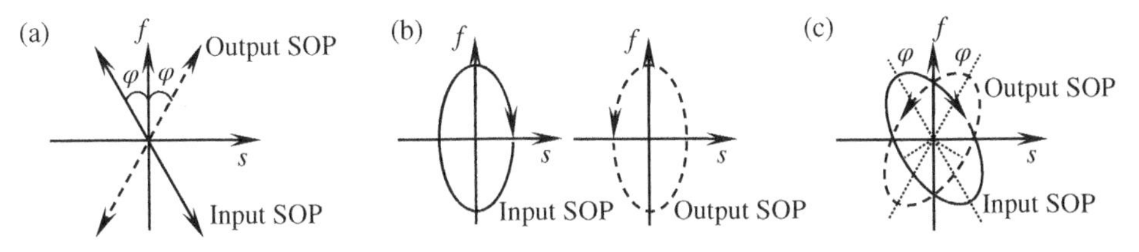

Figure 6.10 (a) Rotation of a linear SOP by an angle of 2φ if the input SOP is orientated φ degrees from the fast axis of the HWP. (b) Handedness reversal of an elliptical SOP. (c) Handedness reversal and rotation of the major axis of an input polarization ellipse by an angle of 2φ if the input major axis of the polarization ellipse is orientated φ degrees from the fast axis of the half-wave plate.

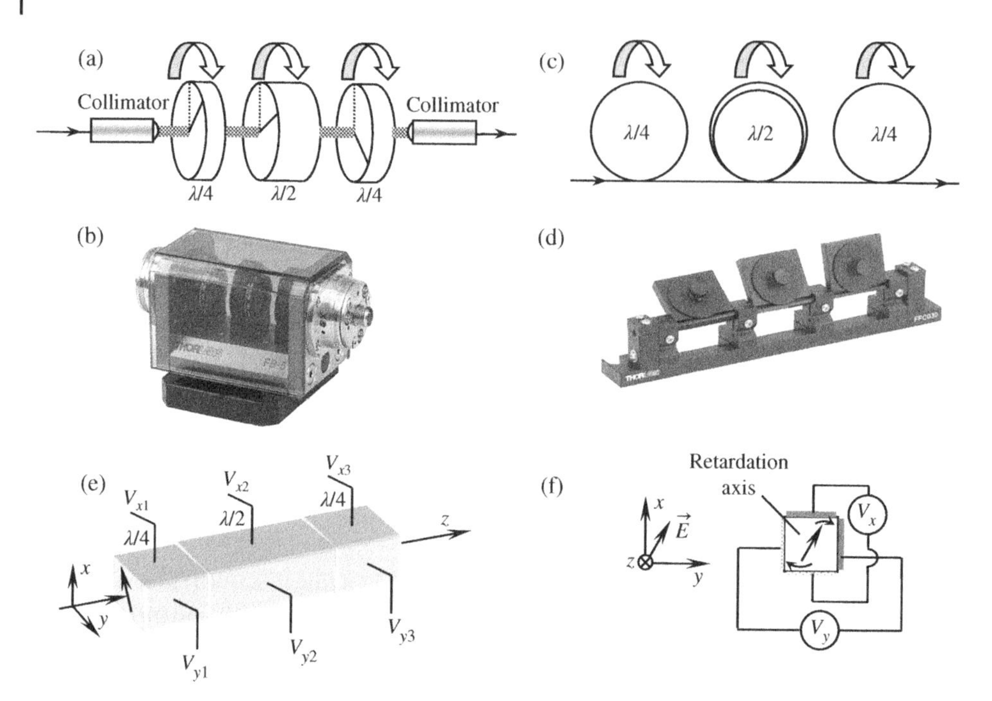

Figure 6.11 Illustration of three different implementations of the tri-plate polarization controller. (a) Free-space implementation; (b) A real product photo of polarization controller of (a) (Source: Courtesy of Thorlabs); (c) An all-fiber implementation (the Lefevre controller): (d) A real product photo of the Lefevre controller (Source: Courtesy of Thorlabs); (e) A tri-plate polarization controller made with electro-optic crystals for high speed operation; (f) The crystal cross section showing the directions of the voltages applied on the device in (e) and the retardation axis.

The reversal of sign indicates that the handedness is reversed.

In general, an arbitrary input SOP can be expressed in the (f, s) coordinates as

$$\varepsilon_{in} = a_f \hat{\mathbf{f}} \pm a_s e^{i\delta} \hat{\mathbf{s}} \tag{6.11e}$$

where a_f and a_s are the field amplitudes and δ is the relative phase, respectively. The output SOP from the HWP can be expressed as

$$\varepsilon_{out} = a_f \hat{\mathbf{f}} \pm a_s e^{i\delta \pm i\pi} \hat{\mathbf{s}} = a_f \hat{\mathbf{f}} \mp a_s e^{i\delta} \hat{\mathbf{s}} \tag{6.11f}$$

Equation (6.11f) indicates that an HWP will rotate the major axis of an arbitrary elliptical SOP by 2φ degrees with respect to its fast axis, together with a handedness reversal, as shown in Figure 6.11c. Equations (6.11b) and (6.11d) are just special cases for Eq. (6.11f).

6.5 Polarization Control with Linear Birefringence

A polarization controller is defined as a device that is capable of converting an arbitrary input SOP into any desired output SOP without a significant transmission loss. In other words, it is capable of transforming any point on a Poincaré sphere into any other point on the sphere. A single wave plate described in Section 6.4 cannot be a general purpose polarization controller because it is only able to convert an input SOP into a limited number of output SOPs by adjusting the angle of the wave plate with respect to the input SOP. In order to transform an arbitrary input SOP into an arbitrary output SOP, multiple wave plates may be used.

Birefringence plate based polarization controllers may be classified into the following three groups: multiple wave plates with fixed retardations but variable orientation angles; a single wave plate with both variable retardation and orientation; and multiple wave plates with fixed orientation but variable retardation.

Controllers based on wave plates of fixed retardation are wavelength-sensitive. Those that rely on physical rotation are generally slow. Other than these fundamental limitations, all three approaches function reasonably well in principle. However, implementation of these approaches in practice determines the performance, cost, and reliability of the devices, as will be discussed in the following text.

6.5.1 Polarization Control with Multiple Wave Plates of Fixed Retardation but Variable Orientation

A classic polarization controller consisting of three rotatable wave plates is illustrated in Figure 6.11a. In this configuration, an HWP is sandwiched between two QWPs and the retardation plates are free to rotate around the optical beam with respect to each other. The first QWP converts any arbitrary input polarization into a linear polarization, as described in Section 6.4.2. The HWP then rotates the linear polarization to a desired angle between $\pm 90°$, as described in Section 6.4.3. Finally by rotating the second QWP between $\pm 90°$, one can convert the linear SOP into any desired elliptical SOP, including linear SOPs of any orientation and circular SOPs of any handedness. In this approach, the retardation of each plate is fixed, but the relative angles of the retardation plates are variable. If the desired output SOP is linear, then the second QWP is not required because the first QWP and the HWP are sufficient to convert and arbitrary elliptical SOP into any linear SOP.

Commercial applications of this approach have produced respectable results for traditional free-space applications. However, for applications in fiber optics, this technique requires collimating light from the input fiber into a beam to pass through the three wave plates and then refocus the beam into the output fiber. The processes of collimating, aligning, and refocusing are time consuming and labor intensive. In addition, the wave plates and micro-lenses are not cheap and need antireflection coating or angle polishing to prevent back reflection. Insertion loss is high because the optical beam has to be coupled out of one fiber and refocused into another. Furthermore, the wave plates are inherently wavelength-sensitive (any fractional wave plate is always specified with respect to a particular wavelength), making the device sensitive to wavelength variations. Finally, electrical motors or other mechanical devices rotate the wave plates, limiting the controller speed. Figure 6.11b shows such a commercial tri-plate polarization controller with three wave plates mounting on three rotation stages and an input and output fixtures on both ends of the device for accepting input and output fiber connectors and for aligning the optical beam.

In order to overcome some of the drawbacks of the tri-plate controllers previously, an all-fiber controller based on this mechanism was devised for reducing the insertion loss and cost, as shown in Figure 6.11c. In this device, three fiber loops replace the three free-space retardation plates. This all-fiber device was first reported by H.C. Lefevre then at Stanford University in 1980 and has been often referred to as the Lefevre controller (Lefevre 1980a, 1980b). Some people also nick name the device Mickey ear controller from its appearance.

As described in Chapter 3, coiling the fiber induces stress, producing birefringence inversely proportional to the square of the coils' diameters. Adjusting the diameters and number of turns can create any desired fiber wave plate. Specifically, the bending induced birefringence Δn_g can be described by Eq. (3.10b) and the total retardation $\Delta\varphi$ of a fiber loop with a radius of R at a wavelength of 1550 nm can be expressed as

$$\Delta\varphi = \left(\frac{2\pi}{\lambda}\right)(2\pi R)\Delta n_g = \left(\frac{2\pi}{\lambda}\right)2\pi R \cdot 6.61 \times 10^{-10}\left(\frac{1}{R}\right)^2 \tag{6.12}$$

Let $\Delta\varphi = \pi/2$, the radius R for making a QWP with a single turn fiber loop is then 2.14 cm. By simply adding a second turn, the retardation can be doubled and an HWP can be made. Like in the free-space tri-plate polarization controller, by rotating the two all-fiber QWPs and the single half-wave plate (HWP) about the straight fiber, as shown in Figure 6.11c, the Lefevre controller is capable of converting an arbitrary input SOP into any desired output SOP. The only difference is that physically rotating a wave plate by θ degrees, the actual relative angle between the input SOP and the wave plate is $\theta' = (1-\rho)\theta$ due to the twist induced optical activity described by Eq. (3.14), where $\rho = g/2$ and $g = 0.16$ from Eq. (3.13c). In other words, in order to achieve a rotation angle θ' of 10°, a physical rotation angle θ must be 10.87°, slightly more than rotating a free-space wave plate.

Despite the reduced loss and cost, the device still suffers from wavelength sensitivity and low speed. In addition, the fiber coils are quite large and the resulting device is generally bulky, as shown in Figure 6.11d for a commercial device. Therefore, the use of these Lefevre controllers is primarily limited to laboratories. To improve the speed of the tri-plate polarization controller and reduce the size for field applications, the electro-optic (E-O) effect can be used to make electro-optic wave plates whose retardation or birefringence axis can be rotated by applied voltages, as shown in Figure 6.11e. Such an electro-optic (E-O) wave plate can be made using $LiNbO_3$ (Campbell and Steier 1971). However, it requires about 1000 V for operation at the 1.55 pm wavelength. To reduce the high voltage, lead lanthanum zirconate titanate (PLZT) crystal with a large quadratic electro-optic effect has been deployed for demonstrating such rotating-wave plates for making a fast polarization controller (Shimizu and Kaede 1988).

As shown in Figure 6.11f, when two voltages (V_x and V_y) 90° out of phase are applied along x- and y-directions across a PLZT crystal:

$$V_{xi} = V_{0i} sin\varphi_i \tag{6.13a}$$

$$V_{yi} = V_{0i} cos\varphi_i \tag{6.13b}$$

where V_{0i} is the amplitude of the voltage applied to the ith E-O plate and φ_i is its phase, the electric field in PLZT is rotated in a plane normal to the light propagation axis z, thus inducing a birefringence rotation in PLZT as φ increases or decreases. As the rotation angle φ_i has no limitation in its working range, the principal axis of the E-O wave plate can be rotated endlessly. The E-O wave plate operates as an HWP and a QWP plate when V_{0i} is a half-wave voltage (the voltage for inducing π retardation) and a quarter-wave voltage (the voltage for inducing $\pi/2$ retardation), respectively. Similar to other tri-plate controllers, any SOP can be transformed into any desired SOP state using an HWP sandwiched between two QWPs. For transforming an arbitrary SOP to any desired linear SOP, only a QWP and an HWP are required. In practice, the length of the HWP is twice as long as that of the QWP. Alternatively, the two QWPs and the HWP can be on the same crystal, but having separate electrodes, with the length of the electrode for the HWP being twice as long as that of the QWPs. In the first demonstration of such a polarization controller using rotatable E-O wave plates, a PLZT with a size of 1 mm × 1 mm × 26 mm with a QWP and HWP was used. Unfortunately, such E-O plate polarization controllers have never been commercialized.

6.5.2 Polarization Controller with a Single Wave Plate of Variable Retardation and Orientation

In an alternative approach, a Babinet–Soleil compensator can convert any input polarization state into any desired output polarization state. The heart of such a device (Figure 6.12a) is a composite wave plate made from two birefringent crystal wedges. The thickness (and therefore the total retardation) of the wave plate varies by sliding the two wedges against one another. The orientation of the composite wave plate can also rotate around the optical beam. Compared with the previous device (Figure 6.11a), this controller has the advantage of insensitivity to wavelength because precision retardation can be achieved for any wavelength. However, it also suffers from high cost, high insertion loss, and low speed for the same reasons for the tri-plate controller (Figure 6.11a).

To reduce the cost and insertion loss, an all-fiber polarization controller based on the Babinet–Soleil compensator principle was developed (Figure 6.12b) in 1996 by Yao with the trade name PolaRITE™ (Yao 1995). Some people simply call it the Yao controller. The device comprises a fiber squeezer that rotates around the optical fiber. Applying a pressure to the fiber produces a linear birefringence, effectively creating a fiber wave plate whose retardation varies with the pressure. A combination of screw and spring is used to precisely control the pressure on the fiber, as shown in Figure 6.12c. As will be explained next, simple squeeze-and-turn operations can generate any desired polarization state from any arbitrary input polarization. In addition to low insertion loss and low cost, the device is compact and wavelength insensitive, especially

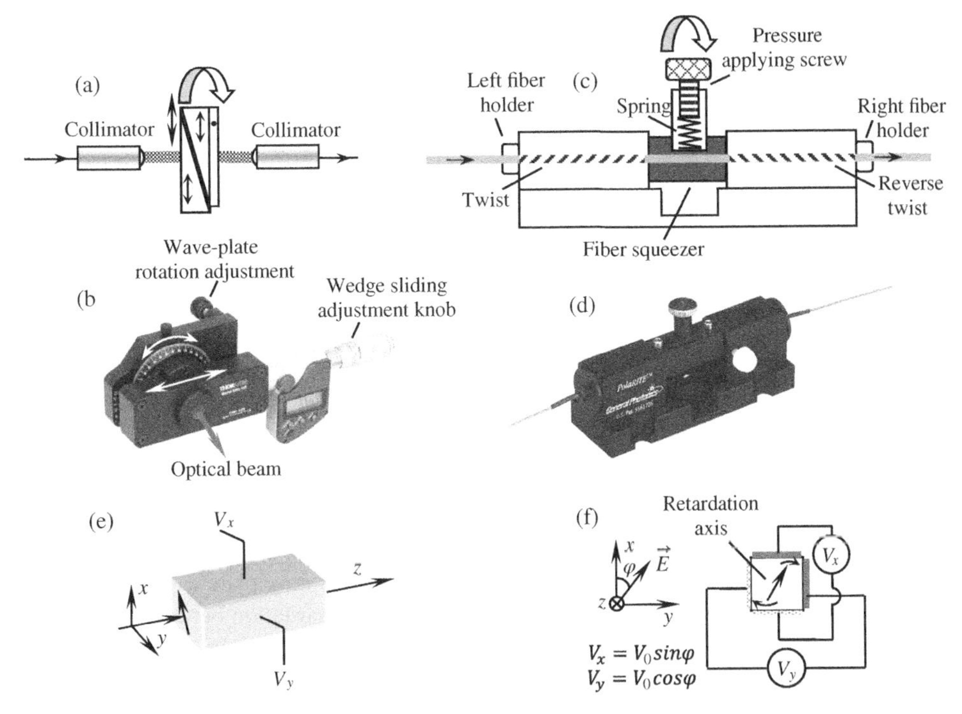

Figure 6.12 (a) Illustration of a polarization controller based on Babinet–Soleil compensator principle; (b) A commercial free-space Babinet–Soleil compensator (Source: Courtesy of Thorlabs); (c) Illustration of the all fiber Babinet–Soleil compensator based polarization controller (the Yao controller); (d) A photo of the first commercial Yao controller (Source: Courtesy of General Photonics/Luna Innovations); (e) An electro-optic Babinet–Soleil compensator made with an E-O wave plate of both variable retardation and orientation angle; (f) Diagraph showing how the voltages are applied and the direction of the resultant electrical field.

compared with the Lefevre controllers. This makes the Yao controllers attractive for integration into different products, such as in the optical coherence tomography (OCT) systems for controlling the polarization in interferometers. However, similar to the controllers that rely on physical rotation, this device is too slow to be used for dynamic polarization control in fiber optic systems where the SOP varies very fast, as discussed in Chapter 3.

As illustrated in Figure 6.12c, the device consists of a strand of SM fiber, a rotatable fiber squeezer (the center), and two fiber holding blocks (left and right). The center portion of the fiber strand is sandwiched in the fiber squeezer and a combination of a screw and a spring is used to precisely control the pressure on the fiber. The spring is cleverly used to regulate the force applied, which translates the compression Δx of the spring from the screw into a precise force F via Hooke's law $F = k\Delta x$, where k is the spring constant or Hooke constant. Turning the screw knob on the fiber squeezer clockwise (CW) will increase the compression of the spring and hence the pressure onto the fiber center portion to increase the linear birefringence in this portion of fiber, and vice versa. As discussed in Chapter 3, the amount of birefringence Δn induced by the force F via the photo-elastic effect is given by Eqs. (3.5) and (3.8):

$$\Delta n = 4n_0^3(p_{11} - p_{12})(1 + v_p)\frac{F/L}{\pi Ed} = \frac{\eta F}{L} \tag{6.14a}$$

where L is the length of the fiber subject to squeezing, n_0 is the refractive index of the fiber, p_{11} and p_{12} are the components of the strain-optical tensor, v_p and the E are Poisson's ratio and Young's modulus of the fiber material (SiO_2), d is the diameter of the fiber, and $\eta = 4n_0^3(p_{11} - p_{12})(1 + v_p)/(\pi Ed)$. The corresponding retardation can be expressed as

$$\Delta\varphi = \frac{2\pi\Delta nL}{\lambda} = \frac{2\pi\eta F}{\lambda} = \frac{2\pi\eta(k\Delta x)}{\lambda} \tag{6.14b}$$

where λ is the wavelength of light. For an SM silica fiber with $d = 125\,\mu\text{m}$, $n_0 = 1.46$, $p_{11} = 0.12$, $p_{12} = 0.27$, and $v_p = 0.17$, one obtains $\eta \approx -8.56\times10^{-8}$ m/N. The minus sign indicates that the fast axis is in the direction of the applied force F. This value of η is close to our measured result of 9.223×10^{-8} m/N (Feng, Zhou, and Yao 2020) to be discussed in Chapter 10. It is interesting to note that the retardation generated by squeezing the fiber is independent of the fiber length L under squeezing for a given applied force. However, the shorter the fiber, the larger the line force (force per unit length) on the fiber, resulting in more stress on the fiber, which may reduce the life time of the fiber under the stress.

At a wavelength of $\lambda = 1.55\,\mu\text{m}$, $\Delta\varphi \approx 0.35\times F$ (rad). For generating a phase shift of $\pi/2$, an applied force F of 4.19 N is sufficient. As will be seen next, for generating any output SOP from an arbitrary input SOP, a retardation of 2π is sufficient, corresponding to a maximum applied force $F_{max} = 16.75$ N. Using $F_{max} = k\Delta x_{max}$, one can select a spring with a proper Hooke constant k and the maximum compression Δx_{max}.

With applied pressure, the fiber center portion acts as a birefringent wave plate with its fast axis in the direction of applied pressure, as shown in Figure 6.13a. The retardation between slow axis and fast axis can be varied between 0 and 2π by changing the applied pressure. When the fiber squeezer is rotated while pressure is applied, the relative angle between the major axis of the incident SOP and the slow axis of the fiber squeezer is changed. Because the rotation will also cause the segments of fiber at the left and right sides of the fiber squeezer to twist in the opposite senses (Figure 6.12c), the actual angle change between axes of the input SOP and the fiber squeezer is different from the physical rotation angle (Ulrich and Simon 1979). This twist induced optical activity will rotate the major axis of the incident SOP by an angle of $\alpha' = \rho\theta$ in the direction of twist from Eq. (3.14), where θ is the physical rotation angle shown in Figure 6.13b, $\rho = g/2$ is a coefficient of

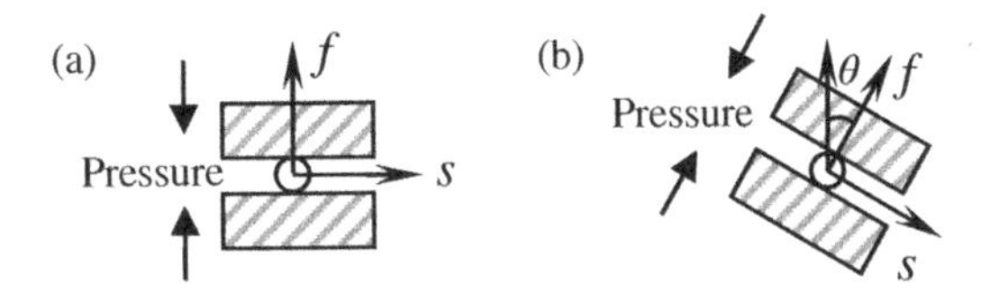

Figure 6.13 Cross-section of a fiber squeezer in which a fiber is sandwiched between two pressure blocks to produce a rotatable wave plate of variable retardation, realizing an all fiber polarization controller based on the Babinet–Soleil compensator mechanism. (a) No fiber twist. (b) The fiber is twisted θ degrees while being squeezed.

twist induced optical activity, and g is given by Eq. (3.13c). For SM silica fibers, ρ is on the order of 0.08, similar to the situation of the Lefevre described earlier. Consequently, for a physical rotation of θ degrees, the net change of the relative angle between the slow axis of the fiber squeezer and the major axis of the input polarization is

$$\theta' = (1 - \rho)\theta \tag{6.15}$$

For example, for a fiber squeezer with a physical rotation angle $\theta = 10°$, the relative angle of the input SOPs major axis with respect to the pressure direction of the fiber squeezer is $\theta' = 9.2°$.

Preferably, one may rotate the fiber squeezer without causing the left segment and the right segment to twist, by first releasing the pressure on the fiber squeezer, then rotating the squeezer, and finally applying pressure to the fiber squeezer again. In this way, for a physical rotation of θ degrees, the net change of the incident angle between the slow axis of the fiber center portion and the input SOP is also θ degrees. This rotate-without twist procedure can be applied for coarse adjustment of polarization. When the output polarization is close to the desired state, the rotate-with-twist procedure can be used to fine tune the output polarization.

As described earlier, by applying pressure to the fiber center portion, the rotatable fiber squeezer causes the fiber center portion to act as a wave plate of variable retardation and rotatable birefringence axes, or as a Babinet–Soleil compensator. Let's first examine how the rotatable fiber squeezer converts any linear SOP into any desired SOP. If one chooses the slow and fast axes of the squeezed fiber center portion as a coordinate system, as shown in Figure 6.13, an input linear SOP to the wave plate can be written as

$$\varepsilon_{in} = cos\theta'\hat{\mathbf{f}} + sin\theta'\hat{\mathbf{s}} \tag{6.16a}$$

where θ' relates to the orientation angle θ of the squeeze induced wave plate by Eq. (6.14). The output SOP immediately after the fiber squeezer can then be written as

$$\varepsilon_{out} = cos\theta'\hat{\mathbf{f}} + sin\theta' e^{i\Delta\varphi}\hat{\mathbf{s}} \tag{6.16b}$$

where $\Delta\varphi$ is the variable retardation induced by the fiber squeezer expressed by Eq. (6.14b). Because θ' can be varied from 0° to 180° by rotating the fiber squeezer and $\Delta\varphi$ can be changed from 0 to 2π by changing the pressure on the fiber center portion, the rotatable fiber squeezer is capable of generating an elliptical SOP with any orientation angle, ellipticity, and handedness from any linear SOP. Note that to achieve a relative 180° rotation angle θ', the fiber squeezer must be rotated 192° from Eq. (6.15). Due to the reverse twist in the fiber from the end of the fiber squeezer to the right fiber holder shown in Figure 6.12c, the major axis of the output SOP will automatically rotate by an angle of $-\rho\theta$ degrees at the right fiber holder. Although the twist induced optical activity $\rho\theta$ should be considered in designing the Yao controller to ensure the maximum rotation angle of the device is sufficient, it does not affect the operation of the device because the rotation angle difference between θ and θ' cannot be noticed in practice.

For an arbitrary elliptical input SOP, it is difficult to show mathematically that the rotatable fiber squeezer can generate any desired elliptical SOP. We now turn to Poincaré sphere to illustrate that it is capable of transforming any point on the sphere to any other point on the sphere, as shown in Figure 6.14a. From the discussion in Chapter 4, one can see that the action of changing the retardation is to cause the SOP changing around an axis perpendicular to s_3 while the action of wave plate rotation is to change orientation angle of this axis in the equator plane (s_1, s_2). Therefore, by the actions of "squeeze and turn," any desired SOP can be generated. Note that in practice, one may notice that the actual SOP trace caused by changing the retardation via squeezing the fiber may not be normal to s_3 in general viewed via a polarimeter or polarization analyzer. This is because the random residual birefringence in the fiber section after the fiber squeezer changes the relative orientation between the fiber squeezer and the coordinate defined by the polarization analyzer. One may re-align the fiber squeezer with respect to the coordinate either mathematically or via another polarization controller after the fiber squeezer so that the SOP trace normal is perpendicular to s_3.

In operation, the two "control knobs" on the Yao controller, one is for adjusting the retardation of the fiber in the fiber squeezer ("squeeze") and the other for adjusting the orientation angle of the fiber squeezer ("turn"), should be adjusted iteratively. For example, let's consider connecting an in-line polarizer described in Figure 6.7a at the output of the device and using an optical power meter to measure the optical power passing through the polarizer, as shown in Figure 6.14b. When the polarization controller is adjusted correctly, the maximum optical power is received by the power meter. Referring to Figure 6.14b,c, the following procedures are recommended for the adjustments: (i) Starting from point "a" in Figure 6.14c, apply a pressure to the center portion of the fiber strand by tightening the knob on the fiber squeezer while monitoring the optical power. If applying a pressure causes a significant increase in monitored optical power, then keep on going until the optical power starts to decrease before stopping at point "b," otherwise loosening the knob. (ii) Rotate the fiber squeezer while maintaining the pressure. If turning the squeezer causes little change in monitored optical power, or causes the optical power to decrease, then turn the squeezer in an opposite direction until the power reaches another local maximum before stopping at point "c." (iii) Repeat step (i) by increasing or decreasing the pressure to reach a new local maximum at point "d." (iv) Repeat step (ii) by turning the rotatable fiber squeezer until a maximum monitored optical power is obtained at point "e."

Despite the advantages (low insertion loss, low cost, and relatively compact size), the Yao controller still suffers from the slow speed issue because mechanical rotation is required, which

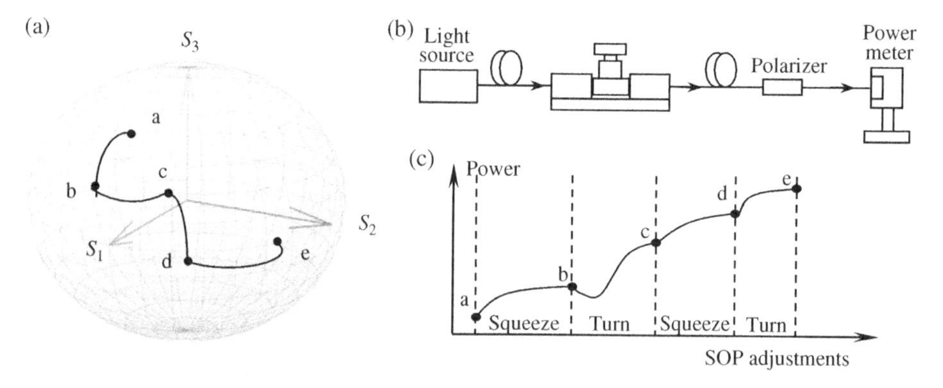

Figure 6.14 (a) Poincaré Sphere illustration of iteratively using the squeeze-and-turn adjustments of a Yao controller for converting an arbitrary SOP to any desired SOP; (b) A setup for optimizing the power output from a polarizer: (c) Illustration of the power maximization process of the squeeze-and-turn adjustments.

prevents it being used for high speed polarization control in real fiber optic communication and sensing systems where the SOP variation speed is high, as discussed in Chapter 3. To overcome this shortcoming, an E-O wave plate similar to that described in Figure 6.11e can be utilized, as shown in Figure 6.12e. The only difference is that in addition to rotating the retardation axis by changing the phase angle φ between V_x and V_y, the amount of retardation is also made to be variable by changing the amplitude V_0 of the applied voltage, as shown in Figure 6.12f. This electro-optic Babinet–Soleil compensator only requires a single E-O wave plate and therefore is simpler than that described in Figure 6.11e, which can be made with bulk $LiNbO_3$ or PLZT crystals. Alternatively, it can be made with $LiNbO_3$ waveguide for reduced operation voltage (Walker et al. 1987a), as will be discussed in Figure 6.16a. Unfortunately, no polarization controller of this kind has been available commercially probably due to different engineering challenges, such as high voltage or temperature instability.

6.5.3 Polarization Control with Multiple Wave Plates of Variable Retardation but Fixed Orientation

Polarization controllers also can be made with multiple free-space wave plates oriented 45° from each other (Figure 6.15a). The retardation of each wave plate varies with an applied voltage; however, the orientation angles are fixed. These variable retardation wave plates can be made with liquid crystals, electro-optical crystals, or electro-optical ceramics. The disadvantage of the liquid crystal device is low speed and the electro-optical one generally requires high operation voltages. Such a polarization controller generally has high insertion loss and high cost due to beam coupling in and out of the fibers.

Figure 6.15b shows a commercial product of such a polarization controller made with an electro-ceramic material having quadratic electro-optic effect, in which the speed of each wave plate for changing the retardation from 0 to π is about 30 μs. A major drawback of the electro-ceramic wave plate is the high temperature sensitivity of the retardation, which complicates system integration because temperature control or close-loop feedback is generally required.

An all-fiber device based on the same operation principle can be realized by using multiple fiber squeezers as variable retarders, as shown in Figure 6.15c, which can reduce the insertion loss and

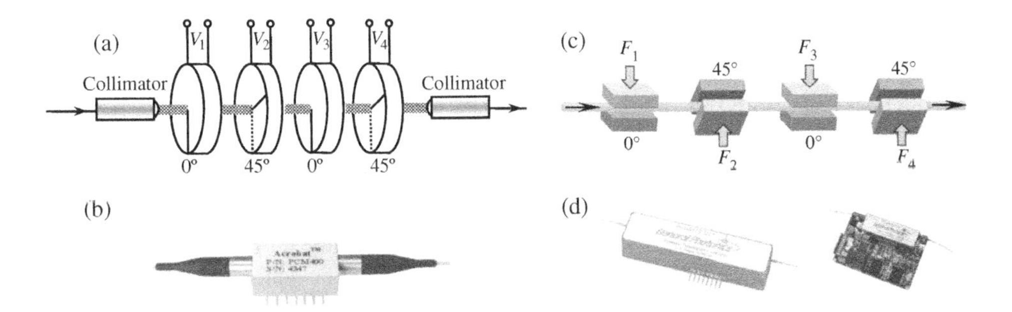

Figure 6.15 Illustration of polarization controllers made with multiple wave plate of fixed orientations but variable retardations. (a) A controller made with four electro-optic wave plates oriented 45° from one another. (b) A photo of a commercial polarization controller of (a) implemented with an electro-ceramic material having quadratic electro-optic effect. (c) A polarization controller made with four fiber squeezers oriented 45° from one another. (d) A photo of a fiber squeezer polarization controller commercialized by the authors of this book (Source: Courtesy of General Photonics/Luna Innovations, Inc.) under the trade name PolaRITE III™.

cost. The retardation of each wave plate varies with the pressure of each fiber squeezer, as discussed in Eq. (6.14b). In practice, the challenge is making the device reliable, compact, and cost-effective. In a commercial dynamic polarization controller, piezoelectric actuators can be used to drive the fiber squeezers for high speed (Yao 2000a). Because it is an all-fiber device, it has no back reflection and has extremely low insertion loss (IL) and PDL. Its response time of 30 μs is fast enough to track the most polarization fluctuations in field-installed fiber links. With an activation-induced loss (the maximum residual insertion loss associated with changing the retardation during operation) of less than 0.003 dB, it is also useful in high-precision PDL instrumentation and in feedback loops for compensating polarization-induced penalties. As a comparison, the activation loss (AL) of other polarization controllers is around 0.1 dB, much larger than what can be achieved with the fiber squeezer approach. Finally, it is also independent of wavelength, working equally well for signals with wavelengths ranging from 1280 to 1650 nm.

The fiber squeezer polarization controller concept was first proposed and demonstrated by Mark Johnson (1979) using electro-magnet to produce squeezing actions and validated by R. Ulrich using such a polarization controller for polarization stabilization (Ulrich 1979) in an optical fiber system, followed by several other researchers (Walker and Walker 1987b; Noe, Heidrich, and Hoffmann 1988). However, it was not commercialized until 2001 by Yao (2000a) at General Photonics Corporation (now part of Luna Innovations, Inc.) under a trade name PolaRITE II™. The main efforts for the commercialization involved the following:

(1) The prevention of fiber breakage from repeated stress. This was achieved by coating the fiber under squeezing with polyimide to fill-in micro-cracks on the fiber surface (Yao 2000b). Such an effort significantly increased the operation life time of the device and made it feasible to be deployed in telecommunication systems which generally required an operation life time of over 20 years. Otherwise, fiber breakages may develop from micro-cracks on the fiber, similar to the glass breakage developed from an inscription on a glass surface by a diamond inscriber.
(2) The reduction of fiber squeezer's operation voltage. This also was achieved by replacing the thick buffer coating on the fiber with the thin polyimide coating. SM fibers for telecommunication and sensor generally have a cladding diameter of 125 μm and a buffer diameter of 250 μm made with soft acrylate material as a protective coating. This 62.5 μm thick buffer would tend to dampen the force exerting on the fiber from the piezo-electric or electro-magnetic actuators. On the other hand, the polyimide coating only has a thickness of 17.5 μm, which is also much harder than the acrylate buffer coating and therefore much effective in transferring the applied force to the fiber core. With this thin coating, the total fiber diameter was reduced to 160 μm (a 36% reduction) and the birefringence induced by an applied force increased by 36% according to Eq. (6.14a) because the induced birefringence is inversely proportional to the fiber diameter. In addition, piezo-electric stacks were used as the force applying actuator (Yao 2002a) in the fiber squeezers with properly designed mechanics. In the final product, the half-wave voltage V_π was reduced to less than 30 V and the total retardation could be generated by each fiber squeezer reached over 5π with a maximum applied voltage of 150 V. Such a large retardation range was important for achieving reset-free polarization control using these fiber squeezer polarization controllers, as will be discussed in Chapter 7.
(3) Reducing the PDL and activation loss (AL) by super-polishing the two surfaces of the fiber squeezer (Yao 2002b). During the product development, it was realized that the unevenness of the fiber squeezer surfaces introduced relatively large PDL and AL and super-polishing the surfaces could effectively reduce the PDL and AL to less than 0.005 and 0.003 dB, respectively.

In principle, three retardation plates are sufficient to convert an arbitrary input SOP into any desired output SOP, as will be shown in Chapter 7. However, in practice, four or more wave plates are required for endless (or reset-free) polarization control (Noe, Heidrich, and Hoffmann 1988). With proper control procedures, reset-free polarization control can be achieved without disruption, as will be discussed later in Chapter 7.

6.5.4 Polarization Controller with $LiNbO_3$-Based Integrated Optical Circuit (IOC)

In the polarizations controllers described in Figures 6.11e, 6.12e, and 6.15b, bulk electro-optic crystals were deployed for achieving high speed operation. Unfortunately, because very high operation voltages were required, the operation speed was actually limited by the high voltage electronics, though the electro-optic effect was much faster. In addition, the high voltage electronics were expensive, further limiting the adoption of such devices in practical systems. To overcome such problems, great efforts were made in using integrated optical circuit (IOC) with electro-optical crystals, particularly lithium niobate crystals ($LiNbO_3$), to make polarization controllers in the past 30 years. Because the dimension d of a $LiNbO_3$ waveguide is much smaller (up to 100 times) than that of a bulk crystal, the electrical field E inside the waveguide is proportionally enhanced for the same applied voltage V ($E = V/d$) and therefore the required voltage can be reduced by up to 100 times.

Figure 6.16a shows an electro-optic Babinet–Soleil compensator made with $LiNbO_3$ IOC (Walker et al. 1987a), which is equivalent in principle to the devices shown in Figure 6.12. The $LiNbO_3$ substrate is of x-cut and the light propagates in the z-direction, where (x, y, z) are the crystal coordinates. Unlike the case with a bulk E-O crystal shown in Figure 6.12e in which an electrical field along the y-direction can be directly applied by using a pair of electrodes along the y-direction, here with the

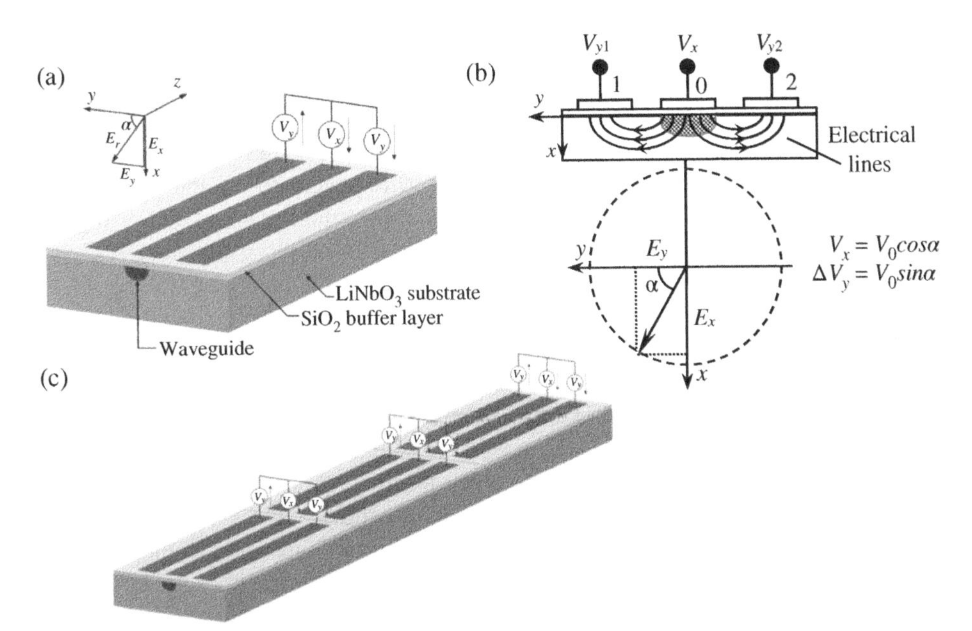

Figure 6.16 (a) Illustration of a Babinet–Soleil polarization controller made with $LiNbO_3$ integrated optical circuit. (b) Diagraph showing how electrical fields of different directions can be generated via three electrodes. (c) Illustration of cascading three rotatable $LiNbO_3$ wave plates for realizing different types of polarization.

waveguide three electrodes are cleverly used to generate electrical fields in both x- and y-directions, as shown in Figure 6.16b. Normally, V_{y2} is set to 0 V. If V_{y1} is also set to zero, the electrical fields in the y-direction generated by electrode pairs (0, 1) and (0, 2) cancel out with each other and the final net electrical field is in the x-direction. On the other hand, if V_{y1} is not zero, a net electrical field along the y-direction will arise. By adjusting the relative strengths of V_x and V_{y1}, the resultant electrical field can be in any desired direction in the (x, y) plane, as shown in Figure 6.16b. The electrical field angle can be expressed as

$$\alpha = ctan^{-1}\left(\frac{E_y}{E_x}\right) = ctan^{-1}\left(\frac{V_{y1}}{V_x}\right) \tag{6.17a}$$

Similar to the case of Figure 6.12f, if the applied voltages follow the sine and cosine relations as follows:

$$V_x = V_r cos\alpha \tag{6.17b}$$

$$\Delta V_y = V_r sin\alpha \tag{6.17c}$$

where $\Delta V_y = V_{y1} - V_{y2}$ and V_r is the resultant applied voltage. In fact, the electrical field can be made to rotate endlessly in the (x, y) plane by recognizing that Eqs. (6.17b) and (6.17c) represent a circle. Note that a bias voltage V_{yB} can be added to electrode 1 to cancel out the modal birefringence of the waveguide (Thaniyavarn 1986) so that refractive index difference between x and y directions is zero when $V_x = 0$.

From the discussion in Chapter 5, the index ellipsoid of $LiNbO_3$ under the transversal electrical field ($E_z = 0$) can be expressed as

$$\left(\frac{1}{n_0^2} - r_{22}E_y\right)x^2 + \left(\frac{1}{n_0^2} + r_{22}E_y\right)y^2 + \left(\frac{1}{n_e^2}\right)z^2 + r_{51}E_y yz + r_{51}E_x zx - r_{22}E_x xy = 1 \tag{6.18}$$

Through the electro-optic coefficients, r_{22}, r_{12} $(= - r_{22})$, and r_{61} $(= - r_{22})$, the resultant electrical field induces birefringence between two principal linear polarization states aligned at $\alpha/2$ to the crystallographic axes (Walker et al. 1987a), with the magnitude of the birefringence proportional to V_r. By adjusting V_r, any desired retardation between 0 and 2π can be achieved. Therefore, with properly applied voltages on the three electrodes, a rotatable wave plate with variable retardation can be realized, which is just an electro-optic Babinet–Soleil compensator capable of converting an arbitrary input SOP into any desired SOP.

If one adjusts and sets V_r to a value (V_π) such that the induced retardation is π, a rotating HWP can be realized. Similarly, by setting V_r such that the induced retardation is $\pi/2$, a rotating QWP can be realized. If one cascades such a rotatable HWP with two such rotatable QWPs, as shown in Figure 6.16c, a $LiNbO_3$ polarization controller of the type described in Figure 6.11 can be made, which is much faster than those in Figure 6.11. In addition, a V_π on the order of few volts can be achieved.

Similarly, if one cascades three such $LiNbO_3$ wave plates together, with the first one having its birefringence axis along the x-direction by setting $\Delta V_y = 0$, the second one having its birefringence axis 45° from the x-direction by setting $\Delta V_y = V_x$, and finally the third one again having its birefringence axis along the x-direction by setting $\Delta V_y = 0$, as shown in Figure 6.16c, a polarization controller of the type described in Figure 6.15 is made. The retardation of each wave plate can be adjusted by varying V_r on each wave plate. Therefore, the $LiNbO_3$ configuration described in Figure 6.16a is quite flexible to configure all different types of polarization controllers. Although this simple picture is modified by the intrinsic TE–TM modal birefringence, it should nevertheless be possible to control both the magnitude and orientation of the net waveguide birefringence.

6.5.5 Minimum-Element Polarization Controllers

The polarization controllers described earlier often have multiple polarization control elements (PCEs), such as wave plates, cascaded together to perform polarization control functions. A minimum-element polarization controller is defined as a controller with just enough PCEs to convert an arbitrary input SOP into any desired output SOP. For example, the minimum number of PCEs in controllers of Figures 6.11, 6.12, 6.15, and 6.16a,b is three, one, three, one, and three, respectively. More PCEs can be added in each controller to overcome imperfections or limitations in some of the PCEs. For example, a fourth PCE is often added in the controller of Figure 6.15 to compensate for the angular inaccuracy of the wave-plate orientations. As will be discussed in Chapter 7, more PCEs can be added to overcome the reset issues for continuously tracking polarization variations in real optical fiber systems.

6.6 Polarization Control with Circular Birefringence

Circular birefringence, such as the Faraday effect and optical activity, can also be used to control polarization. As described in Chapter 4, the eigen polarizations of a medium with pure circular birefringence are of right-hand and left-hand circular and they accumulate a relative phase delay between them when passing through the medium. Because an arbitrary input elliptical polarization can be expressed as the superposition of the two eigen polarizations, the net effect is that the major axis of the polarization ellipse rotates. Unlike the HWP discussed in Section 6.4.3, there is no handedness reversal when the ellipse is rotated. In general, the polarization rotation angle θ can be expressed as

$$\theta = \rho L \tag{6.19}$$

where ρ is the rotation rate per unit length. For example, for a quartz crystal with light propagating in the direction of its optical axis, $\rho = 21.7°/\text{mm}$ at a wavelength of 589.4 nm. For a Faraday rotator (FR), $\rho = V_d B$, where V_d is the Verdet constant and B is the magnetic field. The Verdet constant can be either positive or negative, depending on the properties of the magneto-optic (MO) material. A positive Verdet constant corresponds to L-rotation (counter-clockwise, CCW) when the light beam propagates in the direction of the magnetic field, while a negative Verdet constant corresponds to R-rotation (clockwise) when the light beam travels in the direction of the magnetic field.

As discussed in Chapter 4, Faraday effect and optical activity are quite different in that the former is nonreciprocal while the latter is reciprocal. Such a difference can be manifested by the examples as follows: when a light beam of linear SOP passing through an optically active material is reflected by a mirror, the SOP rotations from the forward and backward passes cancel out, resulting in a zero net SOP rotation. In contrast, the same light beam passing through a Faraday material is reflected by a mirror, the net SOP rotation doubles. Such a nonreciprocal property makes the Faraday effect extremely important in optical systems for controlling reflections, as will be discussed shortly.

The Faraday rotation angle can be easily controlled by changing the strength and direction of the magnetic field applied to the MO or Faraday material, making the Faraday effect attractive for polarization control and measurement applications, as will be discussed in detail in the following text. Likewise, some of the optically active materials, such as liquid crystals, can also be controlled by changing applied voltage across the materials, although in the majority of the optically active materials, the SOP cannot easily be controlled with electrical or magnetic fields, such as the case of a quartz crystal. As mentioned in Chapter 2, the optical activity is largely utilized in LCD screens of electronics products, such as computers, televisions, smart phones, and instruments, although it can also be used for polarization control and measurement.

6.6.1 Magneto-optic or Faraday Materials

There are many different MO materials, ranging from vapors, liquids, and solids. However, for polarization control and measurement applications in optical fiber systems, solid materials with large Verdet constants or large MO effects are preferred. Most glasses have the Faraday effect, although their Verdet constants may be small. In fact, the fused silica for making optical fibers is a good Faraday material though it has relatively small Verdet constant. Because it has extremely low transmission loss, a long interaction length L in Eq. (6.19) is allowed to achieve required polarization rotation. As will be discussed later, such an optical fiber based Faraday material is excellent for electrical current sensing applications, although it is too bulky for making polarization control devices.

Some special glasses, such as MOS-4 and MOS-10 (Molecular Technology GmbH 2022) (www.mt-berlin.com), possess relatively large Verdet constants in visible and near IR, however, dropping rapidly as wavelength increases. For example, the Verdet constants for MOS-4 and MOS-10 are 73 and 87 rad/Tm at 633 nm, respectively, but drop to 21 and 26 rad/Tm at 1060 nm.

The MO crystal Terbium Gallium Garnet ($Tb_3Ga_5O_{12}$, TGG) is a good MO material in the range of 400–1100 nm, excluding 470–500 nm, with large Verdet constants of −134 and −40 rad/Tm at 632 and 1064 nm, respectively. Unfortunately, this material may not be good for optical fiber communication and sensing band from 1260 to 1620 nm.

The most commonly used MO materials for telecommunication wavelengths are various rare-earth iron garnets (RIGs), with the chemical composition of $R_3Fe_nO_{12}$, where R represents a rare-earth element and n is a number. The examples of RIG materials include Yttrium Iron garnet ($Y_3Fe_5O_{12}$ or YIG) and Bismuth Iron garnet ($Bi_3Fe_5O_{12}$ or BIG), with the structure shown in Figure 6.17. Bismuth-substituted RIGs have been found to have greatly enhanced Faraday effect and a variety of optical properties can be adjusted through material composition. Such garnets with large Verdet constants include groups of $Bi_1Tb_2Fe_{4.5}Ga_{0.5}O_{12}$, $(BiTb)_3(FeGa)_5O_{12}$, $(BiYbTb)_3Fe_5O_{12}$, $(BiDy)_3(FeGa)_5O_{12}$, and $(BiLu)_3(Fe, Al)_5O_{12}$.

These RIG MO materials can be produced with common crystal growth methods as bulk crystals, however, most commonly are obtained as thin or thick films on a substrate, such as Gadolinium Gallium garnet ($Gd_3Ga_5O_{12}$, GGG), using Liquid Phase Epitaxial (LPE) or Pulsed Laser Deposition (PLD) methods for large scale productions. Since some of the MO thin films were once used for computer storage memories (the so called magnetic bubble memory) (Scott and Lacklison 1976), the manufacturing processes have been relatively matured by large international corporations, including Mitsubishi Gas Chemical and Sumitomo Metal Mining companies.

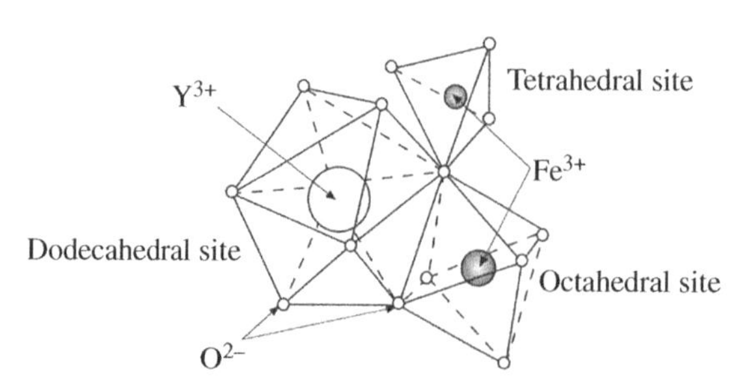

Figure 6.17 The structure of a YIG crystal showing the relative locations of the rare-earth, the iron, and the oxygen ions (Ohta, Hosoe, and Sugita 1983).

Table 6.3 Properties of some commercial magneto-optic rare-earth iron garnets with fixed 45° rotation.

Material[a]	YIG	YIG:Ga	$Bi_1Tb_2Fe_{4.5}Ga_{0.5}O_{12}$	$(BiYbTb)_3Fe_5O_{12}$	$(BiR)_3Fe_5O_{12}$
Rotation (°/mm)	21.4@1310 nm	14.5@1310 nm	−136@1310 nm −92@1550 nm	−155@1310 nm −100@1550 nm	120@1310 nm 90@1550 nm
Saturation field	<1780 Oe	<400 Oe	<225 Oe	<750 (Oe)	Magnet free
Thickness for 45° at saturation	2.1 mm	3.1 mm	330 μm@1310 nm 500 μm@1550 nm	290 μm@1310 nm 450 μm@1550 nm	370 μm@1310 nm 500 μm@1550 nm
Faraday angle θ	45 ± 0.5°	45 ± 0.5°	45 ± 0.5°	45 ± 0.5°	45 ± 0.5°
Insertion loss	<0.2 dB	<0.2 dB		<0.1 dB	<0.1 dB
Extinction ratio			>40 dB	>40 dB	>40 dB
Temperature coefficient $d\|\theta\|/dT$ (°/°C)	−0.023	Data not available	−0.06	−0.04@1310 nm −0.045@1550 nm	−0.09@1310 nm −0.09@1550 nm
Wavelength coefficient $d\|\theta\|/d\lambda$ (°/nm)	Data not available	Data not available	−0.08@1310 nm −0.054@1550 nm	−0.085@1310 nm −0.065@1550 nm	−0.08@1310 nm −0.06@1550 nm
Thermal expansion (°/C)	1.04×10^{-5}	Data not available	1.1×10^{-5}	1.1×10^{-5}	1.1×10^{-5}
Coercive force (Oe)	Not applicable	Not applicable	Not applicable	Not applicable	1000@25 °C
Refractive index	2.2@1310 nm 2.19@1550 nm	2.2@1310 nm 2.19@1550 nm	2.356@1310 nm 2.344@1550 nm	2.33–2.4@1310 nm 2.3–2.37@1550 nm	2.33–2.4@1310 nm 2.3–2.37@1550 nm

a) Data for YIG and YIG:Ga are obtained from the data sheet of Deltronic Crystal Industries, Inc., USA. Data for $Bi_1Tb_2Fe_{4.5}Ga_{0.5}O_{12}$ are from Integrated Photonics, Inc., USA (Now part of II-VI Inc., USA). Data for $(BiYbTb)_3Fe_5O_{12}$ and $(BiR)_3Fe_5O_{12}$ are obtained from the data sheet of Sumitomo Metal Mining Co., Ltd., Japan. Here R represents a rare-earth element not disclosed by the manufacturer.

Table 6.3 lists the properties of some commercially available rare-earth garnets frequently used in optics industry, in which YIG and YIG:Ga are bulk crystals, while $Bi_1Tb_2Fe_{4.5}Ga_{0.5}O_{12}$, $(BiYbTb)_3Fe_5O_{12}$, and $(BiR)_3Fe_5O_{12}$ are thick films fabricated using the LPE process. Note that the manufacturer of the RIG films may offer different models with different specifications, such as lower temperature dependence or lower insertion loss, by slightly altering the compositions. For example, $(BiR)_3Fe_5O_{12}$ can be made magnet free by "freezing" the magnetic domain alignment (to be discussed next) so that no external magnetic field is required to enable the desired 45° Faraday rotation. Such a magnet-free MO garnet can simplify the design, and reduce the cost and size of the devices incorporating the garnet.

Note that the Faraday rotation of some MO crystals has a positive sign, meaning that the polarization rotation is CCW when the light beam propagates in the direction of the magnetic field, while others have a negative sign with the polarization rotates clockwise when the light beam propagates in the direction of the magnetic field.

The Faraday rotation angle or the Verdet constant of an MO crystal has a strong dependence on wavelength, generally decreases rapidly with wavelength. This Faraday rotation dispersion at saturation can be expressed as

$$\rho(\lambda) \approx \rho_0 \frac{\lambda^2 \lambda_0^2}{\left(\lambda^2 - \lambda_0^2\right)^2} \tag{6.20a}$$

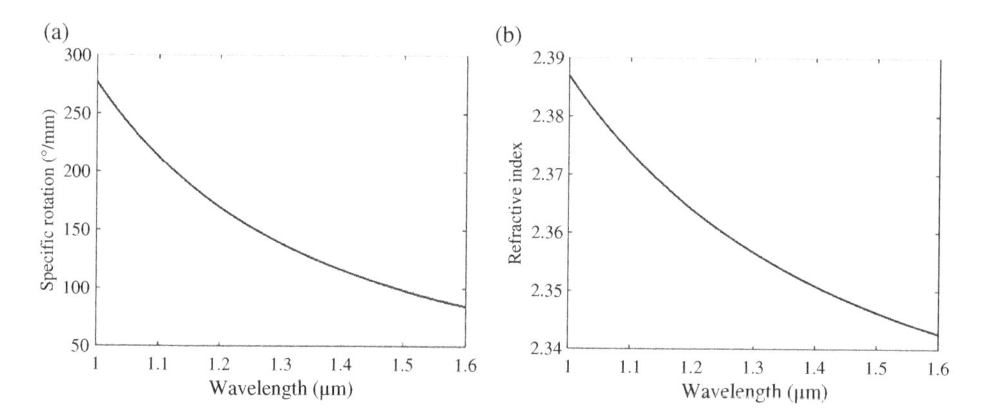

Figure 6.18 (a) Absolute value of specific Faraday rotation of a $Bi_1Tb_2Fe_{4.5}Ga_{0.5}O_{12}$ film at saturation (H_{bias} = 700 oe) as a function of wavelength. As indicated in Table 6.1, the sign of the rotation is negative. (b) The refractive index as a function of wavelength.

where λ is the wavelength in μm. For $Bi_1Tb_2Fe_{4.5}Ga_{0.5}O_{12}$, $\rho_0 = -1065°$ /mm and $\lambda_0 = 0.42$ μm. Figure 6.18a shows the Faraday rotation angle as a function of wavelength obtained using Eq. (6.20a). Such a wavelength dependence is important for designing and using fiber optic devices for optical fiber. The corresponding refractive index n can be expressed as

$$n = \left(1 + \frac{1}{0.5334 - 0.06756/\lambda^2} + \frac{1}{0.401 - 0.008944/\lambda^2}\right)^{1/2} \tag{6.20b}$$

which is the chromatic dispersion expression for this garnet crystal. The corresponding plot of n as a function of wavelength is shown in Figure 6.18b. Other bismuth-substituted rare-earth garnets have similar wavelength responses. As indicated in Table 6.1, the absolute values of the specific rotation of the bismuth-substituted rare-earth garnets also decrease with temperature at a rate of 0.04–0.09°/°C. Note that Eqs. (6.20a) and (6.20b) are extracted from product sheet of Integrated Photonics, Inc. (now part of II-VI, Inc.).

6.6.2 Magneto-optic Properties of Rare-Earth Iron Garnet Films

As mentioned in Section 6.6.1, most MO RIG thick films were originally developed for magnetic bubble memory applications with relatively mature processes. The bismuth-doped versions generally are well accepted in the optical fiber communication industry for making isolators, circulators, switches, Faraday mirrors (FRMs), polarization controllers, and other non-reciprocal devices, as will be discussed in more detail in the following text. The size of the commercial RIG is typically 11 mm × 11 mm and can be diced into small squares with the size of 1 mm × 1 mm or 2 mm × 2 mm. The thickness of the film is on the order of few hundred microns and the total thickness with the substrate is on the order of 500 μm.

6.6.2.1 Perpendicular Anisotropy Thick Films

Magnetic Domains The RIG films are ferromagnetic materials with regions in which the magnetization is oriented in the same direction and in other regions the magnetization is oriented in different directions. These regions are referred to as magnetic domains.

Most MO RIG thick films have perpendicular anisotropy. In particular, they have a positive growth-induced uni-axial magnetic anisotropy with the magnetization vectors being perpendicular to the plane of the film to form serpentine stripe domains in the demagnetized state when no

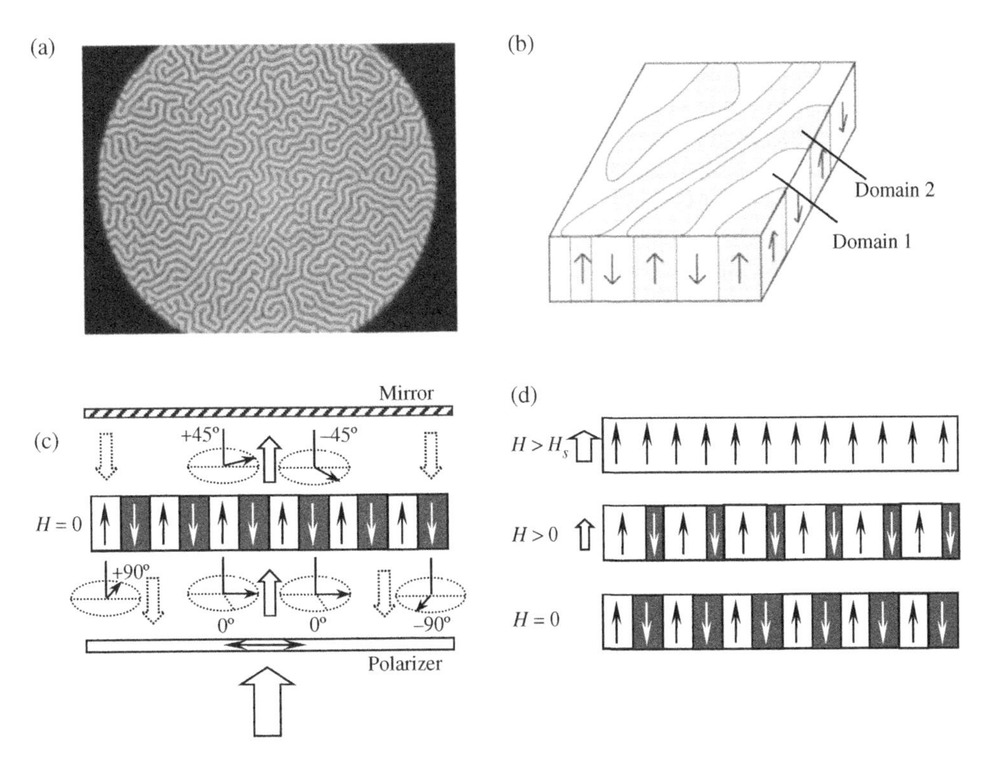

Figure 6.19 (a) A top-view photo of an MO thick film with perpendicular anisotropy showing serpentine stripe domains. (b) The 3D illustration of the domains in the MO film. (c) Polarization rotation of light passing through different domains when the MO film is in the totally demagnetized state ($H = 0$). (d) Comparison of the perpendicular domains in the film under magnetic fields of different strengths.

magnetic field is applied (Fratello and Wolfe 2000; Fratello et al. 2004), as shown in Figure 6.19a. The 3-dimensional (3D) illustration of the domains in the garnet films with perpendicular anisotropy is shown in Figure 6.19b.

As shown in Figure 6.19c (Integrated Photonics, Inc. 2012), when the applied magnetic field $H = 0$, the demagnetized perpendicular Faraday rotator film is of a multi-domain structure with equal proportions of "up" and "down" domains. When a light of linear polarization passes through the film, each individual domain rotates the polarization by $\pm\theta°$, where θ is the Faraday rotation angle typically set at 45° by choosing a proper film thickness. For some special applications to be discussed later, this angle θ is set at 22.5°. Consequently, half of the light beam undergoes a $+\theta°$ rotation and the other half undergoes a $-\theta°$ rotation. The apparent Faraday rotation from all the domains is an average so that the net polarization rotation is 0°. However, if analyzed by a 0° polarizer, only 50% of light from each domain can pass through when $\theta = 45°$, assuming no light scattering. As will be discussed later, the strong diffractions by the domains will further diminish the light transmission. If a mirror is placed after the film, the polarization will be rotated by 90° by each domain after the double pass and will be totally blocked by the input horizontal polarizer, as shown in Figure 6.19c.

When a perpendicular magnetic field is applied to the film (Integrated Photonics, Inc. 2012), as shown in Figure 6.19d, more domains will be aligned with the external magnetic field, resulting in more light with polarization rotating in the +45° direction (assuming $\theta = 45°$) and therefore a net polarization rotation. If the applied magnetic field is larger than the saturation field H_s of the MO film, all the domains are aligned with the magnetic field and the domain walls disappear, resulting in a net +45° polarization rotation. In the meantime, the scattering from the domain walls also disappears. Almost all the isolators, circulators, and FRMs are made with MO films with the applied magnetic field larger than saturation field, as will be discussed later.

If the magnetic field reverses the direction, all the domains will also reverse the direction, resulting in a net −45° polarization rotation. Therefore, one may make a 90° polarization switch with polarization switching between +45° and −45° by changing the direction of the applied magnetic field above the saturation. Alternatively, one may choose the film thickness such that $\theta = 22.5°$ to make ±22.5° polarization rotation switches. In other words, the Faraday rotators made with rare-earth garnet films of perpendicular anisotropy are of binary in nature with $\pm\theta$ polarization rotation angles determined by the film thickness when applied magnetic fields are above a saturation field with opposite signs. Note that such a change of rotation by lateral domain wall motion is damped by the effective friction of the domain walls, limiting the attainable polarization switching speed on the order of microseconds.

In application, the following properties of the Faraday rotators made with rare-earth garnet films of perpendicular anisotropy must be taken into account (Mitsubishi Gas Chemical Company 2004).

Domain Diffractions When a light beam passes through a magnetically unsaturated MO garnet film shown in Figure 6.20a, it will be diffracted by the perpendicular magnetic domains, as shown in Figure 6.20b. For a beam polarized in the x-direction, the un-diffracted beam (the zero order diffraction) experiences a polarization rotation of θ degrees, while the higher order diffractions all experience a 90° polarization rotation around the optical beam. For a totally demagnetized MO film with zero applied magnetic field, the polarization of the un-diffracted beam is orthogonal from that of the diffracted beams.

More than half of the optical power can be in the higher diffraction orders, resulting in high transmission loss of the zero-order beam useful for our intended applications. In general, the larger the applied magnetic field, the smaller the higher order diffractions are. Only when the film is totally magnetized with an applied magnetic field larger than the saturation filed, the higher order diffractions vanish due to the disappearance of the perpendicular magnetic domains. Therefore, such MO films are generally used under a magnetic field above the saturation applied perpendicular to the film surface, with a fixed polarization rotation angle determined by the thickness of the film.

Saturation Field $\mathbf{H_s}$ The magnetic response of the Faraday rotation angle can be measured with the experimental setup shown in Figure 6.21a, in which the PBS is aligned to the input SOP to block the all diffracted beam from entering PD1 because they are orthogonal from the input SOP. As

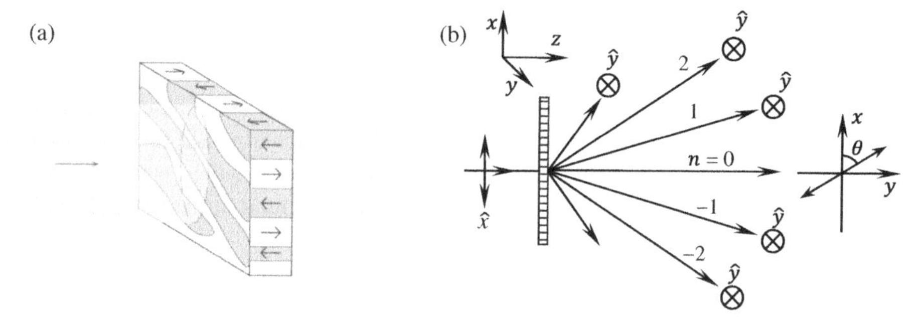

Figure 6.20 (a) Illustration of a light beam passing through a thick film of perpendicular anisotropy with magnetic domains. (b) Diagraph showing the beam is diffracted by the magnetic domains into different diffraction orders with the SOP of each diffraction order labeled on each diffracted beam.

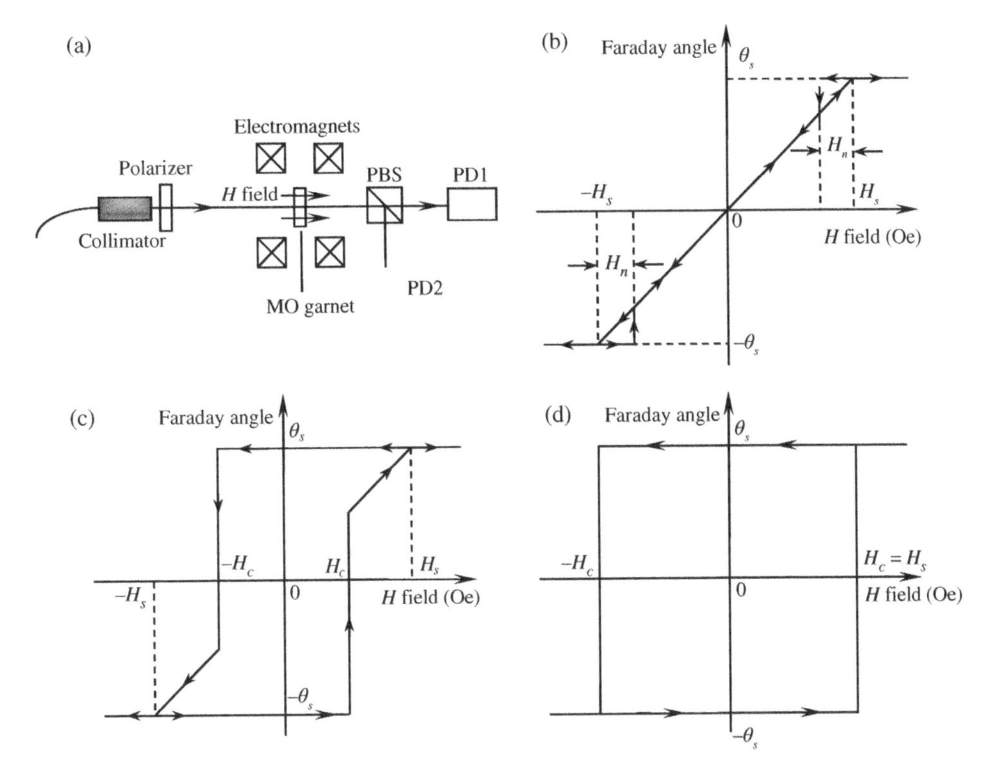

Figure 6.21 Magnetic properties of the Faraday rotation. (a) The measurement setup for measuring the SOP rotation as a function of the applied magnetic field. (b) Faraday rotation angle as a function of the applied magnetic field of a typical MO film, showing the saturation magnetic field H_s, the saturation Faraday rotation angle θ_s, and the nucleation magnetic field H_n. (c) Faraday rotation angle as a function of the applied magnetic field of a special MO film showing the coercive force H_c, in addition to H_s and θ_S. (d) Faraday rotation angle as a function of the applied magnetic field of a magnet free MO film showing the coercive force H_c.

mentioned earlier, when the MO film is totally demagnetized, half of the magnetic domains are "up" and the other half are "down," resulting in a net 0° SOP rotation for the zero order diffraction beam (or un-diffracted beam). The corresponding optical power detected by PD1 is at a maximum value. As the external magnetic field H increases in the direction of light propagation, more magnetic domains become "up," resulting in a net $\theta°$ SOP rotation for the un-diffracted beam until the external magnetic field H reaches a saturation value H_S such that all magnetic domains are aligned with the magnetic field. Under this magnetic field, the total polarization rotation reaches a maximum angle θ_S. The corresponding power detected by PD1 also gradually decreases following a sinusoidal function until it reaches a minimum. Therefore, H_S is defined as the saturation magnetic field and can be easily determined with the measurement setup of Figure 6.21b.

***Nucleation Magnetic Field* H$_n$** As the external magnetic field is reduced from the saturation H_S, the measured power in PD1 does not drop immediately, because the domains still remain aligned. When the magnetic field is reduced to a certain value H_n from H_S, some domains start to flip directions, resulting in the measured power in PD1 starting to increase. The corresponding magnetic field value is called the nucleation magnetic field, as shown in Figure 6.21b. As the magnetic field continues to decrease, more domains are flipped backed to their original direction until the field becomes zero and the corresponding SOP rotation is zero again. As the magnetic field reverses the direction, more domains will be flipped "down," resulting in the SOP rotating in the opposite direction until the magnetic field reaches the negative saturation with the same behavior as for the case with positive H field.

***Coercive Force* H_c** In some MO garnet films, the magnetic domains remain aligned even after the external magnetic field is removed so that no magnet is required in applications. Such garnet films are often referred to as "magnet free" Faraday rotators and are widely used for making miniature fiber optic components, such as isolators, circulators, and FRMs. Even with a reversed magnetic field below a certain value H_c, all the magnet domains still remain locked to the original direction. Only a reverse magnetic field above H_c is able to coerce some of the domains to flip directions, as shown in Figure 6.21c. Therefore, the field strength H_c is often referred to as the coercive force field. In some specially engineered MO films, the coercive force H_c is made to be same as the saturation field H_s, as shown in Figure 6.21d, which means all the domains flip together after the reverse magnetic field reaches H_c.

Overcoming Domain Diffraction The domain diffraction discussed in Figure 6.20 prevents the MO garnet films with perpendicular anisotropy to be used as polarization controllers or used for magnetic field sensing in which the SOP needs to be continuously rotated with the external magnetic field, although these MO films are perfect for making isolators, circulators, and FRMs because only fixed Faraday rotation of ±45° under saturation magnetic field is required. One way to overcome this shortcoming is to apply a transverse magnetic field across the garnet film to ensure the total magnetic field is always above the saturation field H_s (Saitoh and Kinugawa 2003), as shown in Figure 6.22.

As shown in Figure 6.22a, if the magnetic fields along x and z directions satisfy the following equations:

$$H_x = H_0 sin\chi \tag{6.21a}$$

$$H_z = H_0 cos\chi \tag{6.21b}$$

where $H_0 = \sqrt{H_x^2 + H_z^2}$ is the composite magnetic field in a direction having an angle of χ degrees from z, which tries to align the magnetic domains in the MO garnet along this direction. When this

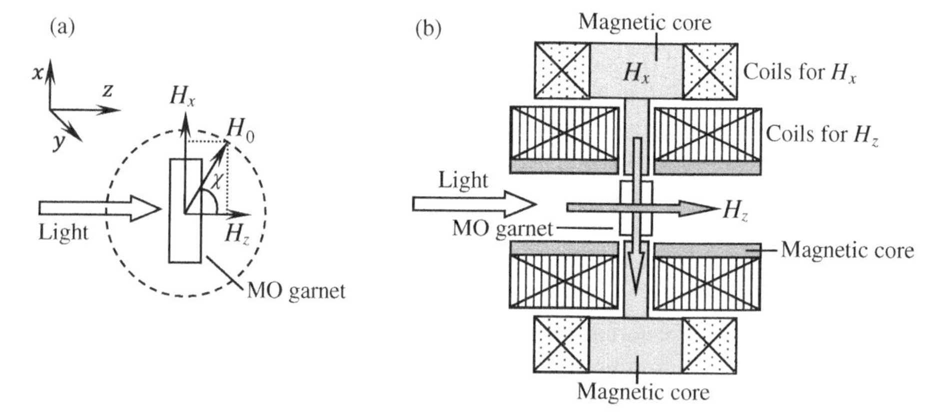

Figure 6.22 Overcoming the domain diffraction by applying a transversal magnetic field. (a) Schematic showing the longitudinal and transversal magnetic fields applied on to an MO garnet. (b) Electromagnet arrangement for generating the transversal (along x) and longitudinal (along z) magnetic fields.

composite magnetic field is larger than the saturation field H_s ($H_0 > H_s$), all domain walls disappear and so do the diffractions. The SOP rotation angle can be expressed as

$$\theta = \frac{H_z}{H_0}\theta_s = cos\chi \cdot \theta_s \tag{6.21c}$$

where θ_s is the saturation Faraday rotation angle of the garnet. Therefore, by changing the magnetic fields along the x and z directions following Eqs. (6.21b) and (6.21c), any polarization rotation angle between $-\theta_s$ and $+\theta_s$ can be achieved.

In Figure 6.22b, the magnetic core is made with a soft magnetic material having low coercivity and high permeability, which is easy to magnetize and demagnetize and is used to conduct magnetic and electromagnetic energy conversion and transmission. Such materials are widely used in various electric energy conversion equipment. Soft magnetic materials mainly include metal soft magnetic materials, ferrite soft magnetic materials, and other soft magnetic materials. The most widely used soft magnetic materials are iron–silicon alloys (silicon steel sheets) and various soft ferrites.

6.6.2.2 Planar Anisotropy Thick Films

For active polarization control and magnetic field/electrical current sensing applications, continuous polarization rotation at high speed is required. Unfortunately, due to the diffractions from the magnetic domains, the thick MO films with perpendicular anisotropy described earlier have difficulties for obtaining continuous polarization rotation with low transmission loss. Although it is possible to overcome such a difficulty by applying a transverse magnetic field, as described in Figure 6.22, the complexity, size, and cost associated with the approach make it less attractive for real world applications. In addition, the polarization rotation speed of the MO films of perpendicular anisotropy is limited because the friction of the domain walls damps lateral domain wall motion responsible for the polarization rotation.

For active polarization control and sensing applications, Faraday rotator or MO garnet films with planar anisotropy are developed (Integrated Photonics, Inc. 2012). In such a thick film, magnetization vectors lie in the plane of the film with large planar magnetic domains in the demagnetized state (Greschishkin et al. 1996; Fratello, Mnushkina, and Licht 2005; Tkachuk et al. 2009), as shown in Figure 6.23a. For the ease of comparison, the photo of magnetic domains of a thick MO film with perpendicular anisotropy is re-shown in Figure 6.23b.

As shown in Figure 6.23a, the in-plane domains are quite large, much larger than the wavelength of the light, and therefore they do not induce any noticeable diffraction of the light entering perpendicularly from the flat surface of the film. Unlike in an MO film with perpendicular domains in which the domain walls move in the plane while changing their sizes in response to an applied magnetic field, as shown in Figure 6.19d, here all the in-plane magnetization vectors rotate toward the direction of the applied magnetic field by η degrees, as shown in Figure 6.23c. This portion of the perpendicular magnetization is in the same direction of the light beam and rotates the SOP of the light beam. The resulting Faraday rotation angle as a function of the magnetic field is shown in Figure 6.23d, in which H_s and θ_s are the saturation magnetic field and saturation Faraday angle, respectively. When the applied magnetic field is equal or larger than the saturation field, all the domain magnetization vectors are aligned parallel with the applied field (Figure 6.23c) so that the Faraday rotation angle reaches the saturation Faraday angle θ_s. The relation between the Faraday rotation angle and the applied magnetic field can be approximately expressed as

$$\theta = sin\eta \cdot \theta_s \approx \frac{H}{H_s}\theta_s \quad \text{for } H < H_s \tag{6.22a}$$

$$\theta = \theta_s \quad \text{for } H > H_s \tag{6.22b}$$

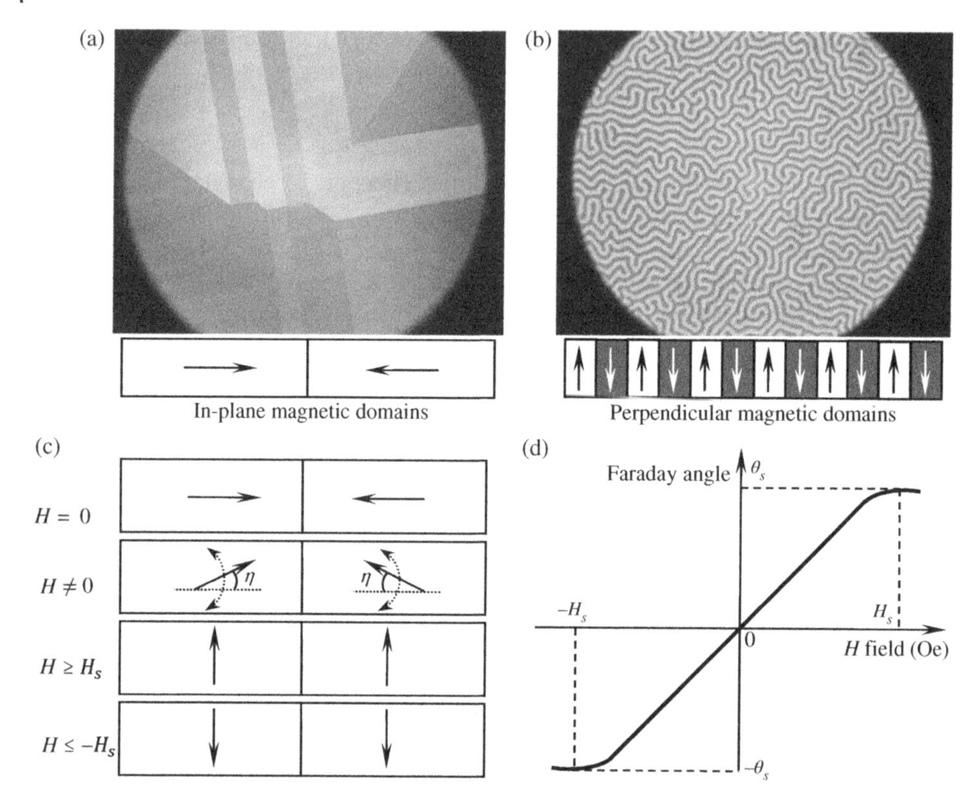

Figure 6.23 Thick film planar Faraday rotator crystal. The top view photo of an MO film with in-plane magnetic domains (a) is compared with that of an MO film with perpendicular magnetic domains (b). (c) Illustration of the in-plane domains under different magnetic fields. (d) The Faraday rotation angle as a function of the applied magnetic field showing the saturation magnetic field H_s and saturation Faraday rotation angle θ_s.

Comparing with the MO films of perpendicular anisotropy, the advantages of this in-plane anisotropy MO thick film are (i) the elimination of the domain diffractions for low transmission or insertion loss, (ii) uniform and continuous polarization rotation in response to applied magnetic field, and (iii) high polarization rotation speed with a frequency response up to 1 GHz. These unique properties make such an MO film ideal for making polarization controllers, variable optical attenuators, and optical current and magnetic sensors. The only drawback is that it has a reduced specific Faraday rotation (°/mm) about 75% that of the perpendicular anisotropy MO films. Therefore, this in-plane anisotropy MO film is about 25% thicker than the perpendicular anisotropy MO films to compensate for this slightly reduced parameter. Table 6.4 lists some typical performance parameters of such a thick MO film.

6.6.3 Faraday Rotator Based Simple Polarization Management Devices

6.6.3.1 Faraday Mirror

A FRM with a total 90° polarization rotation (round trip) is extremely useful for getting rid of polarization fluctuations in interferometers (Kersey, Marrone, and Davis 1991; Ferreira, Santos, and Farahi 1995) and optical delay lines (Yao, Yan, and Chen 2008) made with SM optical fibers and has been widely adopted in various fiber optic systems. A FRM is generally made with a fiber

Table 6.4 Typical performance parameters of an in-plane anisotropy thick MO film.

Specific Faraday rotation (°/mm)	−61@1550 nm, −93@1310 nm
Rotation sensitivity $d|\theta|/dH$@50% saturation (/Oe)	0.06 ± 0.01
Faraday rotation angle (°)	±45°
Insertion loss (dB)	<0.1 dB@1310 nm and 1550 nm
Polarization extinction ratio (dB)	>40 dB
Saturation field H_s (Oe)	~800
Thickness for 45° rotation at H_s (μm)	485@1310 nm, 740 (2 pieces)@1550 nm
Temperature coefficient $d|\theta|/dT$ (°/°C)	−0.09
Wavelength coefficient $d|\theta|/d\lambda$ (°/nm)	−0.09@1310 nm, −0.07@1550 nm
Thermal expansion coefficient (°C)	1.1×10^{-5}
Refractive index n	2.31@1310 nm, 2.29@1550 nm
Curie temperature T_c (°C)	180
Rotation angle tolerance (°)	±1.0@1310 nm, ±2.0@1550 nm (all λ)

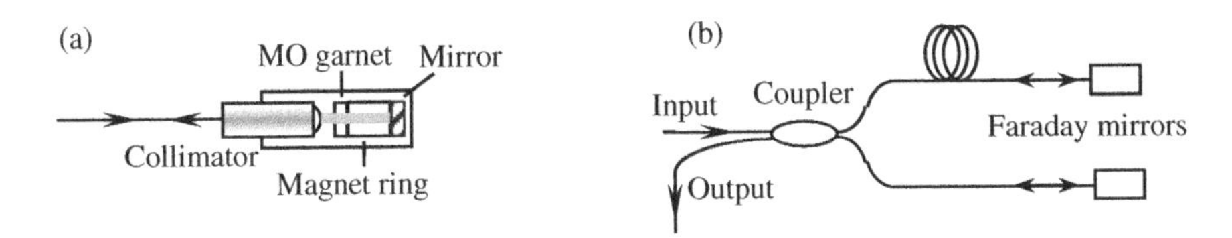

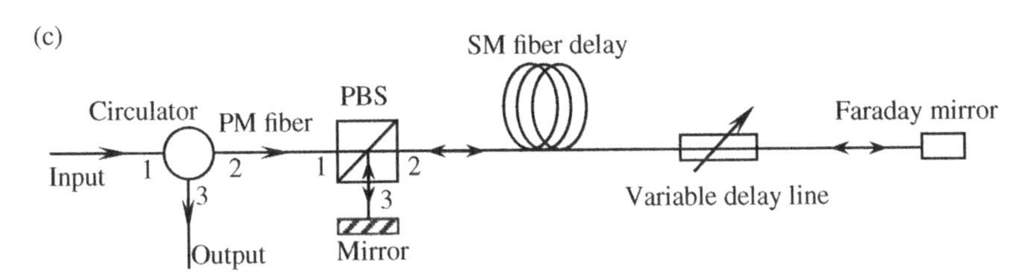

Figure 6.24 Faraday rotator and its application examples. (a) The construction of a Faraday mirror with an MO garnet having a Faraday rotation of 45° at saturation. (b) A Michelson interferometer deploying two Faraday rotators to eliminate polarization fading. (c) An optical delay quadrupler enabled by a Faraday mirror.

collimator, a 45° Faraday rotator, and a mirror, as shown in Figure 6.24a. Both types of Faraday rotators made with perpendicular anisotropy and in-plane anisotropy can be used to make such FRMs; however in practice most FRMs are made with MO films of perpendicular anisotropy described in Section 6.6.2.1. A permanent magnet ring is generally used to magnetically saturate the MO film to reach the saturation Faraday rotation angle of 45° and eliminate the unwanted diffraction. The magnet-free Faraday rotators discussed in Section 6.6.1 are sometimes used to eliminate the magnetic ring for miniaturizing the device.

As described in Chapter 3, the SOP of a light beam generally changes as it propagates in an optical fiber due to residual birefringence in the fiber which causes retardation between the two orthogonal polarization components along the slow and fast axes of the fiber. Upon reflection by a FRM, the

SOP rotated 90°, causing the polarization component in the slow axis in the forward direction to be re-aligned in the fast axis in the backward direction, and vice versa and therefore undoing the retardation everywhere along the optical fiber. Consequently, no SOP fluctuation will be observed after the light double passes a length of optical fiber caused by a FRM with a 90° polarization rotation. The Jones matrix and Mueller matrixes describing a FRM can be found in Table 4.3.

Figure 6.24b shows an optical fiber Michelson interferometer having two FRMs to replace the regular mirrors for eliminating polarization fluctuation caused by varying birefringence inside the fiber. Light from left is input into the coupler to be split into two output beams having the same SOP inside and at the output ends of the coupler because they are originated from the same input light. The two beams propagate in their respective fibers to accumulate different retardation and therefore have different SOPs before being reflected by the FRM. Upon reflection, the SOP of each beam rotates by 90° before propagating back toward the coupler and remains orthogonal with the forward propagating beam at each point along the fiber, including at the coupler. Therefore, upon arriving at the coupler again, the SOPs of the two beams become the same again because they are both orthogonal to that of the forward beams, which ensures that the two beams can fully interfere without polarization fading.

Figure 6.24c illustrates an optical delay quadrupler enabled by a FRM. The device consists of an optical circulator, a PBS with a regular mirror at one of the port, fiber delay line, and a FRM. Light from the input PM fiber is directed via the optical circular to pass the PBS with the polarization aligned with the passing axis of the PBS and propagates through the fiber delay line made with SM fiber toward the FRM. Upon reflection by the FRM, its SOP rotates 90° and is orthogonal everywhere along the SM fiber with respect to the SOP of the forward going beam. At the PBS, the beam is directed into Port 3 and then is reflected by a regular mirror to propagate toward the FRM the second time. Upon reflection by the FRM the second time, the light beam is again rotated 90° before propagating toward the PBS the third time with the polarization aligned with the passing axis of the PBS. After passing through the PBS, it is then directed into the output port of the device via Port 3 of the circulator. Although the delay fiber is of SM, the whole device is PM, which helps to further cut down the cost of the device, in addition to reduce the length of the fiber by 75% for more compact size. If a variable optical delay line is inserted in the optical path between the PBS and the FRM, the range of the variable delay can also be quadrupled.

6.6.3.2 Polarization Switch

For some applications, rotating the SOP by a known amount is desirable. For example, rotating the SOP in one of the Mach–Zehnder interferometer arms by 45° or 90° can effectively remove polarization fading. Figure 6.25a illustrates a Faraday rotator core made with an MO garnet of either perpendicular anisotropy or in-plane anisotropy described previously in Section 6.6.2, together with a coil made with copper wires for generating an axial magnetic field by injecting an electrical current I. A ring made with a semi-hard magnetic core material can be used to enhance the magnetic field and to enable magnetic field latching (meaning that even when the electrical current is off, the semi-hard magnetic core can still hold a magnetic field beyond saturation field for the MO garnet) (en.wikipedia.org/wiki/Ferrite_(magnet)). An MO garnet with a rotation angle of $\pm 22.5°$ and $\pm 45°$ at saturation can be used for switching the SOP by 45° and 90°, respectively. The minus sign of the rotation angle can be realized by reversing the direction of the current flow.

Figure 6.25b shows the basic construction of a fiber pigtailed polarization switch made with an MO garnet, such as a perpendicular anisotropy thick film with a rotation angle of $\pm 22.5°$ or $\pm 45°$ at saturation discussed in Section 6.6.2.1. The input and output fibers can be either SM or PM, depending on the application. Changing the direction of current flow alters the direction

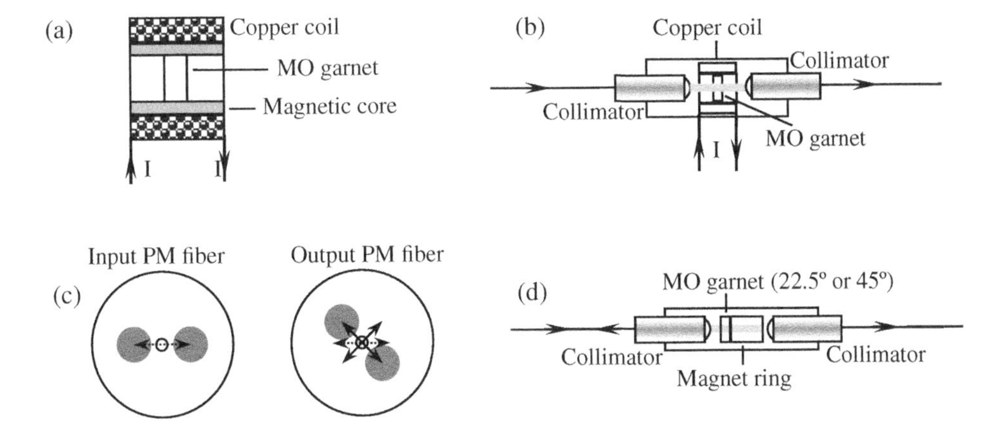

Figure 6.25 (a) A Faraday rotator made with a copper coil, a magnetic core, and an MO garnet. (b) The construction of a fiber pigtailed polarization switch. (c) Illustration of PM fiber orientations of the input and output fibers for a PM polarization switch. (d) The construction of a fiber pigtailed polarization rotator with a fixed 22.5° or 45° SOP rotation angle.

of magnetic field and therefore changes the major axis of an elliptical SOP from +22.5° or +45° to −22.5° or −45° and vice versa, resulting in a net SOP rotation angle of 45° or 90°.

For some applications, it is desirable to switch light from the slow axis of a PM fiber to the fast axis and vice versa. For such an application, a ±45° MO garnet film can be used, with the input and output fibers oriented 45° from each other when fabricating the device, as shown in Figure 6.25c. Assuming the light from the input PM fiber is in the slow axis (along the line between the "Panda eyes"), when a positive magnetic field is applied to the MO garnet for a +45° SOP rotation (clockwise rotation), the SOP is aligned with the slow axis in the output PM fiber. When a negative magnetic field in applied for a −45° SOP rotation (CCW rotation), the SOP is aligned with the fast axis of the output PM fiber, as shown in Figure 6.25c.

6.6.3.3 Variable and Fixed Polarization Rotators

If a continuous SOP rotation is required, an MO garnet with planar anisotropy described in Section 6.6.2.2 can be used in Figure 6.25a, in which the electrical current can be varied to continuously change the magnetic field and hence the Faraday rotation angle.

One may also construct a polarization rotator with a fixed SOP rotation angle of 22.5° or 45°, in which an MO garnet film of perpendicular anisotropy described in Section 6.6.2.1 with a saturation rotation angle of 22.5° or 45° can be used, as shown in Figure 6.25d. Similar to the FRM described in Figure 5.24a, a permanent magnet ring can be used to magnetically saturate the MO film and eliminate the unwanted diffraction. The magnet-free MO garnet discussed in Section 6.6.1 can also be used to eliminate the magnetic ring for reducing the device size.

6.6.4 Variable Faraday Rotator Based Polarization Controllers

Figure 6.26a shows a polarization controller made with three variable Faraday rotators (VFRs) using MO garnets of planar anisotropy described in Section 6.6.2.2 and two QWPs, which is capable of converting an arbitrary input SOP into any desired output SOP (Ikeda 2002). Alternatively, more complicated VFRs using MO garnets of perpendicular anisotropy described in Figure 6.22 can also be used. In the polarization controller, each VFR must at least be able to rotate SOP by 180° or ±90°. Because commercial MO garnets are generally made with a rotation angle of ±45° at saturation, at

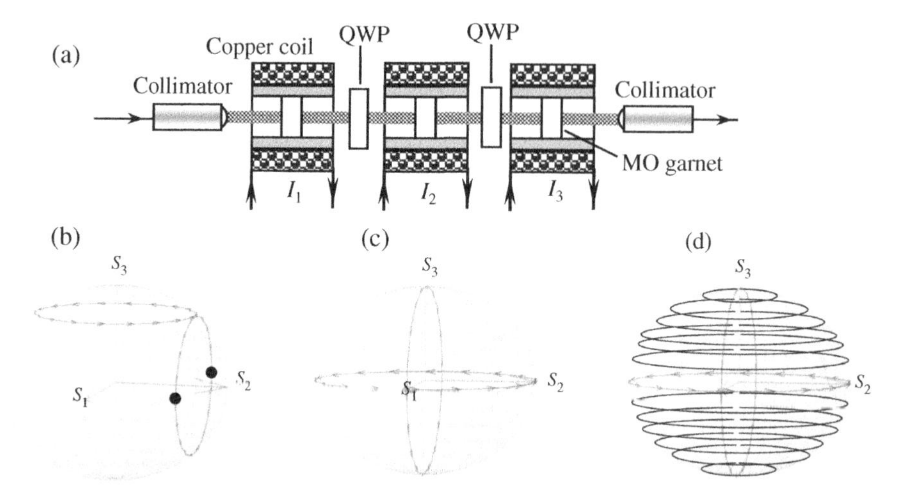

Figure 6.26 (a) Construction of a polarization controller made with three Faraday rotators and two quarter wave plates. (b) Horizontal circle: SOP evolution trace on a Poincaré Sphere as the SOP being rotated by the first VFR; Vertical circle: SOP evolution trace after the light beam passing through the first QWP showing two intersecting points with the equator. (c) Horizontal circle: SOP trace as one of the intersecting SOPs is rotated by the second VFR; Vertical circle: SOP trace after the beam passes through the second QWP. (d) Every SOP point on the vertical circle can be rotated to form a circle parallel to the equator, which enables every SOP point on the Poincaré Sphere to be reached.

least two such MO garnet chips must be used in each VFR to achieve the required ±90° Faraday rotation.

Now let's examine how such a combination of VFR and QWPs can convert an arbitrary input SOP into any output SOP. As illustrated in Figure 6.26b, a VFR rotates an arbitrary elliptical SOP around s_3 axis, with the SOP tracing out a circle in a plane parallel to the equator plane of the Poincaré Sphere. After the beam passes through the first QWP, the SOP rotation trace is converted to be around s_2 axis of the Poincaré Sphere with two points intersecting the equator, as shown by the vertical circle in Figure 6.26b, indicating that the combination of the VFR and the QWP can always convert an arbitrary SOP into one of two linear SOPs. This picture is consistent with the description in Section 6.4.2 that a QWP can always transform an arbitrary input elliptical SOP into a linear output SOP by aligning its slow or fast axis with one of the major axes of the polarization ellipse of the input beam.

The second VFR then rotates one of the linear SOPs around s_3 axis again, assuring the SOP tracing out a large circle on the equator. The second QWP then converts the SOP trace to a large circle around s_2 axis, encompassing north and south poles of the Poincaré sphere, as shown by the vertical circle in Figure 6.26c. Finally, the third VFR can rotate any point on the circle made by the second pair of VFR and QWP around s_3 axis to an arbitrary angle. In particular, by continuously rotating the SOP ±90° with the third VFR from every point on the large circle around s_2 axis, the whole Poincaré sphere can be filled with circles parallel to the equator, as shown in Figure 6.26d, indicating that any output SOP can be generated with an arbitrary input SOP. Therefore, the minimum elements for such a Faraday rotator based polarization controller are three VFRs and two QWPs.

Mathematically, the output SOP $\mathbf{S}_{out}$ can be related to the input SOP $\mathbf{S}_{in}$ as the Muller matrixes of the VFR and QWP as

$$\mathbf{S}_{out} = R(\theta_{F3})M_{QWP}(\alpha_2)R(\theta_{F2})M_{QWP}(\alpha_1)R(\theta_{F1})\mathbf{S}_{in} \tag{6.23a}$$

where $R(\theta_{Fi})$ and $M_{QWP}(\alpha_j)$ are the Mueller matrixes of the ith VFR ($i = 1, 2, 3$) and jth QWP ($j = 1, 2$), respectively, θ_{Fi} is the rotation angle of the ith VFR, and α_j is the orientation angle of the jth QWP. One may always choose $\alpha_j = 0$, then Eq. (6.23a) can be simplified as

$$\mathbf{S}_{out} = R(\theta_{F3})M_{QWP}(0)R(\theta_{F2})M_{QWP}(0)R(\theta_{F1})\mathbf{S}_{in} \tag{6.23b}$$

The expressions for $R(\theta)$ and $M_{QWP}(0)$ can be found in Table 4.3.

6.6.5 Non-reciprocal Fiber Optic Devices Made with MO Garnets

Non-reciprocal devices are critical to optical fiber systems. The most commonly used nonreciprocal devices include optical isolators and circulators. The word nonreciprocity simply means that the optical transmission property in one direction is different from that in the opposite direction. For example, an optical isolator allows light to transmit with a minimum loss in the forward direction from Port 1 to Port 2, however, blocks light from Port 2 to Port 1, whether the backward light is reflected or backscattered light in the fiber system or injected purposely in the opposite direction, as shown in Figure 6.27a. Similarly, a 3-port optical circulator routes light enters Port 1 to Port 2, while routes light enter Port 2 to Port 3, as shown in Figure 6.27b. Finally, a 4-port circulator directs light entering Port 1 to Port 2, light entering Port 2 to Port 3, and light entering Port 3 to Port 4, as shown in Figure 6.27c. As will be shown next, such non-reciprocal optical devices are generally made with Faraday rotators with a fixed saturation angle of 45°. The devices to be described in this section are intended for clear illustration of the physics process involved, which may not be the best configurations for commercial products. Both Jay Damask (2005b) and Yale Cheng (2003) have excellent descriptions of non-reciprocal optical devices in their books and interested readers are encouraged to read them. In the Further reading of this chapter, we also listed some patents for those devices, which are closely related to the commercial products, for serious product-design oriented readers.

6.6.5.1 Polarization Sensitive Isolator

The construction of a polarization sensitive isolator is shown in Figure 6.28a. Here the phrase "polarization sensitive" simply means that device operation depends on the SOP of input and output light. In general, the SOP of a light beam is well defined when propagating in free-space and therefore an optical isolator for a free-space optical beam is only required to work for a specific SOP. Such free-space isolator configuration can be adopted for making an isolator for PM fibers, as shown in Figure 6.28a. The device is made of an input PM fiber collimator, a polarizer, an MO garnet for generating a 45° Faraday rotation at saturation, and ring magnet for generating the saturation magnetic field for the MO garnet, and an output PM fiber collimator. The MO garnet and the ring magnet are collectively called the 45° Faraday rotator (45° FR). As shown in Figure 6.28b, the

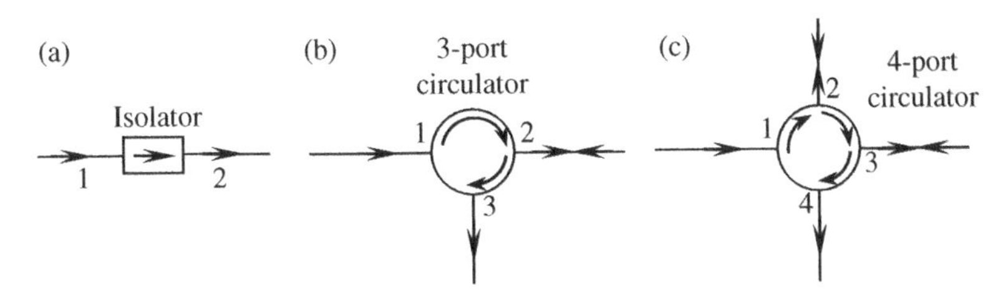

Figure 6.27 Illustration of the functionalities of an optical isolator (a), a 3-port circulator (b), and a 4-port circulator.

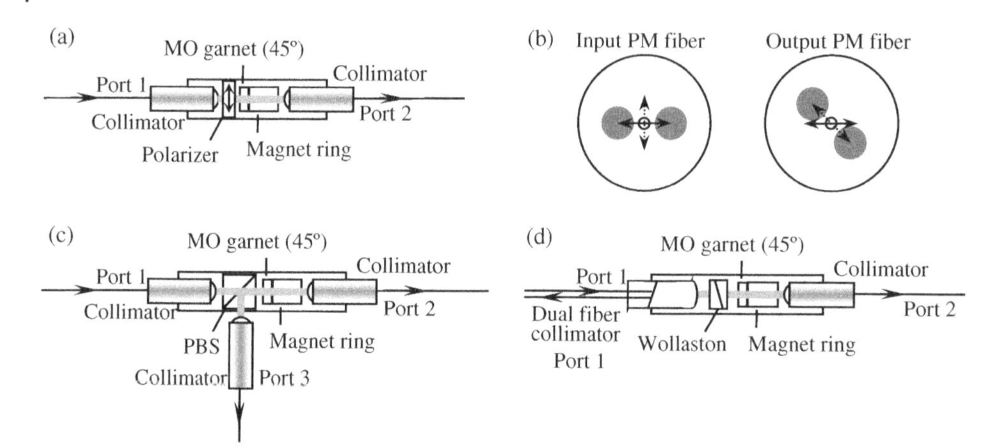

Figure 6.28 Constructions of polarization sensitive or PM fiber pigtailed isolator and circulators. (a) PM fiber pigtailed optical isolator. (b) Slow axis orientation of the input and output PM fibers. (c) PM fiber pigtailed optical circulator made with a PBS cube. (d) PM fiber pigtailed optical circulator made with a Wollaston prism and a dual fiber PM collimator.

input light polarized in the direction of the slow axis of the input PM fiber at Port 1 is aligned with the passing axis of the polarizer before entering the MO garnet to have its SOP rotated 45°. Because the slow axis of the output PM at Port 2 is oriented 45° with respect to the slow axis of the input PM fiber, the light beam at Port 2 will have its SOP aligned with the slow axis of the output PM fiber. Such a forward going optical beam will have a minimum insertion loss. On the other hand, for a light beam going backward from Port 2 toward Port 1, its SOP will be rotated 45° by the FR, which is now 90° from the passing axis of the polarizer and therefore will be totally blocked by the polarizer.

6.6.5.2 Polarization Sensitive 3-Port Circulator

A polarization sensitive circulator is shown in Figure 6.28c, which is almost identical to the PM fiber pigtailed isolator, except that here a PBS is used to replace the polarizer in the isolator, with the slow axis of the input PM fiber being aligned with the passing axis of the PBS. Because the SOP of the backward light from Port 3 toward Port 1 is 90° from the passing axis of the PBS, it will be reflected by the PBS toward Port 3.

Alternatively, a dual fiber collimator described in Figure 6.7b can be used to substitute collimators at Port 1 and Port 3 and a Wollaston prism to replace the PBS, as shown in Figure 6.28d. This way, the size of the device can be greatly minimized.

6.6.5.3 Polarization Independent Isolator

In optical fibers, the SOP of a light beam generally varies randomly; the polarization sensitive isolators and circulators described earlier will not work and polarization insensitive configurations are required.

Figure 6.29 shows a construction of a polarization independent isolator. In the forward direction from left to right, the light beam with an arbitrary elliptical SOP is decomposed into an *e-ray* (the x-polarization) and an *o-ray* (y-polarization) by the left PBD described in Figure 6.4a. The FR rotates the SOPs of the two rays CCW by 45° before they enter the HWP, which further rotates the SOPs of the two rays CCW by another 45°. Finally, the two rays are recombined by the right PBD.

In the backward direction, the beam is also decomposed into two rays of x- and y-polarizations by the right PBD, as shown in Figure 6.29b. The HWP then rotates the SOPs of the two rays clockwise. Because of the non-reciprocity, the FR rotates the SOPs CCW before the two rays enter the left PBD.

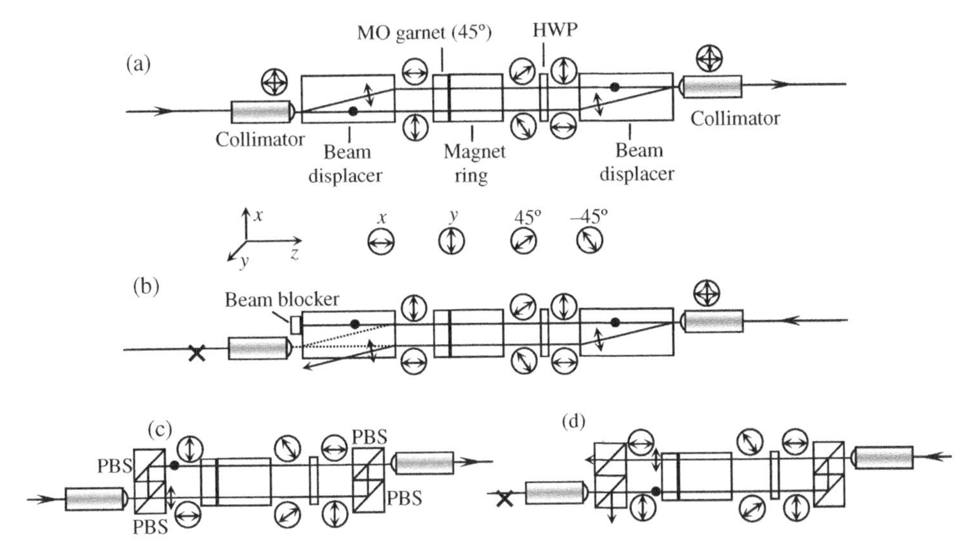

Figure 6.29 (a) A polarization independent isolator made with polarization beam displacers, showing the optical beams and the corresponding SOPs at different locations in the device in the forward direction. (b) The optical beams and the corresponding SOPs at different locations in the device of (a) in the backward direction. (c) A polarization independent isolator made with double PBS cubes, showing SOPs in the forward direction. (d) The device of (c) showing the SOPs in the backward direction.

Because now the SOP of each ray is orthogonal to that of the corresponding forward going beam, the left PBD will direct the two rays into different locations, therefore preventing the backward going rays getting into the input fiber collimator on the left.

Alternatively, each PBD in Figure 6.29a,b can be replaced by a pair of cube PBS, as shown in Figure 6.29c,d, which can be made by depositing a thin film of multiple layers of alternating high- and low-refractive index materials, as shown in Figure 6.6. Unlike the PBD in Figure 6.29a,b, here the *p-ray* passes through the PBS while the *s-ray* is reflected. Again, due to the non-reciprocity of the Faraday rotator, the forward going rays can be recombined and focused back into the fiber at the right-hand side, while the backward going rays cannot be recombined into the fiber at the left-hand side, hence achieving isolation.

6.6.5.4 Polarization Independent Circulator

There are several configurations for making a polarization insensitive optical circulator; here we describe one for the purpose of easy explanation, although it may not be adopted for making the commercial devices we see on the market.

As illustrated in Figure 6.30a, the circulator consisted of a first PBD (labeled 1), a first pair of HWP (labeled 2), a first 45° FR (labeled 3), a second PBD (labeled 4), a second 45° FR (labeled 5), a second pair of HWP (labeled 6), and finally a third PBD (labeled 7). The first PBD is to separate the light from Port 1 into two beams (*o-* and *e-rays*) in the up-down direction, as shown in Figure 6.30b. The first pair of HWPs with their optical axes (OAs) oriented 45° from each other rotates the top beam +45° (clockwise, CW) and the bottom beam −45° (CCW) so that both beams are of +45° linear SOP (L + 45P), as shown in Inset b1. The first FR rotates the L + 45P clockwise by 45° before both beams enter the second PBD as the o-rays (linear horizontal polarization [LHP]) to pass the second PBD without deviation. The SOP of the beams will be further rotated 45° CCW to be of L − 45P before entering the second pair of HWP. The OAs of the top and bottom HWPs are again

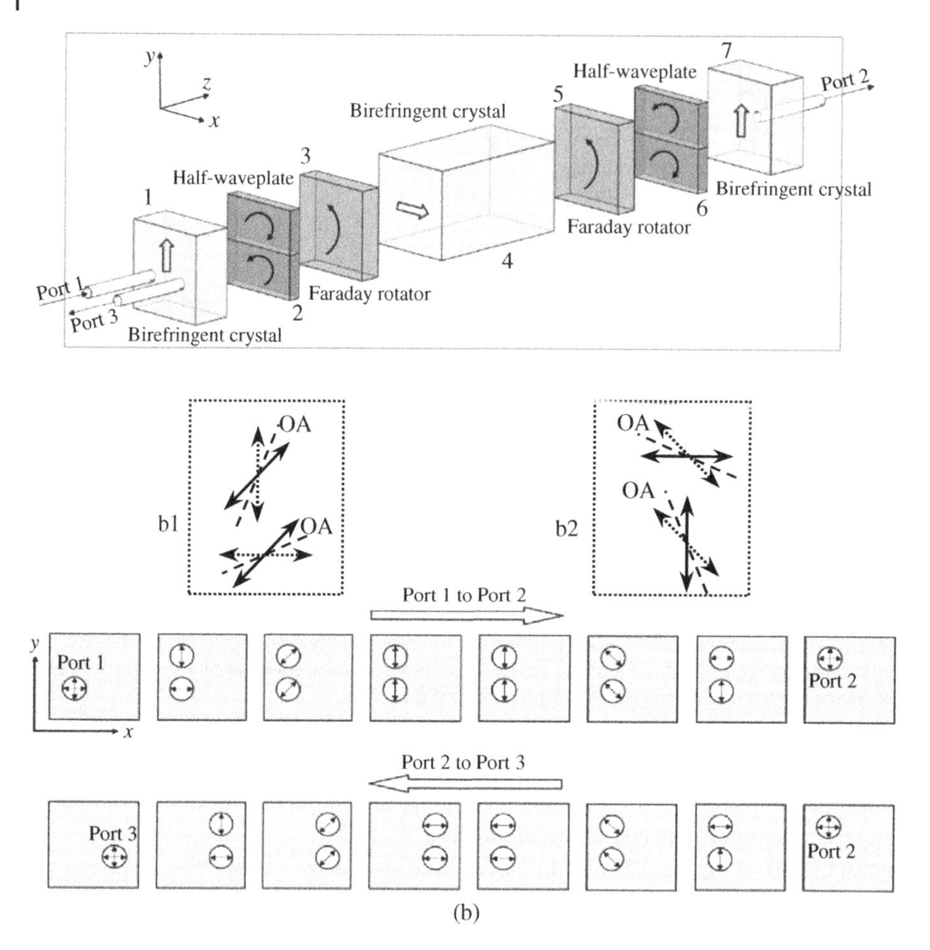

Figure 6.30 (a) The construction of a polarization insensitive circulator. (b) Diagraph showing the positions and SOPs of the optical beams in the device. OA: optical axis of the HWP.

oriented 45° from each other, as shown in Inset b2, to rotate the SOP of the top beam CCW (−45°) and the SOP of the bottom beam CW (+45°). Finally, the third PBD recombines the top and the bottom beams at Port 2.

For a light beam entering Port 2 toward Port 3, it will be separated into the top and bottom beams by the third PBD before their SOPs are rotated by the second pair of the HWP by 45° in opposite directions, as shown in Inset b2 of Figure 6.30b, to be aligned in the same direction (L − 45P). Their SOPs are further rotated CCW 45° after passing through the second FR to be the *e-rays*. These e-rays will be deviated horizontally by the second PBD, as shown in Figure 6.30b. The first FR then rotates their SOP CCW to be L + 45P. After passing through the first pairs of HWP, the SOPs of the top and bottom beams are rotated −45° (CCW) and +45° (CW) to be the *o-ray* and *e-ray*, respectively, for the first PBD to recombine them at Port 3.

6.7 PMD and PDL Artifacts

The PMD and PDL values of optical components described in Section 3.4 are two important specifications to be measured for every component because they directly affect the performance of an

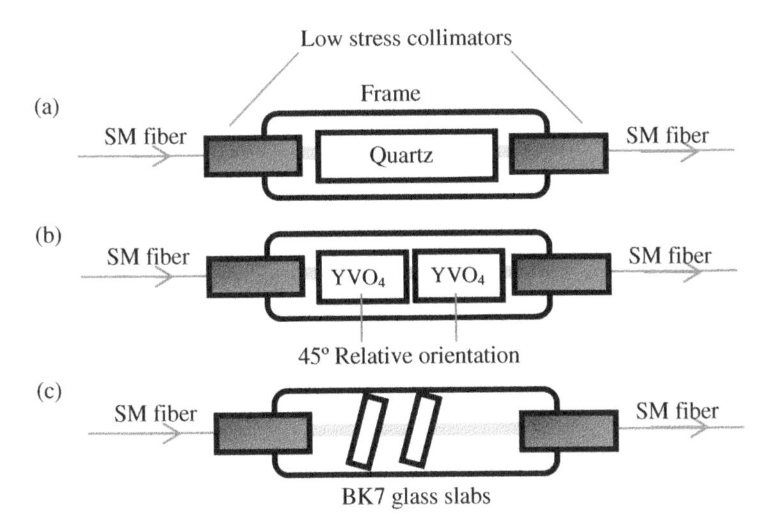

Figure 6.31 Schematic design of DGD, SOPMD, and PDL artifacts. (a) Precision wavelength independent DGD artifact. (b) Wavelength independent combined DGD and SOPMD artifact. The birefringence axes of two birefringence crystals are oriented 45° from each other. (c) Wavelength independent PDL artifact. Fused silica, BK7, or other types of glass can be used to make the artifact.

optical fiber system incorporating them. For the calibration or performance validation of measurement instruments, PMD and PDL artifacts with precise PMD and PDL values are required. Figure 6.31 shows the design and construction of such PMD and PDL standards, with PMD and PDL values defined by quantities that are precisely known, as described in the following text (Yao, Chen, and Liu 2010).

6.7.1 Differential Group Delay (DGD) Artifacts

First-order PMD (differential group delay, DGD) artifacts were constructed using birefringence crystals of precise lengths, such as quartz crystals, whose DGD values can be obtained by

$$DGD = \frac{\Delta n_g L}{c} \tag{6.24a}$$

where L is the length of the crystal, c is the speed of light in vacuum, and Δn_g is the group birefringence defined as

$$\Delta n_g = \Delta n - \lambda \frac{d\Delta n}{d\lambda} \tag{6.24b}$$

In Eq. (6.24a), Δn is the phase birefringence and λ is the wavelength in vacuum.

Quartz is the most well-known birefringence crystal, with birefringences at different wavelengths precisely known to an accuracy of about 0.25% (Williams et al. 2002). In addition, quartz crystals have excellent temperature stability with respect to both dimension and birefringence, with an extremely low temperature coefficient of 1.232×10^{-4}/°C (Williams 1999; Etzel, Rose, and Wang 2000) for total retardation. Furthermore, the processes for cutting and polishing quartz crystals are mature, and the dimensional tolerance is better than 10 μm or even 1 μm. Finally, quartz crystals are environmentally stable and their physical properties are generally not affected by aging. An exemplary quartz crystal of 20 mm length is enclosed in a 27 mm × 6 mm × 6 mm stainless steel package, pigtailed using a pair of low-stress fiber collimators, as shown in Figure 6.31a. The collimators are soldered or glued with special adhesives to the stainless steel package using a standard

Table 6.5 Summary of sources of DGD uncertainty in a quartz DGD artifact.[a)]

Uncertainty source	Uncertainty	Artifact DGD uncertainty
Δn_g	0.000 026 (Williams et al. 2002)	$\sigma_{\Delta ng} = 1.7$ fs
Length of quartz crystal	$\pm 10\,\mu m$	$\sigma_L = 0.3$ fs
Temperature of crystal	$\pm 5\,°C$	$\sigma_T = 0.4$ fs
Pigtail birefringence	± 1 fs	$\sigma_{Pigtail} = 1.0$ fs
Total uncertainty $\sigma_{DGD} = \sqrt{\sigma^2_{\Delta n_g} + \sigma^2_L + \sigma^2_T + \sigma^2_{pigtail}} = \mathbf{2.03}\boldsymbol{fs}$		

a) The artifact is made with a 20 mm long quartz crystal with a DGD value of 627.7 fs at 1550 nm (Yao, Chen, and Liu 2010).

process for making telecom grade fiber optic components with high environmental and long-term stabilities. The use of low-stress fiber collimators is necessary to minimize stress-induced birefringence and its effect on the overall DGD accuracy. As a result, the overall DGD accuracy at 1550 nm is better than 2.03 fs within a temperature range of $23 \pm 5\,°C$ at all times, as shown in Table 6.5. This DGD artifact can therefore be used as a "golden" calibration standard for calibrating different PMD measurement instruments.

6.7.2 Second Order Polarization Mode Dispersion (SOPMD) Artifacts

Wavelength independent second order polarization mode dispersion (SOPMD) artifacts can be constructed by cascading two birefringent crystals, such as YVO_4, with their birefringent axes oriented 45° from each other, as shown in Figure 6.31b. The SOPMD can be calculated from Eq. (4.385) as (Gordon and Kogelnik 2000):

$$SOPMD = \tau_1 \tau_2 sin(2\theta) \tag{6.25a}$$

The corresponding *DGD* from Eq. (4.384) is

$$DGD = \sqrt{\tau_1^2 + \tau_2^2 + 2\tau_1\tau_2 cos(2\theta)} \tag{6.25b}$$

where θ is the relative orientation angle between the two crystal sections. Like quartz, YVO_4 is a commonly used birefringent crystal with precisely known birefringences at different wavelengths. However, with a birefringence about 23 times larger than that of a quartz crystal, it is a better choice for making compact SOPMD artifacts with practical SOPMD values. YVO_4 also has reasonable temperature stability with respect to both dimensions and birefringence, with a fairly low thermal optic coefficient ($<1 \times 10^{-5}/°C$) (Kalisky 2005, p. 168). Like quartz crystals, YVO_4 crystal elements can be made with a dimensional tolerance better than 10 μm or even 1 μm and are environmentally stable, with their physical properties unaffected by aging. The same packaging materials and process used for the DGD artifacts can be used to assemble and pigtail SOPMD artifacts of different SOPMD values. The relative orientation angle can be readily controlled to within $\pm 1°$. The resulting accuracy is 0.86 ps^2 for an artifact made by cascading two 16 mm YVO_4 crystals (see Table 6.6); its anticipated DGD and SOPMD values are 16.15 ps and 130.2 ps^2, respectively (Yao, Chen, and Liu 2010).

Table 6.6 Summary of sources of uncertainty in a YVO_4 SOPMD artifact.[a)]

Uncertainty source	Uncertainty	Artifact SOPMD uncertainty
DGD of crystal 1	0.04 ps	$\sigma_{DGD1} = 0.46\ ps^2$
DGD of crystal 2	0.04 ps	$\sigma_{DGD2} = 0.46 ps^2$
Temperature (±5 °C)	$5 \times 10^{-5}/°C$	$\sigma_T = 0.04\ ps^2$
Pigtail birefringence	0.002 ps	$\sigma_{Pigtail} = 0.032\ ps^2$
Relative orientation angle	±1°	$\sigma_{Orientation} = 0.32\ ps^2$
Total uncertainty= $\sqrt{\sigma^2_{\Delta DGD1} + \sigma^2_{DGD2} + \sigma^2_T + \sigma^2_{pigtail} + \sigma^2_{Orientation}} = \mathbf{0.73ps^2}$		

a) The artifact is constructed with two 16 mm YVO_4 crystals cascaded with a 45° orientation with respect to each other. The anticipated DGD and SOPMD values are 16.15 ps and 130.2ps^2, respectively.

6.7.3 Polarization Dependent Loss (PDL) Artifacts

PDL artifacts can be designed and fabricated using standard optical glass slabs, such as those made with BK7 or fused silica, oriented at different angles with respect to the direction of incident light, as shown in Figure 6.31c. Because the transmittances of the *p*- and *s*-components are different and are wavelength independent, the transmitted light will have a wavelength independent PDL (diattenuation). The transmittance of the *p*- and *s*-polarization components at an interface of an optical slab can be readily calculated using Fresnel Eqs. (2.28a) and (2.28b):

$$T_p \equiv |t_p|^2 = \frac{sin2\theta_i sin2\theta_t}{sin(\theta_i + \theta_t)^2 cos^2(\theta_i - \theta_t)} \tag{6.26a}$$

$$T_s \equiv |t_s|^2 = \frac{sin2\theta_i sin2\theta_t}{sin^2(\theta_i + \theta_t)} \tag{6.26b}$$

where θ_i and θ_t are the incidence and refraction angles, respectively, related by Snell's law: $n_i\, sin\,\theta_i = n_t\, sin\,\theta_t$. The corresponding PDL can be calculated using

$$PDL = \left|10log\left(\frac{T_p}{T_s}\right)\right| = -10log[cos^2(\theta_i - \theta_t)] \tag{6.26c}$$

Because each slab has front and rear surfaces, the PDL value of a PDL artifact containing a single slab of BK7 is the summation of the PDL values from both surfaces. Similarly, the PDL value of a PDL artifact made with multiple slabs is the summation of the PDL values of all slabs with the same orientation used in the artifact. BK7 glass can be used for its temperature stability, dimensional stability, and ease of fabrication. In order to minimize the interference of light reflected from the front and rear surfaces, the rear surface is polished at a 0.5° tilt angle with respect to the front surface. Table 6.7 summarizes the PDL uncertainties caused by different factors. Clearly, the largest uncertainty factor is the incidence angle to the slab.

As shown in Table 6.7, the largest contributions to the total uncertainty are caused by the tolerances of the incidence angle and the angle between the slab's two surfaces. These are fixed once the fiber collimators are soldered to the case, as shown in Figure 6.31c. The exact PDL values can be

Table 6.7 Summary of uncertainty sources in a PDL artifact.

		Artifact PDL uncertainty	
Uncertainty source	**Uncertainty**	**0.4-dB artifact[a)]**	**2-dB artifact[a)]**
Incidence angle tolerance	0.5°	0.013 dB	0.050 dB
Tolerance of angle between front and rear surfaces	0.1°	0.002 dB	0.014 dB
Tolerance of index of BK7 slab	<0.001	0.0012 dB	0.004 dB
Temperature	$\pm 5\,°C$ ($\Delta n = \pm 1.2 \times 10^{-5}$)	$<1.6 \times 10^{-5}$	0.000 05 dB
Total uncertainty:		**0.013 dB**	**0.052 dB**

a) The nominal incidence angles are 33.92° and 62.93°, respectively, for the 0.4-dB and 2-dB artifacts.

accurately measured with a calibration-free method, such as the maximum and minimum search method to be discussed in Chapter 8. The final results are 0.387 ± 0.005 dB for a nominally 0.4-dB artifact and 1.959 ± 0.010 dB for a nominally 2-dB artifact. For the measurement, a highly stable laser source, a high quality polarization controller with extremely low activation loss (such as General Photonics Model PCS-4X with activation loss less than 0.005 dB), and a polarization insensitive power meter built with an integrated optical sphere (such as ILX Lightwave FPM-8210) are used. Once calibrated, the PDL variation due to temperature variation is negligible, as shown in Table 6.7.

6.8 Depolarizer

A depolarizer is a device to convert an input light signal with a high DOP into an output light signal with a low DOP, opposite to a polarizer. Generally, there are two basic approaches to make a depolarizer; one is called space domain depolarizer (SDD), and the other one time domain depolarizer (TDD), as will be discussed in detail next.

6.8.1 Space Domain Depolarizer

An SDD depolarizes a light beam by introducing different polarization rotations equally distributed from 0° to 180° at different locations across the beam, so that collectively the DOP of the whole beam is zero or close to zero. An SDD requiring the input beam having a specific SOP for an optimized performance is called polarization sensitive SDD, while an SDD that does not have the requirement is called polarization insensitive SDD.

6.8.1.1 Polarization Sensitive Space Domain Depolarizer

Figure 6.32a illustrates a side view of a spatial domain depolarizer made with a wedged birefringent crystal, such as quartz, calcite, or YVO_4 (Yao 1999). The optical axis (OA) of the crystal is oriented either along the x-axis or along the y-axis, although that along the x-axis is shown. A light beam propagates along z-axis with an x-polarization component and a y-polarization component. The x-polarization component experiences a refractive index n_e and the y-polarization component experiences a refractive index of n_o. The phase difference between the two polarization components is

$$\Delta\varphi(x) = 2\pi \cdot \Delta n \cdot tan\frac{\alpha}{\lambda} \tag{6.27}$$

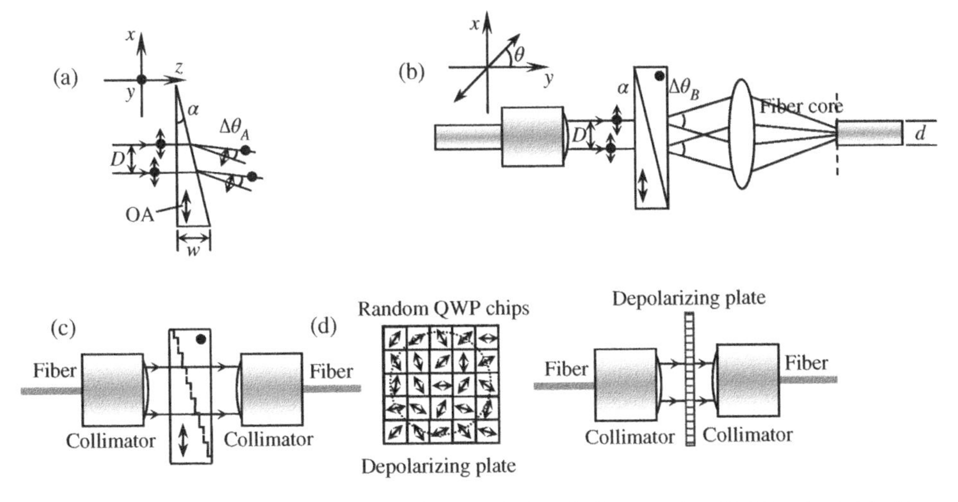

Figure 6.32 Illustrations of different spatial domain depolarizers made with (a) a single birefringent wedge; (b) a composite birefringent wedge; (c) Composite stair step wedge; and (d) Depolarizing plate made with randomly oriented QWPs. For configurations (a–c), the input beam SOP should be either L ± 45P or a circular (LCP or RCP) and an optional polarizer with its passing axis oriented 45° from the OA can be used to define the input SOP.

where x is the position of an optical ray on the x-coordinate, α is the wedge angle, λ is the optical wavelength of the incident light, and $\Delta n = n_e - n_0$ is the birefringence of the crystal. When an input polarization is oriented θ degrees from x-axis, as shown in Figure 6.32b, the output light field can be expressed as

$$\mathbf{E}(x) = E_0[sin\theta\hat{\mathbf{x}} + cos\theta e^{\Delta\varphi(x)}\hat{\mathbf{y}}] \tag{6.28a}$$

where $\Delta\varphi(x)$ is given in Eq. (6.27), E_0 is the amplitude of the input field, $\hat{\mathbf{x}}$ and $\hat{\mathbf{y}}$ are the unit vectors of the x-axis and y-axis, respectively. When $\theta = 45°$, Eq. (6.28a) becomes

$$\mathbf{E}(x) = \frac{E_0[\hat{\mathbf{x}} + e^{\Delta\varphi(x)}\hat{\mathbf{y}}]}{\sqrt{2}} \tag{6.28b}$$

It is evident from Eq. (6.28b) that optical rays incident at different location along the x-axis have different polarization states. For example, for two optical rays with $\Delta\varphi(x_1) = 2m\pi$ and $\Delta\varphi(x_2) = (2m+1)\pi$ (where m is an integer), the corresponding outputs from the depolarizer are orthogonally polarized linear polarization states. For two optical rays with $\Delta\varphi(x_1) = (2m \pm 1/2)\pi$, the corresponding outputs are circularly polarized with opposite sense of rotation, Therefore, the birefringent crystal wedge imparts different polarization states to different portions of an optical beam, resulting in a spatially depolarized beam. By focusing the spatially depolarized light beam into an optical fiber, we obtain a depolarized guided signal.

For an optical beam with a diameter D, the maximum difference in phase angle $\Delta\varphi$ between two rays within the beam is

$$\delta_{max} = max[\Delta\varphi(x_1) - \Delta\varphi(x_2)] = 2\pi \cdot \Delta n \cdot D \cdot tan\frac{\alpha}{\lambda} \tag{6.29a}$$

In order to completely depolarize light, δ_{max} preferably equals to or greater than 2π or

$$\Delta n \cdot tan\alpha \geq \frac{\lambda}{D} \tag{6.29b}$$

In an alternative arrangement, two wedged birefringent crystals are glued together to form a composite wedge, as shown in Figure 6.32b. The birefringence axis of the first birefringence wedge is oriented 90° from birefringence axis of the second wedge. The optical phase difference between the x-polarization component and the y-polarization component of a light ray passing through the composite wedge can be obtained by

$$\Delta\varphi(x) = 2\pi \cdot \Delta n \cdot \frac{(w - 2x \cdot tan\alpha)}{\lambda} \tag{6.30a}$$

where w is the thickness of the polarization depolarizer, as shown in Figure 6.32b. For an optical beam with a diameter of D, the maximum difference δ_{max} between two rays within the beam is

$$\delta_{max} = max[\Delta\varphi(x_1) - \Delta\varphi(x_2)] = 2\pi \cdot \Delta n \cdot 2D \cdot tan\frac{\alpha}{\lambda} \tag{6.30b}$$

where δ_{max} is twice as that of a single wedge. The condition for complete polarization randomization across the beam is

$$\Delta n \cdot tan\alpha \geq \frac{\lambda}{(2D)} \tag{6.30c}$$

As an example, to depolarize an optical beam with a wavelength of 1550 nm and a diameter of 0.5 mm using quartz crystal wedge having a Δn of 0.009, the wedge angle α should be larger than 5°.

The single-wedge and composite-wedge depolarizers described earlier yield an optical beam with spatially randomized SOP; however, as illustrated in Figure 6.32a,b, the x-polarization component and the y-polarization component exit the birefringent wedge at different angles caused by the different refractive indexes experienced by the orthogonal polarization components. For example, the angle differences between two polarization components for the cases of Figure 6.32a,b, respectively, are

$$\Delta\theta_A \approx \Delta n \cdot \frac{\alpha}{n} \tag{6.31a}$$

$$\Delta\theta_B \approx 2\Delta n \cdot \frac{\alpha}{n} \tag{6.31b}$$

where n is the refraction index of the medium after the crystal wedge. Substituting $\Delta n \cdot \alpha$ in Eqs. (6.31a) and (6.31b), respectively, with λ/D in Eq. (6.29b) and $\lambda/(2D)$ in (6.30c) yields

$$\Delta\theta_{min} = \Delta\theta_{Amin} = \Delta\theta_{Bmin} = \frac{\lambda}{(nD)} \tag{6.31c}$$

where $\Delta\theta_{min}$ is the minimum angle difference between the orthogonal polarization components.

The slightly different exit angles may not be a problem for a free Space beam with a sufficiently large beam diameter D. However, the different exit angles make it difficult to couple the x- and y-components exiting from the crystal wedge into an optical fiber with equal coupling efficiency, as shown in Figure 6.32b, which is important for producing a truly depolarized light inside the fiber. Figure 6.32b also illustrates coupling a double wedged depolarizer with an input optical fiber and an output optical fiber. The collimated beam from the input collimator passes through a depolarizer and then is focused into output fiber by a focusing lens. The input fiber can either be a non-polarization maintaining fiber, a PM fiber, or a polarizing (PZ) fiber. When a non-polarization maintaining input fiber is used, a polarization controller may be incorporated to adjust the SOP incident on the crystal wedge to be either ±45° linearly polarized with respect to the optical axis (OA) or to be circularly polarized (RCP or LCP) so that the x- (*e-ray*) and the y-components (*o-ray*) have approximately the same power. When the input fiber is a PM or PZ fiber, the slow or fast axis of the fiber is oriented approximately 45° from the crystal OA. An optional polarizer may be positioned at input side of the crystal wedge with its passing axis oriented 45° from the OA. The incorporation of the optional polarizer facilitates alignment of the input polarization to crystal OA by maximizing power output from depolarizer wedge. Because the x-polarization component and the y-polarization component exit depolarizer wedge with slightly different angles, the focusing lens focuses the orthogonal polarization components to slightly different locations, as shown in Figure 6.32b. The resulting lateral difference Δd_{min} between the orthogonal polarization components may be expressed as

$$\Delta d_{min} = \Delta\theta_{min} \cdot f \approx \frac{\lambda f}{(nD)} \tag{6.32a}$$

where f is the back focal length of the lens and $\Delta\theta_{min}$ is defined in Eq. (6.30c). For an output fiber with a numerical aperture of NA, the focal length f should satisfy the following relationship for a reasonable power coupling into the fiber:

$$\frac{D}{f} \leq NA \tag{6.32b}$$

Substitution of Eq. (6.31b) in Eq. (6.31a) yields

$$\Delta d_{min} = \frac{1}{n}\frac{\lambda}{NA} \tag{6.32c}$$

For optimum polarization randomization, both polarization components should be focused into the output fiber with approximately the same coupling efficiency. On the other hand, to focus both polarization components into the fiber with minimum loss, Δd is preferably less than the fiber core diameter d. For a standard SM fiber with $NA = 0.12$, $\lambda = 1.55\,\mu m$, and $n = 1.5$, $\Delta d_{min} = 8.33\,\mu m$ can be obtained from Eq. (6.32c), which is close to a typical fiber core diameter of 9 μm. Therefore, it is possible to focus both polarization components into an SM fiber; however, the insertion loss may be somewhat high.

Figure 6.32c illustrates a depolarizer which eliminates the angle difference $\Delta\theta_B$. Here the birefringent crystal wedge is formed by many small steps. Each step includes a subsurface oriented parallel to the input surface of the crystal wedge. The normal incidence of the collimated beam on each step surface results in no beam deviation for either the x- or y-polarization components. In addition, within each step, the optical rays experience the same amount of retardation and thus have the same output SOP. However, an optical ray in a different step experiences a different amount of retardation, resulting in a different output SOP. Thus each step corresponds to an output SOP. Because the optical beam covers a sufficient number of steps, the SOPs are randomized across the beam. Note that the step size should be much larger than the wavelength to minimize the beam diffraction or scattering and thus optimize the coupling efficiency into the output fiber. For example, for a light beam at a wavelength of 1.5 μm, the step size can be chosen to be 15 μm. Therefore a beam with a diameter of 1 mm contains approximately 66 sub-beams with different SOP to effectively depolarize the beam.

6.8.1.2 Polarization Insensitive Space Domain Depolarizer

The depolarizers described in Figure 6.32a–c are polarization sensitive; thus, the performance of the device depends on an input beam's polarization state. However, in some fiber optic applications, polarization insensitive devices are preferred. Figure 6.32d illustrates a polarization insensitive depolarizer, which includes many randomly oriented micro QWP chips. As discussed in Section 6.4.2, a QWP is capable of converting a linear SOP into any desired SOP if the relative orientation between the linear SOP and the optical axis of the plate can be arbitrarily rotated. In addition, it can convert an arbitrary elliptical SOP into a linear SOP or another elliptical SOP with an orientation determined by the relative angle between the QWP's OA and the major axis of the input elliptical SOP.

The size of each chip is significantly smaller than the beam size, however, sufficiently larger than the wavelength to minimize diffraction of scattering. Assuming a chip size of 15 μm × 15 μm, a beam with a diameter of 1 mm × 1 mm would contain 3489 QWP chips, sufficiently large to randomize the input SOP and thus produce a depolarized beam with sufficiently small DOP. One method of fabricating depolarizer is to sandwich a plurality of mica chips between two glass plates. Optical cement may be used to hold the mica chips and the glass plates together. The refractive index of the optical cement is preferably close to that of the glass and that of the mica chips. The optical axes of the mica chips may either be randomly oriented or be arranged in such a way that their orientation angles are evenly distributed from 0° to 180°. Alternatively, femtosecond laser processing can be used to make the micro-sized QWP on a glass or fused quartz substrate with randomly distributed orientations (Sakakura et al. 2020).

6.8.2 Time Domain Depolarizer

A TDD depolarizes a light beam by introducing time delays between different polarization components beyond coherence length of the light beam. A TDD requiring the input beam having a specific SOP for an optimized performance is called polarization sensitive TDD, while a TDD does not have the requirement is called polarization insensitive TDD.

6.8.2.1 Polarization Sensitive Time Domain Depolarizer

Figure 6.33a shows the construction of a simplest polarization sensitive TDD. Light from a PM fiber polarized along its slow axis is first collimated with a collimator before entering a birefringence crystal with its birefringence axis oriented 45° from the slow (or fast) axis of the PM fiber.

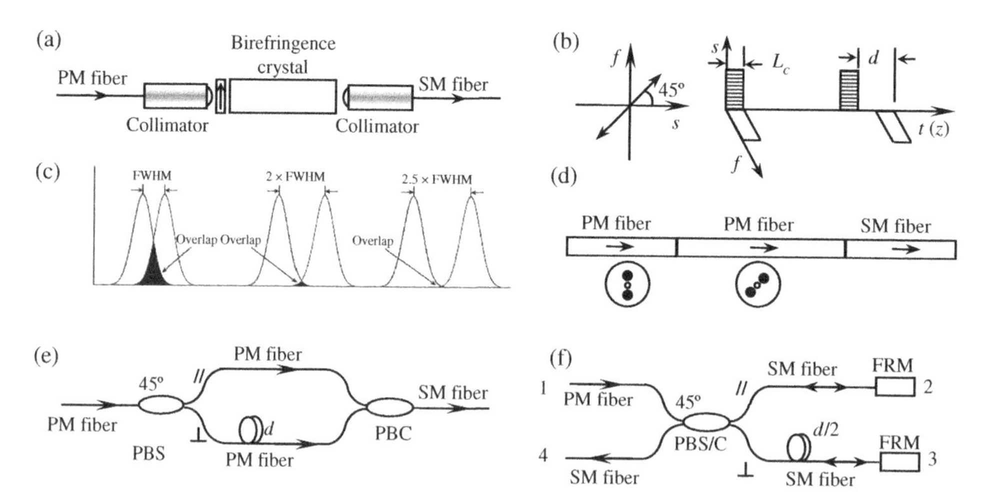

Figure 6.33 Illustration of different time domain depolarizers. (a) Depolarizer constructed with a birefringence crystal of a length L, with its slow axis oriented 45° from the slow axis of the input PM fiber. (b) Left: the relative orientation between the input SOP and crystal's axes. Right: Relative delay between lights in the slow and fast axes of the crystal. The rectangle represents the coherence function in space. (c) Gaussian function overlaps with different separations. (d) A depolarizer similar to (a) but using a piece of PM fiber to replace the birefringence crystal. (e) An all fiber Mach–Zehnder type depolarizer made with a PBS and a PBC. (f) An all fiber Michelson type depolarizer made with a PBS and two Faraday mirrors.

An optional polarizer with its passing axis precisely aligned 45° from the slow (or fast) axis of the crystal to filter out any light with orthogonal polarization component, which can also be used to assist the input PM fiber alignment. The relative orientation between the input SOP and the crystal's slow (s) and fast (f) axes is shown in Figure 6.33b so that 50% of the light is polarized along the slow axis and the other 50% is along the fast axis of the crystal. At the input end of the crystal, both polarization components overlap in time (or space). As they traverse the crystal with a length of L, the light polarized along the fast axis will be ahead of that along the slow axis by $d = \Delta n \cdot L$, where Δn is the birefringence of the crystal and d is called differential group delay (DGD). If d is much larger than the coherence length L_c ($d \gg L_c$) such that the two polarization components are totally separated in space, the DOP at the output of the crystal is

$$DOP = \frac{P_{max} - P_{min}}{P_{max} + P_{min}} = \frac{|P_s - P_f|}{P_s + P_f} = \frac{|cos^2\theta - sin^2\theta|}{cos^2\theta + sin^2\theta} = |cos(2\theta)| \tag{6.33a}$$

where θ is the relative angle between the input SOP and the crystal slow axis. When $\theta = 45°$, $DOP = 0$. If the polarization angle is misaligned by $\pm\Delta\theta$ from 45°, the resulting DOP is $|sin(2\Delta\theta)|$. For a DOP less than 1%, the misalignment angle must be less than $\pm 0.3°$. The length requirement for such a depolarizer must satisfy

$$L = \frac{d}{\Delta n} > \frac{L_c}{\Delta n} \tag{6.33b}$$

Considering that the coherence functions of light sources are generally Gaussian or Lorentzian shapes, with the coherence length L_c defined as the full width at half maximum (FWHM) of the corresponding coherence function, the crystal length L should be larger than $L_c/\Delta n$. As shown in Figure 6.33c, for two Gaussian functions separated by FWHM ($d = L_c$), the overlap portion is 12% of the total area of the two functions, which means that the smallest DOP can be achieved with such separation is 12%. For a separation d of 2 and 2.5 times of L_c, the overlap ratio is 0.93% and 0.16%, resulting in a smallest achievable DOP of 0.93% and 0.16%, respectively. Therefore, $d = 2L_c$ or $L = 2L_c/\Delta n$ is recommended for designing such a depolarizer. For example, for depolarizing a light source with a coherence length of 2 mm, the length requirement of the crystal with a birefringence of 0.2 is 20 mm. Such a depolarizer is often used for depolarizing light sources with short coherence lengths around 2 mm or less; otherwise, the device will be too long to be practical.

Alternatively, one may construct an all fiber depolarizer by replacing the birefringence crystal with a piece of PM fiber, as shown in Figure 6.33d. In particular, one may simply fusion splice the input PM fiber with the piece of PM fiber with a 45° relative orientation between their slow axes. As an example, to depolarize a light source with a coherence length of 2 mm, the length of the PM fiber is required to be more than 8 m, assuming the PM fiber has a typical birefringence of 5×10^{-4}. For light sources of much longer coherence lengths, proportionally longer PM fiber must be used, which can be very bulky and costly considering that PM fibers are much more expensive than SM fibers.

To depolarize light source of much longer coherence length, one may construct an all fiber depolarizer similar to a Mach–Zehnder interferometer in structure, as shown in Figure 6.33e. First equally split the input light using a PBS into two optical fibers of orthogonal polarizations by aligning the slow axis of the input PM fiber 45° from the PBS' passing axis. Next introduce a relative delay d between the orthogonal polarizations by adding an extra length L of PM fiber in one of the arms. Finally recombine them with a PBC before coupling into an output SM fiber. Additional attenuation may be introduced in one of the arms to balance the powers of the orthogonal polarizations to minimize the DOP of the output light in the SM fiber. The relative delay d is about 1.5 times of the extra fiber length because the refractive index of the fiber core is about 1.5. With an extra fiber length of L m, this configuration is capable of depolarizing a light source with a coherence length of $1.5L/2$ m, significantly longer than the configurations of Figure 6.33a,d.

To reduce the size and cost of the configuration in Figure 6.33e, one may use the configuration similar to a Michelson interferometer, as shown in Figure 6.33f, in which the PM fibers in the two fiber arms are replaced SM fibers. In addition, two FRMs described in Figure 6.24a are used to get rid of polarization fluctuations in the SM fiber. The reflected lights from the FRMs are then recombined by the PBS into the fourth ports of the PBS before coupling into an output SM fiber. Because the light double passes in the delay fiber, only half of the fiber length is required compared with that used in Figure 6.33e, which not only reduces the size but also the cost significantly because SM fibers are more than 10 times less expensive than PM fibers. Such a configuration is feasible for depolarizing light sources with coherence lengths of tens of kilometers.

The descriptions earlier are of time domain for easy visualization. These depolarizers can also be described in the spectral domain by considering different wavelengths λ in a light source of a finite bandwidth accumulate different retardations $\Delta\varphi$:

$$\Delta\varphi(\lambda) = \frac{2\pi \Delta n d}{\lambda} \tag{6.34}$$

Consequently, different wavelengths will have different SOPs sitting on a large circle passing through the north and south poles of the Poincaré sphere when the input SOP is aligned 45° from the slow and fast axes of the birefringence material (Figure 6.33a,d) or is equally split into two different optical paths (Figure 6.33e,f). If the spectral width of the light source is sufficiently large such that the SOPs of different spectral components cover the full circle, the light can be considered depolarized. More detail of spectral domain treatment can be found in (Burns 1983).

6.8.2.2 Polarization Insensitive Time Domain Depolarizer – Lyot Depolarizers

Invented by French astronomer Bernard Lyot, the Lyot depolarizer is probably the most discussed and utilized depolarizer. It is similar in construction to that of Figure 6.33a, however, has a second birefringence crystal added for overcoming the polarization sensitivity, as shown in Figure 6.34a. In particular, the second birefringence crystal is at least twice as long as the first one, with slow axis oriented 45° from that of the first crystal. Because of the polarization insensitivity, the input can be of a SM fiber. As an example, considering a beam with an input linear SOP aligned with the slow or fast axis of the first crystal, it will pass through the first crystal without depolarization and without SOP change. However, upon entering the second crystal oriented 45° from the first, depolarization occurs because a differential delay will be generated between the two polarization components along the slow and fast axes of the second crystal.

The inset of Figure 6.33a explains in general how the Lyot depolarizer works: a light beam of an arbitrary SOP enters the first crystal, is decomposed into a slow and fast component of different amplitudes with respect to the first crystal and accumulates a relative delay d equaling at least the coherence length of the light source L_c (the rectangle represents the coherence function of the light source with its width the coherence length). However, it is not completely depolarized because of the different amplitudes. After the beam enters the second crystal with its slow axis oriented 45° from that of the first crystal, each of the two polarization components with the relative delay d further decomposes into a fast and slow component of equal amplitude with respect to the slow and fast axes of the second crystal, as shown in the left diagraph of the inset, resulting in two pairs of wave pockets. It can be seen that in order to effectively separate the two wave pockets, the relative delay generated by the second crystal must be at least twice that of the first crystal. Such a depolarizer can be used for depolarizing light sources with a coherence length on the order of few millimeters.

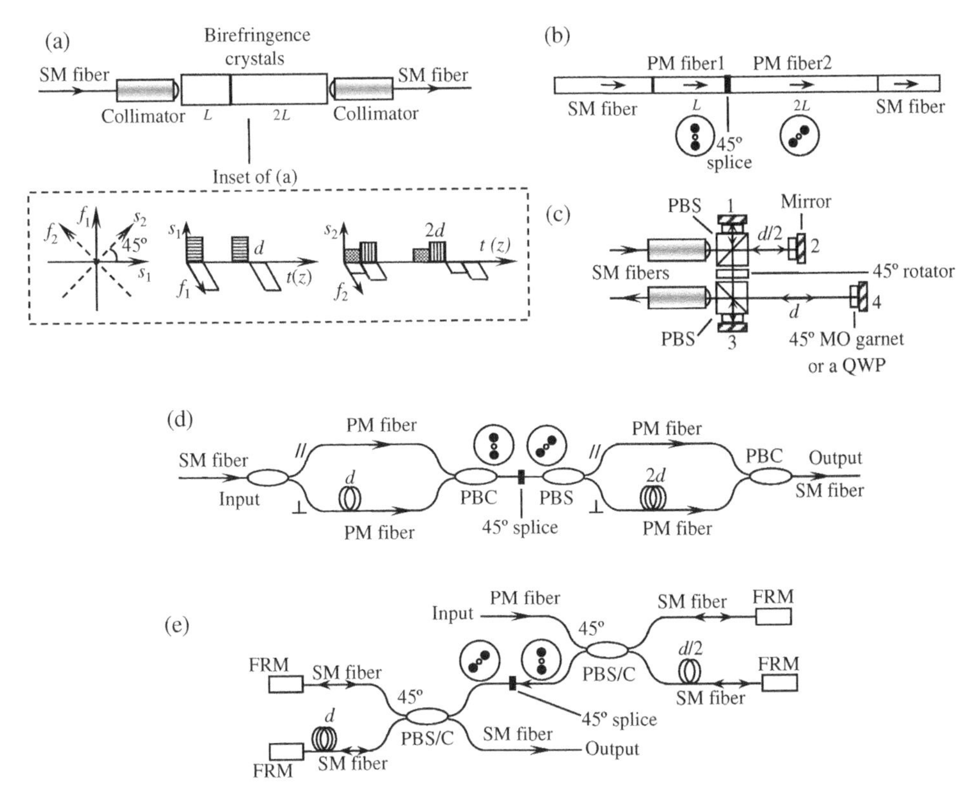

Figure 6.34 The Lyot and Lyot equivalent depolarizers. (a) The fiber pigtailed Lyot depolarizer constructed with two birefringence crystals, with the second one twice of the first one in length. Inset left: the relative orientation between the slow axes of the two crystals. Middle: the decomposition of the input light into the slow and fast axes of the first crystal and their relative delay. Right: the decomposition of the wave pockets in the slow and fast axes of the second crystal and their relative delays. Rectangle shaped coherence function of the light source is assumed for illustration purpose. (b) A Lyot depolarizer made with two PM fibers fusion spliced together with a 45° relative axis orientation. (c) An equivalent Lyot depolarizer made without birefringence materials. (d) An equivalent Lyot depolarizer made with two Mach–Zehnder style polarization sensitive depolarizers for light sources of long coherence length. (e) An equivalent Lyot depolarizer made with two Michelson style polarization sensitive depolarizers for light sources of even longer coherence lengths.

Figure 6.34b shows a Lyot depolarizer made with two pieces of PM fibers, with the second one more than twice as long as the first one. They are fusion spliced together with a 45° relative angle between their slow axes. This all fiber Lyot depolarizer is simplest to construct and can be readily made in any laboratories with PM fiber fusion splicers; however, it is relatively bulky due to a long length PM fiber that must be used. For example, for a light source with a coherence length of 2 mm, the first piece of PM fiber is required to be 8 m and the second piece is required to be at least 16 m, resulting in a total fiber length of 24 m, assuming the PM fiber has a typical birefringence of 5×10^{-4}. Again, such a depolarizer is effective in depolarizing light sources with coherence lengths on the order of few millimeters.

Figure 6.34c illustrates a free-space equivalent Lyot depolarizer made with micro-optical parts. In particular, four 90° polarization rotation reflectors are used, each reflects an incoming beam while rotating its SOP 90°. It can be constructed with either a 45° MO chip or a QWP followed by a mirror, as described in Chapter 4. The optical axis of the QWP is oriented 45° from the linear SOP of the incoming beam. The upper PBS, together with polarization rotation reflectors 1 and 2, acts as the first birefringence crystal in the Lyot depolarizer, while the lower PBS, together with polarization rotation reflectors 3 and 4, acts as the second birefringence crystal in the Lyot depolarizer. The 45° polarization rotator between the upper and the lower PBSs is to rotate the SOP of the light beam from the upper PBS by 45°, thus achieving the same functionality as for physically rotating the two birefringence crystals in a Lyot depolarizer. It can be made either with a 45° MO chip or an HWP oriented 22.5° from the axis of the upper PBS.

A collimated light with an arbitrary SOP from a SM fiber is split by the upper PBS into two beams of orthogonal linear polarizations. The "*s-ray*" reflected from the upper PBS is directly reflected by 90° polarization rotation reflector 1 to become a "*p-ray*" so that it can pass through the upper PBS toward the lower PBS. Similarly, the "*p-ray*" transmitted from the upper PBS travels a distance $d/2$ before being reflected by 90° polarization rotation reflector 2 to become an "*s-ray*," which is further reflected by the upper PBS toward the second PBS. The two orthogonally polarized beams have a relative delay of d (double pass) between the corresponding coherence functions, as shown in the inset of Figure 6.34a (middle graph). Each of the orthogonally polarized beams is then rotated 45° by the 45° rotator such that it can be equally split by the lower PBS to generate a pair of wave pockets of equal power, similar to the illustration in the inset of Figure 6.34a (right graph). Finally, the two orthogonally polarized beams from the lower PBS travel toward their respective polarization rotation reflectors 3 and 4 and are reflected back toward the lower PBS to be recombined before being coupled into the output SM fiber, with a relative delay of $2d$ (double pass), as shown in the inset of Figure 6.34a. Effective depolarization can therefore realize if the delay $2d$ is more than twice of the coherent length of the light source, exactly the same as the Lyot depolarizer described in Figure 6.34a. Comparing with the Lyot depolarizer made with birefringence crystals, this depolarizer can achieve much larger DGDs in a tight space and therefore is capable of depolarizing light sources of much longer coherence lengths. For example, for a given differential delay D, the crystal length required is $L_1 = D/\Delta n$, where Δn is the crystal birefringence, for the configuration of Figure 6.34a. However, for the depolarizer of Figure 6.34c, the air space length required is only $L_2 = D/2$, resulting in a space saving of $2/\Delta n$ times. For example, with a typical Δn of 0.2, the space saving is 10 times. One may further reduce the size of the device by inserting a material with a high refractive index n_h, such as silicon (n_h~3.8), in the air space between the upper PBS and reflector 2, as well as between the lower PBS and reflector 4, resulting in a $L_2 = D/(2n_h)$ and a space saving ratio of $2n_h/\Delta n$. Using n_h of 3.8 and Δn of 0.2, the spacing saving ratio is 38; comparing with birefringence crystal based Lyot depolarizers or the coherence length the amount the depolarizer can handle is increased 38 times. Such a depolarizer is effective in depolarizing light sources with coherence lengths around tens of centimeters.

To depolarize light sources of longer coherence lengths, the Mach–Zehnder configuration of Figure 6.34d can be used, which consists of two polarization sensitive depolarizers of Figure 6.33e made with PM fibers. They are spliced together with a 45° angle difference between the two slow axes of the PM fiber pigtails. The first one acts as the first birefringence crystal in the Lyot depolarizer, while the second one functions as the second birefringence crystal. As in Figure 6.33e, one may simply increase the fiber length difference between the two arms in the Mach–Zehnder structures to depolarize light sources of longer coherence lengths. Such a depolarizer is for depolarizing light sources with coherence lengths up to hundred meters.

For depolarizing light sources of even longer coherence lengths while reducing cost, the Michelson configuration of Figure 6.34e can be used, which consists of two polarization sensitive depolarizers of Figure 6.33f with the delay arms made with SM fibers. They are spliced together with a 45° angle difference between the two slow axes of the PM fiber pigtails. Again, the first one acts as the first birefringence crystal in the Lyot depolarizer, while the second one functions as the second birefringence crystal. As in Figure 6.33f, one may simply increase the fiber length difference between the two arms in the Michelson structures to match the increased coherence lengths of the light sources. Because in this configuration, low cost SM fiber is used for generating the required large differential delays, the total device cost can be significantly reduced compared with the configuration of Figure 6.34e. Such a depolarizer is effective in depolarizing light sources with coherence lengths up to tens of kilometers.

6.8.2.3 Polarization Insensitive Time Domain Depolarizer – Parallel Configurations

In essence, the various Lyot depolarizers described earlier require two polarization sensitive depolarizers cascaded together with a relative 45° polarization rotation, with the second depolarizer having a DGD at least twice that of the first one. The second depolarizer is used to compensate for the polarization sensitivity of the first one for achieving the required polarization insensitivity.

It is possible to achieve the polarization insensitivity without using the idea of cascading two depolarizers, as described in Figure 6.35, which was first proposed by X. Steve Yao (2003, 2006, 2009) and is called Yao depolarizer. As shown in Figure 6.35a, the light from the input fiber is collimated before entering a PBD of Figure 6.4a to be separated into two beams of orthogonal polarizations, the "*o-ray*" and the "*e-ray*," as shown in Figure 6.35b1. The SOP of the *e-ray* is rotated 90° by an HWP to be aligned with the *o-ray*, as shown in b2 separated in space. Now there are two *o-rays*, which are referred to as upper and lower rays. The SOPs of both rays are further rotated 45° relative to the passing axis of the first PBS by the 45° rotator to be equally split by the first PBS into

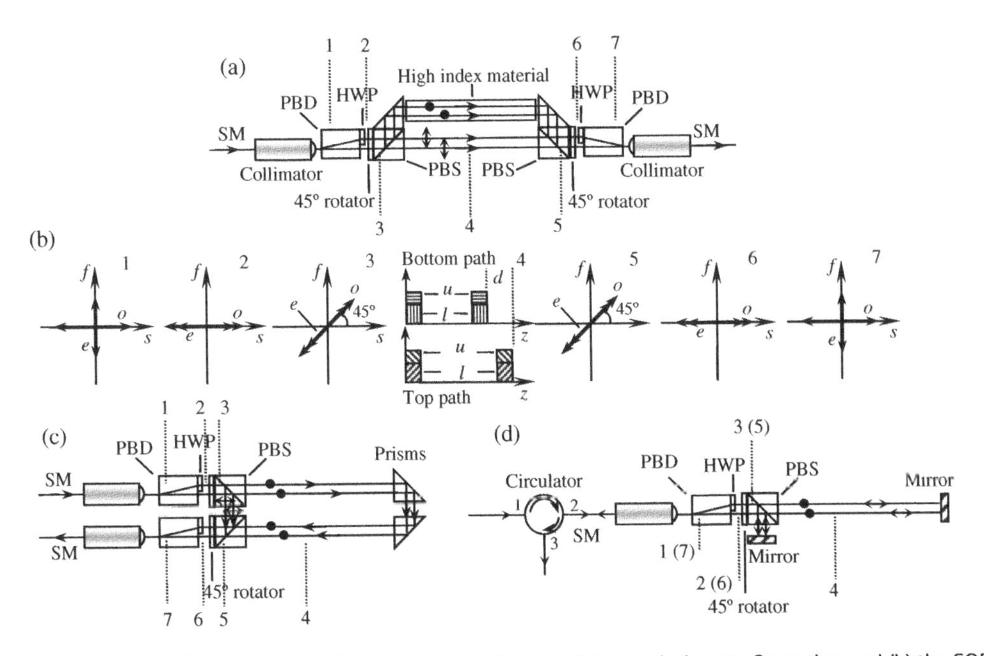

Figure 6.35 Different Yao-depolarizer configurations. (a) A transmission configuration and (b) the SOP at different locations (labeled 1–7) inside the device. *f* and *s* are slow and fast axes of the PBD, which are aligned with the passing and reflecting axes of the two PBSs, and *u* and *l* represent the upper and lower rays, respectively. Rectangle shaped coherence function is assumed for illustration purpose in b4. (c) and (d) Illustrations of two reflections configurations, with the location labels 1–7 corresponding to the drawing labels in (b). The 45° rotator can be eliminated if the PBDs are rotated 45° with respect to the passing axis of the PBSs.

top and bottom paths, as shown in Figures 6.35a, b3. The corresponding coherence functions of the upper and lower rays (labeled "*u*" and "*l*") in the top and bottom paths are shown in b4. The 45° rotator can either be a Faraday rotator or an HWP. Because there is a high refractive index material placed in the top path, the rays in the top path are delayed by a distance of *d* with respect to those in the bottom path. The rays in the top and bottom paths are finally recombined by the second PBS, before passing through the second 45° rotator (at 5), the second HWP (at 6), and the second PBD (at 7) with the corresponding SOPs shown in b5–b7. The 45° rotators can be eliminated by physically rotating each PBD 45° with respect to the passing axis of the corresponding PBS. Similar to the previous discussions, the light can be effectively depolarized when the differential delay *d* is larger than the coherence length of the light source (assuming rectangle coherence function) or twice the coherence length (assuming Gaussian coherence function) of the light source.

Figure 6.35c shows a reflective configuration of the Yao depolarizer for reducing the size of the device for a light source of a given coherence length. The operation mechanism is exactly the same as that of the transmissive configuration, with the SOPs of the light beam at different locations (labeled 1–7) and the differential delay between the PBS reflected and the PBS transmitted paths illustrated in Figure 6.35b, identical to those in Figure 6.35a. As in Figure 6.34c, the differential delay *d* can be greatly enlarged by inserting an optical material with a high refractive index, such as Si ($n \sim 3.8$).

Figure 6.35d shows another reflective configuration of the Yao depolarizer, in which light beams transmitted through and reflected from the PBS are directly reflected back by the corresponding mirrors, therefore eliminating the bottom optical path. The SOPs of the light beam at different locations (labeled 1–7) and the differential delay between the PBS reflected and the PBS transmitted paths can be shown in Figure 6.35b, also identical to those in Figure 6.35a. However, an optical circulator should be used at the left hand side to separate the input and the output light beams.

References

Barlow, A.J., Ramskov-Hansen, J.J., and Payne, D.N. (1982). Anisotropy in spun single-mode fibers. *Electron. Lett.* 18 (5): 200–202.

Bohnert, K. (2002). Temperature and vibration insensitive fiber-optic current sensor. *J. Lightwave Technol.* 20 (2): 267–276.

Burns, W. (1983). Degree of polarization in the Lyot depolarizer. *J. Lightwave Technol.* 1 (3): 475–479.

Campbell, J.P. and Steier, W.H. (1971). Rotating-waveplate optical-frequency shifting in lithium niobate. *IEEE J. Quantum Electron.* QE-7: 450457.

Cheng, Y. (2003). Optical isolators, circulators. In: *Encyclopedia of Physical Science and Technology*, 3e (ed. R.A. Meyers), 381–394. ScienceDirect.

Clarke, I.G. (1993). Temperature-stable spun elliptical-core optical-fiber current transducer. *Opt. Lett.* 18 (2): 158–160.

Damask, J.N. (2005a). Collimator technologies, Chapter 5. In: *Polarization Optics in Telecommunications* (ed. W.T. Rhodes), 212–246. Springer Science + Business Media, Inc.

Damask, J.N. (2005b). Isolators and circulators, Chapters 6 and 7. In: *Polarization Optics in Telecommunications* (ed. W.T. Rhodes), 245–296. Springer Science + Business Media, Inc.

Etzel, S.M., Rose, A.H., and Wang, C.M. (2000). Dispersion of the temperature dependence of the retardance in SiO_2 and MgF_2. *Appl. Opt.* 39: 5796–5800.

Feng, T., Zhou, J., and Yao, X.S. (2020). Distributed transverse-force sensing along a single-mode fiber using polarization-analyzing OFDR. *Opt. Express* 28 (21): 31253–31271.

Ferreira, L.A., Santos, J.L., and Farahi, F. (1995). Polarization insensitive fibre-optic white-light interferometry. *Opt. Commun.* 114: 386–392.

Fratello, V.J. and Wolfe, R. (2000). Epitaxial garnet films for non-reciprocal magnetooptic devices Chapter 3, book chapter in Magnetic film devices. In: *Handbook of Thin Film Devices: Frontiers of Research, Technology and Applications*, vol. 4 (ed. M.H. Francombe and J.D. Adam), 93–141. Academic Press.

Fratello, V.J., Mnushkina, I., and Licht, S.J. (2004). *Magneto-optical Imaging* (ed. T.H. Johnsen and T.H. Shantsev), 311. Dordrecht, The Netherlands: Kluwer Academic Publishers.

Fratello, V.J., Mnushkina, I., and Licht, S.J. (2005). Anisotropy effects in the growth of magneto-optic indicator films. *Mater. Res. Soc. Symp. Proc.* 834, paper J6.2.2: 311–318.

Gordon, J.P. and Kogelnik, H. (2000). PMD fundamentals: polarization-mode dispersion in optical fibers. *Proc. Nat. Acad. Sci.* 97: 4541–4550.

Greschishkin, R.M., Goosev, M.Y., Ilyashenko, S.E., and Neustroev, N.S. (1996). High-resolution sensitive magneto-optic ferrite-garnet films with planar anisotropy. *J. Magn. Magn. Mater.* 157/158: 305–306.

Huang, Y. and Xie, P. (1999). Optical polarization beam combiner/splitter. US Patent 6,282,025, filed 2 August 1999 and issued 28 August 2001.

Ikeda, K. (2002). Arbitrary polarization controller with variable Faraday rotator. *National Fiber Optic Engineers Conference (NFOEC),* 15–19 September 2002, Dallas, TX. p. 1965.

Integrated Photonics, Inc. (2012). Application Note F7.1.

Jaecklin, A. and Lietz, M. (1972). Elimination of disturbing birefringence effects on Faraday rotation. *Appl. Opt.* 11 (3): 617–621.

Johnson, M. (1979). In-line fiber-optical polarization transformer. *Appl. Opt.* 18 (9): 1288–1289.

Kalisky, Y. (2005). *The Physics and Engineering of Solid State Lasers*, 168. Bellingham, Washington, DC: SPIE Press.

Kersey, A.D., Marrone, M.J., and Davis, M.A. (1991). Polarization-insensitive fiber optic Michelson interferometer. *Electron. Lett.* 27 (6): 518–519.

Kurosawa, K., Yoshida, S., and Sakamoto, K. (1995). Polarization properties of the flint glass fibre. *J. Lightwave Technol.* 13 (7): 1378–1384.

Laming, R.I. and Payne, D.N. (1989). Electric current sensors employing spun highly birefringent optical fibers. *J. Lightwave Technol.* 7 (12): 2084–2094.

Land, E.H. (1951). Some aspects on the development of sheet polarizers. *JOSA* 41 (12): 957–963.

Lefevre, H.C. (1980a). Single-mode fiber fractional wave devices and polarization controllers. *Electron. Lett.* 16 (20): 778–780.

Lefevre, H.C. (1980b). Fiber optic polarization controller. US Patent 4,389,090, filed 1980 and issued 1983.

Li, J., Li, S., and Zhu, J. (2014). An adjustable polarization beam displacer (in Chinese). Chinese Patent CN 102981268 B, filed 23 November 2012 and issue 05 November 2014.

MacNeille, S.M. (1943). Beam splitter. US2403731, filed 1 April 1943 and granted 9 July 1946.

Mitsubishi Gas Chemical Company Inc. (2004). Magnetic garnet single crystal (Faraday rotator). Technical Manual: No. G00-05-E, updated 2008.8.1.

Molecular Technology GmbH. (2022). Faraday crystals TGG and glasses MOS-4 and MOS-10. http://www.mt-berlin.com/frames_cryst/descriptions/faraday.htm (accessed 19 February 2022).

Noe, R., Heidrich, H., and Hoffmann, D. (1988). Endless polarization control systems for coherent optic. *J. Lightwave Technol.* 6 (7): 1199–1208.

Norman, S.R., Payne, D.N., Adam, M.J., and Smith, A.M. (1979). Fabrication of single-mode fibers exhibiting extremely low polarization birefringence. *Electron. Lett.* 15 (11): 309–311.

Ohta, N., Hosoe, Y., and Sugita, Y. (1983). Submicron magnetic bubble garnets. In: *Recent Magnetics for Electronics* (ed. Y. Sakurai). Amsterdam: North Holland.

Okoshi, T. (1981). Single-polarization single-mode optical Fibers. *IEEE J. Quantum Electron.* QE17 (6): 879–884.

Peng, N., Huang, Y., Wang, S. et al. (2013). Fiber optic current sensor based on special spun highly birefringent fiber. *IEEE Photonics Technol. Lett.* 25 (17): 1668–1671.

Polynkin, P. and Blake, J. (2005). Polarization evolution in bent spun fiber. *J. Lightwave Technol.* 23 (11): 3815–3820.

Rajneesh, K., Pal, S., and Das, B. (2009). Fabrication and characterization of integrated optical TE-pass polarizer in $LiNbO_3$. *International Conference on Optics and Photonics*, Chandigarh, India (30 October to 1 November 2009).

Saitoh, T. and Kinugawa, S. (2003). Magnetic field rotating-type Faraday polarization controller. *IEEE Photonics Technol. Lett.* 15 (10): 1404–1406.

Sakakura, M., Lei, Y., Wang, L. et al. (2020). Ultralow-loss geometric phase and polarization shaping by ultrafast laser wring in silica glass. *Light Sci. Appl.* 9: 15.

Schneider, H., Harms, H., Papp, A., and Aulich, H. (1978). Low-birefringence single-mode optical fibers: preparation and polarization characteristics. *Appl. Opt.* 17 (19): 3035–3037.

Schreiber, T., Roser, F., Schmidt, O. et al. (2005). Stress-induced single-polarization single transverse mode photonic crystal fiber with low nonlinearity. *Opt. Express* 13 (19): 7621–7630.

Scott, G.B. and Lacklison, D.E. (1976). Magnetooptic properties and applications of bismuth substituted iron garnets. *IEEE Trans. Magn.* MAG-12 (4): 292–311.

Shimizu, H. and Kaede, K. (1988). Endless polarization controller using electro-optic waveplates. *Electron. Lett.* 24 (7): 412–413.

Smith, A.M. (1978). Polarization and magnetooptic properties of single-mode optical fiber. *Appl. Opt.* 7 (1): 52–56.

Stolen, R.H., Ramaswamy, V., Kaiser, P., and Pleibel, W. (1978). Linear polarization in birefringent single-mode fibers. *Appl. Phys. Lett.* 33 (8): 699–701.

Tang, D., Rose, A.H., Day, G.W., and Etzel, S.M. (1991). Annealing of linear birefringence in single-mode fiber coils: application to optical fiber current sensors. *J. Lightwave Technol.* 9 (8): 1031–1037.

Thaniyavarn, S. (1986). Wavelength-independent, optical-damage-immune $LiNbO_3$ TE–TM mode converter. *Opt. Lett.* 11 (1): 39–41.

Tkachuk, S., Fratello, V.J., Krafft, C. et al. (2009). Imaging capabilities of bismuth iron garnet films with low growth induced uniaxial anisotropy. *IEEE Trans. Magn.* 45 (10): 4238–4241.

Ulrich, R. (1979). Polarization stabilization on single-mode fiber. *Appl. Phys. Lett.* 35 (11): 840–842.

Ulrich, R. and Simon, A. (1979). Polarization optics of twisted single-mode fibers. *Appl. Opt.* 18 (13): 2241–2251.

Walker, N.G. and Walker, G.R. (1987b). Endless polarization control using four fiber squeezers. *Electron. Lett.* 23 (6): 290–292.

Walker, N.G., Walker, R.G., Davison, J. et al. (1987a). Lithium niobate waveguide polarization convertor. *Electron. Lett.* 24 (2): 103–104.

Williams, P.A. (1999). Rotating-wave-plate stokes polarimeter for differential group delay measurements of polarization-mode dispersion. *Appl. Opt.* 38: 6508–6515.

Williams, P.A, Etzel, S.M., Kofler, J.D., and Wang, C.M. (2002). Standard Reference Material 2538 for Polarization-mode dispersion (Non-mode coupled). *NIST Special Publication* 260-145.

Wu, S., Xiao, J., Feng, T., and Yao, X.S. (2020). Broadband and high extinction ratio hybrid plasmonic waveguide-based TE-pass polarizer using multimode interference. *J. Opt. Soc. Am. B: Opt. Phys.* 37: 2968–2975.

Wu, S., Hao, J., Zhao, Z., and Yao, X.S. (2021). Low loss and high extinction ratio all-silicon TM-pass polarizer with reflection removal enabled by contra-mode conversion Bragg-gratings. *Opt. Express* 29: 27640–27652.

Yao, X.S. (1995). Apparatus and method for connecting polarization sensitive devices. US Patent 5,561,726, filed 1995 and issued 1996.

Yao, X.S. (1999). Devices for depolarizing polarized light. US Patent 6,498,869, filed 24 July 1999 and issued 24 December 2002.

Yao, X.S. (2000a). Fiber devices based on fiber squeezer polarization controllers. US Patent 6,493,474, filed 30 September 2000 and issued 10 December 2002.

Yao, X.S. (2000b). Fiber squeezer polarization controller with low activation loss. US Patent 6,480,637, filed 30 September 2000 and issued 12 November 2002.

Yao, X.S. (2002a). Dynamic control of polarization of an optical signal. US Patent 6,576,886, filed 20 February 2002 and issued 10 June 2003.

Yao, X.S. (2002b). Transverse-pressure-controlled fiber devices. US Patent 6,754,404, filed 10 December 2002 and issued 22 June 2004.

Yao, X.S., Chen, J., Shi, Y.Q., et al. (2003). Optical depolarizers and DGD generators based on optical delay. US Patent 7,154,659, filed 2003 and issued 2006.

Yao, X.S., Chen, J., Shi, Y.Q. et al. (2006). Optical depolarizers and DGD generators based on optical delay. US Patent 7,535,639, filed 2006 and issued 2009.

Yao, X.S., Yan, L., and Chen, X. (2008). Quadrupling optical delay range using polarization properties. *IEEE Photonics Technol. Lett.* 20 (21): 1775–1777.

Yao, X.S., et al. (2009). Optical depolarizers and DGD generators based on optical delay. US Patent 8,164,831, filed 2009 and issued 2012.

Yao, X.S., Chen, X., and Liu, T. (2010). High accuracy polarization measurement using binary polarization rotators. *Opt. Express* 18 (7): 6667–6685.

Yao, P., Chen, X., Hao, P. et al. (2021). Introduction and measurement of the effective Verdet constant of spun optical fibers. *Opt. Express* 29 (15): 23315–23330.

Further Reading

Collett, E. (2003). *Polarized Light in Fiber Optics*. The PolaWave Group.

Collett, E. (2005). Field guide to polarization. In: *SPIE Field Guides*, vol. FG 05, Greivenkamp, J., Series Editor. SPIE Press.

Goldstein, D. (2003). *Polarized Light*, 2e. Marcel Dekker, Inc.

Huang, Y., Xie, P., Luo, X., and Du, L. (2002). Optical isolator with reduced insertion loss and minimized polarization mode dispersion. US Patent Publication 2002/0060843, filed 5 July 2001.

Kumar, A. and Ghatak, A. (2011). Polarization of light with applications in optical fibers. In: *Tutorial Texts in Optical Engineering*, vol. TT90. SPIE Press.

Rogers, A. (2008). *Polarization in Optical Fibers*, ISBN-13: 978-1-58053-534-2. Artech House, Inc.

Ulrich, R. and Johnson, M. (1979). Single-mode fiber-optical polarization rotator. *Appl. Opt.* 18 (11): 1857–1861.

Xie, P. and Huang, Y. (1998a). Compact polarization insensitive circulators with simplified structure and low polarization mode dispersion. US Patent 6,049,426, filed 17 August 1998 and issued 11 April 2000.

Xie, P. and Huang, Y. (1999a). Compact polarization insensitive circulators with simplified structure and low polarization mode dispersion. US Patent 6,052,228, filed 26 August 1999 and issued 18 April 2000.

Xie, P. and Huang, Y. (1999b). Compact polarization insensitive circulators with simplified structure and low polarization mode dispersion. US Patent 6,212,008, filed 30 March 1999 and issued 3 April 2001.

Xie, P. and Huang, Y. (1999c). Compact polarization insensitive circulators with simplified structure and low polarization mode dispersion. US Patent 6,285,499, filed 29 October 1999 and issued 4 September 2001.

Xie, P. and Huang, Y. (1998b). Optical circulators using beam angle tuners. US Patent 6,175,448, filed 5 November 1998 and issued 16 January 2001.

7

Active Polarization Management Modules and Instruments

The basic components and devices for managing the polarization in fiber optic systems are discussed in Chapter 6. Most of them are passive components, requiring no electronic control, while others are active devices consisting of passive and active components, capable of altering the polarization (state of polarization [SOP] and/or degree of polarization [DOP]) of light by applying voltages or currents onto the device. However, how to achieve a desired polarization or a polarization variation for different applications has not been discussed. In this chapter, we focus on the integration of these components and devices into different modules or instruments with electronics and algorithms for some particular applications.

7.1 Polarization Stabilization and Tracking

7.1.1 Reset-Free Polarization Control

In a practical optical fiber system, a polarization controller (PC) can be used to stabilize the output SOP against the input SOP fluctuations, which consists of multiple polarization control elements (PCEs) as discussed in Chapter 6 (Figures 6.11, 6.15, 6.16, and 6.26). This can be done by inserting a polarization beam splitter (PBS) at the output port of the PC and monitoring the optical power derived from the PBS with a photodetector (PD), as shown in Figure 7.1a. The PD converts the optical power into a photocurrent, which is then amplified by a transimpedance amplifier (TIA). The output photovoltage is then digitized by an analog-to-digital converter (ADC) before being processed by a digital signal processing (DSP) circuit, which is programmed to control the polarization rotation in each PCE by applying a suitable electrical voltage or current to minimize the received photocurrent. Equivalently, the output optical power passing through the PBS into the output fiber is maximized against polarization fluctuations at the input.

Although the polarization controllers described in Chapter 6 are capable of converting an arbitrary input SOP into any desired output SOP, it may not be able to track and undo the polarization variations continuously because each PCE has a limited control range (Walker and Walker 1987; Walker et al. 1988). In other words, when a PCE reaches its maximum range, it cannot continue to change the SOP in a desired direction and a reset of the PCE to its midrange is required in order for the PCE to continue to function. Such a reset will cause an abrupt SOP change and thus disrupt the normal operation of the system. Therefore, a reset free PC is desired (Noe et al. 1988; Bradley et al. 2001).

Figure 7.1b shows a reset-free PC made with six variable Faraday rotators (VFRs) and five quarter wave plate (QWPs) (Ikeda et al. 2003), as compared with the minimum-element configuration of

Polarization Measurement and Control in Optical Fiber Communication and Sensor Systems, First Edition.
X. Steve Yao and Xiaojun (James) Chen.

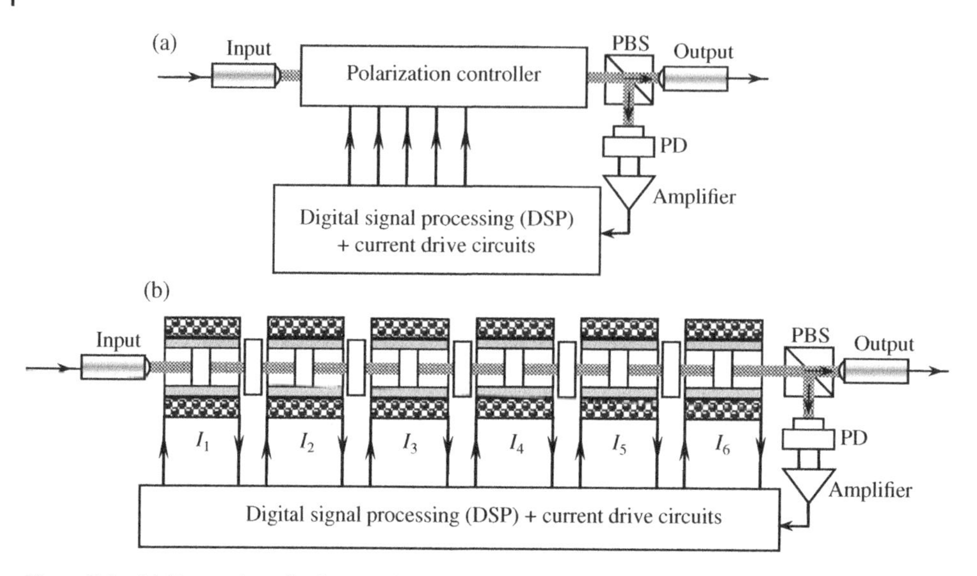

Figure 7.1 (a) Illustration of using a polarization controller to stabilize the output SOP against input SOP fluctuations. A PBS is placed at the output of the controller to analyze the SOP with a photodetector (PD). A digital signal processing circuit is programmed to control the polarization rotation with different current into each Faraday rotator for minimizing the detected photocurrent in PD and hence maximize the power output into the output fiber. (b) A polarization stabilizer made with a reset-free polarization controller consisting of six Faraday rotators and five QWPs. The range of polarization rotations for each Faraday rotator is at least 2π.

three VFRs and two QWPs. Each VFR must have an SOP rotation range of more than 2π, with the nominal rotator setting at 0°, such that the VFR has the room to rotate the SOP from $-\pi$ to $+\pi$ degrees. If one of the VFR reaches its range limit of either $+\pi$ or $-\pi$, the DSP will instruct it to rewind toward zero rotation angle. Because a minimum three consecutive rotators with a QWP sandwiched in between are required to change any SOP to any other SOP, as described in Section 6.6.4, no matter which VFR is about to be out of range, it can always be rewound while other VFR can work together to undo the SOP change caused by the rewind. It is possible to use five VFRs and four QWPs to realize reset-free operation (Ikeda et al. 2003); however, more complicated control algorithm must be used.

Alternatively, one may also realize such a reset free PC consisting of six variable retarders (fiber squeezers or voltage controlled wave plates) as PCEs (Bradley et al. 2001), as compared with the minimum-element configuration of three fiber squeezer or retarders illustrated in Figure 6.15. Each of the PCEs has a retardation variation range larger than 4π. The reason for using six PCEs is that no matter which PCE needs to be rewind, one can always find at least three consecutively connected PCEs (the minimum-element configuration capable of converting any SOP into any other SOP) to compensate for the SOP variation caused by the rewinding process. Such a reset free PC has been successfully commercialized, as shown in Figure 7.2a. Figure 7.2b shows the time response of the stabilizer against a step SOP change, with the upper curve showing the abrupt change of the input SOP and the lower curve showing the output power after the signal passing through the output PBS shown in Figure 7.1a, with the signal recovery time of 1 ms. Finally, Figure 7.2c shows the output power comparison with and without polarization stabilization against a continuous SOP rotation at 60π rad/s caused by an input linear SOP passing through a rotating half wave plate (HWP) at 30 rps. No reset can be seen when the input SOP is continuously rotated 270 full revolutions during the nine seconds shown. Note that for verifying the reset-free operation, it is

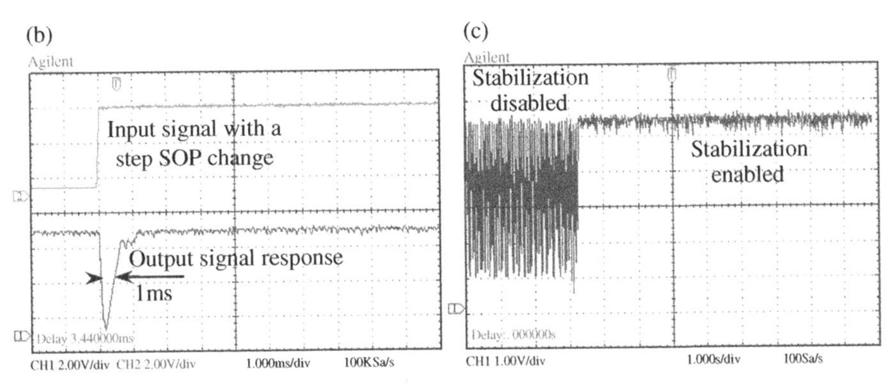

Figure 7.2 (a) A commercial polarization stabilizer made with fiber squeezers developed by the authors of this book (Source: Courtesy of General Photonics/Luna Innovations); (b) The output signal of the stabilizer (lower curve) in response to a step change of input SOP (upper curve). (c) The comparison of the output power with stabilization disabled and enabled against the continuous rotation of an input linear SOP at 60π rad/s.

important to use such input SOP continuously rotating in one direction (unidirectional rotating SOP) generated either by passing the light through a rotating HWP or by a polarization synthesizer to be discussed in Section 7.1.3; otherwise, falsely optimistic conclusion may result.

It is possible to use a minimum of four PCEs to realize reset free operation in the laboratory (Walker and Walker 1987); however, its robustness for real world applications still needs to be proved. Another approach is to cascade two stages of stabilizers together, each has two PCEs, to realize endless polarization stabilization (Martinelli et al. 2006), reducing the total number of PCEs to four, however, at the expense of two detection channels and higher insertion loss.

7.1.2 Polarization Monitoring for Active Polarization Control

As demonstrated in Figure 7.1, in order to actively control and stabilize the polarization, a polarization monitor is required. In Figure 7.1, a PBS is used to monitor the deviation from a linear SOP, which is further used as the feedback signal to maintain the output to a stable linear SOP. Other polarization monitoring parameters can also be used. Figure 7.3 shows different polarization monitoring schemes, including Figure 7.3a for monitoring SOP deviation from a linear SOP used in Figure 7.1.

One may also monitor the deviation from a circular SOP (either right-hand circular [RCP] or left-hand circular [LCP]), as shown in Figure 7.3b. The monitored deviation signal is then used as the feedback signal to control the polarization controller so as to output a stabilized circular SOP against input SOP fluctuations. Figure 7.3c shows an in-line polarimeter (to be described in detail in Chapter 8) is used to monitor the deviation from an arbitrary intended SOP for generating the

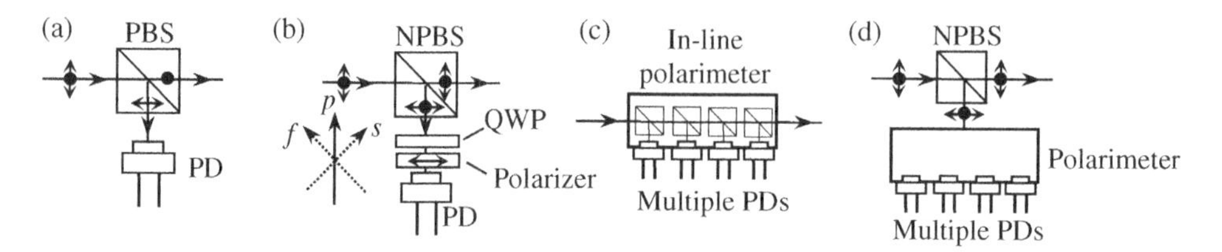

Figure 7.3 Different polarization monitoring schemes for active polarization control. (a) Simple PBS monitor for monitoring SOP deviation from a linear SOP; (b) Circular SOP monitor for monitor SOP deviation from a circular SOP; (c) In-line polarimeter for monitoring SOP (or DOP) deviation from any intended SOP (or DOP); (d) A termination type polarimeter for monitoring SOP (or DOP) deviation from any intended SOP (or DOP). *p*: passing axis axes of the polarizer; *s* and *f*: the slow and fast axes of the QWP; NPBS: polarization insensitive beam splitter.

SOP and maintaining it against input SOP fluctuations. As will be shown next, it can also be used to monitor the DOP of the signal for polarization mode dispersion (PMD) compensation applications. Finally, Figure 7.3d shows a termination type polarimeter for monitoring both SOP and DOP by tapping out a small amount of light using a NPBS to perform the same functionalities as the in-line polarimeter described in Figure 7.3c. In practice, the amplitude-division or wave-front division polarimeters to be discussed in detail in Chapter 8 can be used as the polarization monitor in Figure 7.3d.

7.1.3 Polarization Synthesizer

It is possible to use a polarimeter shown in Figure 7.3c or d to replace the PBS in Figure 7.1 as a polarization monitor. Such polarimeters will be discussed in Chapter 8. Any desired SOP (represented as a set of Stokes parameters or a Stokes vector) can be used as the reference SOP with a normalized Stokes vector $\mathbf{S}_r$. The deviation from the reference, represented as the 3D angle Φ between actual SOP with a normalized Stokes vector **S** and the reference SOP, can be used as the error signal to control the polarization controller to minimize the error signal, therefore, generate and stabilize the output SOP to the desired SOP. Specifically, the 3D angle Φ between an arbitrary SOP and the reference SOP can be expressed as the dot product between $\mathbf{S}_r$ and $\mathbf{S}$: $cos\,\Phi = \mathbf{S}_r \cdot \boldsymbol{S}$. The synthesizer is thus programmed to control the SOP for minimizing the 3D angle Φ. Therefore, a polarization synthesizer is capable of generating any SOP on the Poincaré sphere. Figure 7.4a shows a commercial synthesizer developed and marketed by General Photonics Corporation, with the instantaneous SOP measured at the output of the device by the polarimeter displayed on the Poincaré sphere with a computer using the Stokes data from the synthesizer.

One may use the polarization synthesizer to generate a set of special polarization states, such as LHP, linear vertical polarization (LVP), L + 45P, L − 45P, RCP, and LCP, in a sequential order as shown in Figure 7.4b, and hold a specific SOP against the input SOP fluctuations in a period defined by the user. Such a capability is important for obtaining the Mueller matrix elements of an optical component for the quantitative measurement of the polarization properties of the optical component.

If one programs the synthesizer to change the reference SOP along a particular path on the Poincaré sphere, such as along the equator representing a continuous linear SOP rotation, the actual SOP trace following the programmed reference SOP trace can also be generated, regardless of the input SOP fluctuation. The SOP therefore is able to be rotated along the trace endlessly. Figure 7.4c shows three such SOP traces, one is along the equator (in s_1 and s_2 planes) and the

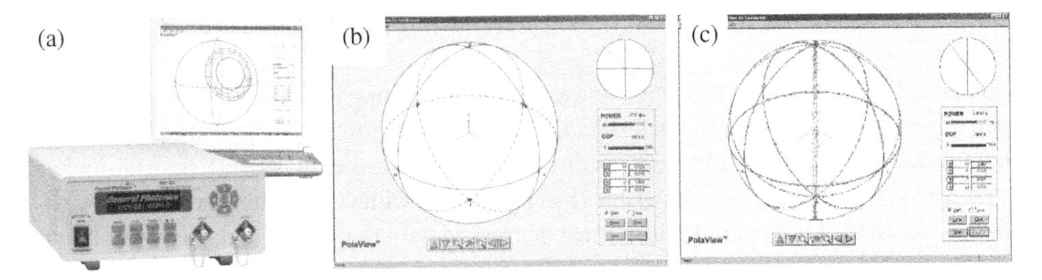

Figure 7.4 (a) A commercial polarization synthesizer with a computer display of generated SOP on a Poincaré sphere. (b) Six special polarization states generated by the synthesizer displayed on the Poincaré sphere. (c) Three special SOP traces generated by the synthesizer displayed on the Poincaré sphere.

other two are along large circles enclosing north and south poles in (s_1, s_3) and (s_2, s_3) planes, respectively. Any other SOP traces can also be generated by first generating a set of SOP points on the Poincaré sphere (represented by Stokes parameters) and then enable the synthesizer to generate these SOP points sequentially in time. This programmable SOP-trace generation capability is useful for testing the reset-free tracking capability of an optical polarization stabilizer described earlier or the tracking capability a digital DSP circuit in a coherent receiver for overcoming polarization related impairments, such as PMD, polarization dependent loss (PDL), and polarization fading, as mentioned in Section 7.1.1.

7.1.4 General Purpose Polarization Tracker

The polarization stabilizers and synthesizer described earlier use an internal monitoring signal as the feedback for tracking to a particular polarization state against input SOP variations. That is to say, the targeted physical quantity to be maintained is a polarization state. For many other applications, the targeted physical quantity to be tracked or optimized may be something else, such as the bit error rate (BER) of a high data rate signal, or the visibility of an interferometer. Therefore, other quantities may be monitored for different purposes. Figure 7.5a shows the block diagram of a general purpose polarization tracker, which not only includes an internal polarization monitoring device, such as those described in Figure 7.3, but also can accept external monitoring signals obtained using different means, as will be discussed next. In this polarization tracker, either the internal feedback signal or the external feedback signal is first digitized by an ADC or A/D before feeding into the microprocessor or DSP circuit with a control algorithm to process the feedback signal and generate a control signal, which is further converted into an analog signal by a digital-to-analog converter (DAC or D/A) and amplified by an electrical driving circuit

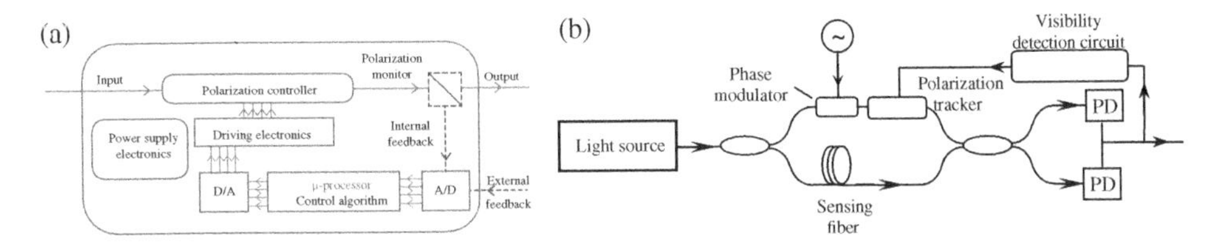

Figure 7.5 (a) Diagram of a general purpose polarization tracker with an internal and an external polarization monitors, enabling users to derive their own feedback signals for different applications. The external feedback signal is analog voltage signal in the range between 0 and 5 V. (b) An example of using a polarization tracker for overcoming the polarization fading problem in a Mach–Zehnder interferometer by taking the visibility as the feedback signal.

before finally being used to control the polarization controller to either maximize or minimize the feedback signal.

A typical example of using an external feedback signal for tracking the polarization is shown in Figure 7.5b, in which the SOP in an interferometric sensor system must be optimized and maintained against environmental perturbations, as discussed in Section 3.6. In general, interferometer based fiber sensor systems are polarization sensitive. In order to achieve maximum sensitivity without fluctuation, the SOP of the reference arm must be tracked with that of the signal arm, as shown in Figure 7.5b. In this application, a phase modulator is used in the reference arm to induce a sinusoidal phase modulation at a frequency much higher than the sensing signal to produce a sinusoidal interference signal with maximum and minimum signal levels of V_{max} and V_{min}, respectively. The visibility of the interference signal (defined as $(V_{max} - V_{min})/(V_{max} + V_{min})$) is detected and converted to an analog signal proportional to the visibility in a range between 0 and 5 V. This analog signal is then used as the external feedback for the polarization tracker, which automatically adjusts the SOP in the reference arm to obtain the maximum visibility for the interferometer.

The following are some specific applications of such a general purpose polarization tracker.

7.1.5 PMD Compensation with a Polarization Tracker

As discussed in Section 3.4, as the bit rate of fiber optic communication systems increases from 10 to 40, 100 Gbps, and beyond, PMD has more and more impact on system performance (Nelson and Jopson 2005). PMD generally causes two principle polarization components of a light signal to travel at different speeds and hence spreads the bit-width of the signal. Consequently, it causes the increase of BER and service outage. Unlike other system impairments, such as chromatic dispersion (CD), PMD effect on the system is random in nature and changes rapidly with time, making it difficult to mitigate (Nagel et al. 2000). The rapid varying PMD effect is mainly due to polarization changes caused by external disturbances on the fiber, such as vibration, wind, and temperature changes. Therefore, PMD mitigation must be dynamic in nature and involve dynamic polarization tracking and control. Before the deployment of coherent detection with polarization multiplexing (Xie 2016), PMD mitigations in the optical domain were extensively studied (Kogelinik et al. 2002; Bulow and Lanne 2005), as will be described next.

Scheme 1: Using a polarimeter to obtain feedback (DOP feedback): As shown in Figure 7.6a, there are two principal states of polarization (PSP) traveling at different speeds in a length of optical fiber with PMD. At the receiving side, they will be split in time due to the PMD effect. When the two PSP components have no relative delay, the *DOP* is 1. If they have the same power and are separated with no overlap, the *DOP* is 0. In fact, the effect of PMD on the optical signal

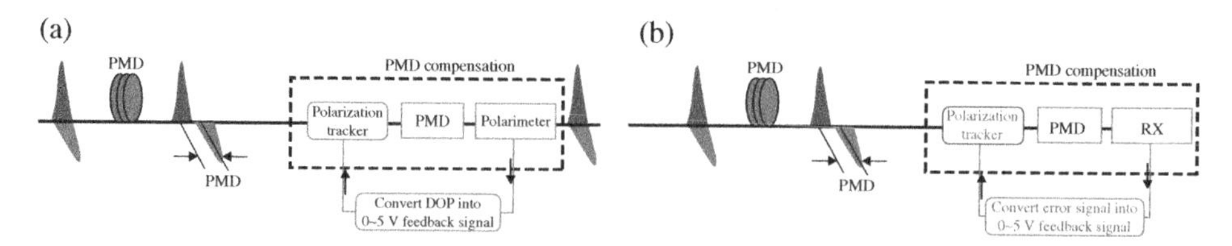

Figure 7.6 PMD compensation using a polarization tracker. (a) Using a polarimeter to derive DOP information of the transmission data signal as the feedback signal for the polarization tracker. The DOP will be maximized when the PMD is properly compensated. (b) Using the BER at the receiver before FEC as the feedback signal for the polarization tracker. The polarization tracker then either maximizes or minimizes the error signal to achieve PMD compensation. RX: Receiver.

is similar to that of the depolarizer described in Figure 6.33b. In general, the bigger the PMD effect on the signal, the smaller the DOP. Therefore, one can use DOP as an indicator of PMD effect on the signal (Kikuchi 2001). In this scheme, one uses a polarimeter to detect the DOP and then feedback to the polarization tracker with the internal feedback option. The user must first convert the DOP information into an analog signal proportional to the DOP, in a range between 0 and 5 V. The polarization tracker then adjusts the polarization of the light signal for the maximum DOP. This action is effectively to align the slow component of the signal with the fast axis of the PMD generating element inside the PMD compensator and maximize the overlap between two polarization components.

Scheme 2: Using the receiver to obtain the feedback signal (BER feedback): The PMD caused signal distortion will cause BER increase. Therefore, one can use BER as the feedback signal for PMD compensation. To get fast feedback, the BER should be obtained before the signal goes into the forward error correction (FEC) circuit or other error correction chips. For a BER larger than 10^{-4}, the BER detection time is less than 1 μs for a data stream with a bit rate of 10 Gb/s, which is sufficiently fast for the feedback to the polarization tracker. The user simply needs to convert the BER information into an analog voltage inversely proportional to BER (the bigger the BER, the smaller the voltage), in a range of 0–5 V. The polarization tracker will then adjust the polarization to maximize the feedback voltage, hence minimizing the BER, to effect PMD compensation.

7.1.6 Polarization Demultiplexing with a Polarization Tracker

Polarization-division multiplexing (PDMux) and coherent detection have emerged as the key technology enablers for 40-Gbps, 100-Gbps, and 400-Gbps networks in recent years because they can significantly increase the spectral efficiency of each channel and allow each channel to transmit high bit rate with a relatively small optical bandwidth (Zhou and Xie 2016). Consequently 40-Gbps, 100-Gbps, or even 160-Gbps channels can be transmitted with existing 10-Gbps wavelength-division multiplexing (WDM) infrastructure of 50-GHz channel spacing. PDMux doubles the spectral efficiency by combining two polarization channels of same bit rate and same wavelength, as discussed in Section 3.5. On the other hand, coherent detection allows multiple levels of phase and amplitude modulations of lower rates to be multiplexed on the same wavelength channel. For example, PDMux, together with quadrature-phase-shifted-keying (QPSK) modulation, enables the reduction of the data bandwidth by four times compared with that of a direct-modulated link (Laperle et al. 2007). At such a small bandwidth, the impairments due to PMD and CD are greatly reduced.

Polarization multiplexing requires extensive polarization management, especially at the receiving end to separate or demultiplex two orthogonal polarization channels, which can be accomplished either in the electronic domain with coherent receiving schemes or in the optical domain using a polarization tracker. We will focus on the later in this section.

In a typical non-coherent PDMux system shown in Figure 7.7a, dynamic polarization control is generally required at the receiver side to effectively demultiplex the two orthogonal polarization states. Different optical domain approaches have been demonstrated for such purpose, all of them requiring a polarization tracker, however with a major difference on how to extract the feedback signal, as will be discussed next. Polarization demultiplexing can also be done in electronic domain using coherent detection schemes; however, it requires extensive development of ASIC level digital signal processing electronics with extremely high up-front cost. In addition, some extremely fast SOP variation events, such as those induced by lightning strikes discussed in Chapter 3, may still not be contained with the electronic domain polarization demultiplexing, causing service interruptions.

The major requirements for the optical dynamic polarization control include high-speed response and reset-free operation (Hidayat et al. 2008; Koch et al. 2011). Due to the polarization perturbations along the transmission links, the typical response time for polarization controllers should be less than tens of microseconds, and the reset-free operation should be completed within milliseconds once there are dramatic polarization jumps. However, for systems that may suffer from the extremely fast SOP variations induced by the lightning strike, the speed requirement on the polarization controllers has to be increased by more than 10 times in order to overcome the abrupt SOP variations in the system. As will be discussed next, the polarization tracker implemented with high-speed polarization controllers can be readily used to accomplish polarization demultiplexing with various feedback schemes.

Scheme 1. Using a pilot tone as the feedback: In this scheme, a low frequency pilot tone around 100 kHz is injected in one of the polarization channels at the transmitter side to distinguish the two channels, as shown in Figure 7.7b. At the receiving side before the ADC, a pilot tone extraction circuit is included in one of the receivers, with which the low frequency pilot tone is separated from the high speed data and its strength is measured. The circuit then converts the pilot tone strength into an analog voltage between 0 and 5 V as the external feedback signal to the polarization tracker. The tracker then automatically controls the polarization before the PBS by maximizing the pilot tone voltage, and hence effectively separates the two polarization channels. Alternately, the feedback signal can be obtained with a separate low speed photodetector, which receives a small portion of optical signal tapped off from one of the two output of the PBS. This way, the receiver does not have to be modified. Note that scheme 1 requires hardware modification at least on one of the transmitters.

Scheme 2. Using signal BER as the feedback: The cross-talk between the two polarization channels will cause BER increase. Therefore, one can use BER as the feedback signal for polarization demultiplexing, as shown in Figure 7.7c. To get fast feedback, the BER should be obtained before the signal goes into the FEC or other error correction chips. For example, for a BER larger than 10^{-4}, the BER detection time is less than 1 μs for a data stream with a bit rate of 10 Gb/s, which is sufficiently fast for the feedback to the polarization tracker. The user simply needs to convert the BER information into an analog voltage inversely proportional to BER (the bigger the BER, the smaller the voltage), in a range of 0–5 V. The polarization tracker will then adjust the polarization to maximize the feedback voltage, hence minimizing the BER, to effectively separate two polarization channels. Note that scheme 2 also requires hardware modification on one of the receivers.

Scheme 3. Using low frequency noise as the feedback: The low frequency correlation noise between two polarization channels is an indication of the channel crosstalk. The noise will be minimized when the two polarization channels are properly separated. Therefore, the crosstalk induced noise can be used as the feedback for polarization demultiplexing, as shown in Figure 7.7d. In one of receiving arms after the PBS, a small portion of signal may be tapped off and detected by a low frequency photodetector, followed by a bandpass filter at certain center frequency with a properly chosen bandwidth. A simple circuit can be used to convert the noise strength into an analog voltage signal as the external feedback for the polarization tracker. This feedback voltage should be in the range of 0–5 V and be inversely proportional to the noise strength. The polarization tracker then automatically separates the two polarization channels by maximizing

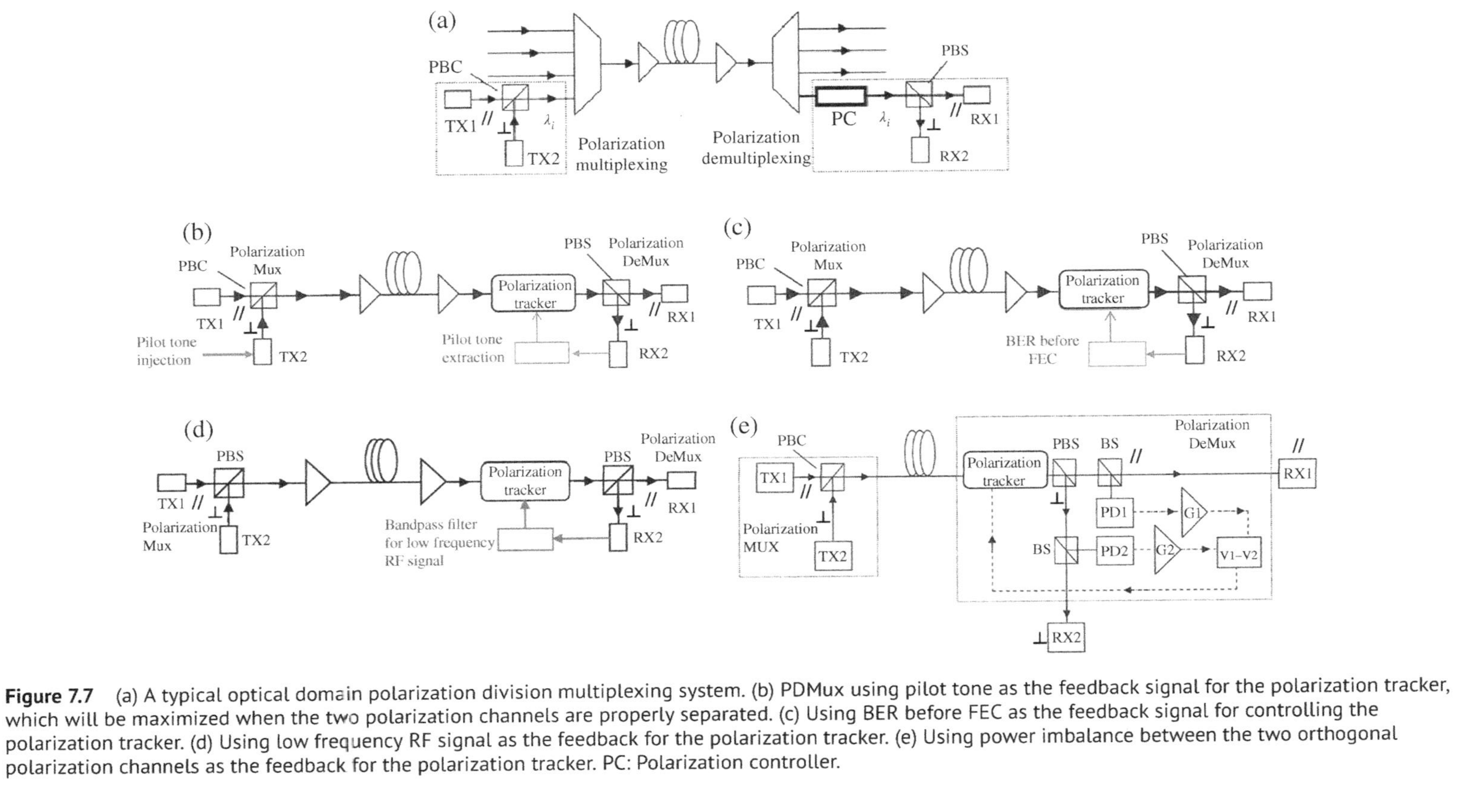

Figure 7.7 (a) A typical optical domain polarization division multiplexing system. (b) PDMux using pilot tone as the feedback signal for the polarization tracker, which will be maximized when the two polarization channels are properly separated. (c) Using BER before FEC as the feedback signal for controlling the polarization tracker. (d) Using low frequency RF signal as the feedback for the polarization tracker. (e) Using power imbalance between the two orthogonal polarization channels as the feedback for the polarization tracker. PC: Polarization controller.

the feedback signal. Alternatively, the low frequency noise signal can be derived from the high speed photodetector inside the receiver without the need of a separate low frequency detector. Note that with a separate low speed detector for the feedback signal detection, this scheme does not require hardware modification on the receivers nor the transmitters.

Scheme 4. Using power imbalance between polarization channels as the feedback: As will be discussed in the following text (Yao et al. 2007), if the optical powers of the two polarization channels at the transmitter side are different by more than 0.5 dB, the two polarization channels can be effectively separated at the receiving side by detecting the power difference and using it as the feedback signal. As shown in Figure 7.7e, after the optical signal is separated by a PBS into two outputs, a small portion of the signal in each output is tapped off by a coupler or beam splitter (PS) and is converted into a voltage signal. The voltage difference between the two tapped signals is used as the feedback to the polarization tracker, which automatically separates the two polarization channels by maximizing the voltage difference. As a requirement, the voltage difference should be in a range between 0 and 5 V. Note that this scheme does not require hardware modification of either the transmitter or the receiver. This scheme was successfully demonstrated using a fiber squeezer based polarization tracker in a dense wavelength division multiplexing or multiplexer (DWDM) system containing 14 WDM channels of 40-Gb/s per channel over 62-km corning LEAF® submarine (LS) fiber transmission with an aggregated data transmission rate of 1.12-Tb/s (14 × 2 × 40-Gb/s).

Now let's describe Figure 7.7e in detail. Optical data streams with orthogonal SOPs (TX1 and TX2) are generated from the same or different light sources at the same wavelength and then multiplexed through a polarization beam combiner (PBC). During signal transmission over a fiber link, the two linear SOPs evolve into elliptical SOPs but still maintain their relative orthogonality, assuming the fiber link has no PDL and PMD. A polarization tracker followed by a PBS is used to demultiplex the two data streams of orthogonal SOPs. Choosing the transmission axes of the PBS as reference coordinates x and y for convenience, the corresponding Muller matrix of the PBS can be expressed as

$$\mathbf{M}_x = \frac{1}{2}\begin{pmatrix} 1 & 1 & 0 & 0 \\ 1 & 1 & 0 & 0 \\ 0 & 0 & 0 & 0 \\ 0 & 0 & 0 & 0 \end{pmatrix} \tag{7.1a}$$

$$\mathbf{M}_y = \frac{1}{2}\begin{pmatrix} 1 & -1 & 0 & 0 \\ -1 & 1 & 0 & 0 \\ 0 & 0 & 0 & 0 \\ 0 & 0 & 0 & 0 \end{pmatrix} \tag{7.1b}$$

Optical data stream i with an arbitrary polarization state S_i can be expressed in Stokes space as

$$\mathbf{S}_i = \begin{pmatrix} S_{i0} \\ S_{i1} \\ S_{i2} \\ S_{i3} \end{pmatrix} = \begin{pmatrix} P_i \\ P_i cos2\varepsilon_i cos2\theta_i \\ P_i cos2\varepsilon_i sin2\theta_i \\ P_i sin2\varepsilon_i \end{pmatrix} \tag{7.2}$$

where P_i is the optical power, ε_i is the ellipticity angle, and θ_i is the orientation angle of the ith optical data stream. After passing through the PBS, the Stokes vectors along the x and y directions are $\mathbf{M}_x\mathbf{S}_i$ and $\mathbf{M}_y\mathbf{S}_i$, respectively, where the first row of each Stokes vector represents the optical power along the corresponding direction. In particular,

$$P_{ix} = \frac{1}{2}P_i(1 + cos2\varepsilon_i cos2\theta_i) \tag{7.3}$$

$$P_{iy} = \frac{1}{2}P_i(1 - cos2\varepsilon_i cos2\theta_i) \tag{7.4}$$

Since the two data streams are incoherent from each other (either from two independent laser sources or made to be incoherent by other means), the emerging optical powers along the x and y axes of the PBS are

$$P_x = P_{1x} + P_{2x} = \frac{1}{2}P_1(1 + cos2\varepsilon_1 cos2\theta_1) + \frac{1}{2}P_2(1 + cos2\varepsilon_2 cos2\theta_2) \tag{7.5a}$$

$$P_y = P_{1y} + P_{2y} = \frac{1}{2}P_1(1 - cos2\varepsilon_1 cos2\theta_1) + \frac{1}{2}P_2(1 - cos2\varepsilon_2 cos2\theta_2) \tag{7.5b}$$

For the two optical data streams with orthogonal SOPs $\mathbf{S}_1$ and $\mathbf{S}_2$, the following relationships hold

$$\varepsilon_2 = -\varepsilon_1 \tag{7.6a}$$

$$\theta_2 = \begin{cases} \theta_1 + \frac{1}{2}\pi, & 0 \le \theta_1 \le \frac{1}{2}\pi \\ \theta_1 - \frac{1}{2}\pi, & \frac{1}{2}\pi < \theta_1 \le \pi \end{cases} \tag{7.6b}$$

Substitution of Eqs. (7.6a) and (7.6b) into Eqs. (7.5a) and (7.5b) yields

$$P_x = \frac{1}{2}(P_1 + P_2) + \frac{1}{2}(P_1 - P_2)cos2\varepsilon_1 cos2\theta_1 \tag{7.7a}$$

$$P_y = \frac{1}{2}(P_1 + P_2) - \frac{1}{2}(P_1 - P_2)cos2\varepsilon_1 cos2\theta_1 \tag{7.7b}$$

To automatically and effectively separate the two orthogonal data channels, we monitor the relative optical power levels of the two channels using two low-speed photodetectors (PD1 and PD2) by coupling a small amount of the signal power. Through low-noise electronic circuits (G_1 and G_2), the power difference between the two polarization states ($P_x - P_y$) is translated into the voltage difference ($V_1 - V_2$), expressed as

$$V_1 - V_2 = \alpha_1 P_x - \alpha_2 P_y = \frac{1}{2}(\alpha_1 - \alpha_1)(P_1 + P_2) + \frac{1}{2}(\alpha_1 + \alpha_1)(P_1 - P_2)cos2\varepsilon_1 cos2\theta_1 \tag{7.8a}$$

where α_1 and α_2 are the response coefficient of the photodetectors and their corresponding amplification circuits, respectively. Through electronic gain balancing or software calibration, they can be adjusted to be equal ($\alpha_1 = \alpha_2 = \alpha$), and therefore the voltage difference of Eq. (7.8a) becomes

$$\Delta V = V_1 - V_2 = \alpha(P_1 - P_2)cos2\varepsilon_1 cos2\theta_1 \tag{7.8b}$$

As can be seen from Eq. (7.8b), as long as there is a power difference between the two channels ($P_1 - P_2 \neq 0$), the voltage difference between the two power monitors depends on the orientation angle θ and ellipticity angle ε_1, which can be changed by the dynamic polarization controller. Therefore, by maximizing the calibrated voltage difference ΔV, we can effectively minimize the crosstalk and readily separate the two orthogonal channels by forcing either a positive maximum $[(\theta = 0°, \varepsilon = 0°)$ or $(\theta = \pm 90°, \varepsilon = \pm 90°)]$ or a negative maximum $[(\theta = \pm 90°, \varepsilon = 0°)$ or $(\theta = 0°, \varepsilon = \pm 90°)]$. The positive maximum corresponds to a Stokes vector of (1, 0, 0), while the negative maximum corresponds to a Stokes vector of (−1, 0, 0). Specifically, when using the polarization tracker for the dynamic polarization control, one may detect and convert the ΔV in a range of 0–5 V and use it as the external feedback signal for the polarization tracker to enable the automatic demultiplexing of the two orthogonal polarization channels.

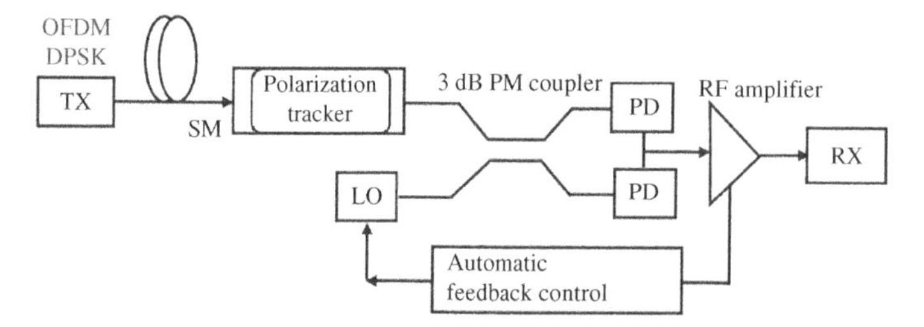

Figure 7.8 A polarization tracker with internal feedback can be used to stabilize the SOP of the signal after it propagates through the transmission fiber. The SOP at the output of the polarization tracker is linear and aligned with the slow axis of the PM fiber pigtail and beat with the local oscillator laser. Coherence detection also needs to actively control and maintain the relative state of polarization between the signal and the local oscillator laser for the maximum signal-to-noise ratio. The "Automatic Feedback Control" in the figure is for controlling the frequency drift and noise of the LO laser.

7.1.7 Polarization Tracking for Coherent Detection

As shown in Figure 7.8, in a coherent detection system, after propagating through a long distance, the signal will be made to interfere with a local oscillator laser for obtaining both the phase and amplitude information with significantly increased sensitivity. In order for the beat signal to have the maximum strength, the SOP of the signal must be the same as that of the local oscillator laser. Unfortunately, the SOP of the signal changes rapidly in the fiber, resulting in fluctuating beat signals. Therefore, the SOP of the signal must be stabilized before it beats with the local oscillator. In this application, a polarization tracker with internal feedback converts the fluctuating SOP into a fixed linear SOP aligned with the slow axis of the output polarization maintaining (PM) fiber (Wree et al. 2007). The signal output from the polarization tracker then combines with the local oscillator (LO) laser via a 3-dB PM coupler to obtain a stable beat signal. The relative frequency between the signal and the LO can be locked by a feedback loop shown in Figure 7.8.

In summary, polarization tracker is very useful in various fiber optical systems to combat polarization related impairments. In particular, the external feedback option makes the polarization tracker especially flexible in fitting into different application schemes of user's choice. With well-designed hardware, fine tuned reset-free control algorithm, and the flexibility of external feedback, this general purpose polarization tracker makes it easier to track and control the SOP in almost all fiber optic systems.

7.2 Polarization Scrambling and Emulation

Polarization stabilization and tracking described earlier is one way to combat the polarization impairments in optical fiber systems discussed in Section 3.4, which convert a random varying SOP into a fixed SOP. Alternatively, one may use an opposite approach, namely, the polarization scrambling, to mitigate some polarization induced impairments, such as PDL in measurement systems, polarization dependent gain (PDG) in optical amplifiers, polarization dependent response (PDR) in receivers, and polarization dependent sensitivity (PDS) in optical fiber sensor systems.

Perhaps a more important application for polarization scrambling and emulation is to test an optical fiber link's tolerance to polarization related impairments. As the appetite for bandwidth continues to sky rocket, PDMux and coherent detection (Derr 1992; Ip et al. 2008) are being used for increasing the transmission speed to reach 100 Gb/s and beyond. In principle, the combination of PDMux and coherent detection scheme is capable of mitigating polarization related issues, i.e. time-varying PMD, PDL, PDG, and SOP in the digital domain (Taylor 2004). Therefore, performance tests of a system against these polarization-related impairments are extremely important for assuring the healthy operation of such high speed fiber optic communication systems.

One of the tests is for obtaining the failure rate of a system due to the polarization variation in the system, requiring a polarization emulator to randomly generate all possible SOPs while mimicking the rate distribution of the SOP variations, i.e. the Rayleigh distribution (Peterson et al. 2004), in a real system described in Figure 3.1.

Early adjustable-rate polarization scramblers are generally made with different types of polarization controllers and are programmed so that SOP traces uniformly covering the whole Poincaré sphere (Lize et al. 2007). Some polarization scramblers are made with an SOP changing rate following Rayleigh Distribution (Krummrich et al. 2005), mainly for emulating SOP variations in a real fiber optic transmission system for statistical system testing (Leo et al. 2003; Boroditsky et al. 2005). Other scramblers are made to change SOP as fast and randomly as possible for mitigating polarization related transmission impairments, with unspecified scrambling rate distributions (Yan et al. 2005).

Another important test is to see the ability of the fiber link for handling fast SOP variations requiring a polarization scrambler capable of generating time-varying random SOPs uniformly covering the whole Poincaré sphere at an adjustable rate, with the highest rate larger than the fastest SOP variations occurring in real fiber communication systems, for example the SOP variations induced by lightning strikes discussed in Section 3.1. For the deterministic test of the polarization response of coherent receivers, a uniform rate polarization scrambling is desired, because non-uniform rate polarization scrambling introduces large test uncertainty and less test repeatability. Because the orthogonal SOP channels are used in the PDMux systems, the ideal polarization scrambler/emulator should work equally well for all polarizations.

7.2.1 Polarization Scrambling Basics

For unpolarized light, the SOP fluctuates randomly around the direction of light beam propagation. Therefore, on average, no direction is favored. The rate of the fluctuation is so fast that an "observer" or a detector cannot tell the SOP at any instant in time. For example, natural light (sunlight, firelight, etc.) is unpolarized. In any other case, the light beam can be considered to consist of partially polarized or fully polarized light. DOP is used to describe how much of the total light intensity is polarized, as defined by Eq. (2.18c). For totally polarized light, the *DOP* is 1. For completely unpolarized light, the *DOP* is 0. Most high performance lasers used in long-haul communication and sensor systems are polarized light sources. However, after the optical signal is transmitted through the optical system, it is somewhat depolarized by the PMD and amplified spontaneous emission (ASE) noises in optical amplifiers.

The natural unpolarized light can be considered as coming from a light source, such as the sun, consisting of many tiny light emitters emitting light with randomly different SOPs, resulting in a *DOP* of 0. As discussed in Chapter 6, unpolarized light can also be obtained by depolarization, which occurs when two orthogonally polarized light beams with equally powers are combined incoherently. The scrambled light is somewhat different.

Light is called "scrambled" if the SOP of totally polarized light is made to vary randomly at a relatively low rate compared with the oscillation frequency of light, as shown in Figure 7.9. At any instant of time, the SOP is well defined, and *DOP* is close to 1. However, on a time average, *DOP* is close to 0. Therefore, the DOP of scrambled light depends on the averaging time or the detection bandwidth of the observer. The faster the scrambling, the less the averaging time is required.

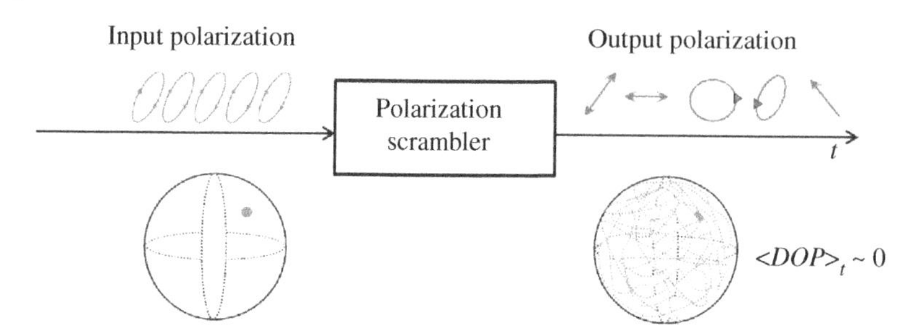

Figure 7.9 Illustration of polarization scrambling. A stable input SOP is converted to a fast time varying SOP uniformly covering the whole Poincaré sphere. $\langle\ \rangle_t$ denotes for time averaging.

7.2.2 Polarization Scrambling Simulation

A polarization scrambler actively changes the SOP randomly and any of the polarization controllers described in Chapter 6 can be used as a polarization scrambler to generate scrambled light; however, one must make sure the generated SOP should uniformly cover the Poincaré sphere with equally probability, as shown in Figure 7.9. Therefore, how to program a polarization controller for generating the randomly distributed SOP is important.

The polarization controllers using multiple variable phase retarders, particularly fiber squeezers described in Figure 6.15, have shown to enable effective polarization scrambling commercially. In order to quickly verify the feasibility of new polarization scrambling ideas and learn how different parameters affect scrambling, it is necessary to simulate how SOP varies when a specific scrambling scheme is applied to the phase retarders. The simulation program can be used to calculate SOP variations caused by N retarders of different orientations when they are driven by N electrical signals of different waveform, frequencies, and amplitude, and display SOP traces on Poincaré sphere. In addition, the program can calculate the SOP variation rate and plot the rate distribution histogram. Moreover, the time or point averaged DOP can also be obtained with the program to reflect SOP coverage uniformity of a particular scrambling scheme (Yan et al. 2005).

In the simulation program, the ith phase retarder is represented by a Mueller matrix from Chapter 4 (also see Collett 2003):

$$\mathbf{M}_i = \begin{pmatrix} 1 & 0 & 0 & 0 \\ 0 & cos4\theta_i sin^2(\phi_i/2) + cos^2(\phi_i/2) & sin4\theta_i sin^2(\phi_i/2) & 0 \\ 0 & sin4\theta_i sin^2(\phi_i/2) & -cos4\theta_i sin^2(\phi_i/2) + cos^2(\phi_i/2) & -cos2\theta_i sin\,\phi_i \\ 0 & 0 & cos2\theta_i sin\,\phi_i & cos\,\phi_i \end{pmatrix} \tag{7.9}$$

where θ_i is the orientation angle and ϕ_i is the retardation of the ith retarder. As described in Chapter 4, the Stokes vector $\mathbf{S}_i$ of the output SOP from the ith retarder is obtained by multiplying its Mueller matrix $\mathbf{M}_i$ with the previous Stokes vector $\mathbf{S}_{i-1}$:

$$\mathbf{S}_i = \mathbf{M}_i \mathbf{S}_{i-1} \tag{7.10}$$

The SOP variation rate is obtained by first calculating the angle between two consecutive points $\mathbf{S}_m$ and $\mathbf{S}_n$ on the SOP trace using:

$$cos\,\theta_{mn} = \frac{\mathbf{S}_m \cdot \mathbf{S}_n}{|\mathbf{S}_m||\mathbf{S}_n|} \tag{7.11}$$

where $\mathbf{S}_m \cdot \mathbf{S}_n$ is the dot product of the two SOP vectors; and dividing the angle by the time interval between the two points. The rate distribution can be obtained by calculating all the rates between two adjacent SOP points and displaying them on a histogram.

Finally, the averaged DOP of the scrambled polarization can be calculated using

$$\langle DOP\rangle = \frac{\sqrt{\langle S_1\rangle^2 + \langle S_2\rangle^2 + \langle S_3\rangle^2}}{S_0} \tag{7.12}$$

where $\langle\ \rangle$ denotes for either time or point average. With this simulation program, one may easily analyze scrambling schemes with different combinations of driving frequencies and determine which scheme has the required SOP coverage uniformity and scrambling rate distribution.

7.2.3 Variable Rate Polarization Scrambling and Emulation

As will be discussed next, there are three types of polarization scrambling methods, classified according to the algorithms for programming polarization controllers: the discrete random polarization scrambling, the continuous random polarization controlling, and the tornado-pattern polarization scrambling.

Discrete random polarization scrambling can be implemented by using random number generators with a computer or microcontroller to generate a set of control signals to control a polarization controller. For example, one may use the polarization controller with three or four variable retarders of fixed orientation, such as the fiber squeezer polarization controller with three or four fiber squeezers, as described in Figure 6.15. Three voltages (V_1, V_2, V_3) are used to drive the three individual squeezers in the controller, each having a half-wave voltage of $V_{\pi i}$ with $i = 1, 2, 3$. By simultaneously driving all fiber squeezers, each with a random voltage between 0 and $2V_{\pi i}$, random and uniform SOPs can be generated, as shown in Figure 7.10a. In practice, four or more variable retarders can be used for more SOP distribution uniformity.

The discrete polarization scrambling causes the SOP changes abruptly, which do not happen often naturally in an optical fiber system. To emulate SOP variations in optical fibers systems, continuous random polarization scrambling is required, which can be implemented by driving three variable retarders, such as fiber squeezers, with three saw waves or triangle waves of different frequencies, each having an amplitude of $2V_{\pi i}$. The ratio of any two frequencies is an irrational number with an infinite decimal, such as π, e, and $\sqrt{3}$ such that the SOP traces on the Poincaré sphere never repeat. For example, the frequencies of the three driving signals can be chosen as $(f_0, \sqrt{2}f_0, \sqrt{5}f_0)$ to avoid repeated SOP traces and uniform distributed SOP on the Poincaré sphere, as shown in Figure 7.10b. The associated distribution of SOP variation rate is shown in Figure 7.10c, which is defined as the 3D angle between two adjacent points on the Poincaré sphere divided by the time interval between the two points.

It is also possible to use sinusoidal wave functions applied to each fiber squeezer to enable polarization scrambling. Using the sinusoidal wave has the advantage to operate the fiber squeezers at the resonate frequencies of the piezo-electric transducers (PZTs) acting on the fiber squeezers for low driving voltage and low power consumption. For example, the half-wave voltage V_π for a PZT driven fiber squeezer is around 30 V with a capacitance of 0.18 μF, and therefore it is difficult to modulate it at frequencies above 20 kHz because of the power and voltage constraints of the electrical circuit and the response bandwidth of the PZT. Fortunately, there exist several resonant frequencies around 30, 60, 100, and 130 kHz of the PZT driven fiber squeezers where the half-wave voltages are reduced to a few volts.

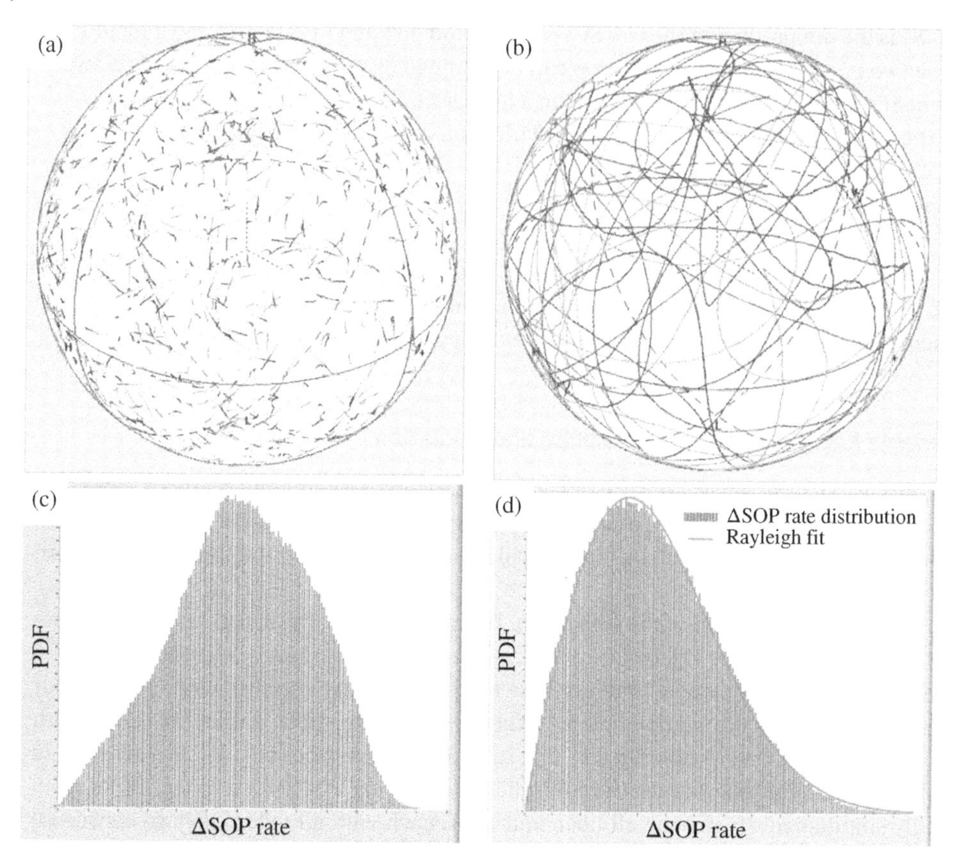

Figure 7.10 Typical Poincaré sphere SOP traces of discrete scrambling (a) and continuous scrambling with triangle waves (b). Typical SOP variation rate distribution of an SOP scrambler with triangle wave scrambling (c) and Rayleigh SOP variation rate distribution implemented with four triangle waves of random frequencies (d). The darker and lighter lines in (a) and (b) represent the traces being in the back side and front side of the Poincaré sphere, respectively.

As discussed in Section 3.1, in real optical fiber communication systems, the SOP variation rates follow a Rayleigh distribution. Therefore, to obtain the failure rate of an optical fiber link due to the SOP variations, an SOP emulator is required, which not only enables the SOP variation traces uniformly cover the whole Poincaré sphere, but also can assure the SOP variation rates follow a Rayleigh distribution. One way to achieve such a Rayleigh distribution, the four variable retarders in Figure 6.15a or c are driven by four saw waves each with a voltage amplitude of $2V_{\pi i}$. In addition, the frequency of the ith saw wave is $f_i(t_j) = RN_i(t_j) \cdot if_0$, where $i = 1, 2, 3, 4, j = 1, 2 \ldots N, f_0$ is a nominal frequency, and $RN_i(t_j)$ is a random number generated by the ith random number generator associated with the ith variable retarder at time t_j. To make sure the SOP trace moves smoothly on the Poincaré sphere, the random numbers from each random number generator can be resorted so that the difference between the adjacent numbers is sufficiently small. Figure 7.10d shows the measured SOP variation rate distribution of a polarization scrambler/emulator made with four fiber squeezers of Figure 6.15c. Clearly, it closely follows a Rayleigh distribution.

7.2.4 Quasi-uniform Rate Polarization Scrambling

The non-uniform rate polarization scrambling schemes described earlier introduce large test uncertainty and less test repeatability for determining the SOP variation tolerance of coherent receivers because of the large spread of the SOP variation rates shown in Figure 7.10. Therefore an alternative scheme is required to overcome this issue, with which not only the generated SOP traces are able to uniformly cover the whole Poincaré sphere, but also the rates of the SOP variation also have a narrow spread (Yao et al. 2012). The tornado scrambling scheme to be described next is such a scheme. In contrast to the random polarization scrambling patterns described earlier, it generates rapid evolving circular SOP traces moving up and down along an axis on the Poincaré sphere, as will be shown shortly. Although the traces do not appear random, they can uniformly cover the entire Poincaré sphere with an averaging DOP approaching zero and in the meantime have a relatively uniform SOP variation rate.

7.2.4.1 Device Construction

The polarization controllers made with multiple variable retarders described in Figure 6.15 can be used to realize such tornado scrambling scheme. In practice, fiber squeezers are used in a commercial polarization scrambler due to its low cost, low insertion loss, and low PDL. In order to increase the scrambling speed, more fiber squeezers are cascaded in series, as shown in Figure 7.11. Each squeezer is driven by an amplified electrical control signal. The first three fiber squeezers are oriented 45° from one another, and the last three fiber squeezers have the same orientation as the third squeezer for overcoming the speed limitation of the fiber squeezers, as will be explained in the following text.

7.2.4.2 SOP Variation Rates Induced by Fiber Squeezers

As described in Chapter 6, the retardation of each fiber squeezer is varied linearly by applying a driving voltage to the piezo-electric actuator on the fiber squeezer. As the retardation is varied from 0 to 2π, the output SOP from the fiber squeezer traces out a complete circle on the Poincaré sphere. The radius of the circle depends on the angle between the input SOP with respect to birefringence

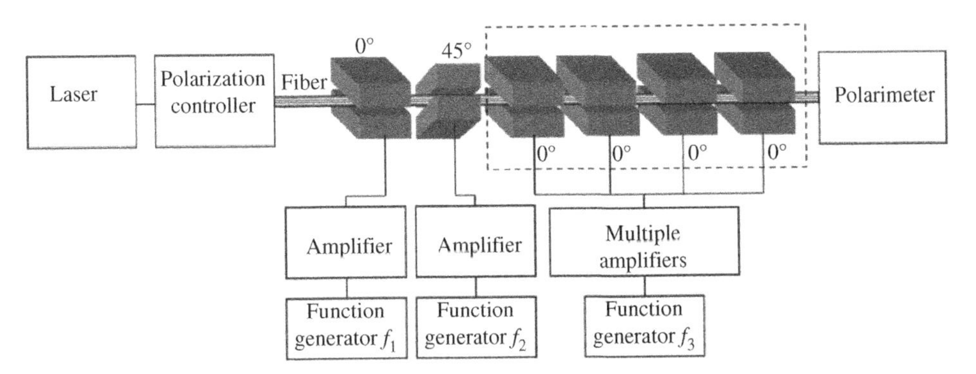

Figure 7.11 Fiber squeezer arrangement and experiment setup for demonstrating the tornado scrambling (Yao et al. 2012). The first three fiber squeezers are oriented 45° from one another and the last three are oriented the same as the third squeezer. The polarization controller in front of the fiber squeezers is for adjusting the input SOP into the fiber squeezers, and the polarimeter is for observing SOP changes on a Poincaré sphere.

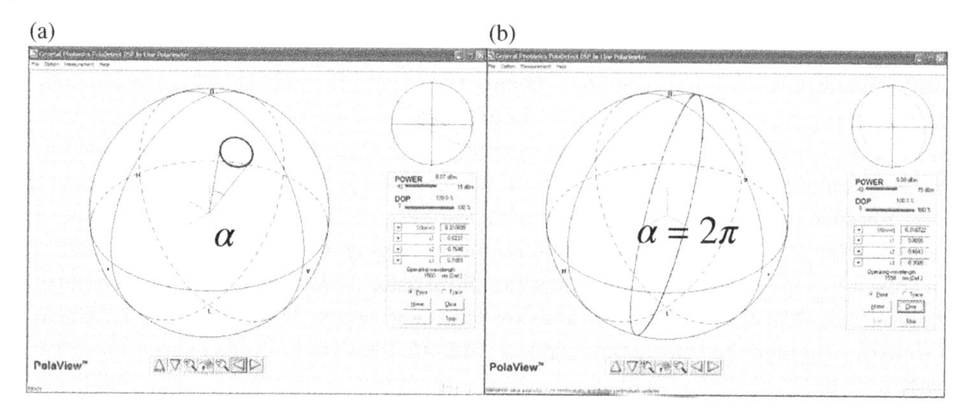

Figure 7.12 (a) When SOP traces out a small circle on Poincaré sphere, the SOP variation angle α is small (a small fraction of 2π) and so is the SOP changing rate; (b) when SOP traces out a great circle on Poincaré sphere, corresponding to the case that input SOP to the wave plate is 45° from its birefringence axis, the angular change per circle reaches its largest value of 2π. The corresponding SOP changing rate is also maximized. In general, for a given retardation variation, the smaller the radius of curvature of the SOP trace, the smaller the rate of SOP variation.

axis of the variable phase plate, i.e. 45° angle results in the largest circle on the sphere while a small or large angle produces small SOP changes and hence small radius, as shown in Figure 7.12. In addition, for a given retardation variation of the fiber squeezer, the SOP variation rate is proportional to the radius of curvature of the SOP trace. The largest radius of curvature produces the largest rate of SOP variation for a given retardation variation.

Two circular traces generated by two adjacent fiber squeezers with a relative orientation of 45° are orthogonal from each other. It can be shown that at least three fiber squeezers are required to generate SOPs to cover the whole Poincaré sphere from any input SOP (Noe et al. 1988; Aarts and Khoe 1989). To program the fiber squeezers for effective polarization scrambling, four parameters on the driving signals can be selected: waveform, frequency, amplitude, and phase. The most important considerations of a polarization scrambling scheme include (i) SOP coverage uniformity, (ii) maximum scrambling rate, and (iii) scrambling rate distribution.

As mentioned previously, SOP coverage uniformity describes how uniform SOPs are distributed on Poincaré sphere after a certain time, and is generally characterized by observing SOP distribution on Poincaré sphere and average DOP over time or over SOP points (Krummrich and Kotten 2004). The point-averaged DOP is a good indicator for uniform SOP coverage, i.e. the smaller the averaged DOP, the better the SOP uniformity is. In particular, if the scrambled SOP can reach the entire sphere with equal probability, the averaged DOP will be zero.

7.2.4.3 Random Polarization Scrambling

Random polarization scrambling makes the SOP trace move randomly on Poincaré sphere with time and covers the whole sphere uniformly, as shown Figure 7.10b and again here in Figure 7.13a for convenience. However, because of the randomly varying trace, the corresponding radius of curvature of the SOP trace also changes randomly, resulting in a wide spread of SOP variation rates, as shown in Figure 7.10c,d, and again in Figure 7.13b. As mentioned previously, such a wide rate distribution is not desirable for testing the polarization response of a system due to the lack of repeatability. As described in Section 7.2.3, at least three variable retarders or fiber squeezers are required.

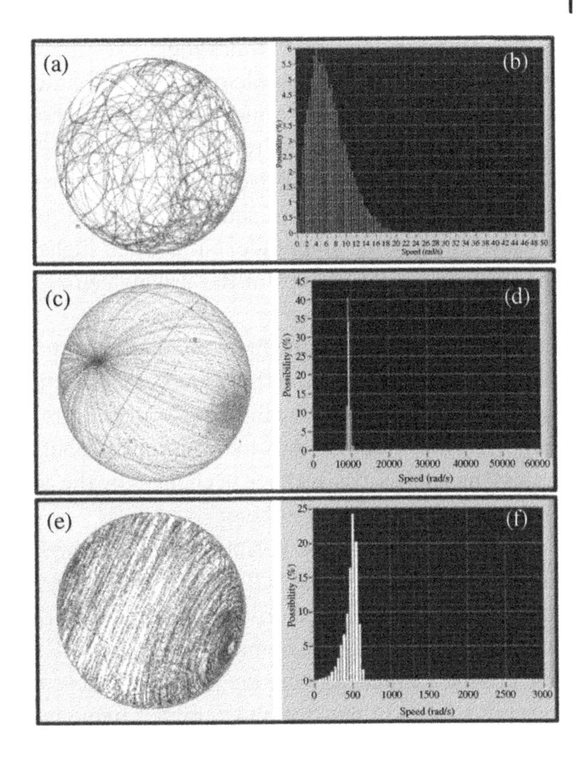

Figure 7.13 (a) Random SOP variation of a Rayleigh scrambling scheme; (b) The corresponding SOP variation rate histogram following Rayleigh distribution; (c) Measured SOP trace of the uniform rate scrambling approach, where the trace evolves like a circle spinning around a diametric axis; (d) Polarization scrambling rate histogram showing a single scrambling rate; (e) Measured SOP trace of the quasi-uniform rate scrambling approach. SOP rotates around Poincaré sphere at a high speed to form a circle, which moves back and forth along the rotation axis of the circles to cover the whole sphere. (f) Polarization scrambling rate histogram showing a quasi-uniform scrambling rate.

7.2.4.4 Uniform Rate Scrambling

One simple method to achieve uniform rate polarization scrambling is only using the first two fiber squeezers in Figure 7.11. The input SOP has to be aligned to 45° from the first fiber squeezer. Both the first and the second squeezers are modulated by a triangle wave with an amplitude of $2V_\pi$ (2π phase retardation). However, the frequency of the driving signal on the first squeezer (f_1) is much higher than that on the second (f_2). In this way, the first squeezer causes the SOP go around the Poincaré sphere in a great circle at a high speed, while the second squeezer rotates the great circle in one of its diametric axis, as shown in Figure 7.13c, to cover the whole sphere completely and uniformly. In the experiment for obtaining Figure 7.13c, the two driving frequencies are $f_1 = 180$ Hz and $f_2 = 1.1459$ Hz. The scrambling rate using this approach is uniform, as shown in Figure 7.13d, because SOP always moves around the great circles of equal radius. The small rate spread is cause by limited sampling point in the polarimeter. The measured 10 000-point averaged DOP is about 6.8%, indicating a reasonable uniform SOP coverage over entire Poincaré sphere is achieved. The non-zero DOP is due to the high sensitivity to the input polarization as it is difficult to align the input SOP at perfect 45° from the first fiber squeezer. Even low DOP is expected to be achieved with a better alignment of the input SOP. The experimental results in Figure 7.13c,d also agree with the simulation results using Eqs. (7.9)–(7.12) under the same parameters.

Although the uniform rate polarization scrambling can be achieved, this method requires that the SOP of the input optical signal has to maintain 45° with the first fiber squeezer. However, the SOP in a real fiber system is not stationary and varies with time. Therefore, in order to keep the SOP input to the first fiber squeezer at 45°, a polarization stabilizer or tracker must be used. As discussed in Section 7.1, a polarization stabilizer generally uses a polarizer or PBS to derive a feedback signal for polarization stabilization and therefore only one polarization state is supported. This not only adds cost, but also causes problems with polarization multiplexed signals in most coherent systems.

7.2.4.5 Quasi-uniform Rate Scrambling

In order to overcome the problems described earlier, one may use the quasi-uniform rate scrambling scheme with a nearly uniform rate to be described in the following text. The scheme requires at least three variable phase retarders oriented 45° from one another, as shown in Figure 7.11. The first two squeezers are driven by two triangle waves of different frequencies but with the same amplitude of $2V_\pi$. The third squeezer, oriented 45° from the second one, is driven by a triangle wave of the same amplitude, but a much higher frequency than those of first two squeezers. The frequency relationships of the driving signals on the three squeezers are: $f_3 \gg f_2 \gg f_1$ (or $f_3 \gg f_1 \gg f_2$).

One may visualize that the first two squeezers cause SOP to vary along certain paths on Poincaré sphere in the absence of the third squeezer. With the third squeezer driving at much higher rate, each SOP point generated by the first two squeezers becomes the starting point for a complete SOP circle. All the circles have the same rotation axis, but different diameters. The circle moves back and forth along circle's axis as the SOP changes by the action of the first two squeezers, and eventually covers the whole Poincaré sphere.

Figure 7.13e shows the experimental result of SOP coverage using the quasi-uniform rate scrambling. The evolution of the SOP trace with time agrees with our reasoning and simulation using Eqs. (7.9)–(7.12). The three driving signals are all triangle waves with 60 V amplitude for inducing 2π phase retardation, with frequencies of $f_1 = 0.1$ Hz, $f_2 = 1.414$ Hz, and $f_3 = 34.6$ Hz, respectively. Low driving frequencies here are purposely chosen for the easy observation of SOP evolution on Poincaré sphere. In practice, much higher frequencies with similar proportionalities can be used for high scrambling rate. The measured DOP (10 000 point average) is only about 3.2%, indicating uniform SOP coverage over the entire sphere. Finally, as shown in Figure 7.13f, the scrambling rate concentrates heavily at the highest end of the rate distribution histogram with a narrow spread. It is important to note that the SOP coverage and scrambling rate distribution of this scheme are not sensitive to the input polarization variations, and therefore no stabilizer is required for implementing the scheme in real systems. Most importantly, the scheme works equally well for both single polarization and polarization-multiplexed signals. Although the scrambling rate in this scheme is not ideally uniform as the scheme in Figure 7.13d, it is sufficient for the high speed polarization-related testing of coherent systems with high repeatability.

Figure 7.14a shows a photo of commercial polarization control instrument with the Tornado polarization scrambling functions described in Figure 7.13e. The Poincaré sphere display of the

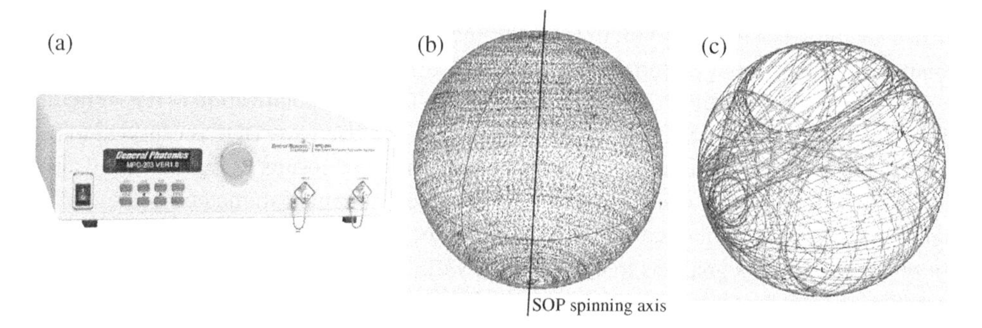

Figure 7.14 (a) A commercial polarization control instrument with Tornado SOP scrambling functions developed by the authors of this book (Source: Courtesy of General Photonics/Luna Innovations); (b) The SOP trace of the Tornado scrambling with fixed spinning axis generated by the commercial instrument; (c) The SOP trace of the Tornado scrambling with varying spinning axis.

SOP traces of this commercial unit is shown in Figure 7.14b in which the SOP spinning axis is fixed and is determined by the N parallel fiber squeezers in Figure 7.11. The spinning axis can be made to vary by including an additional fiber squeezer oriented 45° from the last N parallel squeezers with a retardation modulation frequency f_4 much slower than that of the N squeezers ($f_4 \ll f_3$), with typical SOP traces shown in Figure 7.14c.

7.2.4.6 Rate Multiplication Method for Overcoming Fiber Squeezer Speed Limitations

Fiber squeezers generally have a speed limit around 30π krad/s. Such a speed is not sufficient for testing the performance of coherent receivers. Therefore it is important to find a way to extend the speed limit of the fiber squeezers.

It can be shown mathematically that when N squeezers are placed in succession with the same relative angle, as shown in Figure 7.11 for the last four squeezers, the phase retardations add up. If all the fiber squeezers are driven by a same triangle signal with the same frequency, the phase variation rate will be increased by a factor of N. Consequently, if this composite fiber squeezer is used as the third squeezer in the quasi-uniform rate scrambling scheme in Figure 7.11, the total scrambling rate will also be increased by a factor of N. The scheme is scalable to higher scrambling rates by adding more fiber squeezers and will still work if the fiber squeezers used are replaced with different types of variable phase retarders.

7.2.5 Factors Degrading the Performance of the Polarization Scramblers

The imperfections described in Section 6.5.3, may compromise the performance of the polarization scrambler. For example, the PDL of the polarization controller causes some SOP to have higher transmission loss while the others have less, resulting in increased DOP, despite the SOP traces uniformly covering the whole Poincaré sphere. The minimum DOP achievable from a scrambler with a PDL can be expressed from the definition of DOP in Eq. (2.18c):

$$DOP_{min} = \frac{P_{max} - P_{min}}{P_{max} + P_{min}} = \frac{1 - P_{min}/P_{max}}{1 + P_{min}/P_{max}} = \frac{1 - 10^{-PDL/10}}{1 + 10^{-PDL/10}} \tag{7.13a}$$

where the definition $PDL = 10\,log\,(P_{max}/P_{min})$ is used. For a $PDL = 0.5$ dB, $DOP_{min} = 5.7\%$.

Activation loss (AL) of a polarization controller, which is associated with the loss induced by the driving signal for scrambling the SOP described in Section 6.5.3, may also degrade the performance of the scrambler in a similar way as what the PDL does. Specifically, the minimum DOP achievable using a polarization controller with an AL can be expressed as

$$DOP_{min} = \frac{1 - 10^{-AL/10}}{1 + 10^{-AL/10}} \tag{7.13b}$$

For an AL = 0.2 dB, $DOP_{min} = 2.3\%$. Therefore, for achieving a low DOP, polarization controllers with low PDL and AL should be used. Although $LiNbO_3$ based polarization controllers are of higher speed compared with the fiber squeezer polarization controllers for higher scrambling rates, their high PDL and AL may prevent them for some particular applications, such as for high precision measurements of optical components' insertion loss (IL) and PDL.

7.2.6 Polarization Scrambler Applications

Polarization scramblers have numerous applications in optical communication networks, fiber sensor systems, and test and measurement systems. An important application is for the tolerance tests of polarization related impairments in coherent detection systems of WDM networks, which will be

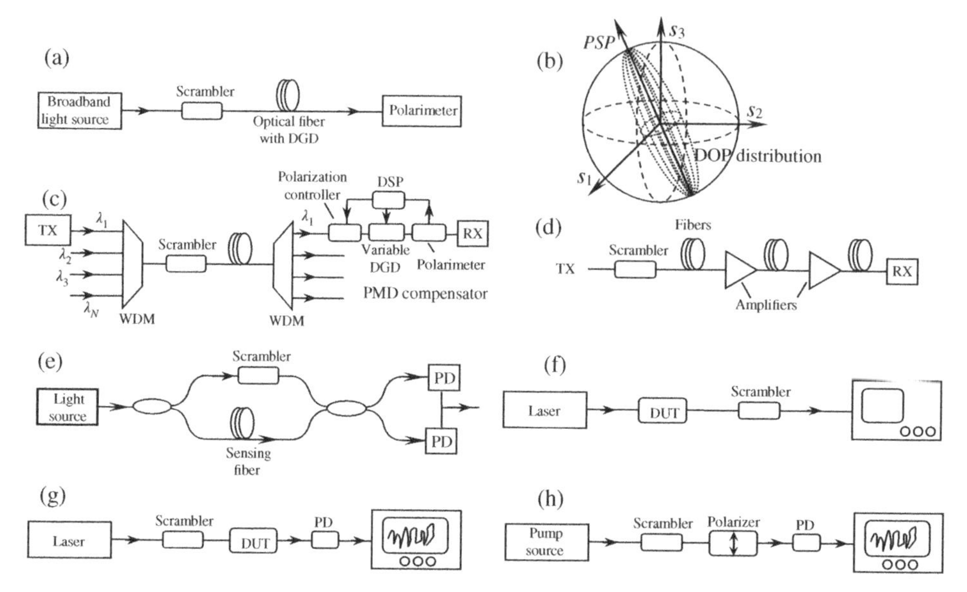

Figure 7.15 Polarization scrambler application examples. (a) Determining the PSP of an optical fiber with DGD or PMD; (b) Illustration of the DOP distributions corresponding to different SOP; (c) Removing the effect polarization state generator (PSG) in an optical fiber communication link. (d) Facilitating PMD compensation. (e) Eliminating polarization fading in an interferometer based sensor system. (f) Removing polarization sensitivity of an optical instrument, such as a power meter. (g) PDL measurement of a device under test (DUT). (h) DOP measurement of a light source. TX: transmitter, RX: Receiver.

separately discussed in Section 7.5. Here we focus on other applications. As shown in Figure 7.15a, a polarization scrambler together with a polarimeter can be used to determine the PSP in an optical fiber. A PSP is in fact the direction of the slow or fast axis of the fiber with residual birefringence or differential group delay (DGD). As described in Chapter 6, when a broad band light signal passes through a device or a piece of optical fiber with DGD, the two orthogonal components of the signal aligned with the slow and fast axes will propagate at different speeds and therefore will have a relative delay between each other, resulting in depolarization of the light signal. When the relative delay is larger than the coherence length of the light source, maximum depolarization occurs and the DOP of the signal will be minimized. However, when the SOP of the light signal is aligned with one of the PSPs (the fast or slow axis of the residual birefringence), no depolarization occurs and the DOP remains at the maximum. On the other hand, when the SOP of the light signal is aligned 45° from a PSP, the maximum depolarization occurs and the *DOP* may approach to zero if the DGD is larger than the coherence length of the light signal. Consequently, when plotting the DOP of all SOP points on the Poincaré sphere, as shown in Figure 7.15b, one sees an ellipsoidal distribution of DOP corresponding to the random distribution of SOP on the Poincaré sphere, with the two end points of the long axis being the slow and fast axis or the two PSP points, and the two end points along the short axis being the linear SOP with an angle of ±45° from the PSP.

Before coherent detection was deployed, scramblers were used to assist in the monitoring of DGD (first-order PMD) in a WDM system (Rosenfeldt et al. 2001), as illustrated in Figure 7.15c. Generally speaking, PMD can be monitored by measuring the DOP of the optical data stream propagated through the fiber. Low DOP usually indicates a large PMD effect. However, such a measurement may be erroneous if the input SOP to the transmission fiber is substantially aligned with the PSP of the fiber. In such a situation, the measured DOP is always large no matter how large the DGD between the two principal states of polarization. A scrambler at the transmitter side can be used to effectively eliminate such an anomaly. Furthermore, it enables a polarimeter in the PMD compensator at the receiver side to identify the PSP, which in turn speeds up PMD compensation. Other optical network applications include signal-to-noise ratio (SNR) monitoring of WDM channels with a polarizer placed after the scrambler.

As shown in Figure 7.15d, a polarization scrambler can be used at the transmitter side to minimize PDG or polarization hole burning of Erbium doped fiber amplifiers (EDFAs) in ultra-long haul systems. For this application, the scrambling rate should be significantly faster than the inverse of the gain recovery time constant of the fiber amplifiers (on the order of 10 kHz).

Polarization scramblers can also be used to eliminate the polarization fading of a fiber sensor, as shown in Figure 7.15e. In such a system, the envelope of the response curve is independent of the polarization fluctuation.

Placing a scrambler in front of a polarization-sensitive instrument, such as a diffraction grating based optical spectrum analyzer, can effectively eliminate the polarization dependence of the instrument, as shown in Figure 7.14f, if the scrambling rate is sufficiently faster than the detector speed in the instrument. In this case, the detector averages the signals corresponding to different SOPs to reduce the polarization sensitivity.

In addition, as illustrated in Figure 7.14g, polarization scramblers can be used to measure the PDL of a device under test (DUT) with the help of a digital scope or a data acquisition card. The resulting PDL of the device can be calculated as

$$PDL = 10\,log\left(\frac{V_{max}}{V_{min}}\right) \tag{7.14a}$$

where V_{max} and V_{min} are the maximum and minimum signal displayed by the digital scope, respectively.

Raman amplifiers generally exhibit strong PDG if the pump laser is highly polarized. To minimize the PDG, a depolarized pump source must be used. The DOP of the pump source directly relates to the PDG of the amplifier, and therefore, must be carefully characterized. Polarization scramblers, again, can be used to accurately measure the DOP with a digital scope, as illustrated in Figure 7.14h. Assuming that the maximum and minimum voltages measured with the digital scope are V_{max} and V_{min}, respectively, the DOP of the light source can be calculated using the following formula:

$$DOP = \frac{(V_{max} - V_{min})}{(V_{max} + V_{min})} \tag{7.14b}$$

In summary, a polarization scrambler is an important device for fiber optic communications, fiber sensors, and fiber optic test and measurement applications. Different applications may pay attention to different set of specifications and capabilities. For test and measurement applications, low PDL and low AL are most important in addition to the scrambling uniformity. For network SOP emulation applications, the capabilities for generating the Rayleigh distribution for SOP variation rates, as well as the high SOP scrambling speed and uniformity, are important. For transceiver SOP variation tolerance test, high scrambling speed and uniform SOP variation rates are preferred. Polarization scramblers made with fiber squeezers have been widely adopted due to their overall performance characteristics, although their scrambling speed cannot reach that of $LiNbO_3$ based devices required for applications requiring unusually high speeds.

7.3 PDL Emulator

In order to evaluate how a certain PDL affects the performance of an optical fiber system, a PDL emulator which is capable of generating different PDL values at different rates is required (Yan and Yao 2002). As shown in Figure 7.16a, a PDL emulator can be constructed with a PBS to separate the incoming optical signal into two paths of orthogonal SOPs, one of which goes through a voltage controlled variable optical attenuator (VOA) and the other one goes through an optical variable delay line (VDL), before they are recombined by a PBC. The VDL is for balancing the optical path lengths between the two paths to eliminate any optical path length difference and the VOA is for attenuating one of the beams to generate the desired PDL value. Figure 7.16b shows the photo of a commercial PDL emulator. By controlling the voltage applied to the VOA with a random number generator, random PDL values can be generated, as shown in Figure 7.16c. By modulating the VOA with a voltage triangle waveform, a PDL ramp with any rate can be generated, as shown in Figure 7.16d. Finally, by modulating the VOA with a square waveform, alternating PDL values with any periodicity can be generated, as shown in Figure 7.16e.

7.4 PMD Generation and Emulation

For evaluating the performance of an optical fiber system against PMD, it is often desirable to generate different PMD values in a defined range to see whether the system can perform satisfactorily against impairments of the PMD imposed onto the system (Willner and Hauer 2005). A PMD generator is a device or instrument capable of generating any desired PMD value in a defined range while a PMD emulator is a device or instrument for generating a series PMD values with the probability density following a specific function, such as the Maxwellian function, mimicking the probability distribution of a real optical fiber communication link. In general, PMD generator is deterministic, which can be controlled to dial in any PMD value at any time, while PMD emulator

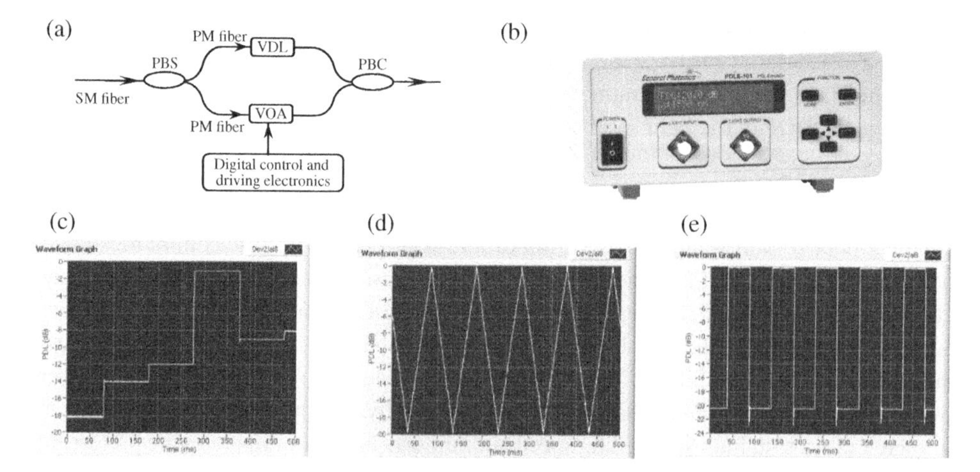

Figure 7.16 (a) The construction of a PDL emulator; (b) The photo of a commercial PDL emulator (Source: Courtesy of General Photonics/Luna Innovations); (c) Random PDL generation with time by driving the VOA with a voltage controlled by a random number generator; (d) PDL ramp generation by the triangle wave modulation of the VOA; (e) Periodic PDL value alternation by the square wave modulation of the VOA. VOA: voltage controlled variable optical attenuator; VDL: Optical variable delay line.

may not be deterministic for obtaining a desired PMD value although it has a certain probability for getting to the PMD value. In this section, we describe some practical techniques or devices for accomplishing this task, one of which is based on polarization beam splitting and recombining and we call it PBS/C technique. The other is based on polarization switching with birefringence crystals, and it is often referred to as polarization switching technique. There are also many other types of PMD emulators, such as that based on cascading multiple sections of birefringent crystals or PM fibers, with the relative orientation angle between any two sections controlled by a motor or a rotatable half-wave plate. J. Damask has an excellent description of such a PMD emulator in his book *Polarization Optics in Telecommunications* (Damask 2005), which therefore will not be discussed in this book.

7.4.1 PMD Generator and Emulator Based on Polarization Splitting and Combining

7.4.1.1 First-Order PMD Generator and Emulator

Figure 7.17a shows the basic hardware configuration of a first-order PMD generator or DGD generator, which consists of an input PBS to split the incoming signal from a single-mode fiber (SM) fiber into two PM fibers (PM fiber 1 and PM fiber 2) of orthogonal polarizations. The PBS can be made with the configurations of Figure 6.7d,e, in which the slow (or fast) axis of each PM fiber is aligned with one of the SOPs exiting from the PBS cube (Figure 6.7d) or the Wollaston prism (Figure 6.7e). PM fiber 1 is connected to an VDL for changing the relative delay ΔL with respect to PM fiber 2. A variable optical attenuator (VOA) is placed in PM fiber 2 to eliminate the PDL of the emulator by balancing the optical powers in the two signal arms. The digital control circuit can be programmed to generate any desired relative time delay $\Delta\tau$ between the two PM fibers, which correspond to the DGD:

$$DGD = \Delta\tau = \frac{n_c \Delta L}{c} \tag{7.15}$$

where n_c is the refractive index of the PM fiber core and c is the speed of light in vacuum. To emulate the DGD (or first-order PMD) in a real optical communication system, one may also program the VDL to generate a sequence of relative delays $\Delta\tau$ with the probabilities following the Maxwellian distribution shown in Figure 3.6 (Kogelinik et al. 2002), which is representative of the probability distribution of DGD, as described in Section 3.4.

Note that the SOP of the recombined light in the output SM may fluctuate around a circle on the Poincaré sphere because the relative phase between the orthogonal SOPs of the signal in the two PM fibers is likely to vary rapidly due to the temperature variation or vibration.

7.4.1.2 All-Order PMD Generator and Emulator

Figure 7.17b shows a configuration capable of generating and emulating both the first-order PMD (DGD) and the second order polarization mode dispersion (SOPMD) values by cascading two DGD emulators together. A magneto-optic polarization rotator (MO rotator) described in Chapter 6 is used to rotate the SOP angle θ between the two first-order emulators. The first and second order PMD can be expressed by Eqs. (6.25b) and (6.25a), and here are rewritten as

$$DGD = \sqrt{\Delta\tau_1^2 + \Delta\tau_2^2 + 2\Delta\tau_1\Delta\tau_2 cos\theta} \tag{7.16a}$$

The corresponding SOPMD is

$$SOPMD = \Delta\tau_1\Delta\tau_2 sin(2\theta) \tag{7.16b}$$

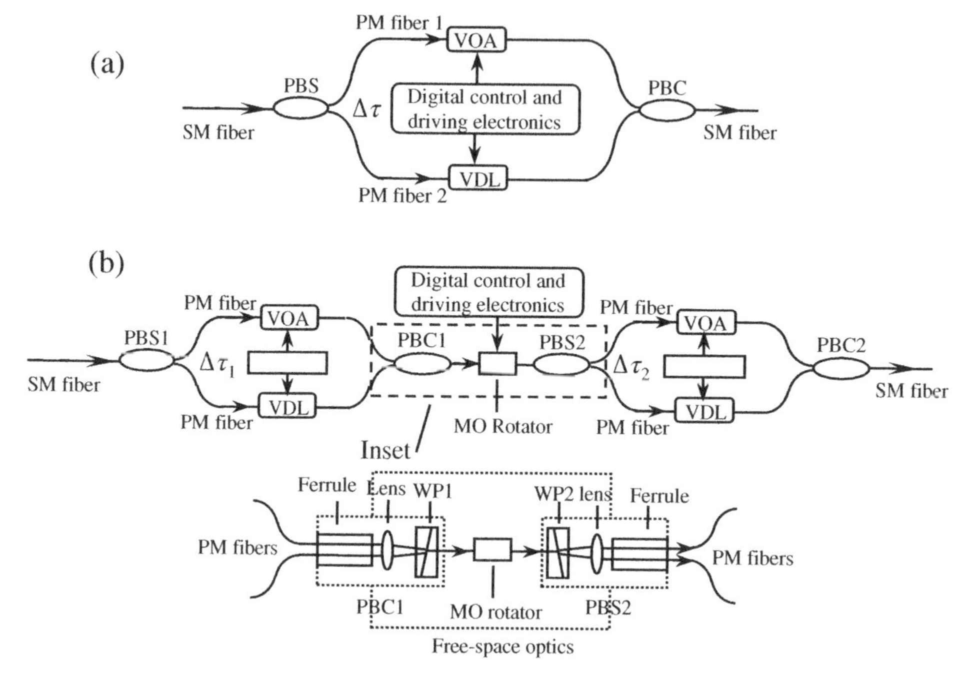

Figure 7.17 Configurations of PMD emulators made with PBS and PBC. (a) A first-order PMD (DGD) emulator. (b) A PMD emulator capable of generating first and second order PMD values by cascading two first-order emulators together. Inset: Details of the interconnection between the two first-order emulators. A magneto-optic polarization rotator (MO rotator) is used to rotate the SOP going into the second DGD. WP1 and WP2: Wollaston prisms.

Similar to the configuration in Figure 7.17a, the SOP of light at the output of the first DGD emulator may drift rapidly due to the vibration and the temperature variations, which may cause undesired changes in the SOP angle θ and unpredictability of the generated SOPMD (Möller et al. 2003).

7.4.2 PMD Generator and Emulator Based on Polarization Switching

The PMD generators and emulators described earlier involved splitting a light signal into two different optical paths of orthogonal polarizations, delaying one of them with respect to the other, before recombining them. Because the phases of the light signals in different optical paths are sensitive to temperature and vibration, the SOP of the combined signal at the output of each PBC may fluctuate, causing uncontrollable SOP and SOPMD outputs. In addition, because the DGD and PMD are generated by controlling the optical delay with a motorized variable delay line, the speed for generating different DGD and PMD values is very slow, on the order of seconds, making them unpractical for generating a large number of PMD values for emulating a Maxwellian PMD distribution mimicking the PMD variations in a real long haul optical fiber communication system. The PMD emulators described in the following text can overcome the issues described earlier.

7.4.2.1 Binary DGD Generator Using MO Polarization Switching

Device Construction As shown in Figure 7.18a, the device is composed of multiple switch/delay sections (Yao 1995; Yan et al. 2003). Each switch/delay section consists of a birefringent crystal to generate a fixed amount of delay and a MO polarization rotator/switch described in Figure 6.25a.

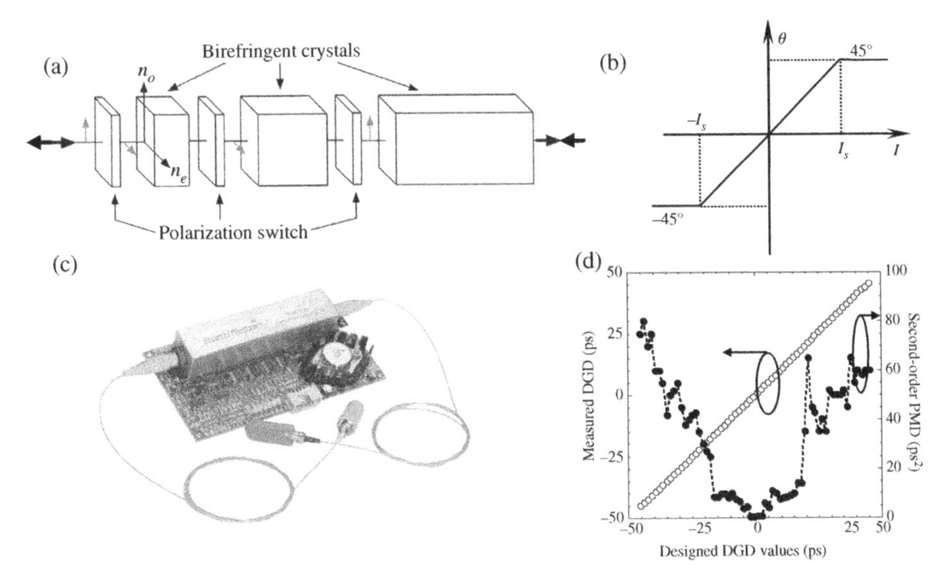

Figure 7.18 Description of a binary programmable DGD generator. (a) The structure showing the device consisting of multiple MO polarization rotation switches and birefringent crystals. (b) A curve showing the SOP rotation angle as a function of electric current applied to a MO rotator. (c) A photo of commercial binary programmable DGD generator developed by the authors of this book (Source: Courtesy of General Photonics/Luna Innovations.). (d) The experimental data showing the measured DGD and SOPMD vs. DGD settings.

The lengths of the birefringent crystals are arranged in a binary power series, increasing by a factor of 2 in each section. Such a binary arrangement requires a minimum number of crystal sections and results in the highest possible delay resolution. Light signal can be input either from left at the shortest delay section or from right at the longest delay section; however, as will be discussed, inputting light from right has certain advantages.

The typical response curve illustrating the binary nature of the MO rotator is shown in Figure 7.18b. By driving the rotator with a current above the saturation current $\pm I_S$, $\pm 45°$ SOP rotation (or 0–90° if the −45° is taken as the reference) can be achieved. Therefore, at any switch/delay section, the input polarized beam can be either switched along the slow or fast axes of the birefringent crystal, corresponding to longer or shorter delay. The total delay is the summation of the delays in each crystal segment and can be varied by the MO polarization switches.

Assuming that the smallest birefringent crystal length is ℓ, a minimum delay time $\delta\tau$ can be defined to represent the delay generated by the shortest delay section, which can be expressed as

$$\delta\tau = \left|(n_e - n_o)\frac{\ell}{c}\right| \tag{7.17a}$$

where n_o and n_e are crystal's ordinary and extraordinary refractive indices, and c is the speed of light. Although the delay/switch sections can be placed in any order, we consider a case when the first bit (the least significant bit) in the switching command controls the shortest delay section and the last bit (the most significant bit) controls the longest delay section.

For a device with N switch/delay sections, when the nth bit is switched from 0 to 1 state (0–90°), the DGD of the nth section changes sign. The total differential time delay for an arbitrary binary state becomes

$$DGD = -\delta\tau \sum_{n=1}^{N} (-1)^{b_n} 2^{n-1} \tag{7.17b}$$

where b_n (= 0, 1) is the binary value of the nth bit, determined by the polarization switch associated with the nth bit, and the delay resolution is twice the unit delay time, i.e. $2\delta\tau$, obtained by switching the least significant bit in Eq. (7.17b). In addition, the DGD can be a negative number due to sign flip of the delay time during switching. In a 6-bit module ($N = 6$), a total of 64 *DGD* values (from $-63\delta\tau$ to $+63\delta\tau$) can be generated with a resolution of $2\delta\tau$.

Because the beams of orthogonal polarizations are co-propagating in the same optical path, the differential path variations due to temperature and vibration found in the configurations in Figure 7.17 are largely eliminated. The remaining differential path variation is due to the different thermal coefficients of the refractive indices n_o and n_e, which is about 2 orders of magnitudes smaller than those of each index along. Consequently, the output SOP from the binary SOP switching DGD generator is much more stable than that from the PBS/PBC based DGD generator. In addition, because the MO polarization rotators/switches have a speed on the order of 20 μs, the resulting speed for generating each DGD state can be as high as 20 μs per state, which is 4 orders of magnitude faster than that of the DGD generator based on motorized VDL described in Figure 7.17a.

Based on the principle described in Figure 7.18a, a programmable delay module using six birefringent crystals (i.e. a 6-bit module) has been commercialized, as shown in Figure 7.18c, which is capable of generating tunable *DGD* values from −45 to +45 ps with a resolution ($2\delta\tau$) of 1.40 ps. One of the most promising features of this module is the sub-millisecond switching speed desirable for fast PMD emulation or compensation.

Figure 7.18d shows the measured DGD and SOPMD data with a commercial PMD analyzer as a function of programmed DGD settings corresponding to different logical DGD states. The data show that the measured DGD values agree well with the designed DGD values and each DGD value is exactly reproducible. Equally as important, the second-order PMD is very small, less than 85 ps^2, even at high values of DGD. In addition, in the wavelength range of 1535–1560 (C-band), the insertion loss of the device is ~1.3 dB with a wavelength dependent variation of <0.15 dB, and the *PDL* values range from 0.02 to ~0.28 dB for all the DGD states.

Dynamic Performance To use the DGD generator in an optical fiber link with continuous data traffic, it is critical that the system performance be unaffected during DGD state switching. Therefore, the DGD generator must have a well-controlled dynamic performance and the switching effect on the performance must be first characterized. In principle, when the device is switched from one DGD state to another, the DGD value will change from one value to another precisely. However, because the device has finite switching speed, the DGD value during switching is different from either the starting or ending states. When the device is switched in smallest step, the maximum DGD value excursion from the DGD value of the ending state is defined as the transient DGD. Because it is difficult to directly measure the transient DGD during switching due to the limited response speed of the measurement instrument, a quasi-static measurement can be performed. Instead of having a full 90° polarization rotation in one step to change the DGD value from one state to another, the polarization rotation angles can be incrementally increased in steps of a few degrees while measuring the corresponding DGD with a commercial polarization analyzer. The results are shown in Figure 7.19a for *DGD* values from 0 to 45 ps in steps of 1.4 ps. The inset shows the detailed DGD values between two adjacent DGD states. It is evident that the transient DGD is always less than the step size. The small transient DGD is important for using the device in PMD compensators.

In order to test the system impacts during fast switching, the DGD generator was inserted into a 10-Gb/s non-return-to-zero (NRZ) transmission link modulated at $2^{31} - 1$ pseudorandom binary sequence (PRBS). The input optical-signal-to-noise-ratio (OSNR) to the DGD module

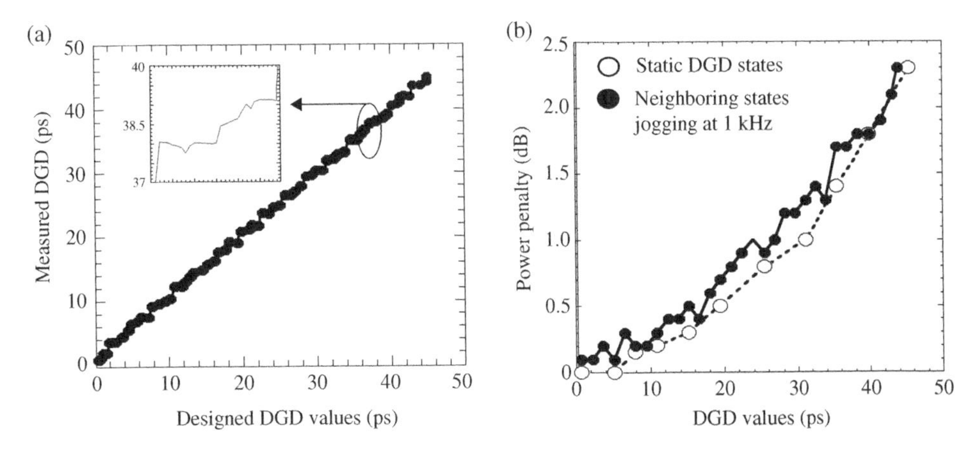

Figure 7.19 Dynamic performance of the DGD generator. (a) Transient DGD between two adjacent DGD states. (b) Power penalties due to dynamic switching (jogging) between neighboring states at 1 kHz (solid circles) at different DGD states, as compared with those due to different static DGD values (open circles).

was set to 30 dB (0.1 nm bandwidth). An optical pre-amplifier before the receiver was used to increase sensitivity. The system's back-to-back sensitivity was measured to be −31 dBm. Power penalties were measured by comparing the receiver sensitivity of the system at a 10^{-9} BER with the back-to-back sensitivity. Two cases were considered: (i) power penalties at static DGD states and (ii) power penalties when neighboring states were jogging back and forth at 1 kHz (1 ms continuous switching). As shown in Figure 7.19b, a negligible power penalty of less than 0.2 dB due to fast polarization switching (jogging) was obtained. Note that such results can only be achieved when the light is input from the crystal with the longest length or the crystal lengths are decreasing along the transmission, as shown in Figure 7.18a. It can be shown both analytically and experimentally that worse power penalty will result if the light is input from the crystal with the shortest length.

7.4.2.2 PMD Emulation with the Binary Programmable DGD Generator

First-Order PMD Emulation The precise and repeatable DGD generation capability of the binary DGD module is ideal for generating a series of DGD values with any statistical distribution for a given number of samples following the Maxwellian, Gaussian, Lorentz, or other statistical functions of a user's choice. However, as described in Chapter 3, the probability density function of the DGD in a typical optical fiber communication link follows Maxwellian distribution. Therefore, to emulate the DGD variations, one may program with a computer or digital circuit to control the DGD module to generate statistical DGD samples with a Maxwellian distribution with a selectable average DGD value $\overline{\Delta\tau}$ using Eq. (3.20a).

Figure 7.20a shows the instantaneous DGD of 500 samples with an average value of 10 ps generated with Eq. (3.20a), while Figure 7.20b shows the corresponding DGD distribution. It is clear that the probability distribution is indeed Maxwellian.

All-Order PMD Emulation By cascading several variable DGD generators with polarization controllers between them, one may emulate not only first-order PMD, but also higher-order PMD statistics (Yan et al. 2006). Figure 7.21a shows the configuration of such emulator, which is constructed from three variable DGD generators separated by two fiber-squeezer-based polarization controllers. Each DGD generator is digitally programmed to generate any *DGD* value from

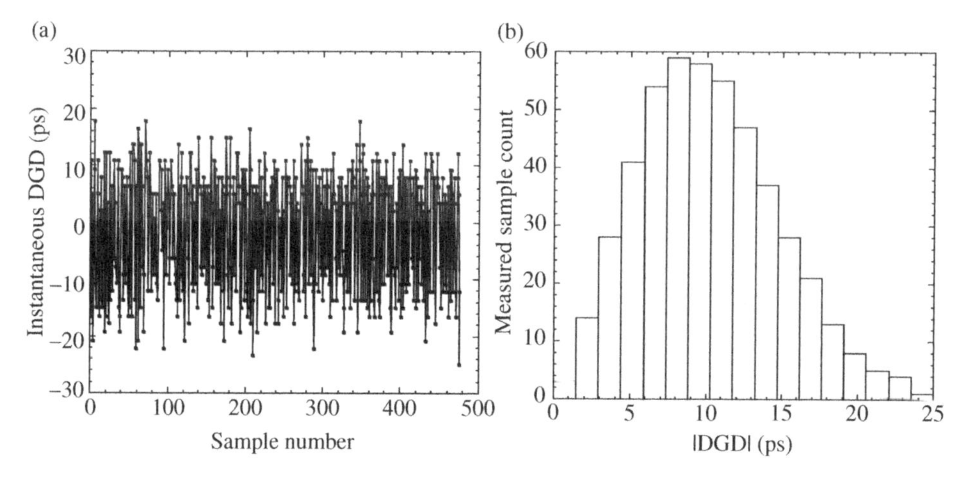

Figure 7.20 (a) Generated DGD samples with an average *DGD* of 10 ps following Eq. (3.20a). (b) The corresponding statistical distribution of the data samples in (a).

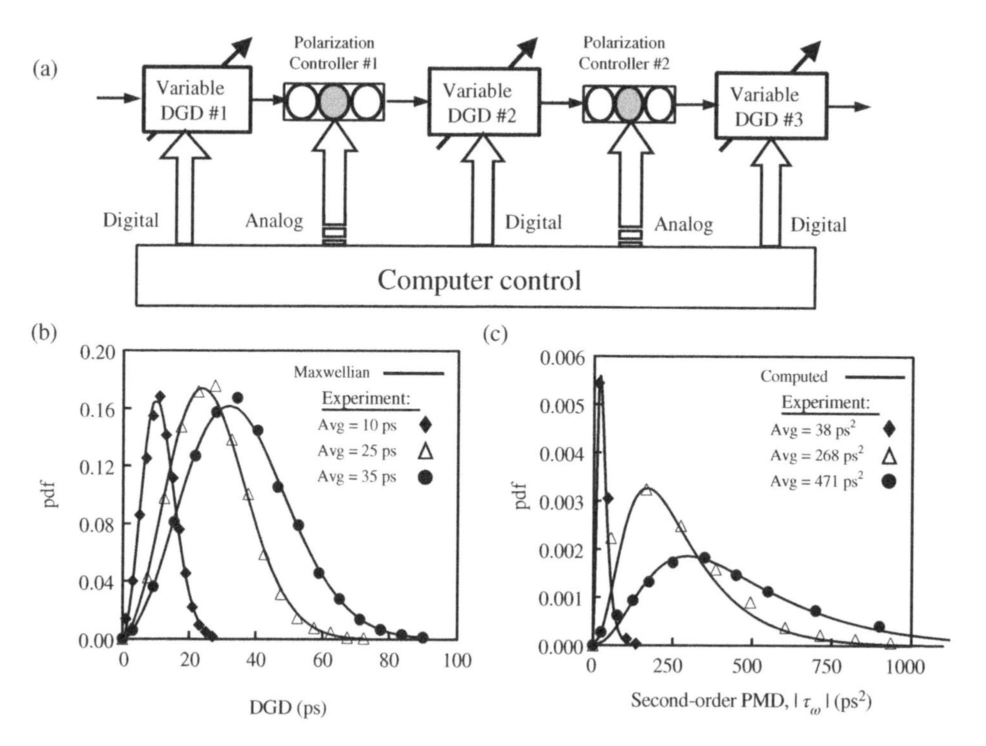

Figure 7.21 All-order PMD emulator with tunable statistics using variable DGD elements (a) and generated (b) first-order and (c) second-order PMD statistics.

−45 to +45 ps with a tuning speed of <1 ms and a resolution of 1.40 ps. A computer is used to control the emulator to randomly generate any desired DGD distribution for each element and to uniformly scatter the polarization between sections. To obtain a Maxwellian DGD distribution at the emulator output, the DGD values of each generator are varied according to a Maxwellian distribution with an average DGD, $\overline{\Delta\tau}$.

It can be shown through simulation (Yan et al. 2006) that such a configuration can yield an average DGD of $3^{1/2}(\Delta\tau)$ for the all-order emulator and an average second-order PMD distribution that has the correct shape but falls slightly short of that expected for a real fiber. Figure 7.21b,c shows generated three statistic distributions of first-order and second-order PMD for averaged *DGD* of 10, 25, and 35 ps, respectively. All of the PMD measurements were performed using the Jones matrix method of a commercial PMD analyzer (Heffner 1992) described in Sections 4.6 and 8.6. As expected, the DGD values closely match the expected Maxwellian distribution. The corresponding SOPMD distributions have averages of 38, 268, and 471 ps^2, which are ~30% lower than expected for a real fiber. 90% agreement with real fiber is expected if two more stages of DGD generator and polarization controller pairs are added. The application of variable DGD generators in PMD emulation may also enable importance sampling (Yan et al. 2004), which is a powerful tool to investigate the rare events with extremely low probability along the Maxwellian tail.

7.4.3 Polarization Optimized PMD Source

7.4.3.1 PMD Emulator vs. PMD Source

The all-order PMD emulator described earlier is not deterministic for generating a particular PMD value because the SOPs between the DGD generators are random, although it may obtain the PMD value with a certain probability. In order to generate deterministic PMD values, a polarimeter may be used after each polarization controller to align two neighboring DGD generators in a certain relative direction; however, this will significantly add cost and complexity. On the other hand, a PMD source to be discussed next is an instrument capable of deterministically dialing in any desired first and second order PMD values, which can also be programmed to generate a series first- and second-order PMD values following a specific probability distribution function (PDF), particularly the Maxwellian function.

7.4.3.2 Polarization Optimization

The PMD effect on a system is highly polarization dependent, as shown in Figure 7.22. When the input SOP is aligned or counter-aligned with the PSP of a fiber link or PMD source, the PMD has no effect on the signal, and therefore no effect on system performance. On the other hand, when the input SOP is either circular or linear with a 45° orientation from the PSP of the fiber link or PMD source, the PMD has its maximum effect on the transmitted optical signal, and hence the worst effect on system performance. Other polarization states experience PMD effects between these two limiting cases. Because SOP can change quickly due to fiber movement or environmental effects, the PMD effect can vary quickly even if the actual amount of PMD in the system does not change.

Because PMD causes depolarization of the optical signal, the DOP of the optical signal can be used to monitor the effect of PMD on the signal. The smaller the DOP, the larger the PMD effect is. DOP is the smallest when PMD impact on signal is maximum, as shown in Figure 7.22. In general, the DOP of the signal passing through a medium with first-order PMD or DGD can be expressed as

$$DOP = \sqrt{1 + 4(\gamma^2 - \gamma)\{1 - [R(\tau)/R(0)]^2\}} \tag{7.18a}$$

where τ is the DGD value of the medium, $R(\tau)$ is the self-correlation function of the optical signal and decreases as τ increases, which relates to the spectrum $F(\omega)$ of the signal by

$$R(\tau) = \int_{-\infty}^{\infty} F(\omega)e^{-i\omega\tau}\,d\omega \tag{7.18b}$$

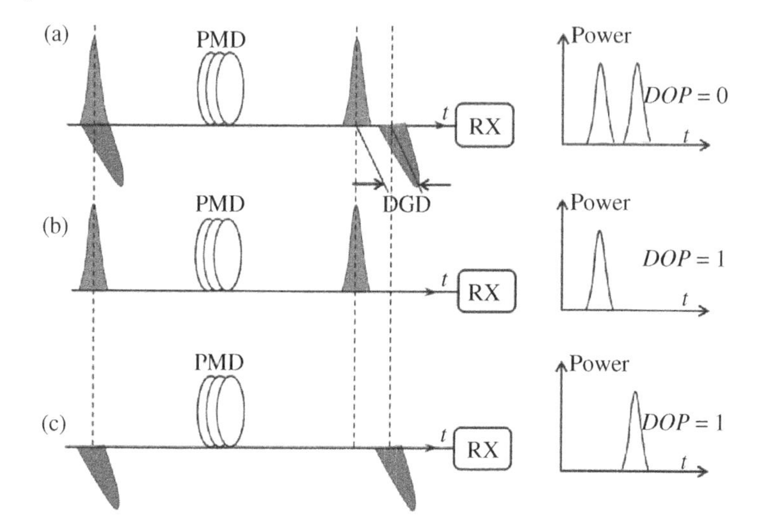

Figure 7.22 Illustration of PMD effect on an optical signal. (a) The SOP of the input signal is aligned 45° from the PSP of the fiber link, causing the worst-case signal distortion. In this case, if the DGD is larger than the width of one bit, then the *DOP* = 0 because the two orthogonal polarization components have the same power, with no phase relationship. (b) and (c) The SOP of the input signal is aligned with the slow or fast PSP axis, respectively. In these cases, no signal distortion occurs except the slight late or early arrival, respectively, of the pulse. The DOP of the signal remains at 1. In all three cases, the PMD of the link remains the same, but the effect on the signal is different due to the different input polarization states.

and γ is the power distribution ratio of the two polarization components with respect to the PSP of the medium with DGD and is defined as

$$\gamma = \frac{P_s}{(P_s + P_f)} \tag{7.18c}$$

In Eq. (7.18c), P_s and P_f are the powers of the polarization components in the slow and fast axes of the DGD element. $\gamma = ½$ when the powers of the two polarization components are equal, corresponding to the case where the input SOP to the medium is 45° from its PSP or where the input SOP is circular.

Clearly, DOP reaches minimum of $R(\tau)/R(0)$ when $\gamma = ½$. At this SOP, the PMD has the worst effect on the quality of the signal. On the other hand, when $\gamma = 1$ or when the input SOP is aligned or counter aligned with the PSP, *DOP* always remains at the maximum value of 1, regardless of the DGD value.

When higher-order PMD is present, the PSP is different for different wavelength components of the signal and Eq. (7.18a) may no longer hold. However, PMD induced signal distortion is always at its worst when the DOP of a signal is minimum and least when it is maximum. Therefore, DOP can be used as an indicator for signal distortion caused by PMD. When performing test of PMD effect on the signal, the input polarization to the PMD source should be optimized to ensure the worst distortion on the signal or the minimized DOP of the signal.

7.4.3.3 Ternary Polarization Rotation Switch

The MO polarization rotators described in Figure 6.25a can be operated at binary mode, with saturated SOP rotation angles of ±22.5° or ±45° when applied electric currents exceed saturation current of $\pm I_s$. Such a binary operation has the advantages of high repeatability and high rotation angle accuracy independent of fluctuations of applied currents. By cascading two such binary

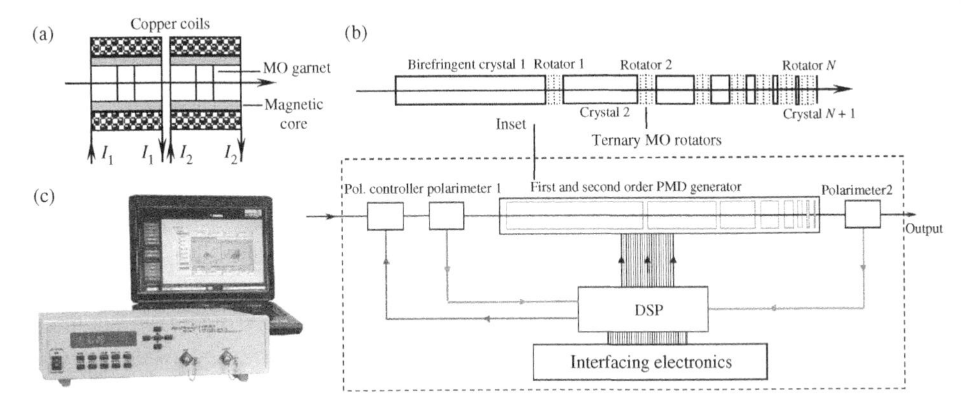

Figure 7.23 Illustration of a ternary polarization rotation switch (a), a polarization optimized PMD source implemented with multiple ternary MO polarization rotators with an inset showing the details of the arrangement of crystals and MO rotators (b), and a photo of such a commercial PMD source developed by the authors of this book (c) (Source: Courtesy of General Photonics/Luna Innovations). DSP: digital signal processing electronics.

SOP rotators together, as shown in Figure 7.23a, ternary SOP rotation switches can be realized: When two MO switches are set to rotate −22.5°, a total of −45° polarization rotation can be achieved. When the two MO switches are set to rotate +22.5°, a total of +45° polarization rotation can be achieved. Finally, when one of the switches is set to rotate −22.5° and the other +22.5°, a net zero polarization rotation can be obtained. Therefore, the ternary polarization rotation switch has three polarization rotation states: (−45°, 0°, 45°). Both the MO thick film of perpendicular anisotropy described in Figure 6.19 and the MO thick film of in-plane anisotropy described in Figure 6.23 can be used in the ternary polarization rotators.

7.4.3.4 Continuous Polarization Rotator

One may also make the MO polarization rotator in Figure 7.23a to rotate the SOP of light by any angle between −45° and 45° by applying a current between $-I_s$ and $+I_s$. In this case, the MO garnet of in-plane anisotropy described in Figure 6.23 is preferred because it does not have the scattering loss at low magnetic field of the MO garnet of perpendicular anisotropy described in Figure 6.19, which causes transient loss during polarization rotation.

7.4.3.5 Polarization Optimized PMD Source Based on Ternary Polarization Switches

PMD sources are generally used to test an optical fiber system's tolerance to PMD or the ability of a transceiver to overcome the adverse PMD effects. Therefore an ideal PMD source for such applications must be able to (i) reliably and accurately generate first- and second-order PMD values over a large range; (ii) automatically optimize polarization state for worst-case PMD effect at any PMD value; (iii) generate continuous PMD traces at variable speed; and (iv) switch PMD states with fast transients between states. The polarization optimized PMD source presented here attempts to satisfy all of these requirements.

Figure 7.23b shows a system diagram of a polarization optimized PMD source, which consists of a polarization controller (PC), a polarimeter at the input end, a PMD generator, a second polarimeter at the output end, and a DSP based electronic circuit. The DSP circuit uses information from the polarimeters to control the polarization controller for automatic polarization optimization. There are two polarization optimization methods. Feedback from the input polarimeter can be used to align the SOP to a state offset by 45° from the PSP of the PMD generator, resulting in the worst-case first-order PMD effect illustrated in Figure 7.22a. Alternatively, the circuit can use the DOP information from the output polarimeter to adjust the input SOP to minimize the DOP of the signal after the PMD generator. This results in the worst-case total PMD effect on the optical signal.

This automatic polarization optimization capability enables fast, accurate PMD tolerance testing of optical fiber links by ensuring the worst-case PMD effect at every PMD value used in the test. Not only does it eliminate the need for an external polarization scrambler, but it also reduces testing time, since the time needed to find and lock to the worst-case SOP with the automatic polarization search is much less than the time typically needed to hit the worst-case SOP using an external polarization scrambler with a low probability.

The PMD generator used in this PMD source design is constructed using birefringent crystals and ternary MO polarization rotators described in Figure 7.23b. The crystals are arranged in descending order of length, with the lengths of adjacent crystals differing by a factor of 2. The crystal lengths are selected to ensure a PMD range large enough to exceed the PMD tolerance of most systems. Each pair of crystals is separated by a ternary polarization rotator with three digitally switched rotation states: +45°, 0°, and −45°. When a rotator is set to its +45° state, the optical axes of the crystals on either side of it are aligned, producing the maximum combined DGD. When it is set to its −45° state, the optical axes of the crystals are counter-aligned, producing the minimum combined DGD. When the rotator is in its 0° state, the optical axes of the crystals are offset by 45°, producing higher-order PMD. Because the rotators are digitally switched, the PMD generator is able to produce highly accurate, repeatable PMD values for PMD tolerance tests. In addition, it is able to switch states in as little as 1 ms, fast enough for PMD response time tests.

The higher-order PMD are not available in the DGD generator described in Figure 7.18 where the polarization rotators can only generate ±45° polarization rotations and cannot generate higher PMD values. The 0° polarization rotation is essential for the generation of higher-order PMD.

The total number of PMD values can be generated with $N+1$ sections of birefringent crystals and N ternary rotators is 3^N and the total values of *DGD* (first-order PMD) is 2^N. It can be shown that for $N = 8$, the PMD source can generate a total of 6561 different PMD states, of which 256 are DGD-only, 2014 have DGD and wavelength independent SOPMD, and the rest are wavelength dependent PMD. For $N = 6$, the total number of PMD can be generated are 729, in which 64 are DGD-only states. Figure 7.24 shows the generated SOPMD and the map of SOPMD vs. DGD for the case of $N = 6$. The distribution of SOPMD may not be sufficiently uniform and dense for certain applications.

In order to overcome this deficiency, rotator 1 between the longest and the second longest crystal can be programmed to rotate SOP continuously between ±45°, while other rotators (rotators

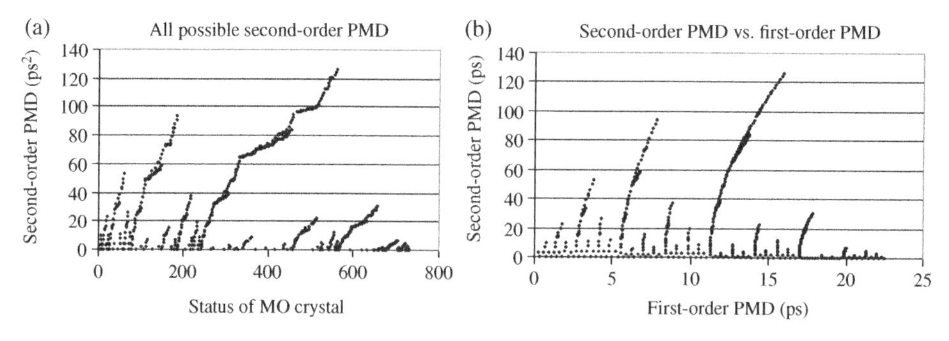

Figure 7.24 (a) SOPMD generated by the ternary PMD source as a function of the polarization status of the eight ternary MO rotators. (b) Distribution of the SOPMD vs. the first-order PMD (DGD).

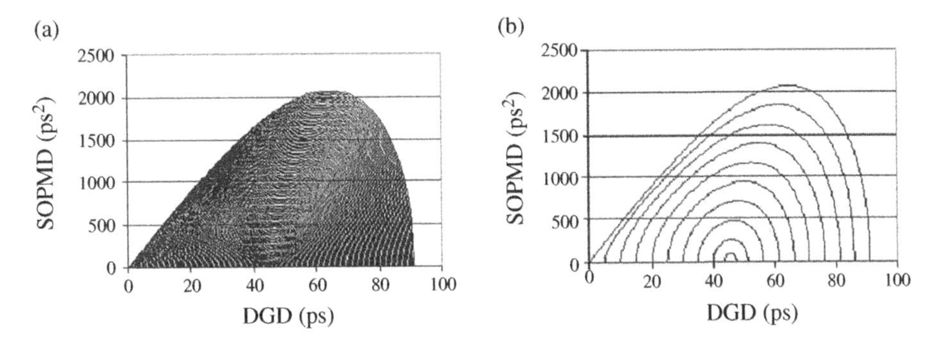

Figure 7.25 The map of SOPMD vs. DGD showing all the PMD values can be generated via the PMD scanning function (a) and a few selected PMD scanning traces generated (b).

2–N) remain to be ternary. This way, quasi-continuous PMD values can be generated. For the case that rotators 2–N are binary with ±45° rotations, the device is then equivalent to the case of two DGD elements cascaded together with a tunable orientation angle θ between them, as described by Eq. (7.16a), while the second DGD element is made to be variable. Such a configuration enables independent DGD and SOPMD control, as well as the generation of continuous PMD traces. Figure 7.25a shows all the states in PMD space that can be generated for the case of $N = 8$, which is too dense to discern, while Figure 7.25b shows a few samples of PMD vs. DGD curves to be viewed clearly. In Figure 7.25b, each PMD trace represents the SOPMD generated by continuously changing the SOP rotation angle θ in Eq. (7.16b) while the different traces correspond to different DGD values of the second DGD element. Figure 7.23c shows a commercial implementation of an 8-bit PMD source, where seven ternary MO rotators, plus one continuous MO rotator ($N = 7 + 1$), and nine sections of birefringent crystals are used.

7.4.3.6 PMD Monitoring by PMD Compensation

The basic concept of PMD measurement by PMD compensation is illustrated in Figure 7.26a (Yao et al. 2010), where an optical signal passes through a fiber link with a certain amount of PMD (labeled as PMD_f herein). At the receiving end, a PMD compensator, consisting of a polarization controller, a variable PMD generator, and a PMD effect monitor, is used to compensate the signal distortion caused by PMD_f. The PMD effect monitor is used to provide the feedback signal to the polarization controller to adjust the polarization of the input signal as to minimize the PMD effect. When properly adjusted, signal's slow polarization component is aligned with the fast axis of the PMD generator, while its fast component is aligned with the slow axis. At the output end of

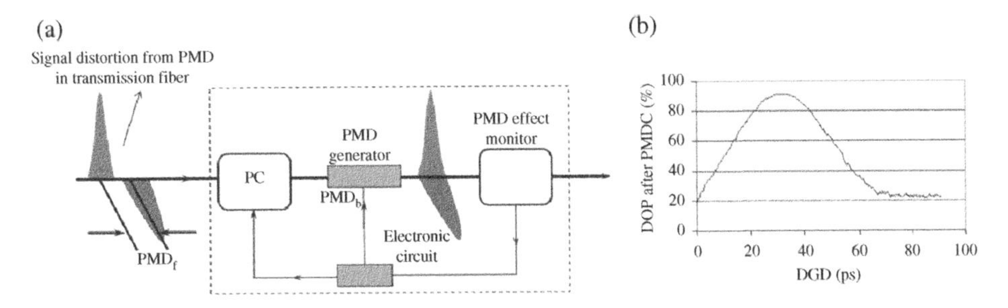

Figure 7.26 (a) Illustration of the concept of PMD measurement by PMD compensation. (b) Measured DOP of the signal by polarimeter 2 as a function of DGD while PMD compensation is performed.

the compensator, the two relatively delayed polarization components are brought closer together. When performing PMD compensation, the compensator will vary the PMD generator and look for a PMD value, PMD_b, which can best compensate the signal distortion caused by PMD_f. Assuming the resolution of the internal PMD generator is sufficiently high, the PMD value for the best PMD compensation, PMD_b, must be the PMD value of the fiber link, PMD_f. Such a compensator is sometimes called PMD nulling compensator (Kogelinik et al. 2002).

The polarization optimized PMD source shown in Figure 7.23 can be used for the implementation of the PMD compensator, with which different PMD values can be generated when the ternary MO polarization rotators rotate signal's polarization between the crystals. In commercial implementations with eight MO rotators ($N = 8$), the PMD generator can generate 256 precise DGD values with a resolution of 0.35 ps and a range of 90 ps, or with a resolution of 0.7 ps and a range of 180 ps.

As discussed in Figure 7.22, the DOP of the signal can be used as the PMD effect monitor and an in-line polarimeter to obtain the DOP (Kikuchi 2001). As described previously, a polarimeter (Polarimeter 1) before the PMD generator can be used to measure the DOP before PMD compensation and a second polarimeter (polarimeter 2) after the PMD generator can be used to provide the feedback DOP signal to the DSP circuit. The DSP circuit changes the PMD setting step by step and controls the polarization controller to maximize the DOP received by polarimeter 2 for each PMD setting. A plot of DOP vs. PMD (DGD) shown in Figure 7.26b can be obtained and the PMD setting corresponding to the maximum DOP value is the PMD of the fiber link at the measured wavelength. It is important to note that the comparison between the two DOP values reveals the effectiveness of PMD compensation and can be used for data analysis, as will be discussed in the next paragraph. Note that with the PMD compensator of Figure 7.23b, one may vary both first- and second-order PMD during the optimization process of PMD compensation. However, because of the large number of first- and second-order PMD values to scan through, the time for finding the optimized PMD is much too long to be practical. Therefore only the first-order PMD (DGD) is varied during the PMD optimization process, even though the PMD in the DWDM channel have both first- and second-order PMD components. The obtained optimized PMD (DGD) value actually contains the contribution of both the first- and second-order PMDs from the fiber because the PMD nulling compensator can also mitigate some higher-order PMD (Karlsson et al. 2001; Lanne et al. 2001), although only first-order PMD (DGD) is used in compensation. The obtained PMD is called the effective PMD and that the average PMD obtained using this method through either wavelength average or time average is expected to be sufficiently close to the real PMD of the fiber link. In the PMD source of Figure 7.23c, the time for obtaining the optimized PMD value among 256 DGD choices is about two seconds.

In operation, the DOP measured by Polarimeter 1 in the PMD source, DOP_1, is an indication whether the SOP of the signal is aligned with PSP of the transmission fiber or not. A value of DOP_1 close to 100% indicates that the input signal's SOP is nearly aligned with fiber's PSP and very little PMD distortion is caused by PMD in the fiber. As a result, the DOP after the compensator measured by Polarimeter 2, DOP_2, is always close to 100% with and without the compensation action. Consequently, the obtained PMD value for the best compensation is less accurate and the corresponding data points can be removed in post data processing.

On the other hand, a small (less than 90%) value of DOP_1 indicates a sufficiently large PMD induced signal distortion and the PMD compensation action is effective in restoring the signal and bringing the DOP_2 back to its maximum. The PMD values obtained for the best compensation for these data points are therefore accurate and can be kept in the data processing.

As expected, this PMD measurement method is less accurate if the PMD in the transmission system is sufficiently small and the corresponding DOP is always larger than 90%, regardless of the polarization launching condition. On the other hand, one usually does not care about small PMD values in the system because their impact on signal transmission is negligible. Therefore this method is very practical for the deployed Long-Haul and Ultra-Long-Haul systems, which has been successfully demonstrated in Verizon's 1500-km ultra-long haul test bed in Richardson, Texas, and in a field trial in a 414-km revenue-generating fiber route.

The PMD source with PMD compensation capability can also be used for system impairment diagnosis. In a fiber link with performance problems, it is sometimes difficult to determine the cause of the problem, whether it is a PMD issue, a chromatic-dispersion issue (CD), a SNR issue, or some other issues. Performing PMD compensation can help to find out whether the problem is caused by PMD or not: If PMD compensation solves the transmission problem, one then knows that PMD is the cause. Otherwise, PMD may not be an issue. With such a diagnosis, not only can one determine whether PMD compensation is required for the fiber link, but also the range of PMD to be compensated, which is helpful for determining the most suitable compensation equipment or parameters to be used, even for coherent detections systems with digital compensation capabilities.

7.5 Polarization Related Tests in Coherent Systems

Coherent detection and PDMux have emerged as key technology enablers for data transmission beyond 40 Gb/s (Ip et al. 2008; Zhou and Xie 2016). The combination can significantly increase the spectral efficiency of a channel and allow each channel to transmit higher bit rates with a relatively small optical bandwidth. Now, 40, 100, or even 160 Gb/s channels can be transmitted using existing 10 Gb/s WDM infrastructure with 50 GHz channel spacing.

PDMux doubles the spectral efficiency by combining two polarization channels of the same bit rate into the same wavelength channel. Coherent detection allows multiple levels of phase and amplitude modulations of lower rates to be multiplexed on the same wavelength channel using advanced modulation schemes. For example, the use of PDMux with QPSK modulation enables 40 Gb/s transmission within the bandwidth of a 10 Gb/s direct channel. The effect of impairments due to PMD and CD is commensurately reduced.

Because coherent detection retains the phase, amplitude, and polarization information of the signal, such systems are able to handle the polarization demultiplexing in the electrical domain using high-speed DSP circuitry and algorithms. Important system impairment correction functions such as CD compensation, polarization mode dispersion compensation (PMDC), and polarization-dependent loss mitigation (PDLM) can also be handled electrically (Xie 2016); however, system performance must always be verified.

7.5.1 Verifying System Performance

For transceiver developers, successful development requires performance evaluation of different DSP circuits and algorithms. For system integrators in the market for transceivers, it is important to compare performance of transceivers from different vendors. Finally, for network operators, evaluation and comparison of the performance of systems from different vendors can help an operator make intelligent purchasing decisions. In all of these cases, the evaluations must include three principal polarization-related functions: polarization demultiplexing, PMD compensation, and PDL mitigation.

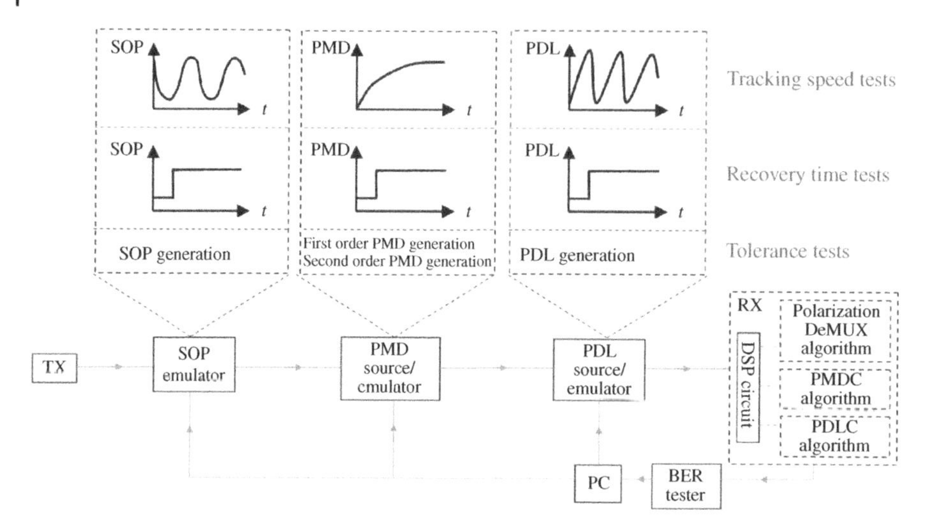

Figure 7.27 A coherent receiver generally includes three polarization-related circuits/algorithms for polarization demultiplexing, PMD compensation, and PDL compensation functions. Three types of emulation equipment are required to generate the different polarization parameters for the complete characterization of these functions, including the PMD/PDL tolerance range; the tracking speed in response to SOP, PMD, or PDL variations; and the recovery time needed to respond to an abrupt change in these parameters.

For a coherent detection system, a general setup can be developed for testing the performances of all polarization-related functions (see Figure 7.27). Note that because polarization impairments vary rapidly with time, the related functions must be tested against variations in different polarization parameters, such as the SOP, PMD, and PDL. To properly test these parameters, it is important to understand the definitions of different polarization-related performance tests for coherent detection systems. These evaluation tests include three basic types: tolerance range, tracking speed, and recovery time (see Figure 7.27). The BER of the system is used as the performance indicator, although other parameters, such as the size of the opening in an eye diagram or the power penalty, can be used as well.

A *tolerance range test* can determine the maximum amount of the impairment under test (PMD or PDL) that can exist in a system before the signal quality degrades beyond an acceptable level. As these impairments increase, the BER increases. The PMD or PDL corresponding to the maximum allowed BER is the system tolerance range as shown in Figure 7.28a.

Unlike CD and SNR, SOP, PMD, and PDL all vary rapidly with time in a real fiber-optic communication system, as discussed in Chapter 3. As shown in Figure 7.28b, a *tracking speed test* determines how rapidly a parameter under test can change before the circuit/algorithm being evaluated can no longer follow the variation, causing the signal quality to degrade beyond a given level. In a tracking speed test using the BER before FEC as the performance indicator, the tracking speed of the circuit/algorithm under test is the variation rate of the parameter under test (such as SOP, PMD, or PDL) corresponding to the maximum allowed BER.

A sudden change in SOP, PMD, or PDL can cause the circuit/algorithm under test to temporarily lose track, resulting in a transient increase in BER. As shown in Figure 7.28c, a *recovery time test* determines how quickly the circuit/algorithm under test can recover from the abrupt change, restoring the BER to an acceptable level. In particular, SOP, PMD, or PDL recovery time is defined as the time t_R required for the DSP to restore BER from an abnormal level caused by a disruptive jump for each of these three parameters.

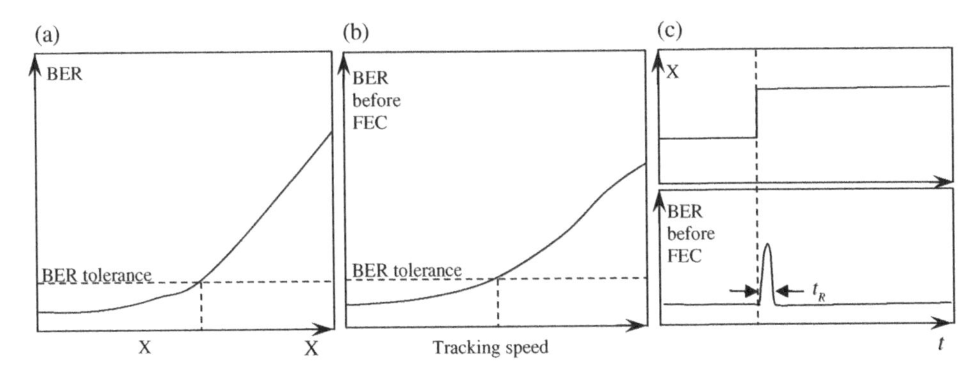

Figure 7.28 Polarization-related performance tests where X represents SOP, PMD, or PDL, including (a) a PMD or PDL tolerance range test; (b) an SOP, PMD, or PDL tracking speed test where the horizontal axis is the changing rate of these three parameters; and (c) recovery time tests where the top graph represents an abrupt change in these parameters. The bottom graph shows the time required for the DSP circuit to regain control of the BER after the step change shown in the top graph.

7.5.2 Polarization Test Instrumentation

The entire ecosystem for coherent detection systems – transceiver designers, system integrators, and network operators – must ensure performance of their solutions, and the proper polarization test instruments must be used to make sure that their system specifications are being quantified and met.

A PMD tolerance range test requires a PMD source and emulator. A personal computer is used to control the source, and a test program can be developed to plot the BER obtained from the BER tester as a function of PMD values, as shown in Figure 7.29a. The source must be able to generate precise, repeatable first- and second-order PMD values with a broad enough range to determine the limits of the compensation circuit/algorithm, as shown in Figure 7.29b,c. Since the effect of PMD on the system is polarization dependent, a polarization scrambler should be used to vary the input polarization to the PMD source in order to make sure that the worst case is covered. The SOP emulator can either be integrated into the PMD source or it can be a separate instrument, and should be able to generate random SOP variations with user-defined speed to uniformly cover the Poincaré sphere. For the polarization optimized PMD source described in Figure 7.23, the polarization controller at the input can be programmed to be such a polarization scrambler. Note that for a polarization multiplexed system, the polarization optimization function of the PMD source cannot be used because of the orthogonal SOP channels involved.

A PDL tolerance range test requires a PDL source and emulator described in Figure 7.16. Analogous to the PMD tolerance range test of Figure 7.29a, the PDL source must be able to generate precise, repeatable PDL values over a range sufficient to determine the PDL limit, as shown in Figure 7.27. A personal computer with a test program can be used to control the generation of the PDL values from low to high and plot the BER as a function of PDL, similar to the PMD tolerance test of Figure 7.29b. Again, an SOP emulator or scrambler should be used in front of the PDL source to ensure that the worst-case SOP is reached for maximum PDL-induced signal distortion.

For an SOP tracking speed test shown in Figure 7.27, the SOP emulator should be able to generate smooth, random SOP variations at a user-defined rate. The BER before FEC is used as the performance indicator to ensure sufficiently high testing speed. Again, a personal computer with a test program can gradually increase the SOP scrambling rate while recording the BER, similar to the

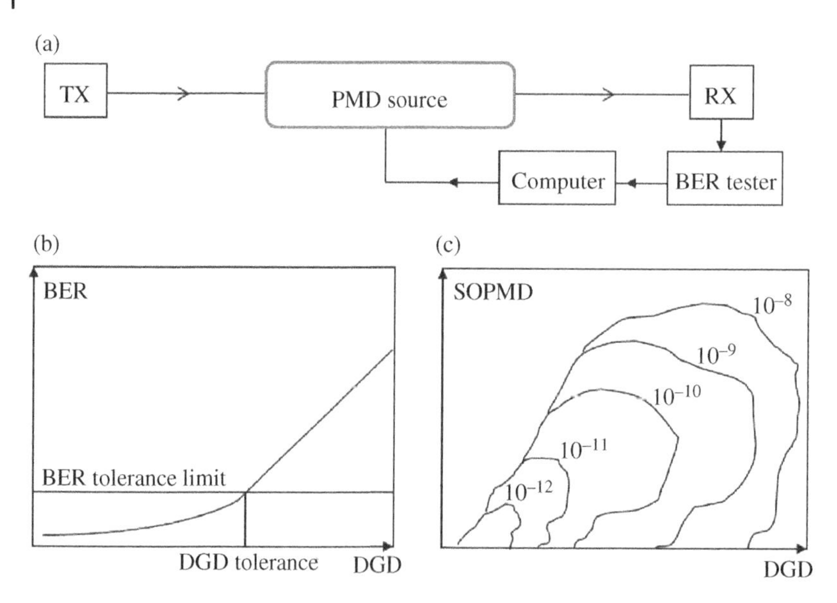

Figure 7.29 (a) Setup for testing the PMD tolerance of a pair of transceiver TX and RX. (b) Exemplary data showing the DGD tolerance limit corresponding to a BER limit. (c) Exemplary data showing the SOPMD tolerance limits corresponding to different BER limits.

PMD tolerance test of Figure 7.29, and finally obtain the plot of BER before FEC vs. SOP scrambling rate and the final SOP tracking speed. The maximum SOP variation rate should be sufficiently high to emulate variations in a real system. As mentioned previously, for PMDC using DSPs in a coherent system, the circuit/algorithm must track SOP variations and rapid PMD changes. To test a PMDC's PMD tracking speed tolerance, the source must be able to generate smooth, fast PMD variation at a user-defined rate and range. A test program can be used to gradually increase the PMD scanning rate while recording the BER before FEC for each rate. The PMD scanning speed required is 100 ps/s or more to ensure statistically realistic coverage of PMD variation speeds in a real system. The continuously rotating MO rotator behind the longest crystal in Figure 7.23b can be used to generate the fast scanning PMD traces shown in Figure 7.25 for the PMD tracking speed test. Finally, polarization scramblers/emulators described in Section 7.2, particularly the quasi-uniform rate polarization scrambler of Figure 7.14, can be used for SOP tracking speed test.

Finally, to test the PDL tracking speed of the PDL compensation circuit/algorithm, a PDL source capable of generating fast PDL variations with user-selectable rate and range, such as that described in Section 7.3, can be used. As with SOP and PMD tracking speed test methods, the PDL tracking speed can be obtained by plotting the BER before FEC vs. PDL scanning rate.

To determine how quickly the circuit/algorithm can recover from an abrupt change in the parameter under test, the emulation instrument must be able to generate such a change significantly faster than the circuit can respond. When the parameter change occurs, the BER may fluctuate as the system loses track. The amount of time the system needs to return the BER to a stable level can then be measured. For an SOP recovery time test, the emulator must generate large SOP discontinuities periodically with rising or falling edges less than the designed response time of the polarization demultiplexing circuits while the BER before FEC is recorded and displayed. Then the SOP recovery time t_R is readily obtained by measuring the duration of the period before BER recovery, as shown in Figure 7.28c. For PMD and PDL recovery time tests, the corresponding sources must similarly be able to generate fast, repeatable PMD or PDL transitions with rising or falling edges less than the specified response time of the PMD or PDL compensation circuits while BER before FEC vs. time is recorded and displayed; PMD and PDL recovery times are obtained from the BER before FEC vs. time plots, as shown in Figure 7.28c.

References

Aarts, W. and Khoe, G. (1989). New endless polarization control method using three fiber squeezers. *J. Lightwave Technol.* 7: 1033–1043.

Boroditsky, M., Brodsky, M., Frigo, N., et al. (2005). Polarization dynamics in installed fiber optic systems. *2005 IEEE LEOS Annual Meeting Conference Proceedings*, 22–28 October, Sydney, Australia, paper TuCC1.

Bradley, E., Miles, E., Loginov, B. et al. (2001). Polarization on the bench. *Photonics Spectra* (November issue): 134–142.

Bulow, H. and Lanne, S. (2005). PMD compensation techniques. In: *Polarization Mode Dispersion* (ed. A. Galtarossa and C. Menyuk), 225–245. Springer Science + Business Media, Inc.

Collett, E. (2003). *Polarized Light in Fiber Optics*. Lincroft, NJ: PolaWave Group.

Damask, J.N. (2005). *Programmable PMD Sources: Polarization Optics in Telecommunications*, 451–477. Springer Science + Business Media, Inc.

Derr, F. (1992). Coherent optical QPSK intradyne system: concept and digital receiver realization. *J. Lightwave Technol.* 10: 1290–1296.

Heffner, B. (1992). Automated PMD measurement using Jones Matrix eigenanalysis. *IEEE Photonics Technol. Lett.* 4 (9): 1066–1069.

Hidayat, A., Koch, B., Zhang, H. et al. (2008). High-speed endless optical polarization stabilization using calibrated waveplates and field-programmable gate array-based digital controller. *Opt. Express* 16 (23): 18984–18991.

Ikeda, K., Takagi, T., Hatano, T. et al. (2003). Endless tracking polarization controller. *Furukawa Rev.* 23: 32–38.

Ip, E., Lau, A., Barros, D., and Kahn, J. (2008). Coherent detection in optical fiber systems. *Opt. Express* 16: 753–791.

Karlsson, M., Xie, C., Sunnerud, H., and Andrekson, P. (2001). Higher-order polarization mode dispersion compensator with three degrees of freedom. *Proceedings of Optical Fiber Communication Conference 2001*, 17–22 March, Anaheim, CA (Optica and IEEE), paper MO1.

Kikuchi, N. (2001). Analysis of signal degree of polarization degradation used as control signal for optical polarization mode dispersion compensation. *J. Lightwave Technol.* 19 (4): 480–486.

Koch, B., Noé, R., Mirvoda, V., and Sandel, D. (2011). 100-krad/s endless polarisation tracking with miniaturised module card. *Electron. Lett.* 47 (14): 813–814.

Kogelinik, H., Jopson, R., and Nelson, L. (2002). Polarization-mode dispersion, Chapter 15. In: *Optical Fiber Communications, IVB* (ed. I. Kaminov and T. Li), 725–861. San Diego: Academic Press.

Krummrich, P.M. and Kotten, K. (2004). Extremely fast (microsecond scale) polarization changes in high speed long hail WDM transmission systems. *Proceedings of Optical Fiber Communication Conference 2004*, 22–27 March, Los Angeles, CA (Optica and IEEE), paper FI3.

Krummrich, P.M., Schmidt, E., Weiershausen, W., and Mattheus, A. (2005). Field trial on statistics of fast polarization changes in long haul WDM transmission systems. *Proceedings of Optical Fiber Communication Conference 2005*, 6–11 March, Anaheim, CA (Optica and IEEE), paper OThT6.

Lanne, S., Idler, W., Thiery, J., and Hamaide, J. (2001). Demonstration of adaptive PMD compensation at 40 Gb/s. *Proceedings of Optical Fiber Communication Conference 2001*, 17–22 March, Anaheim, CA (Optica and IEEE), paper TuP3.

Laperle, C., Villeneuve, B., Zhang, Z., et al. (2007). Wavelength division multiplexing (WDM) and polarization mode dispersion (PMD) performance of a coherent 40 Gbit/s dual-polarization quadrature phase shift keying (DP-QPSK) transceiver. *Proceedings of Optical Fiber Communication Conference 2007*, 25–29 March, Anaheim, CA (Optica and IEEE), paper PDP16.

Leo, P.J., Gray, G.R., Simer, G.J., and Rochford, K.B. (2003). State of polarization changes: classification and measurement. *J. Lightwave Technol.* 21: 2189–2193.

Lize, Y.K., Gomma, R., Kashyap, R. et al. (2007). Fast all-fiber polarization scrambling using re-entrant Lefevre controller. *Opt. Commun.* 279 (1): 50–52.

Martinelli, M., Martelli, P., and Pietralunga, S. (2006). Polarization stabilization in optical communications systems. *J. Lightwave Technol.* 24 (11): 4172–4183.

Möller, L., Sinsky, J., Chandrasekhar, S. et al. (2003). A tunable interferometrically stable three-section higher order PMD emulator. *IEEE Photonics Technol. Lett.* 15 (2): 230–232.

Nagel, J., Chbat, M., Garrett, L. et al. (2000). Long-term PMD mitigation at 10 Gb/s and time dynamics over high-PMD installed fiber. *Proc. ECOC.* 2: 31.

Nelson, L.E. and Jopson, R.M. (2005). Introduction to polarization mode dispersion in optical systems. In: *Polarization Mode Dispersion* (ed. A. Galtarossa and C. Menyuk), 1–33. Springer Science + Business Media, Inc.

Noe, R., Heidrich, H., and Hoffmann, D. (1988). Endless polarization control systems for coherent optic. *J. Lightwave Technol.* 6 (7): 1199–1208.

Peterson, D.L., Leo, P.J., and Rochford, K.B. (2004). Field measurements of state of polarization and PMD from a tier-1 carrier. *Proceedings of Optical Fiber Communication Conference 2004*, 22–27 March, Los Angeles, CA (Optica and IEEE), paper FI1.

Rosenfeldt, H., Knothe, C., Ulrich, R., and Brinkenmeyer, E. (2001). Automatic PMD compensation at 40 Gbits/s and 80 Gbits/s using a 3 dimensional DOP evaluation for the feedback. *Proceedings of Optical Fiber Communication Conference 2001*, 17–22 March, Anaheim, CA (Optica and IEEE).

Taylor, M.G. (2004). Coherent detection method using DSP for demodulation of signal and subsequent equalization of propagation impairments. *IEEE Photonics Technol. Lett.* 16 (2): 674–676.

Walker, N.G. and Walker, G.R. (1987). Endless polarization control using four fiber squeezers. *Electron. Lett.* 23 (6): 290–292.

Walker, N.G., Walker, G.R., and Davison, J. (1988). Endless polarization control using an integrated optical lithium niobate device. *Electron. Lett.* 24 (5): 266–268.

Willner, A.E. and Hauer, M.C. (2005). PMD emulation. In: *Polarization Mode Dispersion* (ed. A. Galtarossa and C. Menyuk), 277–296. Springer Science + Business Media, Inc.

Wree, C., Becker, D., Mohr, D., and Joshi, A. (2007). Coherent receiver for phase-shift keyed transmission. *Proceedings of Optical Fiber Communication Conference 2007*, 25–29 March, Anaheim, CA (Optica and IEEE), paper OMP6.

Xie, C. (2016). Polarization and nonlinear impairments in fiber communication systems. In: *Enabling Technologies for High Spectral-Efficiency Coherent Optical Communication Networks* (ed. X. Zhou and C. Xie), 201–246. Hoboken, NJ: Wiley.

Yan, L. and Yao, X.S. (2002). Variable polarization-dependent-loss source. US Patent 6,975,454, filed 31 July 2002 and issued 13 December 2005.

Yan, L., Yeh, C., Yang, G. et al. (2003). Programmable group delay module using binary polarization switching. *J. Lightwave Technol.* 21 (7): 1676–1684.

Yan, L., Hauer, M., Shi, Y. et al. (2004). Polarization-mode-dispersion emulator using variable differential-group-delay (DGD) elements and its use for experimental importance sampling. *J. Lightwave Technol.* 22 (4): 1051–1058.

Yan, L., Yu, Q., and Willner, A. (2005). Uniformly distributed states of polarization on the Poincare sphere using an improved polarization scrambling scheme. *Opt. Commun.* 249: 43–50.

Yan, L., Yao, X.S., Hauer, M., and Willner, A. (2006). Practical solutions to polarization-mode-dispersion emulation and compensation. *J. Lightwave Technol.* 24 (11): 3992–4005.

Yao, X.S. (1995). Compact programmable photonic variable delay devices. US Patent 5,978,125, filed 30 November 1995 and issued 2 November 1999.

Yao, X.S., Yan, L., Zhang, B. et al. (2007). All-optic scheme for automatic polarization division demultiplexing. *Opt. Express* 15 (12): 7407–7414.

Yao, X.S., Chen, X., Xia, T.J. et al. (2010). In-service light path PMD (polarization mode dispersion) monitoring by PMD compensation. *Opt. Express* 18 (26): 27306–27318.

Yao, L., Huang, H., Chen, X. et al. (2012). A novel scheme for achieving quasi-uniform rate polarization scrambling at 752 krad/s. *Opt. Express* 20 (2): 1691–1699.

Zhou, X. and Xie, C. (2016). Introduction. In: *Enabling Technologies for High Spectral-Efficiency Coherent Optical Communication Networks* (ed. X. Zhou and C. Xie), 1–11. Hoboken, NJ: Wiley.

8

Polarization Related Measurements for Optical Fiber Systems

Polarization measurements generally include the determination of the polarization properties of the light source itself, such as the state of polarization (SOP) and degree of polarization (DOP), and the polarization properties of a medium, such as the birefringence (or polarization mode dispersion [PMD]) and diattenuation (or polarization dependent loss [PDL]). An instrument for measuring the polarization properties of light is called Stokes polarimeter, while an instrument for determining not only the polarization properties of a light source, but also those of an optical medium, is often referred to as the Mueller matrix polarimeter.

A Stokes polarimeter is complete if it is capable of obtaining all four Stokes parameters which fully describe the polarization properties of light, and incomplete if obtaining less than four. Some practical parameters, such as the polarization extinction ratio (PER), can be obtained from incomplete Stokes polarimeters, or be directly measured without the need to deliberately obtain the Stokes parameters.

Similarly, a Mueller matrix polarimeter is called complete if it is capable of measuring all 16 elements in the Mueller matrix of a medium (or device), and incomplete otherwise. A complete Mueller matrix contains all polarization related information of the medium. However, some particular polarization properties of the medium, such as PDL, can also be directly measured without evoking Mueller matrix, as will be discussed in Section 8.4.3.

8.1 Stokes Polarimeters for SOP and DOP Measurements

As described in Chapters 2 and 4, when defining the Stokes parameters of a light beam, four different polarizers with its passing axis set at some specific azimuthal angles are sequentially inserted in the optical beam before the optical powers of the beam corresponding to each polarizer are measured with a photodetector (PD), one at a time, as shown in Figure 8.1. The photocurrents corresponding to these powers are then used to compute the Stokes parameters. Such a polarimeter involving sequentially measuring the optical powers in response to different polarization components in the optical beam is called time-division polarimeter, as will be discussed in detail in Section 8.1.1.

The polarization properties of a light beam can also be measured by first dividing the incoming beam into four sub-beams using two or more beam splitters, and then simultaneously analyzing the optical powers in these beams with different polarizers. Such a polarimeter based on the power splitting is called amplitude-division polarimeter, which will be described in Section 8.1.2.

Polarization Measurement and Control in Optical Fiber Communication and Sensor Systems, First Edition.
X. Steve Yao and Xiaojun (James) Chen.

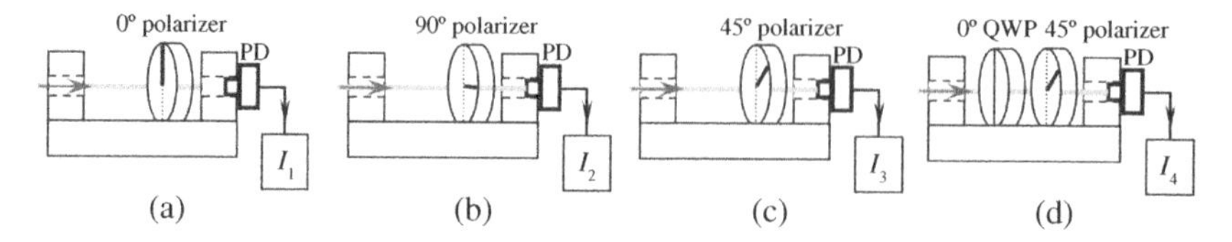

Figure 8.1 The first Stokes polarimeter with a structure for sequentially inserting different polarization elements in the optical beam before measuring the beam powers. (a) A polarizer at 0°; (b) rotate the polarizer 90°; (c) rotate the polarizer 45°; (d) inserting a QWP at 0° before the 45° polarizer. PD: photodetector; I_1, I_2, I_3, I_4: Photocurrent.

One may also spatially divide a light beam into four sub-beams and then simultaneously analyze the powers of the four sub-beams with four different polarizers. The corresponding polarimeter is called wave-front division polarimeters and is discussed in detail in Section 8.1.3.

Let $\mathbf{I} = (I_1, I_2, I_3, I_4)^T$ be the vector of the photocurrent measured with a PD at four different times or with four different PDs at the same time, and $\mathbf{S} = (S_0, S_1, S_2, S_3)^T$ be the Stokes vector of the optical beam to be measured, where T represents the transpose, then we have the following expression relating **I** and **S**:

$$\mathbf{I} = \mathbf{A}\mathbf{S} \tag{8.1a}$$

where **A** is a matrix whose elements depend on the optical system and is called the polarimeter matrix. Once **A** is determined via measurements, calculations, and calibration, as will be described next, the Stokes vector of the optical beam can be uniquely determined by

$$\mathbf{S} = \mathbf{A}^{-1}\mathbf{I} \tag{8.1b}$$

where $\mathbf{A}^{-1}$ is the inverse of **A**.

When the Stokes vector is determined, the DOP can then be obtained by

$$DOP = \frac{\sqrt{S_1^2 + S_2^2 + S_3^2}}{S_0} \tag{8.1c}$$

8.1.1 Time Division Stokes Polarimetry

The first Stokes polarimeter is that due to George Gabriel Stokes, which is also a time-division polarimeter. As shown in Figure 8.1, one makes the polarization measurement of an optical beam by first inserting a linear polarizer in the beam to measure the photocurrent I_1 of light passing through the polarizer with a detector. Second, the polarizer is rotated 90° around the beam with respect to the initial orientation before measuring the photocurrent I_2 with the detector. Third, the polarizer is rotated again to be 45° from its initial orientation to measure the photocurrent I_3. Finally, a quarter-wave plate (QWP) with its optical axis aligned 45° from the polarizer is inserted before the polarizer to form a circular polarizer to measure the photocurrent I_4 of the beam passing through the circular polarizer with the detector. By definition, the Stokes parameters can be expressed as (see related discussion in Chapter 4)

$$\mathbf{S} = \begin{bmatrix} S_0 \\ S_1 \\ S_2 \\ S_3 \end{bmatrix} = \begin{bmatrix} I_1 + I_2 \\ I_1 - I_2 \\ 2I_3 - (I_1 + I_2) \\ 2I_4 - (I_1 + I_2) \end{bmatrix} = \begin{bmatrix} +1 & +1 & 0 & 0 \\ +1 & -1 & 0 & 0 \\ -1 & -1 & 2 & 0 \\ -1 & -1 & 0 & 2 \end{bmatrix} \begin{bmatrix} I_1 \\ I_2 \\ I_3 \\ I_4 \end{bmatrix} \tag{8.2a}$$

Comparison with Eq. (8.1b) obtains

$$\mathbf{A}^{-1} = \begin{bmatrix} +1 & +1 & 0 & 0 \\ +1 & -1 & 0 & 0 \\ -1 & -1 & 2 & 0 \\ -1 & -1 & 0 & 2 \end{bmatrix} \tag{8.2b}$$

As one may notice that such a polarimeter is very slow for polarization measurement, which requires that the SOP and DOP of the optical beam be stable during the measurement.

8.1.1.1 Rotating Element Polarimetry

In order to simplify and speed up the measurement process, different schemes involving rotating one or more polarization elements have been developed. Some of these are incomplete while others are complete polarimeters. Since there have been already good books covering these rotating element polarimeters (Goldstein 2003), only brief reviews are given to such polarimeters in this book. More serious readers are encouraged to read the related books and journal publications listed at the end of this chapter.

Polarimeter with a Rotating Analyzer As shown in Figure 8.2a, this Stokes polarimeter is incomplete, which can only obtain three Stokes parameters S_0, S_1, and S_2, but not S_3 and DOP. To enable fast polarization measurements, a hollow core motor can be used to rotate the polarizer at a constant angular speed ω, with the azimuthal angle θ of the polarizer satisfying:

$$\theta(t) = \theta_0 + \omega t \tag{8.3a}$$

where θ_0 is the initial polarizer azimuthal angle. As the polarizer is rotated, the measured photocurrent $I(\theta)$ with the PD can be expressed as

$$I(\theta) = \frac{1}{2}[a_0 + a_2 cos2\theta(t) + b_2 sin2\theta(t)] \tag{8.3b}$$

One may obtain the Fourier coefficients a_0, a_2, and b_2 from the Fourier analysis of the photocurrent. According to the discussions in Chapter 4 (Table 4.3), the Mueller matrix of a polarizer at an azimuthal angle θ can be expressed as

$$\mathbf{M}_{pol}(\theta) = \frac{1}{2}\begin{bmatrix} 1 & cos2\theta & sin2\theta & 0 \\ cos2\theta & cos^2 2\theta & sin2\theta cos2\theta & 0 \\ sin2\theta & sin2\theta cos2\theta & sin^2 2\theta & 0 \\ 0 & 0 & 0 & 0 \end{bmatrix} \tag{8.4}$$

The SOP of light $\mathbf{S}'$ after the polarizer is therefore

$$\mathbf{S}' = \mathbf{M}_{pol}(\theta)\mathbf{S} \tag{8.5}$$

where $\mathbf{S}$ is the Stokes vector of the light beam to be measured. Substituting Eq. (8.4) in Eq. (8.5), one obtains

$$I(\theta) = S_0' = \frac{1}{2}(S_0 + S_1 cos2\theta + S_2 sin2\theta) \tag{8.6a}$$

By equating Eq. (8.6a) with Eq. (8.3b), one obtains

$$\begin{aligned} S_0 &= a_0 \\ S_1 &= a_2 \\ S_2 &= b_2 \end{aligned} \tag{8.6b}$$

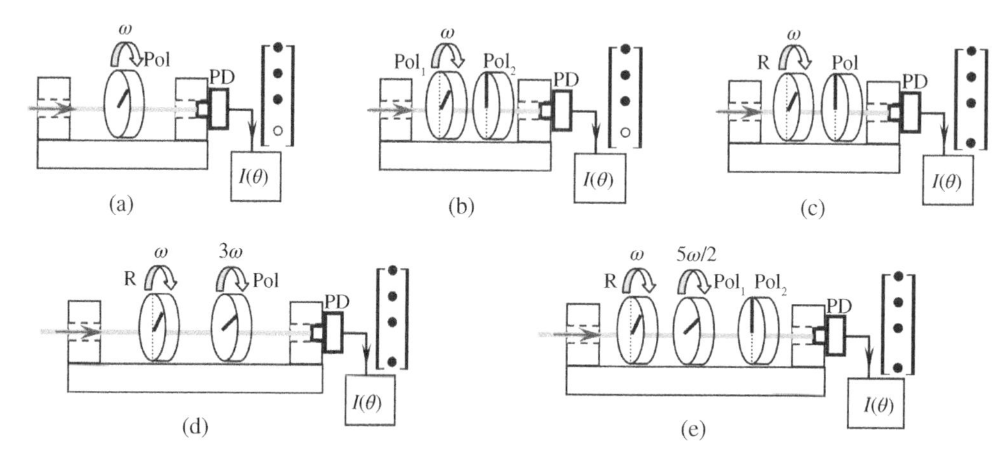

Figure 8.2 Different configurations of rotating element polarimeters. (a) An incomplete Stokes polarimeter with a rotating analyzer; (b) An incomplete Stokes polarimeter with a rotating analyzer and a fixed analyzer; (c) A complete Stokes polarimeter with a rotating retarder and a fixed analyzer; (d) A complete Stokes polarimeter with a rotating retarder and analyzer; (e) A complete Stokes polarimeter with a rotating retarder, a rotating analyzer, and a fixed analyzer. ω: rotation angular speed; R: retarder; Pol, Pol_1, Pol_2: polarizers used as polarization analyzers; PD: photodetector. The vector at the right hand side of each configuration is the Stokes vector that can be obtained. The dark circle represents obtainable vector element, while the clear circle indicates that the element cannot be obtained with the corresponding configuration.

Polarimeter with a Rotating and a Fixed Analyzer As described in Chapter 3, a PD generally has polarization dependent responsivity (PDR), which causes measurement errors. To overcome this short coming, a fixed polarizer is placed in front of the PD, as shown in Figure 8.2b. With this polarizer, the photocurrent by the PD can be expressed as the following Fourier series:

$$I(\theta) = \frac{a_0}{4} + \frac{1}{4}\sum_{n=1}^{2}(a_{2n}cos2n\theta + b_{2n}sin2n\theta) \tag{8.7a}$$

The Stokes vector $\mathbf{S}'$ after the fixed the polarizer can be related with the input Stokes vector $\mathbf{S}$ as

$$\mathbf{S}' = \mathbf{M}_{pol}(0)\mathbf{M}_{pol}(\theta)\mathbf{S} \tag{8.7b}$$

where $\mathbf{M}_{pol}(0)$ and $\mathbf{M}_{pol}(\theta)$ are the Mueller matrices of the rotating and the fixed polarizers at the azimuthal angles of θ and 0 (Table 4.3), respectively. Similar to Eq. (8.6a), by equating $I(\theta)$ with S_0', one obtains the first three elements of the Stokes vector of the light beam in terms of the Fourier coefficients in Eq. (8.7a):

$$\begin{aligned} S_0 &= a_0 - a_4 \\ S_1 &= \frac{2}{3}(a_2 - a_0 + 2a_4) \\ S_2 &= \frac{2}{5}(2b_2 + b_4) \end{aligned} \tag{8.7c}$$

Polarimeter with a Rotating Retarder and a Fixed Analyzer A complete Stokes polarimeter requires at least one retarder, and Figure 8.2c shows such a polarimeter consisting of a rotating retarder, followed by a fixed polarizer before a PD to eliminate the degrading effect of the PDR of the PD. Again, the detected photocurrent can be expressed as a Fourier series:

$$I(\theta) = \frac{a_0}{2} + \frac{1}{2}\sum_{n=1}^{2}(a_{2n}cos2n\theta + b_{2n}sin2n\theta) \tag{8.8a}$$

where θ is the azimuthal angle of the retarder. In most cases, the retarder is a QWP and the Stokes vector $\mathbf{S}'$ after the fixed the polarizer can be expressed as

$$\mathbf{S}' = \mathbf{M}_{pol}(0)\mathbf{M}_{QWP}(\theta)\mathbf{S} \tag{8.8b}$$

where $\mathbf{M}_{QWP}(\theta)$ is the Mueller matrix of the rotating QWP at an arbitrary azimuthal angle θ and $\mathbf{M}_{pol}(0)$ is the Mueller matrix of the fixed polarizer at 0 azimuthal angle (Table 4.3). Again, by equating $I(\theta)$ with S_0', one obtains all four elements of the Stokes vector of the light beam in terms of the Fourier coefficients in Eq. (8.8a):

$$\begin{aligned} S_0 &= a_0 - a_4 \\ S_1 &= 2a_4 \\ S_2 &= 2b_4 \\ S_3 &= b_2 \end{aligned} \tag{8.8c}$$

Polarimeter with a Rotating Retarder and Analyzer In this configuration, as shown in Figure 8.2d, both the retarder and the polarizer are being rotated. For the case where the angular speed of the polarizer is three times that of the retarder, the detected photocurrent is given by

$$I(\theta) = \frac{a_0}{2} + \frac{1}{2}\sum_{n=1}^{3}(a_{2n}cos2n\theta + b_{2n}sin2n\theta) \tag{8.9a}$$

where θ is the retarder azimuthal angle and a_{2n} and b_{2n} ($n = 0, 1, 2, 3$) are the Fourier coefficients, which can be obtained from the Fourier analysis of the PD current. When the retarder is a QWP, the Stokes vector $\mathbf{S}'$ after the rotating polarizer can be expressed as

$$\mathbf{S}' = \mathbf{M}_{pol}(3\theta)\mathbf{M}_{QWP}(\theta)\mathbf{S} \tag{8.9b}$$

where $\mathbf{S}$ is the Stokes vector to be measured. Again, equating $I(\theta)$ with S_0', one obtains all Stokes parameters of $\mathbf{S}$ as

$$\begin{aligned} S_0 &= a_0 \\ S_1 &= a_2 + a_6 \\ S_2 &= b_6 - b_2 \\ S_3 &= b_4 \end{aligned} \tag{8.9c}$$

Polarimeter with a Rotating Retarder and Analyzer plus a Fixed Analyzer This configuration, as shown in Figure 8.2e, while similar to that of Figure 8.2d, however, has another fixed polarizer following the rotating polarizer for eliminating the PDR of the PD. When the rotating polarizer is rotated at an angular speed of 5/2, 5/3, or −3/2 times that of the retarder, the detected photocurrent is then

$$I(\theta) = \frac{a_0}{4} + \frac{1}{4}\sum_{\substack{n=1 \\ n\neq 9}}^{10}(a_n cosn\theta + b_n sinn\theta) \tag{8.10a}$$

where θ is the azimuthal angle of the retarder, and a_n and b_n ($n = 0, 1, \ldots 8, 10$) are the Fourier coefficients of the photocurrent. When the retarder is a QWP, the angular speed factor is 5/2, and the fixed polarizer is at zero azimuthal angle, the Stokes vector $\mathbf{S}'$ after the fixed polarizer can be related to the input Stokes vector $\mathbf{S}$ by

$$\mathbf{S}' = \mathbf{M}_{pol}(0)\mathbf{M}_{pol}\left(\frac{5\theta}{2}\right)\mathbf{M}_{QWP}(\theta)\mathbf{S} \tag{8.10b}$$

Finally, equating $I(\theta)$ with S_0', one obtains all Stokes parameters of **S** as

$$\begin{aligned} S_0 &= a_0 - a_4 \\ S_1 &= 2a_1 \\ S_2 &= 2b_1 \\ S_3 &= b_3 \end{aligned} \tag{8.10c}$$

8.1.1.2 Oscillating Element Stokes Polarimetry

The Stokes polarimeters described earlier involve physically rotating polarization elements with motors, particularly polarizers and wave plates, which is slow and has mechanical wear and tear affecting the measurement accuracy over time. To overcome these shortcomings, a magneto-optic or Faraday polarization rotator described in Figure 6.25a, with MO garnets of planar anisotropy discussed in Section 6.6.2.2, can be used. In particular, an SOP rotation angle of θ by a MO rotator has the effect of mechanically rotating all subsequent elements by an angle of $-\theta$. Because the MO rotators have limited rotation range, continuous rotation in one direction like a motor is not possible. Therefore, one may rotate the SOP back and forth with a sinusoidal or saw-wave modulation signal. With the MO rotators, the SOP rotation rate can be much higher than what a motor can ever achieve, resulting in much higher measurement speed, without suffering the mechanical wear and tear. The only drawback is that with the sinusoidal modulation, the detected signal contains an infinite number of harmonics and therefore complicates the signal analysis. In particular, the amplitudes of the harmonics from a sinusoidal modulation of an angular frequency ω can be expressed as

$$sin(\theta sin\omega t) = 2\sum_{n=0}^{\infty} J_{2n+1}(\theta) sin[(2n+1)\omega t] \tag{8.11a}$$

$$cos(\theta sin\omega t) = J_0(\theta) + 2\sum_{n=0}^{\infty} J_{2n}(\theta) cos 2n\omega t \tag{8.11b}$$

where $J_n(\theta)$ is the nth order Bessel function of the first kind and n is an integer.

Figure 8.3 shows three different configurations of the oscillating polarization elements enabled by MO rotations driven by sinusoidal or saw-wave currents, in which ω, Pol, and R represent angular frequency, polarizer, and retarder, respectively.

Oscillating Analyzer Polarimeter As shown in Figure 8.3a, an MO is placed in front of a polarizer for modulating the SOP entering the polarizer at a frequency of ω, which is equivalent to modulate the polarizer at a frequency of $-\omega$. Alternatively, one may rotate the polarizer back and forth with a hollow shaft motor, however, with the disadvantage of slow speed and mechanical wear and tear. Similar to the rotating polarizer polarimeter described in Figure 8.2a, this polarimeter is incomplete, since it only measures the first three components in the Stokes vector (Azzam 1976).

Modulating an MO rotator rotates the SOP of light entering the polarizer behind it, with an effective polarizer azimuth angle θ of

$$\theta = \theta_0 + \theta_r sin\omega t \tag{8.12}$$

where θ_0 is angle determined by the polarizer azimuth angle and MO cell's DC bias and θ_r is the modulation amplitude produced by the MO rotator cell. Substituting Eqs. (8.11a), (8.11b), and (8.12) into Eq. (8.6a), one obtains

$$I(t) = \frac{1}{2}I_0 + \frac{1}{2}I_1 sin\omega t + \frac{1}{2}I_2 cos2\omega t \tag{8.13a}$$

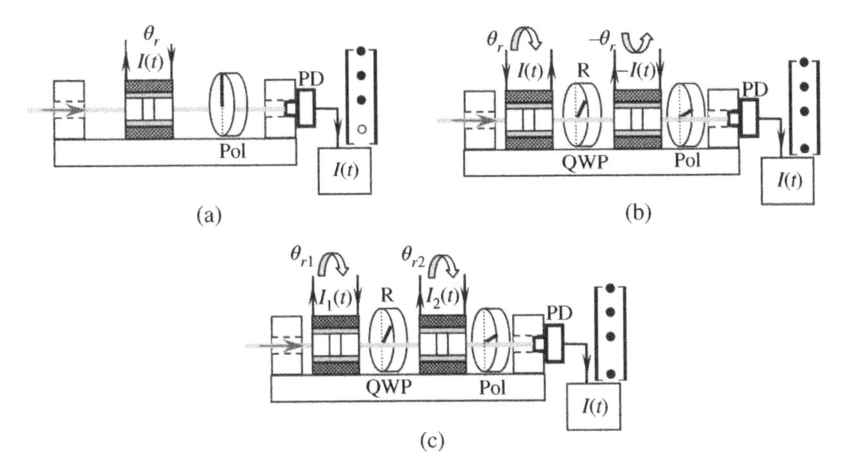

Figure 8.3 Stokes polarimeters with equivalent oscillating polarization elements enabled by MO rotators. (a) An MO rotator being modulated at an angular frequency of ω to effectively oscillate the polarizer (Pol) behind it. (b) An MO rotator being modulated with an amplitude of θ_r in front of a retarder (R) to effectively oscillate the retarder, followed by a second MO rotator in front of a polarizer being modulated with the same amplitude but opposite phase to effectively oscillate the polarizer. (c) An MO rotator being modulated with an amplitude of θ_{r1} in front of a retarder (R) to effectively oscillate the retarder, followed by a second MO rotator being modulated with an amplitude of θ_{r2} in front of a polarizer to effectively oscillate the polarizer. In most cases, the retarder is chosen to be a QWP. In (b) and (c), the modulation frequency ω of the two MO rotators is the same.

where the terms with frequencies higher than 2ω are neglected, and I_0, I_1, and I_2 are the amplitudes of the DC, fundamental frequency, and second harmonic components in the detected signal, respectively:

$$
\begin{aligned}
I_0 &= [S_0 + J_0(2\theta_r)(S_1 cos2\theta_0 + S_2 sin2\theta_0)] \\
I_1 &= 2J_1(2\theta_r)(-S_1 sin2\theta_0 + S_2 cos2\theta_0) \\
I_2 &= 2J_2(2\theta_r)(S_1 cos2\theta_0 + S_2 sin2\theta_0)
\end{aligned}
\tag{8.13b}
$$

Making $\theta_0 = 0$ by adjusting the polarizer azimuth and MO cell's DC bias and setting $2\theta_1 = 137.8°$ so that $J_0(2\theta_r) = 0$ by adjusting the modulation amplitude, one obtains

$$
\begin{aligned}
S_0 &= I_0 \\
S_1 &= \frac{I_2}{[2J_2(2\theta_r)]} \\
S_2 &= \frac{I_1}{[2J_1(2\theta_r)]}
\end{aligned}
\tag{8.13c}
$$

Note that I_0, I_1, and I_2 can be obtained by Fourier analysis of the detected signal $I(t)$.

Polarimeter with an Oscillating Retarder and a Fixed Analyzer As shown in Figure 8.3b, this polarimeter has two MO rotators that sandwich a retarder in between, followed by a fixed angle polarizer. The two MO rotators have equal but opposite SOP rotations. As discussed in Chapter 4, a retarder in between two rotators is equivalent to a rotating retarder, and therefore Figure 8.3b is a complete Stokes polarimeter and is equivalent to that described in Figure 8.2c consisting of a rotation retarder

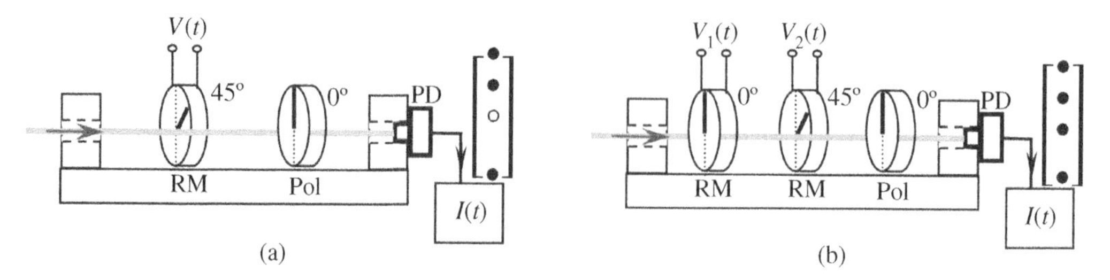

Figure 8.4 Configurations for retardation modulation polarimetry. (a) Incomplete Stokes polarimeter with single modulator and a fixed polarizer; (b) Complete Stokes polarimeter with double modulators and a fixed polarizer. RM: retardation modulator; Pol: polarizer.

followed by a fixed polarizer. The detailed analysis will not be provided in this book and interested readers are encouraged to read (Azzam 1977; Goldstein 2003).

Polarimeter with an Oscillating Retarder and an Analyzer As shown in Figure 8.3c, this complete Stokes polarimeter is almost identical to that described in Figure 8.3b, except that the two MO rotators rotate with different angular frequency ω_1 and ω_2. This configuration is equivalent to physically rotating both the retarder and the following polarizer, as in Figure 8.2d. Again, for detailed analysis, please refer to Azzam (1977) and Goldstein (2003).

8.1.1.3 Retardation Modulation polarimetry

This type of polarimeters is based on modulating the retardations of one or two variable wave plates to periodically alter the SOP passing through the wave plates before being analyzed by a polarizer, as shown in Figure 8.4 (Goldstein 2003). The variable wave plate can be those made with electro-optic crystals described in Figure 6.15a or even the fiber squeezers in Figure 6.15b. Alternatively, liquid crystal and photoelastic variable retardation modulators (RMs) made with quartz crystals can also be used. In Figure 8.4, longitudinal mode retardation modulators (electrical field is along the direction of the light beam) are illustrated, although transverse mode retardation modulators (electrical field is perpendicular to the direction of the light beam) can also be used.

Polarimeter with a Single Retardation Modulator As shown in Figure 8.4a, this polarimeter is consisted of a single retardation modulator (RM) and a fixed angle polarizer, with a 45° relative azimuthal orientation from each other.

The output Stokes vector $\mathbf{S}' = \left(S'_0, S'_1, S'_2, S'_3\right)^T$ from the polarizer can be related to that of the input $\mathbf{S} = (S_0, S_1, S_2, S_3)^T$ by:

$$\mathbf{S}' = \mathbf{M}_{pol}(0^\circ)\mathbf{M}_{RM}(45^\circ)\mathbf{S} \tag{8.14a}$$

where $\mathbf{M}_{RM}(45^\circ)$ is the Mueller matrix of the retardation modulator at a 45° azimuthal angle and $\mathbf{M}_{pol}(\theta_p)$ is the Mueller matrix of at θ_p azimuthal ($\theta_p = 0$). The detected signal is

$$I(t) = S'_0(t) = \frac{1}{2}S_0 + \frac{1}{2}S_1 cos\Delta(t) + S_3 sin\Delta(t) \tag{8.14b}$$

where $\Delta(t)$ is the retardation modulation. For a sinusoidal modulation with amplitude of δ and an angular frequency of ω, $\Delta(t)$ can be expressed as

$$\Delta(t) = \delta sin\omega t \tag{8.14c}$$

From Eqs. (8.11a) and (8.11b), one obtains after neglecting the higher order harmonics

$$cos\Delta(t) = cos(\delta sin\omega t) \approx J_0(\delta) + 2J_2(\delta)cos2\omega t \tag{8.15a}$$

$$sin\Delta(t) = sin(\delta sin\omega t) \approx 2J_1(\delta)sin\omega t \tag{8.15b}$$

The substitution of Eqs. (8.15a) and (8.15b) in Eq. (8.14b) yields

$$I(t) = \frac{1}{2}I_0 + \frac{1}{2}I_1 sin\omega t + \frac{1}{2}I_2 cos2\omega t \tag{8.16a}$$

where I_0, I_1, and I_2 are amplitudes of the DC, first harmonic, and second harmonic terms given by

$$\begin{aligned} I_0 &= S_0 + S_1 J_0(\delta) \\ I_1 &= 2J_1(\delta)S_3 \\ I_2 &= 2S_1 J_2(\delta) \end{aligned} \tag{8.16b}$$

Adjusting the modulation depth $\delta = 137.8°$ to make $J_0(\delta) = 0$, one obtains

$$\begin{aligned} S_0 &= I_0 \\ S_1 &= \frac{I_2}{[2J_2(\delta)]} \\ S_3 &= \frac{I_1}{[2J_1(\delta)]} \end{aligned} \tag{8.16c}$$

Polarimeter with Double Retardation Modulators As shown in Figure 8.4b, this polarimeter consists of two retardation modulators oriented 45° from each other, followed by a polarizer oriented in the same azimuth as the first modulator. The detected signal $I(t)$ can be expressed as

$$I(t) = S'_0(t) = \frac{1}{2}S_0 + \frac{1}{2}S_1 cos\Delta_2(t) + \frac{1}{2}S_2 sin\Delta_2(t) sin\Delta_1(t) - S_3 sin\Delta_2(t) cos\Delta_1(t) \tag{8.17a}$$

where $\Delta_1(t) = \delta_1 sin\,\omega_1 t$ and $\Delta_2(t) = \delta_2 sin\,\omega_2 t$. After setting $\delta_1 = \delta_2 = 137.8°$ to make $J_0(\delta_1) = J_0(\delta_2) = 0$, one obtains

$$I(t) = \frac{1}{2}[I_0 + I_1 cos2\omega_2 t \pm I_2 cos(\omega_2 \pm \omega_1)t + I_3 sin(\omega_2 \pm \omega_1)t] \tag{8.17b}$$

I_0, I_1, I_2, I_3, and I_4 are amplitudes of the DC, $cos2\omega_2 t$, $cos(\omega_2 \pm \omega_1)t$, and $sin(\omega_2 \pm \omega_1)t$ terms, which are related to the Stokes elements:

$$\begin{aligned} S_0 &= I_0 \\ S_1 &= \frac{I_1}{[2J_2(\delta_2)]} \\ S_2 &= \frac{I_2}{[2J_1(\delta_1)J_1(\delta_2)]} \\ S_3 &= \frac{-I_3}{[2J_1(\delta_1)J_1(\delta_2)]} \end{aligned} \tag{8.18}$$

8.1.2 Amplitude Division Polarimeters

In the time-division Stokes polarimeters described in Section 8.1.1, only one PD is used to sequentially analyze the SOP and DOP of a light beam to obtain partial or complete Stokes parameters in the Stokes vector describing the polarization properties of the beam. Such a device requires that the

SOP of light is stationary during the time for completing the measurement and therefore cannot be used to characterize the polarization properties of a light beam with varying polarization.

In this section, we will briefly review amplitude-division polarimeters in which beam splitting components are used to split an input beam into multiple beams to be simultaneously analyzed by multiple PDs (Azzam 1982). Such devices are generally of high speed and are ideal for monitoring fast polarization variations in an optical fiber system.

8.1.2.1 Beam Splitter Based Amplitude Division Polarimeters

PBS Based Polarimeter As shown in Figure 8.5a, this polarimeter comprised a partial polarization beam splitter (PPBS) and two regular polarization beam splitters (PBSs) (Gamiz 1997). The input light beam is first split by the PPBS into two beams, the transmitted beam passes through a QWP oriented at 45° before being split by PBS1 to be detected by PD1 and PD2. The reflected beam from PPBS first passes through a half-wave plate (HWP) oriented at 22.5° before being split by PBS2 to be detected by PD3 and PD4. Let the input SOP be

$$\mathbf{E}_{in} = \sqrt{P_0}(cos\theta\hat{\mathbf{s}} + sin\theta e^{i\delta}\hat{\mathbf{p}}) \tag{8.19a}$$

where P_0 is the optical power, $\hat{\mathbf{s}}$ and $\hat{\mathbf{p}}$ are the unit vectors of the "s" and "p" polarizations, respectively, θ is the azimuthal angle between a principal axis of the input polarization ellipse and $\hat{\mathbf{s}}$, and δ is the phase difference between "s" and "p." The optical powers P_1, P_2, P_3, P_4 detected at the four PDs can be written from Jones analysis (assuming PPBS does not add a differential phase between "s" and "p") as:

$$\begin{aligned}
P_1 &= \frac{1}{2}P_0\left(\frac{4}{5}cos^2\theta + \frac{1}{5}sin^2\theta + \frac{2}{5}sin\delta sin2\theta\right)\\
P_2 &= \frac{1}{2}P_0\left(\frac{4}{5}cos^2\theta + \frac{1}{5}sin^2\theta - \frac{2}{5}sin\delta sin2\theta\right)\\
P_3 &= \frac{1}{2}P_0\left(\frac{1}{5}cos^2\theta + \frac{4}{5}sin^2\theta + \frac{2}{5}cos\delta sin2\theta\right)\\
P_4 &= \frac{1}{2}P_0\left(\frac{1}{5}cos^2\theta + \frac{4}{5}sin^2\theta - \frac{2}{5}cos\delta sin2\theta\right)
\end{aligned} \tag{8.19b}$$

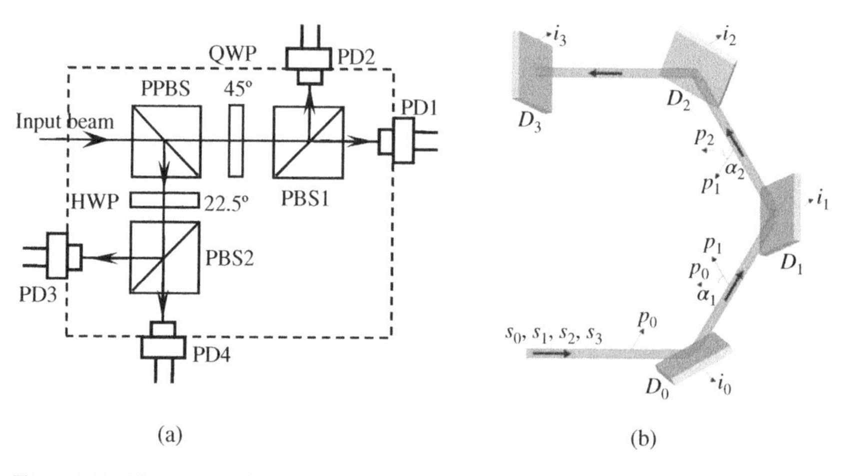

Figure 8.5 Illustration of beam splitter based amplitude division polarimeters. (a) Splitting light with partial PBS (PPBS) and two regular PBS cubes; (b) Splitting light by reflection with photodetector's reception surfaces. For example, the PPBS transmits "p" ray with a coefficient of 80% (T_p = 80%), and "s" ray with a coefficient of 20% (T_s = 20%), while reflects the "p" ray with a coefficient of 20% (R_p = 20%) and the "s" with a coefficient of 80% (R_s = 80%).

One may obtain the SOP information δ and θ from Eq. (8.19b) as

$$tan\delta = \frac{(P_1 - P_2)}{(P_3 - P_4)} \tag{8.20a}$$

$$sin2\theta = \frac{\pm\frac{5}{4}\sqrt{(P_1 - P_2)^2 + (P_3 - P_4)^2}}{P_0} \tag{8.20b}$$

Alternatively, similar to the procedure described in Eqs. (8.1a) and (8.1b), one may also use Mueller matrix analysis to obtain the Stokes vector $\mathbf{S} = (S_0, S_1, S_2, S_3)^T$ of the input beam. Let $\mathbf{P} = (P_1, P_2, P_3, P_4)^T$ be the vector to represent the powers detected with the four PDs; **S** and **P** can be related by a matrix **M**:

$$\mathbf{P} = \mathbf{MS} \tag{8.21a}$$

$$\mathbf{S} = \mathbf{M}^{-1}\mathbf{P} \tag{8.21b}$$

Let $\mathbf{S}'_1, \mathbf{S}'_2, \mathbf{S}'_3, \mathbf{S}'_4$ be the Stokes vectors of the four light beams before entering their corresponding detectors, we have

$$\begin{aligned}
\mathbf{S}'_1 &= \mathbf{M}_{PBS1T}\mathbf{M}_{QWP}(45^\circ)\mathbf{M}_{PPBST}\mathbf{S} \\
\mathbf{S}'_2 &= \mathbf{M}_{PBS1R}\mathbf{M}_{QWP}(45^\circ)\mathbf{M}_{PPBST}\mathbf{S} \\
\mathbf{S}'_3 &= \mathbf{M}_{PBS2R}\mathbf{M}_{HWP}(22.5^\circ)\mathbf{M}_{PPBSR}\mathbf{S} \\
\mathbf{S}'_4 &= \mathbf{M}_{PBS2T}\mathbf{M}_{HWP}(22.5^\circ)\mathbf{M}_{PPBSR}\mathbf{S}
\end{aligned} \tag{8.22}$$

where $\mathbf{M}_{PPBST}$ and $\mathbf{M}_{PPBSR}$ are the transmission and reflection Mueller matrices of the PPBS, respectively, $\mathbf{M}_{QWP}(45^\circ)$ and $\mathbf{M}_{HWP}(22.5^\circ)$ are the Mueller matrices of the QWP at 45° and HWP at 22.5°, respectively, $\mathbf{M}_{PBSiT}$ and $\mathbf{M}_{PBSiR}$ are the transmission and reflection Mueller matrices of PBS*i* ($i = 1, 2$), respectively. Referring to Table 4.3, one finds:

$$\mathbf{M}_{PPBST} = \begin{bmatrix} \frac{1}{2}(\alpha^2 + \beta^2) & \frac{1}{2}(\alpha^2 - \beta^2) & 0 & 0 \\ \frac{1}{2}(\alpha^2 - \beta^2) & \frac{1}{2}(\alpha^2 + \beta^2) & 0 & 0 \\ 0 & 0 & \alpha\beta & 0 \\ 0 & 0 & 0 & \alpha\beta \end{bmatrix} \tag{8.23a}$$

$$\mathbf{M}_{PPBSR} = \begin{bmatrix} \frac{1}{2}(\alpha^2 + \beta^2) & -\frac{1}{2}(\alpha^2 - \beta^2) & 0 & 0 \\ -\frac{1}{2}(\alpha^2 - \beta^2) & \frac{1}{2}(\alpha^2 + \beta^2) & 0 & 0 \\ 0 & 0 & \alpha\beta & 0 \\ 0 & 0 & 0 & \alpha\beta \end{bmatrix} \tag{8.23b}$$

$$\mathbf{M}_{HWP}(22.5^\circ) = \begin{bmatrix} 1 & 0 & 0 & 0 \\ 0 & cos\frac{\pi}{4} & -sin\frac{\pi}{4} & 0 \\ 0 & sin\frac{\pi}{4} & cos\frac{\pi}{4} & 0 \\ 0 & 0 & 0 & 1 \end{bmatrix} \begin{bmatrix} 1 & 0 & 0 & 0 \\ 0 & 1 & 0 & 0 \\ 0 & 0 & -1 & 0 \\ 0 & 0 & 0 & -1 \end{bmatrix} \begin{bmatrix} 1 & 0 & 0 & 0 \\ 0 & cos\frac{\pi}{4} & sin\frac{\pi}{4} & 0 \\ 0 & -sin\frac{\pi}{4} & cos\frac{\pi}{4} & 0 \\ 0 & 0 & 0 & 1 \end{bmatrix} \tag{8.23c}$$

$$\mathbf{M}_{QWP}(45^\circ) = \begin{bmatrix} 1 & 0 & 0 & 0 \\ 0 & cos\frac{\pi}{2} & -sin\frac{\pi}{2} & 0 \\ 0 & sin\frac{\pi}{2} & cos\frac{\pi}{2} & 0 \\ 0 & 0 & 0 & 1 \end{bmatrix} \begin{bmatrix} 1 & 0 & 0 & 0 \\ 0 & 1 & 0 & 0 \\ 0 & 0 & 0 & 1 \\ 0 & 0 & -1 & 0 \end{bmatrix} \begin{bmatrix} 1 & 0 & 0 & 0 \\ 0 & cos\frac{\pi}{2} & sin\frac{\pi}{2} & 0 \\ 0 & -sin\frac{\pi}{2} & cos\frac{\pi}{2} & 0 \\ 0 & 0 & 0 & 1 \end{bmatrix} \tag{8.23d}$$

$$\mathbf{M}_{PBSiT} = \begin{bmatrix} \frac{1}{2} & \frac{1}{2} & 0 & 0 \\ \frac{1}{2} & \frac{1}{2} & 0 & 0 \\ 0 & 0 & 0 & 0 \\ 0 & 0 & 0 & 0 \end{bmatrix} \tag{8.23e}$$

$$\mathbf{M}_{PBSiR} = \begin{bmatrix} \frac{1}{2} & -\frac{1}{2} & 0 & 0 \\ -\frac{1}{2} & \frac{1}{2} & 0 & 0 \\ 0 & 0 & 0 & 0 \\ 0 & 0 & 0 & 0 \end{bmatrix} \tag{8.23f}$$

In (8.23b) and (8.23c), $\alpha^2 = T_p = R_s = 0.8\%$, $\beta^2 = T_s = R_p = 20\%$. Substituting Eqs. (8.23a)–(8.23f) in Eq. (8.22), one obtains

$$\mathbf{S}_1' = \frac{1}{4} \begin{bmatrix} -\frac{4}{5}S_3 + \frac{3}{5}S_1 + S_0 \\ -\frac{4}{5}S_3 + \frac{3}{5}S_1 + S_0 \\ 0 \\ 0 \end{bmatrix}$$

$$\mathbf{S}_2' = \frac{1}{4} \begin{bmatrix} \frac{4}{5}S_3 + \frac{3}{5}S_1 + S_0 \\ -\frac{4}{5}S_3 - \frac{3}{5}S_1 - S_0 \\ 0 \\ 0 \end{bmatrix}$$

$$\mathbf{S}_3' = \frac{1}{4} \begin{bmatrix} -\frac{4}{5}S_2 - \frac{3}{5}S_1 + S_0 \\ \frac{4}{5}S_2 + \frac{3}{5}S_1 - S_0 \\ 0 \\ 0 \end{bmatrix}$$

$$\mathbf{S}'_4 = \frac{1}{4}\begin{bmatrix} \frac{4}{5}S_2 - \frac{3}{5}S_1 + S_0 \\ \frac{4}{5}S_2 - \frac{3}{5}S_1 + S_0 \\ 0 \\ 0 \end{bmatrix} \tag{8.23g}$$

The optical power P_i to be detected by the ith PD is the first row of $\mathbf{S}'_i$ in Eq. (8.23g). Therefore, vector $\mathbf{P} = (P_1, P_2, P_3, P_4)^T$ can be written as

$$\begin{bmatrix} P_1 \\ P_2 \\ P_3 \\ P_4 \end{bmatrix} = \frac{1}{4}\begin{bmatrix} S_0 + \frac{3}{5}S_1 - \frac{4}{5}S_3 \\ S_0 + \frac{3}{5}S_1 + \frac{4}{5}S_3 \\ S_0 - \frac{3}{5}S_1 - \frac{4}{5}S_2 \\ S_0 - \frac{3}{5}S_1 + \frac{4}{5}S_2 \end{bmatrix} = \mathbf{M}\begin{bmatrix} S_0 \\ S_1 \\ S_2 \\ S_3 \end{bmatrix} \tag{8.23h}$$

$$\mathbf{M} = \frac{1}{4}\begin{bmatrix} 1 & +\frac{3}{5} & 0 & -\frac{4}{5} \\ 1 & +\frac{3}{5} & 0 & +\frac{4}{5} \\ 1 & -\frac{3}{5} & -\frac{4}{5} & 0 \\ 1 & -\frac{3}{5} & +\frac{4}{5} & 0 \end{bmatrix} \tag{8.23i}$$

Therefore, with the obtained Mueller matrix $\mathbf{M}$, the Stokes vector $\mathbf{S}$ can be determined with the measurements of the four optical powers.

The Mueller matrix $\mathbf{M}$ in Eq. (8.23i) is obtained by assuming the optical components in Figure 8.5a are ideal, which is used to illustrate conceptually how the Mueller matrix of the four-detector polarimeter can be obtained. In practice, due to the imperfections of the optical components and variations of the responsibility of the PDs, the Mueller matrix representing the actual components can be obtained with a calibration procedure (Azzam et al. 1988a, 1988b, 1989) by inputting light with four precisely known SOPs. As a common practice, one may input linear horizontal polarization (LHP), linear vertical polarization (LVP), L + 45P, and RCP with the same optical power, corresponding to $(1,1,0,0)^T$, $(1,-1,0,0)^T$, $(1,0,1,0)^T$, and $(1,0,0,1)^T$, 16 photocurrent values can be obtained, which can be used to obtain all 16 elements of Mueller matrix $\mathbf{M}'$ relating the four photocurrents $\mathbf{I} = (I_1, I_2, I_3, I_4)^T$ with the Stokes vector $\mathbf{S} = (S_0, S_1, S_2, S_3)^T$. Since the imperfections are temperature and wavelength dependent, $\mathbf{M}'$ should be obtained for different temperature and wavelengths. Corresponding to each temperature T and wavelength λ, the Mueller matrix is $\mathbf{M}'(T, \lambda)$, and the Stokes vectors of an arbitrary SOP and DOP can be obtained with the simultaneous measurements of the four photocurrents by $\mathbf{S} = \mathbf{M}'^{-1}(T, \lambda)\mathbf{I}$.

It is important to note that the calibration procedure described above requires one to input at least four precisely known SOPs to the polarimeter. If the input beam in Figure 8.5a is an optical fiber with a collimator at the end, the precise SOPs are difficult to determine and therefore the calibration procedure is difficult to perform. As will be discussed in Section 8.1.4, a procedure utilizing a known DOP can be performed, which is more flexible, accurate, and fast to complete.

Azzam's Four-Detector Polarimeter This polarimeter invented by Azzam cleverly uses each PD for both light detection and deflection into different directions (Azzam 1985), as shown in Figure 8.5b, and is essentially an amplitude division polarimeter. In particular, when a light beam is incident onto a PD, it undergoes Fresnel refraction and reflection with certain complex coefficients (with a unique phase and amplitude) determined by the angle of incidence, as described in Section 4.2.2, considering that the PDs are highly absorptive. The refracted light is absorbed by the PD while the reflected light continues to the next PD to be absorbed and reflected again. Finally, remaining light incident normally on the last PD, PD4, is completely absorbed. Because the refractive index of a PD is generally much larger than that of the air, the Fresnel reflection can be quite large. By properly adjusting the incident angle of each PD, desired complex reflection coefficients with certain amplitude and phase can be achieved. That is, the PD acts as a beam splitter followed by a retarder. The photocurrent currents are related to the input Stokes vector by

$$\mathbf{I} = \begin{bmatrix} I_0 \\ I_1 \\ I_2 \\ I_3 \end{bmatrix} = \mathbf{A} \begin{bmatrix} S_0 \\ S_1 \\ S_2 \\ S_3 \end{bmatrix} \tag{8.24a}$$

$$\mathbf{S} = \begin{bmatrix} S_0 \\ S_1 \\ S_2 \\ S_3 \end{bmatrix} = \mathbf{A}^{-1} \begin{bmatrix} I_0 \\ I_1 \\ I_2 \\ I_3 \end{bmatrix} \tag{8.24b}$$

By inputting four different special SOPs with the same optical power, such as LHP, LVP, L + 45P, and RCP, corresponding to $(1,1,0,0)^T$, $(1,-1,0,0)^T$, $(1,0,1,0)^T$, and $(1,0,0,1)^T$, 16 photocurrent values can be obtained, which can be used to obtain all 16 elements of matrix $\mathbf{A}$ in Eq. (8.24a). With $\mathbf{A}$ being known, any Stokes vector of an input beam can be obtained from the four photocurrent readings using Eq. (8.24b). Again, since the reflections and responsivities are temperature and wavelength dependent, different matrices at different wavelengths and under different temperatures should be obtained for accurate measurements under all conditions.

There are also many other configurations for the amplitude division polarimeters, such as that based on a diffraction grating (Azzam 1992; Azzam and Giardina 1993), a parallel slab (Jellison 1987; El-Saba and Azzam 1996), or a corner cube (Liu and Azzam 1997). These configurations may be more suitable for free-space optical instruments and less suitable for making fiber optical devices due to the bulky size and wavelength sensitivity. The interested readers are encouraged to read *Polarized Light* by D. Goldstein (2003).

8.1.2.2 In-Line Polarimeter

In fiber optic systems, it is often desirable to monitor the SOP and DOP of the optical signal transmitting in the optical fiber, with a minimum loss to the signal. For this purpose, in-line polarimeters are introduced. Figure 8.6 shows a simple in-line polarimeter design in which four polarization insensitive beam splitters, BS1, BS2, BS3, and BS4, with a small splitting ratio around 1–5% are used to successively split off four beams to be analyzed by different polarizers (Yao 2002).

For the ideal situation that all beam splitters have identical reflectivity and all PDs have the same responsivity, the Stokes vector of the input beam can be related to the four detected photocurrents I_1, I_2, I_3, and I_4 by Eqs. (8.2a) and (8.2b):

$$\mathbf{S} = \begin{bmatrix} S_0 \\ S_1 \\ S_2 \\ S_3 \end{bmatrix} = \begin{bmatrix} I_1 + I_2 \\ I_1 - I_2 \\ 2I_3 - (I_1 + I_2) \\ 2I_4 - (I_1 + I_2) \end{bmatrix} = \mathbf{A}^{-1} \begin{bmatrix} I_1 \\ I_2 \\ I_3 \\ I_4 \end{bmatrix}$$

$$\mathbf{A}^{-1} = \begin{bmatrix} +1 & +1 & 0 & 0 \\ +1 & -1 & 0 & 0 \\ -1 & -1 & 2 & 0 \\ -1 & -1 & 0 & 2 \end{bmatrix}$$

As discussed in Section 8.1.2.1, the beam splitters are imperfect with some certain polarization dependencies, such as the "s" and "p" rays having different splitting ratios and different phases. In addition, the PDs have different responsivities. In order take these factors into account, the matrix **A** must be calibrated. Unlike the calibration of the free-space polarimeters of Figure 8.5, it is difficult here to input precisely known SOPs due to the fluctuating birefringence of the input fiber pigtail, a procedure of using a precisely known DOP for calibration can be adopted, as will be discussed in Section 8.1.4. With such a procedure, different calibrated matrices at different temperatures and with different wavelengths can be quickly obtained. Figure 8.6b shows the picture of a commercial in-line polarimeter with an internal structure similar to that of Figure 8.6a, however, modified to compensate components imperfections.

It is worthy to mention that all in-line polarimeters using four blazed fiber gratings in the optical fiber were also proposed and demonstrated (Bouzid et al. 1995; Westbrook et al. 2000), which can be potentially low cost and small size. However, probably because such devices were sensitive to wavelength and less stable against temperature fluctuations, they were never commercialized.

8.1.2.3 Wave-Front Division Polarimeters

Wave-front division polarimetry relies on analyzing different parts of a beam with different polarization elements, as shown in Figure 8.7a. In particular, the beam's wave front can be divided into four virtual portions, with each portion going through a polarization analyzing element: three linear polarizers at 0°, 90°, 45°, and a circular polarizer consisting of a QWP at 0° followed by a linear polarizer at 45°. Finally, different portions of the beam go into different PDs. Unlike the time-division polarimetry described in Figure 8.1 where the four polarizers are inserted in the

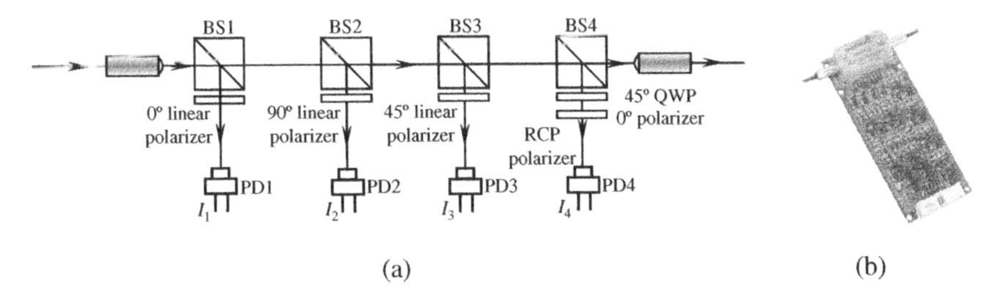

(b)

Figure 8.6 Illustration of an in-line polarimeter. (a) The internal structure showing the polarimeter is constructed with four polarization insensitive beam splitters (BSs), four different polarizers, and four photodetectors (PDs). (b) Image of a commercial in-line polarimeter developed by the authors. Source: Photo courtesy of General Photonics/Luna Innovations.

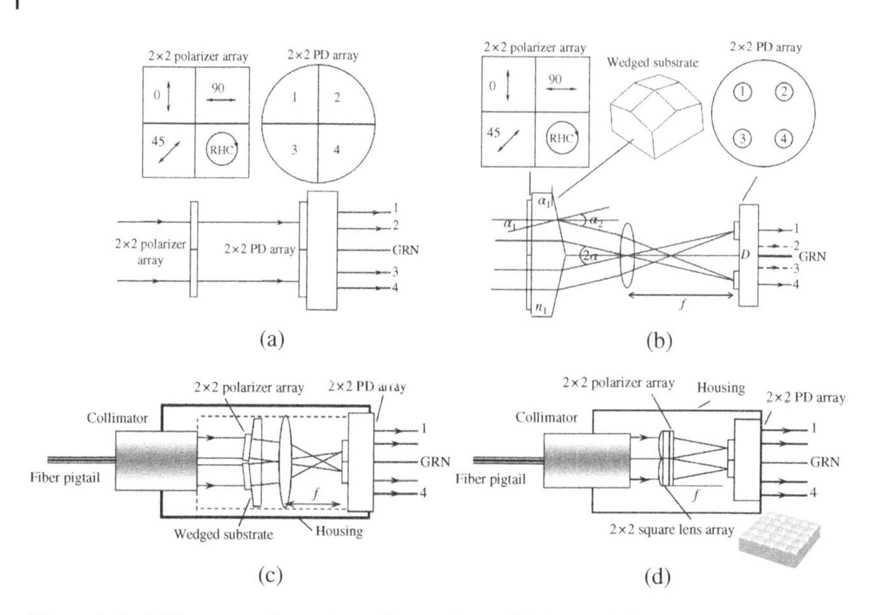

Figure 8.7 Different configurations of wave-front division polarimeters. (a) A simple wave-front division polarimeter made with four polarizers (a 2 × 2 array), followed by a quadrant detector composed of four independent detectors; (b) A wedged substrate is used to divide a wide beam into four sub-beams of different angles before being focused onto four detector chips (2 × 2) by a focusing lens. (c) The fiber pigtailed polarimeter similar to (b), but with the polarizers placed on the slopes of the wedges. (d) A fiber pigtailed wave-front division polarimeter made with a fiber collimator, a 2 × 2 square lens array before a 2 × 2 polarizer array as in (b), followed by a 2 × 2 photodetector array. The inset on the bottom right shows a typical square lens array.

optical beam sequentially followed by a single PD, here they are placed in the beam simultaneously followed by four PDs.

Wedged Substrate Based Wave-Front Division Polarimeter The configuration of Figure 8.7a requires that the PDs have large active areas, which makes them slow, costly, and having higher dark current, in addition to the more serious problem of high cross talk between the four polarization channels. To overcome the issues, a wedged substrate can be used to direct the four sub-beams in four different direction before each of the sub-beam is focused to one of the detectors in the 2 × 2 PD array, as shown in Figure 8.7b (Yao 2006; Yao et al. 2019). Referring to Figure 8.7b, for a wedge angle of α_1, the beam refraction angle α_2 can be obtained by Snell's law as $n_2 sin\,\alpha_2 = n_1 sin\,\alpha_1$, where n_1 and n_2 are the refractive indexes of the substrate and air, respectively. The beam deviation angle α is then

$$\alpha = \alpha_2 - \alpha_1 \approx (n_1 - 1)\alpha_1 \tag{8.25a}$$

where $sin\alpha_i \approx \alpha_i$ $(i = 1, 2)$ and $n_2 = 1$ (in air) are used. The separation D between two focusing spots is

$$D = 2ftan\alpha \approx 2f(n_1 - 1)\alpha_1 \tag{8.25b}$$

For $n_1 = 1.5$, $\alpha_1 = 5°$, and a detection separation $D = 0.5$ mm, the focal length of the lens is

$$f \approx \frac{D}{[2(n_1 - 1)\alpha_1]} = 5.73\text{ mm}$$

Clearly, a very compact polarimeter can be made.

Figure 8.7c shows a fiber pigtailed version of the wave-front division polarimeter of Figure 8.7b. The only difference is that now the 2×2 polarizer array is on the slope side of the wedged substrate.

Lens Array Based Wave-Front Division Polarimeter Figure 8.7d shows another fiber pigtailed wave-front division polarimeter based on a 2×2 square lens array. As can be seen in the figure, a collimated beam entering the array will be focused into four separate spots on the focal plane by four square micro-lenses, where a 2×2 PD array is located with each detector chip on each focal spot. The inset shows the image of a 5×5 square lens array to give readers some ideas of how such a lens array looks like, which can be made by multiple manufacturers around world. A similar polarimeter with a circular lens array and a diverging input beam has also been described by Jay Damask (2005).

Similar to the discussions in the beginning of Section 8.1, photocurrents detected by the four detectors are uniquely related to an SOP $\mathbf{I} = \mathbf{AS}$, as described by Eq. (8.1a). All 16 elements of the calibration matrix $\mathbf{A}$ can be obtained by inputting four known SOPs. After $\mathbf{A}$ is known, the Stokes vector of an arbitrary SOP can be obtained by $\mathbf{S} = \mathbf{A}^{-1}\mathbf{I}$. A DOP based calibration scheme better suited to fiber pigtailed polarimeters will be discussed in Section 8.1.4.

8.1.3 Advantages and Disadvantages of Different Configurations

The time-division polarimeters described in Section 8.1.1 use only one PD to obtain four measurements sequentially, while the amplitude-division described in Section 8.1.2 and wave-front division polarimeters in Section 8.1.3 use four PDs to make required measurements simultaneously. The advantage of using a single PD is that a slight angle deviation of the optical beam into the PD will not affect the calibration matrix as long as the active area of the PD is sufficiently large to receive all the power in the beam. In contrast, for the polarimeters using multiple PDs, a slight beam deviation may change the relative powers into different PDs and therefore mess up the calibration matrix. Consequently, the polarimeter must be re-calibrated each time the beam is re-aligned. Therefore, the single PD time-division polarimeters are more suited to free-space beam applications, while the multiple-PD polarimeters are more suited for fiber pigtailed applications, because the fiber collimator is generally fixed onto the polarimeter body to keep the beam angle into the device fixed, provided that the device is well engineered such that temperature variations will not misalign the beams.

On the other hand, because a time-division polarimeter makes measurement sequentially with a single PD, the SOP and DOP of the beam must be stable before finishing the measurements and therefore is more suited for applications where the SOP variations are much slower than the measurement rates. In contrast, because the amplitude- and wave front-division polarimeters use four PDs to simultaneously make required measurements, they are more suitable for applications for optical fiber systems in which SOP and DOP may change rapidly with time. Measurement rates of few hundred megahertz or higher can be achieved with the polarimeters made with multiple PDs, limited only by the electronics and firmware or software for processing the data.

8.1.4 Polarimeter Calibration with DOP

All optical components have imperfections, as well as temperature and wavelength dependencies, which degrade the Stokes measurement accuracy. In principle, a polarimeter should be calibrated at certain different temperature and wavelength intervals. Specifically, the instrument matrix $\mathbf{M}$ relating the photocurrents and the Stokes vectors must be obtained for different temperature and different wavelengths. The more sensitive the components to the temperature and wavelength, the smaller the calibration intervals must be. For example, the temperature and wavelength intervals can be 5 °C and 10 nm, meaning that one must obtain a calibration matrix every 5 °C and every 10 nm. For a polarimeter with an operating temperature and wavelength ranges of 100 °C and 100 nm, a total of 200 (20 × 10) calibration matrices must be obtained.

As mentioned previously, one may calibrate a polarimeter by inputting at least four known SOPs or Stokes vectors. These calibration SOPs should be evenly distributed on the Poincaré sphere, such as the six distinctive SOPs LHP $(1,1,0,0)^T$, L + 45P $(1,0,1,0)^T$, LVP $(1,-1,0,0)^T$, L − 45P $(1,0,-1,0)^T$, RCP $(1,0,0,1)^T$, and LCP $(1,0,0,-1)^T$, with which the calibration is the simplest to perform. Ideally, every quadrant of the Poincaré sphere should have a calibration SOP point to maximize the accuracy for measuring every SOP on the Sphere. As also mentioned in Section 8.1.2.1, for fiber pigtailed polarimeter, it is difficult to generate the precise input SOPs required for the calibration because a slight motion or temperature variation acting on the fiber pigtail will change the SOP before light entering the measurement optics.

A more attractive approach is to use a known DOP for the calibration, since the DOP is related to all four Stokes parameters: $DOP = \sqrt{S_1^2 + S_2^2 + S_3^2}/S_0$. An inaccuracy of any Stokes parameter can equally affect the accuracy of the DOP. Most importantly, the DOP of a narrow line width laser is not affected by the motion, bending, or temperature variation acting on the fiber pigtail, which remains constant once the laser source is chosen (the DOP invariance).

In practice, a high DOP light source with a DOP at 100% is easy to obtain, such as commercial distributed feedback (DFB) and external feedback semiconductor lasers. In fact, the commercial desk top tunable lasers generally have a DOP of 100%, which is independent of operation temperature and wavelength and therefore are perfect light sources for DOP based polarimeter calibration to be discussed next. Figure 8.8a shows a calibration setup using a high DOP light source, preferably a tunable laser especially when calibration at different wavelengths is required. An optional in-line polarizer with a PER of 35 dB or more can be used in front of the polarization controller to ensure that the DOP of light entering the polarimeter is 100% in case the DOP of the light source is not sufficiently high. The polarization controller or scrambler is used to generate a set of SOPs uniformly cover the Poincaré sphere. The SOP is obtained using the initial instrument **M** matrix

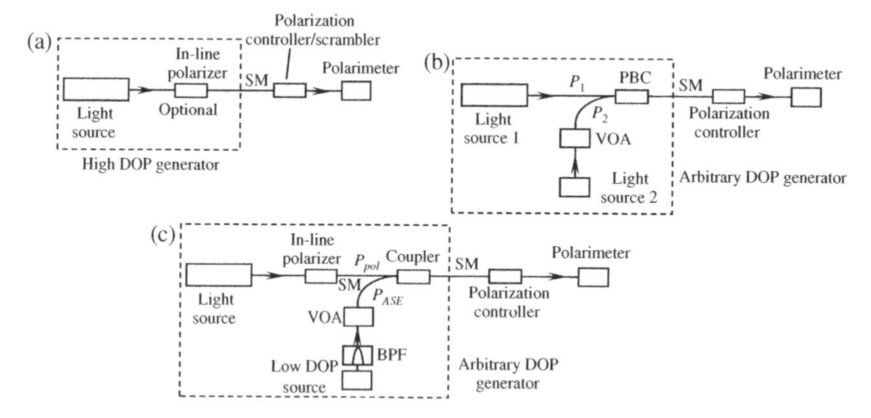

Figure 8.8 (a) Polarimeter calibration setup using a high DOP generator. (b) An arbitrary DOP generator by combining two mutually incoherent beams with orthogonal SOPs. (c) An arbitrary DOP generator by combining two beams of a high DOP and a low DOP. The low DOP source can be an ASE source with a DOP of zero. PBC: polarization beam combiner; VOA: variable optical attenuator; BPF: bandpass filter; SM: single mode fiber.

for each input SOP Stokes vector before the DOP is calculated, which is then compared with the input DOP of unity.

Specifically, when implementing DOP calibration, in addition to assuming the DOP stays constant when the input SOP is varied, we also assume the optical power of the light source stays constant at the input port of polarimeter (the optical power invariance). Such a condition requires that (i) the power of the laser sources is stable during the time of calibration (invariance of power), (ii) no PDL is present in the input optical fiber, and (iii) the polarization controller used has extremely small PDL and activation loss (AL). The first two requirements can be easily satisfied in practice because the laser power is generally extremely stable during the time of calibration (on the order of a few seconds) and the PDL of the input fiber with a length of a few meters or less is also negligible, as shown in Table 3.1. To satisfy requirement (iii), the fiber squeezer polarization controller discussed in Figure 6.15 can be used to generate a set of random SOPs covering the Poincaré sphere, which has negligible PDL and activation loss, as discussed in Chapter 6.

In practice, the photocurrents in the four PDs are first converted to photo-voltages by four transimpedance amplifiers having different gains due to the component non-uniformity, which are then digitized by four analog-to-digital converters (ADCs) for processing and computation. Therefore, the photo-voltages can be related to the input SOP by the instrument matrix $\mathbf{A}_v$ by

$$\mathbf{V} = \mathbf{A}_v\mathbf{S} \tag{8.26a}$$

where $\mathbf{V} = (V_1, V_2, V_3, V_4)^T$ is the photo-voltage vector and $\mathbf{S} = (S_0, S_1, S_2, S_3)^T$ is the Stokes vector. Conversely, if the matrix $\mathbf{A}_v$ is known, an unknown SOP can be obtained by the detected photo-voltages by

$$\mathbf{S} = \mathbf{A}_v^{-1}\mathbf{V} = \mathbf{M}\mathbf{V} \tag{8.26b}$$

For an arbitrary SOP_k generated by the polarization controller, one measures four photo-voltages $\mathbf{V}_k = \left(V_1^k, V_2^k, V_3^k, V_4^k\right)^T$ and obtains a $\mathbf{S}^k$ using a matrix $\mathbf{M}_k \equiv \mathbf{A}_{vk}^{-1}$. The resulting Stokes parameters are

$$S_j^k = \sum_{i=1}^{4} m_{ji}^k V_i^k, \qquad j = 0, 1, 2, 3 \tag{8.26c}$$

where m_{ji}^k is the matrix element corresponding to the kth measurement in the calibration process, as described next. In Eq. (8.26c), S_0^k represents the optical power corresponding to the kth SOP. The corresponding DOP can be expressed as

$$DOP^k = \frac{\sqrt{\left(S_1^k\right)^2 + \left(S_2^k\right)^2 + \left(S_3^k\right)^2}}{S_0^k} \tag{8.26d}$$

In practice, we first generate and select a set of N SOP points uniformly distributed on the Poincaré sphere (the calibration set) with the polarization scrambler and obtains a corresponding set of N photo-voltages $\mathbf{V}$ with the four PDs in the polarimeter, before a set of N SOP points are calculated using Eq. (8.26b) in which the initial matrix $\mathbf{A}_v^{-1}$ can be obtained with the ideal component parameters in the polarimeter, such as that of Eq. (8.2b). To ensure the N SOP points selected are indeed uniformly distributed on the Poincaré sphere, we first divide the whole Poincaré sphere into N sections uniformly covering the whole sphere and then pick one SOP point in each section as one of the calibration SOPs in the calibration set. A set of M SOP points ($M > N$) may be generated, among which N points (typically $N < 50$) are selected into the calibration set.

We first use the property of optical power invariance for calibration and we define a standard deviation function σ_p^l of optical power to evaluate error for the lth iteration:

$$\sigma_p^l = \sqrt{\left\langle \left(S_0^{l,k} - \overline{P}_l\right)^2 \right\rangle_N} = \sqrt{\frac{\left[\sum_{k=1}^{N}\left(\sum_{i=1}^{4} m_{0i}^l V_i^k - \overline{P}_l\right)^2\right]}{N}} \tag{8.26e}$$

where $\langle X \rangle_N$ stands for average over N and $\overline{P}_l = \left(\sum_{k=1}^{N} S_0^{l,k}\right)/N = \sum_{k=1}^{N}\sum_{i=1}^{4} m_{0i}^{l,k} V_i^k/N$ is the calculated average power for the lth iteration. The calibration program adjusts the value of each $m_{0i}^l, i = 0, 1, 2, 3$ by a small step one at a time while observing σ_p^l. If increasing a matrix element m_{01}^l results in a reduced σ_p^l, continue to increase the element in the same direction till a minimum standard deviation σ_p^l is reached. If increasing the matrix element m_{01}^l results in an increased σ_p^l, then decrease the matrix element by a small step till a minimum standard deviation σ_p^l is reached. Repeat the procedure earlier to the next matrix element m_{02}^l till a second minimum standard deviation σ_p^l is reached. Repeat the procedure earlier to all other matrix elements till a 4th minimum standard deviation σ_p^l (corresponding to adjusting m_{04}^l) is reached before starting the $(l+1)$th iteration. Finally, after m iterations till σ_p^m is less than a preset value close to zero, the iteration stops and the matrix elements $m_{0i}^l, i = 0, 1, 2, 3$ are determined.

Similarly, for using the DOP invariance property for calibration, we define a standard deviation function of DOP to evaluate the error during the lth iteration:

$$\sigma_{DOP}^l = \sqrt{\left\langle \left(DOP_0^k - 1\right)^2 \right\rangle_N} \tag{8.26f}$$

where DOP_0^k can be obtained from Eqs. (8.26c) and (8.26d). The same procedures described earlier can be used to adjust the rest of 12 matrix elements $m_{ji}^l, i = 0, 1, 2, 3, j = 1, 2, 3$. If the instrument matrix is perfectly calibrated, σ_{DOP}^l is zero. When σ_{DOP}^l is less than a preset value close to 0, the iteration can then be stopped. Again, for the first calibration iteration ($l = 1$), **M** can be obtained with ideal component parameters, such as that of Eq. (8.2b).

One may repeat the procedures described earlier after tuning the laser to a different wavelength and putting the polarimeter in a temperature chamber at a different temperature to obtain the calibrated instrument matrix at the wavelength and temperature.

After the polarimeter is calibrated with a light source with a DOP of 100%, an arbitrary DOP may be used to verify its performance. Figure 8.8b shows a method to generate an arbitrary DOP for the instrument matrix verification and calibration. From the definition of DOP of Eq. (2.17), the DOP of the combined light after the polarization beam combiner (PBC) is

$$DOP = \frac{|P_1 - P_2|}{(P_1 + P_2)} \tag{8.26g}$$

where P_1 and P_2 are powers from light source 1 and light source 2 with orthogonal SOPs after being combined by the PBC and each can be measured with a power meter by turning the other source off. The two light sources are mutually incoherent. A variable optical attenuator (VOA) can be used to control the power P_2 to achieve any desired DOP values. In Figure 8.8b, the two input fibers to the PBC are polarization maintaining (PM) fibers with the SOP aligned with their slow axis and the output from the PBC is single mode (SM) fiber.

Figure 8.8c shows an alternative method for generating a desired DOP by combining a high DOP light source ($DOP = 1$) with a low DOP light source ($DOP \approx 0$) with a directional coupler. In this configuration, the PM fiber pigtails in Figure 8.8b are all replaced with SM fibers. An ideal low DOP light source is an amplified spontaneous emission (ASE) source. The bandpass filter (BPF) can be used after the low DOP source to restrict the bandwidth of the light source to be around 1–2 nm (less than the calibration interval). The DOP of the combined light can be expressed as

$$DOP = \frac{P_{pol}}{(P_{pol} + P_{ASE})} \tag{8.26h}$$

where P_{pol} and P_{ASE} are powers from the polarized and unpolarized (ASE) light sources, respectively. Again, by adjusting the VOA, different DOP values can be obtained.

8.2 Analog Mueller Matrix Polarimetry

As mentioned at the beginning of this chapter, a Mueller matrix polarimeter is for obtaining the Mueller matrix of an optical device representative of its polarization related properties (Hauge 1980). A complete Mueller matrix polarimeter is capable of obtaining all 16 Mueller matrix elements and incomplete if the Mueller matrix polarimeter can obtain less than 16 matrix elements.

A Mueller matrix polarimeter generally comprises a polarization state generator (PSG) and polarization state analyzer (PSA), as shown in Figure 8.9. Because of the influence of the optical device, the SOP of a light beam after it passes through the device and the PSA is given by

$$\mathbf{S}' = \mathbf{M}_{PSA}\mathbf{M}\mathbf{S}_{PSG} \tag{8.27a}$$

where $\mathbf{S}_{PSG}$ is the Stokes vector representing the SOP generated by the PSG, $\mathbf{S}'$ is the Stokes vector representing the SOP being measured by the detector, $\mathbf{M}$ is a 4×4 Mueller matrix of the optical device, and $\mathbf{M}_{PSA}$ is the 4×4 Mueller matrix representing the PSA. In principle, if one generates four distinctive SOPs $\mathbf{S}_{PSGi}$ $(i = 1, 2, 3, 4)$ with the PSG and measures the corresponding four Stokes vectors $\mathbf{S}'_i$ $(i = 1, 2, 3, 4)$ with the detector, one may obtain a total of 16 equations, which are sufficient to solve for the 16 Mueller matrix elements. However, if the SOPs generated by the PSG cannot uniformly cover the Poincaré sphere, some of the equations may be degenerate, resulting in incomplete Mueller matrix. In practice, more than four SOPs $(i > 4)$ can be generated and analyzed.

Let $\mathbf{S}_{PSGi} = (S_{i0}, S_{i1}, S_{i2}, S_{i3})^T$, $\mathbf{S}'_i = \left(S'_{i0}, S'_{i1}, S'_{i2}, S'_{i3}\right)^T$, and $\mathbf{M}_{PSA} = \begin{bmatrix} a_{11} & a_{12} & a_{13} & a_{14} \\ a_{21} & a_{22} & a_{23} & a_{24} \\ a_{31} & a_{32} & a_{33} & a_{34} \\ a_{41} & a_{42} & a_{43} & a_{44} \end{bmatrix}$, Eq. (8.27a) can be written out as

$$\begin{bmatrix} S'_{i0} \\ S'_{i1} \\ S'_{i2} \\ S'_{i3} \end{bmatrix} = \begin{bmatrix} a_{11} & a_{12} & a_{13} & a_{14} \\ a_{21} & a_{22} & a_{23} & a_{24} \\ a_{31} & a_{32} & a_{33} & a_{34} \\ a_{41} & a_{42} & a_{43} & a_{44} \end{bmatrix} \begin{bmatrix} m_{11} & m_{12} & m_{13} & m_{14} \\ m_{21} & m_{22} & m_{23} & m_{24} \\ m_{31} & m_{32} & m_{33} & m_{34} \\ m_{41} & m_{42} & m_{43} & m_{44} \end{bmatrix} \begin{bmatrix} S_{i0} \\ S_{i1} \\ S_{i2} \\ S_{i3} \end{bmatrix}, \quad i = 1, 2, 3, 4 \tag{8.27b}$$

Because the PD only detects the optical power, or the first term in the Stokes vector $\vec{S}'_{i0}$, one may simplify Eq. (8.27b) as

$$\begin{bmatrix} S'_{i0} \\ \cdot \\ \cdot \\ \cdot \end{bmatrix} = \begin{bmatrix} a_{11} & a_{12} & a_{13} & a_{14} \\ \cdot & \cdot & \cdot & \cdot \\ \cdot & \cdot & \cdot & \cdot \\ \cdot & \cdot & \cdot & \cdot \end{bmatrix} \begin{bmatrix} m_{11} & m_{12} & m_{13} & m_{14} \\ m_{21} & m_{22} & m_{23} & m_{24} \\ m_{31} & m_{32} & m_{33} & m_{34} \\ m_{41} & m_{42} & m_{43} & m_{44} \end{bmatrix} \begin{bmatrix} S_{i0} \\ S_{i1} \\ S_{i2} \\ S_{i3} \end{bmatrix}, \quad i = 1, 2, 3, 4 \tag{8.27c}$$

In this section, we will briefly review some Mueller matrix polarimeters in which the PSG and PSA are made with traditional analog optical devices, such as rotating polarizers or retarders. More

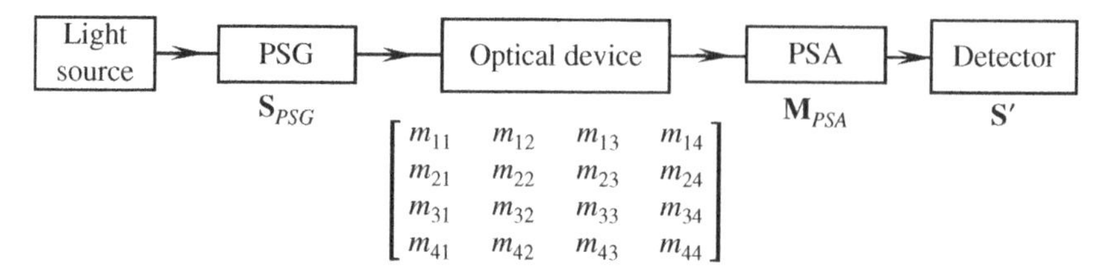

Figure 8.9 Illustration of a Mueller matrix polarimetry system.

advanced binary PSG, PSA, and the associated Mueller matrix polarimeter will be discussed in Chapter 9 in detail.

8.2.1 Rotating Element Mueller Matrix Polarimeters

Figure 8.10 shows four different configurations of Mueller matrix polarimeters made with rotating polarization components in which the first three configurations are incomplete Mueller matrix polarimeters, while the last one is the complete Mueller matrix polarimeter. Figure 8.10a shows an incomplete Mueller matrix polarimeter made with a rotating polarizer as the PSG and another rotating polarizer as the PSA, which rotates three times faster than that of the polarizer in PSG. The detected photocurrent I_d can then be represented as a Fourier series with the form:

$$I_d(\theta) = I_0 \left[\frac{a_0}{4} + \frac{1}{4}\sum_{n=1}^{4}(a_{2n}cos2n\theta + b_{2n}sin2n\theta) \right] \tag{8.28a}$$

where θ is the rotation angle of the first polarizer, a_{2n} and b_{2n} are the Fourier coefficients. The Stokes vector $\mathbf{S}_d$ after the second rotating polarizer can be expressed as

$$\mathbf{S}_d = \mathbf{M}_{PSA}\mathbf{M}\mathbf{M}_{PSG}\mathbf{S}_{in} = \mathbf{P}_2(3\theta)\mathbf{M}\mathbf{P}_1(\theta)\mathbf{S}_{in} \tag{8.28b}$$

where $\mathbf{P}_1(\theta)$, $\mathbf{M}$, and $\mathbf{P}_2(3\theta)$ are the Mueller matrices of the first polarizer, the device under test (DUT), and the second polarizer, respectively. Substituting these Mueller matrices in Eq. (8.28b) (see Table 4.3), one obtains $\mathbf{S}_d$ as a function of Mueller matrix elements m_{ij} of the DUT and different Fourier components, in which the first term is $I_d(\theta)$. The Mueller matrix elements can be determined by relating different Fourier coefficients in the detected signal $I_d(\theta)$ with that in Eq. (8.28a) as (Goldstein 2003)

$$\mathbf{M} = \begin{bmatrix} a_0 & a_2 & b_2 & \square \\ a_6 & a_4 + a_8 & -b_4 + b_8 & \square \\ b_6 & b_4 + b_8 & a_4 - a_8 & \square \\ \square & \square & \square & \square \end{bmatrix} \tag{8.28c}$$

Similarly, one may obtain the Mueller matrix of a DUT using one of other three configurations. The detailed discussions can be found in Goldstein (2003) and Azzam (1978).

8.2.2 Retardation Modulating Mueller Matrix Polarimeters

As shown in Figure 8.11a, one may use the Stokes polarimeter described in Figure 8.4b as the PSA, which includes two retarders with their optical axes oriented 45° from each other for measuring the SOP after light passing through the optical device with a Mueller matrix **M**. An identical PSA can be used in reverse order as the PSG to generate a series of distinctive SOPs required for obtaining

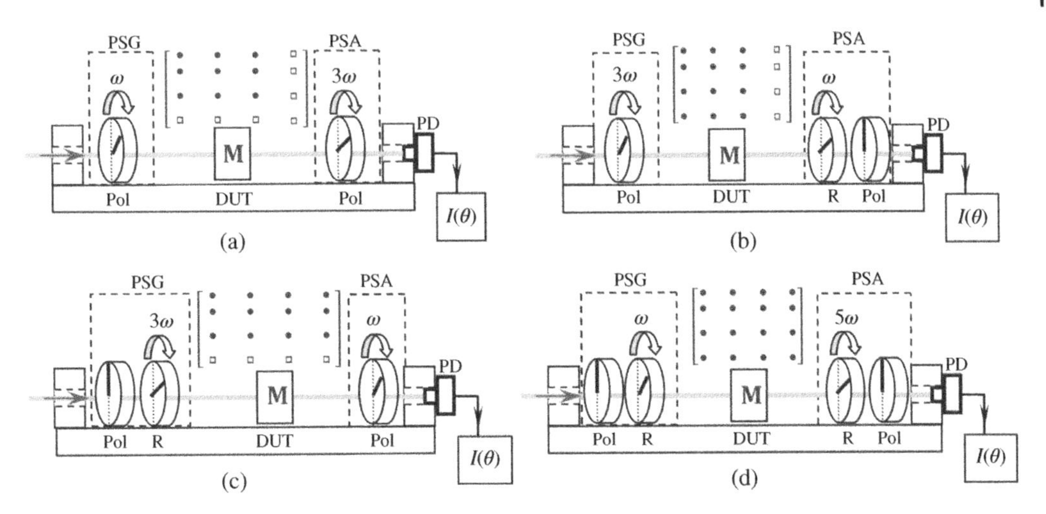

Figure 8.10 Illustration of four different types of Mueller matrix polarimeters based on the rotation of polarization components. (a) A rotating polarizer as PSG and a second rotating polarizer as PSA. (b) A rotating polarizer as PSG with a rotating retarder followed by a polarizer as the PSA. (c) A fixed polarizer followed by a rotating retarder as the PSG, and a rotating polarizer as the PSA. (d) A fixed polarizer followed by a rotating retarder as the PSG and a rotating retarder followed by a fixed polarizer as the PSA. R: retarder, generally a QWP. Pol: polarizer; DUT: device under test; ω: angular speed. The black dots in the matrix indicate the measureable matrix elements and squares indicate the un-measureable matrix elements.

the Mueller matrix of the optical device. In general, the modulation frequencies ω_i and amplitudes δ_i of the four retardation modulators are different and the total retardation $\Delta_i(t)$ for each modulator can be expressed as

$$\Delta_i(t) = \delta_i sin\omega_i t, \quad i = 1, 2, 3, 4 \tag{8.29a}$$

Finally, the Stokes vector $\mathbf{S}_d$ received at the PD can be expressed as

$$\mathbf{S}_d = \mathbf{M}_{PSA}\mathbf{M}\mathbf{M}_{PSG}\mathbf{S}_{in} = \mathbf{P}_2\mathbf{M}_4\mathbf{M}_3\mathbf{M}\mathbf{M}_2\mathbf{M}_1\mathbf{P}_1\mathbf{S}_{in} \tag{8.29b}$$

where $\mathbf{M}_{PSG} = \mathbf{M}_2\mathbf{M}_1\mathbf{P}_1$ and $\mathbf{M}_{PSA} = \mathbf{P}_2\mathbf{M}_4\mathbf{M}_3$ are the Mueller matrices of the PSG and PSA, respectively, and $\mathbf{P}_1$, $\mathbf{M}_1$, $\mathbf{M}_2$, $\mathbf{M}$, $\mathbf{M}_3$, $\mathbf{M}_4$, $\mathbf{P}_2$ are the Mueller matrices of the first polarizer, the first retarder, the second retarder, the DUT, the third retarder, the fourth retarder, and the second polarizer, respectively. Because of the action by the first polarizer $\mathbf{P}_1$, $\mathbf{S}_{in} = I_0(1, 1, 0, 0)^T$, where I_0 is the total optical power after the first polarizer. Since the first term of Stokes parameter $\mathbf{S}_d$ represents the optical power I_d to be detected by the PD, one obtains by substituting Mueller matrices of the retarders and polarizer (Table 4.1) in Eq. (8.29b):

$$\begin{aligned} I_d = \frac{I_0}{2} [&m_{11} + m_{12}cos\Delta_1 + m_{13}sin\Delta_1 sin\Delta_2 - m_{14}sin\Delta_1 cos\Delta_2 + m_{21}cos\Delta_4 + m_{22}cos\Delta_1 cos\Delta_4 \\ &+ m_{23}sin\Delta_1 sin\Delta_2 cos\Delta_4 - m_{24}sin\Delta_1 cos\Delta_2 cos\Delta_4 + m_{31}sin\Delta_3 sin\Delta_4 + m_{32}cos\Delta_1 sin\Delta_3 sin\Delta_4 \\ &+ m_{33}sin\Delta_1 sin\Delta_2 sin\Delta_3 sin\Delta_4 - m_{34}sin\Delta_1 cos\Delta_2 sin\Delta_3 sin\Delta_4 + m_{41}cos\Delta_3 sin\Delta_4 \\ &+ m_{42}cos\Delta_1 cos\Delta_3 \sin\Delta_4 + m_{43}sin\Delta_1 sin\Delta_2 cos\Delta_3 sin\Delta_4 - m_{44}sin\Delta_1 cos\Delta_2 cos\Delta_3 sin\Delta_4] \end{aligned} \tag{8.29c}$$

Substituting Eq. (8.29a) in Eq. (8.29c), one can obtain different terms with 16 different frequencies, including the fundamental frequencies, as well as the sums and the differences of four fundamental frequencies. Therefore, by Fourier analysis of the detected signal, all these terms can be obtained, with which all 16 Mueller matrix elements can be obtained. Interested readers are encouraged to read the book by Goldstein (2003).

8.2.3 Oscillating Element Mueller Matrix Polarimeters

In Figure 8.3b, an oscillating element Stokes polarimeter is described for obtaining the Stokes vector of an arbitrary SOP. The rapid SOP angle oscillation is achieved with MO rotators described in Chapter 6. As shown in Figure 8.11b, the same Stokes polarimeter, which includes a QWP sandwiched between a pair of MO rotators followed by a polarizer, can be used as the PSA in a Mueller matrix polarimeter. An identical PSA can be used in reverse order as the PSG to generate a series of distinctive SOPs required for obtaining the Mueller matrix of the optical device.

As discussed in Figure 8.3b, a retarder (QWP) sandwiched between a pair of MO rotators rotating in opposite directions is equivalent to the rotating retarder in Figure 8.2c. Therefore the MO based PSA and PSG in Figure 8.11b are equivalent to the rotating retarders in Figure 8.10d.

$$\mathbf{S}_d = \mathbf{P}_2(0)\mathbf{M}_4(-5\omega)\mathbf{M}_{QWP}(\theta)\mathbf{M}_3(5\omega)\mathbf{M}\mathbf{M}_2(-\omega)\mathbf{M}_{QWP}(\theta)\mathbf{M}_1(\omega)\mathbf{P}_1(0)\mathbf{S}_{in} \tag{8.30}$$

Substituting the Mueller matrices $\mathbf{P}_1(0)$, $\mathbf{M}_1(\omega)$, $\mathbf{M}_{QWP}(\theta)$, $\mathbf{M}_2(-\omega)$, $\mathbf{M}$, $\mathbf{M}_3(5\omega)$, $\mathbf{M}_4(-5\omega)$, and $\mathbf{P}_2(0)$ in Eq. (8.30), one obtains the optical power I_d (the first element in detected $\mathbf{S}_d$) by the PD as a function of Mueller matrix elements m_{ij} associated with different frequency components, similar to those in Eq. (8.29c). Therefore, by Fourier analysis of the detected signal, all 16 Mueller matrix elements can be obtained.

8.2.4 Imperfections in Mueller Matrix Polarimeters and Instrument Calibration

The components used in the Mueller matrix polarimeters described earlier have many imperfections. For example, the retardation of a QWP is sensitive to wavelength and temperature and it

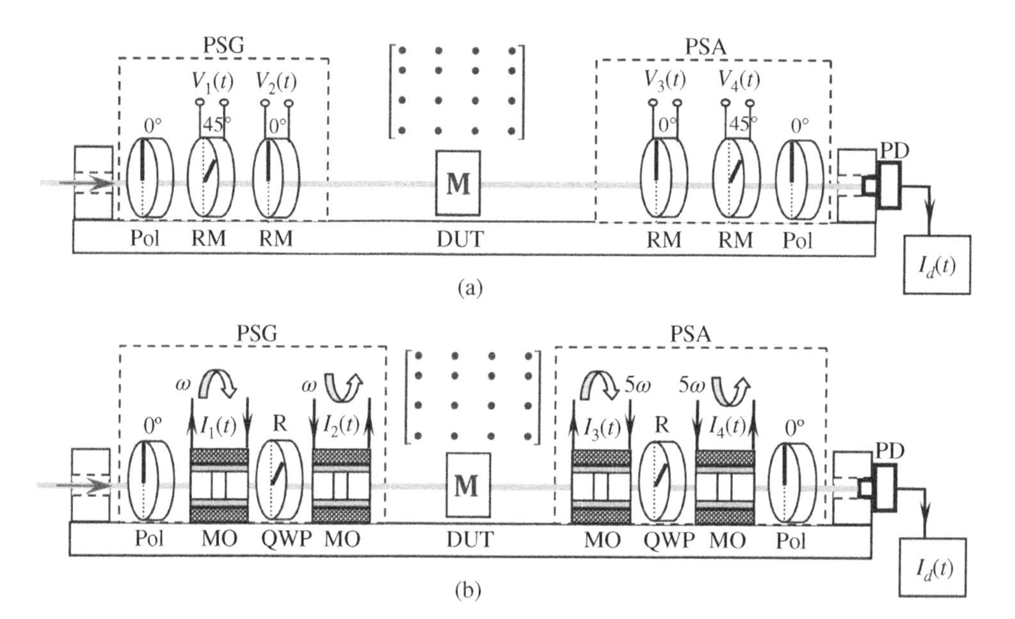

Figure 8.11 (a) Illustration of a complete Mueller matrix polarimeter constructed with two retardation modulators as the PSG and two as the PSA. (b) A complete Mueller matrix polarimeter constructed with a pair of Faraday rotators as the PSG and another pair as the PSA. RM: retardation modulator; MO: Magneto-optic polarization rotator; Pol: polarizer; DUT: device under test.

may not be exactly $\pi/2$ for the wavelength used or under the temperature the measurement is performed. In addition, the relative azimuth angles between the polarizers, between the retarders, and between the polarizers and retarders must be precisely aligned. Slight misalignments may result in significant measurement errors. For example, the angle deviations ε_1, ε_2, ε_3 of the first retarder, the second retarder, and the second polarizer with respect to the first polarizer in the dual-retarder polarimeter of Figure 8.10d should be precisely zero initially before the retarders start to rotate and should remain zero after each retarder completes a 360° rotation. Unfortunately, these deviations are always there and may even vary with aging due to the mechanical wear and tear. Extensive studies have been devoted to the calibration of these types of Mueller matrix polarimeters in attempts to remove the alignment imperfections, as well as the retardation deviations (Azzam et al. 1988a, 1988b, 1989). Interested reader can read (Hauge 1978; Goldstein and Chipman 1990). As will be discussed in Chapter 9, the binary polarization analysis technique can overcome most of the component imperfection issues in the analog Mueller matrix polarimeters discussed in Section 8.2.

8.3 Polarization Extinction Ratio Measurements

To minimize polarization dependent effects described in Section 3.4, it is often desirable to maintain a constant SOP as light propagates through an optical system. With regard to such systems, PER, or polarization crosstalk, is a measure of the degree to which the light is confined in the principal polarization mode as it propagates in the fiber. It is defined as the ratio of the power in the principal polarization mode to the power in the orthogonal polarization mode after propagation through the system, expressed in dB.

$$PER = 10log\left(\frac{P_{principal}}{P_{orthogonal}}\right) \tag{8.31a}$$

As discussed in Section 6.1, a PM optical fiber generally has a strong linear birefringence with a slow axis and a fast axis. Light polarized along the slow axis experiences a large refractive index and therefore travels slower than that polarized in the fast axis. If light input to an ideal PM fiber is polarized along one of the birefringence axes, the SOP will be maintained during propagation through the fiber. For a PM fiber, the PER can be written as

$$PER = 10\left|log\left(\frac{P_{slow}}{P_{fast}}\right)\right| \tag{8.31b}$$

where P_{slow} and P_{fast} are the optical powers in the slow and fast axes of the PM fiber. Large PER generally translates to better polarization stability in the PM fiber system.

While the concept of PER is relatively simple, the parameter can be ambiguous and confusing if the measurement is not done properly. We will discuss how to unambiguously measure or obtain the true PER according to Eqs. (8.31a) and (8.31b).

Four factors generally affect the PER value of a PM fiber. The first is the polarization holding ability of the PM fiber. In practice, when a linearly polarized light is coupled into a PM fiber with its SOP perfectly aligned with the slow axis (or fast axis), a small portion of light can be coupled from the slow axis to the fast axis due to the local imperfections of the PM fiber or the local mechanical stresses on the PM fiber, as will be discussed in detail in Chapter 10. Such a polarization crosstalk will contribute a power, P_{fast}, in the fast axis at the expense of P_{slow}, causing PER to degrade when light travels in the PM fiber. The second factor is the input SOP misalignment. When the input

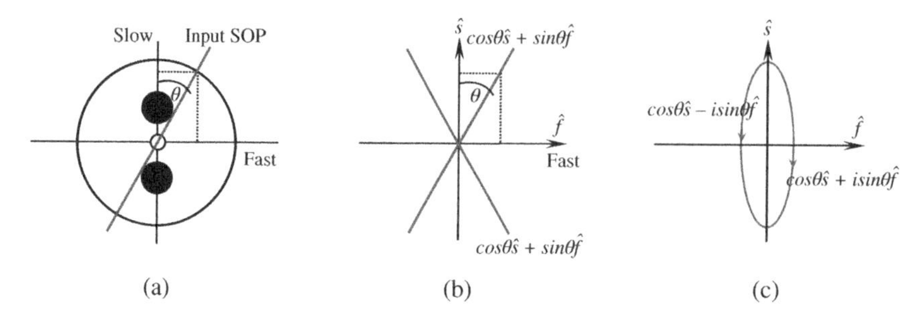

Figure 8.12 (a) Illustration of input SOP with respect to the slow and fast axes of the PM fiber. (b) Output linear SOPs when $\delta = 0, \pi$. (c) Output REP and LEP SOPs when $\delta = \pm\pi/2$.

SOP is not perfectly aligned with the slow axis, as shown in Figure 8.12a, a portion equal to $sin^2\theta$ (corresponding to an electrical field projection of $sin\theta$) will be coupled to the fast axis, while the remaining portion equal to $cos^2\theta$ (corresponding to an electrical field projection of $cos\theta$) is coupled into the slow axis. The PER resulting from such an SOP misalignment is

$$PER = 10\left|log\left(\frac{sin^2\theta}{cos^2\theta}\right)\right| = 10|log(tan^2\theta)| \tag{8.32}$$

The third factor is the ellipticity of the input light, that is, the input SOP is not perfectly linear, but elliptically polarized with a major axis and a minor axis, as described in Chapter 2. In this case, even when the major axis of the polarization ellipse is perfectly aligned with the slow axis of the PM fiber, the PER will also be degraded because the light in the minor axis automatically coupled into the fast axis of the PM fiber. Finally, the fourth factor is the imperfect DOP of the input light. If the input light is not fully polarized ($DOP < 100\%$), there will also be a small portion of light coupled into the fast axis even when the SOP is linear and perfectly aligned with the slow axis, which will also degrade the PER. Therefore, the final PER contains the combined contributions from all four factors.

The polarization component aligned in the slow axis propagates in the PM fiber at a different speed from the polarization component aligned to the fast axis. Therefore, there will be a relative phase difference between the two polarization components:

$$\delta = \frac{2\pi\Delta n_B L}{\lambda} \tag{8.33a}$$

where Δn_B is the birefringence, L is the fiber length, and λ is the wavelength of the light. The resulting SOP at the exit of the fiber will change when the relative phase changes due to temperature variations or mechanical motion on the fiber. The electric field at the exit of the PM fiber can be expressed as

$$\vec{E} = E_0(cos\theta\hat{\mathbf{s}} + sin\theta e^{i\delta}\hat{\mathbf{f}}) \tag{8.33b}$$

where $\hat{\mathbf{s}}$ and $\hat{\mathbf{f}}$ are unit vectors of the slow and fast axes. As shown in Figure 8.12b,c, as δ changes from 0 to π, the output SOP will vary from linear SOP to elliptical SOP with different major axis orientations, which will make it difficult to accurately measure the PER in practice if not done properly.

8.3.1 Rotating Polarizer PER Measurement

The most common method for measuring the PER relies on a rotating polarizer, as shown in Figure 8.13a. Light from a PM fiber first goes through a rotating polarizer before being focused into

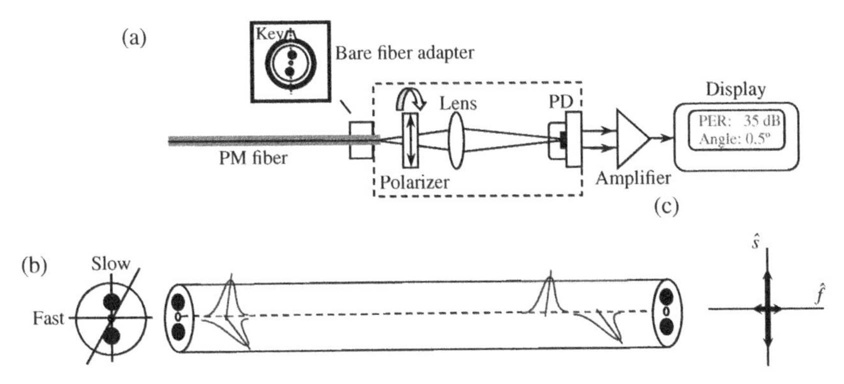

Figure 8.13 (a) The configuration of a PER meter made with a rotating polarizer. (b) Illustration of incoherently decouple the power in the slow and fast axes. In (a), a mechanical reference key can be made on the fiber adapter, which can be used to define the slow axis angle.

a PD. As the polarizer rotating a full circle, a periodical signal with a maximum and a minimum can be obtained. The ratio of the maximum signal V_{max} and minimum signal V_{min} is considered to be the PER of the light exiting from the PM fiber:

$$PER = 10log\left(\frac{V_{max}}{V_{min}}\right) \tag{8.34a}$$

Note that the PER value obtained using Eq. (8.34a) is different from the PER definition of Eqs. (8.31a) and (8.31b) in general. As explained in Figure 8.12, SOP angle corresponding to the maximum power is not along the slow axis, but is determined by the vector summation of the electrical fields from the slow and fast axes of the PM fiber. In addition, V_{max} and V_{min} are not determined by the optical power in the slow and fast axes, but by the major and minor axes of the polarization ellipse of the light field exiting from the PM fiber. Only for the case of Figure 8.12c ($\delta = \pm\pi/2$) for which the major and minor axes are aligned the slow and fast axes of the PM fiber, Eq. (8.34a) is the same as Eq. (8.31b). Unfortunately, such a condition cannot be guaranteed in practice because δ is not controllable.

In order to overcome the difficulty, a light source with a coherence length L_c shorter than differential group delay (DGD) of the PM fiber can be used, as shown in Figure 8.13b, which can be related to the birefringence Δn_B of the PM fiber and fiber length L as

$$L_c < (n_s - n_f)L = \Delta n_B L \tag{8.34b}$$

where n_s and n_f are the refractive indexes of the slow and fast axes, respectively, and $\Delta n_B = n_s - n_f$ is the birefringence of the PM fiber. When this condition is met, the light field in the slow axis no longer overlaps with that in the fast axis in space; therefore, they cannot be coherently added as in Eq. (8.33b). As a result, the maximum signal detected after a rotating polarizer corresponds to the power in the slow axis while the minimum power detected corresponds to the power in the fast axis. Consequently, Eqs. (8.31a) and (8.31b), and the corresponding setup of Figure 8.13a can be used to accurately determine the PER without ambiguity. For a light source with a coherence length of 30 μm and a PM fiber with a typical birefringence Δn_B of 5×10^{-4}, the fiber length required from Eq. (8.34b) is 6 cm. This condition is easy to meet because the fiber pigtails of most devices are about 1 m long. In general, a light source with a band width about 20–30 nm is sufficient to meet the coherence length requirement.

8.3.2 PER Degradation at Fiber Connection

Connecting two pieces of PM fibers is more involved than connecting two pieces of single mode (SM) fibers because the two PM fiber axes must be precisely aligned for minimizing polarization crosstalk, in addition to the precise core alignment for minimum insertion loss. A fusion splice machine can be used to link two PM fibers together, which is capable of automatically rotating the relative angle between the slow axes of the two mating PM fiber while aligning the two fiber cores, as shown in Figure 8.14a.

Another way for interconnecting PM fibers is to pre-install a connector on an end of a PM fiber, as shown in Figure 8.14b, in which the slow (or fast) axis is aligned to a key. The adapter for mating two PM fibers has two key slots on each side to ensure the two keys on the two PM fiber connectors are aligned, which in turn ensures the alignment of the two PM fiber axes.

If the PER of the first PM fiber is PER_1 and the PER of the second fiber is PER_2, as shown in Figure 8.15, the total measured PER can be expressed as (assuming the PERs are sufficiently large)

$$PER_{Total} = -10log\left(10^{-PER_1/10} + 10^{-PER_2/10}\right) \tag{8.35a}$$

The PER of the second fiber can be obtained as

$$PER_2 = -10log\left(10^{-PER_{Total}/10} - 10^{-PER_1/10}\right) \tag{8.35b}$$

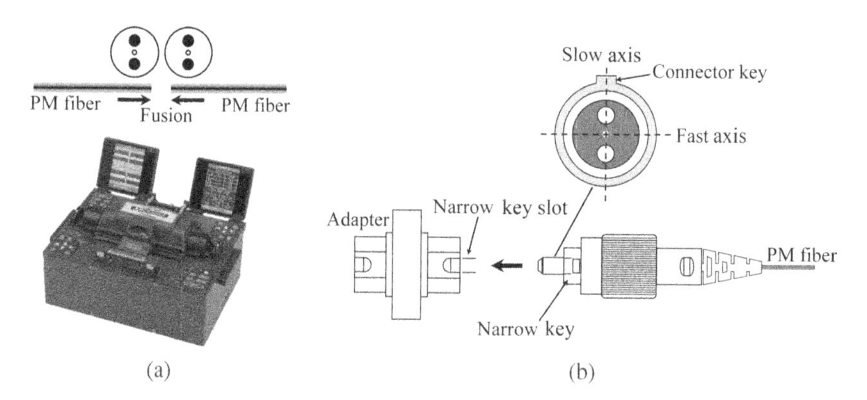

Figure 8.14 (a) Connecting two PM fibers with fusion machine capable of automatically rotating and aligning the fiber slow axes. (b) Connecting two PM fibers with fiber connectors with a key aligned with the fiber slow axis of each fiber. Source: Fujikura Europe Ltd.

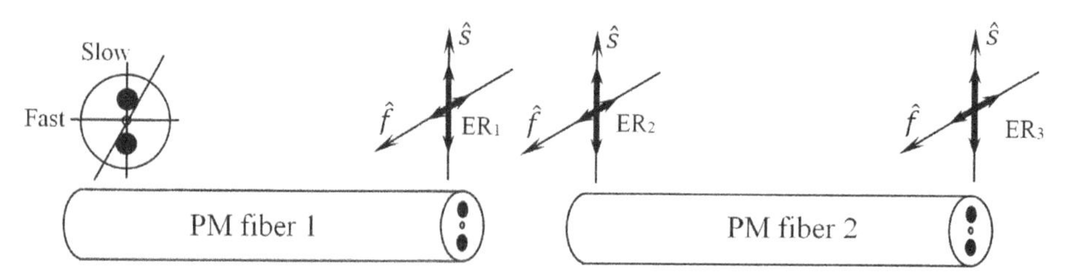

Figure 8.15 PER of two cascaded PM fibers.

8.3.3 Polarization Maximization for Fast PER Measurement

Figure 8.16a shows a common PER measurement setup, in which a fiber pigtailed in-line polarizer is used to maximize the PER from the broadband light source. The output from the polarizer is then fusion spliced to or connected with connectors to the PM fiber pigtail of the DUT. Two error sources are generally encountered in this setup: (i) the alignment between the polarizer and the PM fiber slow axis, as shown by Inset 1 in Figure 8.16a; (ii) the alignment between the slow axes of the polarizer's output PM fiber and the PM fiber pigtail of the DUT, as shown by Inset 2 in Figure 8.16a. The DUT can be a length of PM fiber, or a PM fiber pigtailed device, such as a polarizer, a PBS, or a coupler. To minimize the impact of these error sources, one may first measure the PER of the in-line polarizer before using the setup of Figure 8.16a to measure the PER of the DUT. Finally, Eq. (8.35b) can be used to obtain the PER of the DUT. Such a process is cumbersome and time consuming, with poor repeatability.

To overcome the difficulties described earlier, a device called PER maximizer is introduced by In-line Photonics, Inc., which can be used together with a PER meter to automatically align the linear SOP with the slow axis of DUT's PM fiber pigtail while measuring the PER at high speed. Collectively, they reduce the total PER measurement time (including setup time) to under two minutes while assuring a high measurement accuracy with an uncertainty of less than 0.25 dB, which can be more than 10 times improvement for both the measurement time and accuracy over conventional PER measurement systems.

As shown in Figure 8.16b, the PER maximizer is a motorized linear polarization generator which converts input broadband light to highly linearly polarized light ($PER > 45$ dB) in free space and couples the light directly into the PM fiber pigtail of the DUT. The alignment angle between the linear polarization and the PM fiber axis can be finely adjusted by a motor to achieve over a 45 dB maximum PER. The measured PER is used to control the motorized polarizer for maximizing the PER value until a maximum PER is reached. This setup automatically eliminates the two error sources described earlier, which enables accurate determination of the true PER of the DUT in a

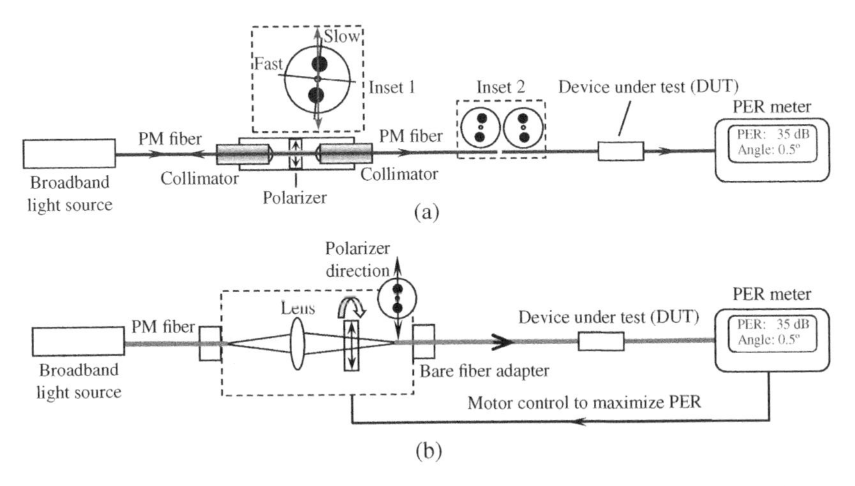

Figure 8.16 (a) PER measurement setup using an in-line polarizer to get a high PER output from the light source. (b) PER measurement setup using a motorized polarization optimizer to automatically get rid of fiber axis alignment errors.

single measurement. Such a motorized polarizer can significantly speed up the measurement with a high repeatability.

8.3.4 PER Measurement with a Stokes Polarimeter

The PER measurements using a rotating polarizer based PER meter require a broadband light source of short coherent length to ensure measurement accuracy. However, in some applications, such as coupling light from a narrow bandwidth DFB laser into a PM fiber pigtail, as shown in Figure 8.17, one does not have the option of having a short coherent length light source. Therefore, alternative solutions are required.

As discussed in Figure 8.12, when the SOP of the laser beam is not precisely aligned with the slow (or fast) axis of the PM fiber, the SOP of the output light from the fiber pigtail will change whenever the retardation δ of PM fiber is changed due to the variations of laser wavelength, temperature, or stress on the PM fiber. The PM fiber is in fact a retarder. As discussed in Section 4.5.2, if δ changes from 0 to 2π, the SOP would trace out a circle on the Poincaré sphere around s_1 axis.

8.3.4.1 PM Fiber PER Measurements

A Stokes polarimeter or a PSA can be used to precisely measure the PER. As shown in Figure 8.18a, if the slow axis of a PM fiber is aligned to the reference plane of the polarimeter at the entrance, the normalized SOP can be calculated as (Born and Wolf 1999; Yao et al. 2010)

$$\mathbf{SOP} = \begin{pmatrix} a_1^2 - a_2^2 \\ 2a_1a_2cos\delta \\ 2a_1a_2sin\delta \end{pmatrix} = \begin{pmatrix} cos2\theta \\ sin2\theta cos\delta \\ sin2\theta sin\delta \end{pmatrix} \tag{8.36a}$$

where θ and δ are the misalignment angle and the phase difference between the slow and fast axes, respectively. Variations in the input wavelength or in the fiber length due to temperature changes or mechanical stress will cause a variation in the relative phase δ, which in turn will cause the SOP to rotate about axis (1,0,0) on the Poincaré sphere when the relative phase δ is changed.

If the slow axis is misaligned to the polarimeter's reference plane by an angle 2ψ, the measured normalized polarization state can be expressed as

$$\mathbf{SOP} = \begin{pmatrix} cos2\psi cos2\theta - sin2\psi sin2\theta cos\delta \\ sin2\psi cos2\theta + cos2\psi sin2\theta cos\delta \\ sin2\theta sin\delta \end{pmatrix} \tag{8.36b}$$

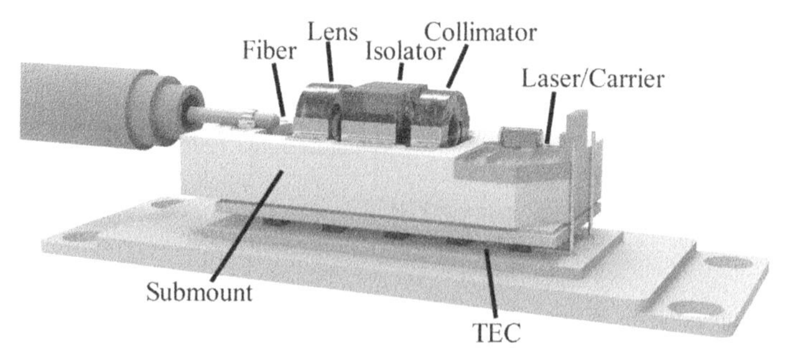

Figure 8.17 Attaching a PM fiber pigtail to a DFB laser requires aligning the SOP from the laser chip to the slow axis of the PM fiber. During the alignment process, the laser wavelength can be modulated by temperature modulation via the thermal electrical cooler (TEC) or by stretching the PM fiber.

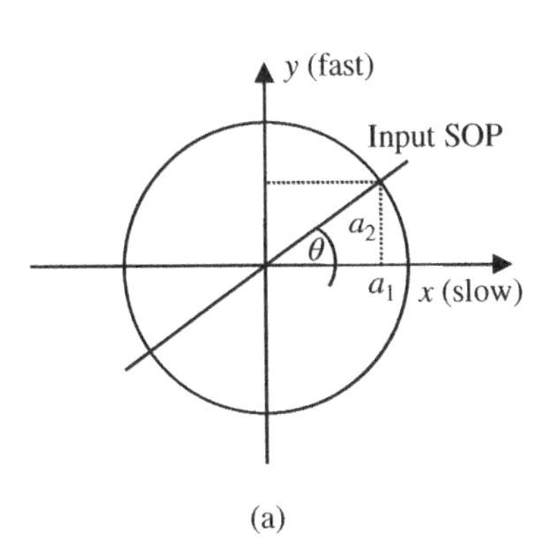

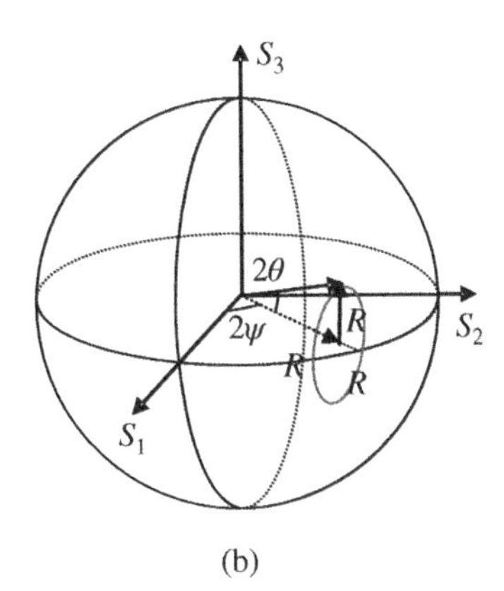

Figure 8.18 (a) Illustration of a linearly polarized light misaligned by an angle θ from the slow axis of a PM fiber. (b) Poincaré sphere illustration of polarization state rotation of output light from a PM fiber due to wavelength variation or to thermal or mechanical stress.

It can be shown that the output SOP rotates about the slow axis $(cos2\psi, sin2\psi, 0)$ to trace out a circle as δ increases or decreases, as illustrated in Figure 8.18b. Note that changes in wavelength, fiber length, or temperature will cause δ to change. The radius of the circle can be calculated to be $R = sin\,2\theta$ so that the PER can be calculated from Eq. (8.32) as

$$PER = -10log\left(\frac{sin^2\theta}{cos^2\theta}\right) = -10log\left(\frac{1 - cos2\theta}{1 + cos2\theta}\right) = -10log\left(\frac{1 - \sqrt{1 - R^2}}{1 + \sqrt{1 - R^2}}\right) \tag{8.37}$$

This result is the same as Eq. (4.342) obtained in Section 4.5.3. Essentially, the PER measured by a Stokes polarimeter, defined in Eq. (8.37), indicates the degree of linear polarization of light inside the PM fiber, or how well the input linear polarization is aligned with the slow (or fast) axis of the PM fiber. For a perfect polarization alignment, R approaches 0 (the circle collapses to a point), and the corresponding PER approaches infinity. At the other extreme, if the light power is equally split between the fast and slow axes, R approaches 1 (the SOP traces a great circle on the circumference of the Poincaré sphere) and the corresponding PER approaches 0.

Note that it only requires a fraction of a circle to calculate the radius R of the circle and therefore the range required for the retardation ($\delta = 2\pi\Delta n_B L/\lambda$) variation is only a fraction of 2π. For example, if only 1/4 of the SOP circle is required, the range of retardation variations required is only $\pi/2$, corresponding to a wavelength scan range of $\Delta\lambda = \lambda_0^2/(4\Delta n_B L)$. For a fiber length of 1 m with a $\Delta n_B = 5\times10^{-4}$, the wavelength scan range required is 1.125 nm around a center wavelength of 1500 nm. Such a wavelength change range may be accomplished by thermally modulating the laser chip with the TEC, as shown in Figure 8.17. Other methods, such as periodically stretching or heating/cooling a section of the PM fiber, can also be applied to change the retardation δ so as to induce an arc or circle on the Poincaré sphere for obtaining the PER if it is difficult to scan the laser wavelength or frequency.

8.3.4.2 PM Fiber Connector Key Orientation Measurements

If light output from the PM fiber is directly coupled into the PSA via free space, there is no alteration of the SOP before the light is analyzed, the rotation axis of the circle is in the plane of the Poincaré sphere's equator, and the angle ψ is the angle between the slow axis and the reference plane of the PSA, as shown in Figure 8.18b. The PSA adapter generally has a reference key slot to accept a keyed PM fiber connector. Because the key slot is perpendicular to the PSA's reference plane, the angle between the slow axis of the PM fiber and the connector key is therefore 90-ψ degrees. For example, $\psi = 90°$ (the circle rotates around S_2 in Figure 8.18b) means that the slow axis of the PM fiber is vertical and aligned to the alignment key direction, and $\psi = 0°$ indicates that the slow axis of the PM fiber is horizontal and perpendicular to the alignment key direction.

8.3.4.3 PM Fiber Connector Stress Measurements

The location of the SOP circle induced by scanning the wavelength of the light source or stretching the PM fiber under test can actually indicate the stress in the fiber tip in the connector, as explained in the following text. Again, if the light output from the PM fiber into the PSA is of free-space, there is no alteration of the SOP before the light is analyzed such that the SOP should always be circled around an axis on the equator. However, in practical measurement, the center of the circle may deviate from the plane of the equator, which is caused by stress induced birefringence in the connector. The effect is equivalent to that of a PM fiber followed by a waveplate. Generally speaking, a greater deviation from the equator indicates a higher stress level. Such a measurement capability is very useful for PM fiber connectorization, because it can reveal information about whether an epoxy or a curing process used for the connectorization induces enough stress on the fiber to cause degradation of the PER. The final degraded PER can be measured with a standard PER meter made with a rotating linear polarizer described in Figure 8.13a.

Figure 8.19 shows PER measurement results for two PM fibers with FC/PC connectors. The effect of stress is clearly shown in one of the PM connectors. In Figure 8.19a, the SOP evolution circle is centered around a point on the equator, indicating negligible stress on the output fiber tip caused by connectorization. Consequently, the difference in PER values measured with a PSA and with a rotating polarizer is negligible. In contrast, in Figure 8.19b, the center of the SOP evolution circle is displaced from the equator, indicating a relatively large stress on the fiber. The stress induced birefringence of the output fiber tip can therefore be expected to degrade the output PER. Indeed, the PER of the light after the output fiber connector is reduced to 31.1 dB measured with a rotating polarizer, although the PER of the light inside the fiber is as high as 39.6 dB measured with a PSA using wavelength scanning method described earlier.

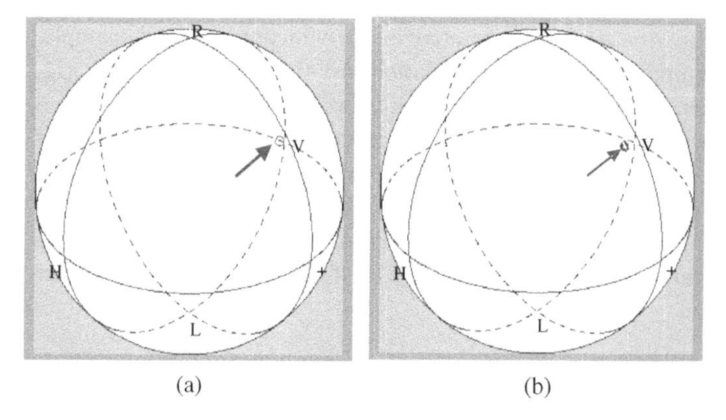

Figure 8.19 PER measurement results: (a) PER measurement result with low stress on output FC/PC connector. (b) PER measurement result with relatively high stress on output FC/PC connector. The corresponding PER of light in the PM fiber is 39.6 dB; however, the PER of the output light is reduced to 31.1 dB due to stress induced birefringence. The deviation of the SOP evolution circle from the equator can be used to indicate stress induced birefringence at the exit end of the PM fiber.

In conclusion, by observing the size and position of the SOP evolution circle on the Poincaré sphere, one can readily distinguish the PER degradation of a PM fiber caused by input SOP misalignment (size of the circle) from that caused by connectorization induced stress (deviation of the circle from the equator). Such a capability is a useful tool for determining the quality of PM fiber connectorization by identifying stress induced birefringence.

8.3.5 Distributed Polarization Crosstalk Measurement Method

An interferometer-based distributed polarization crosstalk analyzer to be discussed in Chapter 10 is able to measure the magnitude and location of all crosstalk events in highly birefringent DUTs such as PM fiber coils or Y-branch waveguides. PER can be calculated by integrating the effects of all crosstalk events in the DUT. This type of measurement provides the most detailed information about a complex PM fiber system. Because it measures the crosstalk caused by individual features of the DUT such as connectors, splices, or sections of stressed or defective PM fiber, it is possible to evaluate the contribution of these individual features to the total PER. It is also possible to exclude sections of the DUT from the PER calculation. Because such a polarization crosstalk analyzer has very high crosstalk measurement sensitivity, it can characterize extremely high PER DUTs.

8.3.6 PER of Free-Space Optical Polarization Components

For measuring the PER of a free-space component, such as a polarizer or PBS, a depolarized light source and a rotating polarizer based PER meter can be used. The PER of the free-space component can be obtained from Eq. (8.34a) as

$$PER = 10log\left(\frac{V_{max}}{V_{min}}\right)$$

where V_{max} and V_{min} are the maximum and minimum detected signal when the polarizer rotates 360°.

8.4 PDL, PDG, and PDR Measurements

As mentioned in Chapters 2 and 3, the PDL of optical components and polarization dependent gain (PDG) of optical amplifiers are critical parameters for optical fiber systems. The presence of PDL and PDG in the transmission link not only causes signal-to-noise ratio (SNR) fluctuation in direct modulated links, but also destroys orthogonality of polarization multiplexed signals in coherent detection systems, making it impossible to fully recover polarization related signal distortion through digital signal processing (DSP). To ensure network quality of service, the PDL of each component should be minimized and the cumulative PDL of the network must be managed within a desired limit. Therefore, accurate PDL measurements are required for almost all optical components, sub-modules, and modules as part of the specified standard parameters (Telcordia 2001). Once the optical component PDL values are known, the system global PDL due to multiple PDL concatenation can be statistically estimated from individual component PDL data (El Amari et al. 1998).

On the other hand, the presence of PDR of PDs greatly affects the detection repeatability and accuracy of optical sensing and measurement systems. Therefore, accurate measurement of PDL, PDG, and PDR is important. As will be seen next, similar methods can be used to measure some of these parameters.

8.4.1 Polarization Scrambling Method for PDL and PDG Measurements

As discussed in Chapter 3, the PDL or PDG of an optical component is the maximum optical power transmittance change over all polarization states, defined as $PDL = 10\,log\,(T_{max}/T_{min})$ in dB, where T_{max} and T_{min} are the maximum and minimum power transmissions. Together, PDL and PDG are sometimes referred to as polarization dependent transmission (PDT). One may use a manual polarization controller to adjust the SOP of light entering the DUT while measuring the transmitted optical power with a power meter. A rough PDL or PDG value can be obtained by manually searching for the maximum and minimum powers via controlling the SOP. For large-scale production environments, automatic PDL measurement is required because manual measurement at a single wavelength can take several minutes or longer. For measurement over a large wavelength range, say, 40 nm at 1 nm interval, it may take one or two hours to measure a single component manually.

As shown in Figure 8.20a, polarization scrambling described in Section 7.2 is the most straight forward method for measuring the PDL and PDG of a component. When the input SOP to the DUT is scrambled by a polarization scrambler, the optical power of light passing through the DUT will fluctuate due to its PDL or PDG. An optical power meter with a response speed faster than the SOP scrambling rate can be used to detect the power fluctuations for determining the maximum and minimum powers P_{max} and P_{min}. The PDL or PDG of the device can be obtained by definition as $10\,log\,(P_{max}/P_{min})$.

For actuate measurement, this method requires that (i) the PDL and activation loss (AL) of the polarization scrambler is much lower than the PDL of the DUT, as discussed in Chapter 6, (ii)

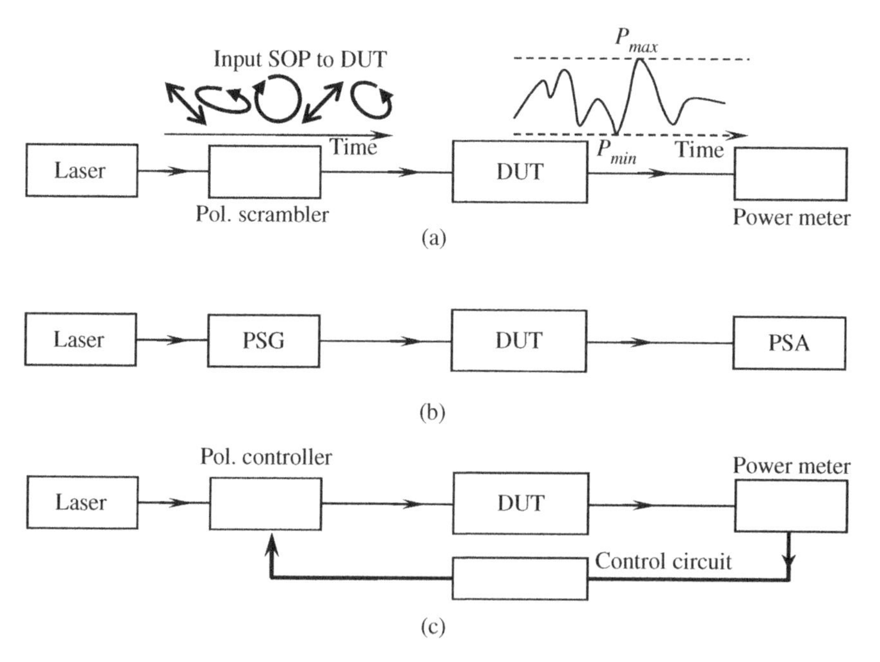

Figure 8.20 Measurement methods for obtaining PDL, PDG, and PDR. (a) Polarization scrambling and the maximum–minimum search methods for measuring PDL and PDG of optical components. (b) Mueller matrix polarimetry method for measuring the PDL and PDG of optical components. (c) The maximum–minimum search methods for measuring the PDL and PDG of optical components.

the polarization scrambler is capable of generating SOPs that uniformly cover the entire Poincaré sphere, (iii) the PDR of the power meter is much smaller than the PDL of the DUT, and (iv) the power fluctuation or drift of the light source is much smaller than the PDL of the DUT. The all-fiber polarization scramblers based on fiber squeezers described in Figure 6.15c are ideal for such an application due to the extremely small PDL (<0.005 dB) and AL (<0.005 dB). The all fiber Lefevre controller of Figure 6.11c or Yao controller of Figure 6.12c can also be used to scramble the SOP; however, their speeds are too slow to be efficient. The polarization scramblers made with $LiNbO_3$ are not suited for this application due to the high PDL and AL, despite their high scrambling speed.

Note that the PDL measurement accuracy and repeatability of the polarization scrambling based method are limited because of the uncertainties for the random SOP to hit the states corresponding to the maximum and minimum transmissions of the DUT in a limited time span, especially for devices with high PDL.

8.4.2 Jones and Mueller Matrix Analysis Methods

These methods sometimes are also called deterministic fixed-states method, which refers to techniques that employ a set of well-defined input SOPs to derive global polarization dependence through Jones (Heffner 1992) or Mueller matrix analysis (Craig et al. 1998; Craig 2003). As shown in Figure 8.20b, a PSG is used to generate the set of SOP and a PSA, such as a polarimeter, is used to analyze the SOP after light passing through the DUT. As discussed in Section 8.2, with the PSG and PSA, the Mueller or Jones matrix containing all the polarization related information of the DUT can be obtained. PDT of the DUT can then be determined.

8.4.2.1 Mueller Matrix Method (MMM)

As detailed in Appendix 8.A, if the Mueller matrix of the DUT is obtained with a Mueller matrix polarimeter described in Section 8.2, the maximum and minimum transmissions can be obtained as

$$T_{max} = m_{00} + \sqrt{m_{01}^2 + m_{02}^2 + m_{03}^2} \tag{8.38a}$$

$$T_{min} = m_{00} - \sqrt{m_{01}^2 + m_{02}^2 + m_{03}^2} \tag{8.38b}$$

where m_{0i} $(i = 0, 1, 2, 3)$ are the first row Mueller matrix elements. The PDL of the DUT can then be obtained as

$$PDL = 10log\left(\frac{T_{max}}{T_{min}}\right) = 10log\frac{m_{00} + \sqrt{m_{01}^2 + m_{02}^2 + m_{03}^2}}{m_{00} - \sqrt{m_{01}^2 + m_{02}^2 + m_{03}^2}} \tag{8.38c}$$

Note that only the first row Mueller matrix elements are involved in the PDL calculation because only the first element S_0 in the Stokes vector is related to the optical transmission power.

8.4.2.2 Jones Matrix Eigenanalysis (JME) Method

PDL is essentially the partial polarizer discussed in Section 4.3.4. If the Jones matrix **M** of the DUT is first measured, we can construct a new matrix **H**:

$$\mathbf{H} \equiv \mathbf{M}^{\dagger}\mathbf{M} = (\mathbf{M}^T)^*\mathbf{M} = \begin{pmatrix} h_{11} & h_{12} \\ h_{21} & h_{22} \end{pmatrix} \tag{8.38d}$$

where the superscript T and the star "*" stands for transpose and conjugate, respectively. With the matrix elements of **H** determined, the PDL of the DUT can be obtained from Eq. (4.135) as

$$PDL = 10log\frac{tr(\mathbf{H}) + \sqrt{(tr(\mathbf{H}))^2 - 4det(\mathbf{H})}}{tr(\mathbf{H}) - \sqrt{(tr(\mathbf{H}))^2 - 4det(\mathbf{H})}}$$
$$= 10log\frac{(h_{11} + h_{22}) + \sqrt{(h_{11} + h_{22})^2 - 4(h_{11}h_{22} - h_{12}h_{21})}}{(h_{11} + h_{22}) - \sqrt{(h_{11} + h_{22})^2 - 4(h_{11}h_{22} - h_{12}h_{21})}} \tag{8.38e}$$

The Jones matrix **M** can be obtained following the procedure described in Section 4.3.5.

The PDL and AL of the PSG, as well as the power fluctuations of the light source, will degrade the PDL measurement accuracy of the DUT. In addition, the measurement system must be calibrated against temperature and wavelength variations to ensure accuracy. Chapter 9 will have more discussions on using Jones and Mueller matrices to obtained PDL and PDG of fiber optic devices utilizing a pair of binary PSG and PSA.

8.4.3 Maximum–Minimum Search Method for Accurate PDL and PDG Measurements

To overcome the measurement uncertainty issues associated with the polarization scrambling method, a fully automatic, deterministic maximum–minimum (max–min) search method for fast, accurate PDL and PDG characterization has been developed (Shi et al. 2006). As shown in Figure 8.20c, the measurement system consists of a polarized light source (DOP ~ 100%), a polarization controller, the DUT, and a power meter or PD. Instead of blindly generating thousands of random SOPs as in Figure 8.20a, here the polarization controller systematically changes the SOP according to the optical power detected to find the maximum and minimum optical power in just tens of steps.

A fiber squeezer based polarization controller described in Figure 6.15c can be used because of its extremely low PDL and AL. In particular, the polarization controller has three fiber squeezers oriented at 0°, 45°, and 0°, respectively. The phase retardation of squeezer is linearly proportional to the voltage applied to the piezoelectric actuator. The phase retardation of the ith squeezer can be expressed as

$$\varphi_i = \frac{\pi V}{V_{\pi i}} + \varphi_{0i} \tag{8.39a}$$

where $V_{\pi i}$ and φ_{0i} are the half-wave voltage and initial retardation bias of squeezer i, respectively. Neglecting IL and PDL, the Mueller matrix **M** of the polarization controller is the multiplication of the matrices of the three fiber squeezer retarders: $\mathbf{M}(\varphi_3, \varphi_2, \varphi_1) = \mathbf{M}_3(\varphi_3, 0)\mathbf{M}_2(\varphi_2, \pi/4)\mathbf{M}_1(\varphi_1, 0)$. Using the retarder matrix from Table 4.3, one obtains

$$\mathbf{M}(\varphi_3, \varphi_2, \varphi_1)$$
$$= \begin{bmatrix} 1 & 0 & 0 & 0 \\ 0 & cos\varphi_2 & -sin\varphi_2 sin\varphi_1 & sin\varphi_2 cos\varphi_1 \\ 0 & -sin\varphi_3 sin\varphi_2 & cos\varphi_3 cos\varphi_1 - sin\varphi_3 cos\varphi_2 sin\varphi_1 & cos\varphi_3 sin\varphi_1 + sin\varphi_3 cos\varphi_2 cos\varphi_1 \\ 0 & -cos\varphi_3 sin\varphi_2 & -sin\varphi_3 cos\varphi_1 - cos\varphi_3 cos\varphi_2 sin\varphi_1 & -sin\varphi_3 sin\varphi_1 + cos\varphi_3 cos\varphi_2 cos\varphi_1 \end{bmatrix} \tag{8.39b}$$

As each φ_i varies between 0 and 2π, any arbitrary output SOP can be obtained from any input SOP. Complete Poincaré sphere coverage ensures that the search results are the true maximum and minimum transmittances.

The simplest automatic maximum–minimum search operation can be achieved using a DC control voltage actuation algorithm. Starting from an arbitrary SOP, the controller takes an initial optical power measurement and steps the DC voltage applied to one fiber squeezer of the polarization controller while holding the voltages at the other squeezers constant. A new optical power or PD output is obtained and is then compared with the previous value. Based on the comparison result of the measured power or photo-voltages, the controller determines the direction and amplitude of the next voltage step. This search process repeats for each squeezer until the optical power change is within the limit of the noise floor or the resolution of the detector circuit. When the maximum is reached, the control circuit records the signal level value P_{max} and starts to search for the other extreme P_{min}. After both P_{max} and P_{min} are obtained, the PDL can be computed as $10\,log\,(P_{max}/P_{min})$. It can be shown that P_{max} and P_{min} are unique, without local maximums and minimums to mislead measurement (Shi et al. 2006).

A different signal processing approach is to combine a small sinusoidal modulation with the DC control voltage to the fiber squeezers. At the maximum and minimum power levels, the ratio of second harmonic to fundamental frequency of the sinusoidal modulation signal is maximized. A sensitive lock-in amplifier or DSP circuit can be implemented to accurately determine the maximum and minimum transmittances using this algorithm.

Similar to the polarization scrambling method, this method also requires that the PDL and AL of the polarization controller, the PDR of the power meter or PD, and the power fluctuation of the light source are much smaller than the PDL of the DUT in order for the measurement to be accurate.

Figure 8.21 shows the performance data of a commercial PDL meter based on the max–min research method described earlier. A tunable laser with a wavelength range of 1520–1620 nm is used as the light source in Figure 8.20c. A fiber-squeezer polarization controller with half-wave voltages around 30 V is used in the measurement system, which has an AL and PDL of less than 0.01 dB. The typical phase retardation range is $0\text{–}5\pi$ for each fiber squeezer. The photo receiver circuit consisted of an InGaAs detector and an adjustable gain transimpedance amplifier and no attempt is made to correct the PDR of the PD.

Three different test devices were selected to cover a large PDL range. An FC/APC connector–air interface was used at the low PDL extreme, while an in-line fiber polarizer represented the high PDL extreme. The third component was a regular 2×2 10-dB fused fiber coupler. The PDL of each component was measured repetitively over a period of more than 30 minutes, and the measurement results were recorded with a personal computer. As shown in Figure 8.21a,b, the maximum–minimum search method performs consistently at both very low and very high PDL extremes, which were challenges to most of the commercial PDL meters using other approaches.

For comparison purposes, the PDL of the 2×2 fused-fiber coupler's 10-dB port was measured using the maximum–minimum search method and the Jones matrix eigenanalysis (JME) method on a commercial instrument. As shown in Figure 8.22a, the maximum–minimum search approach resulted in a much tighter distribution than the commercial instrument's measurement. The standard deviations of the measured PDL using the maximum–minimum search and JME methods were 0.002 and 0.009 dB, respectively.

The measurement repeatability over time and wavelength was investigated by conducting PDL measurements over two weeks at wavelengths from 1520 to 1620 nm in 2 nm steps. A 2×2 directional coupler was used as the DUT sample. The coupler and fiber jumpers were taped to the surface of an optical table to minimize data fluctuations due to photodetector PDR. Except for a few wavelengths near 1590 nm, the maximum–minimum system demonstrated a repeatability of 0.02 dB for all wavelengths over the test period, as shown in Figure 8.22b.

Note that in Figure 8.21a, the measured PDL of the FC/APC connector–air interface is a "vector" sum of connector PDL and photodetector PDR, as discussed in Figure 3.9. Nevertheless, the measurement stability at very low PDL was clearly demonstrated. Based on the experimental results one can see that the maximum–minimum search approach offers many unique advantages. It effectively reduces a two-dimensional pseudo random search over the entire Poincaré sphere to a deterministic quasi-one dimensional line search with unique search results. Therefore, an active and efficient search-and-measure algorithm can be implemented for fast PDL measurements, with a measurement time of less than 30 ms has been achieved commercially. Using a tunable laser as the light source, dense wavelength division multiplexing (DWDM) devices can be quickly characterized in less than 2.4 seconds for over 80 wavelengths.

Like the polarization scrambling approach, the maximum–minimum search method does not require optical power or wavelength calibration matrices. Therefore, its operation wavelength range is very wide, limited only by the light source, fiber bandwidth, and detector response. This approach also offers a large measurement dynamic range for PDL. PDL in the over 30 dB range can be measured accurately, as demonstrated in Figure 8.21b.

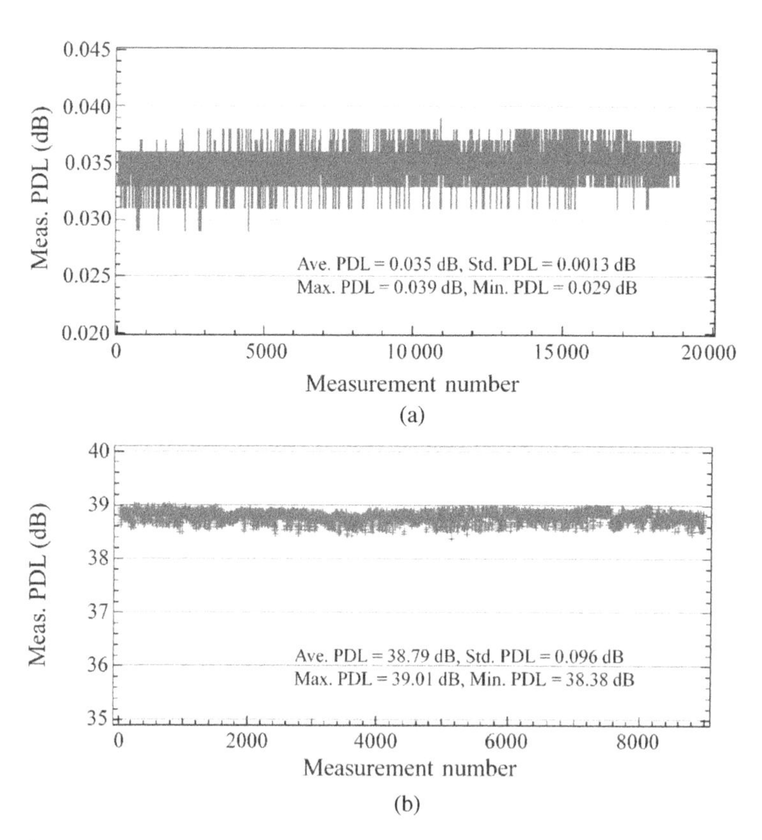

Figure 8.21 PDL measurement data and statistics for (a) an FC/APC–air interface and (b) an in-line fiber polarizer at 1550 nm. FC is a commonly used connector type, while APC stands for angle polished connector whose end face has an angle of 8°. Source: Reproduced from Shi et al. (2006)/IEEE.

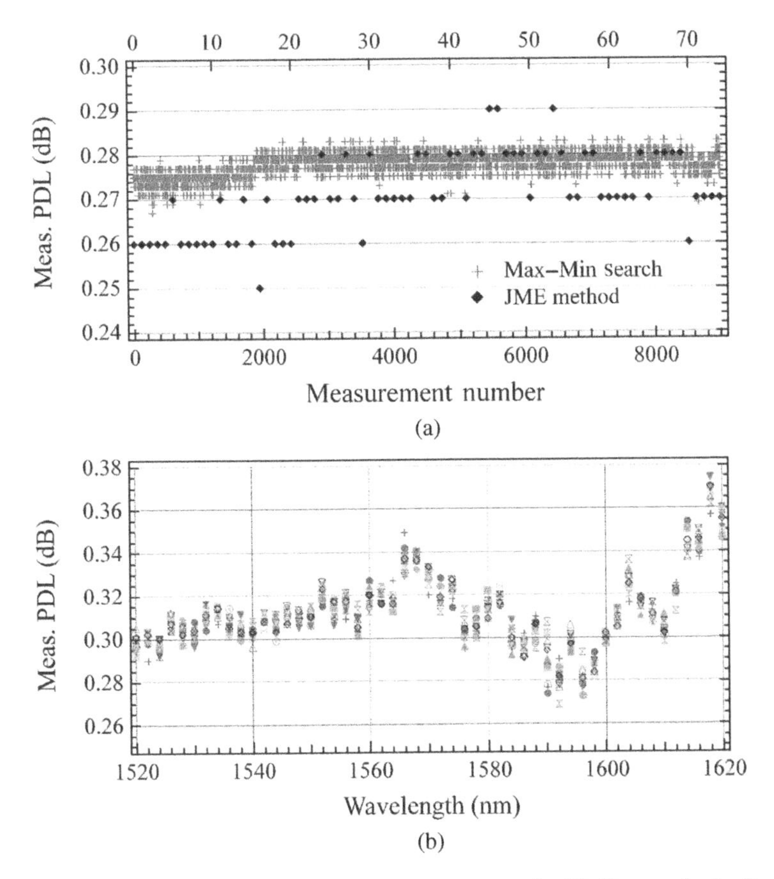

Figure 8.22 (a) PDL repeatability measurement on the 10-dB port of a 2 × 2 coupler using a maximum–minimum search setup, as compared with the results of Jones eigenanalysis (JME) measurement using a commercial instrument (measurement numbers on top axis). (b) Long-term automatic maximum–minimum search PDL measurement repeatability over C- and L-band wavelengths measured for a 2 × 2 fused coupler. Source: Reproduced from Shi et al. (2006)/IEEE.

8.4.4 PDL Measurement Guidelines

PDL measurement is extremely sensitive to unwanted variations in the measurement system, including light source stability, connector back reflections, and even the layout of the test fiber cables. Large measurement inaccuracies or fluctuations may occur if the test setup is not properly arranged even though a high accuracy measurement instrument is used. This section describes precautions for accurate PDL measurements in general and ways to minimize measurement inaccuracies using a maximum–minimum search PDL meter as an example.

8.4.4.1 Comparison of Different Methods

As described earlier, the scrambling method uses a polarization scrambler to generate a pseudo-random subset of all possible SOPs at the input light source; PDL is calculated from the maximum and minimum intensities measured during the SOP scan. The Jones and Mueller matrix polarimetry method measures optical transmission at a set of fixed SOPs. PDL is calculated from the matrix elements obtained from the optical intensity measurements. Both the methods have advantages and limitations in different aspects such as measurement speed, accuracy,

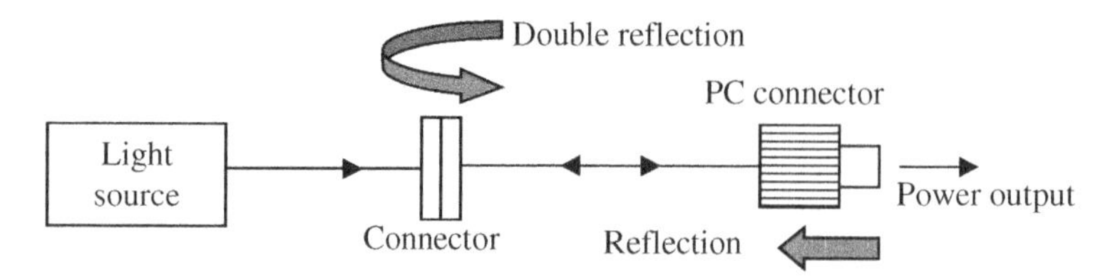

Figure 8.23 Illustration of double reflection.

optical bandwidth, and calibration requirements. In general, these two methods yield accurate PDL measurement values typically between 0.05 and 15 dB. At high and low PDL extremes, the measurement accuracy deteriorates significantly. The main reason for the accuracy deterioration at the high and low PDL extremes is that both methods do not truly measure the maximum and minimum transmittances used for PDL calculation. Small measurement errors or circuit noise can affect the results significantly. In addition, the matrix analysis measurements require system wavelength calibrations. On the other hand, the maximum–minimum search method systematically searches for the maximum and minimum power transmissions through the DUT by feedback control, because it measures PDL and PDG values according to their definitions, without the need for scanning over a large number of polarization states or engaging in extensive intermediate calculations. Therefore, very large PDL measurement over 35 dB can be accurately obtained. In addition, this method provides the attractive features of wavelength insensitivity and calibration free operation, promising high speed, highly accurate PDL and PDG characterization over a wide wavelength band and a wide dynamic range.

8.4.4.2 Error Caused by Light Source Fluctuation

Because $PDL = 10\,log\,(P_{max}/P_{min})$, if the power of the light source fluctuates with time, the measured P_{max} and P_{min} will also fluctuate, resulting in measurement inaccuracies. Therefore, the light source used for PDL measurement must be highly stabilized.

Even if the light source itself is highly stabilized, small back reflections in various positions in the measurement system may feedback to the laser, disturbing its operation and causing output instabilities. Therefore, an isolator is suggested to be used at the input of the instrument to minimize the back reflections, even though the laser source may already have an isolator at its output. In addition, all the connectors from the light source to the PDL meter must be angle polished connector (APC) type to minimize connector back reflection.

8.4.4.3 Error Caused by Double Reflections

There may be small reflections from some components used in the measurement setup, including the connectors and DUT itself. As shown in Figure 8.23, a reflected light from a component may be reflected again by another component. The double reflected light will travel in the same direction as the main input light and therefore interference with the main light. The total output optical power is thus

$$P = P_{in} + P_{dr} + 2\sqrt{P_{in}P_{dr}\hat{\mathbf{e}}_{in}\hat{\mathbf{e}}_{dr}cos\varphi} \tag{8.40a}$$

where P_{in} and P_{dr} are the powers of the main and double reflected lights, $\hat{\mathbf{e}}_{in}$ and $\hat{\mathbf{e}}_{dr}$ are the unit complex vectors for SOP of the main and double reflected lights, and φ is the relative phase between them.

Because the relative phase and SOPs between the double reflected light and the main light will change when the fiber is disturbed, the interference will cause the total output power to fluctuate. The magnitude of the relative fluctuation in dB from Eq. (8.40a) is

$$\Delta = 10log\left[\frac{2\sqrt{P_{in}P_{dr}}}{(P_{in}+P_{dr})}\right] \approx 10log\left(2\sqrt{\frac{P_{dr}}{P_{in}}}\right) \tag{8.40b}$$

Although the double reflected light is very weak, because it interferes with a strong light (the input), its contribution cannot be ignored. For example, if a light is first reflected by an open PC connector with a typical of 4% reflectivity and then is reflected again by a mated PC connector with reflectivity of 0.01% (a return loss of 40 dB), the detected power fluctuation is as large as 0.017 dB. This power fluctuation will cause a PDL measurement fluctuation of 0.017 dB. Such a value cannot be ignored for measuring DUT with PDL of similar values. If an APC connector with a return loss of 60 dB is used, the PDL fluctuation will be reduced to 0.0017 dB, which can be ignored in most cases.

To minimize the contributions from double reflections, APC type connectors with low reflections should be used for connecting light source to the PDL meter if possible. Alternatively, a light source with short coherence length (shorter than the optical path difference of double reflected light and the main light) can also be used. This way, no interference described in Eq. (8.40a) can take place and therefore no interference fluctuations of Eq. (8.40b) to worry about.

8.4.4.4 Error from Connector and Cable Contributions

Beside the DUT, even the fiber cables or connectors used in the PDL measurement also have small PDL values. For example, optical cable itself may have a PDL value on the order of 0.01 dB, and this value may increase when it is bent with a small radius. Connectorized fiber jumpers may also have a small value of PDL on the order of 0.01–0.02 dB. Some poorly connectorized jumper may have even higher PDL values, probably caused by excess stress to the fiber during connectorization. APCs generally have high PDL, especially when not mating with another APC connector. Therefore, it is possible that a connectorized fiber jumper used in the measurement can contribute to a PDL error of 0.02 dB or higher to the PDL measurement of DUT.

8.4.4.5 Variation Caused by System PDL

As shown in Figure 8.24a, in addition to DUT, other optical components in the measurement system, such as connectors A, B, C, and D also have PDLs, which will contribute to the PDL measurement of the DUT and cause measurement errors. According to the discussion in Figure 3.9, the total PDL of the measurement system with multiple PDL components varies between a maximum value and a minimum value, depending on the SOP before each component with PDL. A maximum value corresponds to the case illustrated in Figure 8.24c, while a minimum value corresponds to the case illustrated in Figure 8.24d. Let PDL_i $(i = A, B, C, D)$ be the PDL values of components A, B, C, and D, the maximum and minimum measured PDL values PDL_{max} and PDL_{min} can be written as

$$PDL_{max} = PDL_{DUT} + PDL_A + PDL_B + PDL_C + PDL_D \tag{8.41a}$$

$$PDL_{max} = PDL_{DUT} - (PDL_A + PDL_B + PDL_C + PDL_D) \tag{8.41b}$$

When the fiber before a component is disturbed, the SOP entering the component is also changed, resulting in measurement variations. The maximum PDL value variation is therefore

$$\Delta(PDL) = PDL_{max} - PDL_{min} = 2(PDL_A + PDL_B + PDL_C + PDL_D) \tag{8.41c}$$

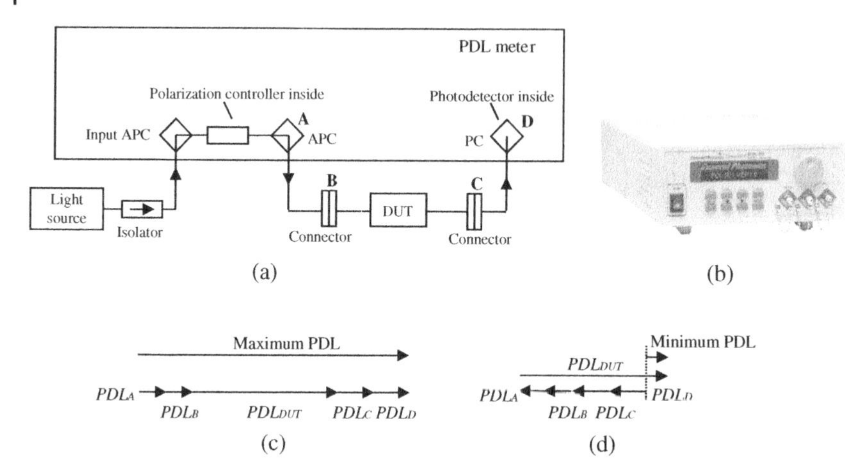

Figure 8.24 (a) Illustration of a PDL measurement setup involving PDLs of four connectors A, B, C, and D using a PDL meter of maximum–minimum search method. (b) Photo of a commercial PDL meter developed by the authors. Source: Courtesy of General Photonics/Luna Innovations. The measured PDL fluctuates between a maximum value (c) and a minimum value (d), depending on the SOP before each PDL component.

If the PDL value of the DUT is much larger than those of the connectors, the relative measurement error is small. Large relative error will result if the DUT has a PDL value similar to those of the connectors. Therefore, for the measurement of DUT with small PDL values (such as a fused coupler), the PDL values of connectors and cables connecting to the DUT must be extremely small because the limiting factor for the measurement accuracy is often not that of the instrument itself, but the residual PDL of the connectors and cables connecting to the DUT.

8.4.5 PDR Measurement

8.4.5.1 The Polarization Scrambling Method

Similar to PDL measurement, this method can also be used for measuring the PDR of PDs, as shown in Figure 8.25a. With a polarization scrambler generating random SOPs uniformly covering the Poincaré sphere, the PD under test detects the optical power P_o and is converted into a photocurrent I_{PD} with a responsivity η dependent on the SOP:

$$I_{PD} = \eta P_o \tag{8.42a}$$

A transimpedance amplifier converts the photocurrents into photo-voltages V_{PD} before being digitized by a DSP circuit, which is instructed to compare the new V_{PD} data with previously kept maximum and minimum photo-voltage data V_{PDmax} and V_{PDmin}. The new V_{PD} is disregarded if $V_{PD} \geq V_{PDmin}$ or $V_{PD} \leq V_{PDmax}$. Only when $V_{PD} \geq V_{PDmax}$ or $V_{PD} \leq V_{PDmin}$, it will be kept as the new V_{PDmax} or V_{PDmin} as the SOP is scrambled. When sufficient numbers of SOPs are generated and the corresponding photo-voltages are measured, the PDR can be obtained as

$$PDR = 10log\left(\frac{V_{PDmax}}{V_{PDmin}}\right) \tag{8.42b}$$

Similar to the PDL measurement, the PDL and AL of the polarization scrambler must be much lower than the PDR of the PDs under test to ensure accurate PDR measurement while the scrambling speed must be sufficiently high to ensure reasonable measurement speed. In general, the PDR of PDs is on the order of 0.05–0.1 dB; only the polarization scramblers made with the fiber-squeezer described in Figure 6.15c can meet this stringent requirement.

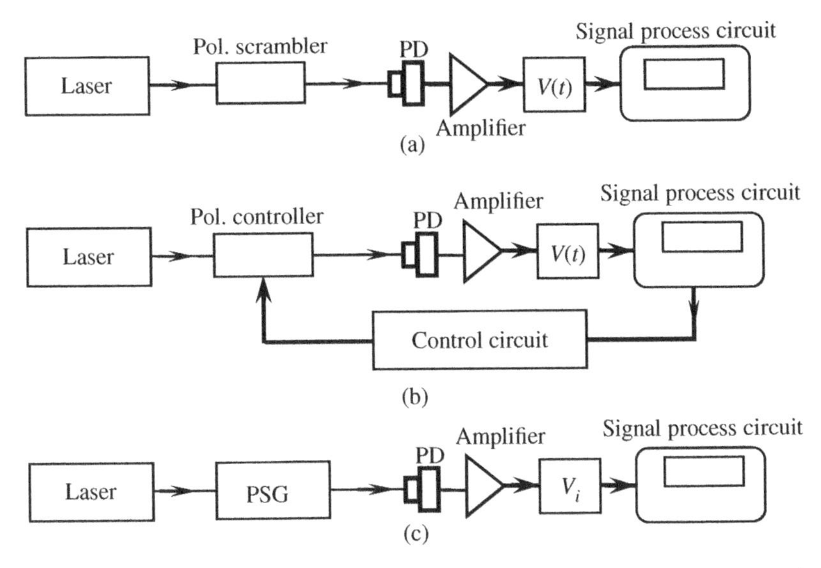

Figure 8.25 Illustration of using the polarization scrambling method (a), the maximum–minimum search method (b), and the Mueller matrix method (c) to accurately measure the PDR of a photodetector.

8.4.5.2 The Maximum and Minimum Search Method

This method can also be used for PDR measurement, as shown in Figure 8.25b, similar to that for the PDL measurement. The PD under test converts an optical power to a photocurrent with a responsivity dependent on the SOP following Eq. (8.42a), which is further converted to a photo-voltage by a trans-impedance amplifier before being digitized by a signal processing circuit. The circuit is programmed to increase or decrease to voltage applied to a fiber squeezer in a fiber squeezer polarization controller. If increase (or decrease) in the voltage results in an increase of detected photo-voltage, continue increasing (decreasing) it until the photo-voltage starts to drop. Hold the voltage to the fiber squeezer while repeating the process to an adjacent squeezer until a maximum is reached. Continue to repeat the process to other fiber squeezers in sequence until the global maximum photo-voltage V_{PDmax} is reached. Apply the same procedure above to find the global minimum photo-voltage V_{PDmin}. Finally, Eq. (8.42b) is used to compute the PDR of the PD. Such a maximum–minimum search routine is easy to implement and fast to execute, and is proven to complete a PDR measurement in less than 30 ms. Comparing with the polarization scrambling method, the maximum–minimum search method has the advantages of better repeatability and higher measurement speed because the process is deterministic. Similar to the polarization scrambling method, the PDR measurement accuracy is limited by the PDL and AL of the polarization controller used.

8.4.5.3 The Mueller Matrix Method

A PD with a PDR is equivalent to a device with PDL (a partial polarizer) in front of a perfect PD without PDR. Therefore, the Mueller matrix analysis method discussed in Section 8.4.2.1 for PDL measurement can be used, as shown in Figure 8.25c. In particular, the detected photo-voltage can be expressed as

$$\begin{bmatrix} V_{i0} \\ \cdot \\ \cdot \\ \cdot \end{bmatrix} = \begin{bmatrix} m_{00} & m_{01} & m_{02} & m_{03} \\ \cdot & \cdot & \cdot & \cdot \\ \cdot & \cdot & \cdot & \cdot \\ \cdot & \cdot & \cdot & \cdot \end{bmatrix} \begin{bmatrix} S_{i0} \\ S_{i1} \\ S_{i2} \\ S_{i3} \end{bmatrix}, \quad i = 1, 2, 3, 4 \tag{8.43a}$$

where the matrix can be considered as the Mueller matrix representing equivalent partial polarizer. Note that in Eq. (8.43a), only the first row elements of the Mueller matrix are shown because other elements are not involved in computing the PDL. By generating four distinctive SOPs and measuring the four photo-voltages, one can obtain four equations for solving for the four Mueller matrix elements, with which the maximum and minimum transmissions T_{max} and T_{min} can be obtained using Eqs. (8.38a) and (8.38b). Finally, the PDR of the PD can be obtained as

$$PDR = 10log\left(\frac{T_{max}}{T_{min}}\right) = 10log\frac{m_{00} + \sqrt{m_{01}^2 + m_{02}^2 + m_{03}^2}}{m_{00} - \sqrt{m_{01}^2 + m_{02}^2 + m_{03}^2}} \tag{8.43b}$$

8.4.6 DOP Measurements

As discussed in Sections 2.5 and 4.7, DOP in an optical beam is defined as

$$DOP = \frac{I_{pol}}{I_{total}} = \frac{I_{pol}}{I_{pol} + I_{unpol}} \tag{2.17}$$

where I_{pol} and I_{unpol} are intensities of polarized and unpolarized portions of a light beam, respectively, and $I_{total} = I_{pol} + I_{unpol}$. As shown in Figure 8.26a, the most straight forward method for measuring the DOP of a signal transmitted through an optical fiber system with different optical devices is using a complete Stokes polarimeter (or PSA) described in Section 8.1 to obtain all four Stokes parameters and calculating the DOP using Eq. (8.1c) as $DOP = \sqrt{S_1^2 + S_2^2 + S_3^2}/S_0$.

As discussed in Section 2.5, the DOP can also be determined by inserting a polarizer in the light beam and measuring the maximum and minimum intensities I_{max} and I_{min} (or powers P_{max} and P_{min}) with a power meter or a PD while rotating the polarizer:

$$DOP = \frac{I_{max} - I_{min}}{I_{max} + I_{min}} = \frac{P_{max} - P_{min}}{P_{max} + P_{min}} \tag{2.18c}$$

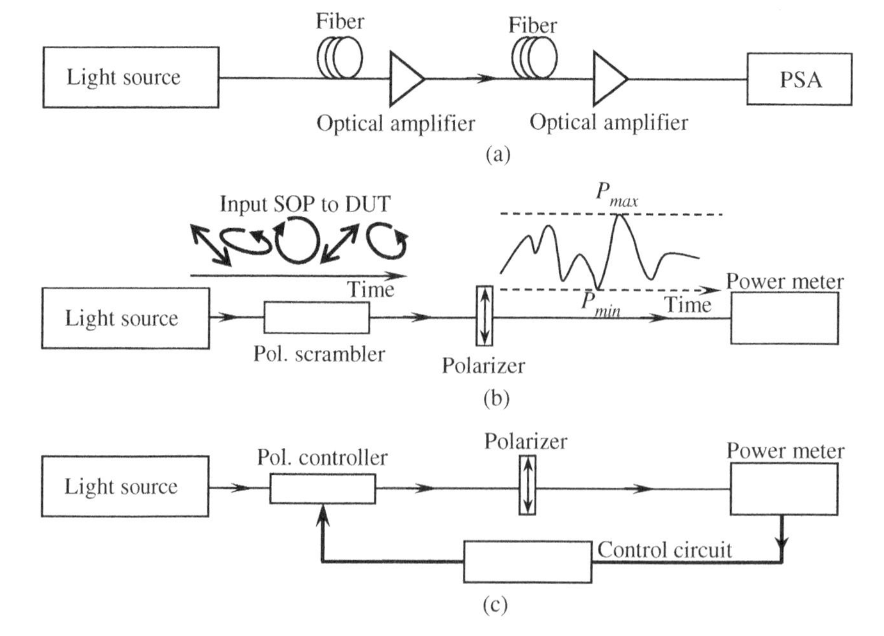

Figure 8.26 Measuring DOP in an optical fiber system using the Stokes polarimetry method (a), the polarization scrambling method (b), and the maximum–minimum search method (c).

Similar to PDL measurement, the polarization scrambling described in Section 8.4.2 can be used to measure the DOP, as shown in Figure 8.26b. In operation, a polarizer is placed before the power meter. The polarization scrambler quickly changes the SOP of the light uniformly covering the whole Poincaré sphere while the optical power is measured, and the maximum and minimum powers are recorded. After a certain amount of time with a sufficient number of SOPs passing through the polarizer, the DOP is computed using Eq. (2.18c).

As shown in Figure 8.26c, the DOP can also be measured using the maximum–minimum search method (Shi et al. 2006), in analogy to the maximum–minimum search method for the PDL measurement. In operation, a polarizer is first inserted in front of the power meter while the optical power passing through the polarizer is monitored. The same maximum and minimum search algorithm similar to that used in PDL and PDR measurements described earlier can be used to obtain P_{max} and P_{min} and the DOP can be obtained using Eq. (2.18c).

Detailed experiments were carried out to verify the performance of the maximum–minimum search method of Figure 8.26c. The setup in Figure 8.8c was used to generate different DOP values in which the ASE source was combined through a 3-dB 2×2 fused coupler with the output of the tunable laser set at 1550 nm. The optical power of each path could be adjusted independently using two VOAs. The DOP of the mixed beam was adjustable from ~3% to 100%. The 3% residual DOP of the ASE source was mainly due to the PDL of the 3-dB fused coupler. Using this light source of variable DOP, DOP values were measured using both the maximum–minimum search method and a commercial Stokes polarimeter (Agilent 8509C Lightwave Polarization Analyzer, discontinued). The measured DOP values agreed very well, as shown in Figure 8.27.

The accuracy of the maximum–minimum search measurement system depends on system noise, circuit architecture, measurement control, and the PDL of the optical components used in the measurement system, including that of the polarization controller. These PDLs add up vectorally in 3D space, as shown in Figure 8.24, and result in measurement error. Therefore, all optical components, including the polarization controller, must have negligible PDL. For example, a small amount of system PDL can directly affect the DOP measurement accuracy, particularly at the low DOP end, as described by Eq. (7.13a). The resulting DOP error can be approximated as

$$\Delta DOP\ (\%) = 11.5 \times PDL\ \ (\text{in dB}) \tag{8.44a}$$

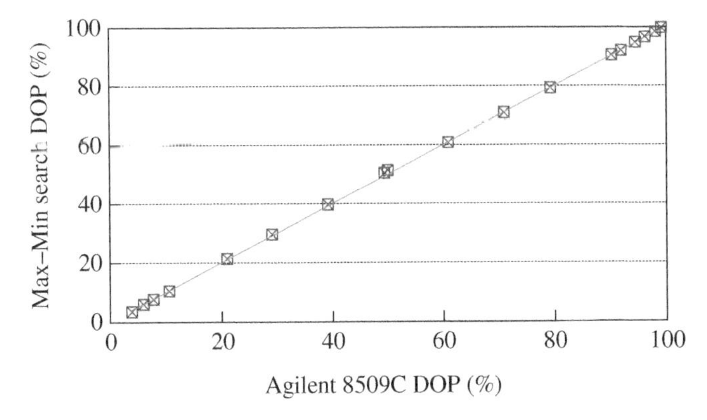

Figure 8.27 Comparison of DOP measured with the maximum–minimum search method and with a commercial Stokes polarimeter (Agilent 8509C). The straight line represents 100% correlation. Source: Reproduced from Shi et al. (2006)/IEEE.

For example, for a measurement system with a combined PDL of 0.1 dB, the DOP measurement error is 1.15%. From Eq. (7.13b), the same approximation can also be used for the DOP measurement error resulting from the activation loss (AL):

$$\Delta DOP\,(\%) = 11.5 \times AL\ (\text{in dB}) \tag{8.44b}$$

8.5 Real-Time Performance Monitoring of a Communication System with DOP Measurement

An optical signal transmitting in optical fiber communication system involving multiple optical amplifiers (such as EDFA) will degrade due to the ASE noise from the amplifiers and polarization related impairments, such as PMD and PDL, from the optical fiber and other optical components, as discussed in Chapter 3. For the early optical fiber networks without deploying coherent detection and polarization multiplexing, the PMD related degradation cannot be ignored when the speed of optical communication network increases to 40 Gb/s and beyond. As shown in Figure 8.28, the SNR of a signal, one of the most important parameters for an optical fiber communication system, will be degraded primarily by the ASE noise each time it passes through an optical amplifier. As will be discussed next, PMD will also contribute to the degradation of the SNR, in addition to the signal distortion discussed in Chapter 3. Therefore, in order to understand the performance of an optical system, it is desirable to simultaneously monitor both the SNR and the PMD effects on the signal while the system is in operation, as well as the channel power (Reza et al. 2004). This way, when a system experiences an outage, the network operator would be able to identify the problem, whether caused by a low SNR, a bad PMD effect, or a low channel power.

It is possible to determine the SNR by using an optical spectrum analyzer to measure the in-band and off-band spectral power, as shown in Figure 8.29, with the assumption that the measured in-band spectral power is the signal power while the off-band spectral power is the noise power. However, such an assumption is not strictly true because the in-band spectral power includes the contributions from both the signal and noise, and therefore may lead to large errors when the SNR is relatively small.

8.5.1 SNR and Channel Power Monitoring via DOP Measurement

The SNR measurement is based on the fact that the signal from a laser source is completely polarized, while the ASE noise from the optical amplifier, such as an EDFA, is totally unpolarized. In the absence of PMD depolarization, if a polarizer is orthogonally aligned with the SOP of the signal, a minimum power $P_{min} = P_n/2$ will be detected because half of the noise power P_n can pass through the polarizer while all the signal power is blocked. If the polarizer is aligned to the SOP of the signal, a maximum power $P_{max} = P_s + P_n/2 = P_s + P_{min}$ will be detected because all the signal power P_s and half of the noise power P_n can pass through. The SNR can thus be expressed as

$$SNR = 10log\frac{P_s}{P_n} = 10log\left[\frac{(P_{max} - P_{min})}{(2P_{min})}\right] = 10log\left[\frac{DOP}{(1 - DOP)}\right] \tag{8.45}$$

where Eq. (2.18c) is used. Therefore, in the absence of PMD depolarization, the SNR of the system can be determined by DOP measurement. The signal power $P_s = P_{max} - P_{min}$ can also be determined during DOP measurement.

As shown in Figure 8.28, a small amount of power from the transmission fiber can be tapped out before entering the RX. An optical BPF centered at a WDM channel wavelength (the intended

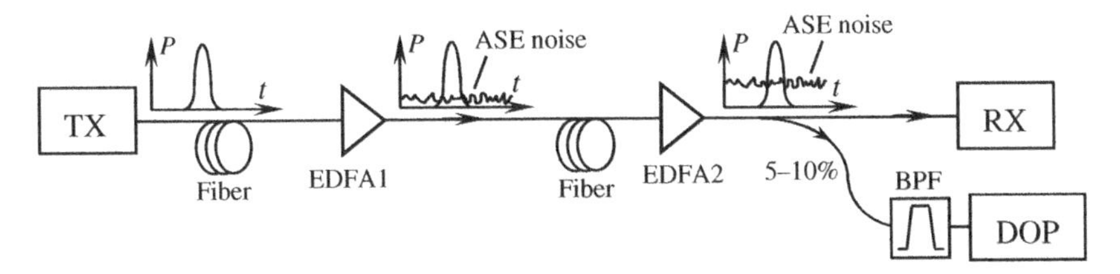

Figure 8.28 Illustration of the SNR degradation due to optical amplifiers' ASE noises as a signal propagating in an optical fiber link and SNR monitoring using a high speed DOP meter. TX: data transmitter; RX: data receiver; EDFA: Erbium doped fiber amplifier; BPF: bandpass filter.

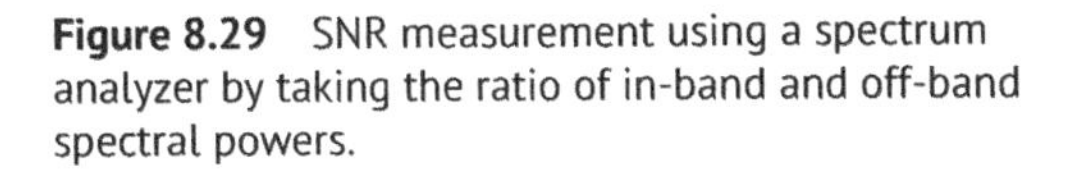

Figure 8.29 SNR measurement using a spectrum analyzer by taking the ratio of in-band and off-band spectral powers.

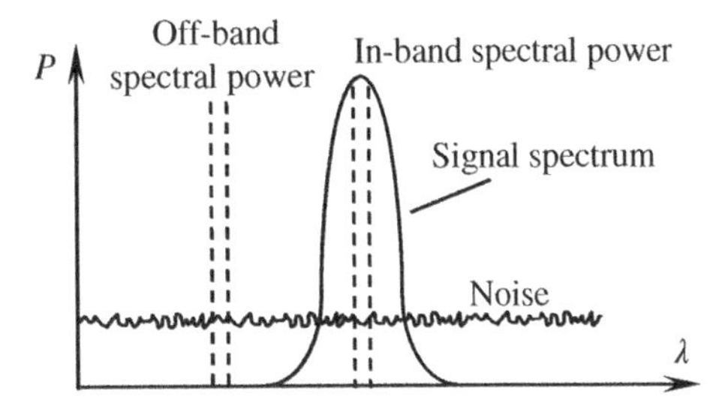

WDM channel) is used to filter out ASE noise in other WDM channels and only allow the signal and noise in the intended WDM channel to pass through. Any of the DOP meters described in Figure 8.26 can be used to monitor the DOP and finally obtain the SNR, as well as the signal power of the WDM channel.

8.5.2 Optical Amplifier Noise Figure Measurement

An accurate DOP meter can also be used to measure the noise figure (NF) of an optical amplifier, as shown in Figure 8.30. One may first measure the SNR of the signal source (SNR_0) without the amplifier by inserting a fiber jumper in the place of the amplifier, and then measure the SNR of the signal after the amplifier (SNR_{amp}). The noise figure NF is simply the difference in dB between the two SNR measurements.

$$NF = SNR_0 - SNR_{amp} \tag{8.46}$$

In practice, the noise figure is generally given within a 0.1 nm bandwidth; therefore the effect of the bandwidth and shape of the filter needs to be taken into account in determining the final SNR

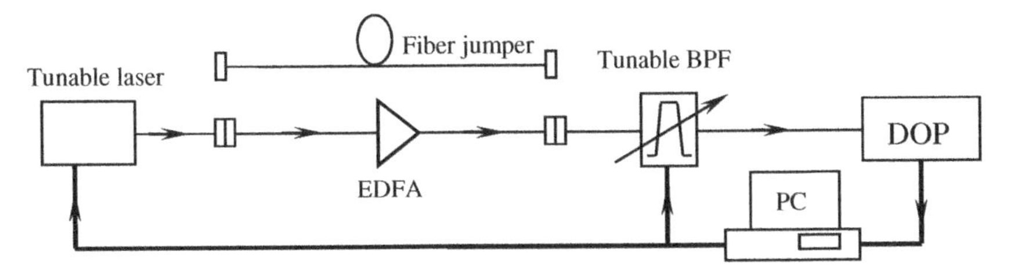

Figure 8.30 Using DOP meter to measure the noise figures of optical amplifiers. The tunable filter is synchronized with the tunable laser to measure the noise figures at different wavelengths.

value. With the aid of a computer, a tunable laser, and a tunable filter, the wavelength dependence and power dependence of the noise figure can also be determined.

8.5.3 SNR, PMD Depolarization, and Channel Power Monitoring via DOP Measurement

As shown in Figure 8.31a, the performance monitor comprised a 1×2 coupler, a tunable bandpass filter (BPF_1) with a bandwidth $\Delta\lambda_1$ in one of the coupler output ports, a second tunable bandpass filter (BPF_2) with a bandwidth $\Delta\lambda_2$ in the other coupler output port, and two DOP meters (Yan et al. 2005; Yao 2013). As shown in Figure 8.31b, the center wavelengths of the two BPFs must be the same as that of the WDM channel to be measured and their bandwidths are sufficiently different from each other. The DOP meter can be a polarimeter or any of those described in Figure 8.26.

The DOP of the light beam received by DOP meter i ($i = 1, 2$) is defined as $DOP_i = P_{pol}/P_{Ti}$, where P_{pol} is the power of polarized portion and P_{Ti} is the total optical power received by DOP meter i consisting of signal power P_s and ASE noise power P_{ni}:

$$P_{Ti} = P_s + P_{ni} \tag{8.47a}$$

Let $\rho(\lambda)$ be the ASE noise power density of light after the tunable filter, the noise powers after the two BPFs are

$$P_{ni} = \int_{-\infty}^{\infty} \rho(\lambda) f_i(\lambda) d\lambda = \overline{\rho}\Delta\lambda_i \tag{8.47b}$$

where $\overline{\rho}$ is the average ASE noise power density, $f_i(\lambda)$ and $\Delta\lambda_i$ are the transmission function and bandwidth of the BPF_i. The total signal power P_s has two portions, one is the polarized portion $P_{pol} = (1 - \alpha)P_s$ and the other is the depolarized portion $P_{dpol} = \alpha P_s$:

$$P_s = P_{pol} + P_{dpol} = (1 - \alpha)P_s + \alpha P_s \tag{8.47c}$$

where α is the depolarization factor to describe the depolarization caused by PMD.

The DOP of the two optical beams into the two DOP meters is therefore

$$DOP_1 = \frac{P_{pol}}{P_{T1}} = \frac{P_{pol}}{(P_{n1} + P_s)} = \frac{(1 - \alpha)P_s}{(\overline{\rho}\Delta\lambda_1 + P_s)} \tag{8.48a}$$

$$DOP_2 = \frac{P_{pol}}{P_{T2}} = \frac{P_{pol}}{(P_{n2} + P_s)} = \frac{(1 - \alpha)P_s}{(\overline{\rho}\Delta\lambda_2 + P_s)} \tag{8.48b}$$

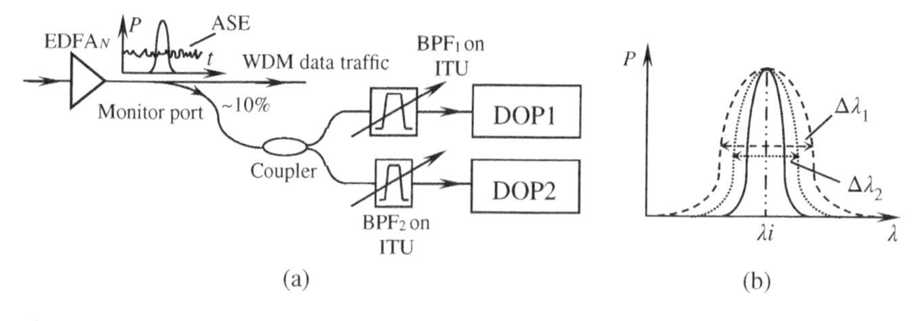

Figure 8.31 (a) Configuration for monitoring SNR, PMD depolarization, and channel power of a WDM channel using two DOP measurements. (b) Spectra of the data in WDM channel λ_i, the bandpass filter 1 (BPF_1), and the bandpass filter 2 (BPF_2). $\Delta\lambda_1$ and $\Delta\lambda_2$: bandwidths of BPF_1 and BPF_2, respectively.

Solving for α and P_s using Eqs. (8.48a) and (8.48b) yields

$$\alpha = 1 - \frac{(\Delta\lambda_1 - \Delta\lambda_2)DOP_1DOP_2}{DOP_1\Delta\lambda_1 - DOP_2\Delta\lambda_2} = 1 - \frac{(1 - \Delta\lambda_2/\Delta\lambda_1)DOP_2}{1 - (\Delta\lambda_2/\Delta\lambda_1)(DOP_2/DOP_1)} \tag{8.49a}$$

$$P_s = \frac{P_{ni}DOP_i}{(1 - \alpha - DOP_i)} = \frac{P_{min}DOP_i}{(1 - \alpha - DOP_i)} \tag{8.49b}$$

The SNR is thus

$$SNR_i = 10log\left(\frac{P_s}{P_{ni}}\right) = 10log\left[\frac{DOP_i}{(1 - \alpha - DOP_i)}\right] \tag{8.49c}$$

In practice, the SNR is generally calculated from the noise in a 0.1 nm bandwidth. Therefore, the SNR of the system is

$$SNR = \frac{\Delta\lambda_i}{0.1}SNR_i \tag{8.49d}$$

where $\Delta\lambda_i$ is expressed in nm.

It is evident from Eqs. (8.49a) and (8.49d) that by measuring the DOPs using the configuration of Figure 8.31, system's SNR and depolarization factor α caused by PMD effect can be simultaneously monitored. Because the DOP meter can also measure P_{min}, the signal channel power P_s can also measured using Eq. (8.49b).

It should be noticed that the value of the depolarization factor α depends on the total PMD of the system and the relative orientation of the polarization with respect to the principle state of polarization (PSP) of the fiber link, as shown in Figure 7.22. It represents the total effect of PMD on the communication link. Since polarization state varies rapidly with time (faster than the PMD variation in general), α may vary accordingly. The difference between the maximum and minimum α values is indicative of the total PMD. Therefore, system's PMD value can be estimated by obtaining the amplitude of α variation in a certain period of time, provided that the polarization variation during the period covers the Poincaré sphere.

8.6 PMD Measurements of Optical Components and Optical Fibers

Many methods have been developed for measuring PMD in optical components and optical fibers up to date, including the time domain, coherence domain, spectrum domain, and frequency domain methods, as summarized in several excellent books, such as those by Paul Hernday (1998), Edward Collett (2003), Paul Williams (2005), and Jay Damask (2005). Interested readers are encouraged to read these books. In this section next, we briefly describe most of the methods and discuss their pros and cons for different application scenarios.

8.6.1 Pulse-Delay Method

As will be discussed in the following text, the time domain method is generally for measuring large PMD in a long optical fiber. Figure 8.32a shows the pulse-delay method which directly measures the DGD between the polarization component along the "fast" and "slow" axes of the fiber illustrated in Figure 7.22. In operation, repetitive optical pulses from a light source, such as a mode-locked laser, are launched into the fiber and received by a fast PD before being displayed on a fast sampling scope. The driving electronic generates signals to drive the pulsed light source while supplying the trigger signals to the sampling scope for synchronization. The polarization controller is adjusted while the pulse arrival time is measured by the sampling scope. When the launched SOP is aligned with the "slow" axis of the fiber, a larger arrival time can be measured and recorded. When aligned with the "fast" axis, a smaller arrival time can be measured and recorded, as shown in Figure 7.22. The difference $\Delta\tau$ is the DGD or the first order PMD. When the launched SOP is aligned 45° from the "slow" or "fast" axis, pulse broadening can be observed. If the DGD of the fiber is sufficiently large, each launched pulse will split into two and the separation $\Delta\tau$ between them can be directly measured to obtain the DGD. As can be seen, this method requires a short pulsed light source, as well as a high speed PD and digital scope, which are all extremely expensive. In addition, the measurement resolution is limited by the sampling speed of the scope.

8.6.2 Modulation Phase-Shift Method

Figure 8.32b shows the modulation phase-shift method in which the light from a narrow bandwidth laser, such as a DFB laser, is modulated by a RF signal from the RF source in a network analyzer via an electro-optical modulator, such as a Mach–Zehnder or electro-absorption modulator. A polarization controller is used to adjust the SOP into the fiber under test before the modulated light signal is detected by the PD and converted back to the RF signal, which is amplified before being mixed with the RF signal directly from the RF source (the local reference RF signal, LO). The relative phase between the detected RF signal and local reference can be directly measured by the network analyzer. Adjustment of the polarization controller will cause the detected phase to vary,

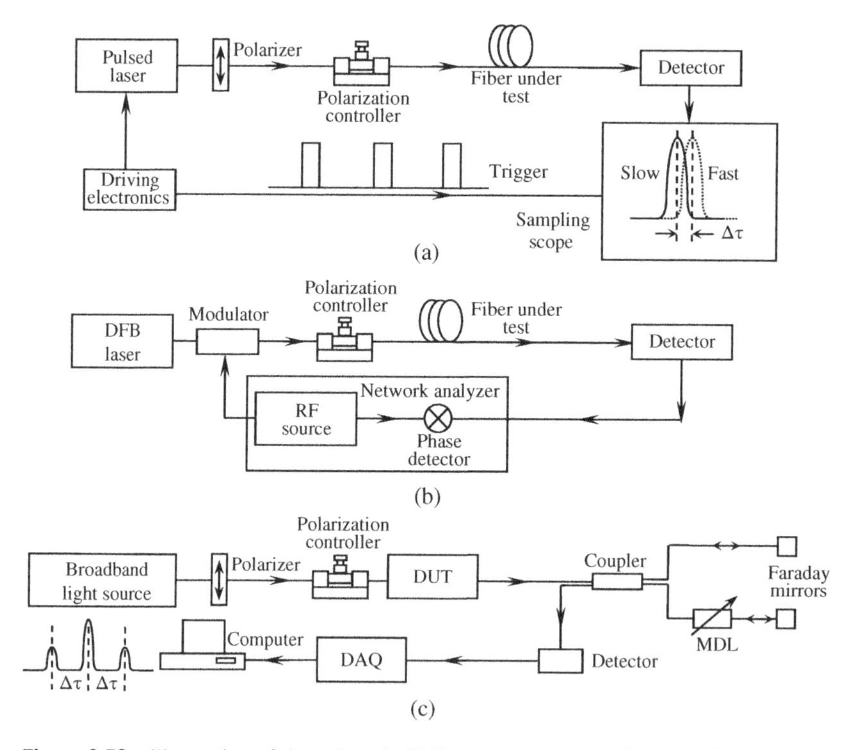

Figure 8.32 Illustration of time-domain PMD measurement methods. (a) Pulse-delay method. (b) Modulation phase shift method. (c) Interferometric method. MDL: motorized delay line; DAQ: digital acquisition card.

with the maximum and minimum phases φ_{max} and φ_{min} corresponding to the launch SOP aligned with the "slow" and "fast" axes of the fiber PMD (the PSP), respectively. The difference between φ_{max} and φ_{min} is related to the DGD or first order PMD $\Delta\tau$ by

$$\Delta\varphi = \varphi_{max} - \varphi_{min} = 2\pi f_m \Delta\tau \tag{8.50a}$$

$$\Delta\tau = \frac{\Delta\varphi}{(2\pi f_m)} \tag{8.50b}$$

In general, the higher the RF modulation frequency f_m, the more accurate the phase measurement is. For example, if the phase detection resolution is 1° at f_m of 10 GHz, the PMD detection resolution $\Delta\tau_{min}$ is 0.28 ps. The maximum detectable PMD corresponds to the differential phase of π, which is $\Delta\tau_{max} = 1/(2f_m)$. At $f_m = 10$ GHz, $\Delta\tau_{max} = 50$ ps.

8.6.3 Interferometric Method

Figure 8.32c shows the measurement setup of the interference method. A broadband light source, such as a super luminescent diode (SLED) or an ASE light source, can be used, which is polarized by a polarizer before entering the DUT. An optional polarization controller described in Chapter 6 is placed before the DUT for adjusting the input SOP. The light exiting from the DUT enters a Michelson interferometer in which two Faraday mirrors described in Chapter 6 are used to eliminate the polarization fluctuations of the interferometer. A motorized delay line (MDL) is used to change the relative delay between the light beams in the two interfering arms. The interference signal is detected by the PD before being digitized by a DAQ card and recorded by a computer as the MDL is scanned. An example of the interferogram envelope of a DUT without polarization mode coupling, such as a piece of PM fiber, is shown at the left hand end of Figure 8.32c. A center interference peak with two side peaks can be seen, in which the relative delay between the main and side peaks is the PMD to be measured. Adjusting the polarization controller may cause the side peaks to disappear, corresponding to the case the input SOP is aligned with one of the PSP axis of the DUT. For strongly mode-coupled DUT, such as a long length of optical fiber, the interferogram has a broad Gaussian-like envelope with a sharp center peak. The width of the Gaussian envelop is determined by the PMD in the fiber. Detailed discussion can be found in Hernday (1998).

8.6.4 PMD Measurement by PMD Compensation

PMD in a field-installed fiber link can be measured by PMD compensation, as described in Figure 7.26. Interested readers may go to Section 7.4.3.6 for details.

8.6.5 Fixed-Analyzer Method

This method can be implemented in two different ways, one is the setup of Figures 8.33a and the other is Figure 8.33b. The former uses an optical spectrum analyzer to detect and analyze the power as a function of wavelength, while the latter uses a tunable laser and a power meter to perform the same measurement.

To better explain how it works, let's first consider the case that the DUT is a pure DGD with a fast and slow axis, without second order PMD or polarization mode-coupling. If the input linear SOP defined by the polarizer is oriented α_1 degrees from the slow axis of the DUT, the input and output SOPs can be expressed as

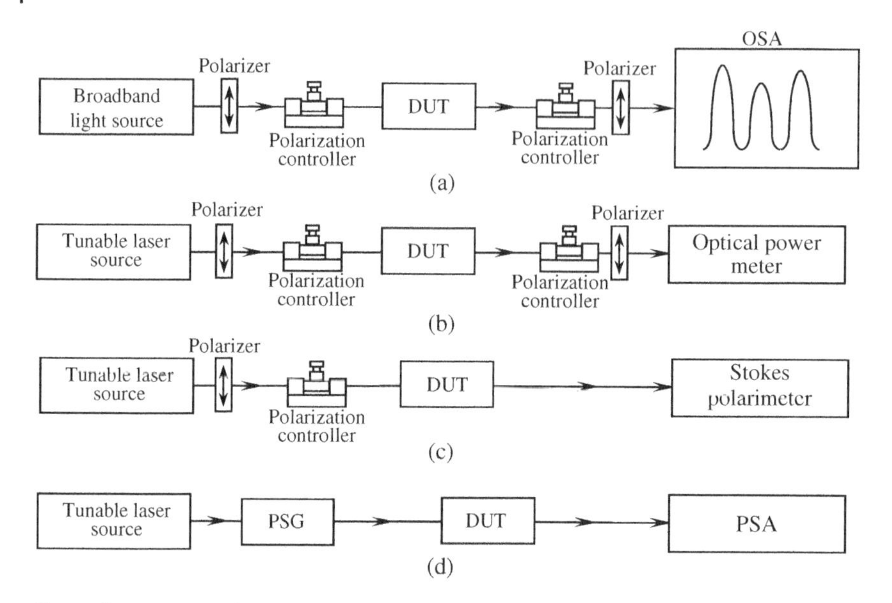

Figure 8.33 (a) The fixed analyzer method with an optical spectrum analyzer. (b) The fixed analyzer method with a tunable laser source. (c) SOP trace analysis method. (d) Setup for PSA, MMM, and JME methods. DUT: device under test; PSA: polarization state analyzer (time-division, amplitude-division, or wave-front division Stokes parameter).

$$\mathbf{E}_{in} = (cos\alpha_1\hat{\mathbf{s}} + sin\alpha_1\hat{\mathbf{f}})e^{i\varphi_0} \tag{8.51a}$$

$$\mathbf{E}_{out} = \left(cos\alpha_1 e^{\frac{i2\pi\Delta\tau c}{\lambda}}\hat{\mathbf{s}} + sin\alpha_1\hat{\mathbf{f}}\right)e^{i\varphi_0} = (cos\alpha_1 e^{i2\pi\Delta\tau f}\hat{\mathbf{s}} + sin\alpha_1\hat{\mathbf{f}})e^{i\varphi_0} \tag{8.51b}$$

where λ is the optical wavelength in vacuum, f is the frequency, and $\Delta\tau$ is the DGD or first order PMD to be measured, and $\hat{\mathbf{s}}$ and $\hat{\mathbf{f}}$ are the unit vectors of the slow and fast axes, respectively. If the analyzing polarizer or the analyzer after the DUT is oriented α_2 degrees from the slow axis,

$$\hat{\mathbf{p}} = (cos\alpha_2\hat{\mathbf{s}} + sin\alpha_2\hat{\mathbf{f}}) \tag{8.51c}$$

The optical power detected after the analyzer is

$$P = |\mathbf{E}_{out} \cdot \hat{\mathbf{p}}|^2 = cos^2\alpha_1 cos^2\alpha_2 + sin^2\alpha_1 sin^2\alpha_2 + \frac{1}{2}sin2\alpha_1 sin2\alpha_2 cos\Delta\varphi \tag{8.51d}$$

where c is the speed of light in vacuum and $\Delta\varphi = 2\pi\Delta\tau \cdot f = 2\pi\Delta\tau c/\lambda$. Clearly, the detected optical power is a periodical function of optical frequency or wavelength, with a periodicity of

$$\Delta\lambda = \frac{\lambda^2}{\Delta\tau c} \quad \text{or} \quad \Delta\tau = \frac{\lambda^2}{\Delta\lambda c} \tag{8.51e}$$

$$\Delta f = \frac{1}{\Delta\tau} \quad \text{or} \quad \Delta\tau = \frac{1}{\Delta f} \tag{8.51f}$$

When $\alpha_1 = \alpha_2 = \pm 45°$, the detected signal has the highest modulation depth of 100%. However, the optimum polarization alignment angle is not critical to the periodicity measurement as long as the signal depth of the periodical signal is larger than 50%. The polarization controller before and after the DUT can be used to adjust for α_1 and α_2, respectively, for the acceptable signal depth. Either Eq. (8.51e) or (8.51f) can be used to determine the $\Delta\tau$ of DUT.

If the DUT has polarization mode coupling or second order PMD, the periodicity of the measured output power spectra may not be uniform in the wavelength range, and therefore the average wavelength or frequency periodicity can be used to compute the PMD $(\Delta\tau)$ (Hernday 1998):

$$\langle \Delta\tau \rangle = \frac{kN\lambda_1\lambda_2}{2|\lambda_1 - \lambda_2|c}$$

where N is the numbers of peaks and valleys over the wavelength scanning range from λ_1 to λ_2 and k is called the polarization mode coupling factor, accounting for the statistical effect of the wavelength dependence of the PSPs, which has a value of 0.824 for randomly-coupled fibers and 2 for non-mode-coupling fibers and devices.

8.6.6 SOP Trace Analysis Method

As shown in Figure 8.34a, for a DUT with a pure DGD (no polarization mode coupling in DUT), the SOP trace on the Poincaré sphere is a circle as the optical frequency scans while the input SOP is fixed by the polarizer, which can be measured with the setup of Figure 8.33c. The PSP, corresponding to the "slow" and "fast" axes of the DUT, is normal to the surface of the circle. The SOP angle variation $\Delta\theta$ around PSP corresponding to a frequency variation $\Delta\omega = 2\pi f$ is simply the differential phase between the "slow" and "fast" polarization components in Eq. (8.51b), which can be obtained as

$$\Delta\theta = 2\pi\Delta\tau\Delta f = \Delta\tau\Delta\omega \tag{8.52a}$$

$$\Delta\tau = \frac{\Delta\theta}{\Delta\omega} \tag{8.52b}$$

This is the same result as Eq. (4.387).

If there is polarization mode coupling in the DUT, the SOP variation trace caused by the optical frequency scan is no longer a perfect circle, but a complicated curve on the Poincaré sphere, as shown in Figure 8.34b in which the SOP is represented by $\mathbf{S}(\omega)$. Any small section of the trace, corresponding to a small frequency variation range $\Delta\omega$ around a frequency ω, can be considered

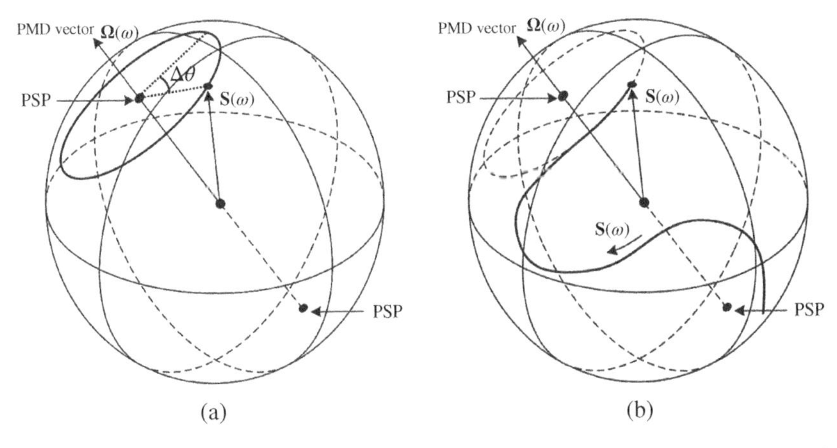

Figure 8.34 (a) The SOP trace of a perfect circle after light from a tunable laser passing through a DUT without polarization mode coupling (pure DGD). (b) The more complicated SOP trace after light from a tunable laser passing through a DUT with polarization mode coupling (including higher order PMD).

an arc on a circle, and the corresponding PMD is a function of frequency ω and can be obtained from Eq. (8.25b):

$$\Delta\tau(\omega) = \frac{d\theta(\omega)}{d\omega} \tag{8.53}$$

Therefore, in principle one may obtain the SOP trace $\mathbf{S}(\omega)$ as a function of laser frequency using a Stokes polarimeter and identify the arcs on the trace at different frequencies to calculate $\Delta\tau(\omega)$ using Eq. (8.53). Unfortunately, such a process is tedious and difficult to implement with software. In order to simplify the computation, let's consider $\mathbf{S}(\omega)$ at three different wavelengths ω_i, $\omega_i + \Delta\omega$, and $\omega_i + 2\Delta\omega$, with the corresponding Stokes vectors $\mathbf{S}_i(\omega_i)$, $\mathbf{S}_{i+1}(\omega_i + \Delta\omega)$, and $\mathbf{S}_{i+2}(\omega_i + 2\Delta\omega)$, respectively, as shown in Figure 4.34c. By replacing $\mathbf{S}_i(\omega_i)$ with $\mathbf{h}_i(\omega_i)$ in Eq. (4.397), the SOP rotation angle between $\mathbf{S}_i(\omega_i)$ and $\mathbf{S}_{i+1}(\omega_i + \Delta\omega)$ can be obtained from Figure 4.35 as

$$sin\left(\frac{\Delta\theta_i}{2}\right) = \frac{\Delta\mathbf{S}_i(\omega_i)/2}{|\mathbf{S}_i(\omega_i) - [\mathbf{S}_i(\omega_i)\cdot\widehat{\mathbf{p}}_i]\widehat{\mathbf{p}}_i|} \tag{8.54a}$$

where $\widehat{\mathbf{p}}_i$ is the unit vector of $\mathbf{\Omega}(\omega_i)$ at ω_i, which can be obtained from Figure 4.34c as

$$\widehat{\mathbf{p}}_i = \frac{\mathbf{\Omega}(\omega_i)}{|\mathbf{\Omega}(\omega_i)|} = \frac{\Delta\mathbf{S}_i \times \Delta\mathbf{S}_{i+1}}{|\Delta\mathbf{S}_i \times \Delta\mathbf{S}_{i+1}|} \tag{8.54b}$$

where $\Delta\mathbf{S}_i$ and $\Delta\mathbf{S}_{i+1}$ are shown in Figure 4.34c and can be expressed as

$$\Delta\mathbf{S}_i = \mathbf{S}_{i+1}(\omega_i + \Delta\omega) - \mathbf{S}_i(\omega_i) \tag{8.54c}$$

$$\Delta\mathbf{S}_{i+1} = \mathbf{S}_{i+2}(\omega_i + 2\Delta\omega) - \mathbf{S}_{i+1}(\omega_i + \Delta\omega) \tag{8.54d}$$

Finally, the PMD at ω_i can be obtained as

$$PMD = \frac{\Delta\theta_i}{\Delta\omega} = \frac{2}{\Delta\omega}sin^{-1}\left[\frac{\Delta\mathbf{S}(\omega_i)/2}{|\mathbf{S}(\omega_i) - [\mathbf{S}(\omega_i)\cdot\widehat{\mathbf{p}}_i]\widehat{\mathbf{p}}_i|}\right] \tag{8.54e}$$

Therefore, if one obtains the Stokes parameters as a function of optical frequency (wavelength) using a Stokes parameter and a tunable laser shown in Figure 8.33c, the PMD at each wavelength can be computed using Eq. (8.54e). In operation, one must make sure the input SOP is not aligned or close to be aligned with the PSP of the DUT by adjusting the polarization controller while observing the SOP trace as a function of wavelength. If the SOP traces out a small circle as the laser wavelength is tuned, then adjust the polarization controller to make the SOP trace sufficiently large.

8.6.7 Poincaré Sphere Arc Method with Two Arbitrary Input SOPs

The operation principle of this method has been discussed in Section 4.6.1, which can be implemented with the setup shown in Figure 8.33c. In operation, one can set the first arbitrary input SOP with the polarization controller before measuring the output SOP of light passing through the DUT as a function of optical frequency (or wavelength) using the Stokes polarimeter to obtain $\mathbf{S}_a(\omega)$. Similarly, one then adjusts the polarization controller to set a second arbitrary input SOP before obtaining the corresponding output SOP as a function of optical frequency, $\mathbf{S}_b(\omega)$. Using Eq. (4.395), one can obtain the PMD $\Delta\tau(\omega_i)$ at any particular frequency ω_i as

$$\Delta\tau(\omega_i) = |\mathbf{\Omega}(\omega_i)| = \frac{1}{\Delta\omega}\left|\frac{\Delta\mathbf{S}_a(\omega_i) \times \Delta\mathbf{S}_b(\omega_i)}{\Delta\mathbf{S}_a(\omega_i)\cdot\mathbf{S}_b(\omega_i)}\right| \tag{8.55}$$

where $\Delta\mathbf{S}_a(\omega_i)$ and $\Delta\mathbf{S}_b(\omega_i)$ are the SOP changes corresponding to a frequency tuning step of $\Delta\omega$ at ω_i. As mentioned in Section 4.6.1, the frequency step $\Delta\omega$ must be sufficiently small to ensure the accuracy of Eq. (8.55).

8.6.8 Poincaré Sphere Analysis (PSA) Method with Three Mutually Orthogonal Output SOPs

As introduced in Section 4.6.2, one may use three mutually orthogonal output SOPs to compute the PMD vector (Williams 2002, 2004). The three orthogonal output SOPs can be obtained with the following two options by (i) generating three mutually orthogonal input SOPs in the Stokes space, such as LHP (0°), L + 45P (+45°), and LVP (+90°), with a PSG before the DUT and measuring the three corresponding output SOPs $(\mathbf{h}, \mathbf{q}, \mathbf{c})$ after the DUT as a function of wavelength (or frequency) with a Stokes polarimeter using the configuration shown in Figure 8.33d; (ii) first generating two distinctive input SOPs with the PSG before measuring the corresponding two output SOPs $\mathbf{a}(\omega_i)$ and $\mathbf{b}(\omega_i)$ as a function of wavelength (or frequency) with the Stokes polarimeter shown in Figure 8.33d, and then constructing the three required mutually orthogonal output SOPs $(\mathbf{h}, \mathbf{q}, \mathbf{c})$ with $\mathbf{a}(\omega_i)$ and $\mathbf{b}(\omega_i)$ using the procedure described in Section 4.6.2. Finally, the magnitude and direction of the PMD (PSP) at a frequency ω_i can be computed as

$$\Delta\tau = \left|\frac{\Delta\theta_i}{\Delta\omega}\right| = \frac{2}{\Delta\omega} sin^{-1}\left\{\frac{1}{2}\sqrt{\frac{1}{2}\left[|\Delta\mathbf{h}(\omega_i)|^2 + |\Delta\mathbf{q}(\omega_i)|^2 + |\Delta\mathbf{c}(\omega_i)|^2\right]}\right\} \tag{8.56a}$$

$$\mathbf{PSP} = \frac{\mathbf{\Omega}(\omega_i)}{|\mathbf{\Omega}(\omega_i)|} = \frac{(\mathbf{c}\cdot\Delta\mathbf{q})\mathbf{h} + (\mathbf{h}\cdot\Delta\mathbf{c})\mathbf{q} + (\mathbf{q}\cdot\Delta\mathbf{h})\mathbf{c}}{|(\mathbf{c}\cdot\Delta\mathbf{q})\mathbf{h} + (\mathbf{h}\cdot\Delta\mathbf{c})\mathbf{q} + (\mathbf{q}\cdot\Delta\mathbf{h})\mathbf{c}|} \tag{8.56b}$$

where $\Delta\mathbf{h}(\omega_i)$, $\Delta\mathbf{q}(\omega_i)$, and $\Delta\mathbf{c}(\omega_i)$ are the differences of $\mathbf{h}$, $\mathbf{q}$, $\mathbf{c}$ between frequencies ω_i and $\omega_i + \Delta\omega$, respectively. As mentioned in Section 4.6.2, the advantage of option 2, as compared with option 1, is that the two input SOPs are not required to be orthogonal, so long as they are sufficiently different from each other, which makes the accuracy of the SOP generated by the PSG in Figure 8.33d less demanding. The PSG in Figure 8.33d can be a polarizer on a rotation stage, or a binary magneto-optic rotator to be described in Chapter 9 for faster and more accurate measurement.

8.6.9 Mueller Matrix Method (MMM)

The operation mechanism of the MMM method has been described in Section 4.6.3, which can be implemented with the Mueller matrix polarimeter of Figure 8.33d. In operation, the Mueller matrix of the DUT as a function of wavelength or optical frequency is obtained with the Mueller matrix polarimeter (described in detail in Section 8.2), while the laser frequency is tuned. Under the assumption that the DUT has no PDL, the 4 × 4 Mueller matrix $\mathbf{M}(\omega_i)$ at each frequency ω_i (or wavelength) can be reduced to a 3 × 3 rotation matrix $\mathbf{M}_{R,3\times3}(\omega_i)$ by removing the first row and the first column of $\mathbf{M}(\omega_i)$. Using the obtained Mueller matrices at different wavelengths, one obtains the transformation matrix $\mathbf{M}_\Delta(\omega_i)$ relating $\mathbf{M}_{R,3\times3}$ at ω_i and $\omega_i + \Delta\omega$:

$$\mathbf{M}_\Delta(\omega_i) = \mathbf{M}_{R,3\times3}(\omega_i + \Delta\omega)\mathbf{M}_{R,3\times3}^{-1}(\omega_i) \tag{8.57a}$$

Finally, the PMD at frequency ω_i can be obtained as

$$\Delta\tau = \left|\frac{cos^{-1}\left[\frac{1}{2}(Tr\mathbf{M}_\Delta(\omega_i) - 1\right]}{\Delta\omega}\right| \tag{8.57b}$$

The PMD direction or the PSP of the PMD vector is the eigenvector of $\mathbf{M}_\Delta$.

Compared with the PSA method described in Section 8.6.8, the MMM method needs to know the input SOP in order to measure the Mueller matrix of DUT while the PSA method does not require the knowledge of the input SOP. Therefore, the PSA method is more convenient and faster than the MMM method. On the other hand, when PDL cannot be ignored, the measurement accuracy of PSA gets worse. On the other hand, the MMM method is more tolerant to the PDL. As will be discussed in Section 9.3.2.2, the matrix $\mathbf{M}_\Delta(\omega_i)$ cannot be represented by a 3×3 rotation matrix when PDL cannot be ignored. It becomes a 4×4 matrix, and the PMD vector becomes a complex vector, but nevertheless can still be used with great accuracy.

8.6.10 Jones Matrix Eigenanalysis (JME) Method

The Jones matrix $\mathbf{M}_J(\omega)$ of the DUT as a function of optical frequency or wavelength can be obtained using the setup of Figure 8.33d, following the procedure of Section 4.3.5. In particular, the PSG in Figure 8.33d generates three distinctive SOPs for each tunable laser frequency and the PSA measures the output SOPs to determine the Jones matrix.

As described in Section 4.6.4, after the Jones matrix $\mathbf{M}_J(\omega_i)$ for each optical frequency ω_i is determined, one may obtain the transformation matrix $\mathbf{\Gamma}(\overline{\omega}_i)$ relating the Jones matrices at optical frequencies ω_i and $\omega_{i+1} = \omega_i + \Delta\omega$:

$$\mathbf{\Gamma}(\overline{\omega}_i) = \mathbf{M}_J(\omega_{i+1})\mathbf{M}_J^{-1}(\omega_i) \tag{8.58a}$$

where $\Delta\omega$ is the frequency tuning step of the tunable laser and $\overline{\omega}_i = (\omega_i + \omega_{i+1})/2 = \omega_i + \Delta\omega/2$. The eigenvectors of $\mathbf{\Gamma}(\overline{\omega}_i)$ are the PSPs of the DUT at the frequency $\overline{\omega}_i$. The corresponding eigenvalues for the "slow" and "fast" eigen polarizations are $\rho_s = e^{i\tau_s\Delta\omega}$ and $\rho_f = e^{i\tau_f\Delta\omega}$, where $\tau_s - \tau_f$ is the group delay between two PSP or the PMD of the DUT at $\overline{\omega}_i$:

$$\Delta\tau(\overline{\omega}_i) = |\tau_s(\overline{\omega}_i) - \tau_f(\overline{\omega}_i)| = \left|\frac{Arg[\rho_s(\overline{\omega}_i)/\rho_f(\overline{\omega}_i)]}{\Delta\omega}\right| \tag{8.58b}$$

where $Arg(\rho_s/\rho_f)$ stands for the phase angle of ρ_s/ρ_f.

8.A Appendix

From Eqs. (4.255) to (4.300), the Stokes vector $\mathbf{S}_{out}$ of the output light exiting from the DUT can be related to the input Stokes vector $\mathbf{S}_{in}$ by the Mueller matrix $\mathbf{M}$:

$$\mathbf{S}_{out} = \mathbf{M}\mathbf{S}_{in} \tag{8.A.1a}$$

$$\begin{pmatrix} S_{out,0} \\ S_{out,1} \\ S_{out,2} \\ S_{out,3} \end{pmatrix} = \begin{pmatrix} m_{00} & m_{01} & m_{02} & m_{03} \\ m_{10} & m_{11} & m_{12} & m_{13} \\ m_{20} & m_{21} & m_{22} & m_{23} \\ m_{30} & m_{31} & m_{32} & m_{33} \end{pmatrix} \begin{pmatrix} S_{in,0} \\ S_{in,1} \\ S_{in,2} \\ S_{in,3} \end{pmatrix} \tag{8.A.1b}$$

From Eqs. (2.14a) to (2.14c), the input Stokes parameters can be expressed as the orientation angle ψ and the ellipticity angle α of a polarization ellipse as

$$S_{in,1} = S_{in,0} cos2\alpha cos2\psi \tag{8.A.2a}$$

$$S_{in,2} = S_{in,0} cos2\alpha sin2\psi \tag{8.A.2b}$$

$$S_{in,3} = S_{in,0} sin2\alpha \tag{8.A.2c}$$

$$S_{in,1}^2 + S_{in,2}^2 + S_{in,3}^2 = S_{in,0}^2 \tag{8.A.2d}$$

Equation (8.A.1b) can be rewritten as

$$\begin{pmatrix} T_0 \\ T_1 \\ T_2 \\ T_3 \end{pmatrix} = \begin{pmatrix} m_{00} & m_{01} & m_{02} & m_{03} \\ m_{10} & m_{11} & m_{12} & m_{13} \\ m_{20} & m_{21} & m_{22} & m_{23} \\ m_{30} & m_{31} & m_{32} & m_{33} \end{pmatrix} \begin{pmatrix} 1 \\ cos2\alpha cos2\psi \\ cos2\alpha sin2\psi \\ sin2\alpha \end{pmatrix} \tag{8.A.3a}$$

where $T_i = S_{out,i}/S_{in,i}$, $i = 0, 1, 2, 3$ and T_0 is the power transmission coefficient, which can be expressed as

$$T_0 = m_{00} + m_{01} cos\beta cos\gamma + m_{02} cos\beta sin\gamma + m_{03} sin\beta \tag{8.A.4a}$$

where $\beta = 2\alpha$ and $\gamma = 2\psi$.

To find the maximum and minimum power transmissions T_{0max} and T_{0min}, one may take the derivatives of T_0 with respect to β and γ and set them to zero, which leads to

$$\frac{\partial T_0}{\partial \gamma} = -m_{01} cos\beta sin\gamma + m_{02} cos\beta cos\gamma = 0 \tag{8.A.4b}$$

$$\frac{\partial T_0}{\partial \beta} = -m_{01} sin\beta cos\gamma - m_{02} sin\beta sin\gamma + m_{03} cos\beta = 0 \tag{8.A.4c}$$

From (8.A.4b) and (8.A.4c), one gets

$$m_{01} sin\gamma = m_{02} cos\gamma \tag{8.A.5a}$$

$$tan\beta = \frac{m_{03}}{m_{01} cos\gamma + m_{02} sin\gamma} = \frac{m_{02} m_{03}}{\left(m_{01}^2 + m_{02}^2\right) sin\gamma} \tag{8.A.5b}$$

From (8.A.5a), using the identity $sin^2\gamma + cos^2\gamma = 1$, one further gets

$$sin\gamma = \pm \frac{m_{02}}{\sqrt{m_{01}^2 + m_{02}^2}} \tag{8.A.5c}$$

$$cos\gamma = \pm \frac{m_{01}}{\sqrt{m_{01}^2 + m_{02}^2}} \tag{8.A.5d}$$

Substitution of (8.A.5c) in (8.A.5b) yields

$$sin\beta = \pm cos\beta \frac{m_{03}}{\sqrt{m_{01}^2 + m_{02}^2}} \tag{8.A.5e}$$

From (8.A.5e), using the identity $sin^2\beta + cos^2\beta = 1$, one obtains

$$sin\beta = \pm \frac{m_{03}}{\sqrt{m_{01}^2 + m_{02}^2 + m_{03}^2}} = \frac{\pm m_{03}}{a} \tag{8.A.5f}$$

$$cos\beta = \pm \frac{\sqrt{m_{01}^2 + m_{02}^2}}{\sqrt{m_{01}^2 + m_{02}^2 + m_{03}^2}} = \frac{\pm\sqrt{m_{01}^2 + m_{02}^2}}{a} \tag{8.A.5g}$$

where $a \equiv \sqrt{m_{01}^2 + m_{02}^2 + m_{03}^2}$.

Substituting (8.A.5c), (8.A.5d), (8.A.5f), and (8.A.5g) in (8.A.4a), we arrive

$$T_0 = m_{00} \pm a = m_{00} \pm \sqrt{m_{01}^2 + m_{02}^2 + m_{03}^2} \tag{8.A.6a}$$

$$T_{0max} = m_{00} + \sqrt{m_{01}^2 + m_{02}^2 + m_{03}^2} \tag{8.A.6b}$$

$$T_{0min} = m_{00} - \sqrt{m_{01}^2 + m_{02}^2 + m_{03}^2} \tag{8.A.6c}$$

$$PDL = 10log\frac{T_{0max}}{T_{0min}} = 10log\frac{m_{00} + \sqrt{m_{01}^2 + m_{02}^2 + m_{03}^2}}{m_{00} - \sqrt{m_{01}^2 + m_{02}^2 + m_{03}^2}} \tag{8.A.6d}$$

References

Azzam, R.M.A. (1976). Oscillating-analyzer ellipsometer. *Rev. Sci. Instrum.* 47 (5): 624–628.

Azzam, R.M.A. (1977). Photopolarimeter using two modulated optical rotators. *Opt. Lett.* 1 (5): 181–183.

Azzam, R.M.A. (1978). Photopolarimetric measurement of the Mueller matrix by Fourier analysis of a single detected signal. *Opt. Lett.* 2 (6): 148–150.

Azzam, R.M.A. (1982). Division-of-amplitude photopolarimeter (DOAP) for the simultaneous measurement of all four Stokes parameters of light. *Optica Acta* 29 (5): 685–689.

Azzam, R.M.A. (1985). Arrangement of four photodetectors for measuring the state of polarization of light. *Opt. Lett.* 10 (7): 309–311.

Azzam, R.M.A. (1992). Division-of-amplitude photopolarimter based on conical diffraction from a metallic grating. *Appl. Opt.* 31 (19): 3574–3576.

Azzam, R.M.A., Elminyawi, I., and El-Saba, A. (1988b). General analysis and optimization of the four-detector photopolarimeter. *J. Opt. Soc. Am. A* 5 (5): 681–689.

Azzam, R.M.A. and Giardina, K.A. (1993). Photopolarimeter based on planar grating diffraction. *J. Opt. Soc. Am. A* 10 (6): 1190–1195.

Azzam, R.M.A. and Lopez, A. (1989). Accurate calibration of the four-detector photopolarimeter with imperfect polarizing optical elements. *J. Opt. Soc. Am. A* 6 (10): 1513–1521.

Azzam, R.M.A., Masetti, E., Elminyawi, I., and Grosz, F. (1988a). Construction, calibration, and testing of a four-detector photopolarimeter. *Rev. Sci. Instrum.* 59 (1): 84–88.

Born, M. and Wolf, E. (1999). Basic properties of the electronmagnetic field. In: *Principles of Optics: Basic Properties of the Electromagnetic Field*, Chapter 1, 6e, 1–66. Cambridge: Cambridge University Press.

Bouzid, A., Abushagur, M., El-Sabae, A., and Azzam, R.M.A. (1995). Fiber-optic four-detector polarimeter. *Opt. Commun.* 118: 329–334.

Collett, E. (2003). *Polarized Light in Fiber Optics.* Lincroft, New Jersey: PolaWave Group.

Craig, R.M. (2003). Accurate spectral characterization of polarization-dependent loss. *J. Lightwave Technol.* 21 (2): 432–437.

Craig, R.M., Gilbert, S., and Hale, P. (1998). High-resolution, nonmechanical approach to polarization-dependent transmission measurements. *J. Lightwave Technol.* 16 (7): 1285–1294.

Damask, J.N. (2005). Review of polarization test and measurement, Chapter 10. In: *Polarization Optics in Telecommunications* (ed. W.T. Rhodes), 429–490. Springer Science + Business Media, Inc.

El Amari, A., Gisin, N., Perny, B. et al. (1998). Statistical prediction and experimental verification of concatenations of fiber optic components with polarization dependent loss. *J. Lightwave Technol.* 16 (3): 332–339.

El-Saba, A.M. and Azzam, R.M.A. (1996). Parallel-slab division-of-amplitude photopolarimeter. *Opt. Lett.* 21 (21): 1709–1771.

Gamiz, V. (1997). Performance of a four-channel polarimeter with low-light-level detection. *SPIE Proc.* 3121: 35–46.

Goldstein, D. (2003). *Polarized Light*, Chapter 27, 2e. New York: Marcel Dekker, Inc.

Goldstein, D.H. and Chipman, R.A. (1990). Error analysis of a Mueller matrix polarimeter. *J. Opt. Soc. Am. A* 7 (4): 693–700.

Hauge, P.S. (1978). Mueller matrix ellipsometry with imperfect compensators. *J. Opt. Soc. Am.* 68 (11): 1519–1528.

Hauge, P.S. (1980). Recent developments in instrumentation in ellipsometry. *Surf. Sci.* 96: 108–140.

Heffner, B.L. (1992). Deterministic, analytically complete measurement of polarization-dependent transmission through optical devices. *IEEE Photonics Technol. Lett.* 4 (5): 451–454.

Hernday, P. (1998). Dispersion measurement, Chapter 12. In: *Fiber Optic Test and Measurement* (ed. D. Derickson), 475–518. New Jersey: Prentice Hall.

Jellison, G.E. Jr., (1987). Four-channel polarimeter for time-resolved ellipsometry. *Opt. Lett.* 12 (10): 766–768.

Liu, J. and Azzam, R.M.A. (1997). Corner-cube four-detector photopolarimeter. *Opt. Laser Technol.* 29 (5): 233–238.

Reza, M., Nezam, M., Yan, L. et al. (2004). Enhancing the dynamic range and DGD monitoring windows in DOP-based DGD monitors using symmetric and asymmetric partial optical filtering. *J. Lightwave Technol.* 22 (4): 1094.

Shi, Y., Yan, L., and Yao, X.S. (2006). Automatic maximum–minimum search method for accurate PDL and DOP characterization. *J. Lightwave Technol.* 24 (11): 4006–4012.

Telcordia. (2001). Generic requirements for passive optical components. GR-1209-CORE. Issue 3.

Westbrook, P., Strasser, T., and Erdogan, T. (2000). In-line polarimeter using blazed fiber gratings. *IEEE Photonics Technol. Lett.* 12 (10): 1352–1354.

Williams, P.A. (2002). PMD measurement techniques – avoiding measurement pitfalls. *Venice Summer School on Polarization Mode Dispersion*, Venice, Italy (24–26 June 2002).

Williams, P.A. (2005). PMD measurement techniques – avoiding measurement pitfalls. In: *Polarization Mode Dispersion* (ed. A. Galtarossa and C. Menyuk), 133–154. Springer Science + Business Media, Inc.

Yan, L., Yao, X.S., Shi, Y., and Willner, A. (2005). Simultaneous monitoring of both optical signal to noise ratio and polarization mode dispersion using polarization scrambling and polarization beam splitting. *J. Lightwave Technol.* 23 (10): 3290–3294.

Yao, X.S. (2002). In-line polarimeter based on integration of free-space optical elements. US patent 6,836,327, filed 18 March 2002 and issued 28 December 2004.

Yao, X.S. (2006). Low-cost polametric detector. US patent 7,372,568, filed 22 June 2006 and issued 13 May 2008.

Yao, X.S. (2013). Monitoring polarization mode dispersion and signal to noise ratio of optical signals based on polarization analysis. US patent 8,787,755, filed 15 April 2013 and issued 22 July 2014.

Yao, X.S., Chen, X., and Liu, T. (2010). High accuracy polarization measurements using binary polarization rotators. *Opt. Express* 18 (7): 6667–6685.

Yao, X.S., Xuan, H., Chen, X. et al. (2019). Polarimetry fiber optic gyroscope. *Opt. Express* 27 (14): 19984–19995.

Further Reading

Chipman, R.A. (1995). Polarimetry, Chapter 22. In: *Handbook of Optics*, 2e, vol. II (ed. M. Bass), 22.1–22.37. New York: McGraw-Hill.

Favin, D., Nyman, B., and Wolter, G. (1993). Systems and method for measuring polarization dependent loss. US5371597, filed 23 November 1993 and issued 6 December 1994.

Galtarossa, A. and Menyuk, C. (2005). *Polarization Mode Dispersion*. Springer Science + Business Media, Inc.

Goldstein, D.H. (1992). Mueller matrix dual-rotating retarder polarimeter. *Appl. Opt.* 31 (31): 6676–6683.

Kogelnik, H., Jopson, R., and Nelson, L. (2002). Polarization mode-dispersion, Chapter 15. In: *Optical Fiber Telecommunication, IVB, Systems and Impairments* (ed. I. Kaminow and T. Li), 725–861. San Diego: Academic Press.

Perlicki, K. (2015). Polarization effects in optical fiber links, Chapter 4. In: *Advances in Optical Fiber Technology: Fundamental Optical Phenomena and Applications* (ed. M. Yasin, H. Arof and S. Harun), 125–158. https://doi.org/10.5772/58517. Intech.

Rochford, K. (2001). *Polarization and polarimetry*. In: *Encyclopedia of Physical Science and Technology*, 3e, vol. 12 (ed. R. Meyers). San Diego: Academic Press.

9

Binary Polarization Generation and Analysis

The polarization state generator (PSG), polarization state analyzer (PSA), and the Mueller matrix polarimeters described in Chapter 8 are analog devices, relying on mechanical rotations or continuous-wave modulation of retardation or polarization angle for achieving desired functions. As one may imagine, devices based on mechanical rotations suffer from relatively poor repeatability due to mechanical wear-and-tear and varied environmental conditions, such as temperature, atmosphere pressure, vibration, dust, lubrication, etc. Many of these changes are not repeatable, and therefore it is difficult to calibrate such devices against these imperfections, which change with time resulting from the changes of environmental conditions. Even for the devices relying on non-mechanical modulations of retardation described in Figure 8.4, and on Faraday rotation described in Figure 8.3, slight variation of the modulating signal's amplitude or phase due to the drift and noise may affect the accuracy of the devices.

In the electronics world, binary or digital electronics have replaced analog electronics for many applications, such as computing, communications, video and audio, and measurements due to their inherent advantages of high repeatability and reproducibility, which are most important for precision measurements because they allow the imperfections in the system to be calibrated out without re-appearing. Analogously, it is not difficult to imagine that if it can be realized, binary polarization analysis can also offer similar advantages and eventually lead to the high measurement accuracy. This chapter focuses on such binary polarization management and measurement techniques.

9.1 Highly Repeatable Magneto-optic Binary PSG

As discussed in Section 8.2, generating at least four distinctive polarization states across the Poincaré sphere with high repeatability, such as at 0°, ±45°, ±90°, right-hand circular polarization (RCP), and left-hand circular polarization (LCP), is important for analyzing polarization properties of lightwave components or systems using Mueller matrix method (Chipman 1995; Goldstein 2003). The information obtained from such analyses can be used to further measure other parameters such as birefringence, polarization mode dispersion (PMD), polarization dependent loss (PDL), degree of polarization (DOP), signal-to-noise ratio (SNR), and state of polarization (SOP) (Willner et al. 2004; Craig et al. 1998; Zadok et al. 2004). Such a PSG can be constructed using rotating quarter-wave plate (QWP) and half-wave plate (HWP), as described in Chapter 8. However, due to its mechanical nature, such a device generally has the disadvantages of slowness, short life-time, low repeatability, and high cost. To overcome these shortcomings, non-mechanical PSGs, such as those made with ferroelectric liquid crystals, are preferred (Craig et al. 1998; Wang and Weiner 2003, 2004; De Martino et al. 2003). In this section, we describe a highly repeatable

Polarization Measurement and Control in Optical Fiber Communication and Sensor Systems, First Edition.
X. Steve Yao and Xiaojun (James) Chen.

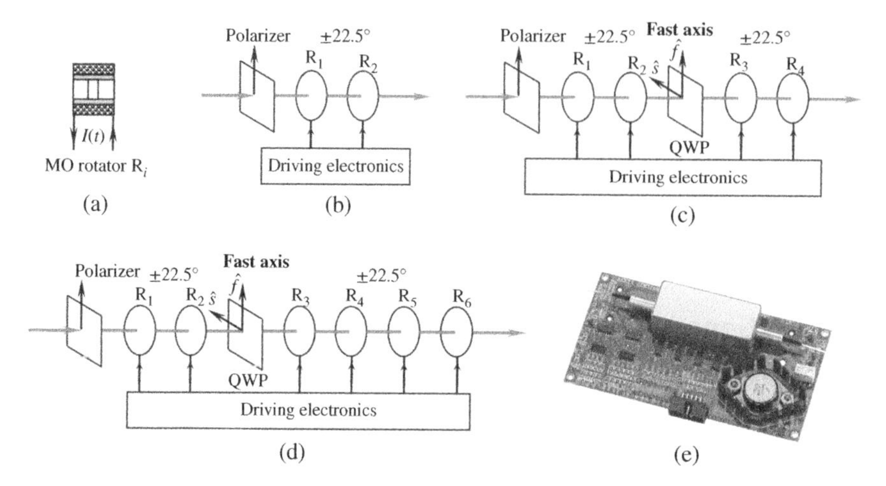

Figure 9.1 (a) Illustration of an MO rotator cell, each is capable of ±22.5° SOP rotation. Device constructions of a 2-bit (b), 4-bit (c), and 6-bit (d) binary polarization state generators (PSG). (e) Photo of the first commercial fiber pigtailed 6-state PSG developed by the authors of this book. Source: Photo courtesy of General Photonics / Luna Innovations.

binary PSG utilizing magneto-optic (MO) polarization rotators described in Figure 6.25a and shown here again in Figure 9.1a (Yao et al. 2005). For a given wavelength, the device can generate highly repeatable polarization states around 0°, ±45°, ±90°, LCP, and RCP across Poincaré sphere. The measured repeatability is better than 0.1° on Poincaré sphere. Another advantage of the device is its predictable wavelength and temperature dependence, typically at −0.067°/nm and 0.1°/C, respectively. The fiber-to-fiber insertion loss of the device is less than 0.9 dB, and its return loss is better than 55 dB. The same device can also be used for Mueller matrix polarization analysis, as will be discussed later in Section 9.3.

9.1.1 Binary PSG Descriptions

9.1.1.1 2-Bit PSG

As shown in Figure 9.1b, a 3-state binary PSG consists of a polarizer and a pair of MO rotators. The polarizer is placed at the input end of the device to define the input SOP. As discussed in Chapter 8, the 3-state PSG is useful for Jones matrix analysis for obtaining PDL and PMD of optical components and optical fibers.

Each MO rotator can be made with a perpendicular anisotropy thick film described in Section 6.6.2.1 or a planar anisotropy thick film described in Section 6.6.2.2, which has the following attractive properties: when applying a positive magnetic field above a saturation field, the rotator rotates SOP by a precise angle around 22.5°. When applying a negative magnetic field beyond saturation, the rotator rotates SOP by a precise angle around −22.5°. Therefore, when two rotators in each pair rotate in the same direction, the net rotation is +45° or −45°. On the other hand, if the two rotators rotate in the opposite direction, the net SOP rotation is zero. Therefore, the 3-state PSG is capable of generating −45°, 0°, +45° (or 0°, 45°, 90°) linear SOPs. Note that the difference for using MO films of perpendicular and planar anisotropy is that the formal has a large transient loss due to the domain scattering while the latter does not. In most cases, this transient loss is not an issue for polarization analysis applications because the data are only taken when the applied magnetic field is above the saturation.

9.1.1.2 4-Bit PSG

As described in Chapter 8, for Mueller matrix analysis, at least four distinctive SOPs are required, with one of them being a LCP or RCP. Figure 9.1c shows the construction of such a 4-bit PSG, consisting of a polarizer, two pair of MO rotators (four pieces total), and a QWP. Again, the polarizer is placed at the input end of the device to define the input SOP and can be either aligned with, orthogonal to, or 45° from the fast (or slow) axis of the QWP, or at other predetermined angles. For the case of aligning to the fast axis of the QWP as in Figure 9.1c, the following SOP can be generated (referenced with respect to the polarizer direction):

(1) Linear SOP at 0°, when rotators in both pairs rotate in opposite directions.
(2) Linear SOP at +45°, when the rotators in the 1st pair rotate in opposite directions, but rotators in the 2nd pair both rotate +22.5° for a total of 45°.
(3) Linear SOP at −45°, when the rotators in the 1st pair rotate in opposite directions, but rotators in the 2nd pair both rotate −22.5° for a total of −45°.
(4) RCP, when rotators in the 1st pair both rotate by 22.5° for a total of 45°, regardless the states of the 2nd pair MO rotators.
(5) LCP, when rotators in the 1st pair both rotate by −22.5° for a total of −45°, regardless the states of the 2nd pair MO rotators.

Note that there are 16 SOP combinations by four bits; however, only five states are distinctive and the rest are degenerate. Table 9.1 is the logic table showing all 16 SOP combinations, in which 1 is to represent +22.5° and 0 to represent −22.5°. In addition, "R_i" stands for rotator #*i*, and "+" and "−" stand for +22.5° and −22.5°, respectively. In the table, the rotation angles of each MO rotator are assumed to be exact ±22.5°. In practice, there are slight deviations due to temperature

Table 9.1 Logic table for a 4-bit PSG.

R_1	R_2	R_3	R_4	Logic	SOP
+22.5°	−22.5°	+22.5°	−22.5°	1010	LVP
+	−	−	+	1001	LVP
−	+	−	+	0101	LVP
−	+	+	−	0110	LVP
+	−	+	+	1011	L+45P
−	+	+	+	0111	L+45P
+	−	−	−	1000	L−45P
−	+	−	−	0100	L−45P
+	+	+	−	1110	RCP
+	+	−	+	1101	RCP
+	+	+	+	1111	RCP
+	+	−	−	1100	RCP
−	−	+	−	0010	LCP
−	−	−	+	0001	LCP
−	−	+	+	0011	LCP
−	−	−	−	0000	LCP

and wavelength variations such that the degenerated SOP states are slightly different, as will be discussed in Section 9.1.3.

9.1.1.3 6-Bit PSG

For Mueller matrix calculations, a minimum of four distinctive SOPs are required. However, some applications may require six distinctive SOPs to have better calibration accuracy. In order to generate such six polarization states, another MO rotator pair can be added to the device (after the second pair) to produce additional +45° and −45° rotations, as shown in Figure 9.1d. Note again that this 6-bit device (with six binary MO rotators) can theoretically generate 64 SOP states; however, only six states are non-degenerate. Table 9.2 lists six typical combinations by this 6-bit device to generate six distinctive SOPs (0° linear, +45° linear, −45° linear, +90° linear [degenerate], RCP, and LCP). Again, here the MO rotators are assumed perfect with exact ±22.5° SOP rotations. Similar to the case of 4-bit PSG, the degenerate SOP states in the 6-bit PSG are slightly different due to the imperfections of MO rotators and alignment errors, which will be discussed in Section 9.1.3. The full logic table for the 64 SOP states without considering the imperfections is listed in Table 9.A.1 in Appendix 9.A.

9.1.2 Experimental Demonstration

The binary PSGs described in Figure 1b–d have been commercialized by the authors of this book at General Photonics Corporation around 2005. To demonstrate their performances, multiple 4-bit and 6-bit PSGs were built and measured with an experimental setup shown in Figure 9.2. Using a commercial polarization analyzer (Agilent 8509C), the generated SOPs on the Poincaré sphere corresponding to each activation combinations have been carefully measured and the results are shown in Figure 9.3a,b for a 4-bit (five states) and a 6-bit (six states) device, respectively (Yao et al. 2005). Points A, B, C, D, and E in Figure 9.3a clearly show five distinctive SOPs generated by the 4-bit PSG, with A being around LCP, C being around RCP, D being around 0° linear, B being around −45° linear, and E being around 45° linear. During the measurement, the polarization

Table 9.2 Logic table for a 6-bit PSG of six distinctive SOPs.

R_1	R_2	R_3	R_4	R_5	R_6	Logic	SOP
+	−	+	−	+	−	101010	LVP
+	−	−	+	+	+	100111	L + 45P
+	−	+	−	−	−	101000	L − 45P
+	−	+	+	+	+	101111	LHP
+	−	−	−	−	−	100000	LHP
+	+	+	+	+	+	111111	RCP
−	−	−	−	−	−	000000	LCP

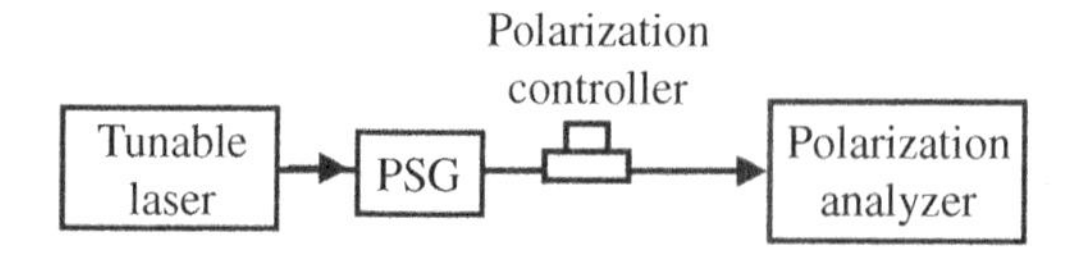

Figure 9.2 Measurement setup for the characterization of the PSGs (Yao et al. 2005).

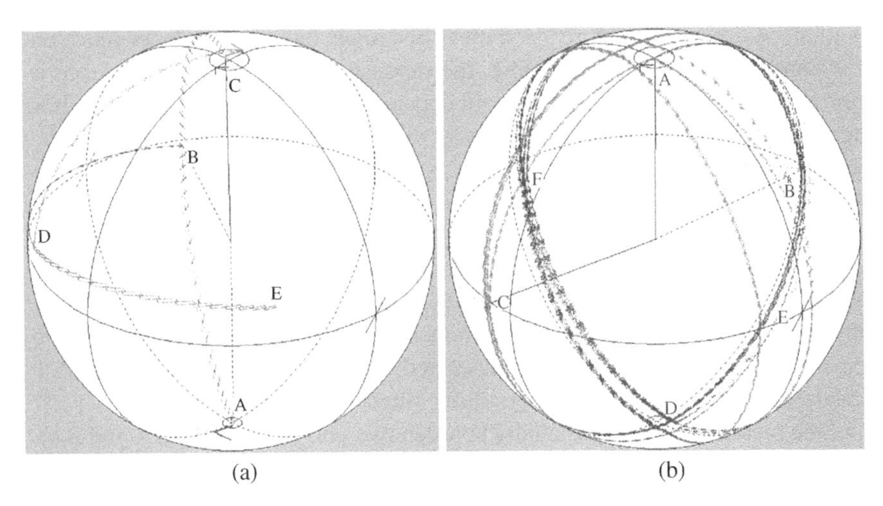

Figure 9.3 Poincaré sphere illustration of SOPs generated by the PSG measured with an Agilent 8509C polarization analyzer. (a) SOPs generated by a 4-bit PSG and (b) SOPs generated by a 6-bit PSG (Yao et al. 2005).

controller in Figure 9.2 was used to align the SOPs with the major axes of the Poincaré sphere. The traces connecting the points are the transient SOP paths when SOPs are switched from one SOP to another.

Similarly, Points A, B, C, D, E, and F in Figure 9.3b are six distinctive SOPs generated by the 6-state PSG. Assuming that a PSG is made with ideal components, many of the bit-combinations will produce the same SOP, resulting in the so-called "degeneracy." However, as will be discussed in Section 9.1.3, because of the imperfection of the components used, these supposed-degeneracy states are slightly off to produce a cluster of SOPs in the vicinity of the targeted SOP in practice. This is why multiple traces of similar orientation, but with some deviation, passing through each targeted SOP are seen. When using the device, one may simply select 6 bit-combinations to generate six consistent SOPs. For example, the bit-combinations corresponding to the six SOPs shown in Figure 9.3b are (000101), (001101), (011101), (011100), (11101), and (111011).

Figure 9.4 shows the measured wavelength dependence of the devices. The vertical axis corresponds to the 3D angle between two SOPs separated by around 90° on the Poincaré sphere (around

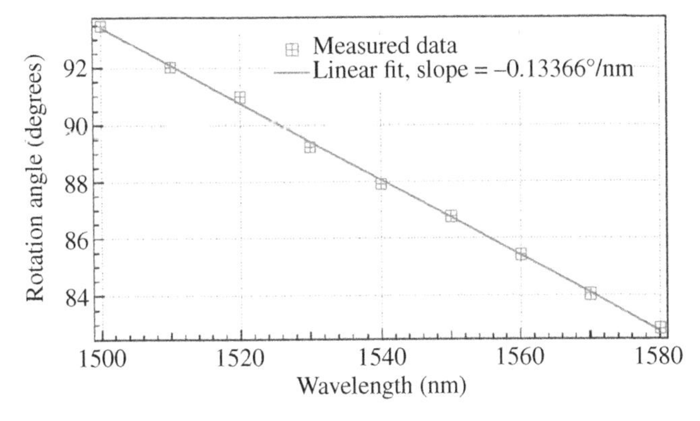

Figure 9.4 Wavelength dependence of the polarization rotation angle of the 6-bit PSG measured on Poincaré sphere. The physical rotation angle is one-half of that measured on the sphere (Yao et al. 2005).

45° of physical rotation). As shown in Figure 9.4, a measured angle has a linear dependence on wavelength, with a slope of −0.134°/nm. Therefore, the physical rotation angle dependence is around −0.067°/nm, in agreement with the wavelength dependence of the MO crystal itself. It is also found that the wavelength dependence is different for different polarization states, as will be discussed in Section 9.1.3.

Note that for Mueller matrix polarization analysis, the inaccuracies of generated polarization states with respect to the targeted polarization states (0°, +45°, −45°, 90°, RCP, and LCP), such as the one caused by wavelength dependence, can be calibrated out in calculations, so long as they are repeatable. In fact, SOP repeatability is a more important specification for Mueller matrix analysis, because it generally involves taking a reference measurement and device under test (DUT) measurements. The repeatability of the generated SOPs between the reference measurement and DUT measurements directly affects the accuracy of the resulting Mueller matrix analysis.

Figure 9.5a shows the measured SOP repeatability between two polarization states A and B on Poincaré sphere with 100 samples. As can be seen, the two states (two end points in the graph) are exactly repeatable with no observable differences. It is interesting to notice that the switching traces from A to B and from B to A are highly repeatable as well. However, they are not exactly the same and are interwoven together, probably due to the hysteresis of the MO rotator. Nevertheless, this trace nonreciprocal property does not affect the performance of the device in any way.

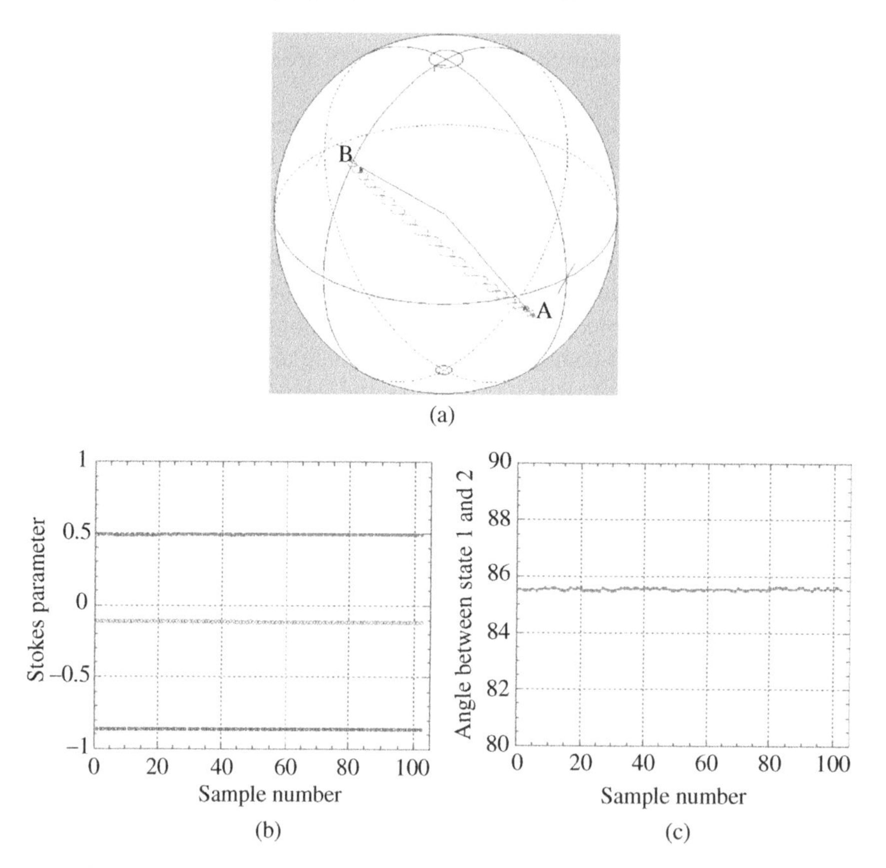

Figure 9.5 Repeatability measurement of a 6-state PSG (100 samples are taken for each graph): (a) Poincaré sphere illustration of the repeatability when the device was switched between two polarization states. (b) Stokes parameters illustration of the repeatability as a function of sample points. (c) 3D angle illustration of the repeatability as a function of sample points. All the measurements show the excellent repeatability of the generated SOPs (Yao et al. 2005).

The SOP repeatability of all six states is shown in Figure 9.3b by repetitively switching between the states 100 times and measuring the corresponding Stokes parameters of each state. The excellent repeatability of the Stokes parameters of a polarization state is shown in Figure 9.5b. All other five states also show the same high repeatability. The 3D angles between the states are also calculated corresponding to each sample point and a typical result is shown in Figure 9.5c, showing a high repeatability. The measured 3D angle repeatability is better than 0.1°, limited by the accuracy of the measurement system.

9.1.3 Imperfections of the Binary PSG

As discussed in Chapter 6, the MO rotators are generally wavelength and temperature dependent. Typical wavelength and temperature coefficients are −0.065°/nm and −0.07°/°C at 1550 nm. In addition, the alignment between the input polarizer and the QWP may not be exact. These imperfections will cause the generated SOPs to deviate from the intended SOPs; however, their effects can be calibrated out when using the SOPs for polarization analysis because of the high repeatability of the each MO rotator. In this section, we discuss how these imperfections affect the generated SOP by a PSG for each logic state.

9.1.3.1 General Expression of the Stokes Vector of Generated PSG

In general, the wavelength and temperature dependences of the Faraday rotation angle of the MO rotators are linear, which are characterized by the wavelength and temperature coefficients k_λ and k_T, respectively. Therefore, the rotation angle θ_j for rotator j can be expressed as

$$\theta_j(\lambda, T) = 22.5^\circ + \Delta\theta_j(\lambda, T) \tag{9.1a}$$

$$\Delta\theta_j(\lambda, T) = \Delta\theta_{0j} + k_{\lambda j}(\lambda - \lambda_0) + k_{Tj}(T - T_0) \tag{9.1b}$$

Here $\Delta\theta_j$ is deviation of rotator j from 22.5° and $\Delta\theta_{0j}$ is the rotation angle bias at the specified operation wavelength and temperature λ_0 and T_0, respectively. For example, for an MO rotator with a specified operating center wavelength of 1550 nm and specified center operation temperature of 23°, $\lambda_0 = 1550$ nm and $T_0 = 23°$. The total SOP rotation angles α and β for the first two and the last four MO rotators can be written as

$$\alpha = \sum_{j=1}^{2} -(-1)^{b_j}\,\theta_j \tag{9.2a}$$

$$\beta = \sum_{j=3}^{6} -(-1)^{b_j}\,\theta_j \tag{9.2b}$$

where b_j is the logic state of each rotator, with $b_j = 1$ standing for +22.5° and $b_j = 0$ standing for −22.5° SOP rotations, respectively. If all the MO rotators are from the same batch of thick films, they are likely to have the same temperature and wavelength coefficients and θ_j in Eqs. (9.2a) and (9.2b) can be replaced with θ.

Referring to Figure 9.1c, the output Stokes vector $\mathbf{S}_{out}$ of a 6-bit PSG in response to an input Stokes vector $\mathbf{S}_{in}$ can be calculated using

$$\mathbf{S}_{out} = \mathbf{M}_{PSG}\mathbf{S}_{in} = \mathbf{M}_{R_6}\mathbf{M}_{R_5}\mathbf{M}_{R_4}\mathbf{M}_{R_3}\mathbf{M}_{QWP}\mathbf{M}_{R_2}\mathbf{M}_{R_1}\mathbf{S}_{in} \tag{9.3a}$$

where $\mathbf{M}_{QWP}$ and $\mathbf{M}_{R_j}$ are the Mueller matrices of the quart-wave plate and the jth rotator, respectively, and $\mathbf{M}_{PSG}$ is the Mueller matrix of the PSG:

$$\mathbf{M}_{PSG} = \mathbf{M}_{R_6}\mathbf{M}_{R_5}\mathbf{M}_{R_4}\mathbf{M}_{R_3}\mathbf{M}_{QWP}\mathbf{M}_{R_2}\mathbf{M}_{R_1} \tag{9.3b}$$

For the PSG described in Figure 9.1, $\mathbf{S}_{in}$ is the Stokes vector of the polarizer. Using the slow and fast axes of the QWP as the coordinate system, if the angle between the polarizer passing axis and the slow axis of the QWP is θ_p, the polarization state of the light after passing through the polarizer can be described by

$$\mathbf{S}_{in} = \begin{pmatrix} 1 \\ cos2\theta_p \\ sin2\theta_p \\ 0 \end{pmatrix} \tag{9.3c}$$

The Mueller matrix of wave plate with retardation of Γ and the Mueller matrix of rotator with a rotation angle ϕ_j can be written as

$$\mathbf{M}_{QWP} = \begin{pmatrix} 1 & 0 & 0 & 0 \\ 0 & 1 & 0 & 0 \\ 0 & 0 & cos\Gamma & sin\Gamma \\ 0 & 0 & -sin\Gamma & cos\Gamma \end{pmatrix} \tag{9.3d}$$

$$\mathbf{M}_{Rj} = \begin{pmatrix} 1 & 0 & 0 & 0 \\ 0 & cos2\phi_j & -sin2\phi_j & 0 \\ 0 & sin2\phi_j & cos2\phi_j & 0 \\ 0 & 0 & 0 & 1 \end{pmatrix} \tag{9.3e}$$

From Eqs. (9.3c) and (9.3d), one obtains

$$\mathbf{M}_{R_6}\mathbf{M}_{R_5}\mathbf{M}_{R_4}\mathbf{M}_{R_3}\mathbf{M}_{QWP} = \begin{pmatrix} 1 & 0 & 0 & 0 \\ 0 & cos2\beta & -sin2\beta & 0 \\ 0 & sin2\beta & cos2\beta & 0 \\ 0 & 0 & 0 & 0 \end{pmatrix} \begin{pmatrix} 1 & 0 & 0 & 0 \\ 0 & 1 & 0 & 0 \\ 0 & 0 & cos\Gamma & sin\Gamma \\ 0 & 0 & -sin\Gamma & cos\Gamma \end{pmatrix}$$

$$= \begin{pmatrix} 1 & 0 & 0 & 0 \\ 0 & cos2\beta & -sin2\beta cos\Gamma & -sin2\beta sin\Gamma \\ 0 & sin2\beta & cos2\beta cos\Gamma & cos2\beta sin\Gamma \\ 0 & 0 & -sin\Gamma & cos\Gamma \end{pmatrix} \tag{9.4a}$$

where β is the total rotation angle of the last four rotators of Eq. (9.2b). Finally,

$$\mathbf{M}_{PSG} = \mathbf{M}_{R_6}\mathbf{M}_{R_5}\mathbf{M}_{R_4}\mathbf{M}_{R_3}\mathbf{M}_{QWP}\mathbf{M}_{R_2}\mathbf{M}_{R_1}$$

$$= \begin{pmatrix} 1 & 0 & 0 & 0 \\ 0 & cos2\beta & -sin2\beta cos\Gamma & -sin2\beta sin\Gamma \\ 0 & sin2\beta & cos2\beta cos\Gamma & cos2\beta sin\Gamma \\ 0 & 0 & -sin\Gamma & cos\Gamma \end{pmatrix} \begin{pmatrix} 1 & 0 & 0 & 0 \\ 0 & cos2\alpha & -sin2\alpha & 0 \\ 0 & sin2\alpha & cos2\alpha & 0 \\ 0 & 0 & 0 & 1 \end{pmatrix}$$

$$= \begin{pmatrix} 1 & 0 & 0 & 0 \\ 0 & cos2\alpha cos2\beta - sin2\alpha sin2\beta cos\Gamma & -sin2\alpha cos2\beta - cos2\alpha sin2\beta cos\Gamma & -sin2\beta sin\Gamma \\ 0 & cos2\alpha sin2\beta + sin2\alpha cos2\beta cos\Gamma & -sin2\alpha sin2\beta + cos2\alpha cos2\beta cos\Gamma & cos2\beta sin\Gamma \\ 0 & -sin2\alpha sin\Gamma & -cos2\alpha sin\Gamma & cos\Gamma \end{pmatrix} \tag{9.4b}$$

where α is the total rotation angle of the first two rotators expressed in Eq. (9.2a). Finally, $\mathbf{S}_{out}$ can be obtained by multiplying Eqs. (9.4b) and (9.3b),

$$\mathbf{S}_{out} = \mathbf{M}_{PSG}\mathbf{S}_{in} = \begin{pmatrix} 1 \\ cos2(\alpha+\theta_p)cos2\beta - sin2(\alpha+\theta_p)sin2\beta cos\Gamma \\ cos2(\alpha+\theta_p)sin2\beta + sin2(\alpha+\theta_p)cos2\beta cos\Gamma \\ -sin2(\alpha+\theta_p)sin\Gamma \end{pmatrix} \tag{9.5}$$

If the polarizer is misaligned with the QWP by an angle of $\Delta\theta_p$, the polarizer angle θ_p can be expressed as:

$$\theta_p = \frac{\pi}{2} + \Delta\theta_p \tag{9.6}$$

9.1.3.2 Stokes Vectors of Some Specific SOPs Generated by PSG

Assuming all the MO rotators are identical such that $\Delta\theta_j(\lambda, T) = \Delta\theta(\lambda, T)$ in Eq. (9.1b), the output distinctive SOPs can be obtained from Eq. (9.5). For example, one may obtain the following Stokes vector.

1. Linear vertical polarization (LVP): For the vertically polarized output, the corresponding possible logic states are: 101100, 101010, 101001, 100110, 100101, 100011, 011100, 011010, 011001, 010110, 010101, and 010011. In essence, the net rotation from all the MO rotators is zero such that $\alpha = \beta = 0$. The Stokes vector of LVP can be simplified from Eq. (9.5) as

$$\mathbf{S}_{LVP} = \begin{pmatrix} 1 \\ cos2\theta_p cos2\beta \\ +sin2\theta_p cos\Gamma \\ -sin2\theta_p sin\Gamma \end{pmatrix} \approx \begin{pmatrix} 1 \\ -cos2\Delta\theta_p \\ 0 \\ sin2\Delta\theta_p \end{pmatrix} \tag{9.7}$$

where $\Gamma = \pi/2$ for QWP is used.

2. Linear horizontal polarization (LHP): For horizontally polarized output, the two MO rotators before the QWP rotate in opposite directions such that $\alpha = 0$, and the last four MO rotators all rotate in the same directions. The corresponding possible logic states are: 101111, 100000, 011111, and 010000. For the case that all rotators rotate $+22.5° + \Delta\theta$ (logic states 101111, 011111), $\beta_{+90°} = 4(22.5° + \Delta\theta) = 90° + 4\Delta\theta$. For the case that all rotators rotate $-22.5° - \Delta\theta$ (logic states 100000, 010000), $\beta_{-90°} = -90° - 4\Delta\theta$. The corresponding Stokes vectors from Eq. (9.5) are

$$\mathbf{S}_{LHP} = \begin{pmatrix} 1 \\ cos2\theta_p cos2\beta - sin2\theta_p sin2\beta cos\Gamma \\ cos2\theta_p sin2\beta + sin2\theta_p cos2\beta cos\Gamma \\ -sin2\theta_p sin\Gamma \end{pmatrix} = \begin{matrix} \begin{pmatrix} 1 \\ -cos2\theta_p cos8\Delta\theta + sin2\theta_p sin8\Delta\theta cos\Gamma \\ -cos2\theta_p sin8\Delta\theta - sin2\theta_p cos8\Delta\theta cos\Gamma \\ -sin2\theta_p sin\Gamma \end{pmatrix} \beta_{+90°} \\ \begin{pmatrix} 1 \\ -cos2\theta_p cos8\Delta\theta - sin2\theta_p sin8\Delta\theta cos\Gamma \\ cos2\theta_p sin8\Delta\theta - sin2\theta_p cos8\Delta\theta cos\Gamma \\ -sin2\theta_p sin\Gamma \end{pmatrix} \beta_{-90°} \end{matrix}$$

$$= \begin{matrix} \begin{pmatrix} 1 \\ cos2\Delta\theta_p cos8\Delta\theta \\ cos2\Delta\theta_p sin8\Delta\theta \\ sin2\Delta\theta_p \end{pmatrix} \beta_{+90°} \\ \begin{pmatrix} 1 \\ cos2\Delta\theta_p cos8\Delta\theta \\ -cos2\Delta\theta_p sin8\Delta\theta \\ sin2\Delta\theta_p \end{pmatrix} \beta_{-90°} \end{matrix} \tag{9.8}$$

Following the same procedure, the Stokes vectors for all six distinctive SOPs can be obtained and are listed in Table 9.3. It can be seen that due to the rotator imperfection, the degeneracy of some of the degenerate logic states is broken, resulting in deviations from the corresponding degenerate states.

It is interesting to notice from Table 9.3 that LVP is independent of wavelength and temperature because it does not contain a term with $\Delta\theta(\lambda, T) = \Delta\theta_0 + k_\lambda(\lambda - \lambda_0) + k_T(T - T_0)$. In addition, some Stokes parameters in certain logic states are also independent of wavelength and temperature, such

Table 9.3 Stokes vectors for all possible logical states with component imperfections.

Target SOP	Possible logic states	Stokes vector	Approximated Stokes vector
Vertical (LVP)	101100, 101010 101001, 100110 100101, 100011 011100, 011010 011001, 010110 010101, 010011	$\begin{pmatrix} 1 \\ \cos 2\theta_p \cos 2\beta \\ +\sin 2\theta_p \cos\Gamma \\ -\sin 2\theta_p \sin\Gamma \end{pmatrix}$	$\begin{pmatrix} 1 \\ -\cos 2\Delta\theta_p \\ 0 \\ \sin 2\Delta\theta_p \end{pmatrix}$
Horizontal (LHP)	101111, 011111	$\begin{pmatrix} 1 \\ -\cos 2\theta_p \cos 8\Delta\theta + \sin 2\theta_p \sin 8\Delta\theta \cos\Gamma \\ -\cos 2\theta_p \sin 8\Delta\theta - \sin 2\theta_p \cos 8\Delta\theta \cos\Gamma \\ -\sin 2\theta_p \sin\Gamma \end{pmatrix}$	$\begin{pmatrix} 1 \\ \cos 2\Delta\theta_p \cos 8\Delta\theta \\ \cos 2\Delta\theta_p \sin 8\Delta\theta \\ \sin 2\Delta\theta_p \end{pmatrix}$
	100000, 010000	$\begin{pmatrix} 1 \\ -\cos 2\theta_p \cos 8\Delta\theta - \sin 2\,\theta_p \sin 8\Delta\theta \cos\Gamma \\ \cos 2\theta_p \sin 8\Delta\theta - \sin 2\,\theta_p \cos 8\Delta\theta \cos\Gamma \\ -\sin 2\,\theta_p \sin\Gamma \end{pmatrix}$	$\begin{pmatrix} 1 \\ \cos 2\Delta\theta_p \cos 8\Delta\theta \\ -\cos 2\Delta\theta_p \sin 8\Delta\theta \\ \sin 2\Delta\theta_p \end{pmatrix}$
Linear at +45° (L + 45P)	101110, 101101 101011, 100111 011110, 011101 011011, 010111	$\begin{pmatrix} 1 \\ -\cos 2\theta_p \sin 4\Delta\theta - \sin 2\theta_p \cos 4\Delta\theta \cos\Gamma \\ \cos 2\theta_p \cos 4\Delta\theta - \sin 2\theta_p \sin 4\Delta\theta \cos\Gamma \\ -\sin 2\theta_p \sin\Gamma \end{pmatrix}$	$\begin{pmatrix} 1 \\ \cos 2\Delta\theta_p \sin 4\Delta\theta \\ -\cos 2\Delta\theta_p \cos 4\Delta\theta \\ \sin 2\Delta\theta_p \end{pmatrix}$
Linear at −45° (L − 45P)	100001, 100010 100100, 101000 010001, 010010 010100, 011000	$\begin{pmatrix} 1 \\ -\cos 2\theta_p \sin 4\Delta\theta + \sin 2\theta_p \cos 4\Delta\theta \cos\Gamma \\ -\cos 2\theta_p \cos 4\Delta\theta - \sin 2\theta_p \sin 4\Delta\theta \cos\Gamma \\ -\sin 2\theta_p \sin\Gamma \end{pmatrix}$	$\begin{pmatrix} 1 \\ \cos 2\Delta\theta_p \sin 4\Delta\theta \\ \cos 2\Delta\theta_p \cos 4\Delta\theta \\ \sin 2\Delta\theta_p \end{pmatrix}$
RCP	111100, 111010 111001, 110110 110101, 110011	$\begin{pmatrix} 1 \\ \sin(4\Delta\theta + 2\Delta\theta_p) \\ -\cos(4\Delta\theta + 2\Delta\theta_p)\cos\Gamma \\ \cos(4\Delta\theta + 2\Delta\theta_p)\sin\Gamma \end{pmatrix}$	$\begin{pmatrix} 1 \\ \sin(4\Delta\theta + 2\Delta\theta_p) \\ 0 \\ \cos(4\Delta\theta + 2\Delta\theta_p) \end{pmatrix}$
	111110, 111101 111011, 110111	$\begin{pmatrix} 1 \\ -\sin(4\Delta\theta + 2\Delta\theta_p)\sin 4\Delta\theta - \cos(4\Delta\theta + \Delta\theta_p)\cos 4\Delta\theta \cos\Gamma \\ \sin(4\Delta\theta + 2\Delta\theta_p)\cos 4\Delta\theta + \cos(4\Delta\theta + 2\Delta\theta_p)\sin 4\Delta\theta \cos\Gamma \\ \cos(4\Delta\theta + 2\Delta\theta_p)\sin\Gamma \end{pmatrix}$	$\begin{pmatrix} 1 \\ -\sin(4\Delta\theta + 2\Delta\theta_p)\sin 4\Delta\theta \\ \sin(4\Delta\theta + 2\Delta\theta_p)\cos 4\Delta\theta \\ \cos(4\Delta\theta + 2\theta_p) \end{pmatrix}$
	111000, 110100 110010, 110001	$\begin{pmatrix} 1 \\ -\sin(4\Delta\theta + 2\Delta\theta_p)\sin 4\Delta\theta + \cos(4\Delta\theta + 2\Delta\theta_p)\cos 4\Delta\theta \cos\Gamma \\ -\sin(4\Delta\theta + 2\Delta\theta_p)\cos 4\Delta\theta + \cos(4\Delta\theta + 2\Delta\theta_p)\sin 4\Delta\theta \cos\Gamma \\ \cos(4\Delta\theta + 2\Delta\theta_p)\sin\Gamma \end{pmatrix}$	$\begin{pmatrix} 1 \\ -\sin(4\Delta\theta + 2\Delta\theta_p)\sin 4\Delta\theta \\ -\sin(4\Delta\theta + 2\Delta\theta_p)\cos 4\Delta\theta \\ \cos(4\Delta\theta + 2\Delta\theta_p) \end{pmatrix}$

(Continued)

Table 9.3 (Continued)

Target SOP	Possible logic states	Stokes vector	Approximated Stokes vector
RCP	111111	$\begin{pmatrix} 1 \\ -sin(4\Delta\theta + 2\Delta\theta_p)cos8\Delta\theta + cos(4\Delta\theta + 2\Delta\theta_p)sin8\Delta\theta cos\Gamma \\ -sin(4\Delta\theta + 2\Delta\theta_p)sin8\Delta\theta + cos(4\Delta\theta + 2\Delta\theta_p)cos8\Delta\theta cos\Gamma \\ cos(4\Delta\theta + 2\Delta\theta_p)sin\Gamma \end{pmatrix}$	$\begin{pmatrix} 1 \\ -sin(4\Delta\theta + 2\Delta\theta_p)cos8\Delta\theta \\ -sin(4\Delta\theta + 2\Delta\theta_p)sin8\Delta\theta \\ cos(4\Delta\theta + 2\Delta\theta_p) \end{pmatrix}$
	110000	$\begin{pmatrix} 1 \\ -sin(4\Delta\theta + 2\Delta\theta_p)cos8\Delta\theta + cos(4\Delta\theta + 2\Delta\theta_p)sin8\Delta\theta cos\Gamma \\ sin(4\Delta\theta + 2\Delta\theta_p)sin8\Delta\theta + cos(4\Delta\theta + 2\Delta\theta_p)cos8\Delta\theta cos\Gamma \\ cos(4\Delta\theta + 2\Delta\theta_p)sin\Gamma \end{pmatrix}$	$\begin{pmatrix} 1 \\ -sin(4\Delta\theta + 2\Delta\theta_p)cos8\Delta\theta \\ sin(4\Delta\theta + 2\Delta\theta_p)sin8\Delta\theta \\ cos(4\Delta\theta + 2\Delta\theta_p) \end{pmatrix}$
LCP	001100, 001010 001001, 000110 000101, 000011	$\begin{pmatrix} 1 \\ sin(4\Delta\theta - 2\Delta\theta_p) \\ cos(4\Delta\theta - 2\Delta\theta_p)cos\Gamma \\ -cos(4\Delta\theta - 2\Delta\theta_p)sin\Gamma \end{pmatrix}$	$\begin{pmatrix} 1 \\ sin(4\Delta\theta - 2\Delta\theta_p) \\ 0 \\ -cos(4\Delta\theta - 2\Delta\theta_p) \end{pmatrix}$
	001110, 001101 001011, 000111	$\begin{pmatrix} 1 \\ -sin(4\Delta\theta - 2\Delta\theta_p)sin4\Delta\theta + cos(4\Delta\theta - 2\Delta\theta_p)cos4\Delta\theta cos\Gamma \\ sin(4\Delta\theta - 2\Delta\theta_p)cos4\Delta\theta - cos(4\Delta\theta - 2\Delta\theta_p)sin4\Delta\theta cos\Gamma \\ -cos(4\Delta\theta - 2\Delta\theta_p)sin\Gamma \end{pmatrix}$	$\begin{pmatrix} 1 \\ -sin(4\Delta\theta - 2\Delta\theta_p)sin4\Delta\theta \\ sin(4\Delta\theta - 2\Delta\theta_p)cos4\Delta\theta \\ -cos(4\Delta\theta - 2\Delta\theta_p) \end{pmatrix}$
	001000, 000100 000010, 000001	$\begin{pmatrix} 1 \\ -sin(4\Delta\theta - 2\Delta\theta_p)sin4\Delta\theta + cos(4\Delta\theta - 2\Delta\theta_p)cos4\Delta\theta cos\Gamma \\ -sin(4\Delta\theta - 2\Delta\theta_p)cos4\Delta\theta - cos(4\Delta\theta - 2\Delta\theta_p)sin4\Delta\theta cos\Gamma \\ -cos(4\Delta\theta - 2\Delta\theta_p)sin\Gamma \end{pmatrix}$	$\begin{pmatrix} 1 \\ -sin(4\Delta\theta - 2\Delta\theta_p)sin4\Delta\theta \\ -sin(4\Delta\theta - 2\Delta\theta_p)cos4\Delta\theta \\ -cos(4\Delta\theta - 2\Delta\theta_p) \end{pmatrix}$
	001111	$\begin{pmatrix} 1 \\ -sin(4\Delta\theta - 2\Delta\theta_p)cos8\Delta\theta - cos(4\Delta\theta - 2\Delta\theta_p)sin8\Delta\theta cos\Gamma \\ -sin(4\Delta\theta - 2\Delta\theta_p)sin8\Delta\theta - cos(4\Delta\theta - 2\Delta\theta_p)cos8\Delta\theta cos\Gamma \\ -cos(4\Delta\theta - 2\Delta\theta_p)sin\Gamma \end{pmatrix}$	$\begin{pmatrix} 1 \\ -sin(4\Delta\theta - 2\Delta\theta_p)cos8\Delta\theta \\ -sin(4\Delta\theta - 2\Delta\theta_p)sin8\Delta\theta \\ -cos(4\Delta\theta - 2\Delta\theta_p) \end{pmatrix}$
	000000	$\begin{pmatrix} 1 \\ -sin(4\Delta\theta - 2\Delta\theta_p)cos8\Delta\theta - cos(4\Delta\theta - 2\Delta\theta_p)sin8\Delta\theta cos\Gamma \\ sin(4\Delta\theta - 2\Delta\theta_p)sin8\Delta\theta - cos(4\Delta\theta - 2\Delta\theta_p)cos8\Delta\theta cos\Gamma \\ -cos(4\Delta\theta - 2\Delta\theta_p)sin\Gamma \end{pmatrix}$	$\begin{pmatrix} 1 \\ -sin(4\Delta\theta - 2\Delta\theta_p)cos8\Delta\theta \\ sin(4\Delta\theta - 2\Delta\theta_p)sin8\Delta\theta \\ -cos(4\Delta\theta - 2\Delta\theta_p) \end{pmatrix}$

as s_2 of RCP with the logic states of (111100, 111010, 111001, 110110, 110101, 110011) and LCP with the logic states of (001100, 001010, 001001, 000110, 000101, 000011). As can be seen from Table 9.3, when the imperfections of the MO rotators are considered, a 6-bit binary PSG with 6 MO rotators can generate 15 different polarization states in which 6 of them are distinctive. Similarly, a 4-bit binary PSG with four MO rotators can generate nine different polarization states, in which five of them are distinctive.

9.1.3.3 Experimental Verification

Figure 9.6 shows the measurement results of the Stokes parameters of some generated SOPs as a function of wavelength compared with the theoretical results of Table 9.3. As expected, all three Stokes parameters of an LVP have no wavelength dependence, as shown in Row 1 of Figure 9.6.

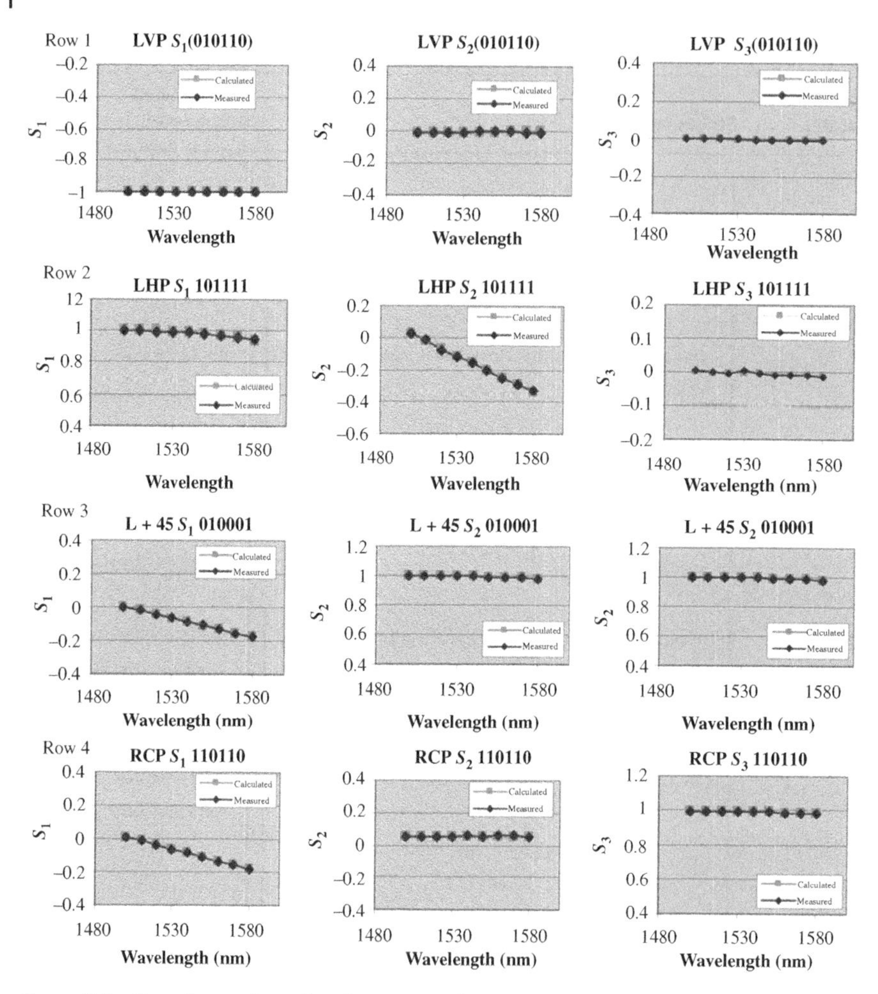

Figure 9.6 Experimental results of measured Stokes parameters of different generated SOPs at some specific logic states. Row 1: LVP at (0101101). Row 2: LHP at (101111); Row 3: L + 45P at (010001); Row 4: RCP at (110110). The results were obtained with a commercial polarimeter.

For an LHP with the logic state of (101111), it can be seen from Row 2 of Figure 9.6 at s_1 and s_3 are less sensitive to the wavelength because they are functions of $cos[\Delta\theta(\lambda, T)]$ ($\Delta\theta \ll 1$), while s_2 is much more sensitive to wavelength variations because it is a function of $sin[\Delta\theta(\lambda, T)]$.

Similarly, for L + 45P with the logic state of (010001) shown in Row 3 of Figure 9.6, s_1 has a much stronger wavelength dependence than s_2 and s_3 because it is proportional to $sin[\Delta\theta(\lambda, T)]$, while s_2 is proportional to $cos[\Delta\theta(\lambda, T)]$ and s_3 is not a function of $\Delta\theta(\lambda, T)$. Finally, for RCP with the logic state of (110110) shown in Row 4 of Figure 9.6, s_1 is wavelength dependent while s_2 and s_3 are not, because s_1 is proportional to $sin[\Delta\theta(\lambda, T)]$ while s_2 is not a function of $\Delta\theta(\lambda, T)$ and s_3 is proportional to $cos[\Delta\theta(\lambda, T)]$ from Table 9.3.

9.1.3.4 Applications

As will be discussed, due to the advantages of compact size, high speed (down to 20 μs SOP switch time), and high repeatability, the binary PSG may find many applications ranging from polarization analysis to network performance monitoring, material birefringence and PMD measurement, swept-wavelength measurement, medical imaging (e.g. polarization resolved coherent topology measurement) (Park et al. 2014; Urbanczyk 1986), and fiber sensor systems. In particular, the PSG described in Figure 9.1 can be used as a polarization analyzer if used in a reverse order, that is, to input an optical signal from the left hand side and detect the signal at the right hand side in Figure 9.1. Thus two identical devices can be used (i.e. one as a PSG and the other as PSA) to measure or monitor the corresponding properties or perturbation of optical medium. As will be discussed in Section 9.5, by connecting a spectrum disperse device, such as a grating or wavelength division demultiplexer, to the PSG's output, many polarization related properties of multiple wavelength channels, such as SOP, DOP, SNR, as well as PMD and PDL, can be simultaneously analyzed in parallel.

9.2 Highly Accurate Binary Magneto-optic Polarization State Analyzer (PSA)

In Section 9.1, binary MO PSGs for generating five or six distinctive polarization states across the Poincaré sphere with a high repeatability of better than 0.1° on the Poincaré sphere are described (Yao et al. 2005). It is mentioned that such a PSG can be configured to analyze the SOP of an optical beam. This section describes in detail such a binary PSA which is highly accurate due to the binary high repeatability nature of the MO crystals. To overcome the wavelength and temperature dependencies of the MO rotators and the QWP described in Section 9.1, a self-calibrating methodology is introduced to automatically extract the effects of wavelength and temperature variations on measurement accuracy (Yao et al. 2006). With the device, a remarkable DOP accuracy of ±0.5%, an SOP accuracy of better than 1.3%, and an angular accuracy and resolution of 0.3° and 0.02° respectively from 1460 to 1580 nm are achieved. This binary MO PSA has the attractive features of low cost, compact size, high repeatability, no moving part, and self-calibration, which is ideal for low cost and high accuracy polarization analysis, system SNR monitoring (Yan et al. 2006), and accurate measurements of PMD (Heffner 1992a; Williams 1999, 2002), PDL (Craig 2013; Heffner 1992b), birefringence and material properties of thin films (Goldstein 2003) when paired with a binary PSG described in Section 9.1. The capability of self-calibration is especially attractive because it enables the device to avoid the cumbersome calibration process (Compain et al. 1999; Goldstein and Chipman 1990) by automatically obtaining the temperature and wavelength dependencies of the optical components used in the device, making it more reliable and easy to operate.

9.2.1 Device Description

9.2.1.1 Device Structure

The binary MO PSA is composed of nine or seven functional components: a QWP, two MO rotators before QWP, four or two MO rotators after QWP, a polarizer P, and a photodetector (PD), as shown in Figure 9.7.

As described in Section 9.1, the MO rotators have the following attractive binary properties: a highly repeatable SOP rotation angle around 22.5° or −22.5° can be obtained with each rotator by applying either a positive or negative magnetic field above a saturation field. Therefore, when two rotators rotate in the same direction, the net rotation is +45° or −45°. On the other hand, if the two rotators rotate in the opposite direction, the net SOP rotation is zero.

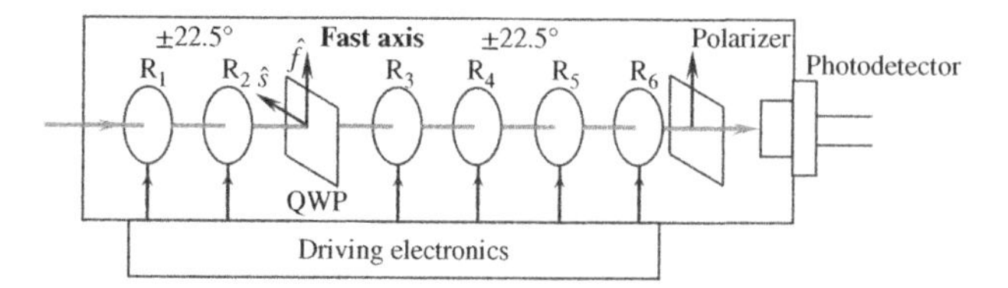

Figure 9.7 Construction of a 6-bit binary MO PSA. R_1, R_2, R_3, R_4, R_5, R_6 are MO rotators, each can be rotated ±22.5° at a saturation current.

9.2.1.2 Operation Theory

The output power P_{out} of the PSA can be calculated by multiplying the Mueller matrices of all components in Figure 9.7 and the corresponding photocurrent I_{out} received in the photodetector is proportional to P_{out}:

$$I_{out} = \eta \mathbf{M}_{PSA}\mathbf{S}_{in} = \eta \mathbf{M}_p \mathbf{M}_{PSG}\mathbf{S}_{in} \tag{9.9a}$$

where $\mathbf{M}_{PSA}$ and $\mathbf{M}_p$ are the Mueller matrices of the PSA and polarizer, respectively, η is the responsivity of the PD, and $\mathbf{S}_{in}$ is the Stokes vector of the input SOP. $\mathbf{M}_{PSG}$ in Eq. (9.9a) is given in Eq. (9.3b) and $\mathbf{M}_{PSA}$ can be written out as

$$\mathbf{M}_{PSA} = \mathbf{M}_p\mathbf{M}_{PSG} = \mathbf{M}_p(\mathbf{M}_{R_6}\mathbf{M}_{R_5}\mathbf{M}_{R_4}\mathbf{M}_{R_3}\mathbf{M}_{QWP}\mathbf{M}_{R_2}\mathbf{M}_{R_1}) \tag{9.9b}$$

where $\mathbf{M}_{R_j}$ and $\mathbf{M}_{QWP}$ are the Mueller matrices of the MO rotator j and the QWP, respectively. The Mueller matrix $\mathbf{M}_p$ for the polarizer can be written as

$$\begin{aligned}
\mathbf{M}_p(\theta_p) &= \mathbf{R}(\theta_p)\mathbf{M}_p(0)\mathbf{R}(-\theta_p)\\
&= \frac{1}{2}\begin{pmatrix} 1 & 0 & 0 & 0\\ 0 & cos2\theta_p & -sin2\theta_p & 0\\ 0 & sin2\theta_p & cos2\theta_p & 0\\ 0 & 0 & 0 & 1\end{pmatrix}\begin{pmatrix} 1 & 1 & 0 & 0\\ 1 & 1 & 0 & 0\\ 0 & 0 & 0 & 0\\ 0 & 0 & 0 & 0\end{pmatrix}\begin{pmatrix} 1 & 0 & 0 & 0\\ 0 & cos2\theta_p & sin2\theta_p & 0\\ 0 & -sin2\theta_p & cos2\theta_p & 0\\ 0 & 0 & 0 & 1\end{pmatrix}\\
&= \frac{1}{2}\begin{pmatrix} 1 & cos2\theta_p & sin2\theta_p & 0\\ cos2\theta_p & cos^2 2\theta_p & sin2\theta_p cos2\theta_p & 0\\ sin2\theta_p & sin2\theta_p cos2\theta_p & sin^2 2\theta_p & 0\\ 0 & 0 & 0 & 0\end{pmatrix}
\end{aligned} \tag{9.9c}$$

where $\mathbf{R}(\theta_p)$ is the rotation transformation matrix to account for the misalignment of the polarizer with respect to the QWP.

Substituting Eqs. (9.4b), (9.9b), and (9.9c) in Eq. (9.9a), one obtains

$$\begin{aligned}
I_{out}(\alpha,\beta) = \eta \Big\{ &\frac{1}{2}S_0 + \frac{1}{2}\left[cos2\alpha cos2\left(\beta-\theta_p\right) - sin2\alpha sin2\left(\beta-\theta_p\right)cos\Gamma\right]S_1\\
&- \frac{1}{2}\left[sin2\alpha cos2\left(\beta-\theta_p\right) + cos2\alpha sin2\left(\beta-\theta_p\right)cos\Gamma\right]S_2\\
&- \frac{1}{2}sin2\left(\beta-\theta_p\right)sin\Gamma S_3\Big\}
\end{aligned} \tag{9.10a}$$

where (S_0, S_1, S_2, S_3) are the Stokes parameters of the input SOP, Γ is the retardation of the QWP, α is the net rotation angle of the two MO rotators before the QWP, and β is the net rotation angle of the MO rotators after the QWP, as described in Eqs. (9.2a) and (9.2b). Assuming all the MO rotators are identical, α, β can be expressed as

$$\alpha = \sum_{j=1}^{2} -(-1)^{bj}\theta(\lambda, T), \quad \beta = \sum_{j=3}^{6} -(-1)^{bj}\theta(\lambda, T) \tag{9.10b}$$

$$\theta(\lambda, T) = 22.5^\circ + \Delta\theta(\lambda, T) \tag{9.10c}$$

$$\Delta\theta(\lambda, T) = \Delta\theta_0 + k_\lambda(\lambda - \lambda_0) + k_T(T - T_0) \tag{9.10d}$$

where b_j, θ, $\Delta\theta_0$, k_λ, k_T, λ_0, and T_0 are all defined in Eqs. (9.1a), (9.1b), (9.2a), and (9.2b). Note that Eq. (9.10a) can also apply to a 4-bit binary PSA made with four MO rotators by reducing the maximum j from 6 to 4 in the expression of β in Eq. (9.10b) such that $\beta = \sum_{j=3}^{4} -(-1)^{b_j}\theta(\lambda, T)$.

9.2.1.3 Degeneracy

Because each MO rotator is binary with two rotation angles, I_{out} has 64 possible values for each input SOP for a device with a total of 6 MO rotators (6-bit PSA) and 16 possible values for a device with 4 MO rotators (4-bit MO) in general. Assuming the rotators are identical and the rotation angles in both directions are the same, one can readily conclude by inspecting Figure 9.7 or Eq. (9.10b) that α only has three possible values ($0, 2\theta, -2\theta$) and that β only has five possible values ($0, 2\theta, 4\theta, -2\theta, -4\theta$) for a 6-bit PSA (four rotators after the QWP). Therefore, I_{out} in Eq. (9.10a) only has $3 \times 5 = 15$ different values, as shown in Table 9.4. The rest are degenerate. Similarly, β has three possible values ($0, 2\theta, -2\theta$) for devices with a total of four MO rotators (two after the QWP) and I_{out} has $3 \times 3 = 9$ different values.

9.2.1.4 Distinctive Logic States (DLS)

Further degeneracy occurs when the MO rotators, the QWP, and the polarizers are perfect, i.e. $\theta = 22.5^\circ$, $\Gamma = \pi/2$, and $\theta_p = 90^\circ$. In this case, I_{out} only have six different values for a device with six rotators and five different values for a device with four rotators. Therefore, from the 15 logic states in Table 9.4, there are only 6 non-degenerate states for a perfect 6-bit PSA. Even in a non-perfect situation, these six states are more distinctive from one another than the rest nearly-degenerate states. We call these six states distinctive logic states (DLS). Similarly, there are only five distinctive logic states for a 4-bit PSA.

Table 9.4 α, β, and logic states of a 6-bit PSA.

I_i	α	β	DLS	I_i	α	β	DLS
I_1	-2θ	-4θ	#1	I_9	2θ	0	#1
I_2	0	-4θ	#2	I_{10}	-2θ	2θ	#6
I_3	2θ	-4θ	#3	I_{11}	0	2θ	#6
I_4	-2θ	-2θ	#4	I_{12}	2θ	2θ	#6
I_5	0	-2θ	#4	I_{13}	-2θ	4θ	#1
I_6	2θ	-2θ	#4	I_{14}	0	4θ	#2
I_7	-2θ	0	#3	I_{15}	2θ	4θ	#3
I_8	0	0	#5				

9.2.1.5 Mueller Matrix Method

Assuming all the parameters in Eqs. (9.10a), (9.10b), (9.10c), and (9.10d), namely, θ_p, Γ, $\Delta\theta_0$, k_λ, k_T, λ_0, and T_0 are known, the output of PSA for the ith logic state can be rewritten as

$$I_i = \begin{pmatrix} M_{i0} & M_{i1} & M_{i2} & M_{i4} \end{pmatrix} \begin{pmatrix} S_0 \\ S_1 \\ S_2 \\ S_3 \end{pmatrix} \quad (i = 1, 2, \ldots 2^N) \tag{9.11a}$$

where N is the total number of rotators, and M_{i0}, M_{i1}, M_{i2}, and M_{i3} can be obtained from Eq. (9.10a) for all MO rotation combinations. For calculating four Stokes parameters of the input light, at least four different equations are required. Therefore, by measuring four output powers (I_a, I_b, I_c, I_d) of four distinctive logic states, one obtains

$$\begin{pmatrix} I_a \\ I_b \\ I_c \\ I_d \end{pmatrix} = \begin{pmatrix} \frac{1}{2} & M_{a1} & M_{a2} & M_{a3} \\ \frac{1}{2} & M_{b1} & M_{b2} & M_{b3} \\ \frac{1}{2} & M_{c1} & M_{c2} & M_{c3} \\ \frac{1}{2} & M_{d1} & M_{d2} & M_{d3} \end{pmatrix} \begin{pmatrix} S_0 \\ S_1 \\ S_2 \\ S_3 \end{pmatrix} = \mathbf{M} \begin{pmatrix} S_0 \\ S_1 \\ S_2 \\ S_3 \end{pmatrix} \tag{9.11b}$$

The four Stokes parameters (S_0, S_1, S_2, S_3) of an unknown optical signal can therefore be obtained by the reverse transform of Eq. (9.11b), and the corresponding DOP can be calculated using

$$DOP = \frac{\sqrt{S_1^2 + S_2^2 + S_3^2}}{S_0}$$

9.2.1.6 Selection of Logic States

The four powers can be selected from the 15 non-degenerate logic states listed in Table 9.4. For calculation accuracies, the four equations chosen should be as distinctive as possible, and therefore should be chosen from the powers of the six distinctive logic states defined previously. For example, we may select the following combinations to be used in Mueller matrix calculations: $[I_1(-2\theta,-4\theta), I_2(0,-4\theta), I_3(2\theta,-4\theta), I_4(-2\theta,-2\theta)]$, $[I_5(0,-2\theta), I_7(-2\theta,0), I_8(0,0), I_{11}(0,2\theta)]$, or $[I_6(2\theta,-2\theta), I_{12}(2\theta,2\theta), I_{14}(0,4\theta), I_{15}(2\theta,4\theta)]$.

9.2.2 Self-Calibrating Binary PSA

9.2.2.1 Self-Calibration Method

The Mueller matrix method described earlier requires the parameters relating to the imperfections of all the components inside a PSA to be known at all wavelengths and all temperatures. However, it is extremely time-consuming to measure the wavelength and temperature dependencies of all the components. Even if the wavelength and temperature dependencies are known, it is often cumbersome to know the exact wavelength and temperature during measurements. To overcome the difficulties, here we describe a self-calibrating methodology to automatically extract the effects of wavelength and temperature variations on measurement accuracy, while in the same time obtaining all the parameters relating to the component imperfections (Yao et al. 2006). One may express Eq. (9.10a) as the function of Stokes parameters, the MO rotator angles, as well as the imperfection parameters:

$$I_i = f[S_0, S_1, S_2, S_3, \alpha_i, \beta_i, \Delta\Gamma(\lambda, T), \Delta\theta(\lambda, T), \theta_p] \tag{9.12}$$

where $i = 1, 2, \ldots 2^N$ and N is the total number of MO rotators in PSA, I_i is the output photocurrent of the PSA for the ith logic state. In practice, there are 15 non-degenerate equations available for the calculation of a 6-bit PSA and 9 for a 4-bit PSA. The rotation angles α and β, the Stokes parameters (S_0, S_1, S_2, S_3) of input light, $\Delta\Gamma(\lambda, T)$ of the QWP, $\Delta\theta(\lambda, T)$, and θ_p of polarizer can be calculated simultaneously by numerically solving Eq. (9.12), without the need of knowing wavelength and temperature, which is the basic concept of self-calibration. A method of solving Eq. (9.12) is basically to numerically search for the optimized values of S_0, S_1, S_2, S_3, $\Delta\theta(\lambda, T)$, $\Delta\Gamma(\lambda, T)$, and θ_p to make $\sum_j (f_j - I_j)^2$ minimum. Note that one should not only use the distinctive states for the calculation because the slight non-degeneracy of the other states actually contains the information of the deviation caused by wavelength and temperature dependencies. Using this method, one can obtain not only the Stokes parameters (S_0, S_1, S_2, S_3) of the input light, but also all the component parameters, including $\Delta\Gamma(\lambda, T)$, $\Delta\theta_0$, k_λ, k_T, θ_p, λ_0, and T_0. In fact, the component parameters obtained at different wavelength and temperature can then be used in Mueller matrix method.

9.2.2.2 Perfect QWP and Polarizer Alignment

Let's examine a special case in which the QWP has a retardation of exact $\pi/2$ without wavelength dependence (achromatic QWP) and the polarizer is perfectly aligned with the fast axis of the QWP, that is, $\Gamma = \pi/2$ and $\theta_p = 90°$. Equation (9.10a) can be simplified to

$$I_{out}(\alpha, \beta) = \frac{1}{2}\eta(S_0 + cos2\alpha cos2\beta S_1 - sin2\alpha cos2\beta S_2 - sin2\beta S_3) \tag{9.13}$$

Table 9.5 lists the PSA output photocurrents corresponding to all possible α and β combinations (logic states) from Eq. (9.13), assuming all the MO rotators is identical.

Define a coefficient C as $C = sin4\theta$, then from Table 9.5 one obtains

$$C = sin4\Delta\theta = \frac{I_{(-45,45)} - I_{(45,45)}}{I_{(45,0)} - I_{(-45,0)}} = \frac{I_{(-45,-45)} - I_{(45,-45)}}{I_{(45,0)} - I_{(-45,0)}} \tag{9.14a}$$

$$= \frac{I_{(-45,45)} + I_{(45,-45)} - I_{(0,45)} - I_{(0,-45)}}{2I_{(0,0)} - I_{(0,45)} - I_{(0,-45)}} \tag{9.14b}$$

$$= \frac{I_{(0,90)} - I_{(0,-90)}}{2[I_{(0,-45)} - I_{(0,45)}]} \tag{9.14c}$$

Table 9.5 All possible PSA output currents under difference logical states ($\theta_p = 90°, \Gamma = \pi/2$).

Rotators' states	Output of PSA
$(\alpha = \beta = 0)$	$I(0,0) = ½\eta(S_0 - S_1)$
$(\alpha = 0, \beta = 45° + 2\Delta\theta)$	$I(0,2\theta) = ½\eta(S_0 + sin4\Delta\theta S_1 + cos4\Delta\theta S_3)$
$(\alpha = 0, \beta = -45° - 2\Delta\theta)$	$I(0,-2\theta) = ½\eta(S_0 + sin4\Delta\theta S_1 - cos4\Delta\theta S_3)$
$(\alpha = 45° + 2\Delta\theta, \beta = 0)$	$I(2\theta,0) = ½\eta(S_0 + sin4\Delta\theta S_1 + cos4\Delta\theta S_2)$
$(\alpha = -45° - 2\Delta\theta, \beta = 0)$	$I(-2\theta,0) = ½\eta(S_0 + sin4\Delta\theta S_1 - cos4\Delta\theta S_2)$
$(\alpha = 45° + 2\Delta\theta, \beta = 45° + 2\Delta\theta)$	$I(2\theta,2\theta) = ½\eta(S_0 - sin4\Delta\theta sin4\Delta\theta S_1 - cos4\Delta\theta sin4\Delta\theta S_2 + cos4\Delta\theta S_3)$
$(\alpha = 45° + 2\Delta\theta, \beta = -45° - 2\Delta\theta)$	$I(2\theta,-2\theta) = ½\eta(S_0 - sin4\Delta\theta sin4\Delta\theta S_1 - cos4\Delta\theta sin4\Delta\theta S_2 - cos4\Delta\theta S_3)$
$(\alpha = -45° - 2\Delta\theta, \beta = 45° + 2\Delta\theta)$	$I(-2\theta,2\theta) = ½\eta(S_0 - sin4\Delta\theta sin4\Delta\theta S_1 + cos4\Delta\theta sin4\Delta\theta S_2 + cos4\Delta\theta S_3)$
$(\alpha = -45° - 2\Delta\theta, \beta = -45° - 2\Delta\theta)$	$I(-2\theta,-2\theta) = ½\eta(S_0 - sin4\Delta\theta sin4\Delta\theta S_1 + cos4\Delta\theta sin4\Delta\theta S_2 - cos4\Delta\theta S_3)$
$(\alpha = 0, \beta = 90° + 4\Delta\theta)$	$I(0,4\theta) = ½\eta(S_0 + cos8\Delta\theta S_1 - sin8\Delta\theta S_3)$
$(\alpha = 0, \beta = -90° - 4\Delta\theta)$	$I(0,-4\theta) = ½\eta(S_0 + cos8\Delta\theta S_1 + sin8\Delta\theta S_3)$

where ±45 are used to represent $\pm 2\theta$ and ±90 are used to represent $\pm 4\theta$ in $I(\alpha,\beta)$. Substituting C in the equations in Table 9.5, one may arrive

$$S_0 = \frac{2CI_{(0,0)} + I_{(0,45)} + I_{(0,-45)}}{1+C} \tag{9.15a}$$

$$S_1 = -\frac{2I_{(0,0)} - I_{(0,45)} - I_{(0,-45)}}{1+C} \tag{9.15b}$$

$$S_2 = \frac{I_{(45,0)} - I_{(-45,0)}}{\sqrt{1-C^2}} \tag{9.15c}$$

$$S_2 = \frac{I_{(-45,45)} - I_{(45,45)}}{C\sqrt{1-C^2}} = \frac{I_{(-45,-45)} - I_{(45,-45)}}{C\sqrt{1-C^2}} \tag{9.15d}$$

$$S_3 = \frac{I_{(45,45)} - I_{(45,-45)}}{\sqrt{1-C^2}} = \frac{I_{(-45,45)} - I_{(-45,-45)}}{\sqrt{1-C^2}} = \frac{I_{(0,45)} - I_{(0,-45)}}{\sqrt{1-C^2}} \tag{9.15e}$$

Generally, the deviation angle $\Delta\theta$ of rotators is small, so C is a small number and Eq. (9.15d) may cause large errors in practice, so that Eq. (9.15c) is recommended to calculate S_2. It should also be noted, when $S_2 \gg 0$ and $S_1 \gg 0$, Eqs. (9.14a) and (9.14b) provide better accuracy to calculate C; If $S_2 \approx 0$ and $S_1 \gg 0$, Eq. (9.14b) gives better accuracy for calculating. If $S_1 \approx 0$, Eq. (9.14c) gives better accuracy.

In summary, if $I(-45,0)$, $I(45,0)$, $I(45,45)$, $I(-45,45)$, $I(0,0)$, $I(0,45)$, $I(0,-45)$, $I(0,90)$, and $I(0,-90)$ are measured, the SOP of input light can be calculated using Eqs. (9.14a)–(9.15e) without the need to know the wavelength of the input light and the temperature of the environment.

9.2.2.3 Experiments

The performance of multiple PSAs has been evaluated with the experimental setup in Figure 9.8. A PSG described in Figure 9.1d is used to generate six distinctive SOPs at different wavelengths and a PSA under test obtains the corresponding Stokes parameters. The results are then compared with those obtained with a commercial high performance reference PSA, which replaces the binary PSA in Figure 9.8.

Figure 9.9a–c shows some typical measurement results of an input SOP using the self-calibration method, and Figure 9.9d shows the obtained wavelength dependences of the rotation angle deviation (from 22.5°) of the MO rotators and the retardation Γ of the QWP. The polarizer angle θ_p obtained is 90.38°. Note that θ, Γ, and θ_p are the averages of six measurements using six distinctive input SOPs (LVP, LHP, L − 45P, L + 45P, RCP, and LCP). Curve fitting in Figure 9.9d yields all the parameters in Eq. (9.10d) for the MO rotators (except k_T): $k_\lambda = -0.035°/\text{nm}$ and $\Delta\theta_0 = -1.359°$ (λ_0 is set at 1550 nm). In Figure 9.9, a small relative alignment error between the PSA and the reference PSA was removed. k_T can also be obtained if similar measurements at different temperatures can be performed.

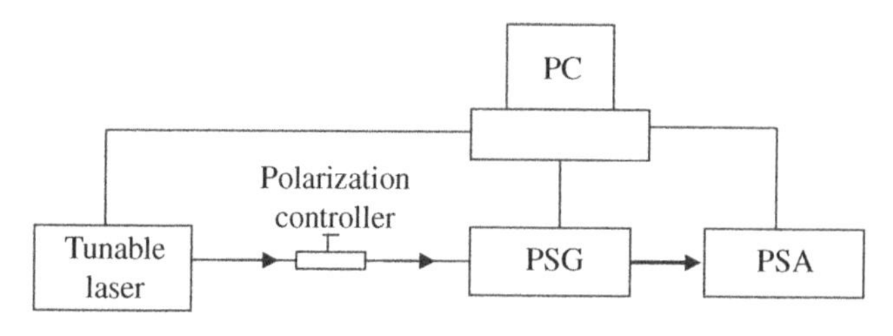

Figure 9.8 Schematic of a binary MO PSA and experiment setup. Light exits a PSG and enters a PSA in free space. A high accuracy reference PSA is put in the place of PSA under test for comparison purpose.

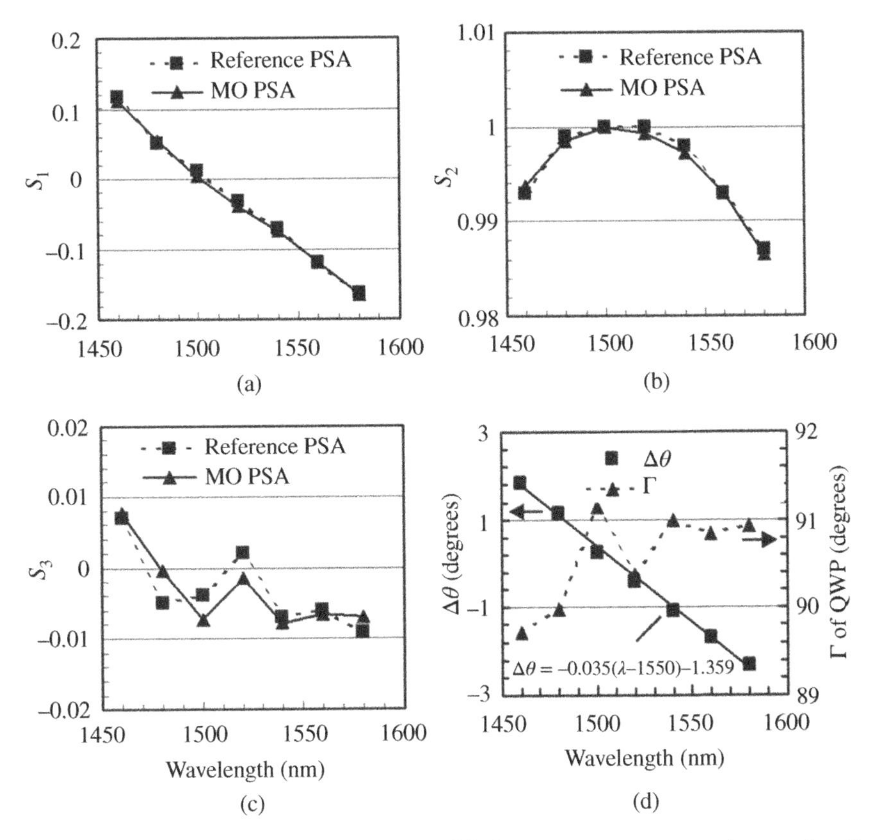

Figure 9.9 Experimental results of measured Stokes parameters s_1, s_2, and s_3 as a function of wavelength as compared with a reference PSA (a–c). The input SOP generated by the PSG is nominally at 45° linear with certain wavelength dependence. (d) Wavelength dependence of the MO rotators and the QWP inside the PSA simultaneously obtained using the self-calibration method (Yao et al. 2006).

The relative SOP error can be obtained by comparing the results with those of the commercial reference PSA using $\sigma = \sqrt{(S_1' - S_1)^2 + (S_2' - S_2)^2 + (S_3' - S_3)^2}$, where S_i' are the Stokes parameters measured using the reference PSA. The DOP accuracies can be obtained by comparing the measurement results with unity because a high extinction ratio (>50 dB) polarizer was placed at the input of the PSG to ensure the DOP of the input light to be 100%. With both the self-calibration and the Mueller matrix methods (after the component parameters are obtained using the self-calibration method), multiple 6-bit PSA units at different wavelengths from 1460 to 1580 nm have been characterized. As shown in Table 9.6 and Figure 9.9, the measurement accuracy remains the same despite the strong wavelength dependencies of MO rotation angle θ and QWP retardation Γ.

Note that the measurement errors presented in Table 9.6 actually include the contributions of PSG fluctuation and inaccuracy of the reference PSA. In order to remove these additional uncertainties, a polarizer was used on a precision rotation stage to replace the PSG. Because the input SOP to the PSA is set by the polarizer and is known, the absolute PSA accuracy can be evaluated. As shown in Table 9.7, the 6-bit PSA has an angular resolution and accuracy of 0.02° and 0.3°, respectively. The DOP accuracy is better than ±0.5% using the self-calibration method. As can be seen, the 6-bit PSA is more accurate than a 4-bit PSA. By taking 50–100 measurements per SOP state, the PSA is found to have a remarkable repeatability of SOP and DOP of less than 0.3%.

Table 9.6 Comparison with a reference PSA (1500–1580 nm).

Method	Self-calibration	Mueller
Maximum SOP error	1.3%	1.5%
Maximum DOP error	±0.35%	±0.65%
DOP standard deviation (STDV)	0.28%	0.4%

Table 9.7 Accuracy measurements with a polarizer (1460–1580 nm).

PSA type	6-bit		4-bit
Method	Self-cal.	Mueller	Self-cal.
Angle resolution	0.02°	>0.02°	0.02°
Maximum angle error	0.30°	0.27°	0.34°
STDV of angle	0.12°	0.09°	0.12°
Maximum DOP error	±0.5%	±0.75%	±1.0%
DOP average	0.999	0.999	1.003
DOP STDV	0.37%	0.46%	0.58%

In summary, because of the high repeatability of the binary MO rotators, a self-calibrating PSA can be realized. The device automatically overcomes wavelength and temperature dependent inaccuracies, and has impressive SOP and DOP accuracies of 0.3° and ±0.5%, respectively, from 1460 to 1580 nm.

9.3 Binary Mueller Matrix Polarimetry

As discussed in Chapter 8, Mueller matrix polarimeters have long been realized with analog PSG and PSA for measuring the polarization related properties of optical components and devices, including optical fibers and fiber links with multiple components. However, as mentioned at the beginning of Chapter 9, these analog devices may suffer from relatively poor repeatability and mechanical wear-and-tear, resulting in compromised performance. With the binary PSG and PSA described in Sections 9.1 and 9.2, one can readily overcome the issues associated with the analog technologies and construct a binary Mueller matrix polarimetry system with the advantages of high measurement speed, accuracy, and repeatability.

In this section, we describe in detail the setup and theory for the use of a binary PSG and PSA for component characterization, and then present experimental data to show the accuracy and repeatability of the technique (Yao et al. 2010).

9.3.1 System Description of Binary Mueller Matrix Polarimetry

As shown in Figure 9.10, the binary Mueller matrix polarimetry system consists of a tunable laser (TLS), a binary MO PSG, a binary MO PSA, and a control computer (PC) (Yao et al. 2010). The

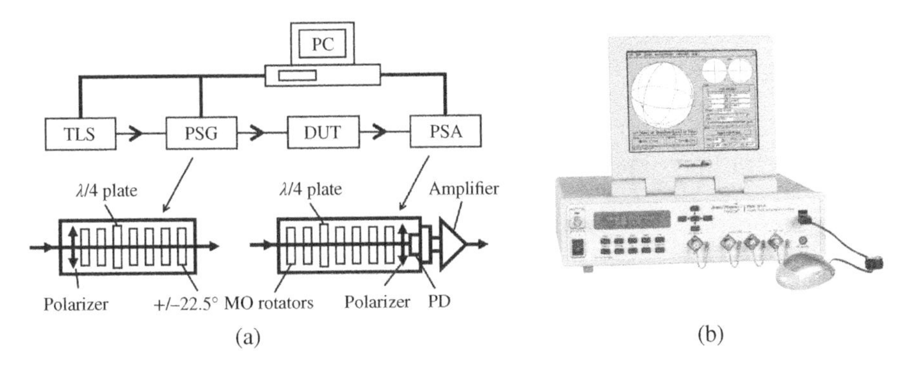

Figure 9.10 The construction diagram (a) and a product photo (b) of a binary polarization measurement system comprising a PSG and a PSA made with binary magneto-optic polarization rotators, a tunable laser, and a computer developed by the authors of this book (Yao et al. 2010). Source: Photo courtesy of General Photonics / Luna Innovations.

DUT is placed between the PSG and PSA. The PSG and PSA each contain six binary MO polarization rotators, a polarizer, and a QWP. The PSA also contains a photodetector (PD) and a signal amplification circuit. As described in Sections 9.1 and 9.2, these MO rotators have the following attractive binary properties: when a positive magnetic field above a saturation level is applied, the rotator rotates the SOP by a precise angle of θ. When a negative magnetic field beyond the saturation level is applied, the rotator rotates the SOP by precisely angle of $-\theta$. For the purposes of the PSG and PSA, θ is designed to be around 22.5°. Therefore, when two rotators rotate the SOP in the same direction, the net rotation is +45° or −45°. Conversely, if the two rotators rotate the SOP in opposite directions, the net SOP rotation is zero. With the rotation directions of each rotator controlled by the control computer, the PSG can generate six distinctive polarization states and the PSA can accurately measure the SOP and the DOP of light entering it by analyzing voltages generated in the photodetector using Mueller matrix analysis. As described in Section 9.2, the binary PSA can also be self-calibrated to remove inaccuracies caused by imperfections in components and workmanship, yielding extremely high measurement accuracies.

Both Jones and Mueller matrix analysis methods can be used to obtain the complete set of polarization related information of a DUT, including PMD (birefringence), PSP (orientation angle of birefringence), PDL (diattenuation), and PDL orientation angle. Other methods, such as the wavelength scanning method, can be used to obtain PMD information only. The system shown in Figure 9.10 can be used to implement all of the measurement methods described in Chapter 8, but in this paper, we will concentrate on its use with the Jones and Mueller matrix methods to measure and calculate the complete polarization information of a DUT, including the validation of the system by using it to measure specially made PMD and PDL artifacts and comparing the results with the theoretical values of the artifacts. Finally, the use of this system for other applications, including the measurements of the polarization extinction ratio (PER) of a polarization maintaining (PM) fiber, the stress in a PM connector, the beat length of a PM fiber and the polarization properties of spun fibers, will be described.

9.3.2 Theoretical Background

9.3.2.1 Jones Matrix Eigenanalysis (JME)

Obtaining the Jones Transfer Matrix As discussed in Section 4.3, generally, the polarization transfer matrix of an optical DUT can be described by a 2×2 complex Jones transfer matrix $\mathbf{\Gamma}$, and the input and output polarization states are related by (Heffner 1992a)

$$\mathbf{J}^{PSA} = \mathbf{\Gamma J}^{PSG} \tag{9.16a}$$

$$\begin{pmatrix} J_x^{PSA} \\ J_y^{PSA} \end{pmatrix} = c^* \begin{pmatrix} \Gamma_{00} & \Gamma_{01} \\ \Gamma_{10} & 1 \end{pmatrix} \begin{pmatrix} J_x^{PSG} \\ J_y^{PSG} \end{pmatrix} \tag{9.16b}$$

where $\mathbf{J}^{PSG}$ is the normalized Jones vector of the light output generated by the PSG, $\mathbf{J}^{PSA}$ is the normalized Jones vector of the light measured by the PSA, and c^* is a complex constant related to the absolute amplitude and absolute phase of the light wave. The ratio of J_x^{PSA}/J_y^{PSA} is independent of c^* and can be expressed as

$$\frac{J_x^{PSA}}{J_y^{PSA}} = \frac{\left(\Gamma_{00}J_x^{PSG} + \Gamma_{01}J_y^{PSG}\right)}{\left(\Gamma_{10}J_x^{PSG} + J_y^{PSG}\right)} \tag{9.16c}$$

which is independent of the constant c^*. If three sets of vectors $\mathbf{J}^{PSG}$ and $\mathbf{J}^{PSA}$ are generated and measured, we can obtain the following equations from Eq. (9.16b):

$$\begin{aligned} &J_{0,x}^{PSG}J_{0,y}^{PSA}\Gamma_{00} + J_{0,y}^{PSG}J_{0,y}^{PSA}\Gamma_{01} - J_{0,x}^{PSG}J_{0,x}^{PSA}\Gamma_{10} = J_{0,y}^{PSG}J_{0,x}^{PSA} \\ &J_{1,x}^{PSG}J_{1,y}^{PSA}\Gamma_{00} + J_{1,y}^{PSG}J_{1,y}^{PSA}\Gamma_{01} - J_{1,x}^{PSG}J_{1,x}^{PSA}\Gamma_{10} = J_{1,y}^{PSG}J_{1,x}^{PSA} \\ &J_{2,x}^{PSG}J_{2,y}^{PSA}\Gamma_{00} + J_{2,y}^{PSG}J_{2,y}^{PSA}\Gamma_{01} - J_{2,x}^{PSG}J_{2,x}^{PSA}\Gamma_{10} = J_{2,y}^{PSG}J_{2,x}^{PSA} \end{aligned} \tag{9.16d}$$

Let $k_{i0} = J_{i,x}^{PSG}J_{i,y}^{PSA}$, $k_{i1} = J_{i,y}^{PSG}J_{i,y}^{PSA}$, $k_{i2} = -J_{i,x}^{PSG}J_{i,x}^{PSA}$, and $k_{i3} = J_{i,y}^{PSG}J_{i,x}^{PSA}$ $(i = 0,1,2)$, then Eq. (9.16d) can be simplified to

$$\begin{aligned} &k_{00}\Gamma_{00} + k_{01}\Gamma_{01} + k_{02}\Gamma_{10} = k_{03} \\ &k_{10}\Gamma_{00} + k_{11}\Gamma_{01} + k_{12}\Gamma_{10} = k_{13} \\ &k_{20}\Gamma_{00} + k_{21}\Gamma_{01} + k_{22}\Gamma_{10} = k_{23} \end{aligned} \tag{9.16e}$$

and the Jones transfer matrix $\mathbf{\Gamma}$ of the DUT can be calculated using Cramer's rule

$$\Gamma_{00} = \frac{\begin{vmatrix} k_{03} & k_{01} & k_{02} \\ k_{13} & k_{11} & k_{12} \\ k_{23} & k_{21} & k_{22} \end{vmatrix}}{\begin{vmatrix} k_{00} & k_{01} & k_{02} \\ k_{10} & k_{11} & k_{12} \\ k_{20} & k_{21} & k_{22} \end{vmatrix}}, \quad \Gamma_{01} = \frac{\begin{vmatrix} k_{00} & k_{03} & k_{02} \\ k_{10} & k_{13} & k_{12} \\ k_{20} & k_{23} & k_{22} \end{vmatrix}}{\begin{vmatrix} k_{00} & k_{01} & k_{02} \\ k_{10} & k_{11} & k_{12} \\ k_{20} & k_{21} & k_{22} \end{vmatrix}}, \quad \Gamma_{10} = \frac{\begin{vmatrix} k_{00} & k_{01} & k_{03} \\ k_{10} & k_{11} & k_{13} \\ k_{20} & k_{21} & k_{23} \end{vmatrix}}{\begin{vmatrix} k_{00} & k_{01} & k_{02} \\ k_{10} & k_{11} & k_{12} \\ k_{20} & k_{21} & k_{22} \end{vmatrix}} \tag{9.16f}$$

where $|\mathbf{K}|$ stands for the determinant of the matrix $\mathbf{K}$. It is important to note that in Eqs. (9.16d)–(9.16f), the three polarization states generated by the PSG are not specified in the analysis. Any three non-degenerate SOPs generated by the PSG can be used to obtain the 2×2 complex Jones transfer matrix $\mathbf{\Gamma}$. The lack of restrictions on the SOPs used eliminates the accuracy requirement for the PSG and makes the system less susceptible to variations caused by the wavelength and temperature dependences of the PSG. By comparison, other Jones matrix analysis implementations (Heffner 1992a) require precise polarization state generation at specific points on the Poincaré sphere (0°, 45°, and 90°), rendering the measurement system more susceptible to imperfections in the PSG.

Obtaining PDL From the discussions in Section 4.3, for a given optical frequency, the PDL of a DUT can be calculated as (Heffner 1992b; Craig et al. 1998):

$$PDL = 10log\left|\frac{r_1}{r_2}\right| \tag{9.17}$$

where $r_{1,2} = \frac{m_{11}+m_{22}}{2} \pm \sqrt{\left(\frac{m_{11}+m_{22}}{2}\right)^2 - m_{11}m_{22} + m_{12}m_{21}}$ are the eigenvalues of the matrix $\mathbf{M} = (\mathbf{\Gamma}^{-1})^* \cdot \mathbf{\Gamma} = \begin{pmatrix} m_{11} & m_{12} \\ m_{21} & m_{22} \end{pmatrix}$, $\mathbf{\Gamma}^{-1}$ is the transpose of Jones transfer matrix $\mathbf{\Gamma}$, and the star * indicates the complex conjugate.

Obtaining PMD As discussed in Section 4.3 (Heffner 1992a), for two adjacent optical frequencies, define a matrix:

$$\mathbf{T}(\Delta\omega) = \mathbf{\Gamma}(\omega_2)\mathbf{\Gamma}(\omega_1)^{-1} \tag{9.18a}$$

The complex eigenvectors ρ_s and ρ_f of matrix $\mathbf{T}(\Delta\omega)$ are the fast and slow principal states of polarization of the DUT. The differential group delay (**DGD**) $\tau(\omega)$ can be calculated from

$$\tau(\omega) = |\tau_s - \tau_f| = \left|\frac{Arg(\rho_s/\rho_f)}{(\omega_1 - \omega_2)}\right| \tag{9.18b}$$

where

$$\omega = \frac{(\omega_1 + \omega_2)}{2} \tag{9.18c}$$

and $Arg(\rho_s/\rho_f)$ stands for the phase angle of ρ_s/ρ_f. The wavelength dependent PMD vector **W** can be defined as

$$\mathbf{W}(\omega) = \tau(\omega)\hat{\mathbf{q}}(\omega) \tag{9.19a}$$

where $\hat{\mathbf{q}}(\omega)$ is the unit vector of the fast principal SOP. As introduced in Section 3.4, the second-order polarization mode dispersion (*SOPMD*), defined as the frequency derivative of the PMD vector $\mathbf{W}(\omega)$, can be calculated as

$$SOPMD \equiv \frac{d\mathbf{W}(\omega)}{d\omega} = \frac{d\boldsymbol{\tau}(\omega)}{d\omega}\hat{\mathbf{q}}(\omega) + \boldsymbol{\tau}(\omega)\frac{d\hat{\mathbf{q}}(\omega)}{d\omega} \tag{9.19b}$$

9.3.2.2 Mueller Matrix Measurement (MMM)

As discussed in Chapters 4 and 8, both PDL and PMD can be obtained using Mueller matrix analysis. Here we summarize the important steps.

Obtaining the Mueller Matrix Let the Stokes vector of the *i*th output of the PSG be

$$\mathbf{S}_i^{PSG} = \begin{pmatrix} S_{0i}^{PSG} \\ S_{1i}^{PSG} \\ S_{2i}^{PSG} \\ S_{3i}^{PSG} \end{pmatrix} \tag{9.20a}$$

The corresponding Stokes vectors measured by the PSA after the light passes through the DUT can be related to Mueller matrix **M** by

$$\mathbf{S}_i^{PSA} = \begin{pmatrix} S_{0i}^{PSA} \\ S_{1i}^{PSA} \\ S_{2i}^{PSA} \\ S_{3i}^{PSA} \end{pmatrix} = \begin{pmatrix} m_{00} & m_{01} & m_{02} & m_{03} \\ m_{10} & m_{11} & m_{12} & m_{13} \\ m_{20} & m_{21} & m_{22} & m_{23} \\ m_{30} & m_{31} & m_{32} & m_{33} \end{pmatrix} \begin{pmatrix} S_{0i}^{PSG} \\ S_{1i}^{PSG} \\ S_{2i}^{PSG} \\ S_{3i}^{PSG} \end{pmatrix} \tag{9.20b}$$

At least four distinctive SOPs must be generated by the PSG and analyzed by the PSA to completely determine Mueller matrix **M** by solving Eq. (9.20b). In such a case, $i = 0, 1, 2, 3$ in Eqs. (9.20a) and (9.20b). However, for higher accuracies, we require that as many as six distinctive SOPs be generated by the PSG and analyzed by the PSA, so that $i = 0, 1, 2, \ldots, 5$. Define a new matrix $\mathbf{S}^{PSA}$ as

$$\mathbf{S}^{PSA} = \begin{pmatrix} S_{00}^{PSA} & S_{01}^{PSA} & S_{02}^{PSA} & S_{03}^{PSA} & S_{04}^{PSA} & S_{05}^{PSA} \\ S_{10}^{PSA} & S_{11}^{PSA} & S_{12}^{PSA} & S_{13}^{PSA} & S_{14}^{PSA} & S_{15}^{PSA} \\ S_{20}^{PSA} & S_{21}^{PSA} & S_{22}^{PSA} & S_{23}^{PSA} & S_{24}^{PSA} & S_{25}^{PSA} \\ S_{30}^{PSA} & S_{31}^{PSA} & S_{32}^{PSA} & S_{33}^{PSA} & S_{34}^{PSA} & S_{35}^{PSA} \end{pmatrix}$$

$$= \begin{pmatrix} m_{00} & m_{01} & m_{02} & m_{03} \\ m_{10} & m_{11} & m_{12} & m_{13} \\ m_{20} & m_{21} & m_{22} & m_{23} \\ m_{30} & m_{31} & m_{32} & m_{33} \end{pmatrix} \begin{pmatrix} S_{00}^{PSG} & S_{01}^{PSG} & S_{02}^{PSG} & S_{03}^{PSG} & S_{04}^{PSG} & S_{05}^{PSG} \\ S_{10}^{PSG} & S_{11}^{PSG} & S_{12}^{PSG} & S_{13}^{PSG} & S_{14}^{PSG} & S_{15}^{PSG} \\ S_{20}^{PSG} & S_{21}^{PSG} & S_{22}^{PSG} & S_{23}^{PSG} & S_{24}^{PSG} & S_{25}^{PSG} \\ S_{30}^{PSG} & S_{31}^{PSG} & S_{32}^{PSG} & S_{33}^{PSG} & S_{34}^{PSG} & S_{35}^{PSG} \end{pmatrix}$$

$$= \mathbf{M} \cdot \mathbf{S}^{PSG} \tag{9.21a}$$

Note that the inverse of $\mathbf{S}^{PSG}$ does not exist because it is not a square matrix, one may first multiply both sides of Eq. (9.21a) by $(\mathbf{S}^{PSG})^T$ to obtain a square matrix $\mathbf{S}^{PSG} \cdot (\mathbf{S}^{PSG})^T$, where $(\mathbf{S}^{PSG})^T$ is the transpose of matrix $\mathbf{S}^{PSG}$. Consequently, the Mueller matrix of the DUT can be obtained by multiplying both sides of the equation with $[\mathbf{S}^{PSG} \cdot (\mathbf{S}^{PSG})^T]^{-1}$:

$$\mathbf{M} = \mathbf{S}^{PSA} \cdot (\mathbf{S}^{PSG})^T \cdot [\mathbf{S}^{PSG} \cdot (\mathbf{S}^{PSG})^T]^{-1} \tag{9.21b}$$

Obtaining PDL As discussed in Section 4.4, for a given optical frequency, the PDL of a DUT can be obtained (Craig et al. 1998) from

$$PDL = 10\, log\left(\frac{T_{max}}{T_{min}}\right) = 10 log \frac{m_{00} + \sqrt{m_{01}^2 + m_{02}^2 + m_{03}^2}}{m_{00} - \sqrt{m_{01}^2 + m_{02}^2 + m_{03}^2}} \tag{9.22}$$

Obtaining PMD For two adjacent optical frequencies, define a matrix

$$\mathbf{M}_\Delta(\overline{\omega}) = \mathbf{M}(\omega_2)\mathbf{M}^{-1}(\omega_1) \tag{9.23a}$$

The complex PMD vector $\mathbf{W} = \boldsymbol{\Omega} + i\boldsymbol{\Lambda}$ can be found from the matrix M_Δ (Dong et al. 2006), where $\boldsymbol{\Omega}$ and $\boldsymbol{\Lambda}$ are the real and complex components of **W**, respectively. The expression for **W** is rather complicated, so we will obtain the result numerically. The DGD and PSP vectors can be calculated using

$$DGD = Re(\sqrt{\mathbf{W} \cdot \mathbf{W}}) \tag{9.23b}$$

$$\hat{\mathbf{q}}_{\pm}(\omega) = \frac{(\pm\mathbf{\Omega} + \mathbf{\Omega} \otimes \mathbf{\Lambda})}{(\mathbf{\Omega} \cdot \mathbf{\Lambda})} \tag{9.23c}$$

where $\hat{\mathbf{q}}_{\pm}$ are the unit vectors of the slow and fast PSP, respectively. The symbol "⊗" in Eq. (9.23c) stands for cross product, and "·" stands for inner product. As with the case using the JME method, the second order PMD can be calculated numerically using Eq. (9.19b), but with $\hat{\mathbf{q}}$ replaced by $\hat{\mathbf{q}}_{-}$.

9.3.3 Experimental Results

9.3.3.1 PMD Measurement Resolution

To characterize the resolution of the binary polarization measurement system, the PMD value of a piece of single mode fiber (SMF) has been carefully measured. A 2-cm section of the bare fiber is placed in a fiber squeezer (General Photonics' fiber squeezer polarization controller, PLC-003, as described in Figure 6.12d) to induce a small amount of birefringence or PMD in the fiber via the photoelastic effect, and then measured the PMD value of the fiber as the pressure on the fiber was increased. As shown in Figure 9.11a, as the pressure on the fiber section increases, the SOP traces out a circle on the Poincaré sphere (Collett 2003; Walker and Walker 1990), as discussed

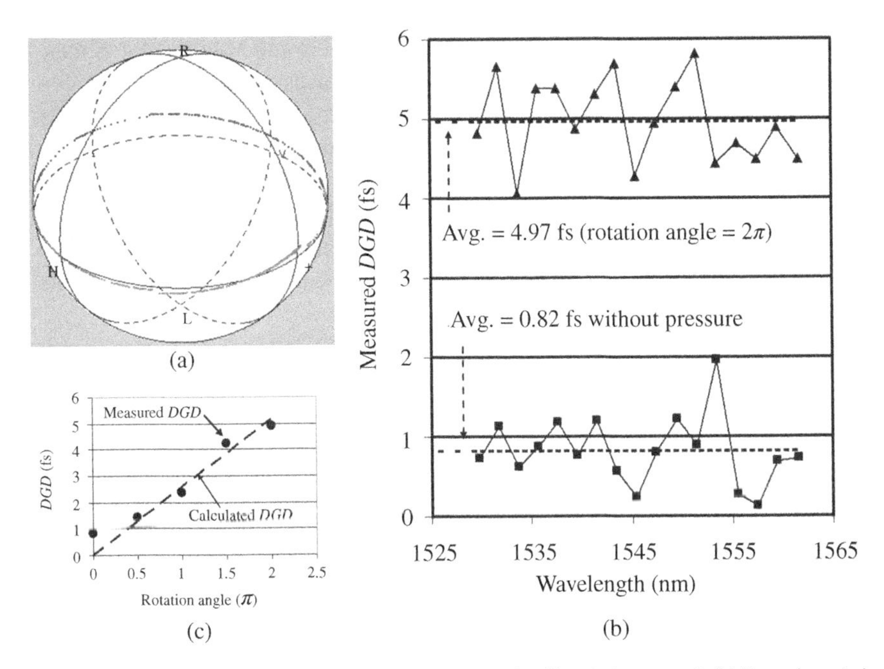

Figure 9.11 (a) SOP evolution as the pressure on the fiber is increased. (b) Experimental results illustrating the PMD measurement resolution of the measurement system using binary MO polarization rotators. (c) Pressure induced DGD as a function of SOP evolution angles at 1550 nm. Black dots: DGD obtained with binary measurement system. Dashed line: DGD obtained by SOP angle counting. A PMD measurement resolution of 1 fs is clearly demonstrated in both (b) and (c). In addition, the measurement accuracy is also confirmed by the rotation angle of the SOP: the 5.2 fs DGD corresponds to a total SOP rotation angle of 2π at 1550 nm (Yao et al. 2010).

in Section 7.2.3. A complete circle corresponds to a 2π retardation caused by pressure induced birefringence or a DGD of 5.2 fs at a wavelength of 1550 nm. Therefore, the pressure induced DGD value can be approximated by counting the SOP revolutions as the pressure on the fiber section is increased using the fiber squeezer. Figure 9.11b shows the DGD as a function of wavelength, measured with the binary system described in Figure 9.10 using the Jones matrix method for two cases: one with no applied pressure and the other with an applied pressure corresponding to an SOP rotation angle of 2π. It is evident that this binary system can easily distinguish PMD value changes as small as 1 fs. The measured DGD of 0.82 fs with no applied pressure is the combination of the DGD of the fiber itself and the noise floor of the measurement system. It can therefore be concluded that the DGD measurement resolution of the binary measurement system is better than 1 fs. The same resolution is also obtained when the Mueller matrix method is used. Figure 9.11c shows the pressure induced DGD as a function of SOP rotation angle. The dashed line is the DGD obtained from the SOP rotation angle, while the black dots represent the DGD obtained with the binary measurement system. The measured DGD shows good agreement with that obtained by SOP angle counting. Again, a measurement resolution of less than 1 fs is clearly demonstrated.

9.3.3.2 DGD and SOPMD Measurements

To validate instrument's measurement accuracy, DGD and SOPMD standards with known values are required. A DGD artifact using a 20 mm long quartz crystal described in Figure 6.31a has been constructed and its DGD value has been measured using the binary system described in Figure 9.10. The DGD value calculated from Eqs. (6.24a) and (6.24b) and birefringence data at 1550 nm from Williams et al. (2002) is 627.7 fs $\pm$ 2.03 fs (see Table 6.1). The measured DGD values, averaged over the C-band (1530–1565 nm), using the Jones matrix eigenanalysis (JME) and Mueller matrix measurement (MMM) methods are 627.0 and 628.1 fs, respectively. The measured DGD as a function of wavelength is shown in Figure 9.12. The wavelength variations of the measured DGD are $\pm$2.75 and $\pm$2.93 fs, respectively, obtained with the JME and MMM methods. As shown in Figure 9.12, the binary measurement system can resolve SOPMD values as small as 0.001 ps^2. However, since the largest measured SOPMD value is 0.004 ps^2 for a DGD artifact with a theoretical SOPMD value of zero ps^2, one can conservatively conclude that this binary measurement system has an SOPMD resolution of 0.005 ps^2. This high resolution reflects the excellent repeatability of the binary PSG and the resolution of the binary PSA used in the system.

Another PMD artifact has been constructed by cascading two YVO_4 crystals of 16 mm length with a 45° relative orientation angle, as described in Figure 6.31b, to validate the performance of

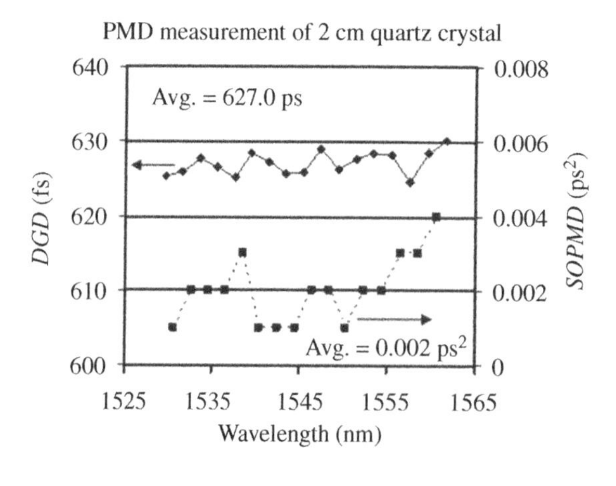

Figure 9.12 1st and 2nd order PMD of a 2-cm quartz crystal DGD artifact (calculated *DGD* = 627.7 fs), measured with Jones Matrix Eigenanalysis method using 2 nm wavelength step. Similar results are obtained with MMM method (Yao et al. 2010).

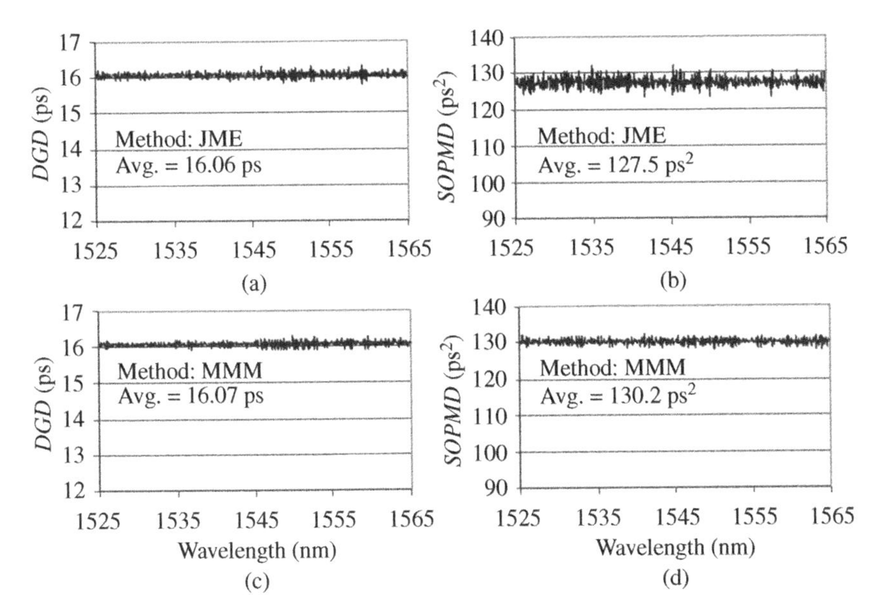

Figure 9.13 DGD and SOPMD measurement results for a PMD artifact made from two 16 mm long YVO_4 crystals with a relative orientation angle of 45°. (a) and (b) DGD values vs. wavelength obtained using JME and MMM methods, respectively. (c) and (d) SOPMD values vs. wavelength obtained using JME and MMM methods, respectively (Yao et al. 2010).

the binary system. The DGD values of the two crystals, measured with the binary measurement system described in Figure 9.10, are 11.410 and 11.416 ps, respectively. The expected DGD and SOPMD values of the PMD artifact are 16.14 ps and 130.2 ps^2, calculated from Eqs. (6.25b) and (6.25a), respectively. As shown in Figure 9.13, using the same binary measurement system, the measured DGD and SOPMD are 16.06 ps and 127.5 ps^2, respectively, using Jones matrix analysis, and 16.07 ps and 130.2 ps^2, respectively, using the Mueller matrix method. The corresponding DGD and SOPMD deviations of the measurement from the theoretical DGD and SOPMD values of the artifacts are 0.08 ps and 2.7 ps^2 using the Jones matrix and 0.07 ps and <0.01 ps^2 using the Mueller matrix method. Clearly, the Mueller matrix method has a smaller PMD measurement deviation than the Jones matrix method for the PMD range tested. In addition, as can be seen in Table 9.8, the Mueller matrix method also has better measurement repeatability than the Jones matrix method.

9.3.3.3 PDL Measurement

Two PDL artifacts described in Section 6.7.3 were constructed. The calibrated PDL values are 1.959 ± 0.01 and 0.387 ± 0.005 dB, respectively, as shown in Table 6.3. Their PDL wavelength dependences, measured using the binary system, are shown in Figure 9.14. The average measured PDL values of the two artifacts are 1.964 and 0.375 dB, respectively, using the Mueller matrix method, and 2.01 and 0.416 dB, respectively, using the Jones matrix method. As can be seen in Figure 9.14, the values measured using the two methods are slightly different, and those measured with the Mueller matrix method show less wavelength variation. As shown in Section 9.3.3.4, the Mueller matrix method also has better measurement repeatability. This is because the Mueller matrix method relies on power measurement, while the JME method is based on SOP angle measurement. As a result, the Mueller matrix method generally has better accuracy and repeatability (as shown in Section 9.3.3.4) than the JME method for low PDL measurement because power measurement generally has better accuracy and repeatability than SOP angle measurement.

Table 9.8 Statistical results of 50 measurements (Yao et al. 2010).

Artifacts		Average	Maximum	Minimum	Maximum − minimum
2-cm quartz DGD artifact (JME)	Average DGD in the wavelength range of 1530–1560 nm with a λ step size of 2 nm	627.02 fs	627.033 fs	627.000 fs	0.033 fs
	@1550 nm with a λ step size of 2 nm	627.757 fs	627.955 fs	627.567 fs	0.388 fs
2 cm quartz DGD artifact (MMM)	Average DGD in the wavelength range of 1530–1560 nm with a λ step size of 2 nm	628.111 fs	628.108 fs	628.107 fs	0.022 fs
	@1550 nm with a λ step size of 2 nm	627.663 fs	627.700 fs	627.511 fs	0.189 fs
SOPMD artifact (JME)	Average SOPMD in the range of 1550–1552 nm with a λ step size of 0.1 nm	128.43 ps^2	129.64 ps^2	128.29 ps^2	1.35 ps^2
	SOPMD@1550 nm with a λ step size of 0.1 nm	129.31 ps^2	132.01 ps^2	126.70 ps^2	5.31 ps^2
SOPMD artifact (MMM)	Average SOPMD in the wavelength range of 1550–1552 nm with a λ step size of 0.1 nm	130.07 ps^2	130.20 ps^2	129.92 ps^2	0.28 ps^2
	SOPMD@1550 nm with a λ step size of 0.1 nm	129.95 ps^2	131.40 ps^2	128.27 ps^2	3.13 ps^2
2-dB PDL artifact (JME)	Measured at 1550 nm with different fiber pigtail positions	1.963 dB	2.04 dB	1.91 dB	0.13 dB
2-dB PDL artifact (MMM)	Measured at 1550 nm with different fiber pigtail positions	1.965 dB	1.982 dB	1.941 dB	0.041 dB
0.4-dB PDL artifact (JME)	Measured at 1550 nm with different fiber pigtail positions	0.389 dB	0.437 dB	0.326 dB	0.11 dB
0.4-dB PDL artifact (MMM)	Measured at 1550 nm with different fiber pigtail positions	0.377 dB	0.395 dB	0.361 dB	0.034 dB

As can be seen from Figure 9.14, the fluctuations in the measured PDL vs. wavelength values are about ±0.03 dB for the Jones matrix method and ±0.005 dB for the Mueller matrix method. Theoretically, the artifact's PDL should be wavelength independent. The observed fluctuation is mainly due to system noise and the wavelength dependence of the fiber pigtail's PDL vector, because the total PDL is the vector summation of the PDL of the artifact and its fiber pigtails as described in Section 8.4.4. Based on these results, the worst-case PDL measurement resolution should be better than 0.01 dB for the Mueller matrix method and 0.06 dB for the Jones matrix method.

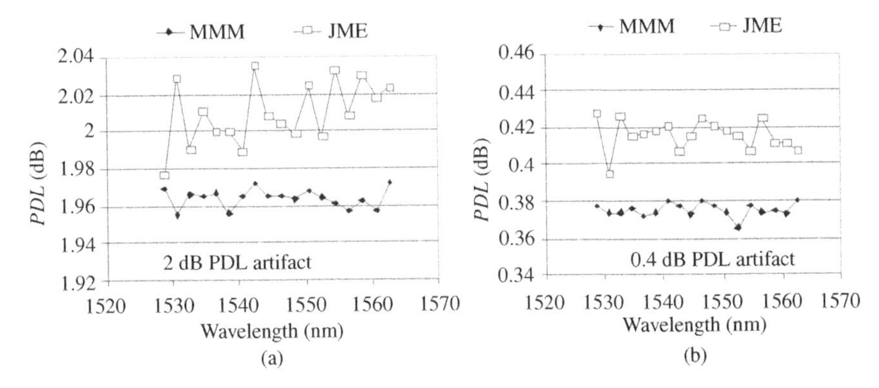

Figure 9.14 Measured PDL as a function of wavelength using the JME and MMM methods. (a) PDL vs. wavelength of a "2 dB" artifact, and (b) PDL vs. wavelength of a "0.4 dB" artifact. PDL obtained using MMM method has much less wavelength variation than that obtained using JME method (Yao et al. 2010).

9.3.3.4 Repeatability Measurements and Accuracy Determination

In order to determine the repeatability of the binary system, the PMD and PDL artifacts have been measured multiple times. The results are listed in Table 9.8. The Mueller matrix method data are shown to have better repeatability and accuracy than the corresponding Jones matrix method data; therefore, Mueller matrix results are used to determine the accuracy and repeatability of the binary measurement system. It is evident that the binary system has a remarkable DGD repeatability of 0.022 fs for a 627.7 fs DGD artifact, an SOPMD repeatability of 0.28 ps^2 for a 130.2 ps^2 SOPMD artifact, a PDL repeatability of 0.041 dB for a 2-dB PDL artifact, and a PDL repeatability of 0.034 dB for a 0.4-dB PDL artifact, measured using the Mueller matrix method. Note that the DGD repeatability of the binary system is more than 100 times better than those of commercially available systems on the market (0.022 vs. 3 fs) (Keysight N7788B/N7788BD Datasheet 2022; Thorlabs PMD5000 Datasheet 2022). No published repeatability data for SOPMD measurement were found for either research or commercial measurement systems.

Note that the total PDL vector of a DUT is the summation of the PDL vectors in 3D space of all components in the measurement setup, including connectors, fiber pigtails, and the DUT itself, as described in Figures 3.9 and 8.24. The positions of the fiber pigtails determine the relative orientation of the PDL vectors involved, and hence affect the total PDL values. During measurement, the fiber pigtails of the PDL artifacts were arranged in different positions to account for PDL variation caused by the residual PDL values of the fiber pigtails and connectors. The measurement results therefore include these residual PDL contributions.

As shown in Table 6.5, the 2-cm quartz DGD artifact has a "standard" DGD value of 627.7 ± 2.03 fs. The worst deviations between the measured and standard DGD values are −0.70 and 0.41 fs, measured using JME and MMM, respectively, as shown in Table 9.8. These deviations are well within the uncertainty range of the DGD artifact. Therefore, the absolute DGD measurement accuracies are determined to be better than $|2.03\ \text{fs}| + |-0.68\ \text{fs}| + |0.39\ \text{fs}| = 3.1$ fs for the JME method, and $|2.03\ \text{fs}| + |0.41\ \text{fs}| + |0.19\ \text{fs}| = 2.6$ fs for the MMM method, where 0.39 and 0.19 fs are the DGD measurement repeatabilities of the binary system when JME and MMM analysis methods are used, respectively, as shown in Table 9.8. Note that the largest contribution to the inaccuracy is not from the measurement system itself, but from the artifact's DGD value uncertainty (2.03 fs). A much higher measurement accuracy might be claimed if an artifact with a much tighter DGD value tolerance were available. Nevertheless, it is also important to point out that the DGD measurement accuracy of the binary measurement system is more than 10 times better than those of commercial PMD measurement systems using the analog PSG and PSA (Keysight N7788B/N7788BD Datasheet 2022).

In order to verify the device uniformity of the MO polarization rotator based binary measurement method, the same DGD artifact using three identical binary measurement systems built according to the design shown in Figure 9.10, has been measured using three different PSG–PSA pairs and three different tunable lasers. The measured DGD values are 628.1, 627.7, and 629.0 fs. The spread is well within the artifact's DGD range of 627.7 ± 2.03 fs.

The SOPMD artifact has a "standard" SOPMD value of $130.3 \pm 0.73\,\text{ps}^2$ (see Table 6.2). The worst measured result deviates by −2.01 and $0.38\,\text{ps}^2$ from the standard value when JME and MMM analysis methods, respectively, are used. Consequently, the worst-case absolute accuracy for SOPMD measurement using the binary system can be estimated to be $|0.73\,\text{ps}^2| + |-2.01\,\text{ps}^2| + |1.35\,\text{ps}^2| = 4.18\,\text{ps}^2$ using the JME method, and $|0.73\,\text{ps}^2| + |0.38\,\text{ps}^2| + |0.28\,\text{ps}^2| = 1.39\,\text{ps}^2$ using the MMM method, where 1.35 and $0.28\,\text{ps}^2$ are the corresponding measurement repeatabilities of the two methods, as shown in Table 9.8. To the best of the authors' knowledge, this SOPMD measurement accuracy of $1.39\,\text{ps}^2$ represents the highest reported SOPMD accuracy to date.

Accuracy estimations similar to those described earlier for DGD and SOPMD yield a PDL measurement accuracy better than 0.223 dB (JME) and 0.074 dB (MMM) for the 2-dB PDL artifact, and 0.17 dB (JME) and 0.06 dB (MMM) for the 0.4-dB artifact.

As previously pointed out, the binary system generally has better accuracy and repeatability using the MMM method; therefore, the MMM analysis method results are used to determine the performance of the binary measurement system. Therefore it can be concluded that the binary measurement system has DGD, SOPMD, and PDL accuracies of 2.6 fs, $1.39\,\text{ps}^2$, and 0.06 dB, respectively.

9.3.3.5 Differences Between Mueller Matrix and Jones Matrix Methods

If different polarization states with the same power pass through an optical device, the output power will generally be polarization dependent because of the PDL of the device. If power variations are measured for at least four non-degenerate polarization states, then the PDL can be calculated using the Mueller matrix method. Because the Mueller matrix method for PDL measurement is based on measurement of power variations, the measurement accuracy is insensitive to polarization changes (for example, because of fiber movement between the PSG and PSA). However, any PSG output power variation (due to fluctuation of laser output power, changes of polarization state along the fiber between the laser and PSG) during testing will produce significant measurement error. Therefore, accurate PDL measurement using the Mueller matrix method requires that the laser source be highly stable and that the fiber between the laser source and the PSG be firmly fixed in place.

If an optical beam with different polarization states but the same power passes through an optical device, the relative angles between the input and the output polarization states will be different because of PDL. If these angle changes are measured for at least three non-degenerate polarization states, then the PDL can be calculated using the Jones matrix method. Because the Jones matrix method is based on measurement of angle variations, the measurement accuracy is insensitive to PSG output power fluctuations. However, any polarization disturbance in the fiber between the PSG and PSA will cause significant measurement error. Therefore, accurate PDL measurement using the Jones matrix method requires that the fiber between the PSG and PSA be fixed.

Because the MMM for PDL measurement is based on power measurement, it can measure small PDL values with high accuracy, but has relatively narrow dynamic range which is limited by the dynamic range of the detector. The JME is based on angle measurement, so it provides high dynamic range, but its resolution is limited by the angle measurement resolution of the PSA. Therefore, for maximum accuracy, the MMM should be used for measurement of small PDL values and the JME for measurement of high PDL values.

9.3.3.6 Summary

The binary Mueller matrix polarimetry system based on binary MO PSG and binary MO PSA has many advantages over the traditional analog systems described in Chapter 8. Because of its binary nature, the system can be easily calibrated against imperfections in the optical components used in the system, caused by wavelength and environmental variations. Consequently, following unprecedented measurement accuracy and sensitivity have been achieved: PMD measurement sensitivity or resolution of 1 fs, PMD measurement accuracy of 2.6 fs, PMD measurement repeatability of 0.022 fs, SOPMD resolution of 0.005 ps^2, SOPMD accuracy of 1.39 ps^2, SOPMD repeatability of 0.28 ps^2, PDL resolution of 0.01 dB, PDL accuracy of 0.06 dB, and PDL repeatability of 0.034 dB. With such high sensitivity and accuracy, the system has proven to be effective in characterizing polarization related parameters in extremely demanding applications, such as fiber optical gyro coils and optical fiber current sensing systems, as will be discussed next.

9.4 Application Examples of Binary Mueller Matrix Polarization Analyzers

9.4.1 PM Fiber Beat Length Measurement

Beat length is important because it measures how well a fiber maintains polarization. It is a measure of how fast the two orthogonal modes become decoupled and thus cease to exchange energy. Fibers with short beat lengths preserve polarization more strongly than those with long beat lengths.

Beat length L_B is defined by the ratio of the wavelength of the transmitted light λ to the fiber's phase birefringence Δn (Rashleigh 1983a; Ritari et al. 2004),

$$L_B = \frac{\lambda}{\Delta n} \tag{9.24a}$$

The beat length of PM fiber can be estimated by measuring its DGD (Rashleigh 1983a; Ritari et al. 2004). The DGD measured by our measurement system is related to the group birefringence Δn_g by

$$DGD = \frac{\Delta n_g L}{c} \tag{9.24b}$$

and the group birefringence is related to the phase birefringence by

$$\Delta n_g = \Delta n - \lambda \frac{d(\Delta n)}{d\lambda} \tag{9.24c}$$

The beat length can be calculated from the measured DGD by

$$L_B = \frac{\lambda}{[DGD * c/L + \lambda d(\Delta n)/d\lambda]} \tag{9.24d}$$

If the chromatic dispersion equation of a fiber is

$$n^2 = 1 + \frac{B_1\lambda^2}{\lambda^2 - C_1} + \frac{B_2\lambda^2}{\lambda^2 - C_2} + \frac{B_3\lambda^2}{\lambda^2 - C_3} \tag{9.24e}$$

The beat length can be calculated using

$$L_B = \frac{\lambda * y}{DGD * c/L} \tag{9.24f}$$

where

$$y = 1 - \frac{1}{n}\left\{n^2 - 1 - \left[\frac{B_1\lambda^4}{(\lambda^2 - C_1)^2} + \frac{B_2\lambda^4}{(\lambda^2 - C_2)^2} + \frac{B_3\lambda^4}{(\lambda^2 - C_3)^2}\right]\right\} \tag{9.24g}$$

As an example, the beat length of a 1.16 m PANDA PM fiber was measured, and a DGD of 1.578 ps was obtained. The corresponding beat length from Eq. (9.24f) is 3.86 mm, consistent with the manufacturer's specified value of 3–5 mm (Corning PANDA PM Data sheet 2004). For a PM fiber with $DGD_{\lambda_2} \approx DGD_{\lambda_1}$, the beat length at λ_2 can be estimated by using the beat length measurement result at λ_1:

$$L_{B\lambda_2} = \frac{\lambda_2}{\lambda_1} L_{B\lambda_1} \tag{9.24h}$$

For example, if we measure the DGD and beat length L_{B1550} of a PM fiber at 1550 nm, its beat length L_{B1310} at 1310 nm is simply $L_{B1310} = (1310/1550)L_{B1550} = 0.845L_{B1550}$.

9.4.2 Characterization of Sensing Coils for Fiber Optic Gyroscopes

Depolarized FOGs that use low-cost SMF coils (Szafraniec and Sanders 1999) are a viable alternative to PM-fiber based FOGs. In addition to low fiber cost, another reason is the higher level of radiation tolerance that can be achieved with simpler SMF designs. How to quantify the performance SMF coils, especially at different temperatures, has also been a challenge. PMD and PDL (Yao 2019) have been identified as the performance indicators of the SMF coils, in addition to the insertion loss induced during the coil winding process.

PMD represents the accumulative retardation or birefringence caused by the stress on the fiber. A well-engineered fiber coil is expected to have extremely low stress on the fiber, and hence low PMD. Excess stress on the various sections of fiber is expected to cause temperature instabilities and local asymmetries. Figure 9.15a shows the measured PMD values of two SM coils made with two different potting adhesives as a function of temperature, using the binary Mueller matrix polarimetry system of Figure 9.10b. Due to the superb measurement resolution and accuracy of the binary

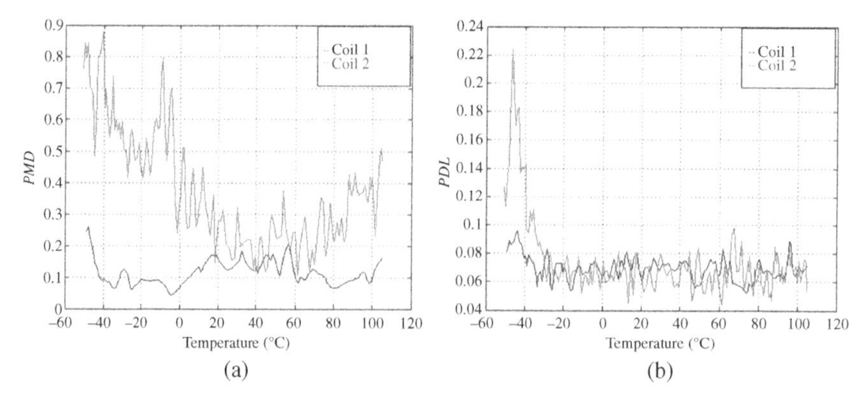

Figure 9.15 (a) Total accumulated PMD and (b) total PDL of coils with different potting adhesives as a function of temperatures. Coil 1 features a specially formulated adhesive for potting fiber gyro coils, whereas coil 2 uses an off-the-shelf adhesive (Yao 2019).

system, very small PMD variations can be clearly obtained. As can be seen, for the coil (Coil 2) made with an off-the-shelf adhesive, a large PMD value increase is observed at low temperatures, indicating that high thermal stress was induced at low temperatures, probably caused by the shrinkage of the adhesive. In contrast, for a coil (Coil 1) made with a specially formulated adhesive, low PMD values are observed at all temperatures, even at the low-temperature setting. Such experimental data indicate that this adhesive is well suited for coil production because of the low shrinkage in the temperature range of interest.

Micro-bending of a fiber can generate PDL, and therefore should be avoided during the fiber-winding process, especially when the fiber is transitioned from one layer to another, because PDL re-polarizes the depolarized light signal and causes instabilities. In addition, micro-bending induced stress on the fiber also induces local asymmetries for the counter-propagating light signals. Figure 9.15b shows the experimental results of PDL measurement of the two different coils made with different potting adhesives, as in Figure 9.15a. For the coil (Coil 2) made with the off-the-shelf adhesive, a large PDL increase is observed at low temperatures, consistent with the PMD measurement shown in Figure 9.15a, whereas the PDL of the coil (Coil 1) made with the specially formulated adhesive shows almost no PDL increase at low temperatures, indicating the superb winding quality of the coil and great suitability of the adhesive for the coil fabrications.

9.4.3 Circular Birefringence Measurement and Spun Fiber Characterization

In optical fiber communication systems, linear birefringence dominates and the PMD effect was assumed to be caused by the linear birefringence. However, as discussed in Chapter 3, circular birefringence induced by fiber-twist or by the Faraday effect also exists in optical fibers, which is important to understand for optical fiber sensing applications.

One particular example is the fiber optic current sensors (FOCS), which is based on Faraday effects for measuring the electric current in power systems and has the advantages of inherent insulation, immunity to electromagnetic interference, and light weight. Unfortunately, variations in stress and temperature of the sensing fiber may change the distribution of linear birefringence in the fiber, which alter the polarization of light in the fiber and degrade sensor system's measurement accuracy (Laming and Payne 1989). In order to reduce the influence of temperature and external stresses, it is desirable to make the sensing fiber with high circular birefringence or low linear birefringence, for example, the spun fiber (Li et al. 1986; Clarke 1993; Peng et al. 2013), the annealed fiber (Tang et al. 1991; Bohnert et al. 2002), the low-stress fiber (Kurosawa et al. 1995), or the helically wrapped fiber (Short et al. 1998). Among the various fibers, the spun fiber is better accepted in the industry (Peng et al. 2013) for its manufacturability and high sensing accuracy over a wide range of environmental conditions, including temperature and vibration, making it suitable for current sensors deployed in both indoors and outdoors in real-life applications.

As discussed in Chapter 3, a spun fiber can be fabricated by first making a preform with two stress rods similar to that for making a linear PM fiber (e.g. a bow-tie fiber or a PANDA fiber), and then spinning the perform during drawing in the molten state (Li et al. 1986). The spinning process induces a large circular birefringence and significantly reduces the average linear birefringence over distance, although the local linear birefringence still remains strong (Li et al. 1986) to overcome the stress and temperature induced birefringence perturbations. It is important to have the detailed knowledge of the circular birefringence Δn_C and the residual total linear birefringence Δn_L of the spun fiber, particularly their temperature dependences, to ensure that the current sensing system incorporating the fiber is insensitive to temperature variations and external stresses and eventually meets its performance requirements.

As described in Chapter 8, previous methods for the birefringence measurements relying on using analog signals to obtain birefringences usually have the disadvantages of low repeatability and require complicated methods to compensate for imperfections in the optical components and in moving mechanical parts. Consequently, it is difficult to use them to detect the small variations of Δn_C and Δn_L in a spun fiber. This section is to describe using binary polarization analysis to accurately measure the circular and residual linear birefringences of spun fibers, which is critical to FOCS applications, especially for meeting the demanding requirement of accuracy on the order of 0.2% or less over the temperature range between −40 and +80 °C (IEC 60044-8 2002; State Grid Corporation of China 2014).

9.4.3.1 Measurement Principle

In general, the measured Mueller matrix $\mathbf{M}(\lambda)$ of an optical medium, such as a spun fiber, at wavelength λ can be decomposed into three Mueller matrices (Manhas et al. 2006; Lu and Chipman 1996; Olivard et al. 1999):

$$\mathbf{M}(\lambda) = \mathbf{M}_\Delta \mathbf{M}_R \mathbf{M}_D \tag{9.25a}$$

where $\mathbf{M}_D$ represents a diattenuator, which measures the differential loss between the orthogonal polarization eigen states (corresponding to PDL, commonly used in fiber optics), $\mathbf{M}_R$ represents a retarder that measures the differential phase between the two eigen states, and $\mathbf{M}_\Delta$ represents a depolarizer that characterizes depolarization. Among them, $\mathbf{M}_R$ relates to the birefringence characteristics and can be separated from $\mathbf{M}(\lambda)$ using the method presented in (Manhas et al. 2006). For the measured fiber, the total retardation is the combined effect of both Δn_L and Δn_C, so the $\mathbf{M}_R$ can be decomposed into two matrices: a linear retarder matrix with a linear retardation of δ and an orientation angle of θ (the axis of the linear birefringence with respect to the horizontal axis) and a circular retarder matrix with an optical rotation of φ:

$$\mathbf{M}_R = \begin{pmatrix} 1 & 0 & 0 & 0 \\ 0 & cos^2 2\theta + sin^2 2\theta cos\delta & cos2\theta sin2\theta\,(1 - cos\delta) & -sin2\theta sin\delta \\ 0 & cos2\theta sin2\theta(1 - cos\delta) & sin^2 2\theta + cos^2 2\theta cos\delta & cos2\theta sin\delta \\ 0 & sin2\theta sin\delta & -cos2\theta sin\delta & cos\delta \end{pmatrix} \begin{pmatrix} 1 & 0 & 0 & 0 \\ 0 & cos2\varphi & sin2\varphi & 0 \\ 0 & -sin2\varphi & cos2\varphi & 0 \\ 0 & 0 & 0 & 0 \end{pmatrix} \tag{9.25b}$$

The values of optical rotation φ and linear retardation δ can be determined from the matrix M_R as (Manhas et al. 2006)

$$\varphi \equiv \frac{\pi L}{\lambda}\Delta n_C = \frac{1}{2} tan^{-1}\left[\frac{M_R(3,2) - M_R(2,3)}{M_R(2,2) + M_R(3,3)}\right] + n\pi \tag{9.25c}$$

$$\delta \equiv \frac{2\pi L}{\lambda}\Delta n_L = cos^{-1}\left(\left\{[M_R(2,2) + M_R(3,3)]^2 + [M_R(3,2) - M_R\,(2,3)]^2\right\}^{\frac{1}{2}} - 1\right) + 2m\pi \tag{9.25d}$$

where $M_R(i, j)$ stands for the matrix element at row i and column j, and m and n are integers to account for phase wrapping. It is difficult to directly use Eqs. (9.25c) and (9.25d) to calculate Δn_C and Δn_L at a wavelength λ because m and n cannot be determined. In practice, it is beneficial to use a differential method to (i) improve measurement accuracy, and (ii) remove the contribution of m

and n, by first measuring φ and δ at two adjacent wavelengths λ_1 and λ_2 and then taking the differences $\Delta\varphi = \varphi(\lambda_1) - \varphi(\lambda_2)$ and $\Delta\delta = \delta(\lambda_1) - \delta(\lambda_2)$, assuming that the wavelength step $\Delta\lambda = \lambda_2 - \lambda_1$ is sufficiently small such that $\Delta\varphi$ and $\Delta\delta$ are less than π and 2π, respectively. The circular birefringence Δn_C and the residual linear birefringence Δn_L can then be obtained from the differential rotation angle $\Delta\varphi$ and the differential retardation $\Delta\delta$, respectively, using Eqs. (9.25c) and (9.25d):

$$\Delta n_C = \frac{\lambda_1 \lambda_2}{\pi L(\lambda_2 - \lambda_1)} \Delta\phi \tag{9.26a}$$

$$\Delta n_L = \frac{\lambda_1 \lambda_2}{2\pi L(\lambda_2 - \lambda_1)} \Delta\delta \tag{9.26b}$$

where Δn_C and Δn_L are assumed constant over wavelength and such an assumption is accurate because of the negligible birefringence dispersion over the small wavelength step.

Appendix 9.B shows an example of using the process described earlier to obtain Δn_C and Δn_L at 20 °C. To further improve the repeatability and accuracy, the laser frequency can be scanned and the values of Δn_C and Δn_L can be calculated at multiple such wavelength pairs for averaging.

9.4.3.2 Experimental Setup

The measurement setup is illustrated in Figure 9.16. The binary polarization analysis system is the same as that in Figure 9.10 and the fiber under test (FUT) is placed between the PSG and PSA. As described in Figure 9.10, the binary PSA and PSG each are composed of a QWP, six binary MO polarization rotators, and a polarizer. The PSA also contains a photodetector and a signal amplification circuit. In PSG, each binary MO rotator controlled by the computer can rotate the SOP ±22.5° to generate six distinctive polarization states. The PSA can accurately measure the SOP of light entering it by analyzing voltages generated in the photodetector and can also be self-calibrated to remove inaccuracies caused by imperfections in components and workmanship, yielding extremely high measurement accuracies. Collectively, the binary PSG and PSA can be used to accurately obtain the Mueller matrix of FUT (Yao et al. 2010), as described in Section 9.3.

Two experiments were carried out using the setup of Figure 9.16. In Experiment 1, 1 m of PM fiber (PM fiber 1550_125-18/250 of YOFC Ltd.) was fusion-spliced with 10 m spun fiber (Spun HiBi

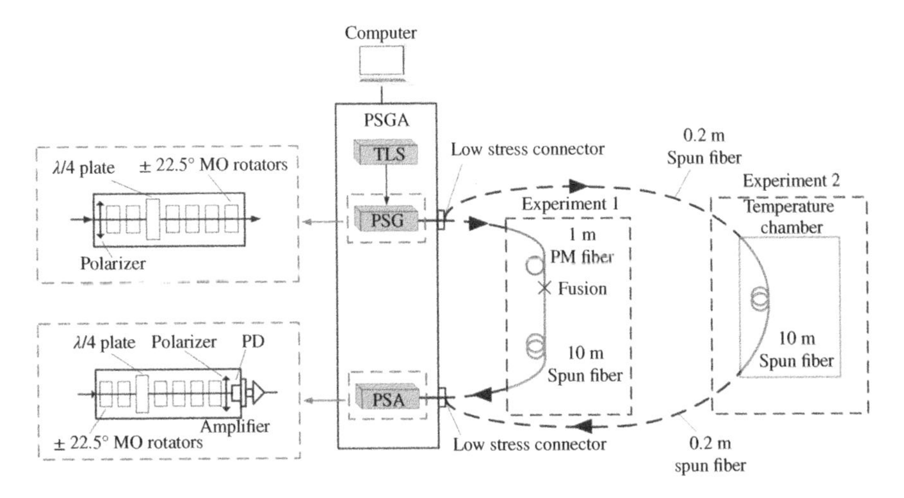

Figure 9.16 Experimental setup. Light from a tunable laser first goes through a PSG before connecting to a fiber under test (FUT). The FUT is placed between the PSG and PSA to obtain the transfer matrix. In Experiment 1, a known length of PM fiber with a known birefringence is fusion-spliced with a 10 m spun fiber. In Experiment 2, a continuous length of 10.4 m spun fiber was used (Xu et al. 2017).

Fiber SHB1250 (7.3/125) of Fibercore Co. Ltd., with a spin pitch of 4.8 mm) to measure Δn_C and Δn_L to validate the binary system and algorithm for the birefringence measurement because the birefringence of the PM fiber is known. The two fibers have well matched mode-field diameters of 6.41 and 6.92 μm, respectively. In Experiment 2, 10 m spun fiber was placed in a temperature chamber to measure the relationship between its circular and the residual linear birefringences and temperature using the same binary measurement system, as shown in Figure 9.16, while having 0.4 m of pigtails outside of the chamber. Assuming the circular and linear birefringences are uniform along the spun fiber, the contribution to the circular and linear birefringences from the 0.4 m outside of the chamber during the data processing was subtracted.

9.4.3.3 Verification of Birefringence Measurements

The binary measurement system and algorithm for obtaining accurate linear birefringence using the binary measurement system has been validated in Section 9.3.3, with artifacts made with quartz crystals having an accurately known linear birefringence. Here the new algorithms for obtaining the circular and the linear birefringences Δn_C and Δn_L from the measured Mueller matrix using Eqs. (9.25a)–(9.26b) are to be validated. The wavelengths $\lambda_1 = 1549.72$ and $\lambda_2 = 1550.52$ nm with a wavelength spacing of 0.8 nm of the internal tunable laser source of the binary system was set. Such a wavelength spacing is sufficiently small to assure $\Delta\varphi < \pi$ and $\Delta\delta < \pi$; however, sufficiently large to allow good measurement repeatability and accuracy.

Δn_C and Δn_L were measured with the following procedures: (i) Fusion-splice 1 m of PM fiber with 10 m of a spun fiber as the FUT, as shown in Figure 9.16, in which the circular and linear birefringences are assumed to be contributed by the spun fiber and the PM fiber, respectively. The linear birefringence Δn_L of the PM fiber before splicing is first accurately measured with the same binary system to be 6.63×10^{-4} at 1550 nm. (ii) Obtaining the Mueller matrix using the binary system and then calculating Δn_C and Δn_L with the new algorithms based on Eqs. (9.25c), (9.25d), (9.26a), and (9.26b). (iii) Cut away 10 cm of the PM fiber and repeat step (ii) to get a second set of measurement results of Δn_C and Δn_L. (iv) Repeat step (iii) to get more sets of measurement results of Δn_C and Δn_L. Figure 9.17 shows the seven sets of measurement results, corresponding to the cases with seven different PM fiber lengths from 1 down to 0.4 m. The linear birefringence of the combined fiber can be obtained with the binary system and the new algorithm to be 6.69×10^{-4}, less than 1% difference from the number obtained using the previously proven algorithm on the PM fiber alone. As expected, the corresponding differential retardation $\Delta\delta$ between two wavelengths of the combined fiber indeed decreases linearly with the length of PM fiber, as shown in Figure 9.17a, with the linear birefringence as the slope of the line. On the other hand, the differential polarization rotation angle $\Delta\varphi$ and the corresponding Δn_C of the combined fiber remain almost constant as the PM fiber is being decreased, as shown in Figure 9.17b, indicating that the new algorithm can effectively isolate the influence of the linear birefringence on the measurement of Δn_C and vice versa. The obtained Δn_C of the combined fiber is 3.47×10^{-5} with a standard deviation of 3.28×10^{-7}, showing again that the linear birefringence of the PM does not affect the measurement of Δn_C of the combined fiber. The experiment earlier therefore validates that the binary system and the new algorithms is capable of simultaneously obtaining the circular and linear birefringences with high accuracies.

9.4.3.4 Thermal Coefficient of Circular Birefringence in Spun Fiber

The temperature dependence of Δn_C is an important parameter for FOCS because it affects the temperature stability of the sensor system and hence its accuracy over temperatures. To obtain the temperature dependency, 10 m of spun fiber coiled with a diameter of 12 cm are put in a

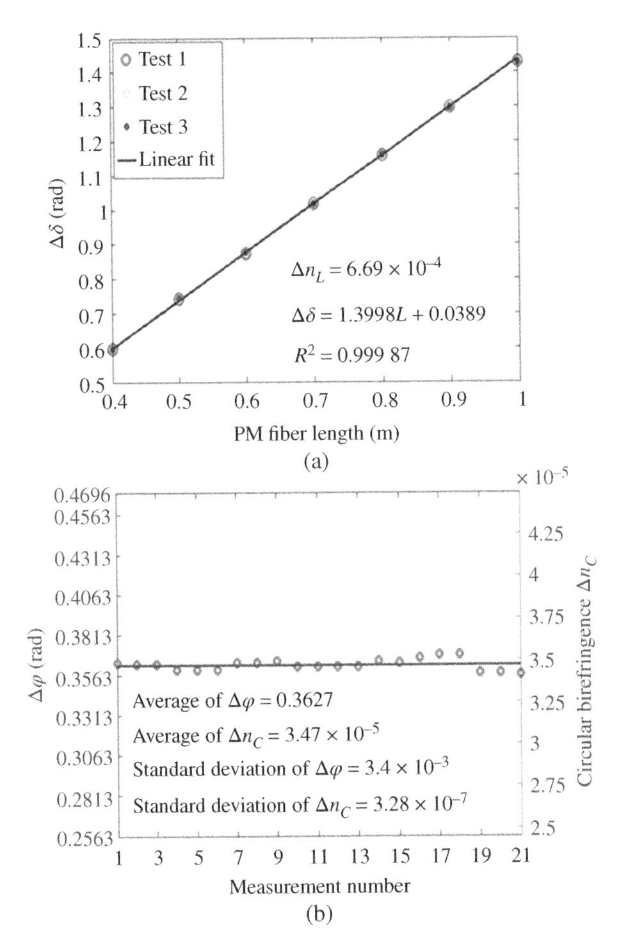

Figure 9.17 (a) Experimental results showing that the differential retardation $\Delta\delta$ between two wavelengths decreases as the length of the PM fiber becomes shorter with a linear birefringence Δn_L of 6.69×10^{-4} representing its slope. (b) Measurement results of differential $\Delta\varphi$ between two wavelengths and the resulting Δn_C of a combined fiber of 10 m spun fiber with varied lengths of PM fiber. In the experiment, the two wavelengths with a spacing of 0.8 nm are set around 1550 nm and three measurements are taken for each PM fiber length with negligible differences. With seven different lengths of PM fiber in the combined fiber, the total number of measurements is 21. Because the length of the spun fiber was unchanged, $\Delta\varphi$ should remain constant and be represented by a horizontal line. Therefore the small deviation of the data from the horizontal line indicates the superb repeatability of the measurements (Xu et al. 2017).

temperature chamber to measure Δn_C in different temperatures from −40 to 80 °C. In the measurements, the wavelength of the tunable laser in the setup is scanned from 1528.77 to 1562.64 nm, with a frequency step of 5×50 GHz (~4 nm). As described previously, one can obtain the birefringence as a function of wavelength and average the multiple sets of birefringence values as the final values. Figure 9.18a shows the measurement results of Δn_C at seven different temperatures (−40, −20, 0, 20, 40, 60, 80 °C). It is evident that Δn_C has a value of 3.24×10^{-5} at 20 °C and decreases linearly with the temperature, with a slope of -5.09×10^{-8}/°C, which is the thermal coefficient of circular birefringence α_C. It is interesting to note that Δn_C of the spun fiber and its thermal coefficient α_C are an order of magnitude less than the linear birefringence Δn_L of a PM

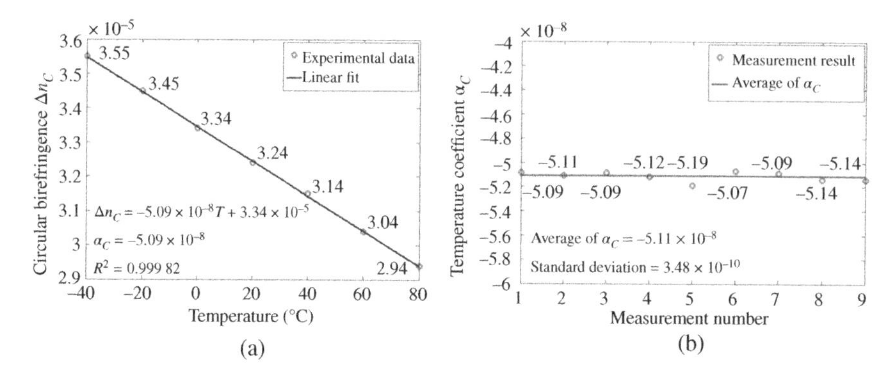

Figure 9.18 (a) Measured Δn_C of spun fiber at seven different temperatures. Curve fitting yields the thermal coefficient α_C of Δn_C to be -5.09×10^{-8}/°C. The relative change of Δn_C per °C, defined as the ratio of α_C and Δn_C at $T=0$, is −0.152%. (b) Repeatability measurements of the thermal coefficient of circular birefringence α_C of a spun fiber under test, with an average of -5.11×10^{-8}/°C and a standard deviation of 3.48×10^{-10}/°C (Xu et al. 2017).

fiber and its thermal coefficient (Ding et al. 2011), respectively, although the relative change of Δn_C of the spun fiber per °C is −0.152%, on the same order of magnitude as that of Δn_L of the PM fiber.

Figure 9.18b shows the measurement repeatability results of nine repeated measurements. The average thermal coefficient α_C of nine measurements is -5.11×10^{-8}/°C and the standard deviation of the nine measurements is 3.48×10^{-10}, which represents less than 1% of measurement fluctuations and demonstrates the superb resolution and accuracy of the binary polarization measurement system.

9.4.3.5 Linear Birefringence Measurement in Spun Fiber with Different Temperatures

In the fabrication of a spun fiber, spinning a fiber with a strong linear birefringence will make the fiber having a strong circular birefringence but greatly reduces its intrinsic linear birefringence (Laming and Payne 1989). For current sensing applications, one prefers Δn_L has a relatively small value that can ensure the accuracy of FOCS and withstand external disturbances. Therefore, it is important to accurately determine Δn_L of a spun fiber and its temperature dependence.

While Δn_C of the 10 m spun fiber being measured, its Δn_L in different temperatures can also be simultaneously obtained, with the results shown in Figure 9.19a. As mentioned previously, the

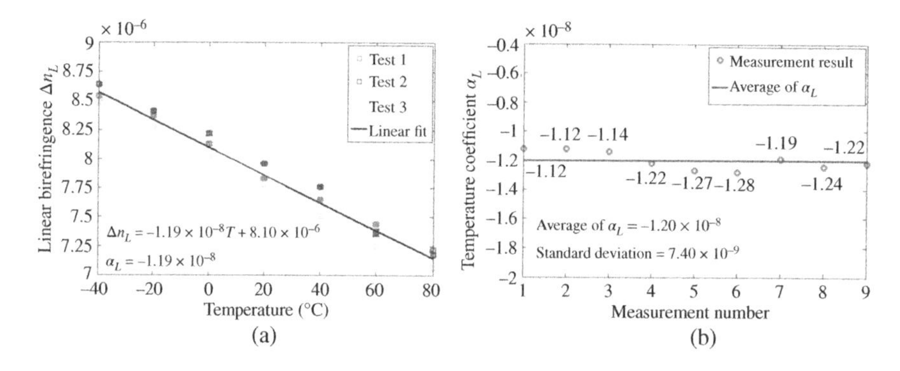

Figure 9.19 (a) Measured Δn_L of 10 m spun fiber at seven different temperatures. The slope of the fitting line represents the thermal coefficient α_L of residual linear birefringence (-1.19×10^{-8}/°C). The relative change per °C, defined as the ratio of α_L and Δn_L at $T=0$, is −0.147%. (b) Repeatability measurements of the thermal coefficient of residual linear birefringence α_L of a spun fiber under test (Xu et al. 2017).

spun fiber is coiled with a diameter of 12 cm and is put in a temperature chamber. Totally three sets of measurements labeled Test 1, Test 2, and Test 3 are performed at seven temperatures from −40 to 80 °C, and the measurements at each temperature are repeated three times. After each set of measurements, the coiled spun fiber in the chamber was purposely re-arranged and re-connected with the measurement system to alter the polarization states inside the fiber. The data fluctuation at each temperature in each test set is represented by an error bar. As one can see from Figure 9.19a that the three measurements at each temperature in each test set is highly repeatable; however, the data at each temperature in different test sets have relatively large fluctuations. Nevertheless, it is evident from Figure 9.19a that Δn_L of the spun fiber is around 7.80×10^{-6} (about four times smaller than its circular birefringence) at room temperature (20 °C) and also decreases with temperature. In Figure 9.19a, nine sets of data at each temperature are averaged and a linear fit of the values is obtained. The slope of the linear fit is -1.19×10^{-8}/°C, which is the thermal coefficient α_L of Δn_L. It is interesting and important to notice that the relative change of Δn_L per °C of the spun fiber is −0.147%, almost identical to the relative change of Δn_C per °C of −0.152%, despite the large difference between the thermal coefficients of the circular and linear birefringences Δn_C and Δn_L. It is also interesting to notice that the temperature dependence of a PANDA PM fiber's birefringence is 6.22×10^{-4}–$5.93 \times 10^{-7}T$ (Ding et al. 2011) and the relative change per °C is −0.095%, also on the same order of magnitude as the relative change of the spun fiber's Δn_C and Δn_L per °C, suggesting that these temperature dependences may all be caused by the anisotropic strain resulting from the differential thermal expansions between the stress rods and the rest of the fiber cladding (Ding et al. 2011; Mochizuki et al. 1982), as discussed in Chapter 3. However, such expectations must be further studied to be conclusive.

To verify the repeatability of the measurement results, just as the circular birefringence measurements, nine repeated measurements were performed with the results shown in Figure 9.19b. The average thermal coefficient α_L of nine measurements is -1.20×10^{-8}/°C and the standard deviation of nine measurements is 7.40×10^{-9}.

Note that the bending induced birefringence in a coiled fiber of 12 cm diameter ($R = 6$ cm) from Eq. (3.10a) (Ulrich et al. 1980) is 5.34×10^{-7}. Such a bending induced linear birefringence is experimentally confirmed to be 6.10×10^{-7} by measuring a length of SMF coiled with a diameter of 12 cm using the binary polarization measurement system. Such a bending induced birefringence is an order of magnitude smaller than the residual linear birefringence of the spun fiber and is expected to have negligible effect on the accuracy of the spun fiber measurement. These experimental results also indicate that the current sensing coils may also be coiled down to a diameter on the order of 12 cm with negligible impact on the temperature stability of the current sensing system.

9.4.4 Effective Verdet Constant Measurement of Spun Optical Fibers

The Verdet constant is a fundamental parameter associated with the Faraday effect describing the polarization rotation of light in certain media caused by a magnetic field (Laming and Payne 1989). For a constant magnetic field B parallel to the path of the light traveling in a Faraday material free of birefringence with a length of L, the Verdet constant $V_{B0}(\lambda)$ at a fixed wavelength λ relates the polarization rotation angle θ_0 by (Tang et al. 1991):

$$\theta_0 = V_{B0}(\lambda)BL = V_{H0}(\lambda)HL \tag{9.27}$$

where B and H are magnetic flux density and magnetic field strength ($B = \mu_0 H$), respectively, $V_{H0} = \mu_0 V_{B0}$ is the Verdet constant of the material relating the H-field to the polarization rotation, and μ_0 is the permeability of vacuum. Note that H has a unit of Ampere per meter; therefore it is easier to use the second term in Eq. (9.27) for current sensing applications. The Verdet constants of different Faraday materials are generally strong functions of wavelength, which have been accurately measured and well tabulated (Smith 1978; Kumari and Chakraborty 2018; Vojna et al. 2019). For fused silica (SiO_2), $V_{H0} = \mu_0 V_{B0} = 5.97 \times 10^{-19}\pi/\lambda^2$ (Cruz et al. 1996).

It is important to point out that Eq. (9.27) assumes that the Faraday material is isotropic free of birefringence. When the light propagates in a Faraday material with birefringence, Eq. (9.27) no longer holds and much more complicated relationships are found between the magnetic field and the polarization rotation angle in general (Harms et al. 1976; Tabor and Chen 1969; Simon and Ulrich 1977). For example, Tabor and Chen (1969) studied the Faraday effect in an Ytterbium Orthoferrite crystal possessing both Faraday rotation and linear birefringence. They showed that the presence of birefringence can drastically affect the behavior of polarization evolution, which is considerably different from pure Faraday rotation of Eq. (9.27). Similarly, as shown in Harms et al. (1976) and Simon and Ulrich (1977), the presence of linear birefringence in optical fibers made the behavior of polarization evolution induced by magnetic field very complicated such that a simple Verdet constant was no longer sufficient to describe the Faraday rotation in general. In other words, the linear relation between the Faraday rotation angle and the magnetic field is no longer preserved with the presence of linear birefringence in the fiber.

As discussed in Chapter 3, the linear birefringence of the fiber generally comes from two sources, one is the imperfection in the circularity of fiber's core and cladding, and the other is from the external and internal mechanical stresses via photoelastic effect (Smith 1980), including the bending induced stress when winding the fiber in a loop around a conductor for current sensing applications. The axis orientation and magnitude of the combined birefringence are sensitive to the temperature and fiber arrangement, which leads to measurement fluctuations of the polarization rotation angle induced by the magnetic field. Consequently, large uncertainties will result when the fiber is used as part of an electrical current sensor for measuring the electrical current flowing in a conductor.

As mentioned in Sections 3.3 and 9.4.3, to overcome the issues caused by the linear birefringence, the spun fibers are adopted (Polynkin and Blake 2005) for electrical current and magnetic field sensing, because it is effective in minimizing the problems caused by the linear birefringence by introducing a large amount of circular birefringence and can be consistently produced by multiple specialty fiber manufacturers (Fibercore Co. Ltd. 2022; iXblue Photonics 2022). It can be shown next that for such a spun fiber with a large circular birefringence and a large local linear birefringence, the desired linear relationship between the Faraday rotation angle and the magnetic field can approximately hold and a parameter called the effective Verdet constant can be defined to relate the polarization rotation angle with the applied magnetic field. To determine the effective Verdet constant, the highly sensitive binary polarization analysis system described Figure 9.10 can be used to measure the magnetic field induced SOP rotation in the fiber. Unfortunately, the wavelength differential method used in Section 9.4.3 is not sensitive enough to measure the polarization change induced by the Faraday effect in the spun fiber and therefore a more sensitive time differential method is adopted, as shown in the following text.

9.4.4.1 Introduction of the Effective Verdet Constant

Theory of the Effective Verdet Constant As discussed in Section 4.3.8, for a polarized light signal launched into a spun fiber, the Jones vector of the signal at the fiber output can be expressed as (Laming and Payne 1989; Barlow et al. 1981; Rashleigh 1983b)

$$\mathbf{E}_{out} = \mathbf{M}\mathbf{E}_{in} \tag{9.28}$$

where $\mathbf{E}_{out}$ and $\mathbf{E}_{in}$ are Jones vectors of the light before and after the fiber. According to Jones equivalent theorem I discussed in Section 4.3.1, the birefringence of a spun fiber of length L can be represented by two lumped birefringent elements, a retarder $\mathbf{M}_\delta$ with a retardation $\delta(L)$ having its principal axis oriented at an angle $\theta(L)$ and a rotator $\mathbf{M}_\varphi$ with a rotation angle $\varphi(L)$ (Laming and Payne 1989):

$$\mathbf{M} = \begin{bmatrix} cos\varphi(L) & -sin\varphi(L) \\ sin\varphi(L) & cos\varphi(L) \end{bmatrix} \begin{bmatrix} cos\frac{\delta(L)}{2} + jsin\frac{\delta(L)}{2}cos2\theta(L) & jsin\frac{\delta(L)}{2}sin2\theta(L) \\ jsin\frac{\delta(L)}{2}sin2\theta(L) & cos\frac{\delta(L)}{2} - jsin\frac{\delta(L)}{2}cos2\theta(L) \end{bmatrix} \tag{9.29a}$$

where the first and the second terms on the right-hand side represent the rotator $\mathbf{M}_\varphi$ and retarder $\mathbf{M}_\delta$, respectively, and $\varphi(L)$, $\delta(L)$, and $\theta(L)$ can be expressed as

$$\varphi(L) = \rho L + tan^{-1}\left\{ \frac{-2(\rho+f)/\Delta\beta}{\sqrt{1+[2(\rho+f)/\Delta\beta]^2}} tan \frac{\sqrt{\Delta\beta^2+4(\rho+f)^2}}{2} L \right\} + n\pi \tag{9.29b}$$

$$\delta(L) = 2\,sin^{-1}\left\{ \frac{1}{\sqrt{1+[2(\rho+f)/\Delta\beta]^2}} sin \frac{\sqrt{\Delta\beta^2+4(\rho+f)^2}}{2} L \right\} \tag{9.29c}$$

$$\theta(L) = \frac{\rho L - \varphi(L)}{2} + \frac{m\pi}{2} \tag{9.29d}$$

$$f = V_{H0}(\lambda)H = \frac{V_{H0}(\lambda)I}{(2\pi r)} \tag{9.29e}$$

In Eqs. (9.29a) to (9.29e), m, n are integers, f is the Faraday rotation angle per unit length of an ideal fiber, ρ is the spin twist rate of the spun fiber, and $\Delta\beta$ is unspun fiber retardation per unit length, which relates to the birefringence Δn of the unspun fiber by $\Delta\beta = 2\pi\Delta n/\lambda$ (the unspun fiber is defined as the PM fiber made with the same preform as the spun fiber, but without spinning the preform during fiber drawing process, as described in Section 3.3). In general, the local Faraday rotation per unit length f for a fiber coil around an electrical conductor depends on coil's radius r, the current I in the conductor, and the Verdet constant $V_{H0}(\lambda)$ of the fiber core, as shown in Eq. (9.29e).

In the absence of electrical current $I = 0$, $\delta(I = 0)$ and $\varphi(I = 0)$ represent the intrinsic linear retardation and circular rotation of a spun fiber of length L, respectively. $\delta(I \neq 0)$ and $\varphi(I \neq 0)$ represent the result of the interaction between the intrinsic linear retardation and circular rotation and Faraday rotation of a spun fiber of length L, respectively. For a spun fiber with a sufficiently large spin twist rate ($2\rho/\Delta\beta \geq 4$, $\rho \gg f$), one obtains

$$\frac{2(\rho+f)/\Delta\beta}{\sqrt{1+[2(\rho+f)/\Delta\beta]^2}} \approx 1 \tag{9.30a}$$

Note that most commercial spun fibers satisfy the conditions $2\rho/\Delta\beta \geq 4$ and $\rho \gg f$. Consequently, one obtains the circular retardation caused by the Faraday effect from Eq. (9.29b):

$$\Delta\varphi_F = \varphi(I) - \varphi(0) = \frac{\varphi^2(I) - \varphi^2(0)}{\varphi(I) + \varphi(0)} \approx \frac{V_{H0}(\lambda)}{[1+(\Delta\beta/2\rho)^2]^{1/2}} HL \tag{9.30b}$$

As will be shown numerically next that Faraday rotation angle of a spun fiber is equivalent to the circular retardation, which can be related to the electrical current by

$$\Delta\varphi_F = V_{eff}(\lambda)HL = \frac{V_{eff}(\lambda)LI}{(2\pi r)} \tag{9.30c}$$

where $V_{eff}(\lambda)$ is defined as the effective Verdet constant of spun fiber. Combining Eqs. (9.29e), (9.30b), and (9.30c) yields

$$V_{eff}(\lambda) = \frac{V_{H0}(\lambda)}{[1+(\Delta\beta/2\rho)^2]^{1/2}} = \frac{V_{H0}(\lambda)}{\left[1+(L_t/2L_p)^2\right]^{1/2}} = \gamma V_{H0}(\lambda) \tag{9.31a}$$

$$\gamma = \frac{1}{[1+(\Delta\beta/2\rho)^2]^{1/2}} = \frac{1}{\left[1+(L_t/2L_p)^2\right]^{1/2}} \tag{9.31b}$$

where γ is called the Faraday effect reduction factor of the spun fiber, and L_t and L_p are the spin pitch and linear beat length of unspun fiber, respectively, with $\Delta\beta = 2\pi/L_p$ and $\rho = 2\pi/L_t$. Note that a different quantity equaling γ^2 was introduced in Laming and Payne (1989) which was defined as the normalized current sensitivity. It can be seen from Eqs. (9.31a) and (9.31b) that the effective Verdet constant of a spun fiber is always less than the original Verdet constant of an ideal fiber made with the same material but free of birefringence, however, approaches to the original Verdet constant as the spin twist rate ρ increases.

Numerical Analysis and Poincaré Sphere Presentation On a Poincaré sphere represented by Stokes parameter (S_1, S_2, S_3), Faraday effect causes the input linear polarization to rotate on the equator of the Poincaré sphere in the absence of linear birefringence in an optical fiber. Similar to the Jones matrix presentation of Eq. (9.29a), in the presence of linear birefringence, the Mueller matrix representing the SOP of light passing through the fiber can also be decomposed into a pure rotator and a pure retarder (Xu et al. 2017), which implies that the SOP trace caused by a variation of circular birefringence circles around S_3 axis in a plane parallel to the equator when the linear birefringence is fixed or changes much slower than the circular birefringence during the measurement (Rashleigh 1983b; Treviño-Martínez et al. 2005). In other words, in the presence of fixed birefringence or extremely slow varying linear birefringence, the polarization rotation angle around S_3 axis corresponds to the Faraday rotation angle. Therefore, one may obtain the Verdet constant by measuring the polarization rotation angle induced by a magnetic field on the (S_1, S_2) plane, where the slow axis of the fiber is assumed to be aligned with the x-axis of the (x, y) coordinates defining (S_1, S_2, S_3).

Figure 9.20a shows the trace of polarization variation on Poincaré sphere obtained using Eqs. (9.28) and (9.29a) through (9.29e) for the case that a linearly polarized light passing through a spun fiber of length 10 m coiled around an electrical conductor with different alternating currents (ACs). In the calculation, the input polarization is assumed to be linear and aligned with the fast axis of spun fiber's residual birefringence ($\theta(L) = 0$) and the following parameters corresponding to a commercial fiber (Fibercore Co. Ltd. 2022) are used: $\Delta\beta$ = 598.5 rad/m, ρ = 1309 rad/m, $V_{H0}(\lambda) = 5.97\times10^{-19}\pi/\lambda^2$ [Verdet constant of fused silica (Cruz et al. 1996)], λ = 1310 nm and r = 70 mm. Figure 9.20b shows the trace in Figure 9.20a on the (S_1, S_2) plane and the corresponding arc angles. Similarly, the polarization variation trace induced by different direct currents (DCs) in the (S_1, S_2) plane is shown in Figure 9.20c. Note that corresponding to a particular current value, the polarization variation angle in Figure 9.20b is $2\sqrt{2}$ times that in Figure 9.20c because the AC swings between negative and positive values (the ACs in Figure 9.20b are the rms values), resulting in a current swing $2\sqrt{2}$ times larger than that in Figure 9.20c with the same current label.

It can be seen from Eqs. (9.28) and (9.29a) that when the linear birefringence $\delta(L)$ is zero, the SOP rotation angle equals to the circular retardation $\varphi(L)$. For spun fibers with a weak linear birefringence, it can be shown numerically using Eqs. (9.28) and (9.29a) through (9.29e) that the current induced SOP rotation angle is still approximately the same as the current induced circular retardation change $\Delta\varphi_F$ calculated using Eq. (9.30b), assuming the input SOP is linear and $\theta = 0$ in

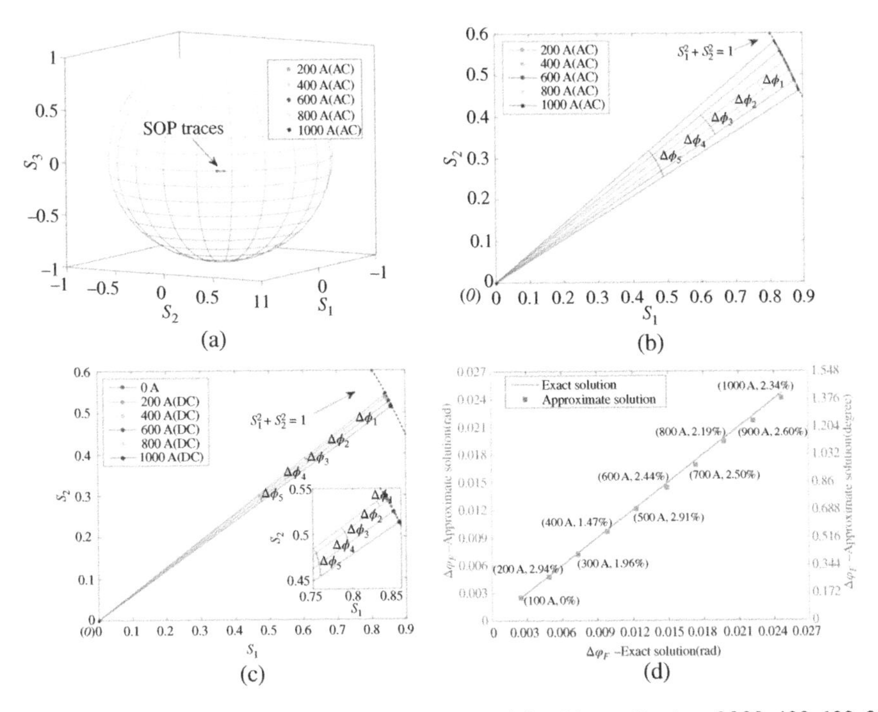

Figure 9.20 (a) Polarization variations induced by ACs with amplitudes of 200, 400, 600, 800, 1000 A (rms), respectively. (b) and (c) The projections of the polarization traces on the equatorial plane of Poincaré sphere induced by different ACs and DCs flowing in the conductor, respectively. (d) Comparison between the circular retardations caused by the Faraday effect of the exact solution of Eq. (9.29b) and the approximate solution of Eq. (9.30b) at different DCs. The relative errors between the approximate and exact solutions are shown at the corresponding data points (Yao et al. 2021).

Eq. (9.30a) with the same spun fiber parameters earlier. Therefore, one can use $\Delta\varphi_F$ to represent the physical SOP rotation angle induced by electrical currents, as shown in Eq. (9.30c).

Now let's compare the circular retardation caused by the Faraday effect $\Delta\varphi_F$ obtained using the exact solution Eq. (9.29b) and the approximate solution of Eq. (9.30b) at different current levels, with the results shown in Figure 9.20d. As can be seen, the maximum relative error of the approximate solution is less than 3%, which is defined as the difference between the exact and approximate solutions divided by the exact solution. And it validates that the approximation of Eq. (9.30b) is sufficiently accurate and the effective Verdet constant is valid for representing the Faraday rotation in a spun fiber. Note that the SOP rotation angle $\Delta\Phi$ on the Poincaré sphere is twice that of the Faraday rotation angle in real space ($\Delta\Phi = 2\Delta\varphi_F$). From the data one can observe that the Faraday rotation angles are quite small, on the order of 0.14° for $\Delta\varphi_F$ or 0.28° for $\Delta\Phi$ per 100 A current. Therefore, care must be taken to eliminate any polarization fluctuations and drifts caused by temperature or fiber motion during the experiment.

9.4.4.2 Experiment

Measurement Setup The experiment setup is shown in Figure 9.21a, which includes a 1310 nm distributed feedback (DFB) laser at the PSG input, a binary MO rotator based polarization analysis system shown in Figure 9.10b, a polarization controller (PC1) at the PSG output, a polarization controller (PC2) at the PSA input, a coiled spun FUT, a current generator capable of generating AC or DC up to 3 kA. A temperature sensor (PT100) is also included to monitor the temperature

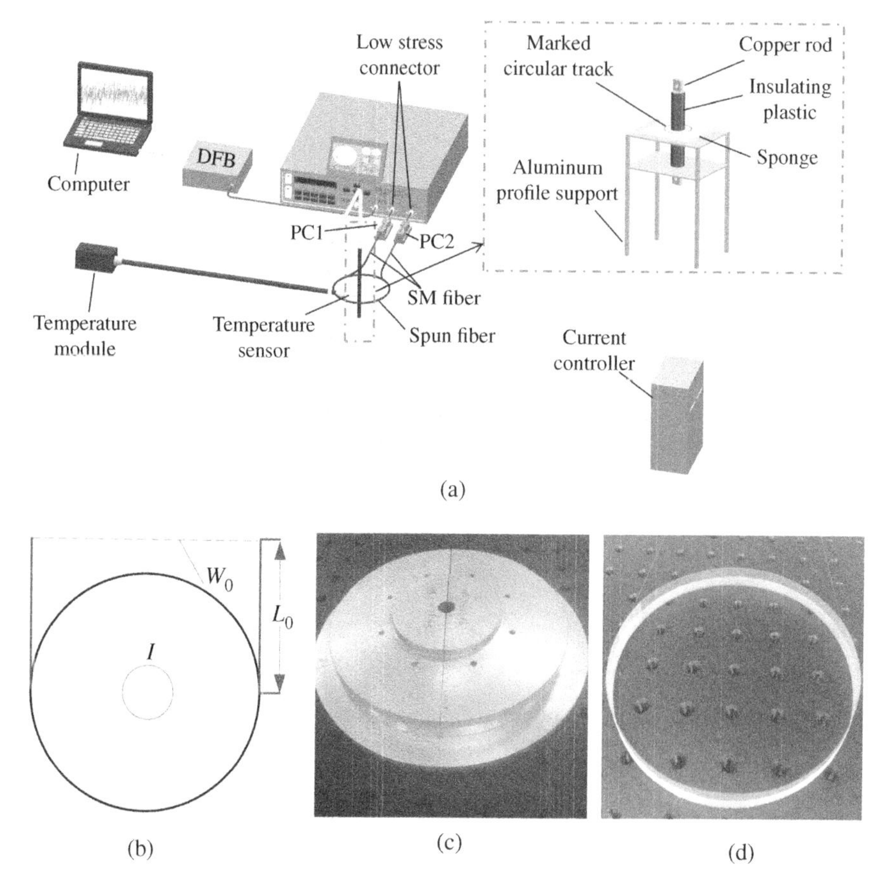

Figure 9.21 (a) Polarization measurement setup for obtaining the Faraday rotation angles induced by electrical currents in a coiled spun fiber. The inset shows the fixture to ensure the conductor rod to be perpendicular to the plane of the spun fiber coil and pass through the center of the coil placed on a sponge with a marked circular track. (b) The sketch of the spun fiber coil with a diameter of 138.8 mm and *N* fiber turns. (c) The photo of the aluminum fixture for making the spun fiber coils with a precise diameter. (d) The coil on an adhesive tape after removed from the aluminum fixture. PC1 and PC2: polarization controllers; PSGA: polarization analysis system constructed with binary polarization rotators in a PSG and a PSA (Yao et al. 2021).

around the coiled fiber. The total length of the FUT is 10 m, with about 9.4 m coiled around the conductor and about 0.3 m fiber at each end for fusion splicing to two SMFs connected to PC1 and PC2. The outputs of PC1 and PC2 are connected to PSG and PSA, respectively, via two low stress connectors. The two fiber pigtails (0.3 m spun fiber each) and two SMFs (0.3 m each) are protected with gooseneck tubes to prevent fiber vibration or fiber motion due to the airflow.

In the setup, PC1 and PC2 are used (i) to re-align the coordinate system of PSG with that of PSA, which were misaligned from the SMF pigtails of the PSG output and the PSA input, as well as the FUT, and (ii) to compensate for the linear birefringence in the SMFs and FUT so that the Faraday rotation is along the equator of the PSA coordinate system to simplify data processing. Note that PC1 and PC2 are of Yao-controllers based on squeeze-and-turn mechanism described in Figure 6.12, which is the optical fiber equivalent of Babinet–Soleil compensator.

As shown in the inset of Figure 9.21a, care is taken to ensure that the copper rod protected with insulating plastic is placed at the center of the fiber coil and is perpendicular to the coil plane for each measurement using a specially made aluminum frame table for minimizing any inconsistency between measurements, which may result from the opening of the coil between the two fiber pigtails, as shown in Figure 9.21b, before connecting the coil to the binary measurement system (no error will result from the Ampere's law if the coil is perfectly closed). The spun fiber coils are securely placed on a circular track marked on a sponge plate for keeping the coils at the same position in different measurements and for dampening acoustic vibrations.

Three different spun fibers from three different commercial vendors (IxF-SPUN-1310-125-EC from iXblue Co. Ltd., SHB1250 (7.3/125) from Fibercore Co. Ltd. and SH 1310/12-5/250 from YOFC Co. Ltd.) for operation at 1310 nm are measured with different electrical currents from 0 to 1 kA. To ensure all spun fiber coils to be tested are identical, a fiber winding fixture was made with a diameter of 138.5 mm, as shown in Figure 9.21c. Before winding the fiber onto the fixture, a layer of adhesive tape with a thickness of 0.03 mm is tightly placed onto the winding surface of the fixture, which helped to hold the fiber together after the coil winding is complete. The resulting single-layer coils therefore all have the same diameter of 138.8 mm, which includes the diameter of the winding fixture (138.5 mm), the thickness of the adhesive tape (0.03 mm × 2 mm), and the fiber diameter of the spun fiber (0.25 mm), as shown in Figure 9.21d. In addition, all fiber coils are ensured to have the same number of turns (21 full turns, single layer) and the same length (10 m) during winding.

Note that coiling the fiber may induce additional linear birefringence; however, the diameter of the fiber coil is sufficiently large (138.8 mm) such that the linear birefringence introduced by the bending is on the order of 10^{-7} (Polynkin and Blake 2005; Feng et al. 2018), which is an order of magnitude smaller than the residual linear birefringence around 7.75×10^{-6} in the spun fiber (Xu et al. 2017). Therefore, the bending induced birefringence from coiling the fiber can be safely ignored in our measurement.

Time Differential Method In Section 9.4.3, the wavelength differential method is used to measure the circular birefringence and residual linear birefringence in some spun optical fibers. In particular, during the measurement, the wavelength of the laser is scanned while the SOP of light after passing through the DUT is analyzed. The birefringence of the DUT can be obtained by analyzing the rate of SOPs change with wavelength or frequency, with the SOPs change in the equator plane corresponding to the circular birefringence and perpendicular to the equator plane corresponding to the linear birefringence. Because the birefringence measured in Xu et al. (2017) are relatively large, on the order of 10^{-7} or more, the wavelength scan induced SOP variation is on the order of 0.10°/nm at the central wavelength of 1310 nm for a fiber with a length of 10 m, sufficiently large to be accurately measured with the binary system with an angular measurement resolution of 0.02°, as shown in Table 9.7.

When the fiber coiled around a conductor is subjected to an axial magnetic field (with a unit of Ampere per meter) induced by a current, the Faraday rotation occurs due to the circular birefringence induced by the Faraday effect, which is extremely weak ($\sim 10^{-13}$) compared with the intrinsic circular birefringence ($\sim 10^{-5}$) and residual linear birefringence (10^{-6}) of the spun fiber (Xu et al. 2017), and thus the corresponding SOPs change resulting from the wavelength change, on the order of 2.2×10^{-4} degrees/nm (assuming a fiber length of 10 m, with a fiber coil radius of 70 mm and under a current of 100 A) on the Poincaré sphere, is too small to be measured with the wavelength differential method. In addition, the wavelength scanning method of the binary system takes about 20 seconds to complete a measurement, which is too slow to avoid polarization drift caused by temperature changes or slight fiber motion due to the airflow. Therefore, here the time differential method is adopted: for Faraday rotation induced by an AC, the SOP angle swing corresponding to the positive current maximum and the negative current maximum can be directly measured, as shown in Figure 9.20b; for Faraday rotation induced by a DC, the current can be turned on until it stabilizes and then off, and the SOP angle difference between the current "on" and "off" states can be measured, as shown in Figure 9.20c. Note that it only takes about 0.1 second to complete a differential SOP angle measurement and the effect of SOP drifts caused by the temperature or fiber motion can be effectively eliminated. As estimated by simulation of Figure 9.20, the Faraday rotation is on the order of 0.28° per 100 A on the Poincaré sphere (assuming a fiber length of 10 m coiled around a conductor with a radius of 70 mm at 1310 nm), which is sufficiently large to be detected by the binary measurement system.

The advantages of the time differential method for measuring the current induced SOP change can be further understood next: the retardation ψ associated with a circular birefringence n_C is $\psi = \pi n_C L/\lambda$, where λ is the wavelength of the laser and L is the fiber length. For a given change Δn_C of n_C, the retardation variation $\Delta\psi$ (corresponding to the SOP change) caused by changing the wavelength and by switching Δn_C on and off are $\Delta\psi_\lambda = \left(\frac{\pi \Delta n_C L}{\lambda}\right)\frac{\Delta\lambda}{\lambda}$ and $\Delta\psi_t = \left(\frac{\pi \Delta n_C L}{\lambda}\right)$, respectively, where $\Delta\lambda$ is the wavelength changing range. Therefore, for a given circular birefringence Δn_C induced by the Faraday effect, the time differential method is $\lambda/\Delta\lambda$ times more sensitive than the wavelength differential method.

SOP Measurement Resolution Before conducting detailed measurements, it is important to understand the minimum detectable SOP rotation angle of the binary polarization measurement system. Figure 9.22a shows the measured SOP fluctuation and drift of the system when a spun fiber (Fibercore) of 10 m is coiled around a conductor without current flow in an experimental setup shown in Figure 9.21a. In principle, temperature variation, fiber motion caused by airflow, and system electronic noise can all contribute to the SOP fluctuation and drift; however, in our experiment, because the temperature was relatively stable during 20 minutes measurement period (brown line in Figure 9.22a) and the fiber pigtails of the FUT were protected with gooseneck tubes, the SOP drift and fluctuation due to these two factors were minimal, and therefore the SOP measurement accuracy and resolution was limited by the system electronic noise (gray line in Figure 9.22a). Taking 3-point average of the data, the electronic noise can be significantly reduced and the corresponding SOP fluctuation is less than 2×10^{-4} rad or 0.012°. Other digital filtering techniques, including taking more averages, can be used to further reduce the system noise floor. In addition, reducing the electronic noise of the amplifier and digital circuit may also be effective in reducing system noise floor.

Figure 9.22b shows the measurement results of SOP change induced by electrical currents. In the measurement, the electrical currents of 10, 20, 50, and 100 A are turned on to reach the

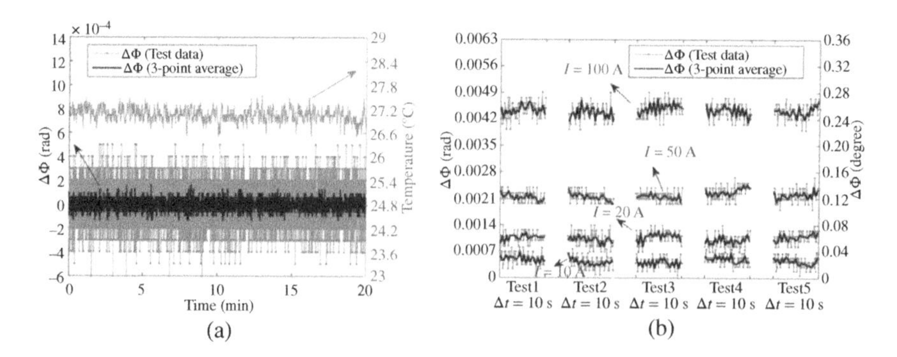

Figure 9.22 (a) System noise floor for differential SOP angle ΔΦ measurement. (b) Five repeated measurements of ΔΦ induced by currents of 10, 20, 50, and 100 A, each having a measurement time of $\Delta t = 10$ seconds. 10 m SHB1250 (7.3/125) fiber from Fibercore was used (Yao et al. 2021).

predetermined value for 10 seconds and then off periodically. The durations for the current off states in between the current on states in Figure 9.22b varied from 10 to 13 seconds, depending on the current levels required by the power equipment operation, although they appeared to be the same in the figure. The differential SOP angle $\Delta\Phi$ on the Poincaré sphere induced by the 10 A current was 4.6×10^{-4} rad or 0.026°, obtained by averaging the data in the 10 seconds period. Therefore a differential angle less than 0.026°, corresponding to a current less than 10 A, can be accurately measured with the binary system.

Experimental Results and Discussions Figure 9.23a shows the measured SOP variation of light propagating through a 10 m spun fiber (Fibercore) coiled around a conductor on the (S_1, S_2) plane caused by different ACs flowing in the conductor, with the measurement setup shown in Figure 9.21a. As described previously, PC1 and PC2 are so adjusted that the SOP traces induced by electrical current are along the equator of the Poincaré sphere. Note that the SOP measurement rate of the binary system is about 30 Hz, while the AC varies at 50 Hz. The SOP traces would move back and forth on the equator in response to the cycles of the AC. It would take multiple cycles of SOP traces to determine the maximum swing of the SOP angle corresponding to the positive and negative maximum currents. In experiment, 250 cycles are taken for each AC current setting, corresponding to a measurement time of five seconds, during which the polarization fluctuations induced by the environmental temperature or fiber motion can be noticed in the experiment. Consequently, the measurement results were less repeatable than those of the DC to be discussed next in Figure 9.23b.

Note that because of the slow SOP drift caused by the environmental temperature or fiber motion, the starting SOP for different current settings was different; however, the differential SOP angle corresponding to each current setting was much less affected by the SOP drift, an advantage of the time differential method. One may notice that the SOP variation traces are not exactly on the equator ($s_1^2 + s_2^2 = 1$), probably because (i) PC1 and PC2 were not perfectly adjusted, and (ii) of the influence of the residual linear birefringence contribution in the spun fiber. This should have minimal impact on the measurement accuracy of Faraday rotation angle, because the deviation is extremely small.

Figure 9.23b shows the differential SOP angles induced by different DCs of light propagating in the same spun fiber as in Figure 9.23a. At each current setting, the current is turned on and off periodically to induce SOPs change in the (S_1, S_2) plane. In particular, the current is first turned on and one must wait 10 seconds for it to stabilize before it is then turned off for measuring the SOP

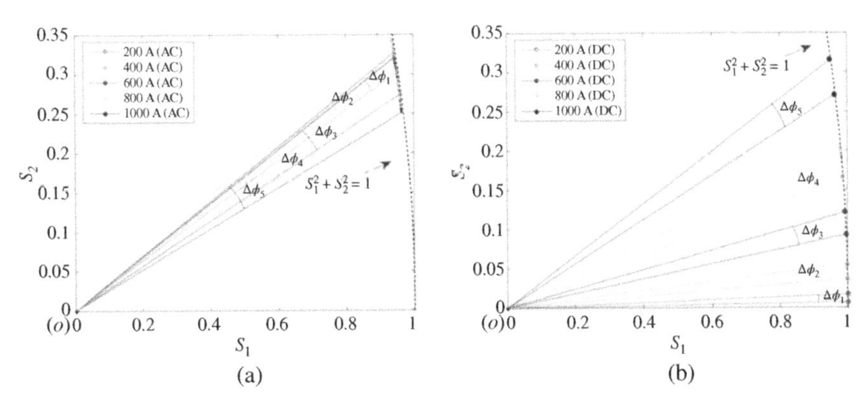

Figure 9.23 Measured differential SOP angles on the equator plane of the Poincaré sphere in response to different electrical currents of 200, 400, 600, 800, 1000 A, respectively. ***O*** represents the center of the equatorial circle. (a) Alternating current (AC, rms value), and (b) direct current (DC). 10 m SHB1250 (7.3/125) fiber from Fibercore was used (Yao et al. 2021).

change between the "on" and "off" states. Because almost no waiting period is required for the current to drop to zero, this differential SOP measurement can be completed in about 0.1 second. Consequently, the effect of SOP drifts caused by the temperature or fiber motion can be effectively eliminated. 3-point averaging was used in processing the data (~40 samples per second). Again, due to the SOP drift caused by the environmental temperature or fiber motion, the starting SOP for different current settings is different; however, the differential SOP angle corresponding to each current setting is not affected by the SOP drift. Note that the currents in Figure 9.23a are rms values so that the differential SOP angle $\Delta\Phi$ for each current setting with AC is $2\sqrt{2}$ times larger than that of DC with the same current label, because the AC swings from minus to plus, while DC changes from 0 to plus.

The differential SOP rotation angle $\Delta\Phi$ as a function of current of three consecutive measurements is shown in Figure 9.24a. At high current setting above 700 A, heat was generated by the current flowing in the conductor, which slightly raised temperature around the spun FUT, as shown by the brown line in the figure monitored by a temperature sensor shown in Figure 9.22a. At the maximum current setting of 1000 A, the temperature is raised by only 0.3 °C, which should have negligible effect on the measurement accuracy. Curve-fitting yields a slope $\alpha = 4.62 \times 10^{-5}$ rad/A, with a goodness-of-fit of 0.9998. The effective Verdet constant can be obtained from this slope and Eq. (9.30c) to be 1.05×10^{-6} rad/A using the following relation:

$$V_{eff} = \frac{\Delta\varphi_F}{HL} = \frac{\alpha\pi r}{L} = \frac{\alpha}{2N} \tag{9.32}$$

where the relations $\Delta\Phi = 2\Delta\varphi_F = \alpha I$, $H = I/(2\pi r)$, and $L = 2\pi rN$ are used. As shown in Eq. (9.32) the accuracy of the effective Verdet constant calculation from measured $\Delta\varphi_F$ can be compromised by the inaccurate measurements of the radius and length of the spun fiber coil. In order for the measurement error of V_{eff} to be less than 0.5%, the errors of the radius and the length of the coil must be controlled to be less than 0.35 and 50 mm, respectively, for fiber coils having a radius r of 69.4 mm and a length L of 10 m.

One may circumvent directly using r and L for V_{eff} calculation by accurately obtaining the number of turns N in the fiber coil from Ampere's law reflected in the last term of Eq. (9.32). However, complication arises due to the existence of two fiber pigtails (as shown in Figure 9.21b,d) of the coil, which fortunately has been successfully accounted for by A.H. Rose and S.M. Etzel (1997). Accordingly, the fiber pigtails with a length of L_0 and a separation of W_0 shown in Figure 9.21b

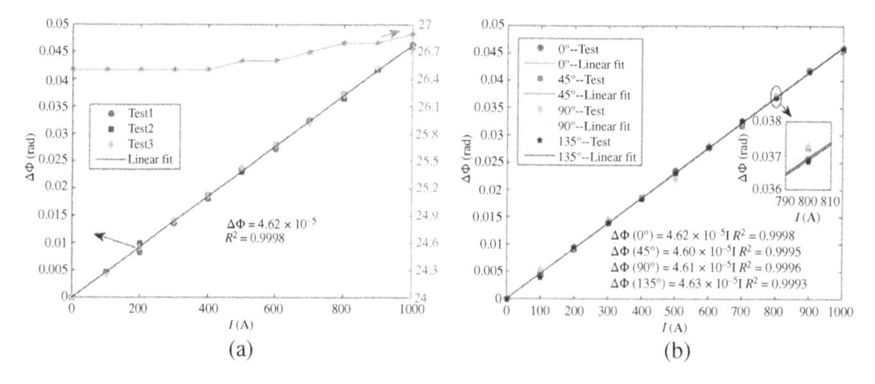

Figure 9.24 (a) Three repeated measurements of the differential SOP $\Delta\Phi$ on the equator plane of the Poincaré sphere induced by different DCs with the 0° linear polarization (1, 0, 0). (b) Measurement results of $\Delta\Phi$ as a function of DCs with four distinctive input linear SOPs: 0° (1, 0, 0), 45° (0, 1, 0), 90° (−1, 0, 0), and 135° (0, −1, 0). 10 m SHB1250 (7.3/125) fiber from Fibercore is used (Yao et al. 2021).

can be counted as effective fractional turns so that the total number of fiber turns can be expressed as

$$N = N_0 + \left(1 - \frac{tan^{-1}(W_0/2L_0)}{\pi}\right) \tag{9.33}$$

where $N_0 = 21$ is the number of complete turns and the second term is the effective fractional turns of the spun fiber coil. With $W_0 = 138.8$ mm and $L_0 = 312.5$ mm, the fractional turns are 0.9 and the total number of turns are 21.9. Therefore, we will use $N = 21.9$ in Eq. (9.32) to obtain the V_{eff} after the slope α is obtained from curve fitting.

The differential SOP rotation angle $\Delta\Phi$ induced by electrical current is expected to be the same with different input linear SOPs to the spun fiber or with different input SOPs on the equator of the Poincaré sphere. Figure 9.24b shows the measurement results of $\Delta\Phi$ for the cases where the input SOPs generated by PSG are 0°, 45°, 90°, and 135°, respectively. Note that although there is an SMF connecting the PSG to the spun fiber, the proper adjustment PC1 can undo the effect of the birefringence in the SMFs to ensure the SOP entering the spun fiber is linear. In between the measurements with different input SOPs, PC1 and PC2 were slightly readjusted to balance out the birefringence offsets in the SMFs connecting the spun fiber to the PSG at the input end, as well as to the PSA at the output end of the spun fiber. One can see from Figure 9.24b that the four curves corresponding to the four input linear SOPs are almost identical, yielding almost the same slope. The average slope obtained from Figure 9.24b is $\overline{\alpha} = 4.61 \times 10^{-5}$ rad/A.

Effective Verdet Constants of Three Different Spun Fibers Different manufacturers choose different preforms with different birefringence or unspun fiber retardation per unit length, and different spin twist rate to produce spun fibers. These differences make different spun fibers having different magnetic sensitivity characterized by their corresponding Verdet constants. Using the published data from different fiber manufacturers, the effective Verdet constants V_{eff} and the corresponding Faraday effect reduction factor γ defined in Eq. (9.31b) can be obtained, as shown in Table 9.9. Note that for the Fibercore fiber, only the parameters of circular beat length L_c of 63–125 mm and spin pitch L_t of 4.8 mm are provided, which can be used to obtain the unspun fiber retardation per unit length $\Delta\beta$ and the spin twist rate ρ using the following relations (Xu et al. 2017): $L_p = \left(\frac{1}{L_c^2} + \frac{1}{L_t L_c}\right)^{-1/2}$, $\Delta\beta = 2\pi/L_p$, and $\rho = 2\pi/L_t$, where L_p is the linear birefringence beat length of unspun fiber.

Figure 9.25 shows the measured differential SOP rotation angles on the equator plane as a function of applied DCs. The effective Verdet constant V_{eff} of each spun fiber can be obtained from its

Table 9.9 Comparison of effective Verdet constants of three different spun fibers (Yao et al. 2021).

	Manufacturer parameters		Theoretical calculation				Measurements	
Fiber	L_p (mm)	L_t (mm)	$\Delta\beta$ (rad/mm)	ρ (rad/mm)	γ	V_{eff} (10^{-6} rad/A)	γ	V_{eff} (10^{-6} rad/A)
iXblue (IxF-SPUN-1310-125-EC)	6.3–9.7	2.5	0.63–1.01	2.51	0.98–0.99	1.07–1.08	0.98	1.07
Fibercore (SHB1250(7.3/125))	8.6–12.2	4.8	0.51–0.79	1.31	0.96–0.98	1.05–1.07	0.96	1.05
YOFC (SH 1310_125-5/250)	7.5–14.0	5.1	0.45–0.84	1.26	0.95–0.98	1.04–1.07	0.95	1.04

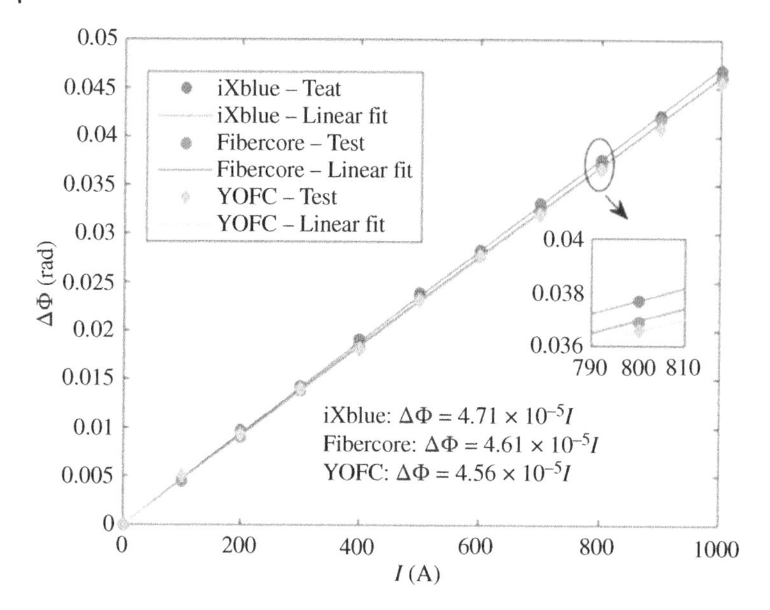

Figure 9.25 Measured differential SOP rotation angles induced by different levels of DC. $\alpha_i = 4.71 \times 10^{-5}$ rad/A, $\alpha_F = 4.61 \times 10^{-5}$ rad/A and $\alpha_Y = 4.56 \times 10^{-5}$ rad/A are the slopes of the fitting curves of the spun fibers from iXblue (IxF-Spun-1310-125-EC), Fibercore (SHB1250 (7.3/125)), and YOFC (SH1310/125/250), respectively (Yao et al. 2021).

curve-fitting slope using Eq. (9.32). The corresponding Faraday effect reduction factor γ of each fiber can be obtained by taking the ratio of the effective Verdet constant and the Verdet constant of an ideal silica fiber free of birefringence, which is 1.09×10^{-6} rad/A. The results are listed in the last two columns in Table 9.9. It can be seen that the spun fiber from iXblue has the highest effective Verdet constant, followed by that from Fibercore and finally YOFC. The corresponding Faraday effect reduction factors of the three fibers from an ideal fiber are 0.98, 0.96, and 0.95, respectively.

The measurement results are consistent with the analytical results of Eq. (9.31a) and validate the theoretical conclusion that the higher the ratio of $\rho/\Delta\beta$, the larger the effective Verdet constant or the more sensitive the Faraday rotation to the applied current or magnetic field. It is expected that the iXblue fiber shall also be less sensitive to local perturbations caused by stress and temperature because of the large local birefringence reflected by the large $\Delta\beta$.

Summary In this section, the term "effective Verdet constant" is introduced to quantitatively characterize the sensitivity of Faraday effect in practical spun optical fibers and is measured with a highly accurate binary polarization rotators based polarization analysis system. In particular, an ultra-sensitive time differential method with an SOP angle resolution better than 0.02° is adopted to measure the Faraday rotation angles induced by different amounts of currents for determining the effective Verdet constants.

9.4.5 Wave-Plate Analyzer Using Binary Magneto-optic Rotators

Optical wave plates for generating linear retardations are one of the most important components for polarization-related analysis and control, as described in Chapter 6. Accurate measurement of retardation and optical axis is therefore important for wave plate manufacturing and quality insurance. To accurately measure the induced retardation, Mueller matrix polarimeters described in Chapter 8 can be used. However, these analog devices have their own short comings:

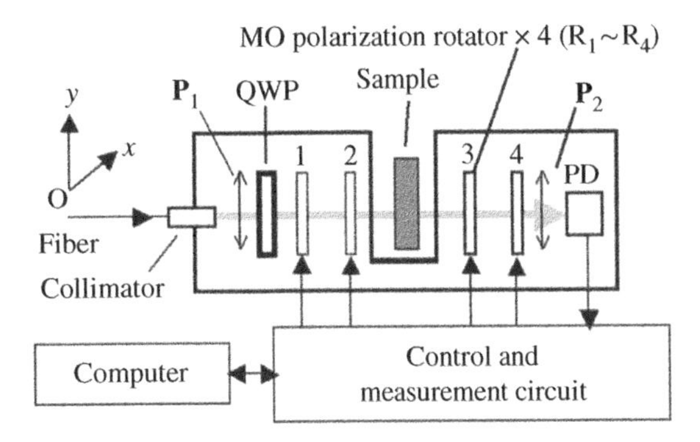

Figure 9.26 The binary wave-plate analyzer using four MO rotators ($R_1 \sim R_4$): Polarizer $\mathbf{P}_1$ is oriented 22.5° from the vertical axis (*y*-axis), polarizer $\mathbf{P}_2$ and the fast-axis of QWP are aligned vertically, and the rotation angles of MO rotators are about ±22.5° at their center wavelength (Chen et al. 2007).

the mechanical rotation based devices suffer from slow speed and mechanical wear-and-tear and the polarization modulation based devices are expensive and complicated.

The binary PSG and PSA based Mueller matrix polarimeter described in Figure 9.10 consisting of a pair of binary PSG and PSA can be used to fully characterize a wave plate; however, such an approach may be excessive, because a PSG/PSA pair not only can measure linear retardation and orientation important to a wave plate, but also other polarization properties, including circular retardation and biattenuation, which are negligible in optical wave plates. In this section, a dedicated scheme for wave plate analysis will be discussed (Chen et al. 2007), which is considerably simpler in construction and therefore costs less than a complete PSG/PSA pair. In addition to advantages intrinsic to MO rotators, including no moving parts, compact, fast, superior repeatability, and stability, this wave-plate analyzer (WPA) can accurately measure the retardation of the wave-plate and the orientation of optical axes simultaneously.

The MO-rotator based binary WPA is shown in Figure 9.26. Polarizer $\mathbf{P}_1$ is placed at the input of the system and is aligned 22.5° with respect to the fast axis of the QWP to generate a right-hand elliptically polarized light. The wave-plate under test (sample) is inserted in the middle slot. Two pairs of MO polarization rotators are placed before and after the sample respectively to rotate light beam's polarization. After passing through the sample, the rotators, and a second polarizer ($\mathbf{P}_2$), the light finally enters a photodetector (PD) for detecting the optical power changes corresponding to different polarization rotation combinations of the MO rotators. A low-noise transimpedance amplifier and 16-bit A/D converter are used to convert the detected photocurrent to digital signals for further data processing.

As discussed previously, each MO rotator can rotate the input SOP by a precise angle around 22.5° or −22.5° when a positive or negative magnetic field above saturation is applied respectively. Therefore, when two rotators rotate in the same direction, the net rotation is +45° or −45°. On the other hand, if two rotators rotate in opposite directions, the net polarization rotation is zero. One may see from Figure 9.26 that the power detected by photodetector will change when the rotation angles of different MO rotators are changed. For different retardation and axis orientation of the wave-plate sample under test, the changes in detected power are different for the same rotation combinations of MO rotators. As will be shown next, the retardation and axis orientation of the wave-plate sample can be uniquely determined from the power measurements corresponding to different rotation combinations of MO rotators.

Following the same procedure leading to Eq. (9.10a), the optical power detected by the photodetector can be written as

$$\begin{aligned} I_{out} = \frac{\kappa I_0}{2} \{ 1 &+ \left[-(\cos 2(\alpha - \theta_{wp})\cos 2(\beta + \theta_{wp}) + \sin 2(\alpha - \theta_{wp})\sin 2(\beta + \theta_{wp})\cos\Gamma \right] S_1 \\ &+ \left[\sin 2(\alpha - \theta_{wp})\cos 2(\beta + \theta_{wp}) + \cos 2(\alpha - \theta_{wp})\sin 2(\beta + \theta_{wp})\cos\Gamma \right] S_2 \\ &+ \sin 2(\beta + \theta_{wp})\sin\Gamma\, S_3 \} \end{aligned} \tag{9.34a}$$

where I_0 is the input optical power entering the analyzer, κ is the attenuation coefficient when taking into account of the insertion loss of all components, (S_1, S_2, S_3) are the normalized Stokes parameters of the light after passing through $\mathbf{P}_1$ and QWP, θ_{wp} is the orientation angle of the fast axis of the wave-plate sample with respect to the horizontal, and Γ is the retardation of the sample. The angles α and β are the total polarization rotation angles of the first pair of rotators (before the sample) and the second pair of rotators (after the sample), respectively, and they can be written as

$$\alpha = \sum_{j=1}^{2} -(-1)^{b_j}\theta(\lambda, T), \quad \beta = \sum_{j=3}^{4} -(-1)^{b_j}\theta(\lambda, T) \tag{9.34b}$$

$$\theta(\lambda, T) = 22.5^\circ + \Delta\theta(\lambda, T) \tag{9.10c}$$

$$\Delta\theta(\lambda, T) = \Delta\theta_0 + k_\lambda(\lambda - \lambda_0) + k_T(T - T_0) \tag{9.10d}$$

where b_j, θ, $\Delta\theta_0$, k_λ, k_T, λ_0, and T_0 are all defined in Eqs. (9.1a), (9.1b), (9.2a), and (9.2b).

Because of the binary nature of the MO rotators, I_{out} has 16 possible values. One can easily find by inspecting Figure 9.26 or Eq. (9.34b) that α and β each has three possible values (0, 2θ, -2θ). Therefore, I_{out} in Eq. (9.34a) has $3 \times 3 = 9$ different values for all 16 rotation combinations or logic states, as shown in Table 9.10. The rest are degenerate.

It is clear by inspecting Eqs. (9.34a) and (9.34b) that power I_{out} is a function of the following seven parameters: I_0, S_1, S_2, S_3, θ, θ_{wp}, and Γ. Therefore Eq. (9.34a) for different non-degenerate states can be rewritten as:

$$I_j = f_j(I_0, S_1, S_2, S_3, \theta, \theta_{wp}, \Gamma), \qquad j = 1, 2, \ldots 9 \tag{9.35}$$

where I_j is the output power of the WPA for the jth non-degenerate state and f_j represents the right hand side of Eq. (9.34a) for the jth non-degenerate state. Assuming the Stokes parameters (S_1, S_2, S_3) generated by $\mathbf{P}_1$ and QWP (Figure 9.26) are known, then input power I_0, rotation angle θ, retardation Γ, and axis orientation θ_{wp} of the sample can be calculated by numerically solving Eq. (9.35). Similar to the self-calibration procedure described in Section 9.2.2, Eq. (9.35) can be solved by numerically searching for the optimized values of I_0, θ, θ_{wp}, and Γ to minimize $\sum_{j=1}^{9}(f_j - I_{j,measured})^2$.

To ensure high accuracy measurement for wave-plates of all possible retardations and orientation angles, the SOP of the input light is chosen to be around (0.5, 0.5, 0.707) by aligning $\mathbf{P}_1$ and QWP with a relative angle of 22.5° to each other. However, because this input SOP will change slightly as

Table 9.10 Relationship of α, β, and logic states of rotators.

I_i	α	β	Logic states ($R_1R_2R_3R_4$)
I_1	0	2θ	0111, 1011
I_2	0	0	0101, 0110, 1001, 1010
I_3	0	-2θ	0100, 1000
I_4	2θ	2θ	1111
I_5	2θ	0	1101, 1110
I_6	2θ	-2θ	1100
I_7	-2θ	2θ	0011
I_8	-2θ	0	0001, 0010
I_9	-2θ	-2θ	0000

a function of wavelength and temperature due to the wavelength and temperature dependences of the QWP, it is necessary to calibrate the SOP of the input light for different wavelengths and temperatures for high accuracy measurements. Such an SOP calibration can be accomplished by simply measuring the I_{out} for all MO rotation combinations before inserting a wave-plate for measurement. When no sample is inserted, Eq. (9.34a) can be rewritten as

$$I_{out} = \frac{\kappa I_0}{2}\{1 + [-(cos2\alpha cos2\beta + sin2\alpha sin2\beta]\ S_1 + [sin2\alpha cos2\beta + cos2\alpha sin2\beta]\ S_2\} \tag{9.36a}$$

Because the light is totally polarized (ensured by polarizer $\mathbf{P}_1$ in Figure 9.26), the following relation holds

$$S_1^2 + S_2^2 + S_3^2 = 1 \quad (S_3 > 0) \tag{9.36b}$$

Therefore, after measuring the output powers under different non-degenerated states of MO rotators, (S_1, S_2, S_3) can be calculated by solving Eqs. (9.36a) and (9.36b) using the least-square-fitting algorithm.

The retardation and the axis orientation of a wave plate sample can be measured using following procedure: (i) The SOP of the input light is determined by measuring the I_{out} under nine non-degenerate logic states before inserting a wave plate sample using Eqs. (9.36a)and (9.36b). (ii) I_{out} under nine non-degenerate logic states are measured after the wave-plate sample is inserted; and (iii) the least-square-fitting algorithm is used to calculate the retardation and orientation angle of the sample using SOP obtained in step (i). The typical measured and fitted data are shown in Figure 9.27. All measurements are taken at 1550 nm and the photodetector outputs have been normalized using the input optical power. The nonlinear least-square-fitting results are shown in

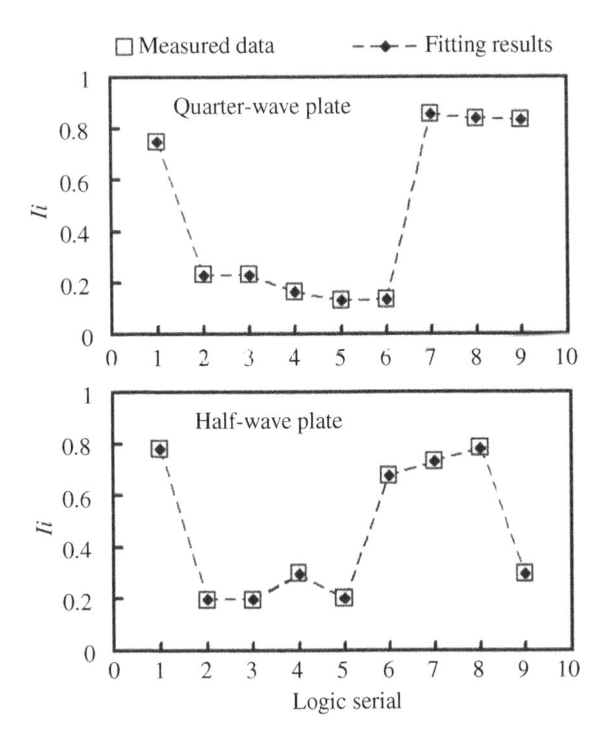

Figure 9.27 Typical measurement results with the normalized intensity for the QWP and HWP. The dash lines are for reference only (Chen et al. 2007).

Table 9.11. The error factor σ between the measured and fitted data is calculated as

$$\sigma = \sum_j \sqrt{\frac{(I_{j,Exp} - I_{j,Fit})^2 / I_0^2}{9}} \tag{9.37}$$

where $I_{j,Exp}$ is the experimentally measured output power of WPA for the jth non-degenerate state and $I_{j,Fit}$ is the calculated output power of WPA for the jth non-degenerate state using least-square-fitting. The low fitting error σ of 0.0016 shows that Eq. (9.34a) can accurately describe the WPA system. The measured retardations of 90.4° and 179.64° of the commercial QWP and HWP are consistent with the datasheet from the vendors (90° ± 0.7° and 180° ± 0.7°, respectively). Using the same setup, the retardation of the air (without any wave-plate sample) is measured to be close to zero (as low as 0.057°, see Table 9.11), which indicates the resolution of the measurement. In addition, one hundred measurements are taken to evaluate the repeatability and stability of the binary WPA system. The standard deviations of the measured retardation are 0.024° and 0.014° for the HWP and QWP, respectively. The corresponding standard deviation of the orientation angles of the optical axes is 0.070° and 0.014°, respectively.

Using a tunable laser, the MO-based binary WPA can easily acquire the wavelength dependence of both the retardation and orientation angles of a wave plate. The typical measured curves are shown in Figure 9.28. The slopes of the retardation are ~−0.064, −0.129, and −2.701°/nm for a zero-order quartz QWP (WP1), zero-order HWP (WP2), and multi-order HWP (WP3), respectively. According to the slopes and the chromatic dispersion of quartz crystal, the order of these wave plates can be calculated to be 0, 0, and 10, which are consistent with the data provided by vendors. The standard deviations of orientation angles are 0.12°, 0.045°, and 0.23° for WP1–WP3 in the wavelength range of 1500–1590 nm, respectively.

The measurement uncertainty of the WPA system is mainly from the non-perfect extinction ratio of polarizer, non-uniformity of the rotation angles of rotators, and uncertainty of SOP generated by $\mathbf{P}_1$ and QWP. Due to the non-linearity of Eq. (9.34a) and curve fitting, it is difficult to give a simple error analysis. Therefore, numerical error analysis according to Eqs. (9.34a) and (9.34b) is performed by assuming a certain device imperfection and the resulting simulated measurement uncertainties are listed in Table 9.12, where the state-dependent-loss is neglected. The simulated data are then analyzed with the same software used in the actual measurements. Therefore the differences between the fitted and theoretical values of the wave-plate represent the uncertainty of the system. The estimated maximum uncertainties of the retardation and the optical axis orientation (for $30° < \Gamma < 330°$) are less than 0.15°.

Table 9.11 Least-square-fitting results of different wave plates.

	HWP	QWP	Air
SOP of the input light	$S_1 = 0.494, S_2 = 0.514, S_3 = 0.701$		
Least-square-fit results			
Retardation of wave-plate Γ	179.68	90.41°	0.057°
Orientation angle of wave-plate θ_{wp}	−2.24°	89.51°	N/A
Rotation angle of rotators θ	21.77°	21.75°	N/A
Fitting error σ	0.0016	0.0015	0.0008

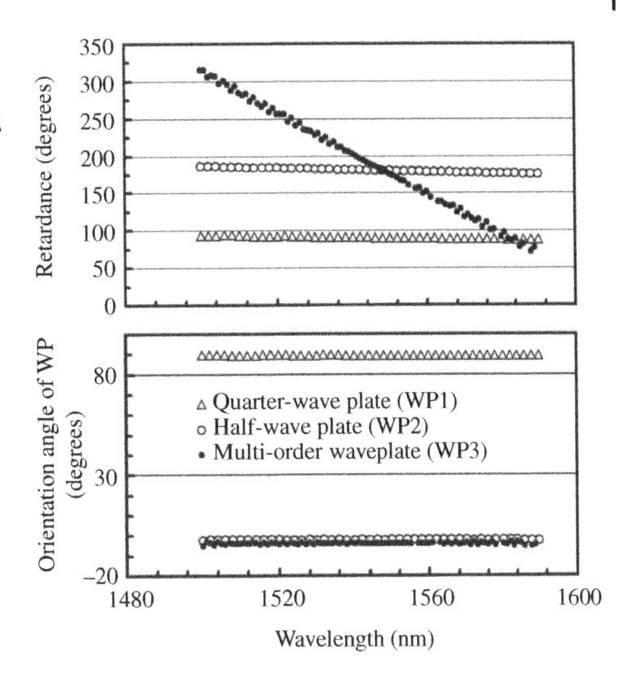

Figure 9.28 Typical wavelength dependence curves of the retardation and orientation angle of three different wave-plates measured with the binary WPA (Chen et al. 2007).

Table 9.12 Uncertainty estimates for the wave-plate analyzer.

Error source	Magnitude of error source	Retardation error	Axis orientation error ($330° > \Gamma > 30°$)
Polarizer (extinction ratio)	>50 dB	<0.01°	<0.01°
Non-uniformity of rotator	0.02°	<0.05°	<0.04°
Uncertainty of SOP generated by $\mathbf{P}_1$ and QWP	0.3%	<0.07°	<0.07°
All of above		<0.15°	<0.15°

In summary, a dedicated MO based binary WPA is described for simultaneously measuring the retardation, the wave-plate order, and the orientation angle of a wave-plate, with high accuracy, repeatability, and stability. The device provides a low cost and accurate alternative for fast wave-plate characterization.

9.5 PDL Measurement of a Multi-port Component Using a Binary PSG

Many optical fiber devices, such as $1 \times N$ couplers or arrayed waveguide gratings (AWGs), have N output ports. It is often time-consuming to measure the PDL values of all the N ports one by one with a single channel PDL meter described in Figure 8.19.

Figure 9.29 shows using a binary PSG described in Section 9.1 to simultaneously measure the PDL values of all the output ports of a multiport device as a function of wavelength. Both the 4-bit and the 6-bit PSG described in Figure 9.1 can be used. For simplicity, here we describe using the

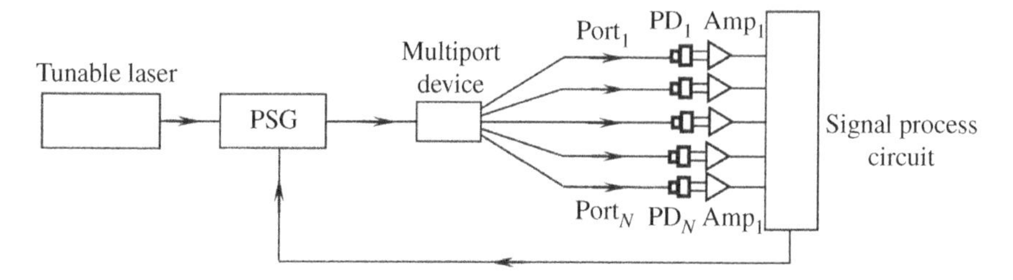

Figure 9.29 Measuring the PDL of a multi-port fiber optic device, such as a $1 \times N$ coupler or an arrayed waveguide grating (AWG) wavelength division multiplexer/demultiplexer (WDM).

4-bit PSG to measure the PDL of the multiport device. Using the logic in Table 9.1, one may set PSG sequentially at four distinctive SOPs: LVP, L+45P, L−45P, and RCP. At each SOP, the tunable laser scans through the wavelength range of interest before the power outputs are measured with the N photodetectors (PD) to get four sets of photocurrents $I_{ij}(\lambda)$, ($j = 1\ldots N$) as a function of wavelength with each PD, where $i = 1,2,3,4$ represents the four SOPs. In general, the Stokes vector $\vec{S}'_j(\lambda) = \left[S'_{j0}(\lambda), S'_{j1}(\lambda), S'_{j2}(\lambda), S'_{j3}(\lambda)\right]^T$ in port j after light passing through the multiport device can be expressed as

$$\mathbf{S}'_j(\lambda) = \mathbf{M}_j \mathbf{S}_i^{PSG}(\lambda) \tag{9.38a}$$

where $\mathbf{M}_j$ is the Mueller matrix of the jth port of the multiport device, and $\mathbf{S}_i^{PSG}$ is the ith Stokes vector generated by the PSG whose expression can be obtained using Eq. (9.5). Since only the first row elements of the Mueller matrix are used in computing the PDL, as expressed in Eq. (8.38c), only the first Stokes parameter $S'_{j0}(\lambda)$ in $\vec{S}'_j(\lambda)$ needs to be measured. The rest can be ignored. Specifically,

$$\begin{bmatrix} S'_{ij0} \\ \cdot \\ \cdot \\ \cdot \end{bmatrix} = \begin{bmatrix} m^j_{00} & m^j_{01} & m^j_{02} & m^j_{03} \\ \cdot & \cdot & \cdot & \cdot \\ \cdot & \cdot & \cdot & \cdot \\ \cdot & \cdot & \cdot & \cdot \end{bmatrix} \begin{bmatrix} S^{PSG}_{i0} \\ S^{PSG}_{i1} \\ S^{PSG}_{i2} \\ S^{PSG}_{i3} \end{bmatrix}, \quad i = 1,2,3,4, \quad j = 1,2,\ldots N \tag{9.38b}$$

The photocurrent generated in PD$_j$ with a responsivity η_j can be written as

$$I_{ij}(\lambda) = \eta_j S'_{ij0} = \eta_j \left(m^j_{00} S^{PSG}_{i0} + m^j_{01} S^{PSG}_{i1} + m^j_{02} S^{PSG}_{i2} + m^j_{03} S^{PSG}_{i3}\right), \quad i = 1,2,3,4 \tag{9.38c}$$

With the four equations corresponding to four distinctive SOPs ($\mathbf{S}_i^{PSG}$, $i = 1,2,3,4$), the Mueller matrix elements ($m^j_{00}, m^j_{01}, m^j_{02}, m^j_{03}$) for the jth port can be obtained. Finally, the PDL of the jth port of the device can be determined as

$$PDL_j = 10 log \frac{m^j_{00} + \sqrt{\left(m^j_{01}\right)^2 + \left(m^j_{02}\right)^2 + \left(m^j_{03}\right)^2}}{m^j_{00} - \sqrt{\left(m^j_{01}\right)^2 + \left(m^j_{02}\right)^2 + \left(m^j_{03}\right)^2}} \tag{9.39}$$

9.6 Multi-channel Binary PSA

The binary PSA described in Figure 9.7 is a single channel device. It is sometimes desirable to measure the SOP and DOP of multiple WDM channels simultaneously (Yan et al. 2006). Figures 9.30 shows a system configuration of such a multi-channel polarimeter consisting of a tunable laser or multiple wavelength source, a fiber pigtailed binary PSA, a wavelength division demultiplexer (AWG), and a photodetector at the end of each WDM channel. One may notice that in the binary PSA of Figure 9.7, a photodetector is directly attached at the end of the PSA package while here a collimator is used to couple light into an optical fiber connecting to an AWG for splitting the light into different WDM channels according to the wavelength. The procedure for obtaining the Stokes parameters $[S_{j0}, S_{j1}, S_{j2}, S_{j3}]$ of each WDM channel labeled j is the same as that for the single channel PSA described in Section 9.2. After the Stokes parameters are obtained, the DOP for channel j can be obtained as $DOP_j = \sqrt{S_{j1}^2 + S_{j2}^2 + S_{j3}^2}/S_{j0}$.

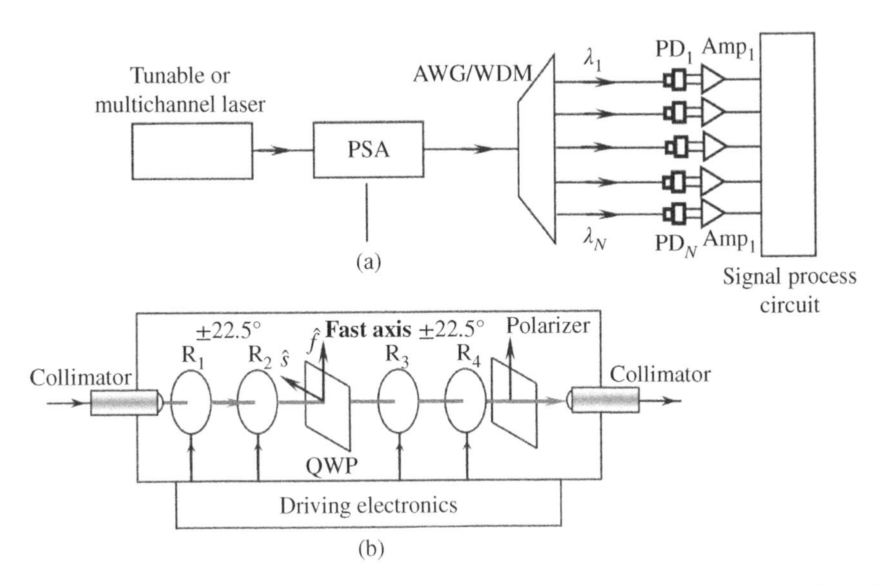

Figure 9.30 A multi-channel polarimeter made with a binary 4-bit PSA and an AWG to split light into different ports according to wavelengths before being detected by a photodetector at the end of each port (a) and the detailed construction of a 4-bit PSA (b).

9.7 WDM System Performance Monitoring Using a Multi-channel Binary PSA

As described in Section 8.5, a DOP meter, including a complete Stokes polarimeter or PSA, can be used to monitor the performance of an optical communication link, including the channel power, the SNR and the PMD induced depolarization (Petersson et al. 2004). However, there are many wavelength channels in a WDM communication system and it is expensive to deploy a PSA for each channel. As shown in Figure 9.31, such an issue can be resolved by using the multi-channel PSA described in Figure 9.30. Similar to the configuration in Figure 8.28, here only the SNR and power of each channel can be monitored.

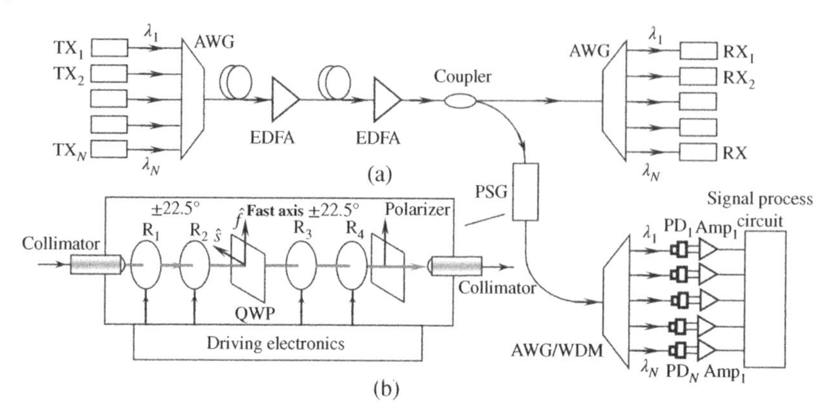

Figure 9.31 A multi-channel PSA is used for monitoring the SNR a WDM communication system (a) and the detailed construction of a 4-bit PSA (b).

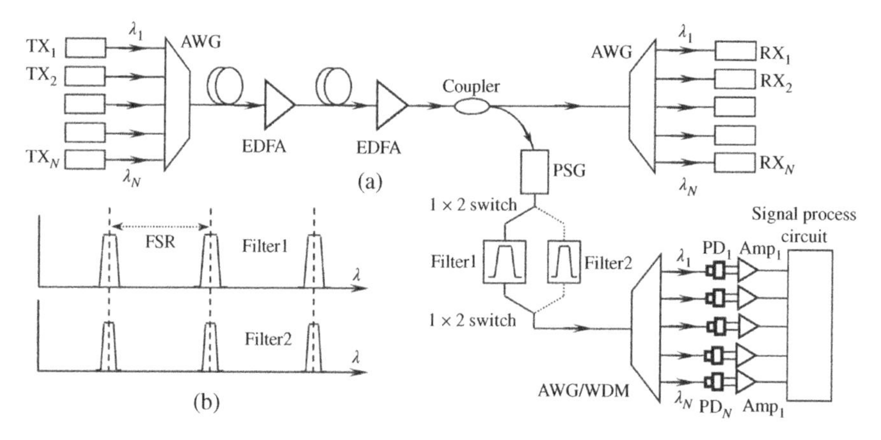

Figure 9.32 (a) Performance monitoring of a WDM system using the multi-channel PSA described in Figure 9.31. (b) Illustration of Filter1 and Filter2 having different bandwidths but the same periodicity equaling to the wavelength spacing of the WDM system to be monitored.

In order to monitor the PMD depolarization (Yan et al. 2005), a configuration of Figure 9.32 can be used (Yao 2013). Here two 1×2 switches are used to alternatively switch the signal into two filters of different bandwidths as shown in Figure 8.31. The filters are periodical with a free spectral range (FSR) equal to the channel spacing of the WDM system, as shown in Figure 9.32b. For each WDM channel, the procedure and algorithm for obtaining the SNR and PMD depolarization factor are the same as described in Figure 8.31. In particular, the switches first direct the optical signal through filter 1 before the MO rotators in the PSA to sequentially generate four distinctive SOPs according to Table 9.1. Afterward, the DOP of each wavelength channel is obtained, as discussed in Section 9.6, and is labeled as $DOP_{j1}, j = 1, 2 \ldots N$. Next, the switches direct the optical signal through filter 2 for the multi-channel PSA system to obtain another DOP for each wavelength channel, which is labeled as $DOP_{j2}, j = 1, 2 \ldots N$. The SNR_j and depolarization factor α_j for the wavelength channel j can be obtained using Eqs. (8.49a) and (8.49c).

9.A Appendix

Table 9.A.1 Logic table of all 64 polarization states of a 6-bit PSG.

Logic states of rotators ($R_6R_5R_4R_3R_2R_1$)	State #	SOP	Logic states of rotators ($R_1R_2R_3R_4R_5R_6$)	State #	SOP
000000	0	LCP	100000	32	LHP
000001	1	LCP	100001	33	L + 45P
000010	2	LCP	100010	34	L + 45P
000011	3	LCP	100011	35	LVP
000100	4	LCP	100100	36	L + 45P
000101	5	LCP	100101	37	LVP
000110	6	LCP	100110	38	LVP
000111	7	LCP	100111	39	L − 45P
001000	8	LCP	101000	40	L + 45P
001001	9	LCP	101001	41	LVP
001010	10	LCP	101010	42	LVP
001011	11	LCP	101011	43	L − 45P
001100	12	LCP	101100	44	LVP
001101	13	LCP	101101	45	L − 45P
001110	14	LCP	101110	46	L − 45P
001111	15	LCP	101111	47	LHP
010000	16	LHP	110000	48	RCP
010001	17	L + 45P	110001	49	RCP
010010	18	L + 45P	110010	50	RCP
010011	19	LVP	110011	51	RCP
010100	20	L + 45P	110100	52	RCP
010101	21	LVP	110101	53	RCP
010110	22	LVP	110110	54	RCP
010111	23	L − 45P	110111	55	RCP
011000	24	L + 45P	111000	56	RCP
011001	25	LVP	111001	57	RCP
011010	26	LVP	111010	58	RCP
011011	27	L − 45P	111011	59	RCP
011100	28	LVP	111100	60	RCP
011101	29	L − 45P	111101	61	RCP
011110	30	L − 45P	111110	62	RCP
011111	31	LHP	111111	63	RCP

9.B Appendix

The values of δ and φ at the two wavelengths λ_1 and λ_2 can be obtained to be $\delta(\lambda_1)$=$(2m\pi+2.1207)$ rad, $\delta(\lambda_2)$=$(2m\pi+1.9516)$ rad, $\varphi(\lambda_1)$=$(n\pi+0.7408)$ rad, and $\varphi(\lambda_2)$=$(n\pi+0.3884)$ rad, respectively, using Eqs. (9.25c), (9.25d), and the matrix M_R at the corresponding wavelength. Finally, the residual linear birefringence Δn_L and the circular birefringence Δn_C of the 10.4 m spun fiber at 20 °C can be obtained to be $\Delta n_L = 7.77 \times 10^{-6}$ and $\Delta n_C = 3.24 \times 10^{-5}$ using Eqs. (9.26a) and (9.26b).

Table 9.B.1 Measured Mueller matrix and its decompositions of the 10.4 m spun fiber (20 °C, $\lambda_1 = 1549.72$ nm).[a)]

$$\mathbf{M} = \begin{pmatrix} 1.0000 & 0.0010 & 0.0000 & -0.0020 \\ 0.0100 & 0.7010 & 0.5200 & 0.5270 \\ -0.0070 & -0.0140 & -0.7350 & 0.6740 \\ 0.0090 & 0.7140 & -0.4530 & -0.5490 \end{pmatrix} \qquad \mathbf{M}_\Delta = \begin{pmatrix} 1.0000 & 0.0000 & 0.0000 & 0.0000 \\ 0.0104 & 1.0193 & -0.0184 & -0.0122 \\ -0.0056 & -0.0184 & 0.9969 & -0.0236 \\ 0.0072 & -0.0122 & -0.0236 & 1.0078 \end{pmatrix}$$

$$\mathbf{M}_R = \begin{pmatrix} 1.0000 & 0.0000 & 0.0000 & 0.0000 \\ 0.0000 & 0.6966 & 0.4913 & 0.5229 \\ 0.0000 & 0.0158 & -0.7391 & 0.6734 \\ 0.0000 & 0.7173 & -0.4608 & -0.5226 \end{pmatrix} \qquad \mathbf{M}_D = \begin{pmatrix} 1.0000 & 0.0010 & 0.0000 & -0.0020 \\ 0.0010 & 1.0000 & 0.0000 & 0.0000 \\ 0.0000 & 0.0000 & 1.0000 & 0.0000 \\ -0.0020 & 0.0000 & 0.0000 & 1.0000 \end{pmatrix}$$

a) Xu et al. (2017).

Table 9.B.2 Measured Mueller matrix and its decompositions of the 10.4 m spun fiber (20 °C, $\lambda_2 = 1550.52$ nm).[a)]

$$\mathbf{M} = \begin{pmatrix} 1.0000 & 0.0030 & 0.0030 & -0.0030 \\ -0.0020 & 0.1540 & -0.3230 & 0.7200 \\ -0.0090 & -0.8130 & -0.5610 & -0.1240 \\ 0.0040 & 0.5660 & -0.6990 & -0.2980 \end{pmatrix} \qquad \mathbf{M}_\Delta = \begin{pmatrix} 1.0000 & 0.0000 & 0.0000 & 0.0000 \\ 0.0007 & 0.8019 & -0.0180 & 0.0561 \\ -0.0053 & -0.0180 & 0.9952 & -0.0155 \\ 0.0035 & 0.0561 & -0.0155 & 0.9457 \end{pmatrix}$$

$$\mathbf{M}_R = \begin{pmatrix} 1.0000 & 0.0000 & 0.0000 & 0.0000 \\ 0.0000 & 0.1335 & -0.3650 & 0.9214 \\ 0.0000 & -0.8055 & -0.5816 & -0.1137 \\ 0.0000 & 0.5774 & -0.7270 & -0.3717 \end{pmatrix} \qquad \mathbf{M}_D = \begin{pmatrix} 1.0000 & 0.0030 & 0.0030 & -0.0030 \\ 0.0030 & 1.0000 & 0.0000 & 0.0000 \\ 0.0030 & 0.0000 & 1.0000 & 0.0000 \\ -0.0030 & 0.0000 & 0.0000 & 1.0000 \end{pmatrix}$$

a) Xu et al. (2017).

References

Barlow, A., Ramskov Hansen, J., and Payne, D. (1981). Birefringence and polarisation mode-dispersion in spun single-mode fibers. *Appl. Opt.* 20: 2962–2968.

Bohnert, K., Gabus, P., Nehring, J., and Brandle, H. (2002). Temperature and vibration insensitive fiber-optic current sensor. *J. Lightwave Technol.* 20 (2): 267–276.

Chen, X., Yan, L., and Yao, X.S. (2007). Waveplate analyzer using binary magneto-optic rotators. *Opt. Express* 15 (20): 12989–12994.

Chipman, R.A. (1995). Polarimetry, Chapter 22. In: *Handbook of Optics*, 2e, vol. II (ed. M. Bass), 22.1–22.37. New York: McGraw-Hill.

Clarke, I.G. (1993). Temperature-stable spun elliptical-core optical-fiber current transducer. *Opt. Lett.* 18 (2): 158–160.

Collett, E. (2003). Chapter 9). *Polarized Light in Fiber Optics*, 183–226. Lincroft, NJ: The PolaWave Group.

Compain, E., Poirier, S., and Drevillon, B. (1999). General and self-consistent method for the calibration of polarization modulators, polarimeters, and Mueller-matrix ellipsometers. *Appl. Opt.* 38 (16): 3490–3502.

Corning Panda PM specialty fibers. (2004). PANDA PM Specialty Fibers. https://www.corning.com/media/worldwide/csm/documents/PANDA%20PM%20and%20RC%20PANDA%20Specialty%20Fiber.pdf (accessed 13 June 2022).

Craig, R.M., Gilbert, S.L., and Hale, P.D. (1998). High-resolution, nonmechanical approach to polarization dependent transmission measurements. *IEEE J. Lightwave Technol.* 16: 1285–1294.

Craig, R.M. (2013). Accurate spectral characterization of polarization-dependent loss. *J. Lightwave Technol.* 21 (2): 432–437.

Cruz, J., Andres, M., and Hernandez, M. (1996). Faraday effect in standard optical fibers: dispersion of the effective Verdet constant. *Appl. Opt.* 35 (6): 922–927.

De Martino, A., Kim, Y., Garcia-Caurel, E. et al. (2003). Optimized Mueller polarimeter with liquid crystals. *Opt. Lett.* 28 (8): 616–618.

Ding, Z., Meng, Z., Yao, X.S. et al. (2011). Accurate method for measuring the thermal coefficient of group birefringence of polarization-maintaining fibers. *Opt. Lett.* 36 (11): 2173–2175.

Dong, H., Shum, P., Yan, M. et al. (2006). Generalized Mueller matrix method for polarization mode dispersion measurement in a system with polarization-dependent loss or gain. *Opt. Express* 14: 5067–5072.

Feng, T., Shang, Y., Wang, X. et al. (2018). Distributed polarization analysis with binary polarization rotators for the accurate measurement of distance-resolved birefringence along a single-mode fiber. *Opt. Express* 26 (20): 25989–26002.

Fibercore Co. Ltd. (2022). Spun HiBi Fiber: Bow-tie spun fiber for Faraday effect current sensors. www.fibercore.com/product/spun-hibi-fiber (accessed 16 March 2022).

Goldstein, D. (2003). Stokes polarimetry. In: *Polarized Light*, 2e, Chapter 27, 533–557. New York: Marcel Dekker, Inc.

Goldstein, D.H. and Chipman, P.A. (1990). Error analysis of a Mueller matrix polarimeter. *J. Opt. Soc. Am. A* 7 (4): 693–700.

Harms, H., Papp, A., and Kempter, K. (1976). Magnetooptical properties of index-gradient optical fibers. *Appl. Opt.* 15 (3): 799–801.

Heffner, B. (1992a). Automated measurement of polarization mode dispersion using Jones matrix eigenanalysis. *IEEE Photonics Technol. Lett.* 4: 1066–1069.

Heffner, B. (1992b). Deterministic, analytically complete measurement of polarization-dependent transmission through optical devices. *IEEE Photonics Technol. Lett.* 4: 451–454.

IEC 60044-8. (2002). Instrument transformers – Part 8: Electronic current transformers, Chapters 4 and 12. International Standard.

iXblue Photonics. (2022). Spun fibers. www.ixblue.com/photonics-space/spun-fibers (accessed 16 March 2022).

Keysight N7788B/N7788BD data sheet. (2022). Optical component analyzer. https://www.keysight.com/us/en/assets/7018-01774/data-sheets/5989-8116.pdf (last accessed 11 March 2022).

Kumari, S. and Chakraborty, S. (2018). Study of different magneto-optic materials for current sensing applications. *J. Sens. Sens. Syst.* 7: 421–431.

Kurosawa, K., Yoshida, S., and Sakamoto, K. (1995). Polarization properties of the flint glass fibre. *J. Lightwave Technol.* 13 (7): 1378–1384.

Laming, R. and Payne, D. (1989). Electric current sensors employing spun highly birefringent optical fibers. *J. Lightwave Technol.* 7 (12): 2084–2094.

Li, L., Qian, J., and Payne, D. (1986). Current sensors using highly birefringent bow-tie fibres. *Electron. Lett* 22 (21): 1142–1144.

Lu, S. and Chipman, R. (1996). Interpretation of Mueller matrices based on polar decomposition. *J. Opt. Soc. Am. A* 13 (5): 1106–1113.

Manhas, S., Swami, M., Buddhiwant, P. et al. (2006). Mueller matrix approach for determination of optical rotation in chiral turbid media in backscattering geometry. *Opt. Express* 14 (1): 190–202.

Mochizuki, K., Namihira, Y., and Ejiri, Y. (1982). Birefringence variation with temperature in elliptically cladded single-mode fibers. *Appl. Opt.* 21 (23): 4223–4228.

Olivard, P., Gerligand, P., Jeune, B. et al. (1999). Measurement of optical fibre parameters using an optical polarimeter and Stokes–Mueller formalism. *J. Phys. D* 32 (14): 1618–1625.

Park, B.H., Pierce, M., Cense, B., and de Boer, J. (2014). Jones matrix analysis for a polarization-sensitive optical coherence tomography system using fiber-optic components. *Opt. Lett.* 29 (21): 2512–2514.

Peng, N., Huang, Y., Wang, S. et al. (2013). Fiber optic current sensor based on special spun highly birefringent fiber. *IEEE Photonics Technol. Lett.* 25 (17): 1668–1671.

Petersson, M., Sunnerud, H., Karlsson, M., and Olsson, B. (2004). Performance monitoring in optical networks using Stokes parameters. *IEEE Photonics Technol. Lett.* 16: 686–688.

Polynkin, P. and Blake, J. (2005). Polarization evolution in bent spun fiber. *J. Lightwave Technol.* 23 (11): 3815–3820.

Rashleigh, S.C. (1983a). Measurement of fiber birefringence by wavelength scanning: effect of dispersion. *Opt. Lett.* 8: 336–338.

Rashleigh, S.C. (1983b). Origins and control of polarization effects in single-mode fibers. *J. Lightwave Technol.* LT-1 (2): 312–331.

Ritari, T., Ludvigsen, H., Wegmuller, M. et al. (2004). Experimental study of polarization properties of highly birefringent photonic crystal fibers. *Opt. Express* 12: 5931–5939.

Rose, A. and Etzel, S. (1997). Verdet constant dispersion in annealed optical fiber current sensors. *J. Lightwave Technol.* 15 (5): 803–807.

Szafraniec, B. and Sanders, G. (1999). Theory of polarization evolution in interferometric fiber-optic depolarized gyros. *J. Lightwave Technol.* 17 (4): 579–590.

Short, S., De Arruda, J., Tselikov, A., and Blake, J. (1998). Elimination of birefringence induced scale factor errors in the in-line Sagnac interferometer current sensor. *J. Lightwave Technol.* 16 (10): 1844–1850.

Simon, A. and Ulrich, R. (1977). Evolution of polarization along a single-mode fiber. *Appl. Phys. Lett.* 31 (8): 517–520.

Smith, A.M. (1980). Birefringence induced by bends and twists in single-mode optical fiber. *Appl. Opt.* 19 (15): 2606–2611.

Smith, A.M. (1978). Polarization and magnetooptic properties of single-mode optical fiber. *Appl. Opt.* 17 (1): 52–56.

State Grid Corporation of China (2014). Performance testing programme of the electronic instrument transformer. Section 3: 5–12.

Tabor, W. and Chen, F. (1969). Electromagnetic propagation through materials possessing both Faraday rotation and birefringence: experiments with ytterbium orthoferrite. *J. Appl. Phys.* 40 (7): 2760–2765.

Tang, D., Rose, A., Day, G., and Etzel, S. (1991). Annealing of linear birefringence in single-mode fiber coils: application to optical fiber current sensors. *J. Lightwave Technol.* 9 (8): 1031–1037.

Thorlabs data sheet (2022). PMD/PDL analysis system PMD5000. https://www.thorlabs.com/catalogpages/obsolete/2016/PMD5000HDR-2.pdf (last accessed 11 March 2022).

Treviño-Martínez, T., Tentori, D., Ayala-díaz, C., and Mendieta-Jiménenz, F. (2005). Birefringence assessment of single-mode optical fibers. *Opt. Express* 13 (7): 2556–2563.

Ulrich, R., Rashleigh, S.C., and Eickhoff, W. (1980). Bending-induced birefringence in single-mode fibers. *Opt. Lett.* 5 (6): 273–275.

Urbanczyk, W. (1986). Optical transfer function for imaging systems which change the state of light polarization. *Opt. Acta* 33 (1): 53–62.

Vojna, D., Slezák, O., Lucianetti, A., and Mocek, T. (2019). Verdet constant of magneto-active materials developed for high-power Faraday devices. *Appl. Sci.* 9: 3160.

Walker, N. and Walker, G. (1990). Polarization control for coherent communications. *J. Lightwave Technol.* 8: 438–458.

Wang, S.X. and Weiner, A. (2003). Fast multi-wavelength polarimeter for polarization mode dispersion compensation systems. Digest of the LEOS Summer Topical Meetings. pp. WB2.4/67–WB2.4/68.

Wang, S.X. and Weiner, A. (2004). Fast wavelength-parallel polarimeter for broad band optical networks. *Opt. Lett.* 29 (9): 923–925.

Williams, P.A. (1999). Rotating-wave-plate Stokes polarimeter for differential group delay measurements of polarization-mode dispersion. *Appl. Opt.* 38 (31): 6508–6515.

Williams, P.A. (2002). PMD measurement techniques – avoiding measurement pitfalls, *Venice Summer School on Polarization Mode Dispersion*, Venice, Italy (24–26 June 2002).

Williams, P.A., Etzel, S., Kofler, J., and Wang, C. (2002). Standard reference material 2538 for polarization-mode dispersion (non-mode coupled). NIST Special Publication 260-145.

Willner, A.E., Motaghian Nezam, S.M.R., Yan, L.S. et al. (2004). Monitoring and control of polarization-related impairments in optical fiber systems. *IEEE J. Lightwave Technol.* 22 (1): 106–125.

Xu, Z., Yao, X.S., Ding, Z. et al. (2017). Accurate measurements of circular and residual linear birefringences of spun fibers using binary polarization rotators. *Opt. Express* 25 (24): 30780–30792.

Yan, L., Yao, X.S., Shi, Y., and Willner, A. (2005). Simultaneous monitoring of both optical signal to noise ratio and polarization mode dispersion using polarization scrambling and polarization beam splitting. *J. Lightwave Technol.* 23 (10): 3290–3294.

Yan, L., Yao, X.S., Yu, C. et al. (2006). High-speed and highly repeatable polarization-state analyzer for 40-Gb/s system performance monitoring. *IEEE Photonics Technol. Lett.* 18 (4): 643–645.

Yao, X.S., Yan, L., and Shi, Y. (2005). Highly repeatable all-solid-state polarization-state generator. *Opt. Lett.* 30 (11): 1324–1327.

Yao, X.S., Chen, X., and Yan, L. (2006). Self-calibrating binary polarization analyzer. *Opt. Lett.* 31 (13): 1948–1950.

Yao, X.S., Chen, X., and Liu, T. (2010). High accuracy polarization measurements using binary polarization rotators. *Opt. Express* 18 (7): 6667–6685.

Yao, X.S. (2013). Monitoring polarization mode dispersion and signal to noise ratio of optical signals based on polarization analysis. US Patent 8,787,755, filed 15 April 2013 and issued 22 July 2014.

Yao, X.S. (2019). Techniques to ensure high quality fiber optic gyro coil production, Chapter 11. In: *Design and Development of Fiber Optic Gyroscopes* (ed. E. Udd and M.J.F. Digonnet), 217–261. Bellingham, Washington, DC: SPIE Press.

Yao, P., Chen, X., Hao, P. et al. (2021). Introduction and measurement of the effective Verdet constant of spun optical fibers. *Opt. Express* 29 (15): 23315–23330.

Zadok, A., Simon, N., and Eyal, A. (2004). The dependence of the output stokes parameters on the state of an arbitrarily located polarization controller in PMD mitigation schemes. *J. Lightwave Technol.* 22: 1533–1538.

10

Distributed Polarization Analysis and Its Applications

The polarization analysis systems described in Chapters 8 and 9 can only measure the accumulative polarization effects of a medium (i.e. a piece of optical fiber) on an optical beam, such as the changes in the state of polarization (SOP) and the degree of polarization (DOP) after it passes through the whole length of the matter, and therefore can only help to obtain the average polarization properties of the medium, such as those characterized by parameters of birefringence, diattenuation, or photoelasticity. Because such parameters are vectors or even tensors with varying orientations along the optical path, the "average" cannot be used to infer the local values of the parameters. For example, in a single mode fiber (SMF), the birefringence (often referred to as differential group delay [DGD]) and the diattenuation (often referred to as polarization dependent loss [PDL]) often vary rapidly along the fiber in both their orientations and magnitudes. The fiber may have strong local birefringence or diattenuation; however, at some moment in time, the average birefringence or diattenuation may be close to zero.

In addition, the polarization extinction ratio (PER) meter discussed in Chapter 8 can only characterize the ratio of the total power accumulated in the fast axis over that in the slow axis in a polarization-maintaining (PM) fiber at the output of the fiber. It is unable to characterize the local PER or polarization crosstalk describing how much light is coupled from the slow axis to the fast axis induced by the local internal or external stresses, as will be discussed later. For some applications, such as fiber optical gyroscopes (FOGs), the ability to identify the location and magnitude of the polarization crosstalk is important for the improvement of the PM fiber coils for FOGs.

Distributed polarization analysis (DPA) refers to methods for measuring some or all polarization related parameters, such as SOP, DOP, PER, birefringence, polarization mode dispersion (PMD), or PDL, as a function of distance. That is, not only a DPA is able to determine the Stokes parameters and Mueller matrix elements at the output end of a medium (such as an optical fiber), but also able to measure the local Stokes parameters and Mueller matrix elements at any section inside the medium or fiber. There are three types of DPA methods in general, categorized by how the distance resolved polarization information is obtained, including the time domain distributed polarization analysis (TD-DPA), the coherence domain distributed polarization analysis (CD-DPA), and the frequency domain distributed polarization analysis (FD-DPA).

As will be described in more detail in the following sections, the polarization sensitive optical time domain reflectometer (P-OTDR) is a time domain partial DPA capable of measuring the local PMD or birefringence along an optical fiber, while the distributed polarization crosstalk analyzer (DPXA) is a coherence domain partial DPA designed for measuring the local polarization crosstalk or PER in a PM fiber. The P-OFDR is a frequency domain partial DPA for measuring the local birefringence in an SMF with high distance resolution, and finally the polarization analyzing

Polarization Measurement and Control in Optical Fiber Communication and Sensor Systems, First Edition.
X. Steve Yao and Xiaojun (James) Chen.

optical frequency domain reflectometer (PA-OFDR) is a complete Mueller matrix distributed polarimeter capable of determining the distance resolved SOP and DOP of the back scattered light by Rayleigh scattering inside an SMF, which in turn can be used to measure the distance resolved birefringence (PMD) and PDL along an SMF. In essence, a DPA is a polarimeter or a polarization analyzer combined with a mechanism or technique to identify the location z and resolve the local polarization information along an optical fiber. The mechanisms include time of flight (ToF), optical frequency domain reflectometry (OFDR), and white light interferometry.

In this chapter, we will concentrate on the CD-DPA and FD-DPA, although the TD-DPA will also be briefly discussed. In addition, we will discuss how each DPA system can be used for different distributed sensing and measurement applications.

10.1 Distributed Polarization Crosstalk Analysis and Its Applications

10.1.1 Polarization Crosstalk in PM Fibers

Polarization crosstalk in a PM fiber arises from three principal causes. First, fiber axis misalignment at fiber connection interfaces, such as connectors or fusion splices shown on the left side of Figure 10.1a, typically causes extremely localized, large-amplitude crosstalk on the order of −35 dB or more, depending on the misalignment angle θ. As shown in Figure 10.1b, for a light wave originally polarized along the slow axis s of the PM fiber, a small portion of amplitude equaling to $sin\theta$ ($sin^2\theta$ of power) will be coupled to the fast axis $\hat{\mathbf{f}}'$ of the connecting fiber:

$$E' = E_0(cos\theta\hat{\mathbf{s}}' + sin\theta\hat{\mathbf{f}}' e^{i\Delta\varphi(L)}] \quad (10.1a)$$

$$\Delta\varphi(L) = \frac{2\pi\Delta nl}{\lambda} = \frac{2\pi l}{L_{b0}} \quad (10.1b)$$

where E_0 is the amplitude of light in the original PM fiber, $\Delta\varphi(L)$ is the retardation, l is the distance traveled in the connecting fiber, Δn is the birefringence of the PM fiber, and $L_{b0} = \lambda/\Delta n$ is the beat length. The power coupling coefficient $h_{s\to f'}$ or crosstalk from $\hat{\mathbf{s}}'$ to $\hat{\mathbf{f}}'$ can be obtained from Eq. (10.1a) as:

$$h_{s\to f'} = sin^2\theta \quad (10.1c)$$

For example, for an angle misalignment of 1°, the crosstalk in dB ($10logsin^2\theta$) is −35 dB.

Second, external mechanical stresses on a sections of the fiber, such as fiber bending, fiber crossing, fiber squeezing, or pressure on the fiber, can cause polarization cross coupling. All such external stresses are equivalent to a transversal force $\mathbf{F}_{ex}$ applied to a section of fiber, as shown in Figure 10.1b, which will be combined with the internal stress $\mathbf{F}_{in}$ induced by the stress rods of the PM fiber to produce a combined stress $\mathbf{F}_c$ rotated from the slow axis by an angle of θ. This combined stress then induces a combined birefringence via the photoelastic effect with a new slow axis in the direction of the combined stress. Therefore, the fiber segment under the transversal force $\mathbf{F}_{ex}$ has a modified birefringence with its slow axis rotated θ degrees from the original slow axis, as shown in Figure 10.1c. This is equivalent to the case that a center piece of PM fiber is sandwiched between two PM fibers, with its slow axis misaligned from the slow axes of the two PM fibers on each side, as shown in Figure 10.1d,e.

Similar to the case of Figure 10.1a, because of the misalignment, light in the slow axis of the fiber on the left is coupled into the fast axis of the center fiber, in which the optical field can be

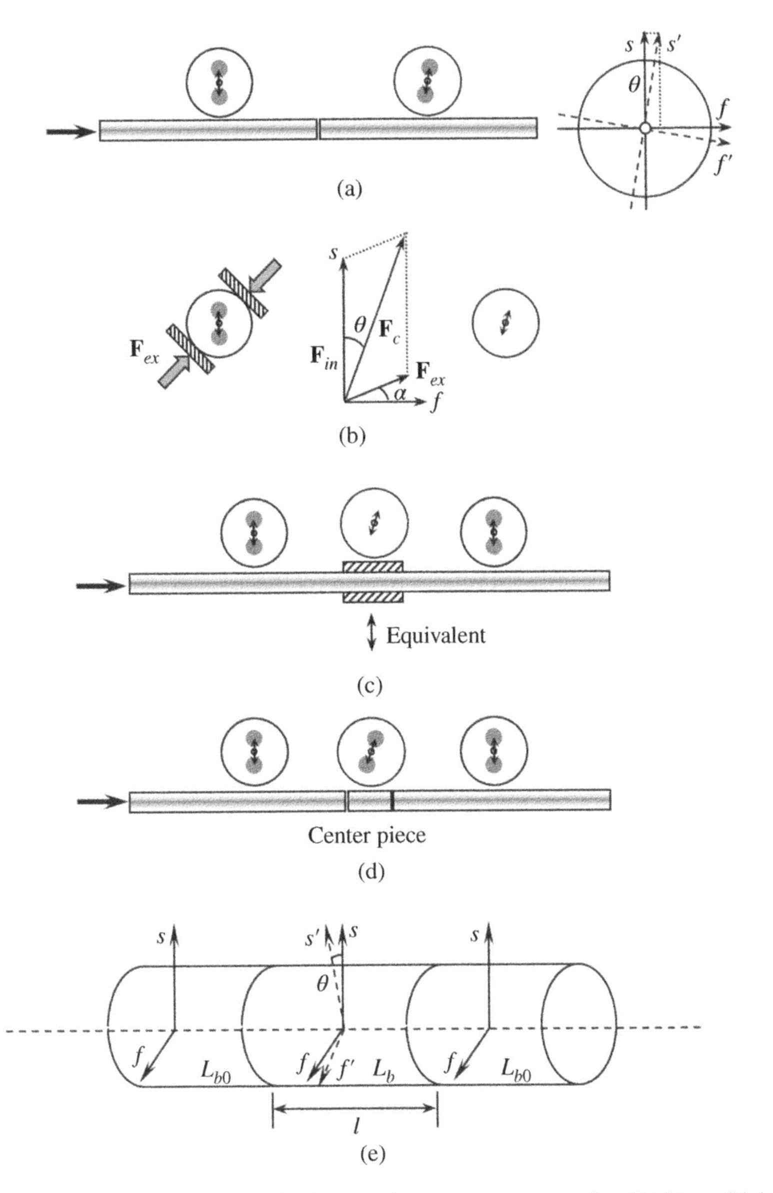

Figure 10.1 (a) Slow axis misalignment between two connecting PM fibers. (b) A transversal force $\mathbf{F}_{ex}$ applied to the PM at an angle from the slow axis of the PM fiber, resulting in the rotation of the combined stress on the fiber and the effective slow axis. (c) Transversal force applied on a PM fiber segment to cause an effective slow axis rotation. (d) The equivalent of (c) with the slow axis of a small segment of PM fiber misaligned with respect to those of two fiber segments on each side. (e) Diagram showing that the slow axis of a center piece of PM fiber is rotated with respective to the left and right pieces for the case of (d).

expressed by Eq. (10.1a). As can be seen, the SOP in the fiber is a periodic function of the distance l of light traveled in the center piece. At the end of the center piece, the light will be coupled into the PM fiber on the right-hand side whose slow axis is misaligned $-\theta$ with respect to the slow axis of the center piece. It can be shown that with this misaligned center piece at the center, polarization crosstalk between the slow and fast axes in the PM fiber on the right-hand side will be induced, with a cross-coupling ratio of (Xu et al. 2009a)

$$h_{s\to f} = \left| sin\theta cos\theta \left(1 - e^{-i2\pi l/L_{b0}}\right)\right|^2 = sin^2(2\theta)sin^2(\pi l/L_{b0}) \tag{10.2a}$$

For the case of Figure 10.1b,c, the misalignment angle θ depends on both the direction and the size of the external force. In general, the transversal force induced coupling ratio or polarization crosstalk h, which is defined as the ratio between the coupled power in the fast axis and the original power in the slow axis, can be expressed as (Chua and Chen 1989; Zhang et al. 2018; Tang 2005)

$$h = F^2 sin^2 2\alpha \cdot \left\{ \frac{sin\left[\pi\sqrt{1+F^2+2Fcos2\alpha}(l/L_{b0})\right]}{\sqrt{1+F^2+2Fcos2\alpha}} \right\}^2 \tag{10.2b}$$

where L_{b0} is the beat length of PM fiber of the unstressed section and F is the normalized force given by

$$F = \frac{2n^3 L_{b0} f(1+\mu)(P_{12} - P_{11})}{\pi\lambda r E} \tag{10.2c}$$

In Eq. (10.2c), r is the radius of the PM fiber, n is the refractive index of the fast axis, f is the magnitude of the force applied to the fiber per unit length ($f = \mathbf{F}_{ex}/l$), α is the angle between the applied force f and the fast axis of the PM fiber (the force angle), μ is the Poisson coefficient, p_{12} and p_{11} are the optical strain coefficients, λ is the wavelength of the light source, and E is Young's modulus of the fiber. Using the parameters for fused silica, Eq. (10.2c) can be simplified to (Tang 2005):

$$F = \frac{5.46 L_{b0}}{r\lambda} f \tag{10.2d}$$

The stress induced crosstalks are mostly complicated, some occurring at sharp points in space, some occurring gradually along a length of fiber, with varied amplitudes, depending on the stress orientations with respect to the slow axis and the stress strengths, as will be shown in Section 10.1.3.

Finally, PM fiber imperfections, such as local birefringence variations, internal shape variations, or internal stress, may cause polarization crosstalk. Such polarization cross-coupling is generally small in amplitude and occurs gradually along a certain length of the PM fiber, as will be discussed in connection with the polarization crosstalk measurement results next.

10.1.2 Description of Distributed Polarization Crosstalk Analyzer (DPXA)

Figure 10.2 illustrates a basic configuration for a DPXA (Li et al. 2015; Yao 2019; Martin et al. 1991). A broadband light source, such as a polarized super luminescent emitting diode source (SLED) with a very short coherence length (~25 μm, corresponding to a 3-dB Gaussian line width of 30 nm) is coupled into the slow axis of a PM fiber under test (FUT) (point A of inset in Figure 10.2). Assume at point B, a polarization crosstalk is induced by an external disturbance and then some lights are coupled into the fast axis of the PM fiber with a coupling coefficient parameter $h = I_1/I_2$, where I_1 and I_2 are the light intensities in the fast and slow axes of the PM fiber, respectively. Because the polarized lights along the fast axis travel faster than those along the slow axis, at output of the fiber the faster light component will be ahead of the slow component by $\Delta Z = \Delta n Z$, where ΔZ is an optical path length difference, Δn is a group birefringence of the PM fiber, and Z is the fiber length between the point where the crosstalk occurs (B) and the output end (point C). A polarizer oriented at 45° to the slow axis of the PM FUT was placed at the end of the fiber. Polarization components from both slow and fast axes were projected onto a same direction of the linear polarizer axis so as

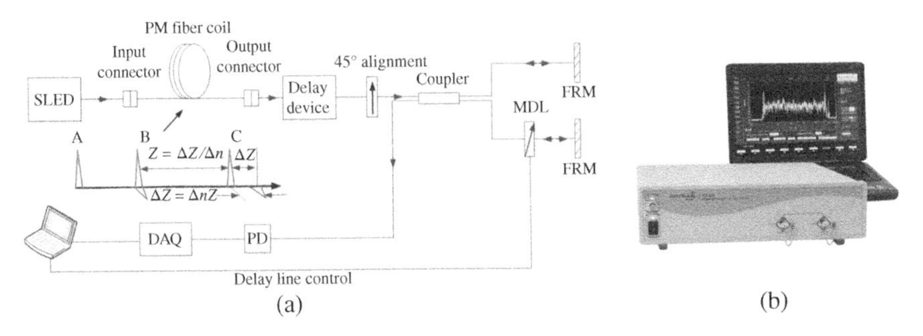

Figure 10.2 (a) Illustration of a ghost-peak-free distributed polarization crosstalk analyzer using a scanning white light Michelson interferometer. The inset shows the delay relation between the original and crosstalk components. Light with a short coherence length traveling in the fiber is polarized along its slow axis at input point A. Crosstalk is induced by a stress at point B where a small portion of light is coupled into fiber's fast axis. A relative delay at the output point C between the two polarization components is ΔZ. The location Z of crosstalk point B can be obtained from a measurement of ΔZ. FRM, MDL, PD, and DAQ are Faraday rotation mirror, motorized delay line, photodetector, and data acquisition card, respectively. (b) A photo of a commercial ghost-peak free DPXA product developed by the authors (Source: Courtesy of General Photonics / Luna Innovations.). With such an instrument, a spatial resolution of 6 cm, a measurement range up to 3.4 km, a crosstalk measurement sensitivity down to −80 dB, and a crosstalk dynamic range of 75 dB can be routinely achieved.

to produce interference pattern between those two components in a scanning Michelson interferometer. When the relative optical path length is scanned, an interference peak appears whenever these two polarization components are overlapped in the space but disappears when they are separated more than a coherence length of light source (i.e. SLED). Then the group birefringence Δn of PM FUT between two positions B and C can be calculated as follows:

$$\Delta n = \Delta Z/Z \tag{10.3}$$

It is evident from Eq. (10.3) that the accuracy of Δn depends on the measurement accuracies of both ΔZ and Z.

Note that the illustration in Figure 10.2 assumes only one polarization crosstalk point along the fiber. If there are multiple polarization crosstalk points, second order interference peaks will occur. That is, the light in the fast axis caused from the coupling at a crosstalk point will couple back to the slow axis at the subsequent crosstalk points down the fiber. As shown in Figure 10.3a, consider a situation where there are three coupling points X_1, X_2, and X_3 along the PM fiber, and the light input to the PM fiber has no fast axis component and is polarized along the slow axis of the PM fiber. At each coupling point, light is coupled not only from the polarization mode along the slow axis to the polarization mode along the fast axis, but also from the polarization mode along the fast axis to the polarization mode along the slow axis. As a result of this coupling, the resulted wave packet series output by the PM fiber includes wave packet caused by multiple couplings. As shown in Figure 10.3b, four wave packets including zero order coupling S_0 (no coupling) and the second order couplings S_{12}, S_{23}, and S_{23} emerging at the output are aligned to the slow axis of the PM fiber and three main packets f_1, f_2, and f_3 emerging at the output are aligned to the fast axis of the PM fiber, where they are generated by coupling from the slow axis to the fast axis (the first order coupling) at points X_1, X_2, and X_3, respectively.

After passing through the 45° oriented polarizer, the wave packets aligned to the slow and fast axes were projected onto a same direction of the linear polarizer axis, as shown in Figure 10.3c. When this mixed light is input to the interferometer, a series of interference peaks can be observed as the delay in one arm of the interferometer is changed and these second order couplings will cause

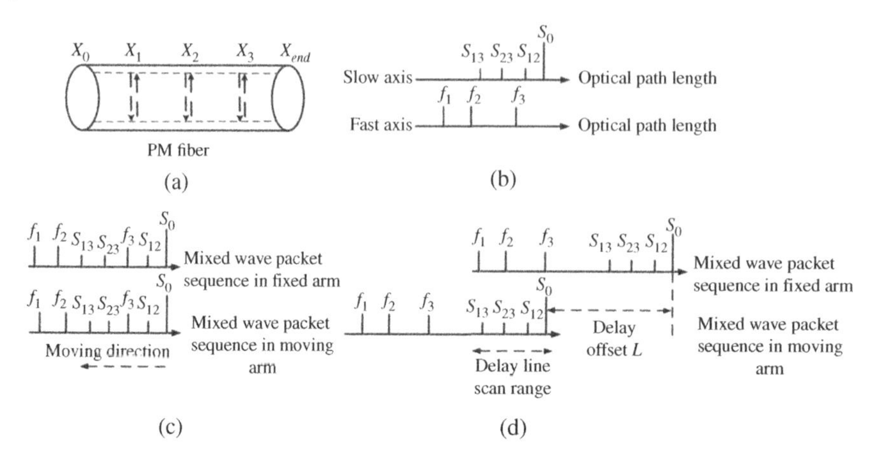

Figure 10.3 (a) Illustration of polarization coupling at locations X_1, X_2, and X_3 along the PM fiber; (b) the wave packet sequences polarized along the slow (denoted by S) and fast axes at output of the PM fiber (denoted by f); (c) Wave pockets in the two interferometer arms after light passing through the 45° oriented analyzer, where the wave packets aligned to the slow and fast axes are mixed together. When this mixed light is input to the interferometer in Figure 10.2, a series of interference peaks, including multiple ghost peaks, will be observed as the delay in one arm of the interferometer is changed; (d) Wave pockets in the two interferometer arms after the Delay device in Figure 10.2 is inserted between the PM fiber's output and the 45° polarizer's input. This differential delay adds an additional delay between the slow axis and the fast axis of the PM fiber. Only the first order interference peaks between S_0 and f_i and third order interference peaks between f_i and S_{ij} can be generated.

ghost crosstalk peaks and result in confusions in simple white light interferometers described in (Martin et al. 1991; Tang et al. 2006, 2007). As shown in Figure 10.2, we use a differential group delay (Delay Device) inside a DPXA to remove all the ghost crosstalk peaks from the second order couplings (Chen and Yao 2010; Yao 2019), making it possible to accurately identify and measure a large numbers of polarization crosstalks along a PM fiber without ambiguity. In particular, as illustrated in Figure 10.3d, the Delay device adds an additional delay L between polarization components in the slow and fast axes, and the delay L in vacuum should be longer than ΔZ where the additional delay L is added to the light polarized along the slow axis of the PM fiber. There, the two wave packet sequences from the fast-axis and slow-axis are separated in time (or space) after the light passes through the analyzer. If we preset the same delay offset between the fixed and moving arms in the interferometer and restrict the range of the variable delay line in the moving arm, the undesired zero order (S_0 with S_0, S_{ij} with S_{ij}, and f_i with f_i) and the second order (S_0 with S_{ij}, and f_i with f_j) interference signals (ghost interference peaks) will not be generated as the delay line scans. Only the desired first order interference signals (S_0 with f_i) and the much weaker third order interference signals (f_i with S_{nm}) will be present. Note that the third order peaks are negligible (less than −75 dB) if the first order coupling f_i is less than −25 dB. More detailed description of such ghost-peak-free DPXA can be found in Chen and Yao (2010).

Figure 10.4 shows a typical polarization crosstalk curve as a function of delay (from the motorized delay line [MDL] in Figure 10.2) measured with a DPXA. The peaks marked Input and Output are due to the birefringence axis misalignments at the input and output connectors. There are also four crosstalk peaks from the axis misalignment at four fusion splice points. The peaks labeled 1–6 are the transversal force induced crosstalks by placing different weights on the fiber at different locations along the PM fiber. No ghost peaks can be seen even with several high crosstalk peaks present.

10.1.3 Identification of Causes for Polarization Crosstalks from Measurement Results

In general, a DPXA can accurately measure the strengths of polarization cross talks occurring at different locations with a spatial resolution of a few center meters. However, it is sometime difficult to interpret the measurement results because in practice a large amount of polarization crosstalk peaks are present along the fiber with different amplitudes and different shapes at different locations. In practice, one may want to identify the causes of the cross talks from the measurement results, or at least get educated guesses of what the causes might be based on the shape and strengths of measured crosstalks at these locations. As will be discussed next, the shape of a crosstalk peak is related to how the crosstalk is induced and therefore can be used to identify the cause for its formation.

10.1.3.1 Crosstalk Caused by Discrete Polarization Coupling Points

It refers to the polarization coupling induced by a sharp stress, a splice point, or multiple stress/splice points separated much larger than the resolution of the measurement instrument, as shown in Figure 10.5a (left section). Such discrete couplings have distinctive peaks in the crosstalk measurements, with the width determined by the spatial resolution of the instrument, as shown in Figure 10.5b, also on the left hand side. The curve in Figure 10.4 is a good example of crosstalk induced by discrete polarization coupling. For this type of coupling, it is meaningful to give a peak crosstalk value for each coupling point, as will be discussed in Figure 10.6.

10.1.3.2 Crosstalk Caused by Continuous Polarization Coupling

It refers to the polarization coupling occurring in a section of fiber induced by a line stress or by fiber internal imperfections, with the section length comparable to or larger than the resolution of the measurement instrument, as shown in Figure 10.5a (middle section). The crosstalk measurement result of such continuous coupling is a broad hump with a width and shape mainly determined by the length of the section stress, as shown in Figure 10.5b. In general, a crosstalk caused by a section of fiber imperfection is very small in amplitude, on the order of −60 dB or lower. Examples of such polarization crosstalk are shown in Figures 10.6 and 10.7. As will be explained, it is not meaningful to give a peak crosstalk value.

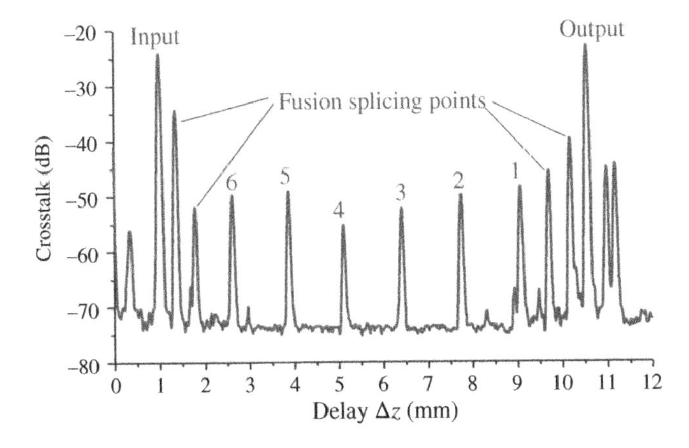

Figure 10.4 A typical polarization crosstalk curve measured with a distributed polarization crosstalk analyzer.

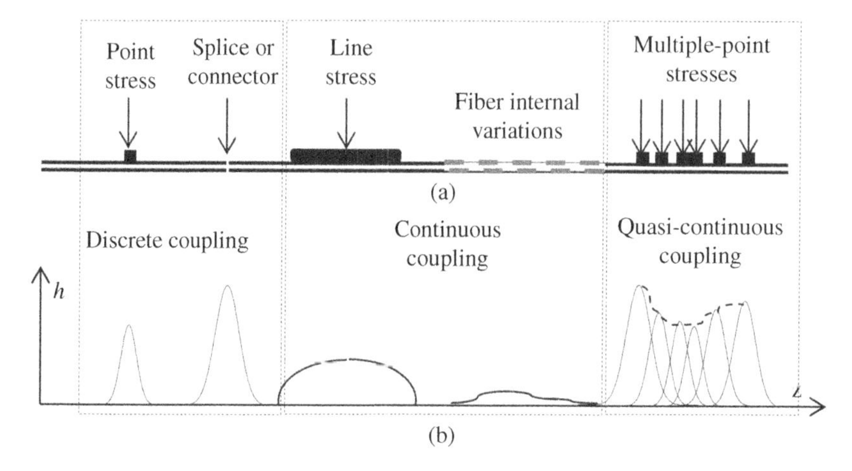

Figure 10.5 Illustration of different types of polarization cross talks. Left: discrete polarization crosstalk peaks induced by a point stress or a splice (a). The shape of peak is a Gaussian, determined by the coherence function of the light source (b). The spatial resolution is also determined by the width of the coherence function. Center: continuous polarization crosstalks induced by a line stress and internal fiber imperfection. Right: quasi-continuous crosstalk induced by densely packed multiple stress points spaced on the order or less than the resolution of the instrument (Yao 2019).

10.1.3.3 Crosstalk Caused by Quasi-Continuous Coupling

It refers to the polarization coupling induced by multiple stress points with spacing less than the resolution of the measurement instrument, as shown in Figure 10.5a (right section). Such a polarization coupling will appear as a broad hump with height variations in polarization crosstalk measurements, with a width and shape determined by the number of stress points, their relative positions, and their relative strengths, as shown in Figure 10.5b. A quasi-continuous coupling cannot be distinguished from continuous coupling. The actual measurement data for such coupling are shown in Figures 10.6 and 10.7. Similar to the continuous coupling, for quasi-continuous coupling, it is not meaningful to give a peak crosstalk value. Accumulative PER between two space points can be used to characterize the polarization coupling, as will be explained in Figure 10.7b.

Figure 10.6a shows the measurement result of a 340 m long PM fiber coil made for fiber optic gyroscope. As one can see that there are many polarization crosstalk spikes distributed along the fiber length, appearing to be noises. Some are high, approaching −50 dB, many are low, below −70 dB. If one expands the view of a section, such as that labeled “A,” one sees many clearly defined polarization crosstalk peaks of different heights. The instrument software enables one to obtain the exact crosstalk peak values above a threshold set by a user, with the exact peak values listed in the crosstalk table on the instrument display screen, as shown in Figure 10.6b. As one can see all the crosstalk peaks above −60 dB are of discrete coupling with clearly defined sharp shape, while some other peaks are due to quasi-continuous coupling because they appear to be formed by two or more peaks closely packed together, with separations less than the resolution of the instrument.

Figure 10.7a shows the expanded view of the section labeled “B” in Figure 10.6a. Similarly, there are many the discrete peaks, as well as some continuous/quasi-continuous coupling peaks shown in the dashed box. For the continuous polarization coupling, it is not meaningful to give a peak crosstalk value. However, one is able to obtain the accumulative coupling occurring in a section of fiber by defining the starting and ending positions along the fiber using two cursors equipped

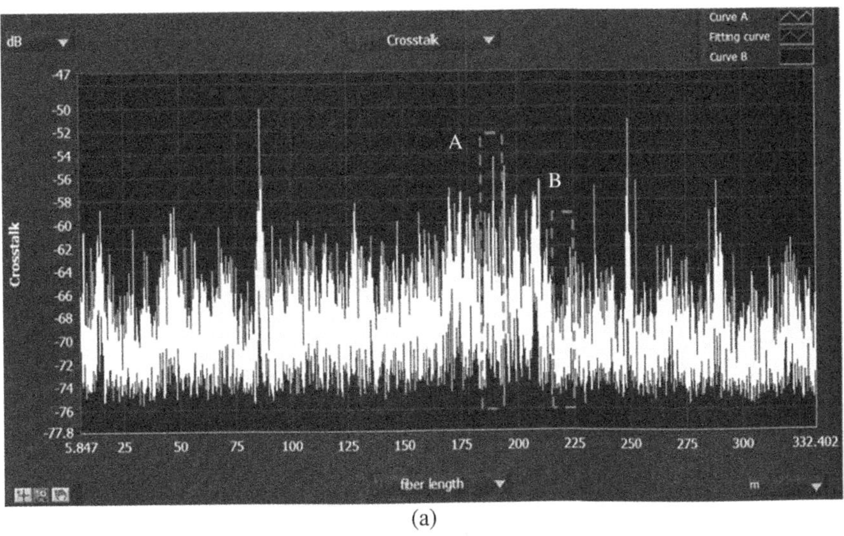

(a)

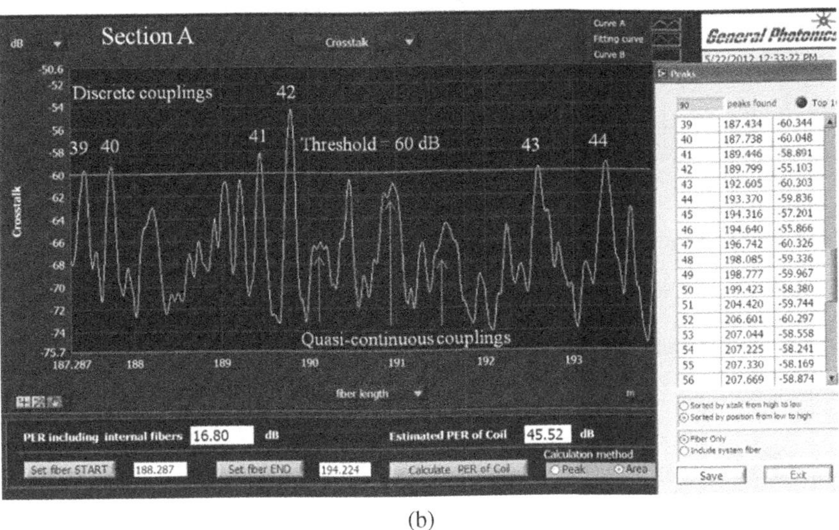

(b)

Figure 10.6 (a) Crosstalk measurement result of a 340 m long PM fiber coil, with the horizontal axis being the fiber length Z. (b) Zoom-in view on section "A" showing discrete and quasi-continuous coupling peaks. Table on the right-hand side lists crosstalk results of discrete peaks larger than −60 dB. Note that the shape and width of discrete crosstalk peaks are determined by the coherence function of the light source. Source: Yao (2019)/SPIE Press.

in the instrument software, as shown in Figure 10.7b, with a result of −61.32 dB. One may use the cursor function to calculate the accumulated PER between any two points in the measurement curve, including the entire length of the PM fiber to get the PER of the whole fiber.

Figure 10.8a shows the polarization crosstalk measurement data of a low quality PM fiber coil made for fiber optic gyro applications (Yao 2019). As one can see, the crosstalk values are very high, some approaching −25 dB, probably due to the high tension when winding the coil, as well as a large number of fiber crossings inside the coil. Because many stress points are closely located, many quasi-continuous crosstalk sections can be observed, as can be seen in Figure 10.8b, which is the

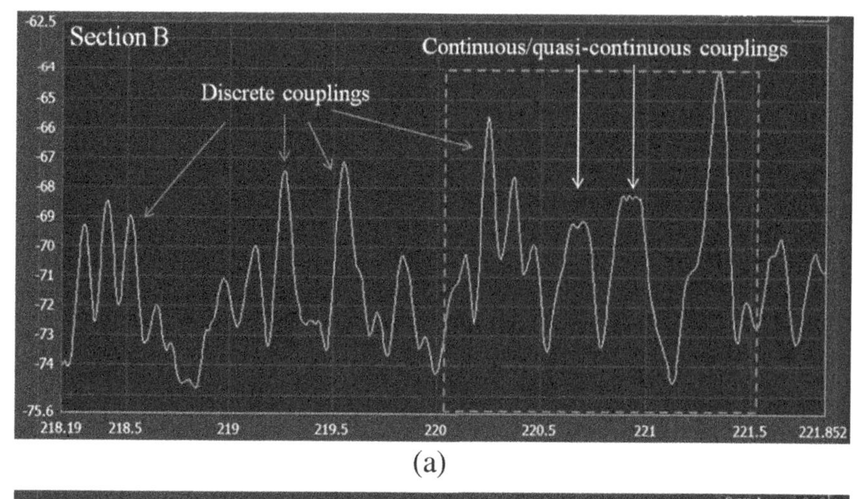

(a)

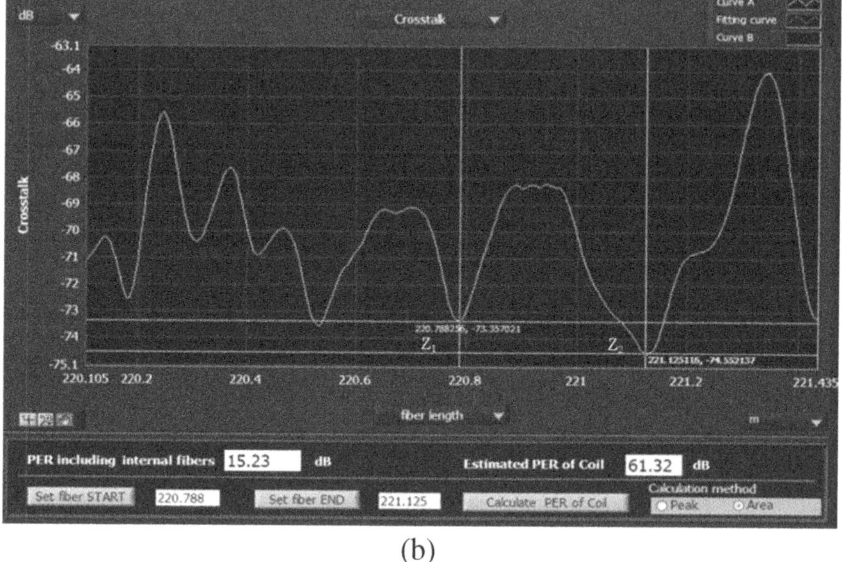

(b)

Figure 10.7 (a) Zoom-in view on section B of the crosstalk measurement of Figure 10.6a, where two continuous or quasi-continuous crosstalk peaks are identified. (b) Accumulative crosstalk value of a continuous/quasi-continuous coupling is obtained by setting Z_1 and Z_2 and calculating integrated PER of the corresponding fiber section. Source: Yao (2019)/SPIE Press.

expanded view of the section in the dashed box in Figure 10.8a. The accumulative PER measured between the two cursors is only −26.46 dB.

10.1.4 Capabilities and Limitations of DPXA

DPXA can give a crosstalk reading about every 4–6 mm, much finer than the crosstalk resolution of the instruments (on the order of 5 cm). The exact spacing between two adjacent data points depends on the birefringence of the fiber and is the ratio of delay resolution of the variable delay line used inside the instrument and fiber birefringence. If the delay resolution of the MDL inside the instrument is Δd, from Eq. (10.3) the corresponding data point resolution δz in the fiber is

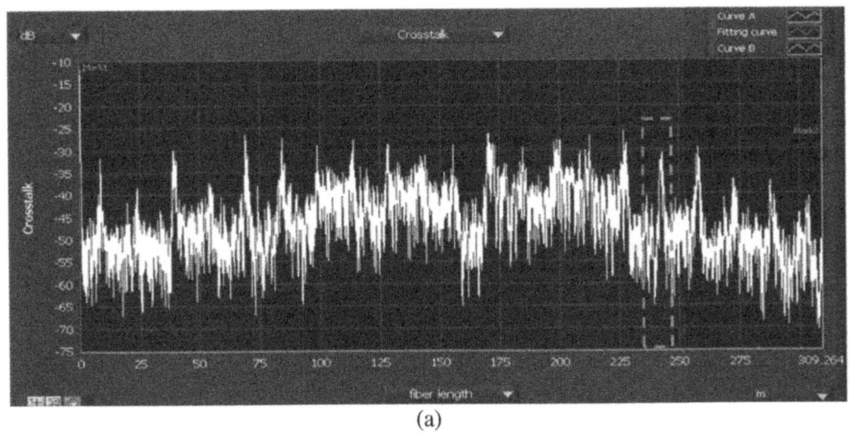

(a)

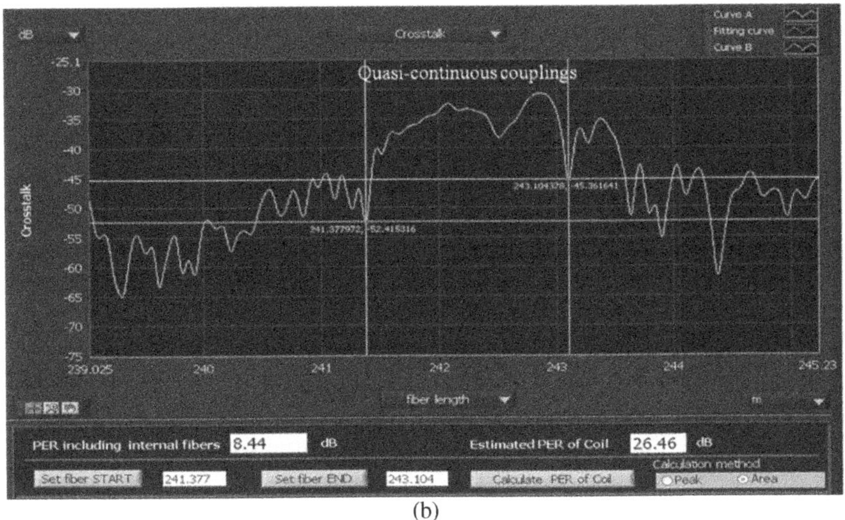

(b)

Figure 10.8 (a) The crosstalk curve of a low quality PM coil of 309 m. (b) The zoom-in view of the boxed section in (a), showing quasi-continuous couplings. The accumulative crosstalk between two cursors is 26.46 dB.

related to the birefringence Δn by

$$\delta z = \frac{2\Delta d}{\Delta n} \tag{10.4a}$$

Similarly, the spatial resolution of a DPXA for measuring polarization crosstalk can be calculated from the coherence length L_c of the light source as

$$\delta l = \frac{2L_c}{\Delta n} \tag{10.4b}$$

The factor of 2 in Eqs. (10.4a) and (10.4b) accounts for the round trip travel of light in the Michelson interferometer. Finally, the fiber measurement range is determined by the MDL travel range l_d by:

$$L_R = \frac{l_d}{\Delta n} \tag{10.4c}$$

For a typical PM fiber with a birefringence Δn of 5×10^{-4}, a typical light source with a coherence length L_c of 25 μm, an MDL with a delay resolution Δd of 1 μm, and a delay range l_d of 0.5 m,

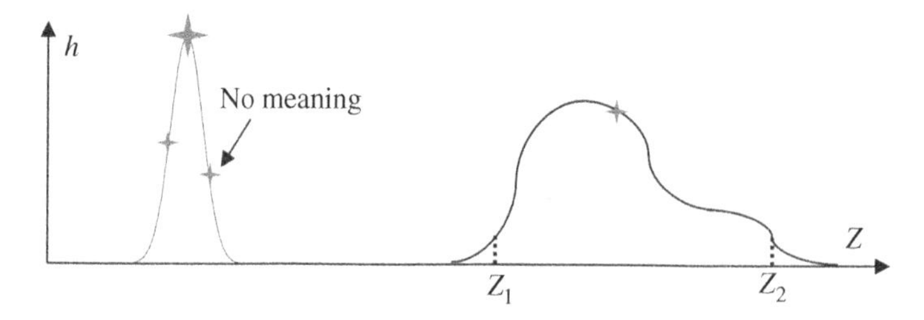

Figure 10.9 Left: illustration of the peak and non-peak values of a discrete crosstalk curve. The non-peak values are the artifacts of light source's coherence function and have no relation to real crosstalk points on the fiber. Right: illustration of a crosstalk hump induced by continuous or quasi-continuous polarization coupling. A point on the hump does not correspond to a crosstalk point on the fiber. Only the integration of the accumulative cross coupling between two points Z_1 and Z_2 is meaningful. One can define points Z_1 and Z_2 in the software interface.

the corresponding data point resolution, the crosstalk measurement resolution, and the crosstalk measurement range are $\delta z = 2$ mm, $\delta l = 50$ mm, and $L_R = 2$ km, respectively.

Assuming the spectrum of the light source has a Gaussian shape, each polarization crosstalk peak induced by a discrete coupling point also has a Gaussian shape because it is essentially the autocorrelation function of the light source, as shown in Figures 10.4 and 10.9 (left). Therefore, only the peak value is meaningful, which represents the crosstalk value of the crosstalk occurring point in space. The other points on the Gaussian curve are just part of light source's coherence function and do not represent meaningful crosstalk values for the corresponding points, as shown in Figure 10.9 (left).

For crosstalks induced by discrete polarization coupling points, a DPXA is able to display the corresponding discrete crosstalk peaks, and gives an accurate crosstalk value of each crosstalk peak and lists all the peak values in the crosstalk table for peaks above a certain value, as shown in Figure 10.6b. For crosstalks induced by continuous or quasi-continuous coupling shown in Figures 10.7b, 10.8b, and 10.9 (right), the crosstalk value of a single point on the broad crosstalk hump is not meaningful. A DPXA is unable to give an accurate crosstalk value for such a point, although a number is given in the crosstalk data file every 4–6 mm (data point resolution). For such crosstalk hump caused by continuous or closely packed quasi-continuous coupling points, it is only meaningful to obtain accumulative crosstalk value, as shown in Figures 10.7b, 10.8b, and 10.9 (right). The instrument can be programmed to calculate such accumulative crosstalk value from point Z_1 to point Z_2, as explained previously. In general, the distance between Z_1 and Z_2 should be much larger than the spatial resolution of the instrument in order to obtain more accurate result.

10.1.5 Applications of Distributed Polarization Crosstalk Analysis

10.1.5.1 Complete Characterization of PM Fibers

In this section, we discuss methods and processes of using the ghost-peak-free DPXA described earlier to accurately obtain all polarization related parameters of PM fibers. We show that by first inducing a series equidistant periodic polarization crosstalk peaks along a PM fiber and then measuring the positions and the widths of these peaks using the analyzer, all birefringence related parameters of the PM fiber, including group birefringence, group birefringence variation along the fiber, group birefringence dispersion, and group birefringence temperature coefficient, can be accurately obtained. We further show that the DPXA has the ability to identify and eliminate polarization crosstalk contributions of connectors or splices in the measurement system and therefore can be used to obtain high accuracy measurement of the PER of PM fibers. Finally, we propose a set of parameters based on the DPA to quantitatively evaluate the quality of PM fibers. The methods and processes described in this section may be applied in the industry for the complete characterization of PM optical fibers.

As described previously, PM fibers having a high internal birefringence that exceeds perturbing birefringence for maintaining a linear polarization along the fiber are important to fiber optic communications and fiber optic sensors, particularly fiber optic gyroscopes. The polarization maintaining ability of a PM fiber is generally characterized by PER or h-parameter (PER per unit length), while the fundamental parameter governing the performance of a PM fiber is actually characterized by its group birefringence. Therefore, it is important for the manufacturers and the users of a PM fiber to know not only the PER, but also the group birefringence and all other group birefringence related parameters, including group birefringence variations with wavelength (group birefringence dispersion) (Tsubokawa et al. 1987; Flavin et al. 2002; Xu et al. 2009b), with temperature (group birefringence thermal coefficient) (Mochizuki et al. 1982; Ding et al. 2011), and along the fiber (group birefringence uniformity) (Zhang et al. 2012).

Measurement fixture To facilitate easy and accurate measurements of group birefringence related parameters, a spool-like fixture to induce periodically spaced polarization crosstalk peaks along a PM FUT is used, as shown in Figure 10.10a. In practice, a standard fiber spool from the fiber manufacturer can be modified by affixing a piece of thin metal rod with a diameter of 2 mm across its width. A single layer of FUT with a length of 280 m is wound on the spool with a 10 g winding tension and each turn of the fiber is placed closely with the previous turn, as shown in Figure 10.10b.

As will be shown next, the accuracy of the length of each fiber turn on the spool is critical to the measurement accuracies of the birefringence parameters of the PM fiber to be measured. As shown in Figure 10.10, the length l of each fiber turn is the circumference l_c of the spool plus an additional length δ_{l1} caused by transitioning each turn of fiber to the next (Figure 10.10b) and an additional length δ_{l2} by the metal rod (Figure 10.10c):

$$l = l_c + \delta_{l1} + \delta_{l2}$$

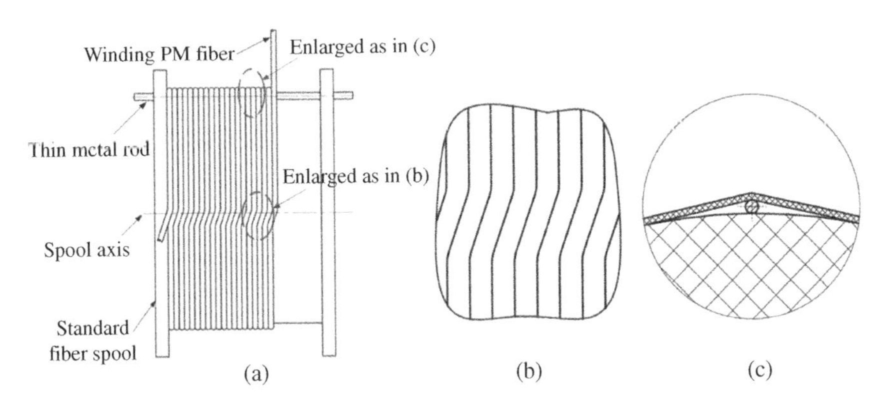

Figure 10.10 (a) Illustration of a length of PM fiber wound on a fiber spool with a thin metal rod to induce periodic transversal stresses on the PM fiber at locations the fiber crossing the rod. (b) Detailed view of the fiber when fiber is transitioning from one turn to the next. (c) Detailed view of the fiber when it crosses the metal rod (Li et al. 2015; Yao 2019).

$$= \pi d_c + d_c \cdot \left(\frac{\sqrt{d_c \cdot d_1}}{d_c - d_1} - \frac{arccos\frac{d_c - d_1}{d_c + d_1}}{2} \right) + 2 \cdot d_c \cdot \left(\frac{\sqrt{d_c \cdot d_2}}{d_c - d_2} - \frac{arccos\frac{d_c - d_2}{d_c + d_2}}{2} \right) \quad (10.5)$$

It can be precisely determined when the diameters of the spool d_c, the fiber d_1, and the metal rod d_2 are known. In practice, one may use l_c to approximate l. For our experiment with $d_c = 0.17$ m, $d_1 = 1.65 \times 10^{-4}$, and $d_2 = 2 \times 10^{-3}$ m, the relative length error $(\delta_{l1} + \delta_{l2})/l_c$ for the approximation is about 0.1%. The relative length error can be reduced to less than 0.015% with the metal rod $d_2 = 5 \times 10^{-4}$ m. Note that the measurement accuracy of the circumference l_c is about 0.006% when a Vernier caliper is used for measuring the diameter of the spool. In comparison, the fiber length measurement accuracy in Ding et al. (2011) is limited by OTDR or a ruler, on the order of 1%.

As expected, "point-like" stresses are automatically applied to the fiber at the points where the fiber goes across the metal rod (Figure 10.10c) to produce multiple periodic polarization crosstalk peaks, with a periodicity precisely defined by Eq. (10.5). These periodic crosstalk peaks act like embedded ruler marks on the fiber, which automatically give out precise length information essential for group birefringence related measurements, as required by Eq. (10.3). Note that if the ghost peaks caused by second order coupling are not eliminated, there would be many false peaks between the true periodic peaks, making it difficult to identify the positions of the true peaks.

In practice, such a spool-like fixture can be machined with a precise predetermined diameter (or circumference) and with a thin slot or bump across its width to induce periodic polarization crosstalk peaks, making the embedded ruler more accurate. It is important to point out that the reasons for making such a fixture are (i) to precisely define the lengths between crosstalk peaks, because the accuracy of group birefringence measurement is proportional to the accuracy of such lengths, as shown in Eq. (10.3), and (ii) to create a distribution of crosstalk peaks to reflect the local group birefringence values along the fiber. Note that the spatial resolution of the local group birefringence is determined by the circumference of the fixture, on the order of the 0.5 m in our experiment. It can be easily improved by reducing the diameter of the fixture, considering that the spatial resolution of the DPXA is around 6 cm (assuming the group birefringence of the PM fiber is 5×10^{-4}), although the 0.5 m resolution is sufficient for most PM fiber characterizations.

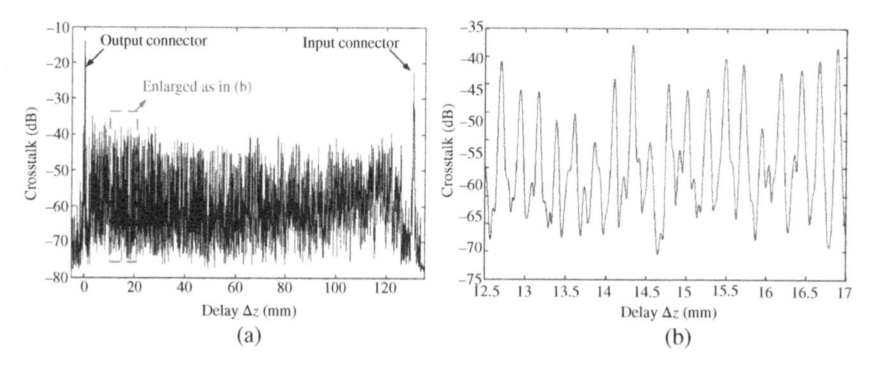

Figure 10.11 (a) Polarization crosstalk curve of 280 m PM fiber wound on the spool as a function of the interferometer delay ΔZ inside the DPXA. The peaks at the far right and left correspond to the crosstalks induced at the input and output connectors, respectively, from slightly axis misalignment between light polarization and PM fiber axis; (b) the equidistant periodic crosstalk peaks are induced by squeezing of the metal cylinder on the PM fiber (Li et al. 2015).

Group Birefringence and Group Birefringence Uniformity Measurements Figure 10.11a is the measured polarization crosstalk curve of a PM PANDA fiber with a diameter of 6 μm, a cladding of 80 μm, and a buffer of 165 μm as a function of the interferometer delay ΔZ, showing the polarization crosstalk peaks induced by the line-pressure from the metal rod on the fiber. The peaks at far left and right correspond to polarization crosstalks induced at the output and input connectors, respectively, due to slight misalignment of lights coupling into the fiber axis. Figure 10.11b shows the detailed view of the equidistant periodic crosstalk peaks caused by transversal pressures induced whenever the fiber crosses the metal rod. As one can see these measured crosstalk amplitudes vary from peak to peak because of the angle variation between the direction of transversal pressure and the fiber's principal axes during winding the fiber onto the spooling wheel; however, such an amplitude variation does not affect the periodicity measurement which is important to the group birefringence measurement. One can readily obtain the spacing between any two stress crosstalk points by simply multiplying the circumference of spool with the number of stress-induced crosstalk peaks between two points. In addition, one can precisely obtain the relative delay ΔZ with the encoder of MDL.

When Eq. (10.3) is used to obtain Δn, the total relative error $\delta_{\Delta n}/\Delta n$ can be expressed as (Ding et al. 2011)

$$\frac{\delta_{\Delta n}}{\Delta n} = \sqrt{\left(\frac{\delta_{\Delta Z}}{\Delta Z}\right)^2 + \left(\frac{\delta_Z}{Z}\right)^2} = \sqrt{\left(\frac{\delta_{\Delta Z}}{\Delta n}\right)^2 + \frac{\delta_Z^2}{Z}} \tag{10.6}$$

where $\delta_{\Delta n}$ is the group birefringence inaccuracy, $\delta_{\Delta Z}$ is the reading error of the delay ΔZ of the variable delay line inside the DPXA, and δ_Z is the measurement error of length Z. Therefore, the absolute length of FUT must be accurately measured in order to obtain an accurate group birefringence Δn according to Eq. (10.6). Any length measurement error will proportionally contribute to the accuracy of Δn. In contrast, here the relative length defined by the circumference of the fiber spool is used to eliminate the need of absolute length measurement and its associated error, and Eq. (10.3) can be rewritten as

$$\Delta n = \Delta_Z/(Nl) \tag{10.7}$$

where l is fiber length corresponding to the period of the crosstalk peaks defined in Eq. (10.5), N is an integer to represent the number of periods chosen in the calculation, and Δ_Z is the corresponding delay in the interferometer for the N periods. The error sources for Δn are from both the relative location inaccuracy $\delta_{\Delta Z}$ between the polarization crosstalk peaks measured with the variable delay line inside the DPXA and the error in the measurement of l. Note that the delay line generally has an error independent of the traveling distance; therefore, multiple periodicities ($N \gg 1$) in the experiment can be used to reduce the effect of delay line error $\delta_{\Delta Z}$, similar to the case of measuring the thickness of a stack of papers in order to accurately determine the thickness of a single paper. In practice, when $N \geq 5$, the measurement uncertainty is sufficiently small. The average Δn obtained is $\Delta n = 4.65 \times 10^{-4}$ when $N = 5$.

Figure 10.12a shows the variation of Δn as a function of distance Z along the fiber for the case of $N = 5$, where Z is obtained by dividing ΔZ with the average Δn obtained in Figure 10.11, as defined in Eq. (10.3). The large data fluctuations at large distances are caused by the dispersion induced peak broadening due to the group birefringence dispersion (Li et al. 2012) to be discussed in the next section, because the broadening increases the uncertainty of $\delta_{\Delta Z}$ in Eq. (10.7). Dispersion compensation procedures described in Li et al. (2012) can be used to further improve the measurement accuracy by multiplying the distributed crosstalk curve with a dispersion compensation function when group birefringence dispersion of the FUT is measured in the next section. Figure 10.12b

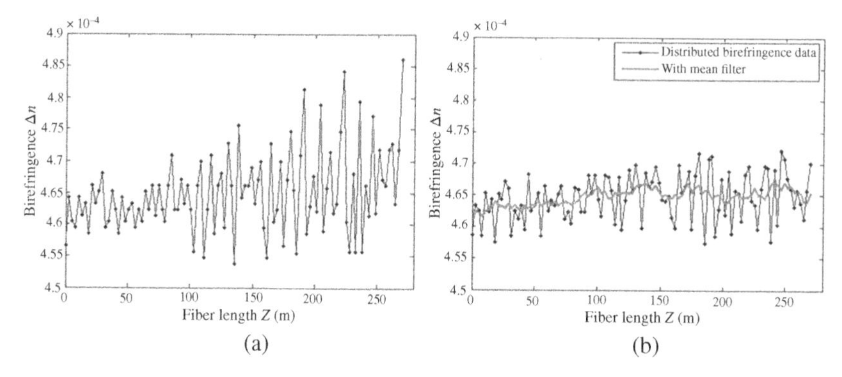

Figure 10.12 (a) Measured group birefringence as a function of distance along the 280 m fiber with $N = 5$ without applying dispersion compensation where the fiber length between any neighboring crosstalk peaks is defined by Eq. (10.5) (0.535 m in the experiment) of the fiber spool. The measurement uncertainty is shown to increase with the distance. (b) The measured group birefringence as in (a), however, the dispersion compensation is applied. The measurement uncertainty at large distances are significantly reduced. The 6-point windowed sweeping average of the group birefringence as a function of distance is shown with the gray line. In both (a) and (b), the distance zero is at the position of FUT's output connector (Li et al. 2015).

is the measured group birefringence as a function of distance along the fiber, showing that the measurement uncertainties are greatly reduced when the dispersion compensation procedure in Li et al. (2012) is applied. It is also evident that the mean Δn slightly varies along the fiber length for the FUT.

Group Birefringence Dispersion Measurement As discussed in Li et al. (2012), the envelope of a measured crosstalk peak (i.e. the interference peak) is influenced by SLED's spectral distribution and group birefringence dispersion ΔD of the PM fiber. In fact, the envelope width increases quadratically with the distance Z due to effect of the group birefringence dispersion, and a relationship between envelop broadening W and group birefringence dispersion ΔD can be expressed as (Li et al. 2012)

$$\frac{W}{W_0} = [1 + (\alpha \Delta D)^2 Z^2]^{1/2} \tag{10.8a}$$

$$\alpha = 2\pi c \left(\frac{\Delta\lambda}{\lambda_0} \right)^2 \tag{10.8b}$$

In Eqs. (10.8a) and (10.8b), c is the speed of light in vacuum, $\Delta\lambda$ is the 3-dB spectral width of the light source with a Gaussian line shape, and λ_0 is center wavelength of the light source used for the measurement, respectively, and W_0 is the $1/e$ width of the interference envelope when the dispersion ΔD or Z equals to zero. One may simply measure the widths of any two polarization crosstalk peaks with a known spacing Z between them to obtain the dispersion ΔD using Eq. (10.8a). However, in order to improve measurement accuracy of ΔD, widths of crosstalk envelops at multiple locations along the PM FUT are measured, and ΔD is then obtained by curve-fitting to Eq. (10.8a).

Figure 10.13 shows the widths of the crosstalk peaks as a function of their locations along the fiber. The distance is measured from the first induced crosstalk peak and evaluated every 20 peaks ($N = 20$ or 10.7 m). As can be seen from Figure 10.6, the widths of crosstalk peak start to show significant broadening at a distance large than 100 m. The group birefringence dispersion ΔD of the PM fiber is accurately obtained by a least-square fitting to Eq. (10.8a) to be $\Delta D = 0.0079$ ps/(km nm).

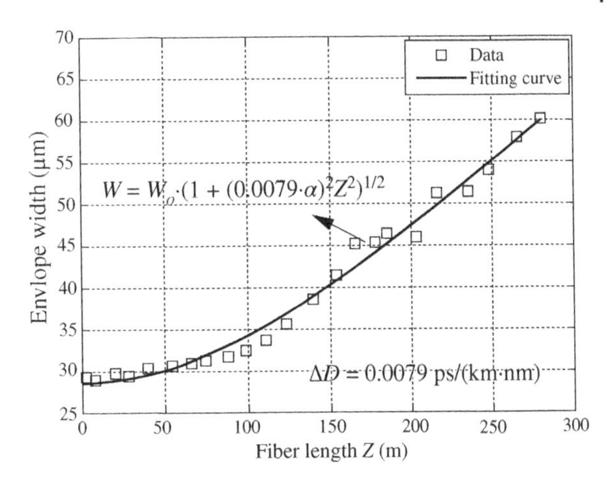

Figure 10.13 Envelope widths of crosstalk peaks induced by stress at various locations along the fiber. The envelope widths of polarization crosstalk peaks broaden as fiber length increases due to group birefringence dispersion. Curve-fitting obtains the group birefringence dispersion of the FUT to be $\Delta D = 0.0079$ ps/(km nm) (Li et al. 2015).

Note that a dispersion compensation function can be obtained once ΔD of the fiber is determined, as described in Li et al. (2012). This dispersion compensation function can be used to remove the broadening of the crosstalk peaks and hence reduce the measurement uncertainties of the group birefringence along the fiber, as described in the section earlier.

Group Birefringence Thermal Coefficient Measurement As discussed in Section 3.3, the birefringence in a PM fiber is expected to be sensitive to the temperature because it resulted from the anisotropic strain due to differential thermal expansion at different regions in the fiber cladding, which varies linearly with temperature in the vicinity of room temperature. The group birefringence Δn can be written from Eqs. (3.4), (3.6), and (3.7) as

$$\Delta n = \gamma(T_0 - T) \tag{10.9}$$

where T is the temperature of FUT, T_0 is the softening temperature of the silica glass with dopants in the stress-inducing region of the cladding, and γ is the thermal coefficient of group birefringence of the PM fiber to be measured.

To measure the temperature coefficient, the fiber spool of Figure 10.10a is put into a temperature chamber, with two fiber pigtails outside of the chamber. Figure 10.14a shows two typical polarization crosstalk curves of PM FUT as a function of relative delay at two different temperatures of 80 and 40 °C. Clearly, the positions of all polarization crosstalk peaks are shifted with the temperature, as predicted by Eqs. (10.3) and (10.9). The thermal coefficient of group birefringence can be obtained by measuring the position of the crosstalk peak induced by the input connector as a function of temperature (Ding et al. 2011). In order to accurately determine the birefringence, the fiber length between the crosstalk peaks must be precisely measured. Because the periodic polarization crosstalk peaks induced by the measurement fixture act as ruler marks along the fiber, the fiber length measurement between any two peaks can be easily obtained with a high precision.

In experiment, the spacing ΔZ between the 1st and 50th peaks is used to obtain the group birefringence Δn as a function of temperature. As shown in Figure 10.14b, the peak positions of the 48th, 49th, and 50th peaks shifted to the left as temperature increases, reducing the spacing ΔZ. The fact that the spacing ΔZ decreases with the temperature indicates that Δn has a negative thermal coefficient. As mentioned previously a thermal coefficient of the group birefringence γ can be obtained by linear-fitting of Δn to Eq. (10.9) by using the least square fitting method at each different temperature. Note that in order to reduce the effect of dispersion, the crosstalk peaks should be chosen close to the output end of the FUT, although dispersion compensation

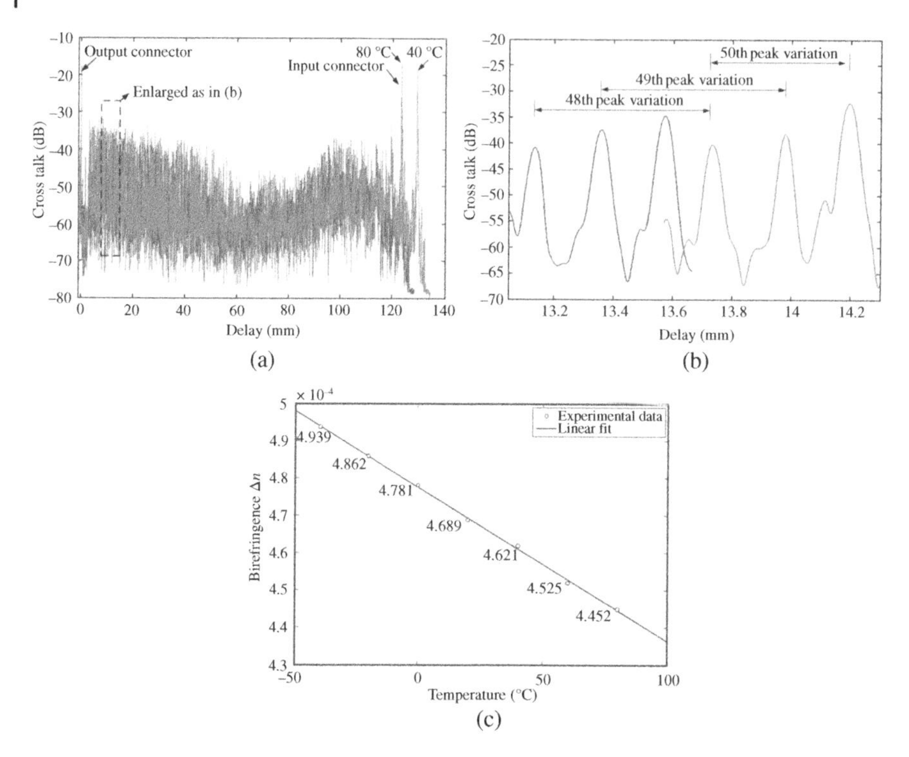

Figure 10.14 (a) Polarization crosstalk curves of a PM fiber as a function of the relative delay at 80 °C (darker line) and 40 °C (lighter line); (b) The expanded view of the positions of the 48th, 49th, and 50th peaks at 80 °C (left 3 peaks) and 40 °C (right 3 peaks); (c) Δn obtained at seven different temperatures by measuring the spacing between the 1st and the 50th crosstalk peaks at different temperatures using Eq. (10.7) (Li et al. 2015).

(Li et al. 2012) may also be used to reduce the peak broadening and improving measurement accuracies for measuring peaks closing to the input connector ($N \gg 50$).

Figure 10.14c shows the measured Δn of a PM fiber at seven different temperatures (i.e. −40, −20, 0, 20, 40, 60, and 80 °C). Linear-fitting Δn to Eq. (10.9) yields a group birefringence thermal coefficient γ of -4.123×10^{-7}.

PER Measurement As discussed in Chapter 8, using a PER meter for measuring the PER of a PM fiber is susceptive to (i) polarization misalignment at the input end of the FUT, and (ii) polarization misalignment between the light source and its fiber pigtail if pigtailed light source is used. Using a DPXA, one can readily identify the crosstalk contributions from the polarization misalignments at the two fiber ends, as well as at the interface between the pigtail and the light source, and eliminate their contributions to the total PER, because the corresponding polarization crosstalk peaks measured with a DPXA are spatially separated. Note that for the PER measurement described in this section has no need to induce the periodic polarization crosstalk peaks as in the previous sections.

Figure 10.15 shows the measured polarization crosstalk curves of a PM fiber jumper with FC/PC connectors and a spool of PM fiber of 250 m directly from a PM fiber vendor, fusion spliced with two FC/PC connectors. When an FUT is connected to the DPXA, the polarization misalignment at the connection points induces significant crosstalk peaks. An auto-search program is implemented in the DPXA software to automatically identify those peaks, because the polarization crosstalk signatures of the fibers inside DPXA are known, as shown in Figure 10.15. In addition,

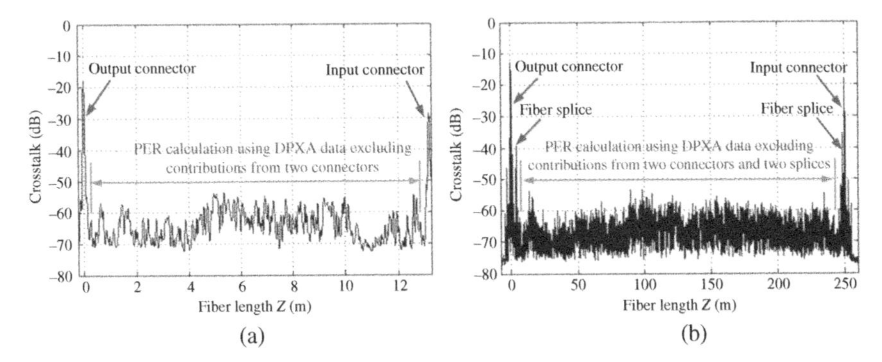

Figure 10.15 (a) Polarization crosstalk curves of a 13 m jumper with two FC/PC connectors and (b) a 250 m PM fiber coil spliced with two FC/PC connectors. PER measurement with a commercial PER always including the contributions of the input connector and two splices, while the DPXA has the ability to identify and eliminate polarization crosstalk contributions of all connectors and splices in the measurement system. Note that fiber length in the horizontal axis is obtained by dividing the fiber delay line distance ΔZ with the average group birefringence obtained using the procedure described in Section 10.1.2 (Li et al. 2015).

the polarization crosstalk peak resulting from the light source and its pigtail is located outside the region defined by the two connectors, and thus is not included for PER calculation. By definition, the PER of the fiber can be calculated as

$$PER = 10\,log\left(\frac{P_f}{P_s}\right) \tag{10.10}$$

where P_f is the total power coupled to the fast axis from the slow axis and can be obtained by integrating of all polarization crosstalks between the two connectors, and P_s is the total power remaining in the slow axis $P_s = P - P_f$, where P is the total received power at the fiber output.

An algorithm in DPXA software has been implemented to automatically calculate the PER excluding the contributions of the two end connectors from the crosstalk measurement curve, as shown in Figure 10.15a. One may also use the DPXA software to calculate the total PER contribution between any two points along the fiber, and therefore to further exclude the contributions from the two fusion splicing points, as shown in Figure 10.15b. Table 10.1 compares multiple PER measurement results of a 13 m fiber jumper and a 250 m PM fiber coil obtained with a commercial PER meter and a DPXA. It is evident that the PER value obtained by PER meter is several dB smaller than that obtained with a DPXA, due to the contributions of crosstalk from the polarization misalignment at the input connector. In addition, as anticipated, the measurement repeatability of a DPXA is much better than that of using a PER meter. Therefore, it is much easier to use a DPXA to obtain more accurate PER measurements than using a PER meter.

Note that a SLED source with a spectral width around 30 nm was used when using a PER meter for the PER measurement, in accordance with the requirement of test standard TIA-544-193 (1999) to avoid measurement fluctuations caused by coherent effect resulting from the use of a narrow line-width laser, as described in Section 8.3.

PM Fiber Quality Evaluation PER or h-parameter (TIA-455-192 1999) only characterizes the accumulative polarization holding performance of a PM fiber, which may not be able to reflect the true polarization performance of the fiber, especially considering that PER measurement using conventional methods may have significant fluctuations as discussed previously. Here we show that a set of parameters from a single DPXA scan can be used to fully describe the performance without ambiguity.

Figure 10.16 shows DPXA scans of three different PM fibers measured directly with the fiber on the fiber spool from the vendor. Four parameters can be used to characterize the quality of a PM fiber for the polarization related performance: (i) the average polarization crosstalk, (ii) the maximum crosstalk, (iii) the number of crosstalk peaks above a certain threshold defined by the manufacturer or the user, and (iv) PER. The average crosstalk is the major contributor to the value of PER and closely relates to h-parameter (PER per fiber length). The maximum crosstalk is an indication whether the PM fiber is degraded or damaged during manufacturing, packing, or shipping

Table 10.1 Comparison of PER measurements of a 13 m PM fiber jumper and a 250 m PM fiber coil obtained with a commercial PER meter (ERM) and a DPXA (Li et al. 2015).

	PER of fiber jumper		PER of fiber coil	
Measurement #	**ERM (dB)**	**DPXA (dB)**	**ERM (dB)**	**DPXA (dB)**
1	22.5	34.81	25.7	30.82
2	24.9	34.27	21.8	30.81
3	23.3	35.06	25.4	30.31
4	26.8	34.64	22.5	30.90
5	25.8	35.14	23.8	31.09
Uncertainty (max–min) (dB)	4.3	0.87	3.9	0.78

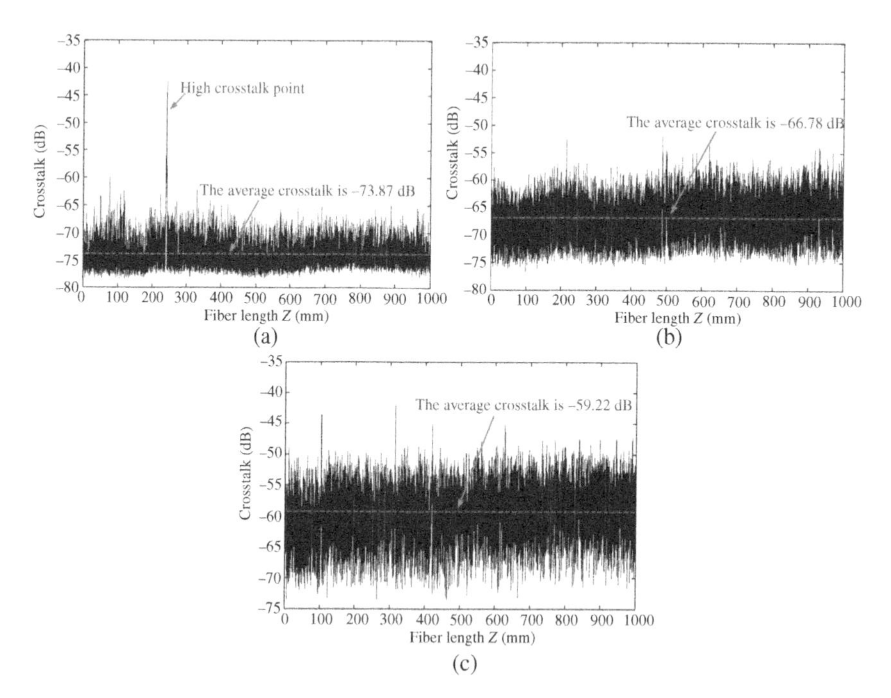

Figure 10.16 Polarization crosstalk curves of three different PM fibers. (a) A PANDA PM fiber at 1310 nm with a buffer diameter of 250 μm. A defect point is seen at around 220 m; (b) a PANDA fiber at 1310 nm of the same core/cladding diameters in (a), however, with a reduced buffer diameter of 169 μm; (c) a third PANDA PM fiber at 1310 nm with the similar cladding diameters as in (a), however, with an even more reduced buffer diameter of 136 μm. Two major defect points with crosstalk more than 45 dB are seen (Li et al. 2015).

Table 10.2 Four parameters to fully characterize the quality of three different PM fibers.

	Average crosstalk (dB)	Maximum crosstalk (dB)	Number of crosstalk peaks above −55 (dB)	PER (dB)
Fiber I	−73.87	−42.36	1	28.8
Fiber II	−66.78	−51.88	23	23.6
Fiber III	−59.22	−42.15	1711	16.5

of the PM fiber, although a single or few high crosstalk peaks contribute insignificantly to the total PER of a long fiber. For some applications, such as fiber gyro coils, the high crosstalk sections must be removed to assure high quality fiber coil production. A large number of high crosstalk peaks present in the fiber may indicate problems in fiber drawing or packaging process. It also makes it impractical to sort out only good fiber sections for demanding applications.

Table 10.2 lists four parameters of the three different fibers under test, obtained from Figure 10.16 (Li et al. 2015). Fiber I is a commercial PANDA fiber at 1310 nm with a beat length of 2.57 mm, a core diameter of 6 μm, a cladding diameter of 125 μm, and a buffer diameter of 250 μm, respectively. Fiber II is a different PANDA fiber at 1310 nm having the same core as Fiber I, but a different beat length of 2.13 mm, a cladding diameter of 80 μm, and a buffer diameter of 169 μm, respectively. Finally, Fiber III is a third type of PM fiber at 1310 nm with a beating length of 2.6 mm, a core diameter of 6.4 μm, a cladding diameter of 80, and a buffer of 136 μm. It is evident from Figure 10.16a that Fiber I has the lowest average crosstalk, resulting in a highest PER of 28.78 dB; however, it has a defect point about 220 m from the output connector with a high crosstalk peak of −42.36 dB, probably caused by mishandling when winding the fiber to the spool. Such a defect cannot be identified with a simple PER measurement and may be permanent, e.g. cannot be recovered even when the corresponding stress is released. On the other hand, Fiber III has the highest average crosstalk of −59.22 dB, corresponding to a low PER of 16.25 dB. It also has a large number of high crosstalk peaks above −55 dB, probably because the thin buffer layer (136 μm) cannot effectively protect the fiber from external stresses. Therefore, all four parameters collectively give a full picture of the quality or performance of the PM FUT.

10.1.5.2 Transversal Force Sensing Using Distributed Polarization Crosstalk Analysis

Equation (10.2b) shows that the polarization crosstalk h is proportional to the square of the normalized force F when $F \ll 1$ and becomes oscillatory as F increases beyond 0.894, as shown in Figure 10.17a. In addition, it is most sensitive to the changes in F when the angle α of the applied force is 45° and the force applying length l is one half of the PM fiber beat length L_{b0} ($l = L_{b0}/2$) when $F \ll 1$. Figure 10.17b shows the plot of h a function of F for the case of $\alpha = 45^\circ$ and $l = L_{b0}/2$ when F is restricted to the peak value of the first oscillatory cycle. In this range, the curve is nonlinear but monotonic, which can be used for force sensing. The inconvenience of the nonlinearity can be overcome if the values of both h and F are on logarithmic scales, as shown in Figure 10.17c. Figure 10.17d is the three-dimensional (3D) plot of h as functions of both l and F. As one can see, as applied force F increases, the optimum l is no longer $L_{b0}/2$, which has shifted toward smaller values. At the peak, $l \approx 0.38L_{b0}$ while the peak h increases to 0.45 approximately.

PM Fiber Sensing Tape As Eq. (10.2b) indicates, the polarization crosstalk h is sensitive to the force applying angle with respect to the birefringence axis of the PM fiber. Therefore, it is important that the forces to be sensed have a fixed angle along the fiber, preferable to be 45° from the fast axis of

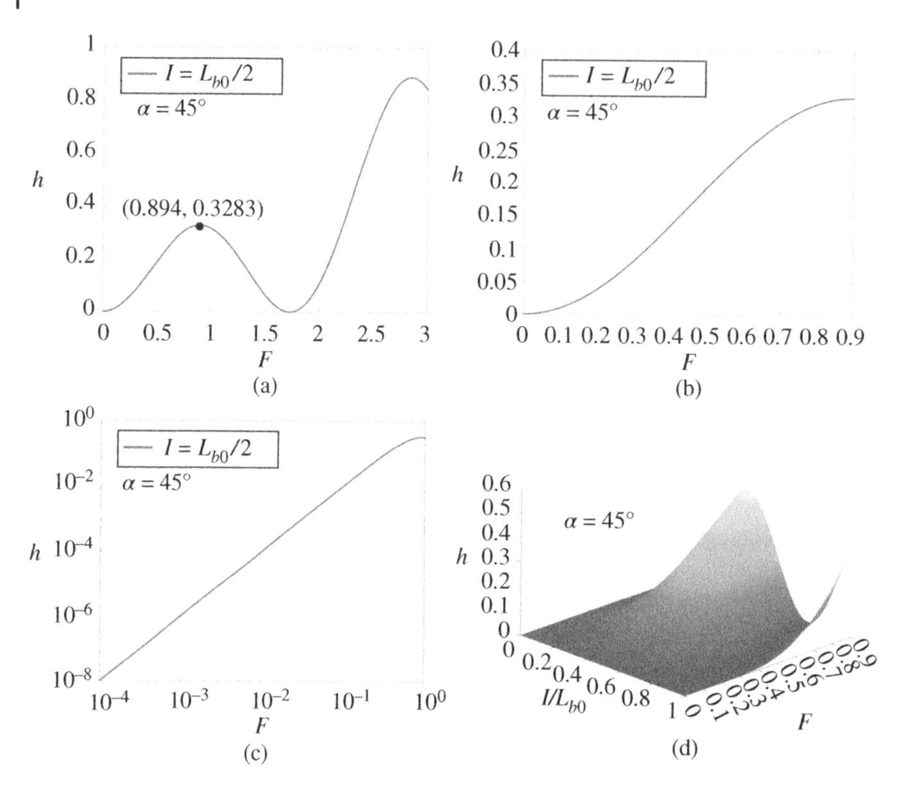

Figure 10.17 (a) Plot of Eq. (10.2b) showing h oscillates as F increases beyond 0.894. (b) Crosstalk h as a function of F when its value is restricted to the peak value of the first cycle. (c) Plot of (b) in logarithmic scale. (d) 3D plot of h as functions of l and F.

the PM fiber. Unfortunately, it is difficult to determine the birefringence axis of a PM fiber in the field, not to mention keeping it 45° from the direction of the pressure/force to be sensed in practice. One way to solve the problem is to first lay the PM fiber on a tape-like strip made with thin metal or synthetic material with its birefringence axis pre-aligned 45° from the strip surface to form a sensing tape, as shown in Figure 10.18a, which can be rolled into a spool for easy transportation if necessary. The sensing tape can then be fixed onto a surface to sense the force or pressure applied perpendicular against the surface, assuring 45° orientations for optimizing measurement sensitivity (Hao et al. 2020).

Figure 10.18b shows the polarization crosstalk (in log scale) as a function of the force applying angle α with respect to the fast axis. It can be observed that (i) the highest crosstalk is obtained when the force angle α is at 45°, and (ii) the crosstalk is least sensitive to the force angle variations around 45°.

In practice, a transparent PET film with a thickness of 0.3 mm and a width of 10 mm is selected as the base material for the sensing tape. The film is waterproof with a smooth surface and having good anti-aging, anti-sticky, and anti-ultraviolet (UV) properties. A length of PANDA PM fiber is bonded on the PET film using a UV curing adhesive at periodical locations, as shown in Figure 10.19, with its birefringence axis aligned 45° against the surface of the PET film using a specially designed equipment described in the following text.

Fabrication Equipment and Machine Vision System Figure 10.20 shows the equipment developed to fabricate the PM fiber based sensing tape described earlier. As illustrated in Figure 10.20a, the

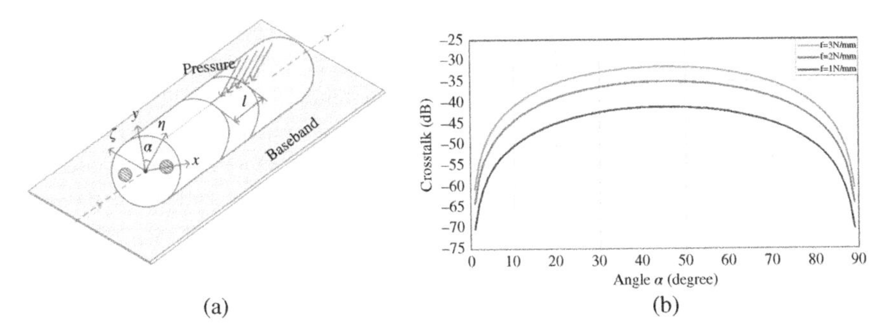

Figure 10.18 (a) Illustration of a sensing tape showing the birefringence axes of the PM fiber and the direction of applied transversal force/pressure. (b) Plot of the relationship between the crosstalk and the force angle, assuming $L_{b0} = 2.32$ mm, $r = 125$ μm, $\lambda = 1550$ nm (Hao et al. 2020).

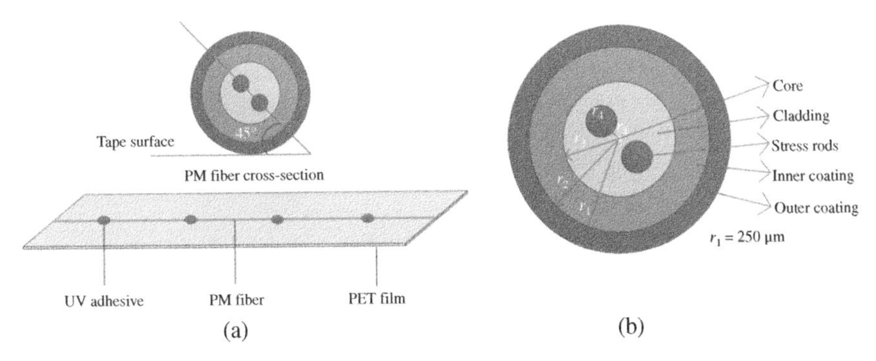

Figure 10.19 (a) Illustration of the PM fiber based sensing tape on which the PM fiber oriented 45° is bonded on the PET film with periodically spaced adhesive drops. (b) Cross-section image of the fiber showing the details of the PANDA PM fiber (Hao et al. 2020).

equipment consists of two synchronized wheels on the right-hand side, one is for supplying the PET film and the other for the PM fiber. The fiber supplying wheel is mounted on a fiber axis rotation apparatus such that it can be rotated back and forth for adjusting the birefringence axis angle of the PM with respect to the PET film surface. The PET film and the PM fiber are moved together passing

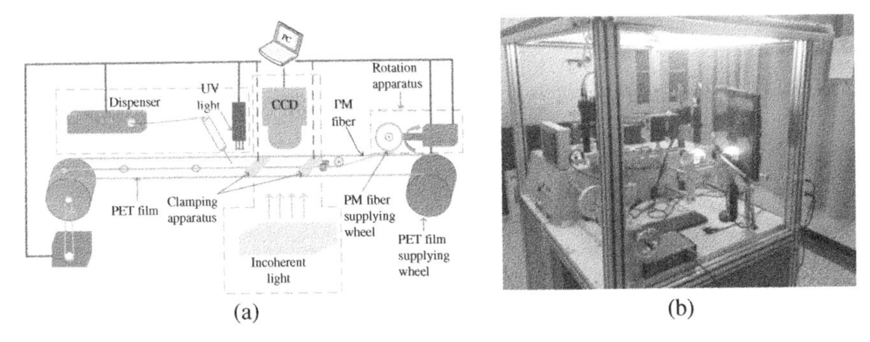

Figure 10.20 (a) Schematic of the equipment and process for fabricating PM fiber based sensing tape. The computer is used to determine the fiber axis orientation from the CCD camera, rotate the fiber supplying wheel, and activate the adhesive disperser and UV light. (b) Photo of the equipment. Source: Hao et al. (2020)/Optica Publishing Group.

through a side-image observation window where there is a machine vision system made of a light source for illuminating the fiber, a charge-coupled device (CCD) camera with a lens system for projecting the fiber side images onto the CCD sensor array. The transparency of the PET film allows for the illumination to pass through and refractive index matching liquid is used between the fiber and PET film to eliminate the influence of the air gap in order to ensure satisfactory image sharpness.

The fiber side-images are analyzed in real time by the machine vision system as the fiber passes through the imaging window to determine the orientation of fiber's birefringence axis, which is then fed back to the fiber rotation apparatus to adjust the fiber axis orientation. After the fiber section with the axis orientation determined passes through the side-image observation window, a UV adhesive drop is applied to the fiber in a preset spacing with a computer controlled adhesive disperser before it is cured by a UV light. Finally, the sensing tape with the fiber bonded on the PET film is wound on the big spool on the left. Two electro-magnetic fiber clamps are also used to clamp down the fiber in the image observation window in the vertical direction, but still allow the fiber to rotate.

The system uses the Polarization Observation by Lens Effect (POL) method (Yan et al. 2019) originally developed for PM fusion splicers for analyzing the fiber side images to determine fiber birefringence axis orientation as the fiber passes through the side-image observation window below the CCD. The axis orientation information then feeds back to the computer for adjusting the fiber rotation apparatus in real time.

Verification of the Theoretical Results As shown in Figure 10.18, when a load or force is applied onto a PM fiber at a particular location, a polarization crosstalk peak will occur at the location and its amplitude is sensitive to the angle between the direction of the force/pressure and the fiber's birefringence axis angle, which can be measured and validated with the DPXA described in Figure 10.2. The instrument has a polarization crosstalk measurement floor of −80 dB, crosstalk repeatability, and accuracy of ±0.5 dB, a spatial resolution of 4 cm, and a spatial measurement uncertainty of ±2.53 cm.

A 2 m-long sensing tape using the equipment and process described earlier was fabricated, with different sections having different fiber axis orientation angles ranging from 0° to 90°, measured again with the POL method described in the previous section after the tape was made. The PM fiber

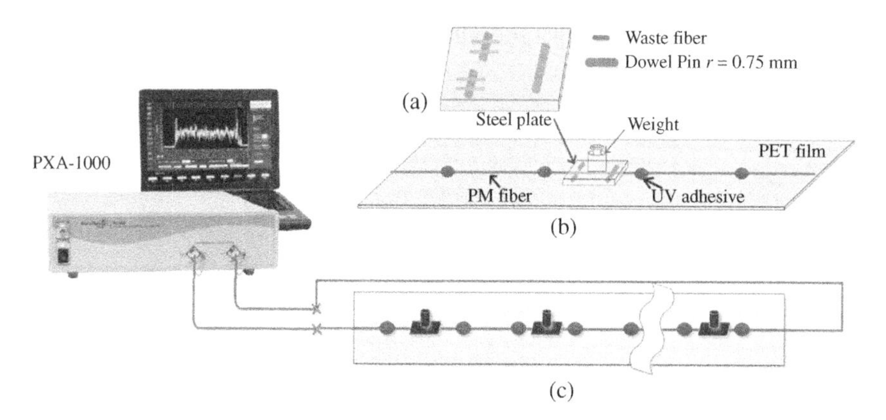

Figure 10.21 (a) The pressure plate design with a dowel pin to act on the sensing fiber and two other dowel pins on each side of the sensing fiber for balancing purpose. (b) The sensing tape with the pressure plate on top. (c) Using a DPXA to measure the polarization crosstalk induced by the pressure plate on the sensing tape. Source: Hao et al. (2020)/Optica Publishing Group.

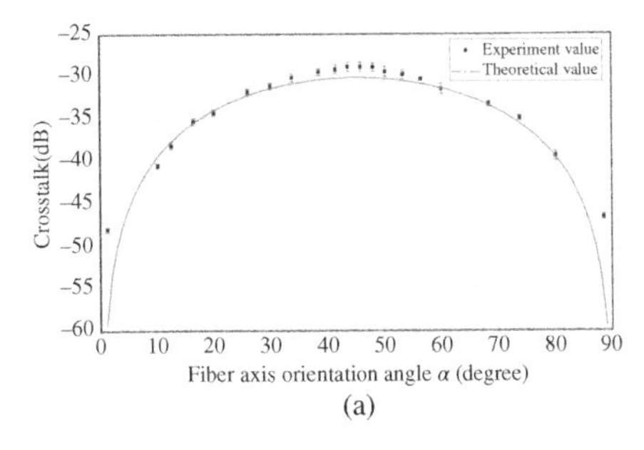

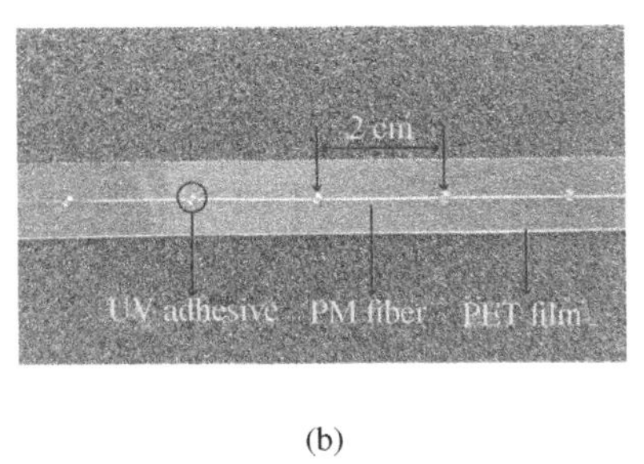

Figure 10.22 (a) Experimental and theoretical results of polarization crosstalk as a function of fiber axis orientation angle induced by a 100 g weight at 1550 nm. (b) Photo of the sensing tape with a thickness of 0.3 mm and width of 10 mm. Source: Hao et al. (2020)/Optica Publishing Group.

is of PANDA for operating at 1550 nm with a core, cladding, and coating diameters of 6.92, 125, and 250 μm, respectively. In order to apply a consistent force on all fiber sections, a pressure plate with a dimension of 15 mm × 10 mm × 0.3 mm and a weight of 5 g was placed onto each plate, as shown in Figure 10.21a. The long dowel pin on the right end was designed to be in contact with the sensing fiber and the two short dowel pins with the same diameter were to be on both sides of the sensing fiber for balancing purpose, as shown in Figure 10.21b. The two short dowel pins were slightly elevated with two waste optical fibers of the same size as the sensing fiber to ensure the pressures plate is leveled when sitting on top of the sensing fiber. All dowel pins and the waste fibers were all glued onto the steel plate with the same UV adhesive for fixing the fiber on the PET film. In the experiment, the pressure plate was put on top of the sensing fiber and a weight of 100 g was placed on the center of the pressure plate for applying a constant force on the sensing fiber. Finally, the DPXA was used to measure the polarization crosstalk value at each point the pressure was applied using the pressure plate, with 21 points all together, one at a time, as shown in Figure 10.21c. The measurement results are shown in Figure 10.22a.

The diameter of the force applying dowel pin on the pressure plate is 1.5 mm and the force acting width was estimated to be 0.1 mm, resulting in an effective stress on the PM fiber of approximately 3.5 N/mm. In the calculation, only 1/3 of the weight was included because the pressure plate had three dowel pins of the same diameter to share the load applied. Using the expression for the crosstalk ratio h given in Eqs. (10.2b) and (10.2d), the polarization crosstalk as a function of the fiber axis orientation angle was obtained at a wavelength of 1550 nm, with the results also shown in Figure 10.22a. The photo of the fabricated sensing tape is shown in Figure 10.22b.

Overall, the experimental and theoretical results agreed reasonably well, except at fiber axis orientation angles less than 5° or larger than 85° where the polarization crosstalk values were too small so that the measurement accuracy were affected by the residual polarization crosstalk (RPC) induced in the PM fiber during manufacturing process.

Validation of 70 m-Long Sensing Tape With the equipment and process described earlier, a 70 m-long PM fiber based sensing tape operating at 1550 nm band was successfully fabricated with the fiber birefringence axis aligned nominally 45° with respect to the tape surface. To verify the accuracy of the fiber axis orientation, 20 points in the 70-m-long PM fiber sensing tape were randomly selected, and the POL method was used to measure the angle between the fiber birefringence axis and the PET film, with the result shown in Figure 10.23. It can be observed that the repeatability of the measured angle is within ±0.6° and the average angles of the 20 points range between 42° and 48°, resulting in an angular accuracy of ±3° around 45°.

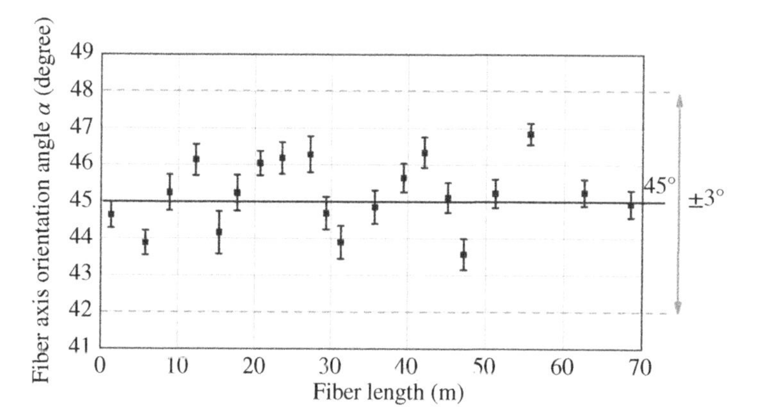

Figure 10.23 Post fabrication measurement of the birefringence orientation angle of the PM fiber based sensing tape (Hao et al. 2020).

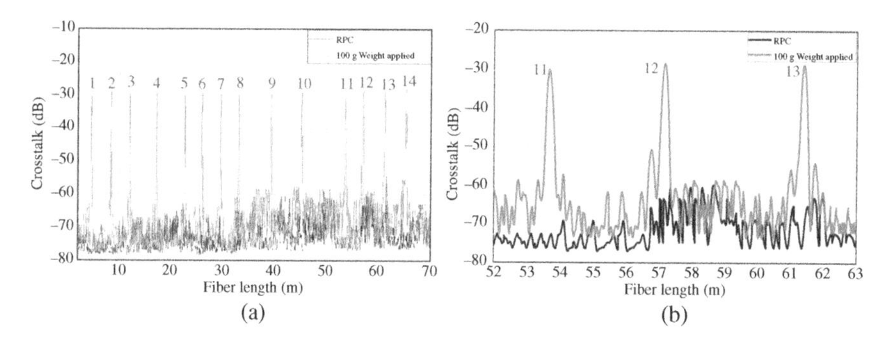

Figure 10.24 (a) Sensing tape uniformity test with 14 identical pressure plates randomly located, each having a 100 g weight applied. (b) Expanded view of the polarization crosstalk peaks from distances 52 to 63 m for more detailed examination (Hao et al. 2020).

Distributed Force Sensing with 70 m-Long Sensing Tape To demonstrate the force sensing uniformity, 14 points on the 70 m-long PM fiber sensing tape were randomly selected, and 14 identical pressure plates described in Figure 10.21a were made and placed onto the 14 points on the sensing tape, with a 100 g weight placed on each pressure plate. The DPXA was used to simultaneously measure all induced polarization crosstalk peaks, with the resulting polarization crosstalk measurement shown in Figure 10.24a,b. It is evident that there appeared 14 polarization crosstalk peaks of similar heights, with a maximum difference less than 2.11 dB and standard deviation of 0.62 dB, indicating an excellent uniformity of fiber orientation angle alignment along the sensing tape. One may notice that in Figure 10.24 the RPC peaks (black curve) of the sensing tape are not consistent at different locations along the tape and appeared to be higher at larger distances. Some of these RPC peaks are caused by unwanted stresses on the PM fiber from the shrinkages of the adhesive at the bonding points. In the experiments, the curing time of UV adhesive, the amount of adhesive, and the irradiation power of the light source all lead to the difference of the stress on the fiber and therefore the variations of the RPC peaks. In addition, due to the limited space in the laboratory, it is difficult to lay the 70 m long sensing tape flat, resulting in bending and twisting of the sensing tape and hence the additional RPC peaks. The unwanted stresses are more severe in the fiber section from 40 to 70 m, which leads to the increased RPC peaks.

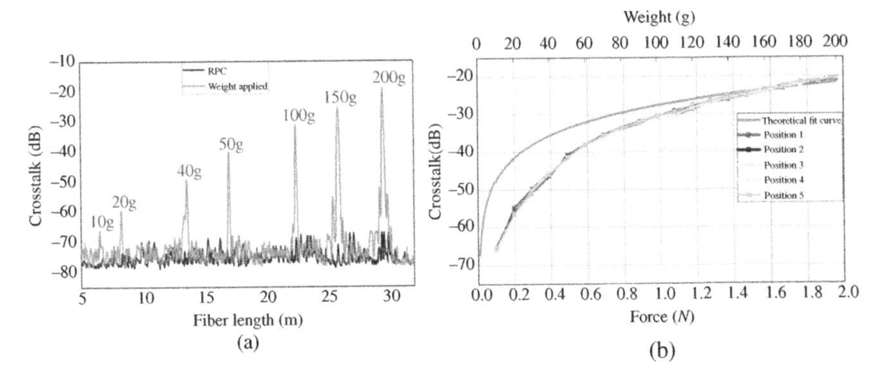

Figure 10.25 (a) Distributed force sensing with different weights applied to the sensing tape at different locations. (b) The calibration curve of crosstalk vs. the loading force (weight) applied to the sensing tape at five randomly selected sensing points, compared with the theoretical fit curve of Eq. (10.2b). The positions 1–5 are at locations 3.36, 12.18, 28.66, 40.37, and 56.62 m, respectively (Hao et al. 2020).

Finally, the distributed transversal force sensing with the 70 m sensing tape was demonstrated using seven pressure plates randomly placed on the sensing tape, with weights ranging from 10 to 200 g, as shown in Figure 10.25a. It is evident that the system can clearly distinguish weights less than 10 g (corresponding to a line force of 0.33 N/mm). The calibration curves of crosstalk vs. weight at five randomly selected sensing points were also obtained, which are compared with the theoretical curve of Eq. (10.2b), as shown in Figure 10.25b. As can be seen that these calibration curves have reasonably good repeatability; however, they deviate significantly from the theoretical curve when the weights applied to the PM fiber are small. This deviation is likely due to the cushion effect of the fiber buffer (250 μm) on the PM fiber used because Eq. (10.2b) did not take the buffer into account. Therefore, in practice, the calibration curves must be used for converting the polarization crosstalk values to the weight values. Note that the nonlinear response may complicate the implementation of the sensing system and compromise the sensing accuracy across the sensing range; however, with a properly implemented calibration procedure using the calibration curves, the inaccuracy caused by the nonlinearity can be minimized digitally. Note from Eq. (10.2b) that the sensing sensitivity can be controlled by changing the acting width l of the force with respect to the PM fiber beat length. The maximum sensitivity occurs when the force acting width is $(2n-1)/2$ of the fiber beat length. Therefore, for small scale weight sensing, one can design the sensing plate with a force acting width close to $(2n-1)/2$ of the beat length. For large scale force sensing, the force acting width can be designed away from $(2n-1)/2$ beat length.

The demonstrations earlier are for direct transversal force sensing. With some clever mechanical fixtures, pressure can be converted to transversal stress or force, which can then be sensed by the DPXA based distributed sensing system.

10.1.5.3 Temperature Sensing Using Distributed Polarization Crosstalk Analysis

As shown in Eq. (10.9) and Figure 10.14c, the birefringence of a PM fiber is linearly dependent on the temperature, and therefore the spacing between any two polarization crosstalk peaks is also linearly dependent on the temperature, as demonstrated in Figure 10.14b. Such a property can be utilized for temperature sensing (Yao et al. 2015; Ding et al. 2016). From Eq. (10.3), the relative delay Δz between any two stress points corresponding two polarization crosstalk peaks i and j is

$$\Delta z_{ij} = \Delta n(Z_j - Z_i) = \gamma(T_0 - T)L_{ij} \tag{10.11a}$$

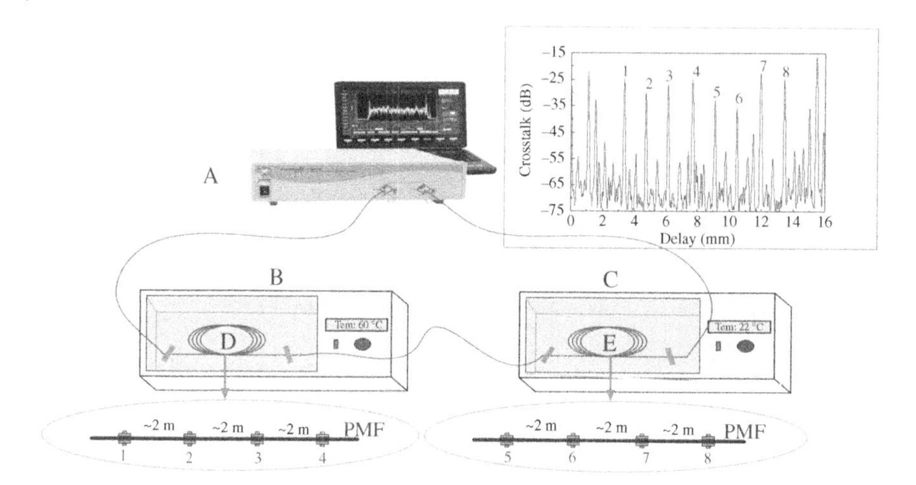

Figure 10.26 Experimental setup with eight preset stress points corresponding to eight crosstalk peaks. Source: Ding et al. (2016)/Optica Publishing Group.

where $L_{ij} = Z_j - Z_i$ is the fiber distance between the two points and Δz_{ij} is the delay of the MDL inside the DPXA, as shown in Figure 10.2. Therefore, if the distance L_{ij} between the two stress points and the temperature coefficient γ of the PM fiber are known, the temperature change can be precisely obtained from the MDL delay variation δz_{ijT} caused by the temperature change:

$$\delta z_{ijT} = \Delta z(T_1) - \Delta z(T_2) = \gamma(T_2 - T_1)L_{ij} = \gamma \Delta T L_{ij} \tag{10.11b}$$

Figure 10.26 shows simple experiment setup demonstrating the distributed fiber temperature sensing. Eight preset stress-induced crosstalk points with a separation of ~2 m are induced by eight clamps in a section of PM fiber, as shown in the inset of Figure 10.26 and the delays Δz_{ij} between any two crosstalk peaks (i, j) can be easily obtained from the horizontal axis readings. One half length of the PM fiber containing four preset stress points (#1–#4) is placed in water tank B with variable temperature for sensing and the other half containing four preset stress points (#5–#8) is placed in water tank C with a constant temperature of 22 °C for comparison.

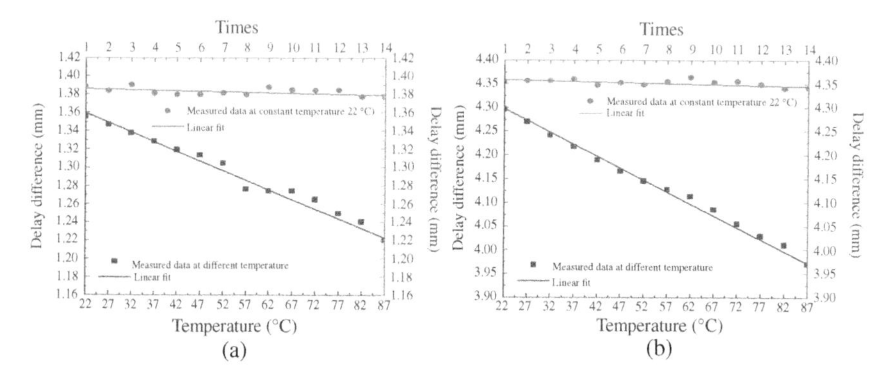

Figure 10.27 (a) Variations of delay difference Δz_{23} between crosstalk peaks 2 and 3 (square) as the temperature increases, compared with the delay difference Δz_{67} between crosstalk peaks 6 and 7 (solid circle) at constant temperature of 22 °C; (b) Variation of delay difference Δz_{14} between crosstalk peaks 1 and 4 (square) as the temperature increases, compared with the delay difference Δz_{58} between crosstalk peaks 5 and 8 (solid circle) at a constant temperature of 22 °C (Ding et al. 2016).

As the temperature changes in water tank B, the spacing between different crosstalk peaks Δz_{ij} varies with the water temperature surrounding the PM fiber, and by measuring the variations of the spacing δz_{ijT} between the crosstalk peaks(i, j), temperature sensing can be achieved. In the experiment, the temperature was increased from 22 to 87 °C with a 5 °C interval in water tank B. The squares and the associated solid line in Figure 10.27a are the experimental data and the linear fit of delay difference Δz_{23} between peaks 2 and 3 as a function of temperature, respectively, while the solid circles and the associated line are the data and linear fit of delay difference Δz_{67} between peaks 6 and 7 at a constant temperature of 22 °C.

Similarly, Figure 10.27b shows the experimental data (squares) and the linear fit (the associated solid line) of delay difference Δz_{14} between peaks 1 and 4 as a function of temperature, as compared with the data (solid circles) and linear fit (the associated line) of delay difference Δz_{58} between crosstalk peaks 5 and 8 at a constant temperature of 22 °C. The temperature sensing coefficient is calculated to be approximately −0.89 μm/(°C m), e.g. for every degree of temperature increase, the spacing between two crosstalk peaks induced at two locations separated by 1 m is 0.89 μm. Therefore, in order for the system to have a temperature resolution of 0.1° at a spatial resolution of 1 m, the MDL delay resolution must be less than 0.089 μm. The MDL used in the DPXA of Figure 10.2 has a resolution about 1 μm, which must be improved to have reasonable temperature measurement sensitivity and accuracies. A linear encoder with resolution and accuracy or the order of 50 nm or better can be used for the improvement, which is available commercially at reasonable costs.

10.1.5.4 Simultaneously Transversal Stress and Temperature Sensing with DPXA

As discussed in Sections 10.1.5.2 and 10.1.5.3, the transversal force sensing measures the magnitude of the crosstalk, which is the quantity on the vertical axis of the polarization crosstalk curve, while the temperature sensing measured the distance between some preset crosstalk peaks, which is the quantity on the horizontal axis of the measurement curve. These two quantities are orthogonal or independent from each other and therefore they can be measured simultaneously without affecting each other, which is advantageous compared with other distributed sensing techniques, such as those based on the fiber Bragg gratings (FBGs), Raman OTDR, and Brillouin OTDR. Utilizing such a unique property, one may devise a distributed sensor system to simultaneously sense the stress and temperature, as described in Yao (2015) and Feng et al. (2017).

10.2 Distributed Mueller Matrix Polarimetry and Its Applications

The distributed polarization crosstalk analysis described in Section 10.1 only applies to PM fibers for measuring a single parameter, namely the distance resolved PER along the fiber. Not only are PM fibers generally much more expensive than SMF, but also more difficult to handle, making it difficult to deploy for many applications. In this section, we describe distributed Mueller matrix polarimetry for the complete polarization analysis of the backscattered and reflected light in a single mode (SM) fiber (Galtarossa and Palmieri 2005). It is capable of measuring not only the distance resolved Stokes parameters of backscattered light, but also the Mueller matrix elements along the fiber, which contains the essential polarization related properties of the fiber, such as the birefringence (PMD) and die-attenuation (PDL).

10.2.1 System Description

In Chapter 9, we described a binary Mueller matrix polarimetry system relying on a polarization state generator (PSG) and a polarization state analyzer (PSA) made with magneto-optic (MO) crystals to achieve high accuracy polarization measurements (Feng et al. 2018). Such a non-distributed polarization analysis (NDPA) system has been proved to achieve exceptionally high accuracy measurements of SOP, DOP, linear and circular birefringence, PDL, and PER. Here we describe a distributed Mueller matrix polarimetry system incorporating the same type PSG and PSA made with binary MO polarization rotators in an OFDR system, which is capable of performing distance resolve polarization measurements. We sometimes call it an OFDR based DPA system or sometimes simply a PA-OFDR. The same concept can be applied to optical coherence tomography (OCT) system, which is essentially an OFDR system for z direction information with a two-dimensional (2D) beam scanner for (x, y) plane information, or simply a three-dimensional (3D) DPA system.

In this system, for each input SOP generated by the PSG, the SOP of the reflected or back scattered light wave at any point in an optical path, such as in an optical fiber or waveguide, can be precisely measured using PSA in the OFDR. By generating four distinctive input SOPs with the PSG and analyzing the corresponding SOPs of reflected or back scattered light at each point along the optical path for each input SOP with the PSA, the Mueller matrix of the optical medium at each point in the optical path can be obtained. Consequently, the birefringence vector of the optical medium at each point along the optical path can be derived. Because of the high speed ($\sim$20 μs) and high repeatability advantages of the binary PSG and PSA, this binary DPA is also of high speed and high accuracy in obtaining both the SOP distribution and the birefringence vector distribution along the optical path. In comparison, a polarization sensitive OFDR (P-OFDR) using analog polarization analysis shows low measuring speed and precision (Galtarossa et al. 2009, 2010), compared with this binary polarization analysis, or only has partial polarization analysis capability (Soller et al. 2005). On the other hand, the non-distributed binary DPA system made with PSG and PSA can only obtain the final SOP and total accumulative birefringence of the whole optical path.

Figure 10.28a shows the basic schematic of the binary DPA system called PA-OFDR, while Figure 10.28b shows its commercial realization. Light from a tunable laser (TL) with a long coherence length is coupled into a PM fiber. Around 5% is coupled out by a first coupler C1 to a k-clock consisting of a first circulator (CIR1) and a Michelson interferometer made with SMF in which two Faraday rotation mirrors (FRMs) are used to eliminate polarization fluctuations. The outputs from the interferometer are detected and amplified by a first balanced photodetector (BPD1) to get the incremental frequency of the TL. The light in the PMF continues to propagate and around 10% is coupled out by a second coupler C2 as a local oscillator (LO) beam. The remaining light first goes through a PSG and is then directed into the single mode fiber under test (SMF-UT) via a second circulator (CIR2). The back scattered and reflected light from the SMF-UT is directed to port 3 of CIR2 and then goes through a PSA before entering a PM fiber to be mixed at a third coupler C3 with the LO beam from C2. The interference signals from two outputs of C3 are detected and amplified by a second balanced detector (BPD2), and the output of BPD2 is finally sent to the analog to digital converter (ADC) in the digital circuit board to be converted to digital signal with 16 bit resolution. The zero crossings of the interference signal from the k-clock interferometer detected by BPD1 are converted to trigger pulses in the digital circuit to trigger the ADC so that the signal from BPD2 is digitized with equal frequency spacing. The working principle of the k-clock and the data processing algorithm can be found in detail in Ding et al. (2012, 2013). Fast Fourier transform (FFT) of the digitized signal then reveals the location information of backscattered and reflected light originated at different locations in the SMF-UT. As in Chapter 9, both PSG and PSA are made with binary MO crystals, as shown in the inset of Figure 10.28a. The PSG is capable of generating four distinctive SOPs, while the PSA is capable of analyzing any SOP with four distinctive logic states or MO settings.

Figure 10.29 shows the data acquisition and processing flow chart for measuring the SOP matrix as a function of z. For each frequency scan of the TL, the PSG generates one of four SOPs and the

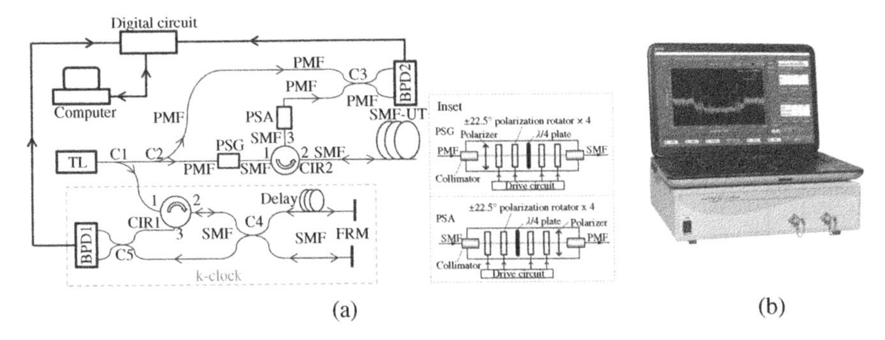

Figure 10.28 (a) Schematic of distributed polarization analysis (DPA) system, PA-OFDR. TL: tunable laser; C1, C2, C3, C4, C5: couplers; PSG: polarization state generator; PSA: polarization state analyzer; PMF: polarization maintaining fiber; SMF: single mode fiber; CIR1, CIR2: circulator; BPD1, BPD2: balanced photodetectors; FRM: Faraday rotation mirror; SMF-UT: SMF under test. (b) Product photo of an OFDR based DPA developed by the authors (Source: Courtesy of General Photonics / Luna Innovations) (Feng et al. 2018).

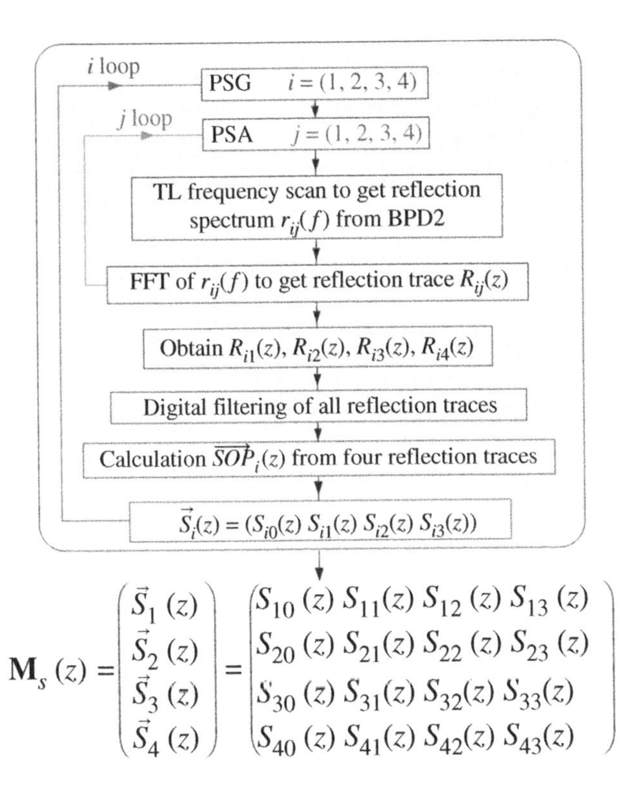

Figure 10.29 Data acquisition and processing flow chart for obtaining the state of polarization (SOP) matrix as a function of z. Note $\mathbf{M}_s(z)$ can be reduced to a 3×3 matrix if polarization dependent loss (PDL) in the fiber can be neglected (Palmieri 2013; Feng et al. 2018).

PSA measures the SOP with four sequential MO settings, and a total of 16 frequency scans and PSG/PSA settings are required to get the full Stokes (or SOP) matrix $\mathbf{M}_s(z)$ of the signal returned from each point along the SMF-UT:

$$\mathbf{M}_s(z) = \begin{bmatrix} \mathbf{S}_1(z) \\ \mathbf{S}_2(z) \\ \mathbf{S}_3(z) \\ \mathbf{S}_4(z) \end{bmatrix} = \begin{bmatrix} S_{10}(z) & S_{11}(z) & S_{12}(z) & S_{13}(z) \\ S_{20}(z) & S_{21}(z) & S_{22}(z) & S_{23}(z) \\ S_{30}(z) & S_{31}(z) & S_{32}(z) & S_{33}(z) \\ S_{40}(z) & S_{41}(z) & S_{42}(z) & S_{43}(z) \end{bmatrix} \tag{10.12}$$

where $\mathbf{S}_i(z) = [S_{i0}(z), S_{i1}(z), S_{i2}(z), S_{i3}(z)]^T$, $(i = 1, 2, 3, 4)$ is the Stokes vector of the reflected or back scattered light at location z measured with the PSA corresponding to the ith SOP generated by the PSG. When a polarized light is propagating in the SMF-UT, the evolution of the light's SOP can be described using the well-known equation of motion of the Stokes vector $\mathbf{S}_i(z)$ discussed in Section 4.5.3 (Wanner et al. 2003; Palmieri 2013) as

$$\frac{\partial \mathbf{S}(z)}{\partial z} = \mathbf{W}_{rt}(z) \times \mathbf{S}(z) \tag{10.13}$$

where z is the distance of light propagated within the fiber and $\mathbf{W}_{rt}(z)$ is the local round-trip birefringence vector. In practice, one may use the following steps to calculate the birefringence (Yao et al. 2010; Wuilpart et al. 2000; Williams 2004; Dong et al. 2006).

For a small fiber segment with a length of Δz, the SOP matrix $\mathbf{M}_s(z+\Delta z)$ at $z+\Delta z$ relates to the SOP matrix at z by $\mathbf{M}_s(z+\Delta z) = \mathbf{M}_\Delta(z)\mathbf{M}_s(z)$, where $\mathbf{M}_\Delta(z)$ is the Mueller matrix of the fiber segment Δz and can be obtained as

$$\mathbf{M}_\Delta(z) = \mathbf{M}_s(z+\Delta z)\mathbf{M}_s^{-1}(z) \tag{10.14}$$

It can be shown that the retardation angle $\theta(z)$ can be expressed as (Wuilpart et al. 2000; Williams 2004; Dong et al. 2006):

$$\theta(z) = cos^{-1}\left(\frac{Tr[\mathbf{M}_\Delta(z)] - 1}{2}\right) = \frac{2 \cdot 2\pi\Delta n(z)\Delta z}{\lambda} \tag{10.15}$$

where the factor of two accounts for the round trip passage of light in the fiber segment. Finally, the local birefringence $\Delta n(z)$ can be calculated from the $\theta(z)$ as

$$\Delta n(z) = \frac{\theta(z)\lambda}{4\pi\Delta z} \tag{10.16}$$

where λ is the wavelength.

10.2.2 Expression of Bending-Induced Birefringence in SMF

As discussed in Chapter 3, the bending-induced birefringence Δn, which is mainly caused by lateral internal stress in a bent SMF, can be represented by the well-known expression as (Ulrich et al. 1980)

$$\Delta n = \frac{n^3}{4}(P_{12} - P_{11})(1+\sigma)\left(\frac{r}{R}\right)^2 \tag{10.17a}$$

where n is the refractive index of the core, P_{11} and P_{12} denote the strain-optical coefficients, r is the cladding radius, R is the bending-radius, and σ is Poisson's ratio. As shown in Eq. (10.16), the local birefringence of the SMF is inversely proportional to the square of the bending radius, and we can define a bending-induced birefringence coefficient k as

$$k = \frac{n^3 r^2}{4}(P_{12} - P_{11})(1+\sigma) \tag{10.17b}$$

Hence, Eq. (10.16) can be rewritten as

$$\Delta n = k\left(\frac{1}{R}\right)^2 \tag{10.17c}$$

Since $P_{12} - P_{11} = 0.15$, $\sigma = 0.17$ for fused silica (Ulrich et al. 1980) and $n = 1.46$, $r = 62.5\ \mu\text{m}$ for SMF, the theoretical value of k for common SMF can be estimated to be $5.334 \times 10^{-10}\ \text{m}^2$ from Eq. (10.17b). Note that such an estimate was calculated by assuming the optical fiber made with fused silica with a uniform cross section and did not consider the contributions from a more complicated fiber cross section with core, cladding, and buffer coating.

10.2.3 Measurement Setup and Results

It is difficult to prove the correctness of spatial SOP distribution measurement inside an optical fiber or an optical waveguide because the SOP is not well defined and changes with time. In order to verify the measurement accuracy of OFDR based DPA system, the birefringence distribution inside an optical fiber is to be measured because it can be well defined and stable over time.

To demonstrate the distributed bending-induced birefringence measurement feasibility of the DPA system (PA-OFDR), and to verify the correctness of the theory as described earlier, a simple experimental system is designed, as shown in Figure 10.30a. A commercial tunable laser (TUNICS T100S-HP) with a linewidth of ~100 kHz and a wavelength tunable range from 1520 to 1600 nm was used as the frequency-swept laser source for the system and output the polarized light to the PA-OFDR through a PM fiber jumper. The clock signal was communicated between the tunable laser (TL) and the PA-OFDR via a general-purpose interface bus (GPIB). Twelve fiber loops with different radii were made along the SMF-UT with a total length of ~4.5 m to produce the bending-induced birefringence variations as a function of bending radius. As shown in Figure 10.30b, the bending radius gradually decreases from ~3.75 to ~1.00 cm with a 0.25 cm interval for the 12 fiber loops along the SMF-UT. The fiber loops were produced by respectively looping 12 different sections of the SMF-UT around 12 glass cylinders with different radii, and then removing the cylinders to relax the fiber after the loops fixed on a paperboard using adhesive tapes as shown in the inset to avoid introducing undesired strain or stress to the fiber loops. The radii of the 12 glass cylinders were measured using a vernier caliper. The experimental data were obtained and processed by a LabVIEW based data acquisition system.

The distributed birefringence variation along the SMF-UT segment (distance from 7.0 to 12.0 m) with the 12 fiber loops is measured using the PA-OFDR, as shown in Figure 10.31a. It can be seen that loops with smaller radii induce higher local birefringence and loops with larger radii have wider peak and low peak value, consistent with the expectations based on Eq. (10.17c). In data processing, the measured SOP data along the SMF-UT are divided into numerous segments, each having a length of Δz and then the birefringence value of each fiber segment is calculated. This fiber segment length Δz is defined as birefringence spatial resolution (BSR). Six (6) different BSRs (0.25, 0.5, 1, 5, 20, and 100 mm) are used to calculate the distributed birefringence along the SMF-UT and the results are plotted using different line shapes with different colors. In general, BSR is much larger than the reflection spatial resolution (RSR) because a minimum distance is required to accumulate sufficient SOP changes above the noise level of the measurement system for the accurate birefringence calculation using Eq. (10.14). The smaller the local birefringence, the larger the minimum distance or BSR is required for a sufficient SOP change; however, the BSR should not be larger than the feature length of the local birefringence to be measured. Here, the feature length corresponds to the length of the fiber having a birefringence induced by the fiber bending/loop. On the other hand, the larger the local birefringence, the smaller BSR should be selected. As can be seen from the curves, an inappropriate selection of BSR produces different measurement errors for the fiber loops with different radii, corresponding to different local birefringence with different feature length. For instance, the 100 mm BSR produces the smallest error for measuring the birefringence induced by the fiber loop No. 1 with 3.75 cm radius while producing the largest error for measuring the birefringence induced by the fiber loop No. 12 with ~1.00 cm radius because the BSR is already larger than the perimeter of loop No. 12. Figure 10.31b shows the zoom-in measurement curves of loop No. 10 in Figure 10.31a, and one can see that all BSR selections, except for BSR of 100 and 20 mm, are adequate for accurate birefringence measurements for this loop. Fortunately, the BSR of 0.25 mm is adequate for the accurate measurement of birefringence produced by all of the fiber loops. In order to avoid the measuring uncertainties, 20 repeated measurements were performed and each curve in Figure 10.31 represents 20 times average. Note that the birefringence distribution inside a perfect circular fiber loop is expected to be uniform in theory. The Gaussian-like birefringence distribution in each fiber loop is possibly caused by the following reasons: (i) low-pass digital filtering during data processing, (ii) averaging of multiple data traces, including multiple frequency scans in each measurement and 20 full-measurements, and (iii) imperfections of the fiber loops, such as non-circularity, birefringence induced adhesive tapes for fixing each fiber loop, fiber bend between two adjacent loops, and fiber twist.

Based on the experimental result measured using the BSR of 0.25 mm as plotted by the black solid line in Figure 10.31a, the values of all the 12 peaks were picked out and plotted in Figure 10.32a to characterize the bending-induced birefringence as a function of bending radius. The peak value was selected as the birefringence of each loop because (i) this value has the maximal probability in Gaussian-like distribution, and (ii) this point is approximately at the center of each loop and therefore is farthest from the birefringence transition regions on both sides of the loop and from the adhesive tapes. Curve-fitting the experimental data to Eq. (10.17c) yields the expression in the

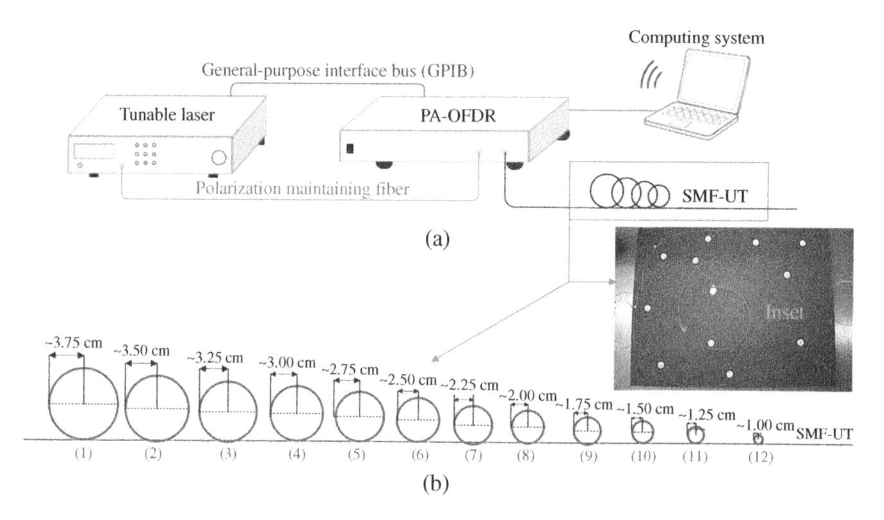

Figure 10.30 (a) Experimental setup of bending-induced birefringence measurement. (b) Illustration of a setup of 12 fiber loops with different bending radii in SMF-UT, with each loop having a single turn. Inset showing the photo of the 12 fiber loops. Source: Feng et al. (2018)/Optica Publishing Group.

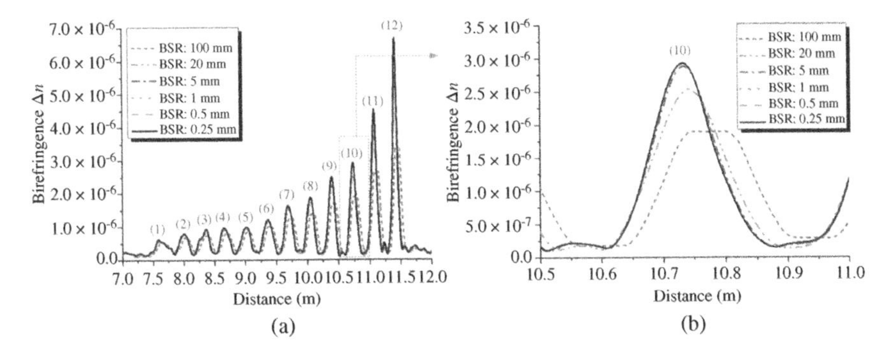

Figure 10.31 (a) Birefringence curves of the SMF-UT segment with 12 fiber loops of different radii measured by PA-OFDR in different birefringence spatial resolutions (BSRs). (b) Zoom-in curves located around loop No. 10 in (a). (Feng et al. 2018).

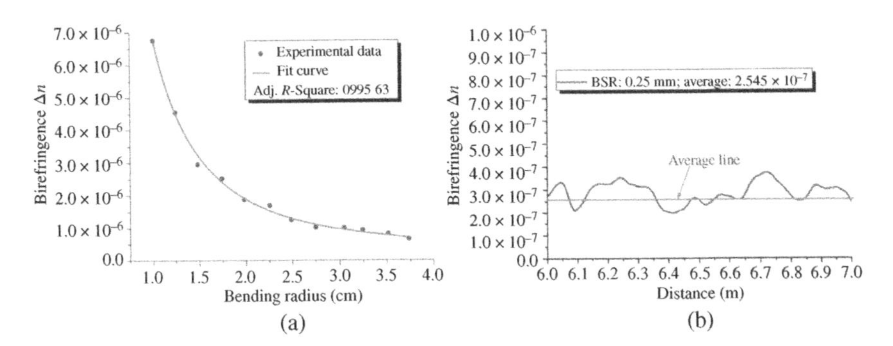

Figure 10.32 (a) Bending-induced birefringence as a function of fiber bending radius measured with a DPA system called PA-OFDR; The residual birefringence (RB) is apparent as the bending radius goes to infinity. (b) Birefringence curve of the SMF-UT section without any loop, measured by the DPA system, showing fiber's distributed RB; the average value of the RB was also shown (Feng et al. 2018).

following text, with an adjusted (Adj.) R-Square (goodness of fit) of 0.995 63.

$$\Delta n = 6.601 \times 10^{-10}\left(\frac{1}{R}\right)^2 + 2.365 \times 10^{-7} \tag{10.18}$$

As can been seen in Eq. (10.18), experimentally obtained value of bending induced birefringence coefficient k was 6.601×10^{-10} m^2, in good agreement with the theoretical value of 5.334×10^{-10} m^2.

Note that, as shown in Figure 10.32a, a residual birefringence (RB) is apparently present in the measurement data as the bending radius goes to infinity, which was not included in Eq. (10.17c), which is believed to be the contribution of the inherent geometric asymmetry and internal stress in the SMF-UT, and a constant term should be added in Eq. (10.17c) to represent it. Consequently, curve fitting yields a RB of 2.365×10^{-7} in Eq. (10.17c). Figure 10.32b shows the birefringence distribution along the SMF-UT section (from 6.0 to 7.0 m) without any loop, measured by the PA-OFDR based DPA system using a BSR of 0.25 mm. The average of 2.545×10^{7} is highly consistent with the value from the curve fitting, indicating that the PA-OFDR is capable of performing distributed RB measurement with high accuracy. In addition, from the measured RB results in Figure 10.32b, one may conclude that the birefringence measurement resolution of the PA-OFDR is better than 2×10^{-7}. The RB measurement capability and accuracy of the PA-OFDR will be further discussed in Section 10.2.4.

It is worth noting that the birefringence of a fiber to be measured by the PA-OFDR cannot be too high to cause the accumulated retardation over the minimum BSR length larger than π, which will cause phase wrapping errors. Another limitation is that a large number of laser wavelength scans are required, which causes the broadening of the birefringence measurement resolution and losing some detailed spatial features. However, these limitations can be resolved by improving the data acquisition and processing algorithms, and using lasers with much faster wavelength scanning rates. An amplitude division or a wave-front division polarimeter described in Chapter 8 may be used as the PSA to significantly reduce the number of wavelength scans, with the drawbacks of increased cost resulting from the increased photodetectors and associated digital signal processing (DSP) channels.

10.2.4 Validations with a Non-distributed Mueller Matrix Polarimetry System

10.2.4.1 Residual Birefringence (RB) Validation

To verify the presence of RB in SMFs and the measurement accuracy of the PA-OFDR, a high precision NDPA system made with binary MO crystal based PSG and PSA, as described in detail in Chapter 9, is used to measure the RB of an FUT with a length of ~4.5 m from the same fiber batch as in the previous experiment in Section 10.2.3, using the experimental setup shown in Figure 10.33, in which such an NDPA system is labeled PSGA. It is important to note that the birefringence measurement accuracy of the PSGA has been validated with a quartz crystal of precisely know birefringence and therefore is traceable to the quartz birefringence standard, as described in Chapter 9. The FUT was connected with the output port of the binary PSG and the input port of the binary PSA with two low-stress connectors, and is loosely laid on an optical table with a minimum bending radius of ~0.5 m along the whole fiber length, as shown in Figure 10.33. Such a large bending radius only contributes a bending-induced birefringence on the order of 10^{-9} from Eq. (10.17c), which is 2 orders of magnitude smaller than the expected RB value and therefore can be safely neglected.

In measurements, the PSG generates six distinctive input SOPs one at a time and the PSA analyzes the output SOPs from the FUT for each input SOP with six distinctive MO logic settings to obtain the Mueller matrix of the FUT as a function of wavelength. The accumulative DGD $\tau(\omega)$ of the FUT then can be obtained from the Mueller matrix analysis as described in Chapter 9. Note that

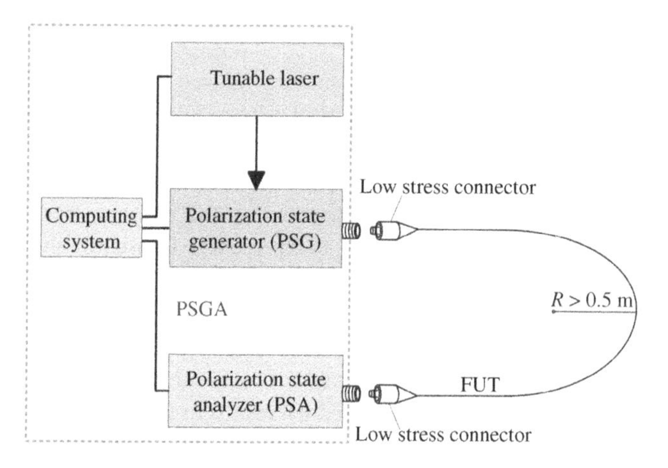

Figure 10.33 Measurement setup of FUT's residual birefringence (RB) based on a binary Mueller matrix polarimetry system (PSGA) described in Chapter 9 (Feng et al. 2018).

Table 10.3 Repeated measurements of FUT's RB with non-distributed polarization analysis system.[a]

Groups	1	2	3	4	5	6
RB	2.546×10^{-7}	2.549×10^{-7}	2.577×10^{-7}	2.586×10^{-7}	2.556×10^{-7}	2.544×10^{-7}
Average	$\mathbf{2.560\times10^{-7}}$					

a) Feng et al. (2018).

in Figure 10.33, six (6) distinctive logic states for polarization generation and analysis are used, as compared with four distinctive logic states in the DPA system described in the last section, for higher measurement accuracy. Finally, the RB Δn_{RB} of the FUT can be obtained as

$$\Delta n_{RB} = \frac{\tau(\omega)c}{L} \tag{10.19}$$

where L is the length of the FUT and ω is the angular frequency. Six groups of repeated measurements were carried out and the results are listed in Table 10.3. In each group of the measurements, 10 repeated experiments were performed and the average DGD $\tau(\omega)$ was calculated. Note that we slightly re-arranged the FUT layout on the optical table before each group of the measurements started. As shown in Table 10.3, the average of six measured RB values is calculated to be 2.560×10^{-7}, which is very close to the average RB of 2.545×10^{-7} measured by the DPA system, with a relative error of only 0.59%, validating the measurement accuracy of the DPA system and the correctness of the fitted Eq. (10.18). Note that unlike the DPA method described in Figure 10.30, NDPA method described here can only obtain the accumulative RB of the FUT along the whole fiber length L. The small deviations of the bending-induced birefringence coefficient k between the theoretical and experimental values and the RB between values obtained with DPA and NDPA may be attributed to the following factors: (i) the slight inaccuracies of the parameters used in the theoretical calculation as compared with those of a real fiber, (ii) the inaccuracies of the bending radius measurements of the 12 fiber loops, (iii) the errors associated with the non-circularity of the fiber loops because the fiber loops may not be perfectly circular, and (iv) the errors associated with the fiber twist.

10.2.4.2 Bending-Induced Birefringence Validation

Because the birefringence measurement accuracy of the NDPA system (PSGA) is traceable to the birefringence standard of a quartz crystal, it will be convincing to use the PSGA to validate the accuracy of the bending-induced birefringence values measured with the DPA system (PA-OFDR). Since the PSGA is of a non-distributed birefringence measurement system, for each bending induced birefringence measurement, only a single fiber loop can be made in the FUT. In the experiment, an FUT of the same length and from the same fiber batch is used as in the experiment of Figure 10.30. As shown in Figure 10.34a, for each measurement, a single fiber loop on the FUT was made and its birefringence was measured with the PSGA. Total 12 sets of measurements were performed, each having a different loop radius gradually decreasing from ~3.75 to ~1.00 cm with a 0.25 cm interval. Each set measurement was repeated 10 times and the average of the 10 measurements was taken as final birefringence value of the corresponding fiber loop. The fiber loops were produced using the same glass cylinders as described in Section 10.2.3. The FUT with the fiber loop was affixed with adhesive tapes on the optical table for measurement stability. The adhesive tape is of soft material to minimize the pressure induced birefringence. Theoretically, the bending-induced birefringence

Δn_{loop} resulting from looping a loop with a perimeter of l_{loop} can be calculated from the DGD $\tau_{loop}(\omega)$ of the fiber loop as

$$\Delta n_{loop} = \frac{\tau_{loop}(\omega)c}{l_{loop}} \tag{10.20}$$

To eliminate the contribution of the RB in the FUT, a reference measurement of the FUT's Mueller matrix $\mathbf{M}_{ref}$ without a loop was first taken before each Mueller matrix $\mathbf{M}_{total}$ measurement of the FUT with the loop was performed. The Mueller matrix of the FUT of the loop without the RB contribution can be obtained with

$$\mathbf{M}_{loop} = \mathbf{M}_{total} \cdot \mathbf{M}_{ref}^{-1} \tag{10.21}$$

Finally, the DGD $\tau_{loop}(\omega)$ can be calculated from the Mueller matrix $\mathbf{M}_{loop}$ of the fiber loop, as described in Chapter 9.

Using the methodology described earlier, the bending-induced birefringence Δn_{PSGA} for all of the 12 fiber loops were measured and plotted in Figure 10.34b using the squares. Note that the birefringence value for each bending radius is the average of 10 repeated measurements. We curve fitted the PSGA measurement data in Figure 10.34b to Eq. (10.17c), and obtained Eq. (10.22) with an Adj. R-Square of 0.997 43 as

$$\Delta n_{PSGA} = 6.49 \times 10^{-10} \left(\frac{1}{R}\right)^2 \tag{10.22}$$

Note that the BBC obtained here is 6.49×10^{-10} m^2, highly consistent with the value of 6.601×10^{-10} m^2 obtained using the DPA system, with a relative error of only 1.68%. For visual comparison, the birefringence data measured by the PA-OFDR in Figure 10.32a were also plotted in Figure 10.34b with solid circles, but with the RB subtracted from every data point. It is evident that the data points obtained with the two different methods agree well with each other, with a maximum relative error of 8.71%. The slight deviation may be attributed to the measurement uncertainties of the perimeters of the 12 fiber loops and the errors resulting from the RB induced by the adhesive tapes for affixing the FUT on the optical table. The measurements earlier with a non-distributed Mueller matrix polarimetry system not only validated the performance of the distance-resolved birefringence measurement with the distributed Mueller matrix polarimetry

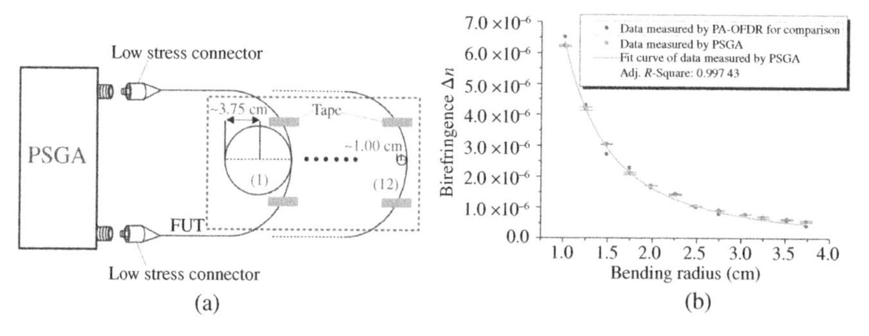

Figure 10.34 (a) Measurement setup of bending-induced birefringence of SMF with only a single loop made on the fiber; 12 loops were successively made and measured, one at a time, with radii gradually decreasing from ~3.75 to ~1.00 cm with a 0.25 cm interval. (b) Bending-induced birefringence as a function of the fiber bending radius measured by the PSGA system (squares). The small error bars indicating the high precision of the PSGA measurements. For comparison, the birefringence values measured with the PA-OFDR are also plotted (solid circles) and note that the residual birefringence has been subtracted from every measured birefringence datum (Feng et al. 2018).

system (PA-OFDR) in Section 10.2.3, but also cross checked the accuracy first bending-induced birefringence coefficient measurement of an SMF.

10.2.5 Distributed Transversal Force Sensing

With the distributed birefringence measurement capability of a DPA system validated, we now describe using it for distributed transversal force/transversal stress (TF/TS) sensing, an important area with multiple applications ranging from industrial and environmental structures to defense constructions (Saida and Hotate 1997; He and Hotate 2002; Maier et al. 2010; Wei et al. 2016; Schenato et al. 2020), especially in pressure monitoring area, including oil/gas downhole and pipeline, geotechnical engineering, water distribution, and sewerage utilities (Schenato et al. 2020; Vanrolleghem and Lee 2003; Sadeghioon et al. 2018; Zhang et al. 2019; Zhu et al. 2019). Compared with the numerous development efforts for distributed fiber optic strain and temperature sensing (Zhou et al. 2013, 2016; Song et al. 2014; Wada et al. 2016; Zhang et al. 2017; Ding et al. 2018; Bassil et al. 2019), the studies on distributed fiber optic TF sensing are limited (Maier et al. 2010; Wei et al. 2016; Schenato et al. 2020). Unlike the TF sensing described in Section 10.1, in which expensive PM fibers and time consuming fiber axis alignment are required, here standard communication fiber can be used for simplicity and reduced cost.

10.2.5.1 Sensing Principles

Birefringence Induced by TF in SMF As described in Section 3.3, when a length of SMF is subject to a transversal line force (TLF) f, the birefringence Δn induced via the photoelastic effect can be expressed as (Namihira et al. 1977; Smith 1980a, 1980b)

$$\Delta n = \frac{4n^3}{\pi E}(1+\sigma)(p_{12}-p_{11})\left(\frac{f}{d}\right) = \zeta f \tag{10.23a}$$

where n is the refractive index of fiber core, E is Young's modulus, σ is Poisson's ratio, d is the diameter, and p_{1j} ($j = 1, 2$) are the strain-optic coefficients, describing the photoelastic effect which relates the change of refractive index to the mechanical strain, for an isotropic material (Bertholds and Dandliker 1988). Finally, the proportion constant ζ relating the TLF and birefringence is defined as the TF measurement sensitivity:

$$\zeta = \frac{4n^3}{\pi dE}(1+\sigma)(p_{12}-p_{11}) \tag{10.23b}$$

Taking the parameters of fused silica: $n \approx 1.46$ at 1550 nm, $E = 6.5 \times 10^{10}$ N/m, $p_{11} = 0.12$, $p_{12} = 0.27$, and $\sigma = 0.17$, for an SMF with a cladding diameter $d = 125$ μm, the proportion constant ζ is estimated to be 8.559×10^{-8} RIU/(N/m). Such a value is calculated by assuming the SMF made with fused silica with uniform stresses inside the core region, without considering the difference between the fiber core and cladding.

Experimental Setup The experimental setup of distributed TF sensing using the PA-OFDR is shown in Figure 10.35 (Feng et al. 2020). The construction and operation of the PA-OFDR are the same as that described in Figure 10.28. Following the same procedure described in Section 10.2.1, the TF induced birefringence can be obtained using Eq. (10.16).

From Eq. (10.23a), the distributed TF fiber sensing using a PA-OFDR system can be achieved with a sensitivity equal to the proportion constant ζ. A simple experiment set up shown in Figure 10.35 was used to demonstrate the distributed TF fiber sensing. The whole SMF-UT, with a length of ~18 m, was placed on a stainless steel optical table, and affixed with soft adhesive tapes in order

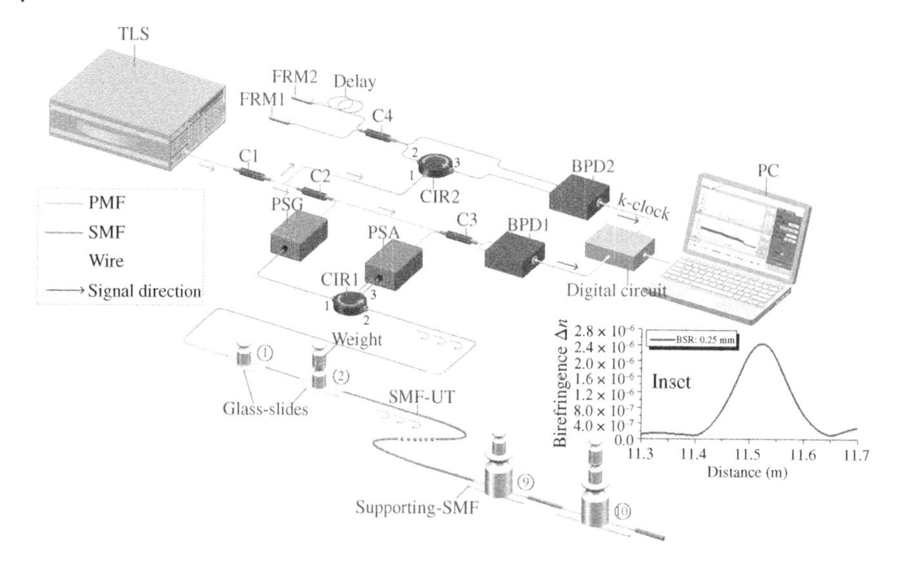

Figure 10.35 Schematic of experimental setup for distributed TF fiber sensing, with TFs simultaneously applied at 10 different positions along the SMF-UT, using a PA-OFDR system. Different weights are placed on the top of a glass-slide for applying different TFs onto the SMF-UT and the length of the glass-slide determines the force-applying length. A supporting-SMF, in parallel with the SMF-UT, is used for maintaining the balance of every glass-slide. The coatings of each fiber segment subject to the TF and the corresponding supporting SMF are removed. Inset shows measured birefringence vs. distance along SMF-UT under a typical TF with a BSR of 0.25 mm for data processing. PMF: polarization maintaining fiber; TLS: Yenista T100S-HP tunable laser source; C1, C2, C3: PMF couplers; C4, C5: SMF couplers; PSG: polarization state generator; PSA: polarization state analyzer; CIR1, CIR2: optical circulators; BPD1, BPD2: balanced photodetectors; FRM1, FRM2: Faraday rotation mirrors; SMF-UT: SMF under test; PC: personal computer. Data acquisition and processing were implemented with a personal computer and a LabVIEW™ based software (Feng et al. 2020).

to avoid inducing undesired stress. Care was also taken to avoid twisting the fiber. In addition, any possible bending along the SMF-UT was assured to have a radius larger than 10 cm, for assuring bending-induced birefringence negligible. Ten (10) fiber segments were selected, as numbered in Figure 10.35, with an interval of ~0.5 m from ~11 to ~16 m along the SMF-UT, to apply transversal force along the SMF-UT. The buffer coatings of the 10 fiber segments were stripped carefully with no damage to the fiber itself. A TF was precisely applied to each fiber segment by putting one or more calibration weights stacked together on a glass-slide pressing on the SMF-UT. To keep each glass-slide balanced, a supporting-SMF, identical with the SMF-UT and with the coating removed, was used, and subsequently only half of the total weight was experienced by the SMF-UT. All the glass-slides had the same width of 2.5 cm but were cut into different lengths for changing the force-applying lengths onto the fiber. Therefore, the TLF f applied to a segment of SMF-UT can be expressed as

$$f = \frac{(m_w + m_s)g}{2L_s} = \frac{m_t g}{2L_s} \tag{10.24}$$

where m_w is the mass of the calibration weights, m_s is the mass of the glass-slide, m_t is the total mass loaded on the fiber segment, $g = 9.8$ N/kg is the gravitational acceleration, and L_s is the contact length between the glass-slide and the fiber segment (the force-applying length).

In the inset of Figure 10.35, a typical measured curve of birefringence along the SMF-UT, induced by a TF with a weight of 400 g and a force-applying length of 7.5 cm, is shown, obtained using a

BSR of 0.25 mm for the data processing. In order to avoid the measuring uncertainties, ten (10) repeated measurements are performed and the curve represents the 10 times average. Note that the birefringence distribution inside the 7.5 cm force-applying length is expected to be uniform. The Gaussian-like distribution is mainly caused by the low-pass digital filtering during data processing and averaging of multiple data traces, including multiple frequency-scans in each measurement and 10 full measurements. Due to the data processing described earlier, the Gaussian-like peak is found to have a full width at half maximum larger than the force-applying length. This issue can be solved by optimizing the data processing algorithm in the future.

10.2.5.2 Calibration of TF Measurement Sensitivity

In order to demonstrate a true direct distributed TF fiber sensing, a calibration should be performed to determine the TF to birefringence conversion coefficient ζ or the TF measurement sensitivity of each SMF batch, although it can be estimated to be 8.559×10^{-8} RIU/(N/m) using Eq. (10.23b) theoretically. In practice ζ may be slightly different from fiber batch to batch due to small variations in fiber dimensions and material properties. Since either the TF itself or the force-applying length can change the TLF f in Eq. (10.23a), two groups of experiments with different force-applying lengths and different TF are conducted to minimize potential errors in the calibration process. Note that one may apply different TLF to a single segment of optical fiber to perform the calibration for obtaining ζ, ten (10) different TLF's are chosen to be simultaneously applied to 10 different fiber sections to perform the calibration, which can average out the slight position dependent variations, as will be shown next.

Calibration by Changing Force-Applying Length By taking advantage of distributed measurement capability of the PA-OFDR system, the first group of experiments was carried out by applying the same amount of calibration weight on 10 fiber segments along an SMF-UT with different force-applying lengths from 8 to 17 cm, with 1 cm increment, respectively, in a single measurement. For instance, a 400 g mass was loaded on each glass-slide by stacking three weights (200, 100, and 100 g) on top of one another to obtain the birefringence curve along the SMF-UT with the PA-OFDR, as shown in Figure 10.36a, averaged over 10 repeated measurements. As can be seen, ten (10) TF-induced birefringence peaks with 10 different force-applying lengths were produced, with the corresponding force-applying length marked on each peak. These peaks are inversely proportional to the force-applying length, as expected from Eq. (10.23a). Figure 10.36a also shows the RB of the fiber on the fiber section from 8 to 11 m without applied force, which is mainly the results of inherent geometric asymmetry and internal stress of the fiber, with an average value of 1.707×10^{-7}. This RB is smaller than the previous measured value in the last section, probably due to the batch differences. In addition, a minimum RB of 9.172×10^{-8} was obtained, indicating that the PA-OFDR's birefringence measurement resolution should be $<1 \times 10^{-7}$ RIU, improved from the previous record. The RB is important because it sets the limit on the minimum measurable TLF, as will be shown shortly.

The peak values of the TF-induced birefringence as a function of TLF is plotted in Figure 10.36b, with the error bars from the 10 repeated measurements shown on each data point with short gray lines. The values of TLF in the horizontal axis were obtained with Eq. (10.24). It is evident that this sensing system has an excellent birefringence measurement repeatability. Curve-fitting of the averaged data point with Eq. (10.23a) yields the TF sensitivity ζ (TLF to birefringence conversion coefficient) to be 9.162×10^{-8} RIU/(N/m), with the goodness of fit (Adj. R-Square) of 0.994 95, in good agreement with the theoretically estimated value of 8.559×10^{-8} RIU/(N/m). The slight deviation is expected, because the theoretical value was calculated with parameters of bulk fused silica, but the experimental data were obtained with real fiber having a core and cladding, in addition to the measurement uncertainties, such as the nonuniformity of the contact length between the fiber segment and the glass-slide. Additionally, the RB of 1.604×10^{-7} obtained from the curve-fitting is very close to the directly measured average value of 1.707×10^{-7}, validating again that this PA-OFDR system is capable of performing high accuracy distributed measurement of SMF's RB for determining the minimum detectable TF.

In addition to loading the fiber segments with 400 g weight, different weights of 200, 300, 500, and 600 g were applied to obtain the TF measurement sensitivity, which are listed in Table 10.4, together with the corresponding RB obtained from curve-fitting. As can be seen, the TF sensitivity values are highly consistent, but the RBs have relatively larger variations.

Calibration by Changing Applied TLF In this group of experiments, different amounts of TF were applied to 10 different fiber sections along the SMF-UT, all with the same length of glass-slide, to measure the induced birefringences in a single measurement. For instance, using a force-applying length of 12 cm, the birefringence distribution along the SMF-UT shown in Figure 10.37a was obtained by taking 10 repeated measurements and averaging them. The corresponding TLF is calculated using Eq. (10.24) and marked on each peak. As expected from Eq. (10.23a), the TF-induced birefringence increases proportionally with the applied TLF along the fiber. The peak values of the TF-induced birefringence as a function of TLF are plotted in Figure 10.37b with the error-bars representing the measurement uncertainty of the 10 measurements. Curve-fitting the experimental data to Eq. (10.23a) yields the expression given in Figure 10.37b, with an Adj. R-Square value of

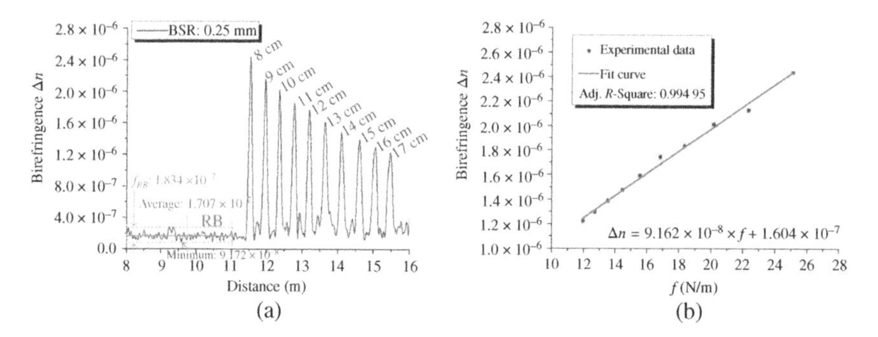

Figure 10.36 (a) Measured birefringence curve along an SMF-UT with a 400 g weight applied onto 10 fiber segments with different force-applying lengths shown on each peak. Averaging of 10 repeated full measurements was performed for curve plotting. The residual birefringence (RB) of the SMF-UT was obtained from fiber section without any applied load from 8 to 11 m with the maximal fluctuation range of RB (f_{RB}) and the minimum RB value shown in the figure. (b) TF-induced birefringence as a function of transverse line-force (or TLF) f. Error-bars are to show the variations of 10 repeated measurements (Feng et al. 2020).

Table 10.4 TF sensitivity and RB values obtained by varying force-applying lengths under different loading weights (Feng et al. 2020).

Loading weights (g)	**200**	**300**	**400**	**500**	**600**
TF sensitivity (RIU/(N/m))	9.419×10^{-8}	9.216×10^{-8}	9.162×10^{-8}	9.211×10^{-8}	9.125×10^{-8}
RB	1.116×10^{-7}	1.052×10^{-7}	1.604×10^{-7}	1.271×10^{-7}	1.287×10^{-7}

Table 10.5 TF sensitivity and RB values obtained by varying TF on different force-applying lengths (Feng et al. 2020).

Force-applying length (cm)	**8**	**10**	**12**	**14**	**16**
TF sensitivity (RIU/(N/m))	9.209×10^{-8}	9.365×10^{-8}	9.132×10^{-8}	9.185×10^{-8}	9.203×10^{-8}
RB	1.262×10^{-7}	1.031×10^{-7}	1.489×10^{-7}	1.636×10^{-7}	1.503×10^{-7}

0.998 53. Using the expression, a TF sensitivity of 9.132×10^{-8} RIU/(N/m) and a RB of 1.489×10^{-7} are obtained and listed in Table 10.5. To verify the measurement repeatability of the PA-OFDR system, similar experiments using different force-applying lengths of 8, 10, 14, and 16 cm, respectively, were performed and the corresponding TF sensitivities and RBs via curve-fitting were obtained, with the results listed in Table 10.5. Similar to those in Table 10.4, the TF sensitivity values are highly consistent with one another, indicating an excellent sensing repeatability.

Using the data in Tables 10.4 and 10.5, the average values of TF sensitivity ζ and RB of 9.223×10^{-8} RIU/(N/m) and 1.325×10^{-7}, respectively, and the corresponding equation relating the TF-induced birefringence with the TLF f are obtained as

$$\Delta n = 9.223\times10^{-8}\times f + 1.325\times10^{-7} \tag{10.25}$$

The averaged TF sensitivity of 9.223×10^{-8} RIU/(N/m) is close to the theoretical estimate of 8.559×10^{-8} RIU/(N/m), with a relative difference of only 7.31%.

10.2.5.3 Validation of Distributed TF Fiber Sensing

With the TF sensitivity ζ obtained in Eq. (10.25), let's now proceed to validate TF sensing by first measuring the birefringence and then calculate the TLF f using

$$f = \frac{\Delta n}{\zeta} = 1.084\times10^{7}\times\Delta n \tag{10.26}$$

Note that the RB in Eq. (10.25) cannot be directly included in the calculation of TLF f because its direction may not be aligned with that of Δn (they are both vectors in practice). It can be considered

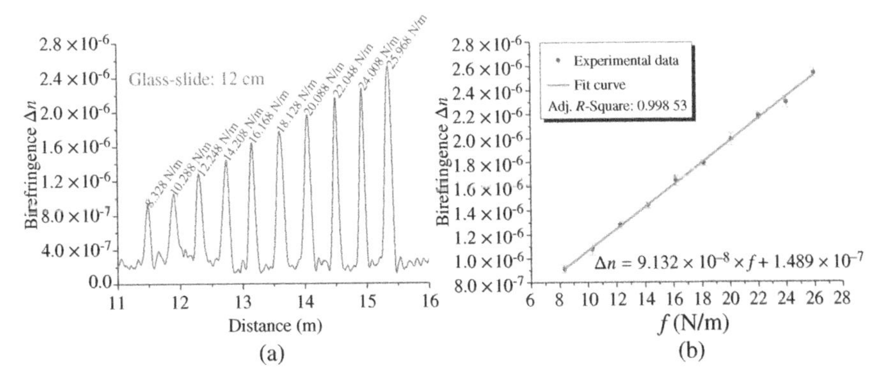

Figure 10.37 (a) Measured birefringence curve along the SMF-UT with 10 different TFs applied onto 10 different fiber segments using the same length of glass-slides (12 cm) obtained with 10 repeated measurements. The TLF f applied to each fiber segment is calculated using Eq (10.24). (b) TF-induced birefringence as a function of TLF f. The error-bars are plotted in gray to show the measurement repeatability of our PA-OFDR system (Feng et al. 2020).

Table 10.6 Comparison of applied and measured TLF with the PA-OFDR sensing system (Feng et al. 2020).

No. of fiber segment	Force-applying length (cm)	m_t (g)	Applied TLF (N/m)	Measured TLF (N/m)	Relative error (%)
①	8	329.50	20.182	21.799	8.012
②	8	329.50	20.182	21.326	5.668
③	9	370.43	20.168	21.810	8.142
④	9	370.43	20.168	22.432	11.226
⑤	10	411.36	20.157	22.093	9.605
⑥	10	411.36	20.157	22.037	9.327
⑦	11	451.64	20.119	22.350	11.089
⑧	11	451.64	20.119	22.406	11.367
⑨	12	491.94	20.088	21.590	7.477
⑩	12	491.94	20.088	22.653	**12.769**
No. of fiber segment	**Force-applying length (cm)**	**m_t (g)**	**Applied TLF (N/m)**	**Measured TLF (N/m)**	**Relative error (%)**
①	8	1609.50	98.582	98.851	0.273
②	8	1609.50	98.582	99.957	1.395
③	9	1810.43	98.568	100.130	1.585
④	9	1810.43	98.568	100.965	**2.432**
⑤	10	2011.36	98.557	98.894	0.342
⑥	10	2011.36	98.557	100.693	2.167
⑦	11	2211.64	98.519	99.902	1.404
⑧	11	2211.64	98.519	100.553	2.065
⑨	12	2411.94	98.488	100.076	1.612
⑩	12	2411.94	98.488	99.328	0.853

as a background noise source, limiting Δn measurement accuracy. If the orientation of both Δn and RB was obtained in the measurement, RB could be subtracted vectorially from Δn to improve the measurement accuracy.

Two TF sensing experiments (Experiment I and Experiment II) were performed, the first one with a set of TLFs of ~20.1 N/m applied onto 10 fiber segments (sequentially numbered ①–⑩) in a sensing SMF-UT (Exp. I) from the same fiber spool as that in Figures 10.36 and 10.37 (identical production batch), and the second one with a set of TLFs of ~98.5 N/m applied onto the 10 fiber segments (No. ①–⑩) in a second sensing SMF-UT (Exp. II) from the same fiber spool. Note that the force-applying lengths for the TF loads on the 10 fiber segments were purposely chosen to have five different lengths of 8, 9, 10, 11, and 12 cm to show that the measurement accuracy is independent of the force-applying lengths. Table 10.6 lists the applied TLF and the corresponding force-applying length on each fiber segment in the two experiments, with the upper portion from Exp. I and the lower portion from Exp. II.

Figure 10.38 shows the measured TLF f as a function of fiber distance by first obtaining the TF induced birefringence Δn and then converting it to f using Eq. (10.26). Ten (10) repeated measurements were averaged for each curve. Table 10.6 lists the measured TLFs for the 10 fiber segments

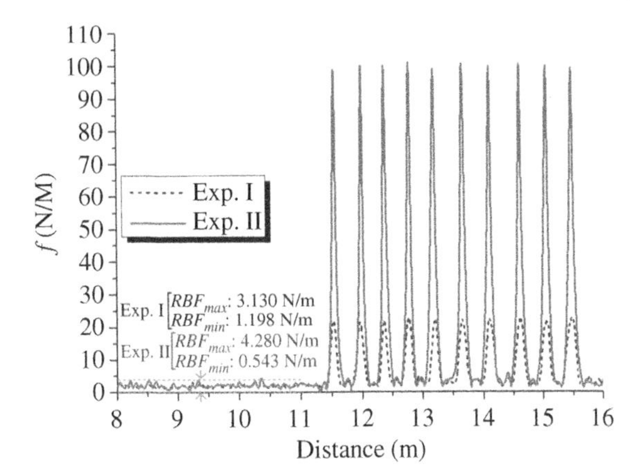

Figure 10.38 Measured TLF curves from two separate experiments (Exp. I and Exp. II) with different TLF applied onto 10 fiber sections along two SMF-UTs with different force-applying lengths. Solid line: ~98.5 N/m TLF applied. Dotted line: ~20.1 N/m TLF applied. Five different force-applying lengths were chosen to apply the TLF on the fiber sections. The RB equivalent TLF (RBF) on the fiber section without the applied TLF are also shown on the left side of the curve from 8 to 11 m, with the maximum RBF_{max} of 3.130 and 4.280 N/m for the two experiments, respectively (Feng et al. 2020).

in the two experiments. The relative errors between the applied TLFs and the measured TLFs are also listed in Table 10.6. As can be seen, the maximum relative errors are 12.769% for the case of ~20.1 N/m applied TLFs in Experiment I and 2.432% for the case of ~98.5 N/m applied TLFs in Experiment II. In addition, the relative errors obtained in Experiment I for the smaller applied TLFs are all much larger than those obtained in Experiment II for the larger applied TLFs. Clearly, the RB has a larger impact on the measurement accuracy for the case of smaller applied TLF, consistent with one's expectations. Other factors affecting the measurement accuracy include system noises and the uncertainty of the contact lengths between the glass and the fiber which affects the accuracy of the force-applying length.

The RB equivalent TLF (RBF), which is the RB converted to TLF using Eq. (10.26), is shown on the left hand-side from 8 to 11 m in Figure 10.38. The maximum values of RBF (RBF_{max}) corresponding to the two fibers in two experiments are 3.310 N/m (Exp. I) and 1.198 N/m (Exp. II), respectively, and the minimum values of RBF (RBF_{min}) are 1.198 N/m (Exp. I) and 0.543 N/m (Exp. II), respectively. In practice, it is feasible to perform a RB measurement of the entire sensing fiber to determine the RBF as a function of distance along the fiber so that the minimum detectable TLF can be determined for every position of the fiber.

Maximum Detectable TLF The maximum TLF detectable value f_{max} is an important parameter of a distributed TF fiber sensing system. Considering that the accumulative retardation induced by the TF over the range of BSR of the DPA system cannot be over π to avoid phase wrapping problem, according to Eq. (10.16), the maximum TF-induced Δn must satisfy:

$$\Delta n \cdot BSR \leq \frac{\lambda}{4} \tag{10.27}$$

With the PA-OFDR based DPA system, the BSR is 0.25 mm and the resulting maximum measurable TF-induced birefringence Δn_{max} =1.55 × 10^{-3} RIU at 1550 nm. The corresponding TLF from Eq. (10.26) is

$$f_{max} = \frac{\Delta n_{max}}{\zeta} = 1.68 \times 10^4 \ \ (\text{N/m}) = 16.8 \ \ (\text{N/mm}) \tag{10.28}$$

Note that BSR is defined as the minimum fiber length for the retardation or birefringence calculation in data processing and can be software changed in the PA-OFDR system.

Minimum Detectable TLF The minimum TLF detectable value f_{min} is another important parameter for distributed TF sensing, which is defined as the applied TLF for inducing a birefringence

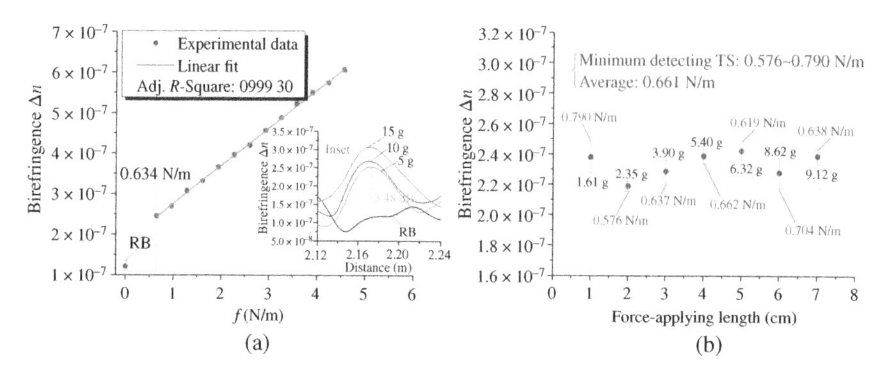

Figure 10.39 (a) TF-induced birefringence as a function of applied TLFs generated by loading different weights from 5 to 65 g with 5 g increment onto a fiber segment via a glass-slide of 7.5 cm for determining f_{min}. Inset shows the birefringence distributions along the fiber segment under test with 0, 5, 10, and 15 g weight on the force-applying glass-slide, respectively. Inset also illustrates the 3-dB criteria for determining f_{min}. (b) Minimum detectable TLF induced birefringence as a function of force-applying length. For both figures, six repeated measurements were averaged for each data point (Feng et al. 2020).

twice of the RB or 3 dB above the RB, as shown in the inset of Figure 10.39a. To determine f_{min}, a fiber segment with low RB was located and weights from 5 to 65 g with an increment of 5 g over a glass-slide having a length of 7.5 cm (the force-applying length) were applied to measure the TF induced birefringence Δn, with the results shown in Figure 10.39a. For each data point, six (6) repeated measurements were averaged, with the error-bars representing the maximum data variations, indicating an excellent repeatability of the measurement system. Curve-fitting of the experimental data obtained a straight line with a superb goodness of fit of 0.999 30. As a reference, the RB is also plotted in the figure. It is evident from the fitted line that the minimum detectable TLF, f_{min}, is 0.634 N/m.

As an alternative, another set of experiments with different force-applying lengths of 1, 2, 3, 4, 5, 6, and 7 cm, respectively, was performed and the minimum detectable TF-induced birefringence as a function of force-applying length was obtained as shown in Figure 10.39b. For each force-applying length, the corresponding total weight on the fiber is also given. As can be seen, the minimum detectable TLFs, varying from 0.576 to 0.790 N/m, are all close to the average of 0.661 N/m (or 6.61×10^{-4} N/mm). The result indicates that the PA-OPDR based DPA system is highly sensitive to the TF. Taking a conservative estimate by assuming a force-applying length of 5 mm, the system is capable of sensing a weight as small as 0.68 g from the f_{min} obtained earlier, which is about 1/4th weight of a US penny, to put in perspective. Additionally, according to the obtained f_{max} of 16.8 N/mm and f_{min} of 6.61×10^{-4} N/mm, a TF sensing dynamic range of this system is calculated to be over 44 dB.

Spatial Resolution and Maximum Sensing Distance Spatial resolution and the maximum measurement distance are other two important parameters for distributed TF sensing. To determine the spatial resolution, a single-point TLF on a fiber segment using a digital force gauge (DFG) (Sundoo Instrument, model SH-50) was applied, as shown in Figure 10.40a. In the experiment, a total length of ~6.3 m SMF-UT was used and the fiber segment with its buffer coating removed was fixed between two micro-translation stages with soft adhesive tapes. In addition, the BSR of the PA-OFDR system is selected to be 0.25 mm, which is the fiber length over which the retardation was accumulated and calculated in data processing. The contact length of the DFG's force-applying head against the fiber was 1.0 mm, resulting in a TF-applying length of 1.0 mm. Ideally, a TF-applying length less than the BSR of 0.25 mm should be used to determine the TLF spatial resolution; unfortunately, 1.0 mm is the smallest force-applying head available for the DFG. Therefore the TLF spatial resolution obtained with the 1.0 mm force-applying head is expected to be the upper limit of the true spatial resolution.

In obtaining the TF measurement spatial resolution, a certain TF was applied to the fiber segment with the DFG while the TLF distribution along the fiber was measured, as shown by the solid line (Test-1) in Figure 10.40b. Then the two translation stages were adjusted to move the contact point of the fiber segment with the force-applying head of the DFG step by step with an increment of 0.1 mm while the corresponding TLF distribution along the fiber after applying the same amount of TLF by the DFG was measured, as shown by the dashed line (Test-2). Figure 10.40b shows the measured two TLF curves with a spatial separation of 3.7 mm, and the combined curve was also plotted. As can be seen that the central dip in the combined curve just disappears with the 3.7 mm separation, which therefore can be considered as the TF measurement spatial resolution of our PA-OFDR system following the Sparrow criterion (Sparrow 1916; Nayyar and Verma 1978), although the 3-dB width of the TLF response curve is ~5 mm. Note that ten (10) repeated measurements were averaged for each curve plotting in Figure 10.40b.

According to the Nyquist sampling theorem, the length of delay line of the k-clock interferometer in Figure 10.27 should be at least two times longer than the maximum fiber length to be measured by the OFDR system. In order for the maximum TF sensing distance longer than 100 m, the delay line used in the k-clock interferometer must be longer than 200 m theoretically. In the PA-OFDR based DPA system, the delay line used in the k-clock interferometer was chosen to be 250 m to ensure a >100 m measurement range, although the coherence length of the tunable laser used is around 600 m. In order to verify the TF measurement range, a TLF was first applied onto a fiber segment in an SMF-UT with a length around 6.3 m using a DFG with the setup in Figure 10.40a and obtained a TLF peak (Peak-1) with the PA-OFDR system at the location of 5.652 m, as shown in Figure 10.41a. An SMF with a length ~97.2 m was then inserted before the first 6.3 m fiber, which would effectively move the location of the TF application point around 97.2 m away from the OFDR exit reference point and extend the total fiber length to 103.5 m. Another measurement was taken next using the PA-OFDR and indeed the TF peak was moved to the location at 102.865 m (Peak-2), as shown in Figure 10.41a. With this operation, the fiber segment and the TF applied to the fiber segment were not changed. Note that the inserted 97.2 m fiber was from the same fiber batch as

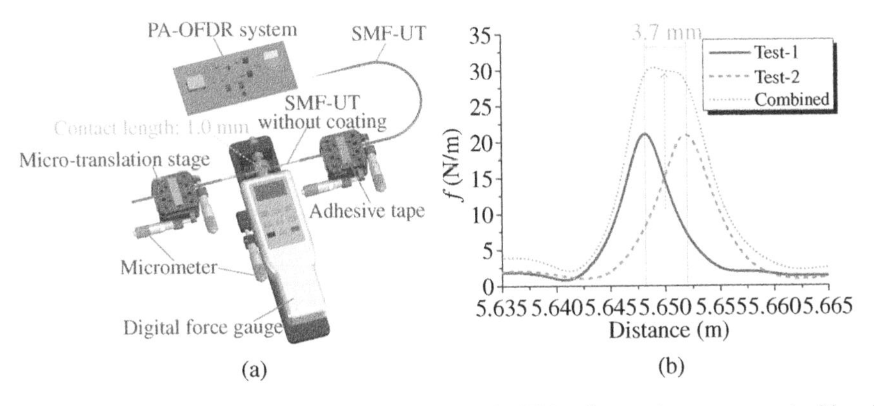

Figure 10.40 (a) Experimental setup of single-point TF loading and measurement with a digital force gauge. (b) Two measured TLF curves along SMF-UT with a spatial separation of 3.7 mm. The combined curve indicates a spatial resolution of 3.7 mm using the Sparrow criterion. Note that the 3-dB width of each TLF curve is about 5 mm (Feng et al. 2020).

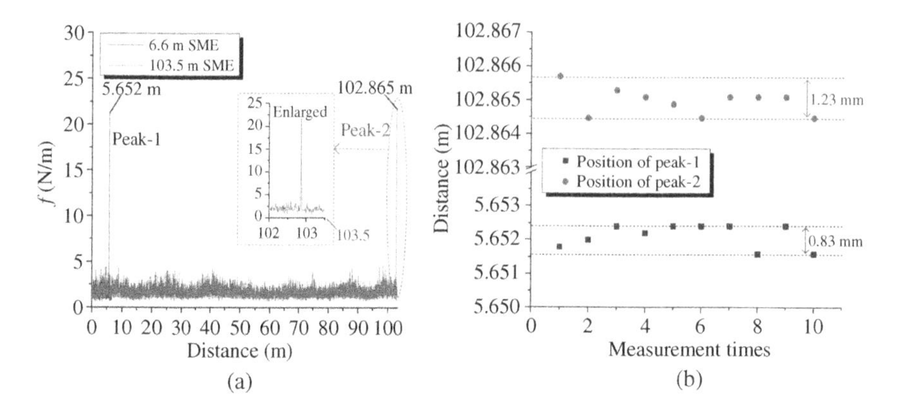

Figure 10.41 (a) TF sensing with a 6.6 m SMF-UT and a 103.5 m SMF-UT. Peak-1 at 5.652 m and peak-2 at 102.862 m were produced with a DFG. Inset displays the enlarged local position of peak-2 to show the maximum measurement distance of 103.5 m. Ten repeated measurements were averaged for each curve. (b) Location uncertainties of peak-1 and peak-2 obtained from 10 repeated measurements.

the first 6.3 m fiber and the data shown in Figure 10.41a are the average of 10 repeated measurements. The average values of Peak-1 at 5.652 m and Peak-2 at 102.865 m are 21.179 and 21.685 N/m, respectively, with excellent agreement.

Figure 10.41b shows the location uncertainties of the measured TLF peak for Peak-1 and Peak-2, respectively, with 10 repeated measurements. As can be seen, a maximum variation of 1.23 mm was obtained for peak-2, as compared with a maximum variation of 0.83 mm for peak-1, much smaller than the validated spatial resolution of 3.7 mm. Therefore, it can be confidently concluded that the TF sensing range of our PA-OFDA is at least 103.5 m, which can be extended by using a longer delay line in the k-clock interferometer and a data acquisition card with higher speed (Feng et al. 2020).

It is worthy to note that the Mueller matrix polarimetry method described earlier requires the SOP to be stable along the SMF-UT during each wavelength scan. Light scattered further down in the fiber is more susceptible to SOP disturbances by temperature and vibration, limiting the accuracy of sensing in longer fibers or in situations with large fiber disturbances.

Influence of SMF Buffer Coating on Distributed TF Sensing A standard SMF for optical communications generally has a buffer coating with a thickness of 62.5 μm made of polyacrylate material with a small Young's modulus to protect it against external stresses, and therefore not suitable for TF sensing. In the experiments earlier, the polyacrylate coating of SMF-UTs at the places for applying TF was removed. However, in practical TF fiber sensing applications, the SMF should have a suitable coating, which allows the efficient force transmission onto the fiber while in the same time protects the fiber. Therefore, it is crucial to investigate the influences of different buffer coatings of the SMF on TF sensing applications.

It was found in a previous work (Zhang et al. 2019) that a thin polyimide coating with a large Young's modulus can efficiently transfer an external TF onto a PM fiber. Here a standard SMF with polyacrylate coating and a polyimide coated SMF (YOFC HT-9/125-14/155) with a core, cladding, and coating diameters of 9, 125, and 155 μm, respectively, are used to measure the TF sensitivity and the RB of the fiber using the same methodology described earlier. The polyimide coated SMF has the same index profile of silica glass as those of the standard SMF, but with a polyimide coating thickness of 15 μm. In each type of fiber, TFs were applied to 10

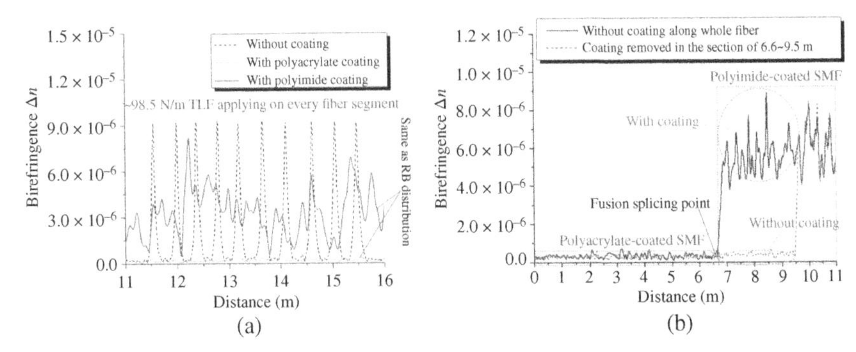

Figure 10.42 (a) Birefringence distributions as a function of fiber distance measured with PA-OFDR system, when TLFs of ~98.5 N/m applied to 10 fiber segments along SMF-UTs without coating (dashed line), with polyacrylate coating (dotted line at bottom) and with polyimide coating (solid line), respectively. (b) RB distributions as a function of fiber distance of a polyacrylate-coated SMF (from 0 to 6.6 m) fusion spiced with a polyimide-coated SMF (from 6.6 to 11 m); Solid line: without polyimide coating removed; Dashed line: polyimide coating was removed in the section from 6.6 to 9.5 m and the coating remained on the rest of the fiber (Feng et al. 2020).

different fiber segments similar to the experiments in Figures 10.36–10.38, but without removing the fiber coating. Figure 10.42a shows the measured birefringence distributions as a function of fiber distance when 10 different segments were subject to TLFs of ~98.5 N/m. For comparison, the corresponding curve measured using a length of fiber with the coating removed as that in Figures 10.36–10.38 was also plotted. As can be seen, with the TLFs of ~98.5 N/m applied to the 10 fiber segments with either the polyacrylate or the polyimide coating, no TF-induced birefringence peaks can be observed, indicating that they were buried under the RB of the corresponding fiber and the applied TLF was not sufficient to induce observable induced birefringence. In addition, one can observe that the standard SMFs with and without the polyacrylate coating have the same RB level; however, the polyimide-coated SMF has much higher RB distribution along the fiber, indicating that the polyimide coating induced much higher internal stress, similar to what was observed in polyimide-coated PMF (Zhang et al. 2019). To further validate the observation, a length of standard polyacrylate-coated SMF was fusion-spliced with a length of polyimide-coated SMF as a new fiber sample, and measured its RB distribution with our PA-OFDR system, with the results shown in Figure 10.42b. As can be seen from the dark-solid line, the RB increased sharply beyond the splicing point and had a much larger fluctuation in the section of the polyimide-coated SMF. After the polyimide coating of the polyimide-coated SMF was removed following the splicing point from 6.6 to 9.5 m, the RB distribution was measured again, with the results displayed by the dashed line in Figure 10.42b, in which the polyimide coating of the fiber from 9.5 to 11 m was not removed for comparison. As expected, the RB of the fiber section with the polyimide coating removed decreased to the same level as that of the polyacrylate-coated SMF. Therefore one can conclude that the polyimide coating is the sole reason for the increased RB and that such a high RB level will significantly limit the minimum detectable TLF. Better coating process can be developed to eliminate the stress on the fiber with the aid of the PA-OFDA.

To obtain and compare the TF measurement sensitivities ζ of SMFs with the polyacrylate coating and polyimide coating, another length of fiber by fusion spliced the two types of fiber together were prepared. Two segments, one located in the polyacrylate-coated SMF section and the other in the polyimide-coated SMF section with a low RB, were chosen for being applied different TLFs on each segment with a same force-applying length of 7.5 cm starting from 100 N/m. Figure 10.43a,b shows the measured birefringence variations as a function of TLF. As can be seen, the minimum

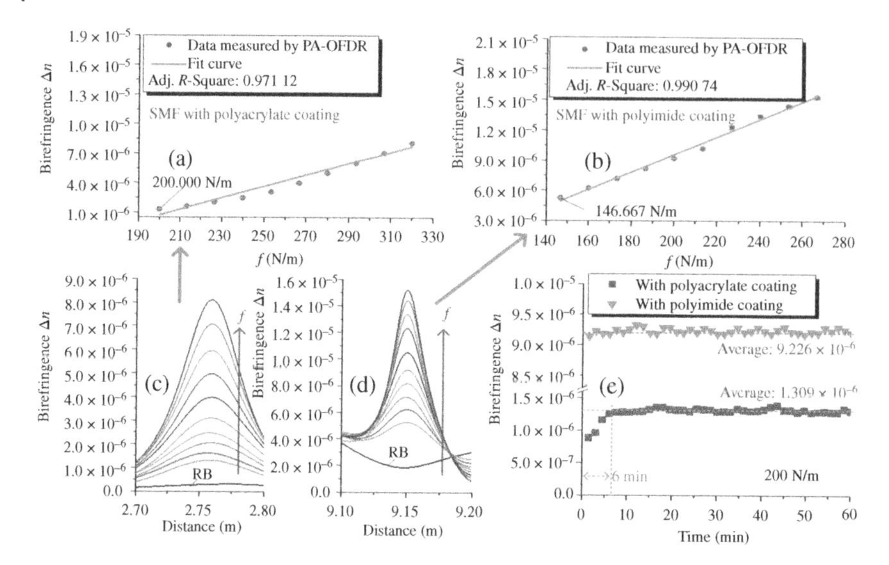

Figure 10.43 TF-induced birefringence variations as a function of TLF using SMFs with polyacrylate coating (a) and polyimide coating (b), respectively. TF sensitivities of both SMFs were obtained via curve-fitting of the experimental data. Variations of TF-induced birefringence against applied TLF for SMFs with polyacrylate coating (c) and polyimide coating (d), respectively, together with the RB of each fiber type. (e) TF-induced birefringence as a function of time when a 200 N/m TLF applied to the SMFs with polyacrylate coating (squares) and polyimide coating (triangles) (Feng et al. 2020).

Table 10.7 Different SMF-UT's TF sensitivity ζ measured with the PA-OFDR system (Feng et al. 2020).

SMF-UT	Without coating	With polyacrylate coating	With polyimide coating	Theoretical value
TF sensitivity (RIU/(N/m))	9.223×10^{-8}	5.839×10^{-8}	8.707×10^{-8}	8.559×10^{-8}

detectable TLFs are 200.000 and 146.667 N/m, for SMFs with polyacrylate coating and polyimide coating, respectively, indicating that the polyimide coating has better force transmission property in comparison. Beyond the minimum detectable TLF, the TF-induced birefringence starts to increase with TLF for both types of fiber, as shown in Figure 10.43c,d respectively. Curve-fitting the experimental data in Figure 10.43a,b yields the TF sensitivities of 5.839×10^{-8} and 8.707×10^{-8} RIU/(N/m) with a goodness of fit of 0.971 12 and 0.990 74, respectively. For comparison, the measured TF sensitivities for different types of SMF are listed in Table 10.7. It is evident that the TF sensitivity of polyimide-coated SMF is much closer to the measured value of the SMF without coating, as well as to the theoretical value. On the other hand, the SMF with polyacrylate coating significantly reduces the TF sensitivity compared with that of a naked SMF.

The response time of the two types of fiber under a same TLF of 200 N/m and a same force-applying length of 7.5 cm was also investigated, as shown in Figure 10.43e. As can be seen, the response time for the polyimide-coated SMF to react to the TF is too fast to be measurable with the PA-OFDR system. In contrast, the time taken for the polyacrylate-coated SMF to produce a stable TF-induced birefringence is ~6 minutes, which is not desired for TF sensing in real time.

From the experimental data one may conclude that the polyimide coated fiber is a good candidate for distributed TF sensing for its large TF sensitivity and fast response time. However, the large RB associated with polyimide coating is a drawback which limits the minimum detectable TLF. Further development is required to reduce the stress of the polyimide coating on the fiber and therefore the RB. As an added advantage, with the capability to measure coating stress induced RB, the PA-OFDR will prove useful for evaluating and improving the coating quality of optical fibers.

10.2.6 Investigation Clamping-Force Induced Birefringence of SM Fibers in V-Grooves

V-grooves are commonly used to hold optical fiber tips for attaching fiber pigtails to micro-optic devices and photonics integrated circuits (PICs), such as silicon photonics and planar-lightwave-circuit (PLC) based devices (Suzuki et al. 2019; Son et al. 2018; Carroll et al. 2016; Dong 2016). They are particularly attractive for holding multiple fiber tips for simultaneously attaching to multiple waveguide ports of a PIC chip because the accuracy of groove spacing can be made extremely high to match the corresponding waveguide spacing on the PIC (Suzuki et al. 2019; Kim et al. 2015). In practice, the fiber tips with a length around a centimeter are affixed in the V-grooves with optical cement and clamped by either a flat-lid or an identical V-groove chip, which inevitably exerts transversal forces on the fiber tips and may induce birefringence via photoelastic effect (Takahashi et al. 2004; Zhang et al. 2021; Wen et al. 2015) in general, as discussed in Chapter 3. Such a birefringence may lead to the degradation of polarization related performance, such as PER and polarization stability. We show in this section that a DPA system described in Sections 10.2.1–10.2.5 can be used to effectively measure the local birefringence in the V-groove induced by the clamping-force to precisely determine the optimum V-groove angles for the two different types of V-grooves (Feng et al. 2022).

10.2.6.1 Theoretical Expressions

Figure 10.44a shows an SMF in a V-groove with a groove-angle of θ clamped with a flat-lid (case 1). The fiber is naked with the buffer coating removed and is pressed by a force f_0 through a glass-slide. A birefringence Δn induced in the fiber for case 1 can be expressed by Eq. (3.9b) for the case free of friction at the V-groove walls. Here it is rewritten with θ replacing 2γ ($\theta = 2\gamma$) and f_0 replacing f ($f_0 = f$) for convenience.

$$\Delta n = \frac{1}{2}\left(1 - \frac{cos\theta}{sin(\theta/2)}\right)\zeta f_0 \tag{10.29a}$$

where ζ is a proportional constant relating to the mechanical properties of the fiber.

Similarly, Figure 10.44b shows a naked SMF clamped by two identical V-grooves with a groove-angle of θ (case 2). A force f_0 is imposed on top of the V-groove. Two V-grooves are parallel and are not in contact with each other. A birefringence Δn induced in the SMF for case 2 can be given as

$$\Delta n = \left(1 - \frac{1}{tan(\theta/2)}\right)\zeta f_0 \tag{10.29b}$$

The detailed derivation for Eqs. (10.29a) and (10.29b) and the definition of coefficient ζ can be found in the Appendix 10.A. For the two cases earlier, the friction between fiber and groove walls is not included, which can be safely ignored in commercial V-grooves made with ZrO_2 or SiO_2, as will be shown later in this section. In addition, for both cases, the weights of the flat-lid in Figure 10.44a and the upper V-groove in Figure 10.44b can be included into the pressing-force f_0 during the actual calculations in experiments, so they are not considered as separated variables in the analysis.

Figure 10.45 shows the plots of Eqs. (10.29a) and (10.29b), the normalized birefringence $\Delta n/(\zeta f_0)$ as a function of the groove-angle θ. As can be seen, the clamping-force induced birefringence in the fiber is zero at groove-angles of $\theta = 60°$ and $\theta = 90°$, respectively, for the two cases, indicating that the clamping-force distribution on the fiber is isotropic and causes only isotropic change of geometrical shape when the fiber is pressed into a 60° V-groove or clamped by two 90° V-grooves. In addition, for each situation, one sees that the values of $\Delta n/(\zeta f_0)$ are opposite in sign when the groove-angles are respectively smaller and larger than the zero-birefringence (ZB) groove-angle (60° or 90°), implying that the fast and slow birefringent axes of a fiber induced by the anisotropic force can be switched by changing groove-angles.

10.2.6.2 V-Grooves and Measurement Setup

Three kinds of V-grooves were investigated, including standard 60° V-grooves made of black ZrO_2 used for holding fibers in commercial fiber fusion splicers, custom-made V-grooves made of silica (SiO_2) glass with different groove-angles, and off-the-shelf multi-channel fiber-array V-grooves made of silica-glass. The product pictures of the three kinds of V-grooves are shown in Figure 10.46a1,b1,c1, respectively. The local details of the V-grooves were examined by an Olympus microscope (Model: BX53MTRF-S) from their cross-sections, as shown in Figure 10.46a2,b2,c2, respectively. With the aid of the CCD pictures captured by the microscope, the groove angle θ of each V-groove can be accurately measured, as illustrated in Figure 10.46d. The measured groove-angle of the ZrO_2 V-groove is 59.99°, and the measured groove-angles of the V-grooves on a typical 4-channel SiO_2 fiber-array substrate are 59.61°, 59.31°, 59.46°, and 59.73°, respectively.

To investigate the birefringence of SMF clamped in different V-grooves, two types of silica-glass V-grooves were custom-made. For the first type (type-1), the depth of each V-groove is chosen such that 20% of the fiber protrudes above the flat-surface as shown in Figure 10.46e1, and the corresponding angles of the custom-made V-grooves are listed in Table 10.8, which are used for experiments of an SMF clamped in a V-groove by a flat-lid.

For the second type (type-2), the depth of each V-groove is made such that 55% of the fiber protrudes above the flat-surface as shown in Figure 10.46e2, with the groove angles of these custom-made V-grooves listed in Table 10.9, which are used for experiments of a fiber clamped by two V-grooves. As can be seen in Tables 10.8 and 10.9, there are slight deviations between the specified groove-angles and corresponding measured values resulting from the production

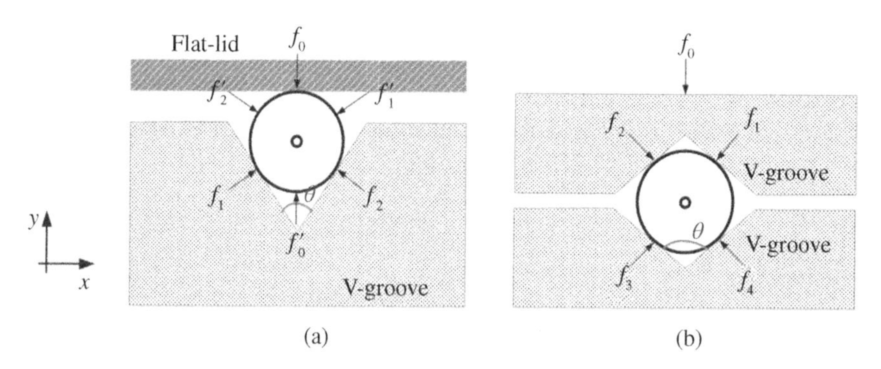

Figure 10.44 Schematic diagrams of fiber showing (a) clamped in a V-groove by a flat-lid (case 1) and (b) clamped by two identical V-grooves (case 2) (Feng et al. 2022).

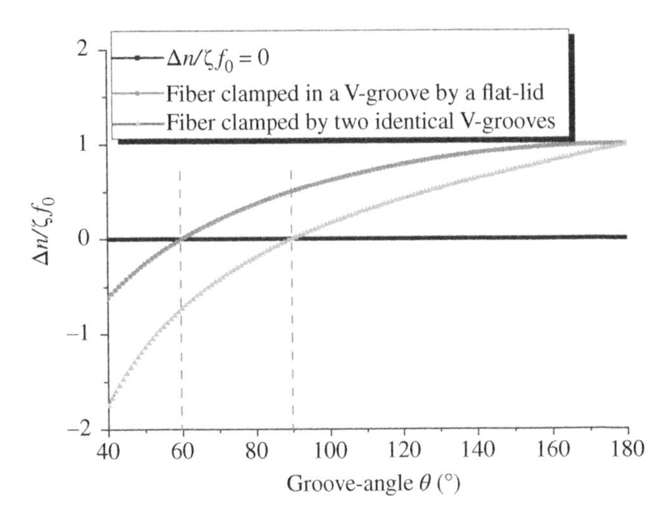

Figure 10.45 Normalized birefringence induced in SM fiber clamped in a V-groove by a flat lid (darker line) and clamped by two identical V-grooves (lighter line) (Feng et al. 2022).

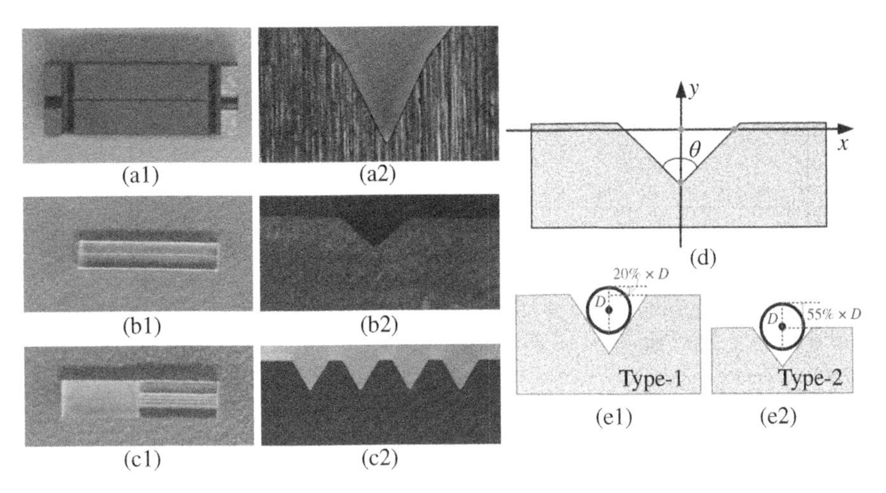

Figure 10.46 Product photos of (a1) a standard 60° V-groove made of ZrO_2 used for holding fiber in a commercial fiber fusion splicer, (b1) a custom-made silica-glass V-groove and (c1) a commercial silica-glass multi-channel fiber-array substrate with four V-grooves; (a2), (b2), and (c2) showing the corresponding micrographs of above V-grooves, respectively. (d) Coordinate system on a V-groove for calculating the groove-angle. (e1) Type-1 V-groove; (e2) Type-2 V-groove; *D* denotes fiber diameter. Source: Feng et al. (2022)/Optica Publishing Group.

imperfections. Note that during production, the V-grooves are first made with a length of 100 mm before each of them is cut into 10 identical pieces with a length of 10 mm.

Figure 10.47 shows the setup for experimentally investigating the birefringence characteristics of fibers clamped in different V-grooves. For type-1 V-grooves, as shown in Figure 10.47a1, a weight is placed on a flat-lid (glass-slide) to press a naked SMF segment (with the buffer coating removed) into a V-groove. Similarly, for type-2 V-grooves, a naked SMF segment is placed into a V-groove to press the fiber segment with an identical V-groove with a weight on top of a glass-slide, as shown in Figure 10.47b1. To maintain the balance of the glass-slide for each case, an identical fiber segment is used to support the glass-slide, as shown in Figure 10.47a1,b1. The size and weight of the

Table 10.8 Custom-made Type-1 V-grooves (Feng et al. 2022).

Customized θ	Measured θ	Customized θ	Measured θ
40°	40.95°	110°	112.04°
50°	52.16°	120°	120.15°
60°	60.92°	130°	129.64°
70°	70.99°	140°	140.63°
80°	82.06°	150°	150.03°
90°	90.35°	160°	159.37°
100°	102.07°		

Table 10.9 Custom-made Type-2 V-grooves (Feng et al. 2022).

Customized θ	Measured θ	Customized θ	Measured θ
40°	40.73°	110°	109.16°
50°	52.30°	120°	119.81°
60°	61.54°	130°	130.55°
70°	70.89°	140°	137.75°
80°	80.65°	150°	149.34°
90°	90.34°	160°	158.22°
100°	101.95°		

glass-slides are $35 \times 25\,mm^2$ and 2.14 g, respectively. As a reference, a fiber segment is put on a flat-surface (corresponding to a groove angle of 180°) directly to be pressed with a weight on top of a glass-slide, as shown in Figure 10.47c1. Here the flat-surface is the optical table surface and another glass-slide with the same length (10 mm) as that of the V-groove is used. The photos of corresponding experiment setups are given in Figure 10.47a2,b2,c2, respectively.

In addition, to validate the theoretical results of Eqs. (10.29a) and (10.29b), and to demonstrate the distributed measurement of V-groove induced birefringence, 14 fiber segments clamped in different V-grooves along a length of SMF for type-1 and type-2 V-grooves are set, respectively, as shown in Figure 10.47d,e. The groove-angle successively increases for segments 1–13 in accordance with Tables 10.1 and 10.9 for type-1 and type-2 V-grooves, respectively. The last segment 14 is the flat-surface reference with an equivalent groove angle of 180°.

10.2.6.3 Experimental Results and Discussions

All experiments were carried out at the room temperature (with air conditioning constantly running) in a quiet environment. As mentioned previously, the birefringence value Δn directly obtained from our DPA system may be affected by the noise filtering algorithms used in data processing, and therefore a calibration should be implemented. Here, a high accuracy NDPA system is used for calibration, whose birefringence measurement accuracy is traceable to a quartz standard, as described in Chapter 9. With this binary Mueller matrix polarimetry system, the absolute birefringence values induced by different weights applied to a glass-slide over a fiber segment (similar to the setup in Figure 10.47c1) are measured, to obtain a calibration equation $\Delta n = 8.914 \times 10^{-8} \times f_0$ relating the birefringence Δn to the applied force f_0. Comparing with the

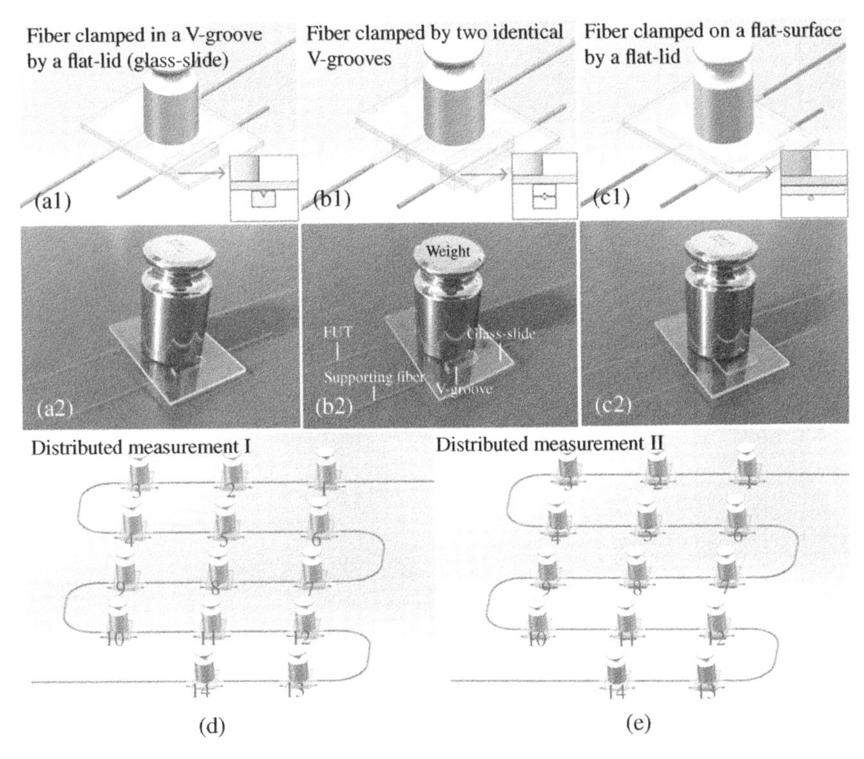

Figure 10.47 Schematic diagrams of SM fiber (a1) clamped in a V-groove by a flat-lid, (b1) clamped by two identical V-grooves and (c1) clamped on a flat-surface by a flat-lid; (a2), (b2), and (c2) showing the corresponding photos respectively. (d) Distributed measurement I: fiber clamped in V-grooves by flat-lids and (e) distributed measurement II: fiber clamped by two identical V-grooves. 14 force-applying segments are measured simultaneously for each type of V-grooves, with the groove-angles successively increasing according to Tables 10.8 and 10.9, respectively. Source: Feng et al. (2022)/Optica Publishing Group.

relation between the birefringence Δn and the applied force f_0 obtained with the PA-OFDR based DPA, a calibration coefficient $\alpha = 0.206$ is obtained, and the Δn values measured with DPA can be directly divided by α to get the calibrated Δn values.

First, the birefringence of an SMF in a V-groove clamped by a flat-lid is investigated experimentally. Figure 10.48a shows the measured birefringence (Δn) distribution along the fiber under different pressing-forces (f_0, with a unit of N/m), when the fiber was clamped in the ZrO_2 V-groove with a groove-angle of 59.99°, at the location of ~2.01 m. The corresponding linear-force values are given in the legend. As shown in the inset, the force-applying length was 10 mm determined by a flat glass lid. In the experiment another identical ZrO_2 V-groove was used for maintaining balance of the glass-slide. As can be seen, with the increase of pressing-force from 0 to 245.0 N/m, no obvious clamping-force induced anisotropic birefringence can be observed, which agrees well with the analytical result in Figure 10.45. For comparison, two similar experiments were carried out, one was clamping the fiber on a flat-surface and the other was clamping it in a type-1 V-groove with a groove-angle of 58.81°, as shown in Figure 10.48b,c respectively. The birefringence increases significantly in response to the pressing-force increase in Figure 10.48b, which is equivalent to clamping the fiber in a 180° V-groove. As can be seen in Figure 10.48c, if the groove-angle slightly deviates from 60°, a birefringence can be induced as the clamping force increases. Figure 10.48d shows the induced birefringence as a function of clamping-force for the three types of V-grooves with data extracted from Figure 10.48a–c, with the slopes of 7.369×10^{-10} (59.99°), 1.088×10^{-7} (180°), and

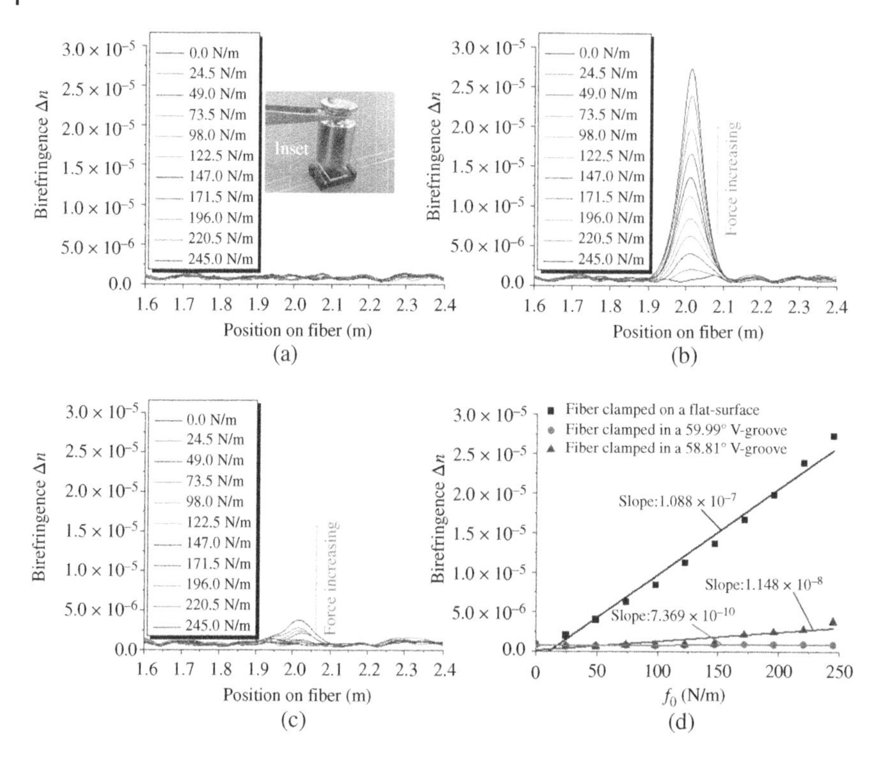

Figure 10.48 Birefringence distribution measured by DPA when an SM fiber is (a) clamped in a 59.99° ZrO_2 V-groove, (b) clamped on a flat-surface and (c) clamped in a type-1 58.81° silica-glass V-groove, by a flat-lid, respectively. (d) Birefringence vs. clamping-force f_0 for three cases earlier, with slopes obtained by linear curve-fitting of the measurement data. Source: Feng et al. (2022)/Optica Publishing Group.

1.148×10^{-8} (58.81°), obtained by linear curve-fitting, respectively. Therefore, the induced birefringence of a fiber in a 59.99° V-groove is more than 100 times less sensitive to the clamping-force than that in a 180° V-groove (flat-surface). Note that the clamping-force induced birefringence in a 59.99° V-groove cannot be distinguished from the background birefringence in an SMF, indicating that the birefringence induced by a 59.99° V-groove is practically zero.

Second, the birefringence of an SMF clamped by two identical type-2 V-grooves is investigated experimentally. Figure 10.49a,b shows the birefringence (Δn) distribution along the SMF under different clamping-forces (f_0) when it is clamped by two type-2 61.54° V-grooves and two type-2 90.34° V-grooves, respectively. The same clamping-forces and force-applying length were applied with those in Figure 10.48. As can be seen in Figure 10.49a, the clamping-force induced birefringence increases significantly as the clamping-force increases from 0 to 245.0 N/m for the fiber clamped by two 60° (61.54°) V-grooves. In contrast, no clamping-force induced birefringence can be observed for the fiber clamped by two 90° (90.34°) V-grooves even under a clamping-force as large as 245.0 N/m, as shown in Figure 10.49b. The clamping-force induced birefringence, if any, is completely buried in the background birefringence of the SMF. For comparison, the fiber in a ~90° (90.34°) V-groove was clamped with a flat glass lid, with the results shown in Figure 10.49c. One can see that the clamping-force induced anisotropic birefringence is still obvious. These results agree well with the theoretical analysis in Figure 10.45.

Similar to that in Figure 10.48d, the slopes of the induced birefringence as a function of clamping-force were also obtained for the three cases in Figure 10.49a,c to be 6.0×10^{-8},

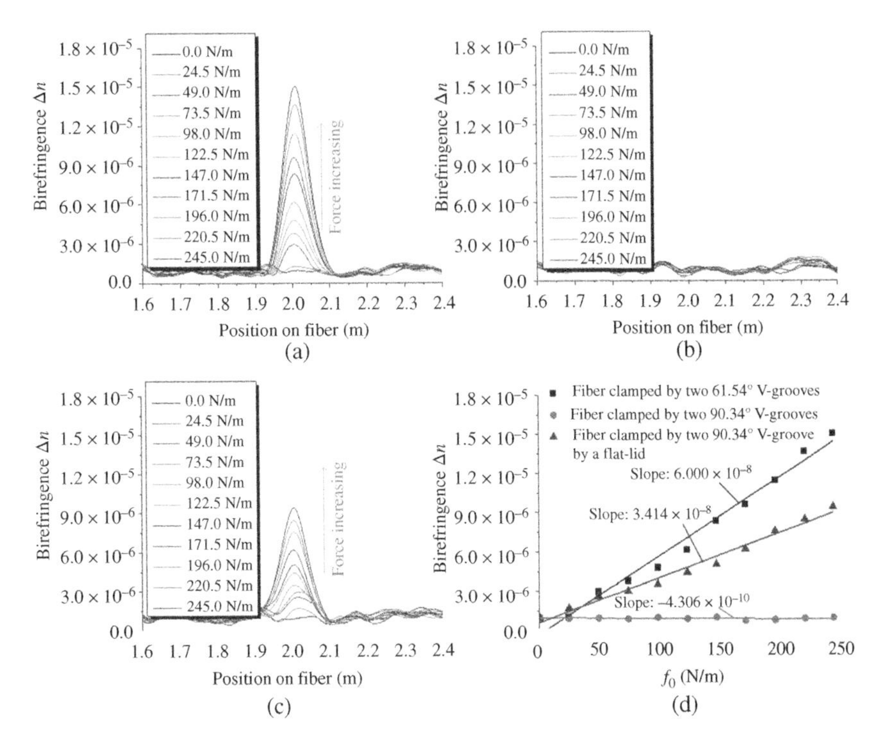

Figure 10.49 Birefringence distribution measured by DPA when a fiber is (a) clamped by two type-2 61.54° V-grooves, (b) clamped by two type-2 90.34° V-grooves, and (c) clamped by a type-2 90.34° V-groove and a flat glass lid. (d) Birefringence vs. clamping-force f_0 for three cases earlier, with slopes obtained by linear curve-fitting of the measurement data (Feng et al. 2022).

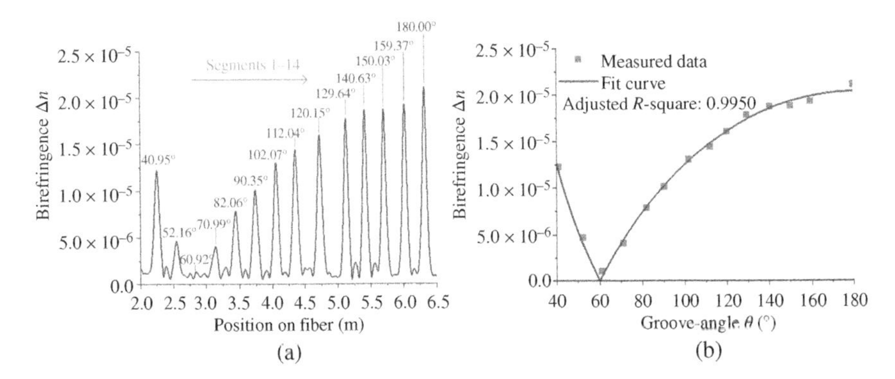

Figure 10.50 (a) Distributed measurement I, with a clamping-force of 245 N/m on each fiber segment. (b) Birefringence Δn as a function of groove-angle plotted with data obtained from (a), and curve fitting Δn to the absolute value of Eq. (10.29a) at different groove-angles θ (Feng et al. 2022).

-4.306×10^{-10}, and 3.414×10^{-8}, respectively, as shown in Figure 10.49d. Note that the negative slope with a magnitude of 4.306×10^{-10} is due to the fact that the clamping-force induced birefringence in Figure 10.49b is too small to be measured because it is buried in the background birefringence of the SMF. From the results in Figures 10.48 and 10.49, one can conclude that not only the improper choice of V-groove angle but also the improper clamping method for an SMF

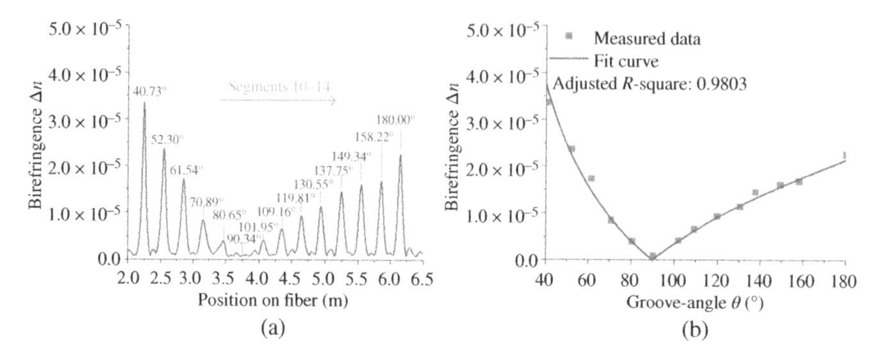

Figure 10.51 (a) Distributed measurement II, with a clamping-force of 245 N/m on each fiber segment. (b) Birefringence Δn as a function of groove-angle plotted with data obtained from (a), and curve fitting Δn to the absolute value of Eq. (10.29b) at different groove-angles θ (Feng et al. 2022).

in a V-groove may cause large birefringence in the fiber and therefore degrade the polarization related performance of the devices whenever V-grooves are used.

Using the group of type-1 V-grooves listed in Table 10.8, the distributed birefringence measurement (Measurement I) was performed with the experimental setup described in Figure 10.47d. A segment of SMF was clamped in each of the V-grooves with a flat glass lid. A clamping-force of 245.0 N/m introduced by a 500 g weight was imposed onto the fiber in each V-groove. The last fiber segment on the flat-surface in Figure 10.47d was clamped by a flat glass slide of the same length as other V-grooves to emulate a 180.00° V-groove. Figure 10.50a shows the birefringence distribution measured by the PA-OFDR based DPA system, with the corresponding groove-angles labeled on top of birefringence peaks at the locations of the V-grooves. Figure 10.50b shows the peak birefringence as a function of groove-angle using the data extracted in Figure 10.50a. Because the present data processing algorithm can only obtain the absolute values of Δn, without knowing its signs ("+" or "−"), therefore curve-fitting the obtained $|\Delta n|$ as a function of groove-angle θ to Eq. (10.29a) was performed, with the result shown with the solid line. In the calculation, $\zeta = 8.559 \times 10^{-8}$ (N/m)$^{-1}$ calculated according to Eq. (10.23b) and $f_0 = 245.0$ N/m (500 g weight) were used. It can be seen that the measurement data agree extremely well with Eq. (10.29a) (with a goodness of fit of 99.50%) and that the ZB groove-angle for an SMF clamped in a V-groove with a flat-lid is exactly 60°, as predicted by the theory, indicating that the friction between fiber and groove-walls can be safely ignored.

Similarly, using the group of type-2 V-grooves listed in Table 10.8, distance resolved local birefringence was measured (Measurement II) with the experimental setup described in Figure 10.47e. A segment of SMF was clamped in each of the V-grooves with another identical V-groove. The measured local birefringence corresponding to each V-groove is shown in Figure 10.51a, and the birefringence $|\Delta n|$ as a function of groove-angle extracted from Figure 10.51a is shown in Figure 10.51b. Again, curve-fitting the measurement data to Eq. (10.29b) yields the solid line in Figure 10.51b, showing that the measurement data agree with the theoretical results extremely well (with a goodness of fit of 98.03%) and that the ZB groove-angle for an SMF clamped with two identical V-grooves is 90°. Because Eq. (10.29b) was obtained by assuming the friction between the SMF and the V-groove walls is zero, this excellent agreement between the theory and experiment indicates that the friction is practically zero, as in the case of type-1 V-grooves.

Finally, the clamping-force induced birefringence of an SMF in commercial multi-channel fiber-array V-grooves made with silica-glass was measured. The off-the-shelf four-channel

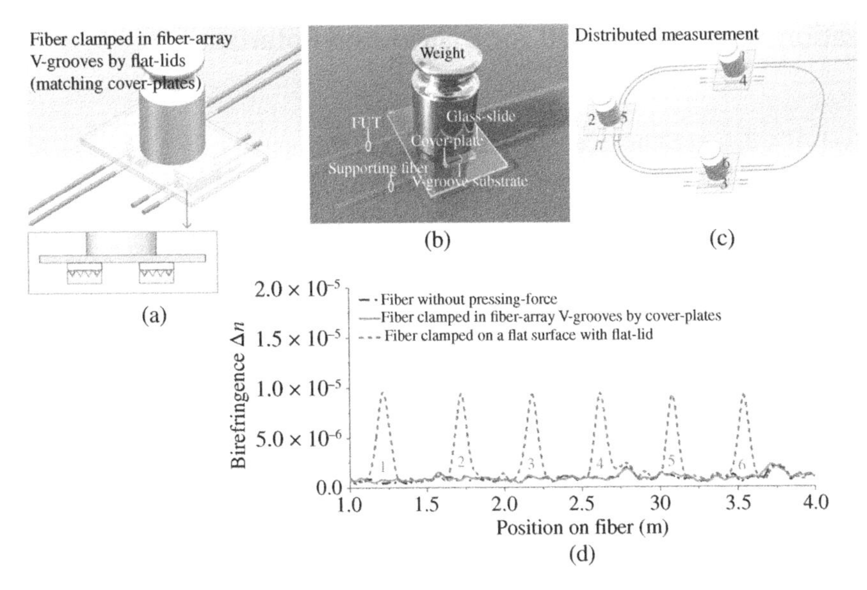

Figure 10.52 (a) Schematic diagram and (b) photo of fiber clamped in V-grooves of commercial multi-channel fiber-array substrates; Each substrate with four ~60° V-grooves, and specialized cover-plates used. (c) Schematic diagram of distributed birefringence measurement by setting six force-applying positions along FUT. (d) Birefringence distribution plotted by solid line measured by DPA for experiment setup in (c), and measurements of fiber without pressing-force and fiber clamped on a flat surface by removing the fiber-array substrates, plotted by black dash-dot line and dash line, respectively, for comparison. Source: Feng et al. (2022)/Optica Publishing Group.

fiber-array V-grooves, each having a matching flat cover-plate with a size of $5 \times 2.5\,\text{mm}^2$, was used to clamp a fiber segment in each V-groove. The groove-angles of the fiber-array V-grooves have been measured in Section 10.2.6.2, which are all very close to 60°. A 500 g weight and a glass-slide were used to press the cover-plate onto the fiber segments, as shown in Figure 10.52a,b. For maintaining the balance of glass-slide, two V-grooves were used to hold the fiber to be measured and the other two were used to hold two dummy fibers for supporting the glass-slide. To measure the distance resolved local birefringence induced by the V-grooves, the experimental scheme shown in Figure 10.52c was implemented. As can be seen, six force-applying segments were deployed along an SMF by using three four-channel fiber-array V-grooves. As expected, there is no any clamping-force induced birefringence observed along the fiber, as shown by the solid line in Figure 10.52d. For comparison, the birefringence distribution along the SMF before exerting the clamping-force (by putting the weight and the glass-slide on the V-grooves) was also measured, as shown by the black dash-dot line, which is almost overlapped with the solid line. Finally, the birefringence distribution was measured after removing all V-grooves and directly putting the fiber on the optical table before pressing it with the same glass-slide and the same weight. It can be seen that six clamping-force induced birefringence peaks were clearly obtained at the exact six forcing-applying locations. The results indicate that the clamping-force induced anisotropic birefringence in the SMF in the commercial 60° fiber-array V-grooves clamped with the flat cover-plate is negligible under a clamping-force up to 245.0 N/m. However, based on the result of Figure 10.48c, if the groove-angle deviates from 60° by as small as 2°, the induced birefringence cannot be ignored.

10.3 Polarization Scrambled OFDR for Distributed Polarization Analysis

The PA-OFDR based DPA described in Section 10.2 requires that the optical fiber for sensing is stable during the time the measurement is taking place, including the time for a total of sixteen (16) wavelength scans, because for each of the four (4) PSG generated SOPs, the PSA must use four (4) binary states to analyze the received SOP, which generally takes about 20 seconds to complete a measurement if the time for each wavelength scan is about 1 second. The options for minimizing the fiber disturbances during the measurement include (i) increasing the laser wavelength scanning speed at the expense of requiring higher speed electronics to acquire and process the data, (ii) using amplitude division or wave-front division polarimeters to replace the binary PSA at the expenses of requiring more photodetection and data acquisition channels, as described in Chapter 8, and finally (iii) developing a different DPA system that does not require stable fiber to keep the SOP stable during the measurement.

In this section, we focus on option (iii) to describe a polarization scrambled OFDR (PS-OFDR) (Wei et al. 2016), which is capable of measuring distance resolve birefringence along an optical fiber and hence can be used for sensing transversal stress in the fiber due to severe bending and pressing, which may shorten the life time of the installed optical fiber systems. The PS-OFDR based DPA system has the advantages of insensitivity to SOP variations that resulted from temperature fluctuations and fiber motions in comparison to the PA-OFDR based DPA system described in Section 10.2.

10.3.1 System and Algorithm Descriptions

Figure 10.53 shows the setup for such a PS-OFDR measurement system, where a tunable laser with a linewidth around 100 kHz is tuned continuously up to 4 nm around 1550 nm. A small fraction of light ~5% from coupler 1 is split and then sent to a FRM based Michelson interferometer with a relative time delay of 10 μs for generating clock signals at each "zero-crossing" position of the optical interference fringe signals. The generated real-time clock signals were then used for the PS-OFDR raw data sampling. The light was then further split by a 90/10 coupler (coupler 3), in which 10% of the light was used as an LO light to coherently probe Rayleigh Back Scattering (RBS) light and 90% of the light was then sent to the FUT via a circulator. The SOP before entering and after exiting the FUT is randomly varied by an all-fiber polarization controller described in Figure 6.15 to uniformly cover the Poincaré Sphere. The Rayleigh backscattered light from different locations of the fiber is brought to interfere with the portion of light directly from the tunable laser via a polarization diversity interferometer (the dotted line box in Figure 10.53). The interference signals of the two polarization components are analyzed after detection by PD2 and PD3, and the local birefringence information is then calculated. The transverse stress related to the birefringence is then obtained using the signal analysis procedures described next.

The optical fibers with non-perfect circular symmetry or with bend, twist, or other transverse stress often exhibit local birefringence. The light passing through a section of optical fiber with birefringence will experience a variation of SOP. Under assumption of no PDL, when a polarized light is propagating in a birefringence fiber, the evolution of the light polarization state can be described using the well-known equation of motion of the Stokes vector, $\mathbf{S}(z)$, expressed in Eq. (10.13) and rewritten here for convenience.

$$\frac{\partial \mathbf{S}(z)}{\partial z} = \mathbf{W}_{rt}(z) \times \mathbf{S}(z)$$

where z is the distance of light propagated within the fiber, $\mathbf{W}_{rt}(z)$ is the local round-trip birefringence vector. If the evolution of light's SOP can be analyzed along the fiber length, then the local birefringence can be deduced. In particular, we found that the $\mathbf{W}_{rt}(z)$ can be determined from a mean-square value of differences of local transmission between two closely-spaced fiber sections, similar to Cyr et al. (2009) in which the local birefringence was obtained by analyzing the SOP evolution between two closely-spaced frequencies of light. Assuming the local birefringence $W(z) = W_{rt}/2$ and following the same procedure as in Cyr et al. (2009), except substituting the change in optical frequency $\Delta\omega$ with Δz, one obtains

$$W(z) = \frac{1}{2\Delta z} \cdot \sqrt{15\langle \Delta T^2(z)\rangle_{I/O-SOP}} \tag{10.30}$$

where $\Delta T(z) = T(z + \Delta z/2) - T(z - \Delta z/2)$ is the normalized light power difference between two positions at $z + \Delta z/2$ and $z - \Delta z/2$, and $T(z) = P_i(z)/[P_2(z) + P_3(z)]$ is the normalized light power at the location z, where $P_i(z)$ is the light power at z measured with detector PD_i ($i = 2, 3$). In Eq. (10.30), $\langle\rangle_{I/O-SOP}$ stands for averaging over random input and output SOP variations. The fiber transverse stress as a function of the distance can be extracted from measured distributed fiber birefringence using Eqs. (10.23a) and (10.23b).

A key aspect of this PS-OFDR approach is that the back-reflected light from the FUT at the two neighboring locations is subsequently detected, and their SOPs are analyzed for many random and uniformly distributed input and output SOPs that is uncorrelated with respect to each other. It is worthy to note that because the input SOPs incident into and the output SOP exited from FUT are random and without specific requirement, the system is independent of the SOP variations from fiber cable movement and therefore is well suited for field applications, where the SOP along the fiber is constantly changing (Cyr et al. 2009).

10.3.2 Experimental Results

An example of the measured reflection trace for an FUT length up to 860 m is shown in Figure 10.54, where the OFDR back-reflection curve was extracted from the sum of two squared light intensities

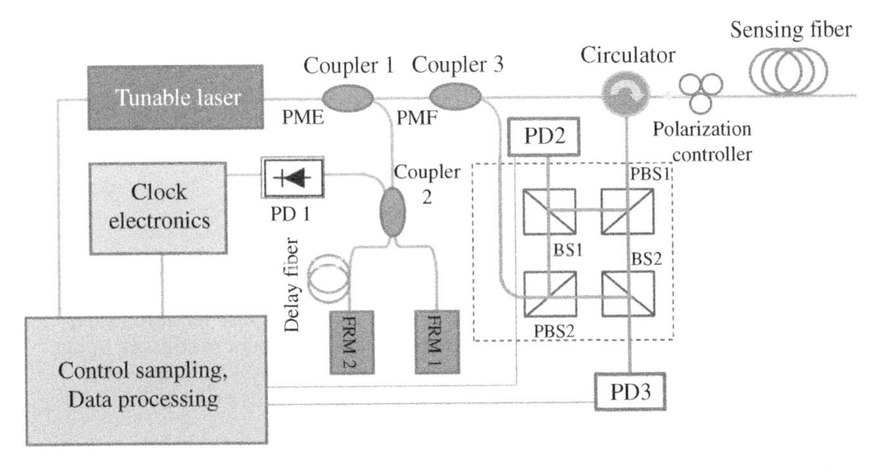

Figure 10.53 Experimental setup of a PS-OFDR based DPA for distributed transversal stress measurement. PBS: polarization beam splitter; PD1–3: photo detectors. The pigtails of couplers 1 and 3 are of polarization maintaining (PM) to avoid polarization fluctuations when light interferes. PBS1 and 2 are free space polarization beam splitters; BS1 and 2 are free space polarization insensitive beam splitters. The PM fiber inputting to PBS2 is oriented such that the power is equally split by PBS2 (Wei et al. 2016).

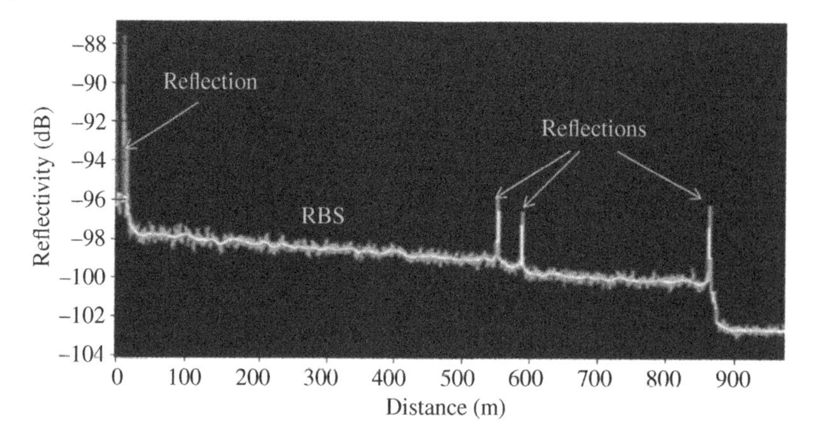

Figure 10.54 A measured OFDR reflection trace with back-reflections for an FUT with a length up to 860 m (Wei et al. 2016).

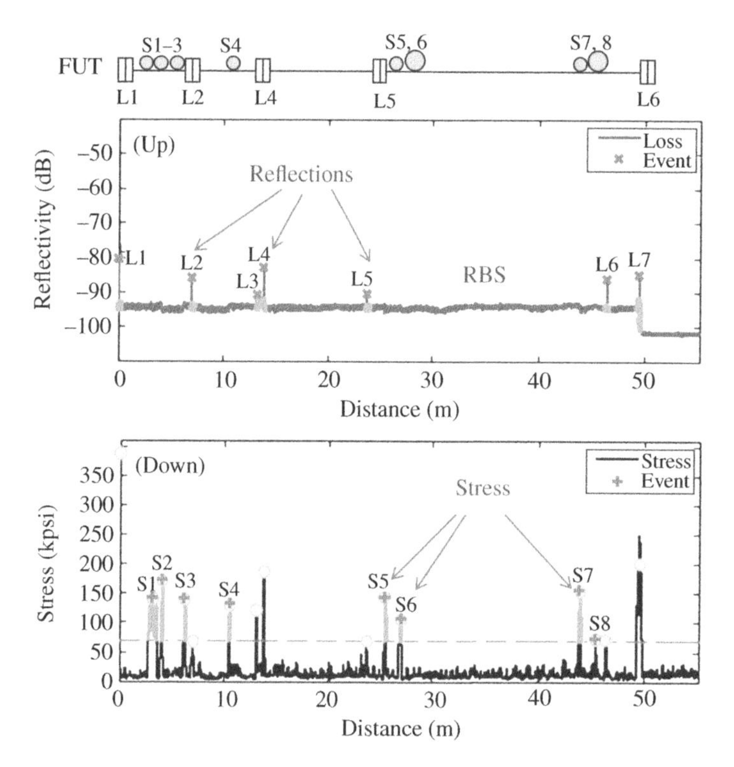

Figure 10.55 PS-OFDR signal from a length of fiber with six connectors and one splice for introducing reflections and eight fiber loops with different diameters for inducing different amount of stresses. Upper curve: measured reflections L1–7; Lower curve: measured fiber bending induced stress S1–8 (Wei et al. 2016).

measured by two detectors (PD2 and PD3) and its reflectivity was calibrated by an internal reference reflector (not shown in Figure 10.53).

In order to demonstrate the measurement principle described earlier, an experimental test was carried out by using bending induced transverse stresses along an FUT made with a bending insensitive SMF (Corning ClearCurve ZB) where eight bending loops at eight different FUT locations were designed. The bending radius is different at different positions and they are typically between 5 and 15 mm. In addition to the eight bends, there are also six FC/APC connectors and a fiber splice. The results are illustrated in Figure 10.55. In the experiment, the polarization controller randomly adjusted its launched SOP and the receiver analyzed the received signal corresponding to 200 different polarization controller settings randomly distributed on the Poincaré Sphere. The tunable laser is scanned in a range of 4 nm with a scanning speed of 10 nm/s. The results as shown in the upper and lower plots of Figure 10.55 were acquired simultaneously by the same data sampling of the raw data but with different signal processing algorithms, where the reflection trace in the upper plot was computed by the sum of squared measured results from two photo-detectors (PD2 and PD3 in the Figure 10.53) and the fiber birefringence trace $W(z)$ was extracted using Eq. (10.30) with an average of over 200 inputs and analyzed SOPs. The displayed stress value was obtained by a calibration procedure with a calibration weight. The measurement and computation take about five minutes to complete.

As can be seen, in the reflection trace (the upper plot of Figure 10.55) there are seven light back-reflections associated with six connectors (L1, 2, 4–7) and a fiber splice (L3). Of particular interest is the stress measurement as shown on the lower plot of Figure 10.55 with features associated with fiber bendings, connectors, and splice. With cross reference of reflection trace, eight fiber bending induced stresses feature S1–8 that can be unambiguously distinguished from those of light back-reflections. The induced transverse stress is higher for smaller bend radii than for larger bend radii, consistent with the bending induced birefringence calculations of Eq. (10.17c). All eight bending induced fiber birefringence events are seen clearly with a signal to noise ratio over 2. The spatial resolution of the system for the stress measurement is of 0.5 mm, determined by

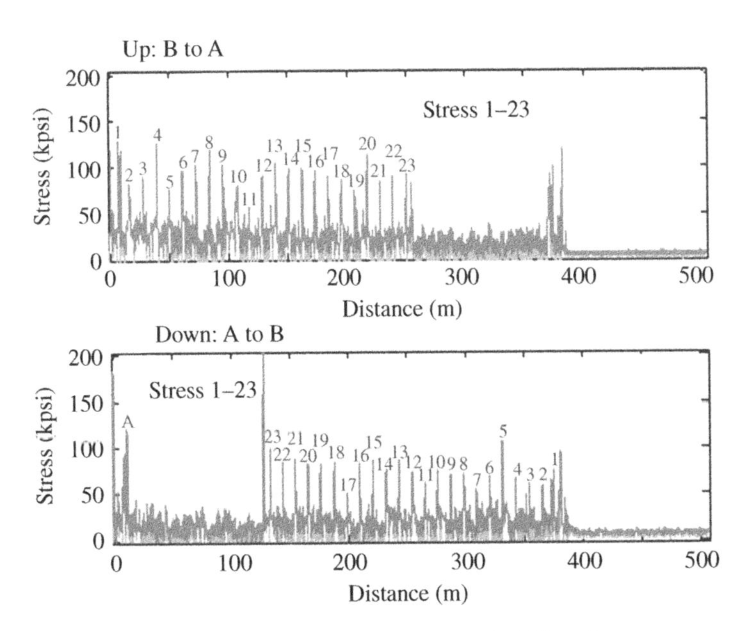

Figure 10.56 Distributed stress measurement of a 250 m fiber coil connected with a loose fiber about 130 m in length. Upper: Light enters the fiber from the coil side B. Lower: Light enters the fiber from the spool side A. The fiber used is the bend insensitive fiber Corning ClearCurve ZBL single-mode optical fiber and the tension for winding the coil is about 10 g. All measurements are conducted at room temperature (Wei et al. 2016).

the wavelength scanning range of the laser. The measurement uncertainty from multiple measurements is ~10% when 200 randomly distributed SOPs were used, and it can be reduced by averaging more SOPs but with longer measurement time. The relative stress levels of all eight fiber loops are consistent with Eq. (10.17c).

As an example for distributed stress sensing, the PS-OFDR was used to identify stress locations and magnitudes along the optical fiber inside a fiber sensing coil. As discussed in Section 9.4.2, the accumulative effect of the stress in a fiber sensing coil can be characterized by PMD and measured by a binary Mueller matrix system; however, no distance resolved information can be obtained by the binary system in Chapter 9. Here the distributed stress information obtained can be used for improving the winding process and minimizing such stresses for making better fiber coils, such as a crossover free coil. To demonstrate its capabilities, a PS-OFDR prototype system is used to measure a fiber coil with an inner diameter of 80 mm and clearly identifies periodic stress peaks, as shown in Figure 10.56. The periodicity of 23 peaks is the same as the length of each layer of the coil, consistent with the expectation that stresses are induced at fiber crossover points when the fiber is transitioning from one layer to the other during the winding process. There are totally 23 peaks, corresponding to 23 layers of fiber in the fiber coil. To further prove that the periodic peaks are indeed from the coil winding, a length of fiber around 130 m loosely wound on a large spool of about 50 cm in diameter was spliced with the coil before the distributed stress tests were performed from both ends of the combined fibers A and B, with A side being the large spool and B side being the fiber coil. Indeed, the periodic stress peaks were only observed in the fiber coil, not in the loosely wound fiber spool, no matter which direction the light was launched. This preliminary test result of transversal stresses on the fiber coil indicates that the PS-OFDR is sufficiently sensitive to detect such a small winding induced stresses and is promising for the production process control to improve the fiber coil quality.

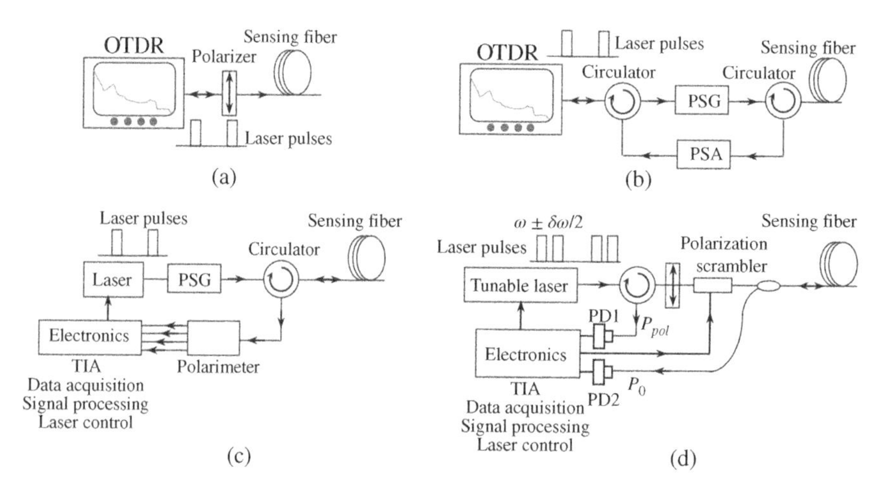

Figure 10.57 Different configurations of polarization OTDR (P-OTDR). (a) The simplest P-OTDR based on a standard OTDR hardware on the market. (b) A complete Mueller matrix polarimetry OTDR combining a time division PSG and PSA described in Chapters 8 and 9 with a standard OTDR hardware. (c) An OTDR configuration using a time-division PSG and amplitude or wave-front division PSA described in Chapter 8 with four photodetectors. (d) A polarization scrambling OTDR (PS-OTDR) insensitive to fiber disturbances.

10.4 P-OTDR Based Distributed Polarization Analysis Systems

The DPA described in Section 10.2 are implemented by incorporating different polarization analysis schemes, such as partial or complete Mueller matrix analysis, inside an OFDR. Similarly, one may also incorporate different polarization analysis in an OTDR (Barnoski et al. 1977) to realize different DPA systems.

Figure 10.57 shows different configurations of commonly used P-OTDR (Galtarossa 2004). The configuration shown in Figure 10.57a is the P-OTDR in its simplest form which can use the hardware of a standard commercial OTDR to implement; however, different data processing algorithm must be developed to obtain the desired polarization properties of the fiber as a function of fiber distance. This configuration is essentially the same as the first P-OTDR proposed and demonstrated by Alan Rogers in 1980 (Rogers 1980; Galtarossa 2004), with the only difference of using a PBS to replace the polarizer. Such a P-OTDR can be classified as an incomplete Mueller matrix distributed polarimeter or DPA. In operation, short laser pulses passing through the polarizer are launched in the sensing fiber and Rayleigh backscattered lights at different locations z_i are reflected back to the OTDR unit to be detected by the avalanche photodetector (APD) inside the OTDR. A polarization controller described in Chapter 6 may be used before the polarizer to maximize the forward going optical power passing through the polarizer. The location of the reflection can be determined by the ToF of the returned optical pulses with respect to their corresponding launching times. The SOPs of the returned light pulses reflected from different locations are different because they experienced different optical paths with different birefringence, which will be converted to different power levels by the polarizer. By analyzing the power level variations, one may obtain partial information about the birefringence distribution along the sensing fiber caused by the transversal stresses on the sensing fiber. In the fiber sections with a large birefringence or PMD, the power variation caused by the SOP variation is more rapid as a function of distance in the OTDR trace. Because only a single SOP is launched into the sensing fiber with different birefringence orientations in different fiber sections, the SOP variations of the returned light pulses do not contain the full information of the local birefringence.

Figure 10.57b shows a complete Mueller matrix DPA system which can be implemented with a standard commercial OTDR (Palmieri 2013). The system contains a PSG to generate different SOP and a PSA for analyzing the SOPs of the returned light pulses corresponding to each PSG generated polarization state. Finally, the returned light pulses passing through the PSA are detected by the APD inside the OTDR. Both the analog and binary PSGs and PSAs described in Chapters 8 and 9 can be used here in principle, although the binary PSG and PSA has the advantages of high speed, high repeatability, compact size, and better controllability for systems integration, similar to that used in the PA-OFDR described in Figure 10.28. In fact, similar algorithms described in PA-OFDR can be applied in the P-OTDR here to extract the complete birefringence information for transversal stress sensing applications. Compared with the PA-OFDR based DPA system, this P-OTDR based DPA system may suffer from degraded spatial resolution and measurement sensitivity.

Figure 10.57c shows another configuration of a complete Mueller matrix DPA which uses a four-port amplitude division or wave front division polarimeter as a PSA for fast polarization analysis (Rogers et al. 2000a; Rogers 2000b; Ellison and Siddiqui 2000). Because four outputs are to be detected simultaneously, a standard commercial OTDR, which only contains a single APD, cannot be used. A four-channel signal detection and processing electronics need to be developed, which may significantly increase the system cost, but having the advantage of much higher detection speed.

Finally, Figure 10.57d shows a polarization scrambled P-OTDR which is not sensitive to the SOP fluctuations caused by temperature variations and fiber motions, similar to the PS-OFDR described in Figure 10.53 (Cyr et al. 2009). In fact, the signal processing algorithm for the PS-OFDR is adopted from that used in this PS-OTDR. Unlike other P-OTDR systems, here the wavelength or frequency of the laser pulses is tunable. As shown in Figure 10.57d, a pair of pulses in each period is launched into the sensing fiber through a polarization scrambler (Cyr et al. 2009). The optical frequencies of the pulse pair differ by the amount of $\delta\omega$. The local birefringence or PMD is obtained by analyzing the SOP variations between the two pulses with different wavelengths, as shown next. The SOP evolution caused by the optical frequency variations in the presence of the local birefringence can also be described by the equation of motion of the Stokes vector $\mathbf{S}(z,\omega)$ (Poole and Favin 1994; Wai and Menyuk 1996)

$$\frac{\Delta\mathbf{S}(z,\omega)}{\Delta\omega} = \mathbf{\Omega}_{rt}(z,\omega) \times \mathbf{S}(z,\omega) \tag{10.31}$$

where z is the distance of light propagated within the fiber, $\mathbf{\Omega}_{rt}(z,\omega)$ is the local round-trip PMD vector with $|\mathbf{\Omega}_{rt}| = \Delta\tau_{rt}$ (local DGD), and ω is the optical frequency. If the evolution of light's SOP can be analyzed with the change of optical frequency, then the local birefringence can be obtained. In particular, according to Cyr et al. (2009), the $\mathbf{\Omega}_{rt}(z,\omega)$ can be determined from a mean-square value of differences of local transmission between two closely-spaced optical frequencies. Assuming the local DGD $\Delta\tau(z,\omega) = \Delta\tau_{rt}/2$, one obtains

$$\Delta\tau(z,\omega) = \frac{1}{2\delta\omega}\sqrt{15\langle\Delta T^2(z,\omega)\rangle_{I/O-SOP}} \tag{10.32}$$

where $\langle\rangle_{I/O-SOP}$ stands for averaging over random input and output SOP variations, $\Delta T(z,\omega) = T(z,\omega+\delta\omega/2) - T(z,\omega-\delta\omega/2)$ is the normalized light power difference between two frequencies at $\omega+\delta\omega/2$ and $\omega-\delta\omega/2$, and $T(z,\omega) = P_{pol}(z,\omega)/P_0(z,\omega)$ is the normalized light power at the location z, where $P_{pol}(z,\omega)$ is the returned light power passing through the polarizer at z measured with PD1 and $P_0(z,\omega)$ is the returned light power without going through the polarizer at z measured with PD2. The local birefringence relates to the local DGD by $\Delta n = \Delta\tau \cdot c/\Delta L$, where ΔL can be considered as the fiber length corresponding to the pulse width in the OTDR system. The fiber transverse stress as a function of the distance can be extracted from measured distributed fiber birefringence using Eqs. (10.23a) and (10.23b).

The OTDR based TD-DPA systems described earlier have been widely used for measuring the PMD distribution in optical fiber communication links. It can also be used to sense the distance resolved transversal stress along fiber via the local birefringence measurement, or local magnetic and electrical fields by measuring the local SOP changes caused by the Faraday Effect and the optical Kerr Effect, respectively. However, because the spatial resolution of TD-DPA or P-OTDR is limited by the pulse width to be on the order of 0.5 m or more in general, which is two or three orders of magnitude larger than that of a FD-DPA (PA-OFDR), the P-OTDR may not be best suited for distributed sensing applications.

10.A Appendix

In this appendix, we briefly describe how to obtain Eqs. (10.29a) and (10.29b). For an SM fiber in a V-groove with a groove-angle of θ clamped with a flat-lid (case 1), as shown in Figure 10.44a, a birefringence Δn induced by three forces, f_0, f_1, and f_2, has its principal axes along the x and y directions, where f_1 and f_2 are the reactive forces of the groove. The birefringence can be calculated by $\Delta n = \Delta\beta/k$, where k is the propagation constant of light in vacuum expressed as $k = 2\pi/\lambda$, and $\Delta\beta = \beta_x - \beta_y$ denotes the difference of propagation constants of lights polarized along x and y principal axis directions of the fiber respectively. λ is the light wavelength. The same birefringence Δn would be induced by three other forces f_i' (where $i = 0, 1, 2$), which are of equal magnitudes as the f_i but whose azimuths are rotated by 180°. Considering the linearity of the elastic equations, a doubled birefringence $2\Delta n$ must result if all six forces are acting simultaneously. Therefore, we can determine the difference of propagation constants $\Delta\beta_i$ by grouping the forces in pairs (f_i, f_i') with opposite azimuths, and subsequently determine the corresponding birefringence Δn_i, where $\Delta\beta_0$ has its fast axis at the azimuth $\alpha_0 = 90^\circ$, whereas $\Delta\beta_1$ and $\Delta\beta_2$ have their fast axes at $\alpha_1 = +\theta/2$ and $\alpha_2 = -\theta/2$, respectively. Considering all three $\Delta\beta_i$ simultaneously and referring to the direction of f_0 as the fast axis (y direction), according to the Poincaré formalism we have (Wang et al. 2009)

$$2\Delta\beta = \sum \Delta\beta_i cos2(\alpha_i - \alpha_0) \tag{10.A.1}$$

Each pair of forces induces a birefringence Δn_i having a magnitude $\Delta n_i = \Delta\beta_i/k = \zeta f_i$ according to Eq. (10.23a), with the coefficient ζ defined in Eq. (10.23b).
Substitution of Eq. (10.23a) in Eq. (10.A.1) yields

$$\Delta\beta = \frac{1}{2}k(f_0 - f_1 cos\theta - f_2 cos\theta)\zeta \tag{10.A.2}$$

By making the force analysis for f_i with the aid of Figure 10.44a, one obtains

$$f_1 = f_2 = \frac{f_0}{2sin(\theta/2)} \tag{10.A.3}$$

Substituting Eq. (10.A.3) in Eq. (10.A.2), Eq. (10.29a) can be obtained.

For a naked SM fiber clamped by two identical V-grooves with a groove-angle of θ (case 2), as shown in Figure 10.43b, the forces, f_i ($i = 1, 2, 3, 4$), are generated on the fiber, under the pressing force f_0. From the force analysis, one can easily obtain $f_1 = f_3$ and $f_2 = f_4$. Referring to the direction of f_0 as the fast axis, a birefringence Δn can be introduced by an anisotropic force f_{ani} derived as

$$f_{ani} = (f_1 + f_2)sin\left(\frac{\theta}{2}\right) - (f_1 + f_4)cos\left(\frac{\theta}{2}\right) \tag{10.A.5}$$

It can be seen from Figure 10.43b that

$$(f_3 + f_4)sin\left(\frac{\theta}{2}\right) = (f_1 + f_2)sin\left(\frac{\theta}{2}\right) = f_0 \tag{10.A.6}$$

Substituting Eq. (10.A.6) in Eq. (10.A.5), one can obtain

$$f_{ani} = f_0\left(1 - \frac{1}{tan(\theta/2)}\right) \tag{10.A.7}$$

Because $\Delta n_{ani} = \zeta f_{ani}$ from Eq. (10.23a), Eq. (10.29b) can be obtained subsequently.

References

Barnoski, M., Rourke, M., Jensen, S., and Melville, R. (1977). Optical time domain reflectometer. *Appl. Opt.* 16 (9): 2375–2379.

Bassil, A., Wang, X., Chapeleau, X. et al. (2019). Distributed fiber optics sensing and coda wave interferometry techniques for damage monitoring in concrete structures. *Sensors* 19 (2): 356.

Bertholds, A. and Dandliker, R. (1988). Determination of the individual strain-optic coefficients in single-mode optical fibres. *J. Lightwave Technol.* 6 (1): 17–20.

Carroll, L., Lee, J., Scarcella, C. et al. (2016). Photonic packaging: transforming silicon photonic integrated circuits into photonic devices. *Appl. Sci.* 6 (12): 426.

Chen, X. and Yao, X.S. (2010). Measuring distributed polarization crosstalk in polarization maintaining fiber and optical birefringence material. US Patent number 8,599,385, filed 14 May 2010 and issued 3 December 2013.

Chua, T.H. and Chen, C.L. (1989). Fiber polarimetric stress sensors. *Appl. Opt.* 28 (15): 3158–3165.

Cyr, N., Chen, H., and Schinn, G. (2009). Random-scrambling tunable POTDR for distributed measurement of cumulative PMD. *J. Lightwave Technol.* 27 (18): 4164–4174.

Ding, Z., Meng, Z., Yao, X.S. et al. (2011). Accurate method for measuring the thermal coefficient of group birefringence of polarization-maintaining fibers. *Opt. Lett.* 36: 2173–2175.

Ding, Z., Liu, T., Meng, Z. et al. (2012). Improving spatial resolution of optical frequency-domain reflectometry against frequency tuning nonlinearity using non-uniform fast Fourier transform. *Rev. Sci. Instrum.* 83 (6): 066110.

Ding, Z., Yao, X.S., Liu, T. et al. (2013). Compensation of laser frequency tuning nonlinearity of a long range OFDR using deskew filter. *Opt. Express* 21 (3): 3826–3834.

Ding, D., Feng, T., Zhao, Z., et al. (2016). Demonstration of distributed fiber optic temperature sensing using polarization crosstalk analysis. Conference on Lasers and Electro-Optics (Optical Society of America, San Jose, California, 2016). p. JTu5A.105.

Ding, Z., Wang, C., Liu, K. et al. (2018). Distributed optical fiber sensors based on optical frequency domain reflectometry: a review. *Sensors* 18 (4): 1072.

Dong, P. (2016). Silicon photonic integrated circuits for wavelength-division multiplexing applications. *IEEE J. Sel. Top. Quantum Electron.* 22 (6): 370–378.

Dong, H., Shum, P., Yan, M. et al. (2006). Generalized Mueller matrix method for polarization mode dispersion measurement in a system with polarization-dependent loss or gain. *Opt. Express* 14 (12): 5067–5072.

Ellison, J. and Siddiqui, A. (2000). Automatic matrix-based analysis method for extraction of optical fiber parameters from polarimetric optical time domain reflectometry data. *J. Lightwave Technol.* 18 (9): 1226–1232.

Feng, T., Ding, D., Li, Z., and Yao, X.S. (2017). First quantitative determination of birefringence variations induced by axial-strain in polarization maintaining fibers. *J. Lightwave Technol.* 35 (22): 4937–4942.

Feng, T., Shang, Y., Wang, X. et al. (2018). Distributed polarization analysis with binary polarization rotators for the accurate measurement of distance-resolved birefringence along a single-mode fiber. *Opt. Express* 26 (20): 25989–26002.

Feng, T., Zhou, J., Shang, Y. et al. (2020). Distributed transverse-force sensing along a single-mode fiber using polarization-analyzing OFDR. *Opt. Express* 28 (21): 31253–31271.

Feng, T., Miao, T., Lu, Z., and Yao, X.S. (2022). Clamping-force induced birefringence in a single-mode fiber in commercial V-grooves investigated with distributed polarization analysis. *Opt. Express* 30 (4): 5347–5359.

Flavin, D., McBride, R., and Jones, J. (2002). Dispersion of birefringence and differential group delay in polarization-maintaining fiber. *Opt. Lett.* 27: 1010–1012.

Galtarossa, A. (2004). Spatially resolved PMD measurements. *J. Lightwave Technol.* 22 (4): 1103–1115.

Galtarossa, A. and Palmieri, L. (2005). Reflectometric measurements of polarization properties in optical-fiber links. In: *Polarization Mode Dispersion* (ed. A. Galtarossa and C. Menyuk), 168–197. Springer Science + Business Media, Inc.

Galtarossa, A., Grosso, D., Palmieri, L., and Rizzo, M. (2009). Spin-profile characterization in randomly birefringent spun fibers by means of frequency-domain reflectometry. *Opt. Lett.* 34 (7): 1078–1080.

Galtarossa, A., Palmieri, L., and Geisler, T. (2010). Distributed characterization of bending effects on the birefringence of single-mode optical fibers. *Opt. Lett.* 35 (14): 2481.

Hao, P., Yu, C., Feng, T. et al. (2020). PM fiber based sensing tapes with automated 45° birefringence axis alignment for distributed force/pressure sensing. *Opt. Express* 28 (13): 18829–18842.

He, Z. and Hotate, K. (2002). Distributed fiber-optic stress-location measurement by arbitrary shaping of optical coherence function. *J. Lightwave Technol.* 20 (9): 1715–1723.

Kim, C., Kim, D., Cheong, Y. et al. (2015). 300-MHz-repetition-rate, all-fiber, femtosecond laser mode-locked by planar lightwave circuit-based saturable absorber. *Opt. Express* 23 (20): 26234–26242.

Li, Z., Meng, Z., Chen, X. et al. (2012). Method for improving the resolution and accuracy against birefringence dispersion in distributed polarization cross-talk measurements. *Opt. Lett.* 37: 2775–2777.

Li, Z., Yao, X.S., Chen, X. et al. (2015). Complete characterization of polarization-maintaining fibers using distributed polarization analysis. *J. Lightwave Technol.* 33 (2): 372–380.

Maier, R., MacPherson, W., Barton, J. et al. (2010). Distributed sensing using Rayleigh scatter in polarization-maintaining fibres for transverse load sensing. *Meas. Sci. Technol.* 21 (9): 094019.

Martin, P., Le Boudec, G., and Lefevre, H. (1991). Test apparatus of distributed polarization coupling in fiber gyro coils using white light interferometry. In: *Proc. SPIE 1585, Fiber Optic Gyros: 15th Anniversary Conference, Boston*, 173–179.

Mochizuki, K., Namihira, Y., and Ejiri, Y. (1982). Birefringence variation with temperature in elliptically cladded single-mode fibers. *Appl. Opt.* 21: 4223–4228.

Namihira, Y., Kudo, M., and Mushiake, Y. (1977). Effect of mechanical stress on the transmission characteristics of optical fiber. *Electron. Commun. Japan* 60 (7): 107–115.

Nayyar, V.P. and Verma, N.K. (1978). Two-point resolution of Gaussian aperture operating in partially coherent light using various resolution criteria. *Appl. Opt.* 17 (14): 2176–2180.

Palmieri, L. (2013). Distributed polarimetric measurements for optical fiber sensing. *Opt. Fiber Technol.* 19: 720–728.

Poole, C.D. and Favin, D.L. (1994). Polarization-mode dispersion measurements based on transmission spectra through a polarizer. *J. Lightwave Technol.* 12 (6): 917–929.

Rogers, A. (1980). Polarization optical time domain reflectometry. *Electron. Lett.* 16 (13): 489–490.

Rogers, A. (2000b). Distributed measurement of strain using optical fibre backscatter polarimetry. *Strain* 36 (3): 135–142.

Rogers, A., Wuilpart, M., and Blondel, M. (2000a). New polarimetry for fully-distributed optical-fibre strain and temperature sensing. *Proc. SPIE* 3986: 302–311.

Sadeghioon, A., Metje, N., Chapman, D., and Anthony, C. (2018). Water pipeline failure detection using distributed relative pressure and temperature measurements and anomaly detection algorithms. *Urban Water J.* 15 (4): 287–295.

Saida, T. and Hotate, K. (1997). Distributed fiber-optic stress sensor by synthesis of the optical coherence function. *IEEE Photonics Technol. Lett.* 9 (4): 484–486.

Schenato, L., Pasuto, A., Galtarossa, A., and Palmieri, L. (2020). An optical fiber distributed pressure sensing cable with Pa-sensitivity and enhanced spatial resolution. *IEEE Sensors J.* 20 (11): 5900–5908.

Smith, A.M. (1980a). Birefringence induced by bends and twists in single-mode optical fiber. *Appl. Opt.* 19 (15): 2606–2611.

Smith, A.M. (1980b). Single-mode fibre pressure sensitivity. *Electron. Lett.* 16 (20): 773–774.

Soller, B., Gifford, D., Wolfe, M., and Froggatt, M. (2005). High resolution optical frequency domain reflectometry for characterization of components and assemblies. *Opt. Express* 13 (2): 666–674.

Son, G., Han, S., Park, J. et al. (2018). High-efficiency broadband light coupling between optical fibers and photonic integrated circuits. *Nanophotonics* 7 (12): 1845–1864.

Song, J., Li, W., Lu, P. et al. (2014). Long-range high spatial resolution distributed temperature and strain sensing based on optical frequency-domain reflectometry. *IEEE Photonics J.* 6 (3): 1–8.

Sparrow, C.M. (1916). On spectroscopic resolving power. *Astrophys. J.* 44: 76–86.

Suzuki, K., Konoike, R., Hasegawa, J. et al. (2019). Low-Insertion-loss and power-efficient 32 × 32 silicon photonics switch with extremely high-delta silica PLC connector. *J. Lightwave Technol.* 37 (1): 116–122.

Takahashi, M., Ido, T., and Nagara, T. (2004). A polymer PLC platform with a fiber-alignment V-groove for a low-cost 10-GbE WWDM transmitter. *IEEE Photonics Technol. Lett.* 16 (1): 266–268.

Tang, F. (2005). Measurement of polarization coupling in polarization-maintaining fiber using white light interferometry and its applications. PhD thesis. Tianjin University (in Chinese).

Tang, F., Wang, X., Zhang, Y., and Jing, W. (2006). Distributed measurement of birefringence dispersion in polarization-maintaining fibers. *Opt. Lett.* 31 (23): 3411–3413.

Tang, F., Wang, X., Zhang, Y. et al. (2007). Characterization of birefringence dispersion in polarization-maintaining fibers by use of white light interferometry. *Appl. Opt.* 46 (19): 4073–4080.

TIA-455-192. (1999). H-Parameter test method for polarization-maintaining optical fiber.

TIA-455-193. (1999). Polarization crosstalk method for polarization-maintaining optical fiber and components.

Tsubokawa, M., Shibata, N., Higashi, T., and Seikai, S. (1987). Loss of longitudinal coherence as a result of the birefringence effect. *J. Opt. Soc. Am.* A4: 1895–1901.

Ulrich, R., Rashleigh, S., and Eickhoff, W. (1980). Bending-induced birefringence in single-mode fibers. *Opt. Lett.* 5 (6): 273–275.

Vanrolleghem, P. and Lee, D. (2003). On-line monitoring equipment for wastewater treatment processes: state of the art. *Water Sci. Technol.* 47 (2): 1–34.

Wada, D., Igawa, H., and Murayama, H. (2016). Simultaneous distributed measurement of the strain and temperature for a four-point bending test using polarization-maintaining fiber Bragg grating interrogated by optical frequency domain reflectometry. *Measurement* 94: 745–752.

Wai, P. and Menyuk, C. (1996). Polarization mode-dispersion, decorrelation, and diffusion in optical fibers with randomly varying birefringence. *J. Lightwave Technol.* 14 (2): 148–157.

Wang, W., Wu, N., Tian, Y. et al. (2009). Optical pressure/acoustic sensor with precise Fabry–Pérot cavity length control using angle polished fiber. *Opt. Express* 17 (19): 16613–16618.

Wanner, T., Marks, B., Menyuk, C., and Zweck, J. (2003). Polarization mode dispersion, decorrelation, and diffusion in optical fibers with randomly varying elliptical birefringence. *J. Lightwave Technol.* 14 (2): 148–157.

Wei, C., Chen, H., Chen, X. et al. (2016). Distributed transverse stress measurement along an optic fiber using polarimetric OFDR. *Opt. Lett.* 41 (12): 2819–2822.

Wen, X., Ning, T., Bai, Y. et al. (2015). Ultrasensitive temperature fiber sensor based on Fabry–Pérot interferometer assisted with iron V-groove. *Opt. Express* 23 (9): 11526–11536.

Williams, P. (2004). PMD measurement techniques and how to avoid the pitfalls. *J. Opt. Fiber Commun. Rep.* 1 (1): 84–105.

Wuilpart, M., Rogers, A., Megret, P., and Blondel, M. (2000). Fully distributed polarization properties of an optical fiber using the backscattering technique. In: *Proc. SPIE 4087, Applications of Photonic Technology*, 396–404.

Xu, T., Tang, F., Jing, W. et al. (2009a). Distributed measurement of mode coupling in birefringent fibers with random polarization modes. *Opt. Appl.* 39 (1): 77–90.

Xu, T., Jing, W., Zhang, H. et al. (2009b). Influence of birefringence dispersion on a distributed stress sensor using birefringence optical fiber. *Opt. Fiber Technol.* 15: 83–89.

Yan, J., Miao, L., Huang, T. et al. (2019). Development of method for polarization alignment of PANDA polarization maintaining fiber. *Opt. Fiber Technol.* 53: 101999.

Yao, X.S. (2015). Measurements of strain, stress, and temperature by using 1-dimensional and 2-dimensional distributed fiber-optic sensors based on sensing by polarization maintaining fiber of distributed polarization crosstalk distribution. US Patent 9,476,699, filed 5 March 2015 and issued 25 October 2016.

Yao, X.S. (2019). Techniques to ensure high quality fiber optic gyro coil production, Chapter 11. In: *Design and Development of Fiber Optic Gyroscopes* (ed. E. Udd and M.J.F. Digonnet), 217–261. Bellingham, Washington DC: SPIE Press.

Yao, X.S., Chen, X., and Liu, T. (2010). High accuracy polarization measurements using binary polarization rotators. *Opt. Express* 18 (7): 6667–6685.

Yao, X.S., Yao, Y., and Chen, X. (2015). 1-Dimensional and 2-dimensional distributed fiber-optic strain and tress sensor based on polarization maintaining fiber using distributed polarization crosstalk analyzer as an interrogator. US Patent 9,689,666, filed 5 March 2015 and issued 27 June 2017.

Zhang, H., Wen, G., Ren, Y. et al. (2012). Measurement of beat length in polarization-maintaining fibers with external forces method. *Opt. Fiber Technol.* 18: 136–139.

Zhang, H., Yuan, Z., Liu, Z. et al. (2017). Simultaneous measurement of strain and temperature using a polarization-maintaining photonic crystal fiber with stimulated Brillouin scattering. *Appl. Phys Express* 10 (1): 012501.

Zhang, H., Wang, Y., Wen, G. et al. (2018). Frequency demodulation of dynamic stress based on distributed polarization coupling system. *J. Lightwave Technol.* 36 (11): 2094–2099.

Zhang, Z., Feng, T., Li, Z. et al. (2019). Experimental study of transversal-stress-induced polarization crosstalk behaviors in polarization maintaining fibers. *Proc. SPIE* 11191: 111910W.

Zhang, Z., Yang, B., Jiang, J. et al. (2021). Side-polished SMS based RI sensor employing macro-bending perfluorinated POF. *Opto-Electron. Adv.* 4 (10): 200041.

Zhou, D., Li, W., Chen, L., and Bao, X. (2013). Distributed temperature and strain discrimination with stimulated Brillouin scattering and Rayleigh backscatter in an optical fiber. *Sensors* 13 (2): 1836–1845.

Zhou, D., Chen, L., and Bao, X. (2016). Distributed dynamic strain measurement using optical frequency-domain reflectometry. *Appl. Opt.* 55 (24): 6735–6739.

Zhu, C., Zhuang, Y., Chen, Y. et al. (2019). Distributed fiber-optic pressure sensor based on Bourdon tubes metered by optical frequency-domain reflectometry. *Opt. Eng.* 58 (7): 072010.

11

Polarization for Optical Frequency Analysis and Optical Sensing Applications

11.1 Optical Frequency Analysis Techniques

Tunable diode laser spectroscopy (TDLS) systems and coherent distributed optical sensing systems, such as frequency modulated continuous wave (FMCW) Lidar, optical coherence tomography (OCT), and optical frequency domain reflectometer (OFDR), generally require a frequency tunable laser scanning at high speed. Ideally, the laser frequency tuning should be linear in order to obtain the accurate instantaneous frequency for TDLS and best spatial resolution for coherent distributed sensing. Unfortunately, in reality, almost all tunable lasers have certain tuning nonlinearity, as shown in Figure 11.1a. In addition, some lasers may have frequency wiggles causing local tuning direction reversal, as shown in the inset. For example, for an up-ramp (or down-ramp) frequency tuning, one expects the frequency increases (or decreases) monotonically with time. However, due to the frequency wiggle, one may see some local frequency drops with time at some instants.

In order to minimize the negative impact of the scan nonlinearity, unbalanced Mach–Zehnder interferometers are often used to get the k-clock or the frequency clock (f-clock) by relating the times at the zero-crossings of the interference signal to the optical frequency increments, as shown in Figure 11.1b. In particular, whenever the interference signal crosses the zero, a frequency increment equal to one half of the free spectral range (FSR) of the interferometer is obtained, as shown in Figure 11.1c. Such a method suffers from two major drawbacks. First, it assumes that the optical frequency changes monotonically during the scan, which may result in wrong frequency measurement whenever a local frequency reversal occurs around a zero-crossing due to the laser frequency wiggles. Second, for obtaining high frequency resolution for some application, such as long range distributed sensing, the fiber delay required in the unbalanced Mach–Zehnder interferometer is very long, on the order of hundred meters or more, which make it difficult to miniaturize the system. In addition, the method only measures incremental frequency changes and no information on the absolute frequency of wavelength of the laser can be obtained.

Wavemeters (Hall and Carlsen 1997; Snyder 1982) based on Fizeau interferometer (Gardner 1985; Gray et al. 1986), scanning Michelson interferometer (Derickson and Stokes 1998), and Fabry–Perot (F–P) interferometer (Derickson and Stokes 1998) can be used to precisely measure laser frequency (wavelength) variations. However, their measurement speed and frequency resolution are generally not sufficient for the demanding distributed sensing applications, although they have exceptionally high wavelength measurement accuracies.

For optical spectrum analyzers (OSAs) (Vobis and Derickson 1998) relying on diffractive gratings or tunable narrow-band filters, the resolution, spectral range, and measurement speed generally counter play with one another (Yao et al. 2008), and therefore good performances of all three

Polarization Measurement and Control in Optical Fiber Communication and Sensor Systems, First Edition.
X. Steve Yao and Xiaojun (James) Chen.

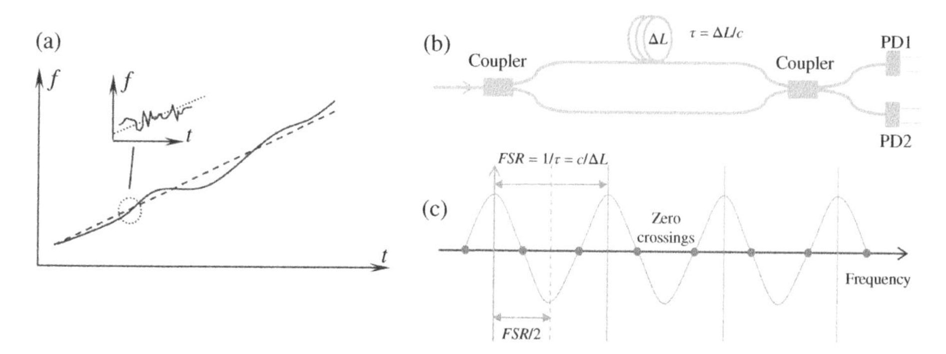

Figure 11.1 (a) Illustration of laser frequency tuning nonlinearity (Solid line) in comparison with a linear reference (Dashed line) and frequency wiggle (Inset) causing local tuning direction reversal. (b) An unbalanced Mach–Zehnder interferometer for incremental frequency measurement. (c) The incremental frequency equaling to one half of FSR can be obtained by counting the zero-crossings of the interference signal.

parameters cannot be achieved simultaneously. For example, the spectral resolution and the measurement range (FSR) of an F–P filter based OSA are inversely proportional to each other. In order to achieve fine resolution, the spectral range tends to be sacrificed. As another example, a diffraction grating based spectrum analyzer with a charge coupled device (CCD) can be reasonably fast; however, its resolution or range is not sufficient due to limitations imposed by the size of the CCD and the appearance of higher diffraction orders from the grating when the spectrum range is too large or when the grating period is too small.

For TDLS and coherent distributed sensing applications, it is important to know the following transient characteristics of the tunable lasers: the frequency tuning rate, nonlinearity, range, and repeatability, in addition to the power variation and frequency wiggle during each scan. For telecom coherent detection and distributed acoustic sensing (DAS) systems, fast detection of the frequency jitter of the narrow band distributed feedback (DFB) lasers (fixed wavelength) is also desirable (Baney and Sorin 1998), with which it is possible to reduce the jitter via feedback or compensate for the jitter effect.

Polarization analysis is generally not associated with obtaining optical spectral information by conventional wisdom. However, in this section, we introduce two polarization related techniques for detecting optical frequency at high speed and with high resolution for the applications described earlier, which cannot be easily accomplished with the conventional spectral analysis techniques.

11.1.1 Polarimeter-Based Optical Frequency Analyzer

In this section, we describe a polarimeter-enabled optical frequency analyzer (P-OFA) (Yao et al. 2008) for the characterization of tunable lasers for TDLS and coherent distributed sensing applications, as well as for measuring the frequency jitter of narrow line-width lasers for coherent detection in telecom and various coherent sensor applications. This method is based on analyzing both the state of polarization (SOP) and the degree of polarization (DOP) information of a single frequency light source after it passes through a variable differential group delay (DGD) module. Thanks to the high-speed amplitude division or wave front division polarimeters described in Chapter 8, this P-OFA can readily achieve a measurement speed up to 500 MHz, and is therefore capable of measuring the center frequency or wavelength of a fast scanning laser as a function of time. The measurement speed is limited only by the bandwidth of the photo-detectors, the transimpedance amplifiers, and digital processing electronics.

A unique and attractive feature of the P-OFA is that it can determine the direction of the frequency change, a capability unobtainable with most of the conventional spectrum analysis methods. Consequently, the P-OFA is capable of having arbitrarily high frequency resolution, yet, with arbitrarily large spectral range, provided that the measurement speed is sufficiently fast compared with the rate of the SOP change caused by the spectral change of the light source and that the signal-to-noise ratio in P-OFA's detection circuit is sufficiently high for accurate SOP measurement. Such a feature opens a door for many spectral related measurements not imaginable with conventional spectrum analysis methods.

Finally, the P-OFA can obtain the information of the instantaneous spectral shape (power vs. frequency) of a swept-wavelength light source as a function of time. With this unique capability, a 3D plot of the spectral shape of a modulated light signal as a function of center wavelength of a tunable light source can be generated. This capability enables detailed spectral characterization of a fast swept-wavelength source that cannot be obtained with other conventional methods.

11.1.1.1 Swept-Frequency (Wavelength) Measurement

Concept The basic concept of the P-OFA for measuring the instantaneous wavelength of a swept laser source is shown in Figure 11.2a, where the input light source first enters a fixed DGD element (e.g. a birefringent crystal or a polarization beam splitter [PBS]/polarization beam combiner [PBC] based device shown in Figure 7.16a) before being analyzed by a high-speed polarimeter. As described in Chapters 4 (Section 4.5) and 8, when a tunable light source passes through a DGD element, its SOP will trace out a circle on the Poincaré sphere when the wavelength of the light source is tuned, as shown in Figure 11.2b. The rate of the SOP change as a function of frequency is

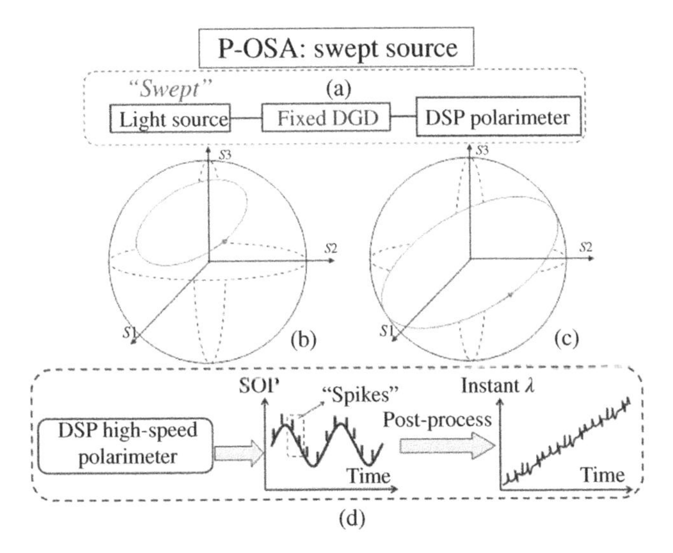

Figure 11.2 Concept of the polarimeter-based optical frequency analyzer used for a swept-wavelength source. (a) Experimental configuration. (b) SOP trace of light at DGD output when input SOP is not aligned 45° from the slow (or fast) axis of the DGD. (c) Output SOP trace when input SOP is aligned 45° from the DGD axis. (d) Obtaining the instantaneous wavelength (frequency) of a tunable light source from SOP rotation angle. The spikes correspond to transient frequency surges not observable with many other spectral measurement techniques (Yao et al. 2008).

determined by the value of the DGD element. Therefore, one can obtain the DGD value from the SOP trace on the Poincaré sphere.

If we choose a precisely known DGD element, by using the reverse effect, the frequency of the light source can be determined from the SOP trace on the Poincaré sphere. Let τ be the DGD value of the DGD element, the complex amplitude of the electrical field of the light after the DGD element can be expressed as

$$\mathbf{E} = E_s(e^{i\theta}\hat{\boldsymbol{s}} + E_f\hat{\boldsymbol{f}})e^{i\varphi_0} \tag{11.1a}$$

$$\theta = 2\pi f\tau \tag{11.1b}$$

where θ is the retardation of the DGD element, E_s and E_f are the amplitudes of the electrical field along the slow and fast axes of the DGD, respectively, $\hat{\boldsymbol{s}}$ and $\hat{\boldsymbol{f}}$ are the unit vectors along the slow and fast directions, respectively, and φ_0 is the common phase term.

If $E_s = E_f$, the SOP will trace a largest circle on the Poincaré sphere, as shown in Figure 11.1c, resulting in the highest frequency measurement resolution or sensitivity. When considering two specific frequency values, f_1 and f_2, during a wavelength sweep, the angular difference, denoted as $\Delta\theta$, between the two polarization states of these two different frequencies is simply the phase difference in Eq. (11.1b), which can be expressed as

$$\Delta\theta = 2\pi(f_1 - f_2)\tau \tag{11.2a}$$

For a known differential delay τ, the frequency difference can be calculated from the SOP angular difference on Poincaré sphere as

$$f_1 - f_2 = \frac{1}{2\pi}\frac{\Delta\theta}{\tau} \tag{11.2b}$$

Figure 11.2d shows the working principle of the swept-wavelength operation mode. The high-speed polarimeter collects the time-resolved SOP traces, which carries the detailed polarization evolution information. Based on the known DGD value, one can translate the SOP traces into the instantaneous frequency evolution by calculating the polarization rotation angle on the Poincaré sphere using Eq. (11.2b).

Spectral Range and Resolution For conventional OSAs, such as those based on F–P interferometer, FSR is defined as the spacing between the periodic pass-bands. The FSR usually determines the spectral measurement range of the analyzer. One of the major tradeoffs for the conventional OSAs is that the FSR is always proportional to the spectral resolution. This means that the high spectral resolution will always result in limited measurement range.

For the P-OFA, if the SOP resolution is 0.036°, one can have 360/0.036 = 10 000 points resolved in a full circle (one-cycle) on the Poincaré Sphere, which can be easily achieved with a polarimeter. From Eq. (11.2b), one can see that for a given τ, the resolution for measuring the center frequency and the one-cycle measurement range of the polarimeter based spectrum analyzer are

$$\text{Frequency resolution: } \delta f = \frac{\Delta f}{10000} = \frac{10^{-4}}{\tau} \tag{11.3a}$$

$$\text{One-cycle measurement range: } \Delta f = \frac{1}{\tau} \tag{11.3b}$$

As an example, for a DGD value of 1000 ps, the frequency resolution will be 100 kHz, and the one-cycle measurement range will be 1 GHz. The tradeoff between the resolution and the range still exists if we consider only one-cycle on the Poincaré sphere.

However, since the SOP circle will be able to repeat itself on the sphere if the frequency variation range Δf is larger than $1/\tau$, one can utilize the valuable information of the direction of the SOP evolution. By combining Eq. (11.3b) with the SOP direction information, one obtains the following equation as the total measurement range by the multiplication of the SOP cycle number N:

$$\text{Total measurement range:} \quad \Delta f = N \times \frac{1}{\tau} \tag{11.3c}$$

where N can be any arbitrarily large integer number.

This results in the unique feature of arbitrarily large spectral range, without any compromise to the spectral resolution. Since only the value of the fixed DGD element determines the spectral resolution, one can simply improve the resolution by introducing arbitrarily large DGD element without sacrificing the measurement range. The P-OFA is therefore not limited by the traditional tradeoff between the spectral range and frequency resolution.

Experimental Setup and Results As shown in Figure 11.3, two different types of tunable frequency light source were used in the experiment. The first one is a spectral spliced amplified spontaneous emission (ASE) source using a high speed F–P tunable filter (LamdaQuest 2022) (up to 40 KHz sweeping speed). The second swept source is the HP 8164A tunable laser source (TLS) with a sweep step size of 0.05 nm and a dwell time of 0.1 second. A 5.7 ps birefringent crystal is applied as a fixed DGD element, with input adjusted by a polarization controller for equal power splitting between two eigen polarization states. The output port of the DGD element is directed to the General Photonics' high-speed digital signal processing (DSP) in-line polarimeter (POD-101D) for real-time Poincaré sphere display and SOP trace recording at a sampling rate near ~1 MHz (Luna Innovations 2022).

Figures 11.4 and 11.5 show the results of the swept-wavelength input using a high speed F–P filter based spectral-slicing source (Yao et al. 2008). A high power EDFA with an output of ~15 dBm is used as the ASE source. The sweeping wavelength range of the high-speed tunable filter is properly adjusted using a function generator, with frequency, amplitude, and offset matched to the EDFA gain bandwidth. Figure 11.4a shows the modulation of one (S1) of the recorded Stokes parameters from the polarimeter when the tunable filter is swept at 1-KHz rate. The amplitude and the offset from the sinusoidal function generator are set to be 10 and 3.5 V, respectively. By utilizing the directional SOP evolution, the accumulated polarization rotation angle is obtained, as shown in the right Y axis of Figure 11.4b. Note that the multiple full circle of the SOP modulation can be correctly interpreted to the accumulated rotation angle. From Eq. (11.2b), one may further derive the swept wavelength as a function of time (in the left Y axis of Figure 11.4b) from the accumulated polarization angle. The starting wavelength is determined using a spectrum analyzer. In practice, one may utilize a tunable DGD element to obtain the reference wavelength using the method described in Section 11.1.1.2. One can see that the swept wavelength curve resembles well with the sinusoidal sweeping function and the time period is determined by the swept frequency of 1 KHz.

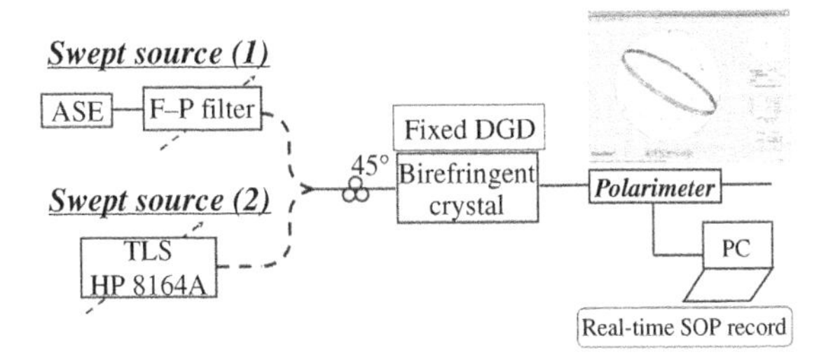

Figure 11.3 P-OFA experimental setup for analyzing two types of swept sources. Note that only one source is connected at a time. The SOP of the input light is adjusted to 45° with respect to the birefringent axis of the DGD. Source: Yao et al. (2008)/Optica Publishing Group.

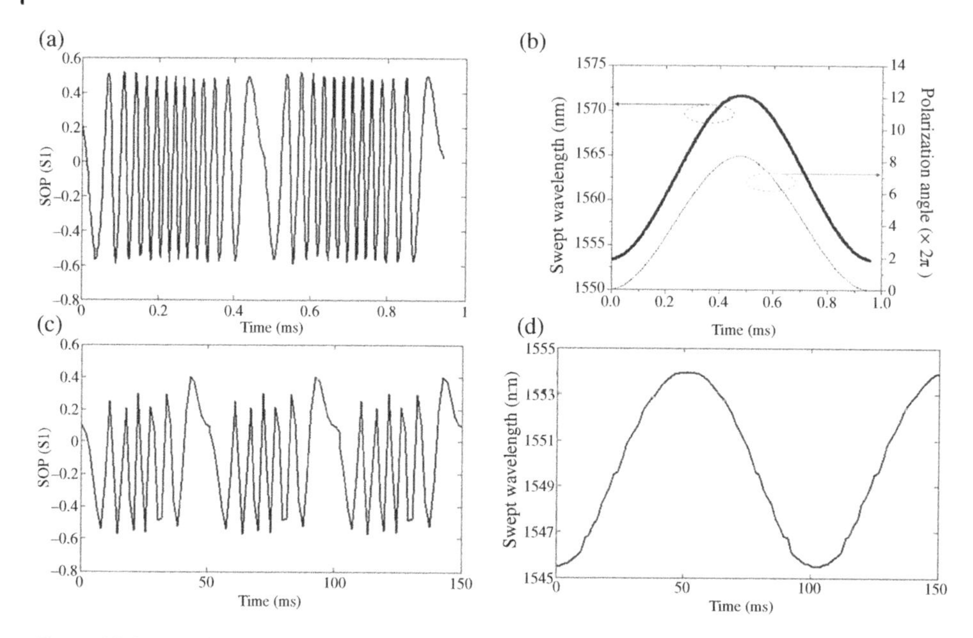

Figure 11.4 (a) SOP (S1) of 1-kHz tuning F–P filter (a). (b) The corresponding swept wavelength. (c) SOP (S1) trace of 10-kHz tuning F–P filter. (d) The corresponding swept-wavelength measured by the P-OFA. Note that the starting wavelength is obtained with a commercial spectrum analyzer, although the P-OFA can also determine the absolute starting wavelength directly, as described in Section 11.1.1.2.

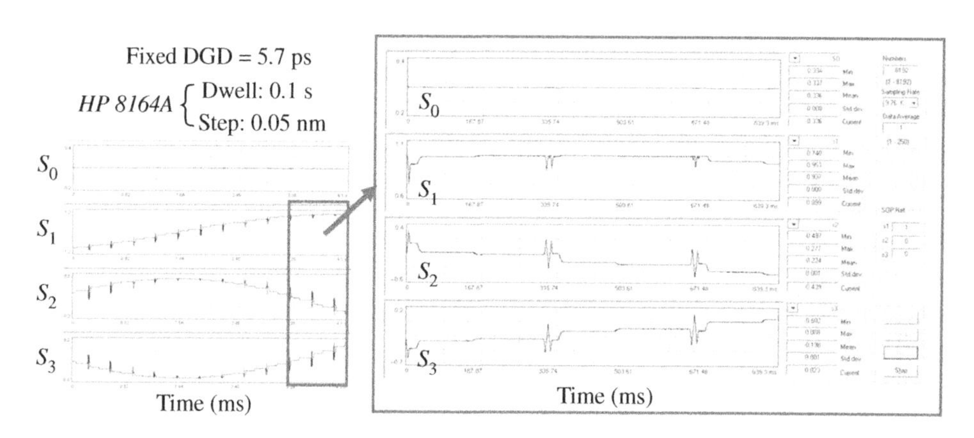

Figure 11.5 Polarimeter oscilloscope mode display. SOP evolutions are recorded when the input is swept at a speed of 0.1 second per 0.05 nm step (Yao et al. 2008).

Figure 11.4c shows one of the recorded Stokes parameters from DSP polarimeter when the tunable filter is swept at a higher rate of 10 kHz. The amplitude and the offset of the function generator are 5 and 4.5 V, respectively. Figure 11.4d shows the derived swept wavelength as a function of time. A period of 100 μs proves the swept frequency of 10 kHz. Due to the limited sampling rate of the DSP polarimeter, the recovered SOP trace cannot be as smooth as that of the 1 kHz case. Improved results are expected if the sampling rate of the polarimeter is increased. The reduced SOP modulation cycle and thus the reduced swept wavelength range is due to the smaller amplitude swing applied to the F–P filter.

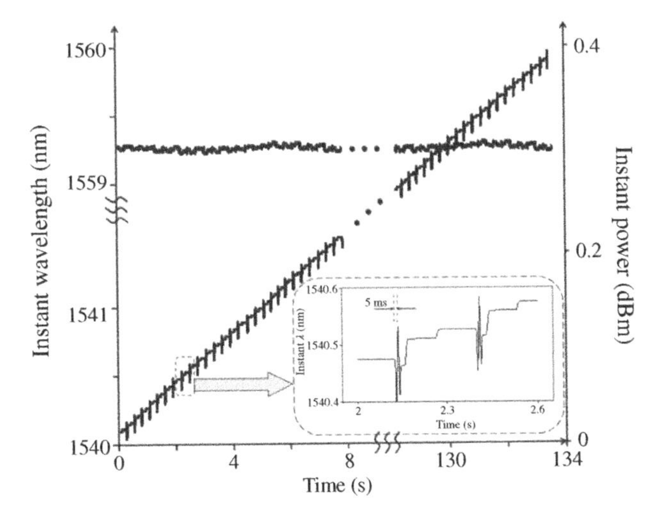

Figure 11.6 Instantaneous wavelength and power as the input light source is swept at a speed of 0.1 second. The starting wavelength is from the setting of the commercial tunable laser. Note that the transient dynamics of the swept laser source can be clearly revealed, as shown in the inset (Yao et al. 2008).

Based on the sampled SOP traces, the recorded SOP data can be post-processed by calculating the accumulated polarization rotation angle, taking into account the direction of the rotation. The polarization rotation angle can be further translated into the time-resolved swept frequency, as shown in Figure 11.6, where the instantaneous wavelength is obtained from 1540 to 1560 nm using the periodic nature of the SOP traces in Figure 11.5. The starting wavelength is the setting of the tunable laser and its value is not obtained from the analysis, although the P-OFA is capable of determining the absolute wavelength as described in Section 11.1.1.2. Note that the range can be further increased by recording more SOP evolution circles. A zoom in the curve in Figure 11.6 shows that a relatively low speed (0.1 second per step) swept source actually has a fast transition time on the order of millisecond. This reveals that when the light source is stepped from one wavelength to the next, it experiences a fast initialization state in which the wavelength is oscillating. It then quickly jumps to the desired value within several tens of milliseconds. However, most of the time is then used for wavelength locking and stabilization by the wavelength locking circuit inside the tunable laser.

From Figures 11.5 and 11.6, one can see that P-OFA exhibits the powerful capability of capturing the transient dynamics of a swept source. This capability greatly surpasses those of the conventional OSAs. The instantaneous power evolution is also measured from the time-resolved S_0 trace.

The direction of the SOP traces has another interesting usage for determining the direction of the wavelength change. As can be seen from the inset of Figure 11.6, the fast oscillation (on the order of millisecond) that occurred during wavelength transitioning can be resolved in terms of the direction of the instantaneous frequency changing, which can be well correlated to the SOP evolutions shown in Figure 11.5. This unique feature is also unobtainable with traditional OSAs and can find interesting applications in the field of swept frequency analysis.

11.1.1.2 Spectral Shape Analysis

Concept Figure 11.7 shows the concept and principle of using the P-OFA for analyzing the spectrum of a fixed wavelength source. In this spectrum analysis mode, a variable DGD element is applied and the spectrum of the light source is analyzed by post-processing the recorded SOP

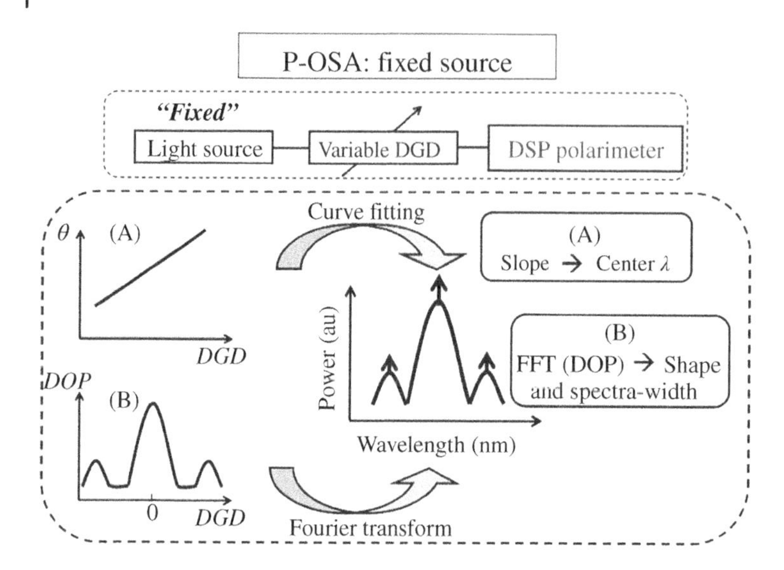

Figure 11.7 Concept and principle of the real-time P-OFA used for spectral shape analysis. Curve fitting of (A) determines the center frequency while Fourier transform of (B) yields the spectral shape and width. The spectral resolution is inversely proportional to the range of the variable DGD (Yao et al. 2008).

and DOP information from the polarimeter as the DGD is tuned. For a fixed wavelength input source, the polarization rotation angle is a linear function of the DGD (τ) value. By curve fitting the measurement data using Eq. (11.4), one can readily obtain the center frequency of the light source.

$$\theta = \theta_0 + 2\pi f\tau \tag{11.4}$$

In order to determine the spectral shape and width of the input source, one utilizes the DOP information of the light as the DGD value is varied from zero to well beyond the coherence length of the source. As discussed in Section 4.7.4, since the DOP is well correlated with the self-correlation function of the light source (Born and Wolf 1999), which in turn relates to the power spectrum by the Fourier transform, one obtains the following expression of the power spectrum for the case of equal power splitting between two principle polarization states of the DGD element.

$$P(\omega) = S_0 \int_{-\infty}^{\infty} DOP(\tau)e^{i\omega\tau}d\tau \tag{11.5}$$

where S_0 is the total received power and ω is the relative angular frequency. From Eqs. (11.4) and (11.5), one can conclude that for a fixed wavelength input, the spectrum of the source can be obtained accurately by measuring both the SOP and the DOP as a function of DGD, as shown in Figure 11.7.

Experimental Setup and Results The experimental setup for the spectrum width analysis of a fixed wavelength source is shown in Figure 11.8. In order to verify the capability of the P-OFA, two interesting spectral features are generated by modulating a narrowband tunable laser using two different on-off-keying (OOK) modulation formats (non-return-to-zero [NRZ] and return-to-zero [RZ]) at 40-Gb/s. The variable DGD module consists of a 2×2 PBS for splitting the input light into two orthogonal polarization states (port 1 → port 2 and 3) and combining them again at the output (port 2 and 3 → port 4). One motorized delay line (MDL), with a tuning range of 560 ps, is inserted in one of the arms. Both arms are path length matched when the MDL is set at its origin. Two

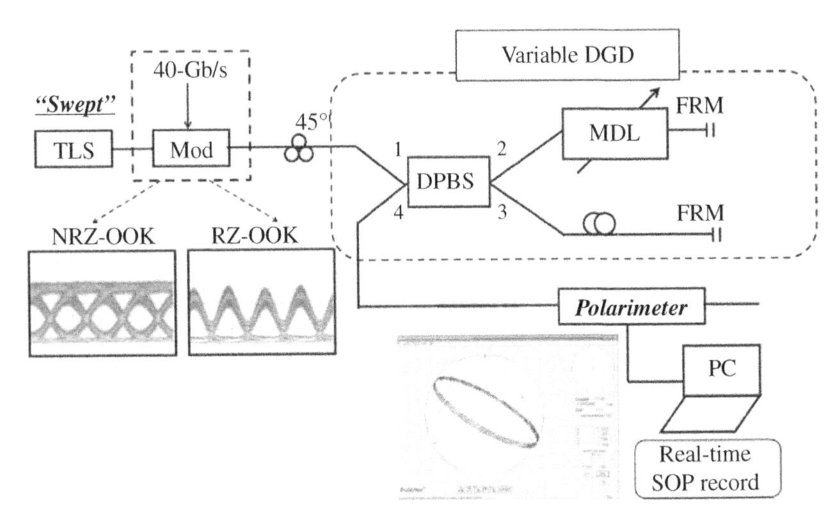

Figure 11.8 Experimental setup for spectrum analysis of fixed wavelength source. Source: Yao et al. (2008)/Optica Publishing Group.

Faraday rotating mirrors (FRMs) described in Chapter 6 are placed at the end of both arms for ensuring polarization stability of the light in the two arms when they recombine at the PBS. The output port of the PBS is directed to the DSP in-line Polarimeter described in Figure 8.6. A polarization controller is placed at the input of the polarimetric interferometer to ensure equal power splitting of the two arms when they recombine at the PBS, and thus the largest SOP circle (shown in the inset) on the Poincaré sphere, resulting in the highest frequency resolution.

Figure 11.9 shows the experimental results of the spectra analysis operation of the P-OFA. DOP of both the 40-Gb/s NRZ-OOK and RZ-OOK are recorded as the MDL values are increased from 0 to 500 ps, as shown in Figure 11.9a. This corresponds to 1000 ps DGD tuning range due to the double pass configuration of the experimental setup. One can see from the DOP vs. DGD curve that the NRZ-OOK curve remains almost constant around 60% when the DGD is beyond 25 ps, while the RZ-OOK curve exhibits periodic DOP natures due to the fact that it has pronounced 40-GHz tones and some residual 20-GHz tones. Figure 11.9b shows the measured spectra of both NRZ-OOK

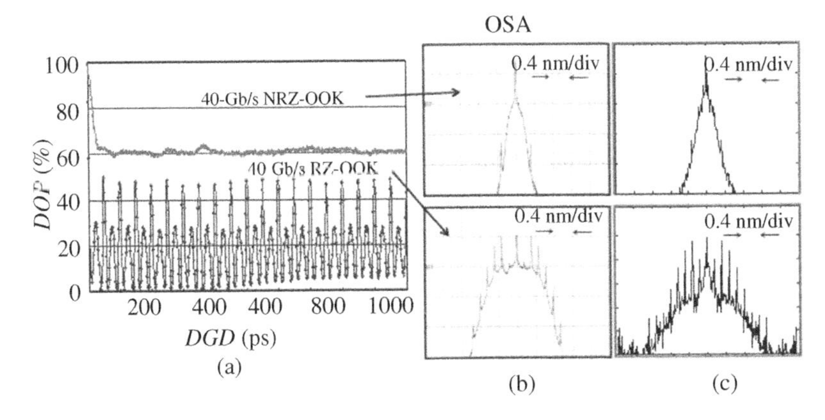

Figure 11.9 (a) Experimental results of DOP values when the DGD is changed from 0 to 1000 ps for both 40-Gb/s NRZ-OOK and RZ-OOK signals. (b) The measured OSA spectra. (c) The derived P-OSA spectra for comparison (Yao et al. 2008).

and RZ-OOK measured with a conventional OSA, which exhibits dominant 40-GHz spaced tones and much wider spectrum width. Using Eqs. (11.4) and (11.5), one obtains the spectra for the two different formats by processing the rotation angle for center frequency as well as the DOP curve for spectral shape and width. Figure 11.9c shows the derived spectra using our P-OFA method. For the same horizontal and vertical scales, one can see that the P-OFA provides very similar spectra width and shapes, with a much better spectral resolution due to the 1000-ps DGD tuning range, which corresponds to a line-width resolution of less than 1 GHz.

Spectral Shape Measurement of Swept-Wavelength Source The spectral shape and the width of a swept-wavelength source at each wavelength is an important parameter, since it contains the coherence length information of such sources for OCT applications (Huang et al. 1991). However, they cannot be directly measured with conventional OSAs (Huber et al. 2006). In this section, we show that the P-OFA can directly measure the spectral shape of a fast swept-wavelength source, in addition to the measurement of the spectral shape (power vs. frequency) of a fixed wavelength source, as described earlier. With such a capability, a unique three dimensional (3D) plot of the spectral shape of a wavelength-swept light source with feature-rich spectrum as a function of its center wavelength (or time) can be obtained.

Figure 11.10 shows the 3D display of the P-OFA with an added dimension of the swept wavelength or time. For the swept-frequency input, every time the DGD module is tuned to a specific value, the SOP and DOP information are recorded when the wavelength of the source is swept a full cycle. By tuning the DGD element gradually from minimum to maximum value, one obtains the whole set of SOP and DOP as two dimensional (2D) matrices with respect to both swept wavelength values as well as tuned DGD points. By rearranging the two dimensional matrices, one can display each spectrum at every swept wavelength using Eqs. (11.4) and (11.5). For Figure 11.10, during each wavelength scan ranging from 1540 to 1560 nm, alternating 40-Gb/s NRZ-OOK or 40-Gb/s RZ-OOK modulation formats are generated so as to obtain feature-rich yet contrasting spectra. The RZ-OOK spectrum shows a better distinguishable and equally spaced carrier tones, as well as a much wider spectrum. This capability enables detailed spectral characterization of a fast swept-wavelength source that cannot be obtained with any conventional methods.

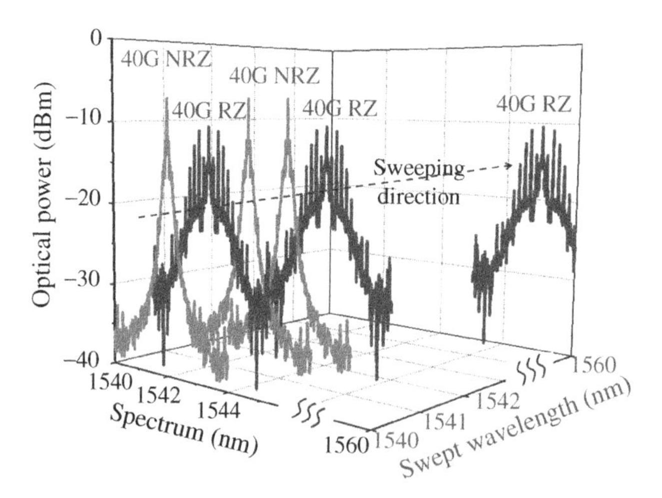

Figure 11.10 A 3D plot using the P-OFA data. Swept wavelength is the added dimension compared with the conventional OSAs. Note that the absolute wavelengths are directly obtained with P-OFA via curving fitting (Yao et al. 2008).

11.1.2 Sine–Cosine Optical Frequency Detection with Polarization Manipulation

The polarimeter based optical frequency analyzer described earlier is capable of detecting optical frequency with high speed, high resolution, and large range, simultaneously. However, it has two major drawbacks. First, the configuration of Figure 11.2 is sensitive to the input polarization and a polarization stabilizer described in Chapter 8 must be used to ensure measurement stability, which adds cost and complications to the measurement system. Second, the SOP in general has an arbitrary orientation on the Poincaré Sphere because the coordinate system of the polarimeter may not be aligned with the axis of the DGD used, which requires complicated data processing that compromises the detection speed.

11.1.2.1 Concept Description of Sine–Cosine Optical Frequency Detection

Figure 11.11a illustrates a new configuration (Yao et al. 2022), which overcomes the drawbacks of the configurations in Section 11.1.1. The input beam of an arbitrary polarization first enters a polarization beam displacer (PBD) (as discussed in Figure 6.4) made with a birefringence crystal in which the two beams 1 and 2 with the orthogonal polarization components are displaced. A 90° polarization rotator is then placed in one of the beams to rotate the SOP of the beam 90° so that the two beams have the same polarization state. A DGD element, such as a birefringence crystal with its slow axis oriented 45° from the linear SOP of the two beams, is placed after the PBD. A polarizer oriented 45° from the slow axis of the DGD element is used to clean up the polarization deviation from the imperfections of the polarization rotator made of either a half-wave plate (HWP) or a

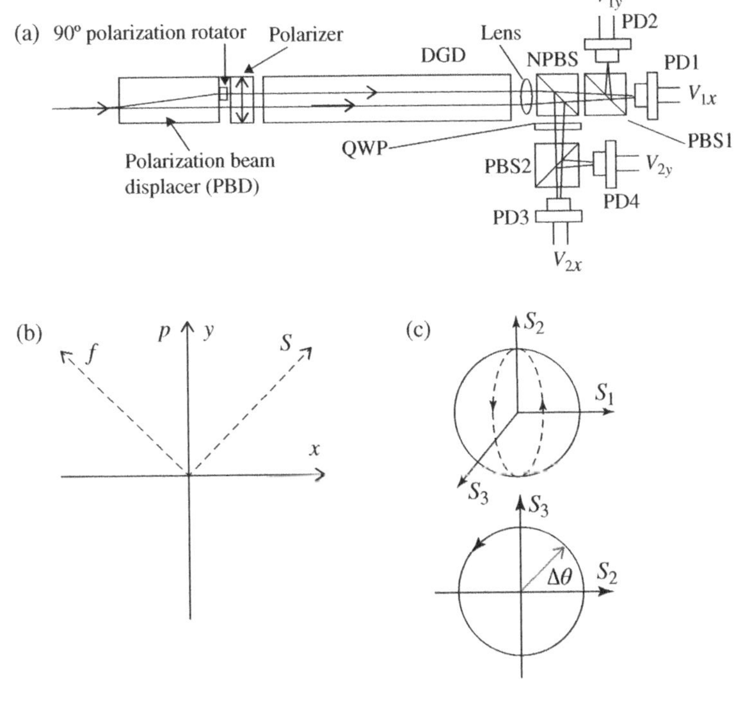

Figure 11.11 (a) Basic configuration of a sine–cosine optical frequency detector based on SOP analysis. (b) Relationship between the birefringence axes $(\hat{s}, \hat{f})$ of the DGD element and the orientation directions $(\hat{x}, \hat{y})$ of the PBD and PBS, as well as the polarizer $\hat{p}$. (c) The trajectory of the SOP on the Poincaré Sphere. NPBS: polarization insensitive beam splitter, PBS: polarization beam splitter, PD: photodetector.

Faraday rotator described in Chapter 6, as shown in Figure 11.11b. A non-polarizing beam splitter (NPBS) is placed at the output end of the DGD element to further split the beams into two optical paths. A quarter wave plate (QWP) ($\lambda/4$ plate) is placed in one of the paths with its optical axis oriented parallel or anti-parallel with the DGD element to introduce a $\pi/2$ phase shift between the slow and fast polarization components. A PBS is placed in each of the two paths to split the beams into two photodetectors (PDs). All PBS' are oriented the same as the polarizer, which is 45° from the birefringence axes $(\hat{\mathbf{s}}, \hat{\boldsymbol{f}})$ of the DGD element. A lens with a properly selected focal length is used to focus two beams onto the two PDs. Figure 11.11c shows the SOP tracing out a circle on in the (s_2, s_3) plane as the optical frequency scans.

The optical field $\mathbf{E}$ of light after passing through the DGD element can be expressed by Eq. (11.1a):

$$\mathbf{E} = (E_s e^{i\theta}\hat{\mathbf{s}} + E_f\hat{\boldsymbol{f}})e^{i\varphi_0}$$

Because the passing axis and the reflecting axis are aligned with $\hat{\boldsymbol{x}}$ and $\hat{\boldsymbol{y}}$ respectively, one obtains from Figure 11.12b as

$$\hat{\boldsymbol{x}} = \frac{1}{\sqrt{2}}(\hat{\mathbf{s}} - \hat{\boldsymbol{f}}) \tag{11.6a}$$

$$\hat{\boldsymbol{y}} = \frac{1}{\sqrt{2}}(\hat{\mathbf{s}} + \hat{\boldsymbol{f}}) \tag{11.6b}$$

The photocurrents detected in PD1 and PD2 as

$$I_{1x} = \alpha_1|\vec{E}\cdot\hat{x}|^2 = \frac{\alpha_1}{2}\left(E_s^2 + E_f^2 - 2E_sE_f cos\theta\right) \tag{11.7a}$$

$$I_{1y} = \alpha_2|\vec{E}\cdot\hat{y}|^2 = \frac{\alpha_2}{2}\left(E_s^2 + E_f^2 + 2E_sE_f cos\theta\right) \tag{11.7b}$$

where α_1 and α_2 are the factors accounting for the transmission losses of the lights in the paths to PD1 and PD2, and responsivities of PD1 and PD2, respectively. The photocurrents in Eqs. (11.7a) and (11.17b) can be converted into photovoltages V_{1x} and V_{1y} with a transimpedance amplifier

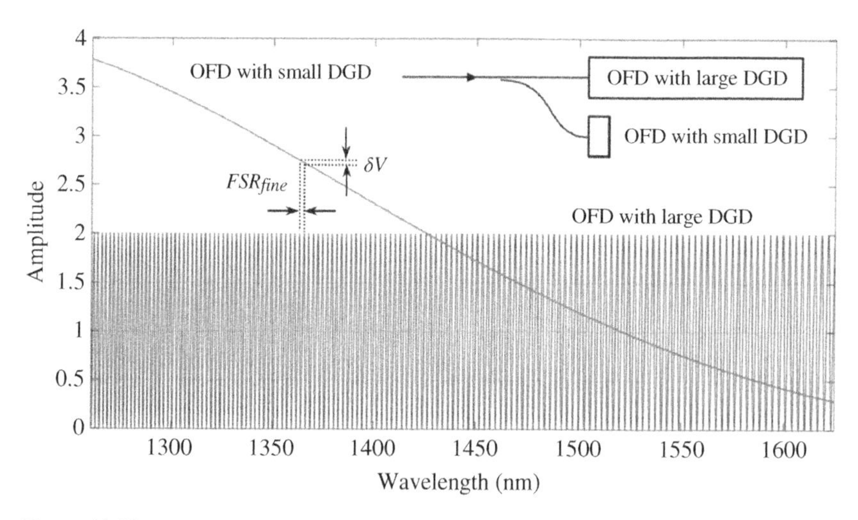

Figure 11.12 Illustration of an OFD with absolute optical frequency (wavelength) detection capability by combining an OFD of large DGD with an OFD of a small OFD, providing that the one is able to resolve δV corresponding to the free spectral range, FSR_{fine}, of the large DGD (Yao and Chen 2018).

after each PD. One may always adjust the gains of the amplifiers such that the peak photovoltage V_{10} from the two PD1 and PD2 is the same to obtain

$$V_{1x} = V_{10}(1 - cos\theta) \tag{11.8a}$$

$$V_{1y} = V_{10}(1 + cos\theta) \tag{11.8b}$$

$$cos\theta = \frac{(V_{1y} - V_{1x})}{(V_{1x} + V_{1y})} \tag{11.8c}$$

where $E_s = E_f = E_0/\sqrt{2}$ is used because the input polarization is oriented 45° from the slow axis of the DGD.

After the light passes through the QWP, the field can be expressed as

$$\mathbf{E} = [E_s e^{i(\theta+\pi/2)}\hat{\mathbf{s}} + E_f\hat{\boldsymbol{f}}]e^{i\varphi_0} \tag{11.9a}$$

Similar to the derivation of (11.8a)–(11.8c), one obtains the photovoltages V_{2x} and V_{2y} from PD3 and PD4, respectively, as

$$V_{2x} = V_{20}(1 - sin\theta) \tag{11.9b}$$

$$V_{2y} = V_{20}(1 + sin\theta) \tag{11.9c}$$

$$sin\theta = \frac{(V_{2y} - V_{2x})}{(V_{2x} + V_{2y})} \tag{11.9d}$$

Combining Eqs. (11.8c) and (11.9d), one obtains the optical frequency increment Δf as

$$\Delta f = \frac{\theta}{2\pi\tau} = \frac{1}{2\pi\tau}tan^{-1}\frac{(V_{2y} - V_{2x})/(V_{2x} + V_{2y})}{(V_{1y} - V_{1x})/(V_{1x} + V_{1y})} \tag{11.10}$$

It is important to notice that the optical frequency variation can be represented as sine and cosine functions of the detected signals, and therefore one can obtain both the magnitude and direction of the frequency change with endless range without ambiguity, similar to a sine–cosine or quadrature decoder for a motor. There are many hardware and software solutions already developed for obtaining the motor rotation angle in real time using the sine and cosine information of the angle. Therefore, the differential delay τ can be made arbitrarily large for achieve high frequency resolution, without scarifying the measurement range because the multiple periods can be counted from the sine–cosine interpretation algorithms.

The discussions earlier are only for obtaining the incremental optical frequency. Sometimes it is important to obtain the absolute frequency or wavelength. Figure 11.12 shows using the combination of an optical frequency detector (OFD) with a small DGD (called coarse OFD) and an OFD with a large DGD (fine OFD) to make the absolute frequency detector. The latter is for achieving high frequency resolution, while the former is for determining the absolute optical frequency. As can be seen in the figure, the value of the small DGD for the coarse OFD is chosen such that in the frequency (or wavelength) range of interest, the detected optical power change is limited in the first half of the cosine function, or one half of the FSR (FSR_{coarse}) of the coarse OFD, for the coarse absolute optical frequency measurement. Within this half period, there many a large amount of periods produced by the fine OFD with the large DGD. So long as the frequency (wavelength) resolution of the coarse OFD is sufficiently fine to resolve a period of the fine OFD or the FSR (FSR_{fine}) of the large DGD, corresponding to the case that the circuit and signal processing is able to resolve δV in Figure 11.12, the absolute frequency of the light source can be obtained unambiguously.

11.1.2.2 Experimental Results of Sine–Cosine Optical Frequency Detection

A sine–cosine OFD was fabricated with a DGD of 2.7 ps (0.8 mm) using an YVO_4 crystal according to Figure 11.11, with which a frequency resolution of less than 10 MHz (0.1 pm) is achieved. Figure 11.13a shows the measured instantaneous frequency (top) of a commercial tunable laser module around 1550 nm (Pure Photonics) and the corresponding tuning rate (bottom) using the sine–cosine OFD. The tuning nonlinearity can be clearly detected. Figure 11.13b shows the measured wavelength of a commercial desk-top tunable laser (Yenista) as a function of time (solid line) and the digitally generated k-clock of equal frequency spacing. Any k-clock spacing can be generated using the embedded field programmable gate array (FPGA) circuit in the OFD module. As can be seen, the k-clock ticks are more dense during the time the wavelength (or frequency) ramp is more steep, while is more sparse whenever the slope becomes less steep, as expected.

Figure 11.14 shows the measurement results of an OCT laser (Thorlabs) operating at 8 kHz repetition rate with a wavelength scanning range of 115 nm. Figure 11.14a is the spectrum of the laser measured with a traditional grating based OSA from Yokogawa (Model #) using the peak detection function. Figure 11.14b shows the instantaneous wavelength (thin line) and optical power (thick line) as a function of time, measured with the sine–cosine OFD displayed in a two dimensional (2D) plot, while Figure 11.14c shows the same data in a 3D plot. Note that the instantaneous wavelength cannot be detected during the time the laser power output is below

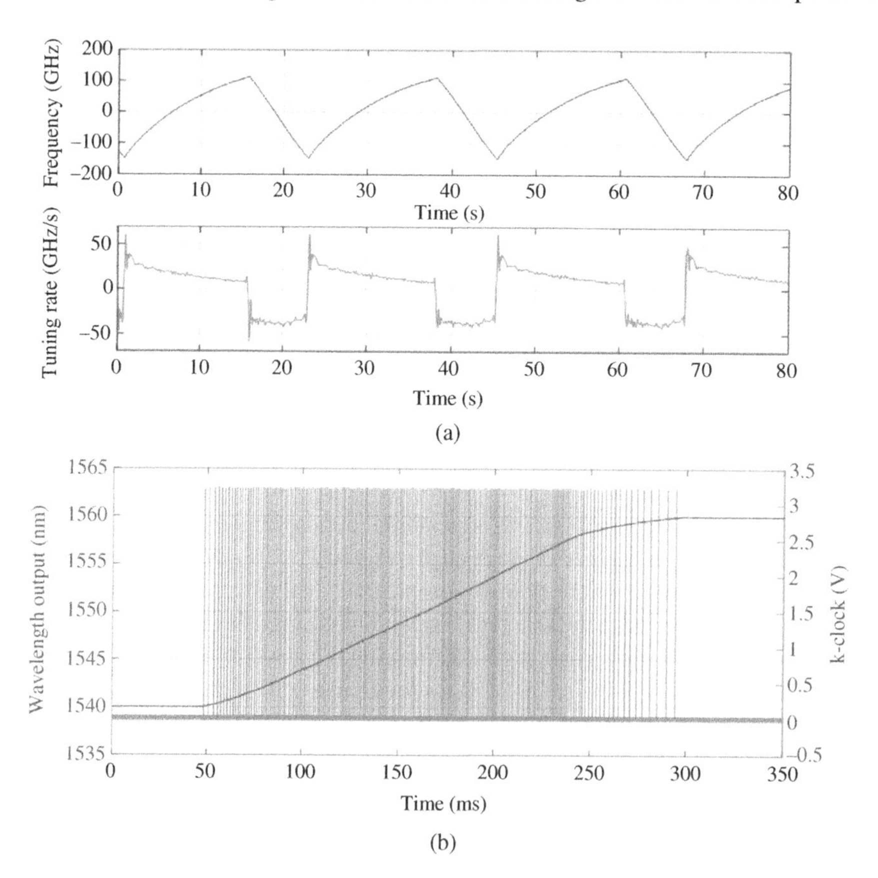

Figure 11.13 (a) Measured optical frequency of a tunable laser module (top) and its tuning rate (bottom). (b) Measured wavelength ramp of a tunable laser (solid line) and digitally generated k-clock ticks (Yao et al. 2022).

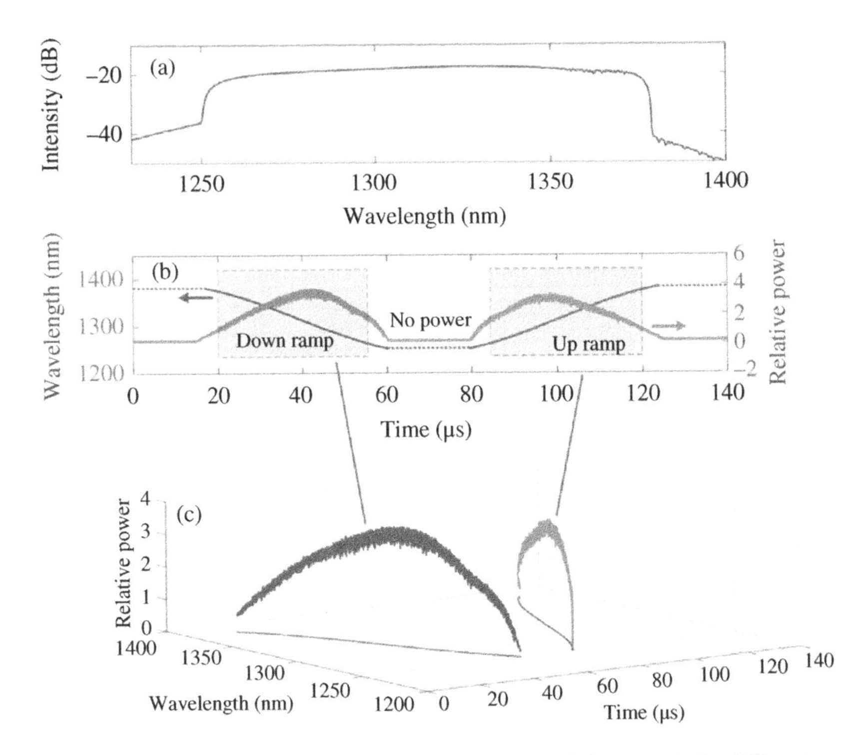

Figure 11.14 Measurement results of swept-wavelength laser made for OCT systems scanning at 8 kHz repetition rate using the sine–cosine OFD. (a) Spectrum measured with a grating based optical spectrum analyzer. (b) Instantaneous wavelength and power when the laser wavelength is scanned. Dotted line indicates the no optical power region during a wavelength scan. (c) 3D plot of the laser output of the shaded regions (Yao et al. 2022).

a certain threshold, which is represented by the dashed line in Figure 11.14b. Clearly, both the wavelength and power as a function of time can be unambiguously detected with the OFD.

Figure 11.15a shows the measured instantaneous frequency (gray line) of the OCT laser earlier, with the dashed line showing the linear fit and solid line the sinusoidal fit, while Figure 11.15b shows the differences between the measurement data and the fitting curves. It can be seen that the frequency scan fits to the sinusoidal curves (solid line) much better than the linear fit (dashed line) for both the up- and down-frequency ramps. Therefore, the frequency can be accurately represented with the sinusoidal function.

Figure 11.16 shows the frequency ramp of another OCT laser (Thorlabs) at 100 kHz repetition rate with a wavelength range of 110 nm. Figure 11.16a is the spectrum of the laser measured with the traditional OSA using the peak detection function. The instantaneous wavelength (thin line) and optical power (thick line), measured with the sine–cosine OFD displayed in a two dimensional (2D) plot and 3D plot, are shown in Figure 11.16b,c, respectively. Note that the instantaneous wavelength cannot be detected during the time the laser power output is below a certain threshold, which is represented by the dashed line in Figure 11.16b.

Figure 11.17a shows the optical frequency up ramp of the laser in the shaded region of Figure 11.16b measured with the OFD. The linear (square) and sinusoidal (solid line) fits of the measurement data (gray line) almost overlap so that only the solid line can be seen. For analyzing the scan nonlinearity, the deviations of the measured frequency from the linear and sinusoidal fits are calculated, as shown in Figure 11.17b.

Figure 11.18 shows some of the applications of the sine–cosine OFD (Yao and Chen 2018). As shown in Figure 11.18a, the OFD can be used as the interrogator for the fiber Bragg grating

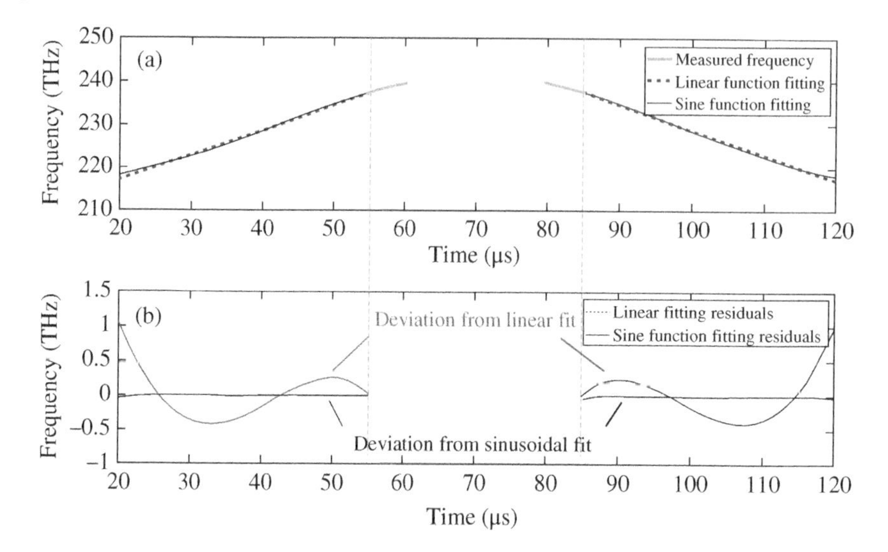

Figure 11.15 (a) Frequency up and down ramps of the OCT laser extracted from the data of Figure 11.14b. (b) Frequency scan nonlinearity analysis of (a).

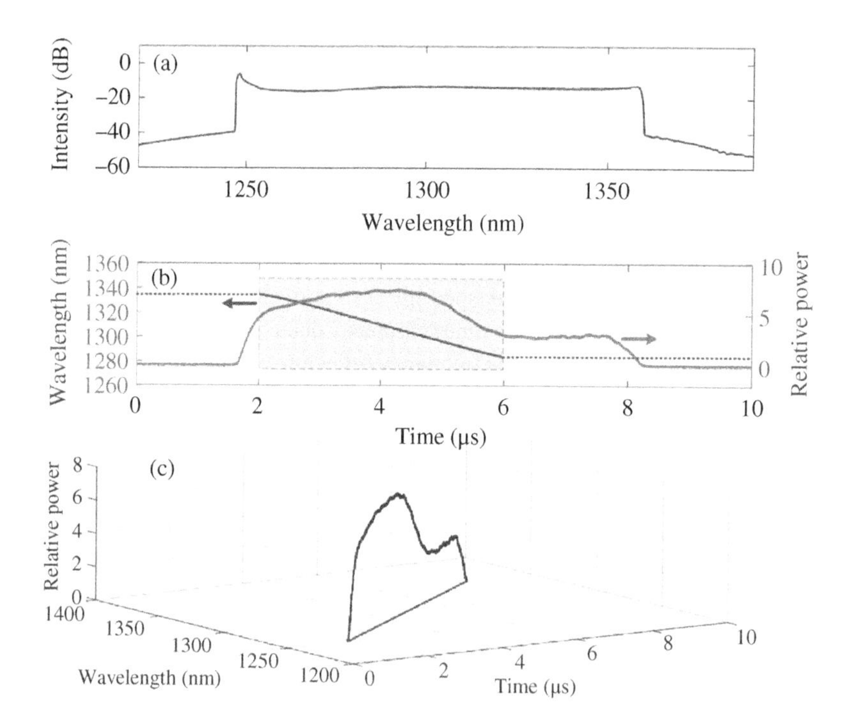

Figure 11.16 Measurement results of swept-wavelength laser made for OCT systems scanning at 100 kHz repetition rate using the sine–cosine OFD. (a) Spectrum measured with a grating based optical spectrum analyzer. (b) Instantaneous wavelength and power when the laser wavelength is scanned. Dotted line indicates the low optical power region during a wavelength scan. (c) 3D plot of the laser output of the shaded region (Yao et al. 2022).

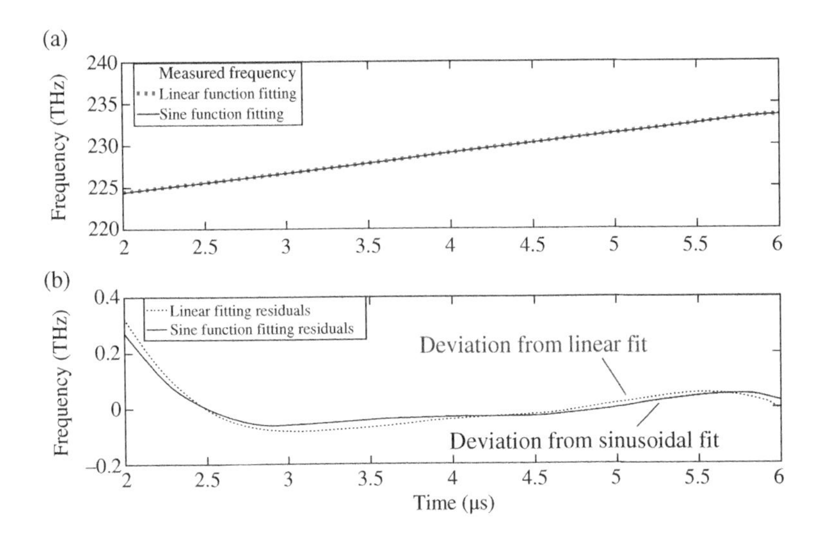

Figure 11.17 Frequency scan nonlinearity analysis for the frequency ramp in Figure 11.16b. (a) Sinusoidal (solid line) and linear (square) fits of the measurement data (gray line). (b) Deviations of the measurement data from the sinusoidal (solid line) and linear (dots) fit (Yao et al. 2022).

(FBG) sensors in which the slight center wavelength shifts of the reflected light from different sensing FBGs can be precisely measured at high speed, with a wavelength resolution better than 0.1 pm and speed up to MHz, significantly better than those of an OSA based FBG interrogator.

In operation, light from a broadband light source, such as ASE or superluminescent light-emitting diode (SLED), is inputted into the fiber with N FBGs for sensing temperature, strain, or

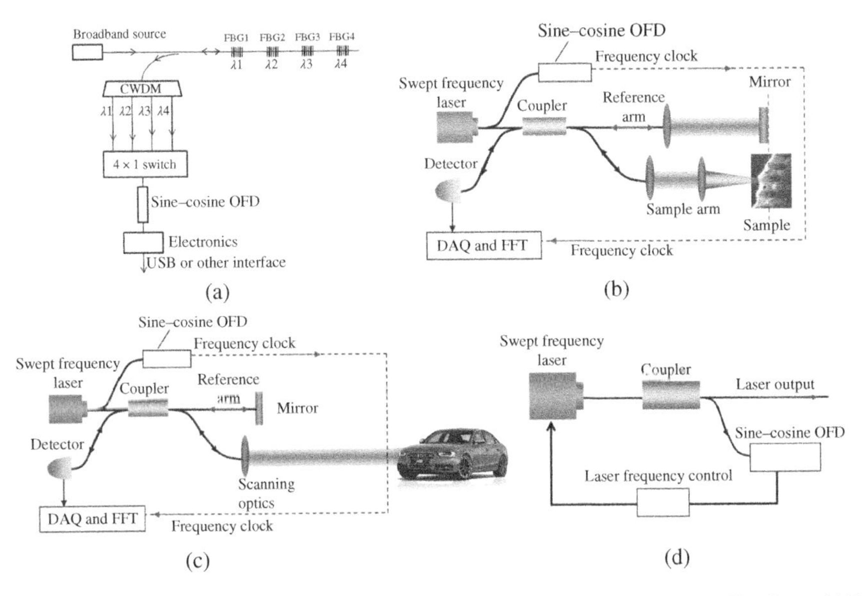

Figure 11.18 Sine–cosine OFD for sensing and laser frequency control applications. (a) The sine–cosine OFD as an interrogator for fiber Bragg grating (FBG) sensor system. (b) The sine–cosine OFD for generating f-clock for an OCT system. (c) The sine–cosine OFD for generating f-clock for a coherent Lidar system. Source: Audi. (d) The sine–cosine OFD real-time measurement of optical frequency for control and synthesizing optical frequencies with a tunable laser.

vibration at different locations. These FBGs have different center wavelengths for the reflection spectra. A coarse wavelength division demultiplexer (CWDM) is used to separate the reflected light from the N FBGs into N channels so that an $N \times 1$ switch can be used to direct one channel at a time to the OFD. The changes in temperature, strains, or stress on the FBGs will cause the reflection center wavelengths to change, which will be precisely measured by the OFD at a high speed. The amount of center wavelength change can be used to determine the temperature, the strain of the stress on each FBG to be sensed.

Figure 11.18b,c shows using the f-clock obtained by the OFD of the frequency-swept laser in an OCT or a coherent Lidar system to trigger data acquisition card for digitizing the detected interference signal obtained, while Figure 11.18d shows using the precise optical frequency detected by the OFD to control the laser frequency of the tunable laser to generate any desired frequency.

11.2 Polarimetry Fiber Optic Gyroscope

11.2.1 Introduction

Gyroscopes are essential for fully autonomous vehicle control. Since their first demonstration by Vali and Shorthill more than 42 years ago (Vali and Shorthill 1976), fiber optic gyroscopes (FOGs) have proven useful in various military and civilian applications (Lefèvre 1997; Sanders et al. 1995), such as precision rotation rate and angle detection for navigation systems in land, air, and sea vehicles, to become the most widely used fiber optic sensors in the world. Perhaps FOGs have the largest market share among all the fiber optic sensors.

Several FOG configurations have been proposed (Zarinetchi et al. 1991; Ezekiel and Balsamo 1997; Ciminelli et al. 2013; Li et al. 2017; Liang et al. 2017); however, as of today, the interferometric fiber optic gyroscope (I-FOG, Figure 11.19a) (Lefèvre 1997), which exploits a Sagnac interferometer, is the only one that has been adopted for real-world applications due to its superior performance. Sagnac interferometers typically operate at a minimum sensitivity point, with the interference signal having a cosine relationship with the rotation-induced phase difference between the two counter-propagating optical waves (Lefèvre 1993), which makes slow rotations difficult to detect. In addition, the nonlinear nature of the signal around this minimum sensitivity point leads to inaccurate measurements. Therefore, the I-FOG requires active biasing of the Sagnac interferometer at the most sensitive operation point (the quadrature point). Measures to mitigate these effects rely on phase modulation as a general biasing method and closed-loop techniques that enhance the dynamic range and detection sensitivity of I-FOGs (Lefèvre 1993, 1997; Cahill and Udd 1979).

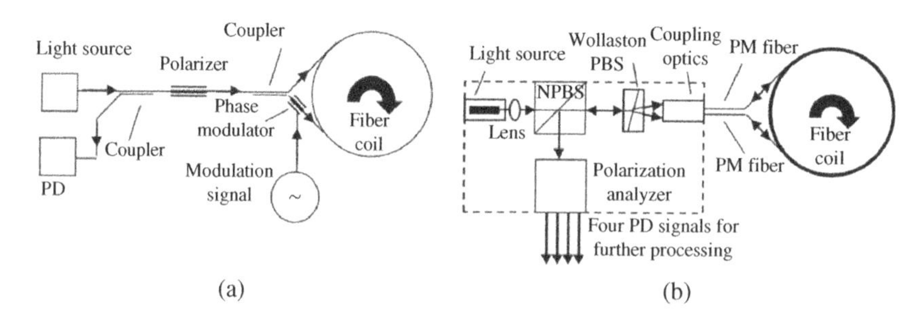

Figure 11.19 Comparison between (a) I-FOG; (b) P-FOG configurations. NPBS: non-polarizing beam splitter; Wollaston PBS: polarizing beam splitter made with Wollaston prisms; PM fiber: polarization-maintaining optical fiber; PD signals: photodetector signals (Yao et al. 2019).

A 3 × 3 coupler-based Sagnac interferometer has also been proposed to solve the bias problem without phase modulation (Sheem 1980), but its inherent nonreciprocity causes bias errors (Bergh et al. 1981), which has stalled its commercialization. Attempts at enhancing the performance of I-FOGs include work on the key gyroscope components, such as refinements to fiber coils (Li et al. 2013, 2015) and the improvement of a $LiNbO_3$-based integrated optical circuit (IOC) incorporating a 1 × 2 coupler and a phase modulator (Tran et al. 2017). Despite these advances, I-FOG systems remain too expensive, which limits their economic viability in cost-sensitive commercial applications, such as robotics and autonomous vehicles.

In this section, we describe a different approach that can acquire rotation information from a Sagnac loop at cost-effective levels for autonomous or robotic vehicle applications. This approach relies on the polarization analysis of the counter-propagating optical waves exiting from the Sagnac loop to detect rotation induced polarization variations, and therefore the resulting device is called a polarimetry fiber optic gyroscope (P-FOG) (Yao 2013; Yao et al. 2019). Specifically, the counter-propagating optical waves entering the Sagnac loop are orthogonally polarized to each other using a PBS and subsequently recombined using the same PBS at the exit of the Sagnac loop without producing an interference signal. The resulting beam is analyzed by polarimetry. We show that the Stokes parameters s_2 and s_3 are simply the cosine and sine functions of the phase difference, which is proportional to the rotation rate of the Sagnac loop, while the parameter s_1 is always equal to 0. Therefore, the phase difference and the corresponding rotation rate can be precisely obtained over an unlimited dynamic range by applying sine and cosine interpretation algorithms commonly used in sine–cosine encoders for motion control applications (Jenkins and Hilkert 2008; Kim et al. 2006), similar to the sine–cosine OFD described in Section 11.1. Interestingly, a similar orthogonal polarization Sagnac loop was been proposed for rotation detection in 1994 (Doerr et al. 1994; Lynch 1999), which relied on incomplete polarization analysis for obtaining the rotation information. Unfortunately, this orthogonal polarization fiber optic gyroscope (OP-FOG) has only provided a signal associated with the sine function of the phase difference, limiting its detection range to slow rotations and, possibly thwarting its commercialization. In contrast, the P-FOG approach here performs a full polarization analysis involving all the Stokes parameters to give both the sine and cosine functions of the phase difference. This enables phase unwrapping and hence rotation rate detection over an unlimited range, which represents a highly desirable feat that has only been attainable using a closed-loop I-FOG incorporating an expensive IOC and related high-speed electronics (Lefèvre 1993; Cahill and Udd 1979).

11.2.2 Operation Principle

Figure 11.19b illustrates an implementation of the P-FOG discussed earlier (Yao 2013; Yao et al. 2019). As shown in the figure, linearly polarized light emitted by the light source first passes through a NPBS to produce the input beam, which is split into two orthogonally polarized components of equal power using a Wollaston PBS (described in Chapter 6) oriented 45° from the SOP of the input beam. These two components are subsequently coupled into both ends of a polarization-maintaining optical fiber (polarization maintaining [PM] fiber) and are aligned with the slow axis (or fast axis) of the fiber. Next, they travel in opposite directions along the PM fiber, forming a closed loop, before recombining at the PBS. As in a Sagnac interferometer, the counter-propagating waves experience a relative delay or phase difference when the system is under rotation. This relative delay is simply a DGD between two orthogonally polarized components. Unlike in an I-FOG, the two orthogonally polarized counter-propagating waves cannot interfere either at the PBS or at the NPBS. It is important to point out that the Sagnac loop of the P-FOG (Figure 11.19b) displays a fully reciprocal optical path and therefore there is absence of nonreciprocal bias, which is critical for gyroscope applications (Merlo et al. 2000). Note that although the two counter propagating waves inside the fiber coil are co-polarized along the slow (or fast) axis of the PM fiber, they always remain orthogonal when entering and exiting the PM fiber coil at the Wollaston PBS, and consequently cannot directly interfere at the exit port (or the Wollaston PBS) of the Sagnac loop. However, as will be described next, the polarization analysis will involve interferences of different polarization components of the combined beam from the exit of the Sagnac loop for extracting the phase difference between the counter propagating waves induced by rotation.

The electric field of the optical input beam entering the PBS (Figure 11.20b) can be written as

$$\mathbf{E}_{in} = \left(\frac{E_0}{\sqrt{2}}\right)(\hat{\boldsymbol{x}} + \hat{\boldsymbol{y}}) \tag{11.11}$$

where E_0 is the amplitude of the input optical beam and $\hat{\boldsymbol{x}}$ and $\hat{\boldsymbol{y}}$ denote the two passing axes, or principal axes, of the PBS. After the two orthogonally polarized beams traverse the optical loop and recombine at the PBS, the electric field becomes

$$\mathbf{E}_{out} = \left(\frac{E_0}{\sqrt{2}}\right)(\hat{\boldsymbol{x}} + \hat{\boldsymbol{y}}e^{i\Delta\varphi}) \tag{11.12}$$

As in an I-FOG, $\Delta\varphi$ is the phase difference between the counter-propagating beams resulting from the physical rotation of the optical loop. It can be expressed as

$$\Delta\varphi = \frac{2\pi DGD}{\lambda_0} = \left[\frac{2\pi LD}{(\lambda_0 c)}\right]\Omega \tag{11.13}$$

where L is the fiber length, D is the mean fiber coil diameter, λ_0 is the center wavelength, c is the speed of light, and Ω is the rotation rate. In Eq. (11.12), we assume that PBS and NPBS do not induce any differential phase shift between the two orthogonally polarized beams so that the polarization rotation of the output light depends solely on the gyroscope rotation.

The recombination of the polarized components at the PBS is expected to alter the output SOP when the fiber loop is rotated. Such a SOP change can be measured using a polarization analyzer

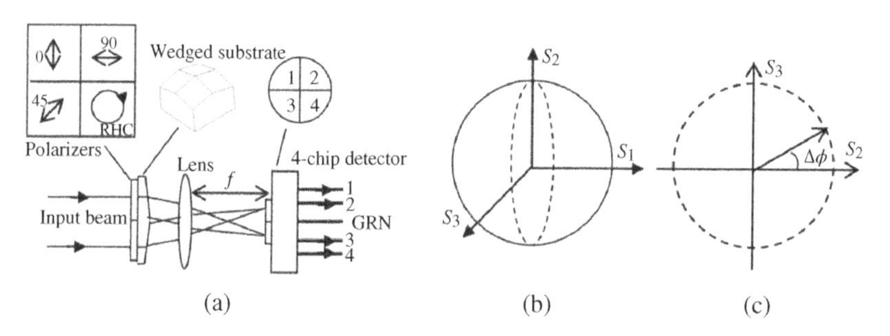

Figure 11.20 Illustration of how the SOP is measured and its behavior. (a) The space-division polarization analyzer for simultaneous polarization measurements; (b) SOP trace on a Poincaré sphere; (c) the SOP trace in s_2 and s_3 planes as the P-FOG rotates. The wedged substrate, which comprises four facets with distinctive angles, divides the input beam into four subbeams propagating in slightly different directions. Each subbeam passes through a polarizer chip with a particular orientation before being focused on its corresponding PD chip, which converts the optical power into a photovoltage. The four photovoltages give the Stokes parameters of the input beam. The 2×2 PD array in (a) is similar to a quadrant photodetector commonly used for position sensing (Yao et al. 2019).

or polarimeter (such as that of Figure 11.20a). In particular, the SOP is expected to outline a circle that passes through both poles (right- and left-hand circular polarizations) of the Poincaré sphere as the phase difference increases (Figure 11.20b,c), as will be described next.

Polarization analysis can be implemented in a number of ways, including those based on rotating wave plate and polarizer described in Chapter 8, and binary magneto-optic rotators described in Chapter 9. However, these methods rely on taking multiple measurements sequentially to obtain the complete polarization information (i.e. Stokes parameters), which require the SOP of the optical beam to be relatively stable and therefore are not suitable for a P-FOG in which the SOP varies dynamically. For such an application, all measurements for obtaining the Stokes parameters must be taken simultaneously, such as the amplitude-division and wave-front division polarimeters described in Chapter 8. Figure 11.20a shows such a polarization analyzer (Yao 2006), which has been described in Figure 8.7 in more detail, capable of performing the simultaneous measurements to determine the SOP of an optical beam. In a P-FOG, the output beam from the NPBS (Figure 11.19b) is expanded and passes through a four-faceted optical wedge to create four spatially divided subbeams. These subbeams propagate in four slightly different directions before being focused by a lens onto four PD chips placed at distinct spots on the focal plane of the lens, similar to a quadrant PD, as shown in Figure 11.20a. These PD chips detect the optical powers of the four subbeams and convert them into proportional photovoltages. Four additional polarizer chips can be placed on the wedge (Figure 11.20a) or in front of the PD chips to analyze the SOP of the optical beam. The first subbeam passes through a polarizer aligned with the $\hat{\boldsymbol{x}}$ axis before entering the first PD to give the photovoltage V_1 as $V_1 = \alpha|\mathbf{E}_{out} \cdot \hat{\boldsymbol{x}}|^2 = \alpha E_0^2/2$, where the coefficient α accounts for optical loss, PD quantum efficiency, and the electronic gain of each channel. The second subbeam passes through an orthogonal polarizer aligned with the $\hat{\boldsymbol{y}}$ axis before entering the second PD to yield the photovoltage V_2 as: $V_2 = \alpha|\mathbf{E}_{out} \cdot \hat{\boldsymbol{y}}|^2 = \alpha E_0^2/2$. The third subbeam passes through a polarizer with a 45° orientation with respect to the $\hat{\boldsymbol{x}}$ axis before entering the third PD to afford the photovoltage V_3 as $V_3 = \alpha|\mathbf{E}_{out} \cdot (\hat{\boldsymbol{x}} + \hat{\boldsymbol{y}})/\sqrt{2}|^2 = \left(\alpha E_0^2/2\right)(1 + cos\Delta\varphi)$. Finally, the fourth subbeam passes through a circular polarizer before entering the fourth PD to provide the photovoltage V_4 as $V_4 = \alpha|\mathbf{E}_{out} \cdot (\hat{\boldsymbol{x}} - i\hat{\boldsymbol{y}})/\sqrt{2}|^2 = \left(\alpha E_0^2/2\right)(1 + sin\Delta\varphi)$. The circular polarizer consists of a quarter-wave plate whose birefringence axis is aligned with the $\hat{\boldsymbol{x}}$ (or $\hat{\boldsymbol{y}}$) axis combined with a polarizer whose axis forms a 45° angle with respect to the $\hat{\boldsymbol{x}}$ (or $\hat{\boldsymbol{y}}$) axis. Optical losses and detector efficiencies vary according to channels, but the electronic gain can always be adjusted in either hardware or software to ensure that α is the same for all channels. As discussed in Chapters 4 and 8, let $S_0 = (V_1 + V_2)$; the Stokes parameters of the recombined light beam can be calculated using $s_1 = (V_1 - V_2)/S_0$, $s_2 = (2V_3 - S_0)/S_0$, and $s_3 = (2V_4 - S_0)/S_0$ as

$$s_1 = 0 \tag{11.14}$$

$$s_2 = cos\Delta\varphi \tag{11.15}$$

$$s_3 = sin\Delta\varphi \tag{11.16}$$

From Eqs. (11.15) and (11.16) one can see that $s_2^2 + s_3^2 = 1$, indicating that the SOP traces out a circle in the (s_2, s_3) plane on the Poincaré sphere (Figure 11.20c) as $\Delta\varphi$ increases. In addition, the Stokes parameters s_3 and s_2 represent the sine and cosine, respectively, of the phase difference $\Delta\varphi$, which is the polarization rotation angle in the (s_2, s_3) plane or the polar angle of the SOP on the Poincaré sphere. This situation is similar to the case where a sine–cosine rotary encoder (or an analog quadrature encoder) is used to obtain the rotation angle in a motor (Jenkins and Hilkert 2008; Kim et al. 2006). Consequently, $\Delta\varphi$ can be accurately evaluated over an unlimited range using the interpretation algorithms commonly used in sine–cosine rotary encoders (Jenkins and Hilkert 2008; Kim et al. 2006). Interestingly, the SOP trace is contained in the (s_2, s_3) plane, which eliminates the need to measure s_1 and can further simplify the detection optics of P-FOG (Figure 11.20). These simplified configurations are described in detail in Yao (2013). Compared with the prior OP-FOG (Doerr et al. 1994; Lynch 1999), this scheme yields an additional $cos\Delta\varphi$ term, which is crucial because it enables the unwrapping of the phase difference, and consequently provides an unlimited dynamic range for rotation rate measurements. Note that although the Sagnac loop in the P-FOG displays a fully reciprocal optical path, the wave plate in the polarization analyzer may cause additional bias drift because wave plate's retardation generally has temperature dependence. Fortunately, it is possible to make a temperature insensitive wave plate using two different birefringence materials with opposite signs of thermal coefficient to minimize such bias drift (Hale and Day 1988). Alternatively, one may use a thermal sensor to monitor the temperature and to measure the temperature induced phase difference at different temperatures when the P-FOG experiences no rotation. This phase difference can then be subtracted out in real measurement to minimize bias drift caused by the temperature sensitivity of the QWP's retardation.

11.2.3 Experimental Validation

There are also some slightly different configurations for the polarization analysis in a P-FOG, as described in detail Yao (2013). In principle, such a P-FOG exhibits following five major advantages. (i) It does not require phase modulation to bias the gyroscope system, resulting in significant cost savings. (ii) The SOP rotation angle is linearly proportional to the system rotation rate, simplifying the extraction of both the amplitude and direction of the rotation rate from the SOP measurement. (iii) The simplicity of the optical circuit facilitates its fabrication with a silicon optical bench. (iv) The P-FOG benefits from simpler and less power-consuming electronics than an I-FOG because it does not need a dedicated signal to drive the phase modulator nor high-speed field-programmable gate array/digital signal processing for the digital closed-loop design. Only a low-power circuit is required to detect the polarization rotation information. (v) Subtraction and division operations substantially reduce relative intensity noise effects (Doerr et al. 1994) from the light source and multipass interferences along the optical paths in the detection circuit during the Stokes parameter computations, which decrease the cost of light sources. It should be pointed out that a P-FOG requires four PDs to simultaneously detect four beams of light, while an I-FOG only requires a single PD, although the signal processing for extracting the rotation information is more straightforward and faster for the P-FOG.

To demonstrate the feasibility of the polarimetric designs described earlier, a P-FOG was constructed using a 585 m-long PM fiber coiled with a quadrupole winding pattern to minimize the Shupe effect (Shupe 1980), which characterizes the phase difference error originated from the fiber coil caused by temperature variations. Figure 11.21a shows the optics module containing all the optical components except the light source and the PM fiber coil. A 1310 nm SLED with a linewidth of 40 nm and an output power of 1 mW acted as the light source. The PD signals were amplified using a four-channel low-noise transimpedance amplifier board developed in house. Data acquisition and processing were performed using a low-cost digital signal processing board equipped with multiple analog-to-digital converters. It was necessary to package the entire P-FOG in an enclosure to fully characterize its performance using a rotation stage and a temperature chamber. The packaged unit comprises the optics module, electronic circuit, light source, and PM fiber coil (Figure 11.21b). It also has a universal asynchronous receiver–transmitter (UART) interface, outputting 200 $\Delta\varphi$ data points per second.

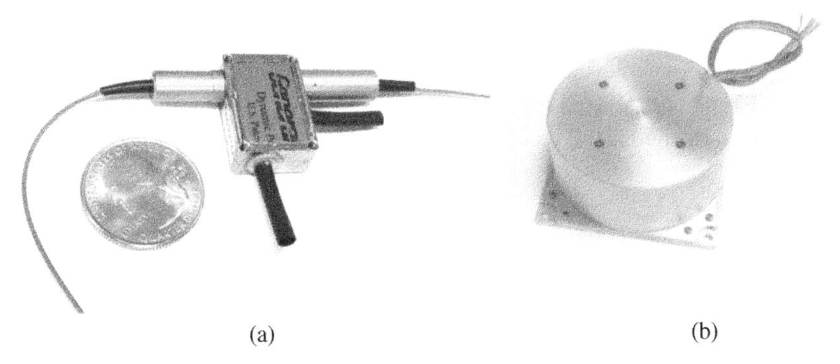

(a) (b)

Figure 11.21 Photos of the proof-of-concept P-FOG developed by the authors. (a) Assembled optical module of the P-FOG configuration with a size of $24.5 \times 17.5 \times 9\,\text{mm}^3$; (b) Fully packaged P-FOG in an aluminum enclosure, including the optical module, an SLED light source, PM fiber, and all of the supporting electronics with a size of $98 \times 98 \times 47.1\,\text{mm}^3$. The coil made with 165 m-diameter PM fiber has a length, inner diameter, outer diameter, and height of 585 m, 83.1, 91.7, and 12.64 mm, respectively. Source: Yao et al. (2019)/Optica Publishing Group.

The ability of the P-FOG to sense rotation was evaluated using a high-precision rotation table that can generate rotation rates between 0.001 and 1000°/s. The rotation table was programmed to rotate back and forth with a gradually increasing rate of 0–1000°/s. Figure 11.22a shows the output of the P-FOG over time. The device correctly detected the varying rotation rates. The scale factor was obtained by plotting the P-FOG output $\Delta\varphi$ as a function of the input rotation rate Ω and linearly fitting the curve (Figure 11.22b). The linear fit gave the scale factor as a slope of 0.0144, in agreement with the value calculated using Eq. (11.13) ($\Delta\varphi = 0.0143\Omega$ (Ω in °/s)) for the PM fiber coil chosen. The output of the P-FOG remained linear even at ±1000°/s, despite the fact that the rotation induced phase difference reached 14.3 rad (~4.55π) at a rotation rate of 1000°/s. This result demonstrates the phase unwrapping capability and, consequently, unlimited dynamic range of the P-FOG. This dynamic range is a clear enhancement over those reported for OP-FOG (Doerr et al. 1994; Lynch 1999) and open-loop I-FOG devices (Lefèvre 1993), which were limited to a fraction of π, and rivals that of a closed-loop I-FOG (Merlo et al. 2000).

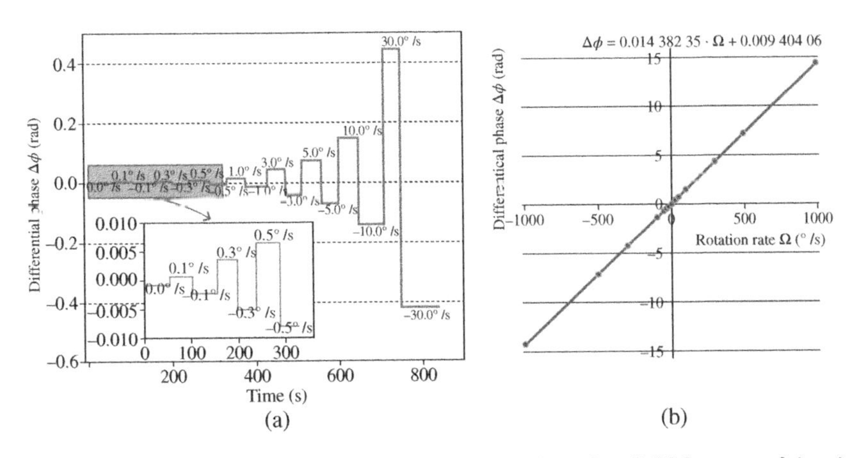

Figure 11.22 P-FOG performance demonstration. (a) Real-time P-FOG output of the phase difference $\Delta\varphi$ as the P-FOG is rotated back and forth at increasing rotation rates; (b) $\Delta\varphi$ as a function of rotation rate Ω. The linear fit (solid line) of the measured data points yields the scale factor (Yao et al. 2019).

To evaluate the detection sensitivity of the P-FOG, it is necessary to generate highly accurate rotation at slow rates ranging within fractions of degrees per hour, which is beyond the capability of the rotation table available in the lab. To achieve these slow rates, the earth's rotation can be explored by vertically mounting the P-FOG on a leveled rotation stage (Tanaka et al. 1994) (Figure 11.23a). The rotation rate Ω experienced by the P-FOG can be expressed as (Tanaka et al. 1994)

$$\Omega = \Omega_0 cos\psi cos\theta \quad (11.17)$$

where Ω_0 is the earth's rotation rate (15°/h), ψ is the latitude angle, and θ is the rotation stage angle. The measurement was conducted in Suzhou China, at 31.3° north latitude. Figure 11.23b shows the measured rotation rate as a function of the rotation stage angle using a $\Omega_0 cos\psi$ value of 12.82° (15*cos31.3°) to calibrate the measurement data. The measured rotation rate varies sinusoidally with respect to angle θ, consistent with Eq. (11.17). Finally, a subdegree per hour angular rate sensitivity is observed; however, bias drift, which mainly results from the temperature dependence of the optical head and the electronic circuit, restricts further sensitivity enhancement. The temperature insensitive QWP or the procedure for extracting the temperature induced differential phase described previously can be implemented to minimize the bias drift.

Bias instability and angular random walk (ARW) are the most important parameters for a gyroscope (Lefèvre 1993), which describe the instability of the bias offset and the average error as a result of random noise of the gyroscope, respectively. To further evaluate the performance of the P-FOG, the bias instability and AWG of the P-FOG at room temperature and in variable temperature environments are measured by placing the whole P-FOG, including the fiber coil and the P-FOG body, in a temperature chamber. In data processing, the measured $\Delta\varphi$ data are first converted to rotation rates using the scale factor obtained in Figure 11.22 and are then further averaged to obtain the bias instability, while the ARW is obtained by processing the data using Allen Variance (Lefèvre 1993) analysis.

As shown in Figure 11.24a, the bias instability at 25 °C averages 0.09°/h over a 100 seconds period and slightly drifts downward over time. Figure 11.24b shows the bias instability under variable temperature conditions at heating/cooling rates of 1 °C/min (dashed line). The maximum peak-to-peak bias instability approximates 0.85°/h, which is considered to be outstanding for a gyroscope without temperature compensation or control. The ARW of the P-FOG measures $0.0015°/\sqrt{h}$ with a 1 second integration time at 25 °C, which is indicative of a low-noise device.

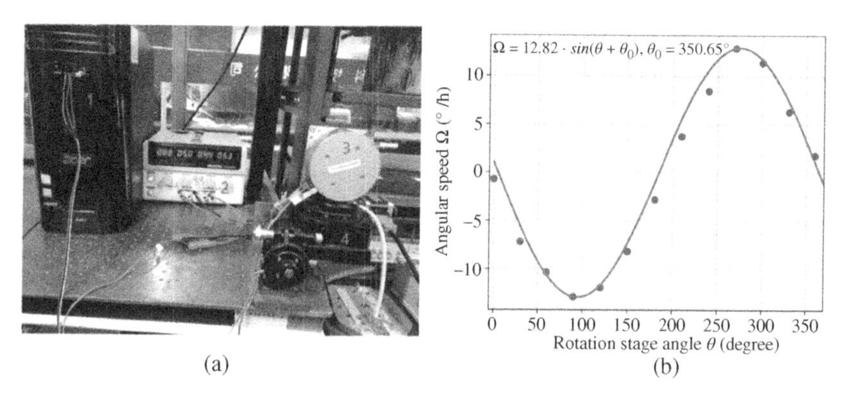

Figure 11.23 Detection sensitivity measurement of the P-FOG. (a) The P-FOG is mounted on a leveled rotation stage to sense a fraction of the earth's rotation rate; (b) P-FOG output as a function of rotation stage angle. The solid line is obtained by fitting the data points with a sine function. Source: Yao et al. (2019)/Optica Publishing Group.

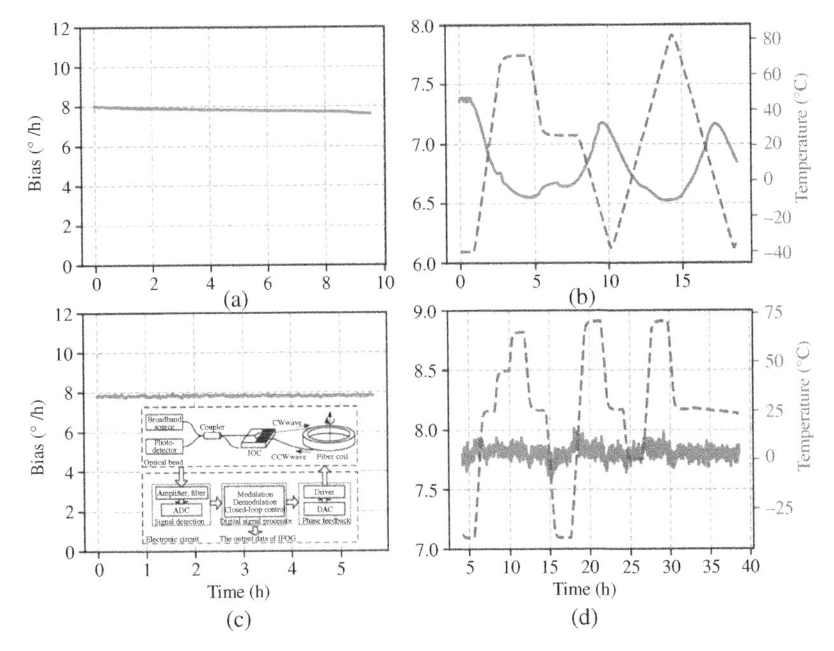

Figure 11.24 Bias instability measurements of P-FOG and I-FOG. (a) Bias instability of P-FOG at 25 °C; (b) Bias instability of P-FOG in a variable temperature environment (solid line); (c) Bias instability of a closed-loop I-FOG at 25 °C. Inset: closed-loop I-FOG configuration; (d) Bias instability of the closed-loop I-FOG (solid line) with only the PM fiber coil placed in the temperature chamber, the rest of the I-FOG, including the light source, PIN-FET detector, the IOC, the electronics, and the data interfaces, were all enclosed in a compact package and placed outside of the chamber. The fiber leads connecting the fiber coil inside the chamber and the I-FOG package outside the chamber were well protected with goose neck tubes to prevent air-flow induced fiber motions. The dashed lines represent the programmed temperature profiles. Both gyroscopes are stationary and the heating and cooling rates amount to 1 °C/min (Yao et al. 2019).

For comparison, a $LiNbO_3$ IOC-based commercial closed-loop I-FOG classified as a high-end tactical-grade gyroscope (Lefèvre 1993) was modified by replacing its original fiber coil with the 585 m-long PM fiber coil used in the P-FOG and measured its bias instability at 25 °C. Measurements were conducted using a 1310 nm SLED as a light source and a PIN photodiode combined with a high-impedance field-effect transistor preamplifier (PIN-field effect transistor [FET]) as a low-noise optical detector. Light source power and spectral width have been optimized by the I-FOG manufacturer along with the detector bandwidth and gain. The modulation frequency used by the I-FOG's electronic circuit to drive the IOC was readjusted to match the eigenfrequency of the fiber coil (Lefèvre 1993). The results are shown in Figure 11.24c. The technically mature I-FOG shows lower bias instability than the proof-of-concept P-FOG (0.037 vs. 0.09°/h). However, its ARW measures $0.0069°/\sqrt{h}$ for a 1 second integration, which is 4.6 times higher than that of the P-FOG ($0.0015°/\sqrt{h}$).

Low bias instability in varying temperature environments is also critical for a gyroscope to achieve in order to assure its accuracy at all times. The bias instability of the P-FOG under variable temperature conditions (Figure 11.24b) includes contributions from the PM fiber coil (Lefèvre 1993; Shupe 1980) and the P-FOG body, which consists of the optical head, electronic circuitry, and mechanical package. We are more interested in the contribution from the P-FOG body because it reflects the temperature behavior of the P-FOG concept. The contribution of the fiber coil, which results from the Shupe effect (Shupe 1980), can be independently determined and be excluded from the total bias instability of the P-FOG to obtain the contribution from the P-FOG body alone. To evaluate fiber coil's contribution under variable temperature conditions, the fiber coil is separated from the closed-loop I-FOG main body containing the optics and electronics components (Figure 11.24c, inset) and is placed alone in a temperature chamber for testing. These measurements show that the peak-to-peak and average bias instabilities of the fiber coil resulting from the temperature ramps (Figure 11.24d, dashed line) approximate 0.4 and 0.056°/h, respectively. The peak–peak bias instability of the coil is 0.45°/h lower than that of the entire P-FOG, which includes contributions from the fiber coil, optical assembly, and electronic circuit (Figure 11.24b).

High data output rate of the P-FOG at 200 points per second is also attractive for vehicle navigation and control, because this low-latency positioning information is critical to high speed vehicles, robots, and drones. For example, at 100 km/h, a vehicle moves 0.14 m between 200 Hz position reports.

Table 11.1 compares the key performance data of the proof-of-concept P-FOG with those of the modified commercial closed-loop I-FOG. The P-FOG displays somewhat worse bias instability but superior ARW performance than the technically mature I-FOG. Nonetheless, according to the gyroscope classification of Table 11.2 (Lefèvre 1993), these performance parameters would qualify the P-FOG device as a high-end tactical-grade instrument. Continued efforts to optimize the optical assembly and electronics can significantly reduce the bias instabilities of the P-FOG for more demanding applications.

Table 11.1 Comparison of the performance data for the first proof-of-concept P-FOG and a modified commercial closed-loop I-FOG using the same fiber coil (Yao et al. 2019).

	Average bias instability (25 °C)	Peak-to-peak bias instability (varying temperature)	Random walk	Maximum rotation rate	Data output rate
P-FOG	0.09°/h	0.85°/h for the whole P-FOG	$0.0015°/\sqrt{h}$	1000°/s[b] demonstrated	200/s
I-FOG	0.037°/h	0.4°/h[a] for the fiber coil only	$0.0069°/\sqrt{h}$	600°/s[c] for the device obtained	1/s

a) Only the fiber coil is placed inside the temperature chamber and subjected to temperature variations.
b) Unlimited in principle; limited in practice only by the speed of the electronics.
c) Limited by the speeds of electronics and phase modulators.

Table 11.2 Performance grade classification of gyroscopes (Lefèvre 1993).

Grade	ARW (Angle random walk)	Bias instability
Inertial	$<0.001°/\sqrt{h}$	<0.01°/h
Tactical	$(0.5–0.05)°/\sqrt{h}$	(0.1–10)°/h
Rate	$>0.5°/\sqrt{h}$	(10–1000)°/h

11.3 Polarimetric Magnetic Field and Electrical Current Sensors

As discussed in Chapters 3, 4, and 6, the magnetic field can induce circular retardation, causing left hand circular (LHC) and right hand circular (RHC) polarized light to have different phases (the Faraday phase shift) when the light propagating in a Faraday material, which in turn causes the major axis of an elliptically polarized light (the Faraday rotation) to rotate. Therefore, by measuring the Faraday rotation or Faraday phase shift, the strength of the responsible magnetic field can be obtained (Papp and Harms 1980; Nguyen et al. 2012). The Faraday materials include different bulk glasses (Barczak et al. 2009), garnet films (magneto-optic [MO] rotators) (Jia et al. 2022), and optical fibers (Smith 1980). Because the garnet films have very large Verdet constant while the optical fibers enable long interaction length with the magnetic field, they are capable of producing relatively large Faraday rotation or circular retardation in response to a magnetic field and therefore are frequently used for magnetic and current sensing applications.

The electrical current sensors made with optical fibers (Kurosawa 1997; Kurosawa et al. 2000) are particularly attractive for the monitoring, control, and protection of substations and power distribution systems in smart grid (Brigida et al. 2013). Comparing with traditional current sensors, they have the advantage of being able to separate the sensor head and electronic processing unit in different locations, and therefore do not require any electrical power at the sensor head, making it safe for high voltage applications. Other advantages include small size, light weight, immune to electromagnetic interferences, low power consumption, and no current saturation.

Accuracy, sensitivity, dynamic range, and environmental stability (Gu et al. 2013) are the main performance indicators for the fiber optic current sensors. An effective way to increase the dynamic range is to increase the magnetic field detection sensitivity, requiring significantly reducing the noise in optoelectronics detection circuitry. The temperature dependence of the Verdet constants of the Faraday glasses or crystals poses challenges for the for the measurement accuracy (Gu et al. 2013; Jia et al. 2022) and must be overcome in the sensor systems design. In addition, the temperature effect on optical signal derived from the sensor head is also critical for achieving environmentally stable operation of the sensor, and must be considered and compensated. Finally, interferences of the magnetic fields from the neighboring current carrying conductors can also affect the current detection accuracy, and measures must be implemented to minimize their impacts.

In this section, we describe several magnetic and current sensor systems which rely on analyzing polarization variations caused by the magnetic field. Specifically, we first describe fiber optic magnetic and current sensor systems utilizing a Faraday rotator as the sensing element operating in transmission and reflection modes, respectively, and finally current sensor systems using optical fiber as the sensing element.

11.3.1 Transmissive Magnetic and Current Sensors Using MO Garnet Films

Figure 11.25a shows the configuration of a transmissive magnetic field and current sensor (Yao 2014, 2017a), in which either a broad bandwidth light source, such as ASE or SLED, with a low DOP or a polarized light source, such as a DFB laser or a SLED, can be used as the light source. The low DOP source is for using low cost single mode (SM) fiber or multimode (MM) fiber to deliver light into the sensor head located remotely at the current sensing site. If a polarized or high DOP light source is used, a more expensive PM fiber must also be used, with its SOP aligned with the slow (or fast) axis of the PM fiber. In the sensor head, light first passes through a polarizer to define

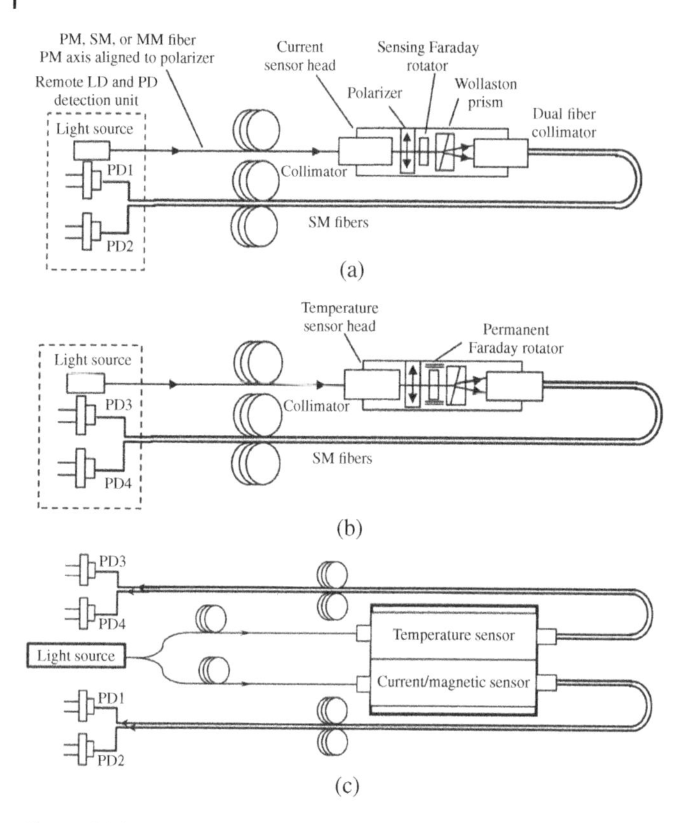

Figure 11.25 (a) Illustration of a transmissive magnetic field sensor relying on a Faraday rotator. (b) A polarimetric temperature sensor with the exact the same configuration as the magnetic sensor using a permanent Faraday rotator. (c) A magnetic or current sensing system with temperature monitoring capability for real time temperature compensation for achieving high accuracy (Yao 2014, 2017a).

the input polarization. For the case of using a low DOP source, there is no requirement of the orientation of the polarizer with respect to the input light. On the other hand, for the case of PM fiber, the polarizer's passing axis should be aligned to the slow (or fast) axis of the PM fiber. After the polarizer, the linearly polarized light then passes through a Faraday material, such as a length of glass, a length of Faraday crystal, or piece of Faraday film capable of rotating the polarization of light in response to a magnetic field parallel to the Faraday material.

A planar anisotropy Faraday thick film described in Figure 6.23 is preferred due to its large rotation sensitivity and low domain scattering. A micro Wollaston prism is used to separate the input linear polarization into two orthogonal polarization components propagating at a small crossing angle, say 3.7°. The orientation of the Wollaston prism is 45° from the polarizer's passing axis such that when there is no magnetic field present, the two orthogonal polarization components have equal powers (50% splitting). A dual fiber collimator with a crossing angle similar to that of the Wollaston prism is then used to focus the two polarization two separate output fibers. The two output fibers can be either SM or MM to transmit the light back to the signal processing unit remotely located from the optical sensing head to be detected by two PDs.

The electric field of the optical beam after the input polarizer can be written as

$$\mathbf{E}_{in} = E_0\hat{\mathbf{y}} \tag{11.18a}$$

where $\hat{\boldsymbol{y}}$ is the passing axis of the input polarizer. After passing through the Faraday rotator with a rotation angle of θ, the electric field becomes

$$\mathbf{E}_F = E_0(cos\theta\hat{\boldsymbol{y}} + sin\theta\hat{\boldsymbol{x}}) \tag{11.18b}$$

The two principle axes of the Wollaston prism can be represented as

$$\hat{\boldsymbol{w}}_1 = \frac{1}{\sqrt{2}}(\hat{\boldsymbol{y}} + \hat{\boldsymbol{x}}), \quad \hat{\boldsymbol{w}}_2 = \frac{1}{\sqrt{2}}(\hat{\boldsymbol{y}} - \hat{\boldsymbol{x}}) \tag{11.18c}$$

where a relative orientation of 45° between the input polarizer and the Wollaston prism is assumed. The powers of the two polarization components in the two output fibers are therefore

$$P_1 = \alpha_1|\mathbf{E}_F \cdot \hat{\boldsymbol{w}}_1|^2 = \frac{1}{2}\alpha_1 E_0^2(1 + sin2\theta) \tag{11.19a}$$

$$P_2 = \alpha_2|\mathbf{E}_F \cdot \hat{\boldsymbol{w}}_2|^2 = \frac{1}{2}\alpha_2 E_0^2(1 - sin2\theta) \tag{11.19b}$$

where α_1 and α_2 are the optical losses caused by the coupling and imperfections of the optical parts. The photocurrents I_1 and I_2 received by the two PDs are therefore

$$I_1 = \frac{1}{2}\rho_1\alpha_1 E_0^2(1 + sin2\theta) = I_{10}(1 + sin2\theta) \tag{11.20a}$$

$$I_2 = \frac{1}{2}\rho_2\alpha_2 E_0^2(1 - sin2\theta) = I_{20}(1 - sin2\theta) \tag{11.20b}$$

where ρ_1 and ρ_2 are responsivities of PD1 and PD2, respectively, and I_{10} and I_{20} are the nominal photocurrent in PD1 and PD2, respectively. If the signals from the two PDs are separately amplified, as in Figure 11.25a, one may always adjust the gains of the two amplifiers to obtain

$$V_1 = V_0(1 + sin2\theta) \tag{11.21a}$$

$$V_2 = V_0(1 - sin2\theta) \tag{11.21b}$$

where $V_0 = \rho_1\alpha_1 G_1 E_0^2/2 = \rho_2\alpha_2 G_2 E_0^2/2$, and G_1 and G_2 are the trans-impedance gains of PD1 and PD2, respectively.

$$sin2\theta = \frac{(V_1 - V_2)}{(V_1 + V_2)} \tag{11.21c}$$

Note that any power fluctuations of the light source can be eliminated with the double outputs. Configuration of Figure 11.25a can be used for both alternating current (AC) and direct current (DC) sensing. If the AC is to be sensed, the configuration can be simplified to have a single output fiber and a single PD in Figure 11.25a because Eq. (11.21a) can be separated into DC and AC components:

$$V_{DC} = V_0 \tag{11.22a}$$

$$V_{AC} = V_0 sin2\theta \tag{11.22b}$$

The Faraday rotation angle can be obtained by

$$sin2\theta = \frac{V_{AC}}{V_{DC}} \tag{11.22c}$$

In practical circuits, the AC and DC components of a signal can easily be separated using a bias tee or high-pass and low-pass filters.

In general, the Faraday material is temperature sensitive, causing magnetic field and current sensing inaccuracies. The current induced rotation angle can be expressed as

$$\theta(H,T) = \theta_0(T) + V_d(T) \cdot L \cdot H(t) \tag{11.23}$$

where $\theta_0(T)$ is the rotation bias, $V_d(T)$ and L are the Verdet constant and the length of the sensing Faraday material, and $H(t)$ is the magnetic strength to be sensed. In order to minimize the error induced by the temperature sensitivity of the Faraday rotator, the local temperature at the sensing head should be monitored so that the polarization rotation caused by the temperature variation can be subtracted.

Figure 11.25b shows a temperature sensor with the same construction as the current sensor in Figure 11.25a, where the variable Faraday rotator in Figure 11.25a is replaced with the permanent Faraday rotator of γ degrees, which is magnetically shielded to prevent the influence of the external magnetic field to be sensed. The Wollaston prism is oriented to allow equal power splitting into the two output fibers at the room temperature during the assembly of the sensor head. The permanent Faraday rotator is preferably set at a nominal rotation angle of 45° ($\gamma = 45°$), which will vary due to the temperature dependence inherent of the permanent Faraday rotator. The temperature induced rotation angle variation will cause the photocurrents detected in PD1 and PD2 to vary, similar to those of Eqs. (11.21a) and (11.21b):

$$V_3 = V_{TO}[1 + sin2\Delta\gamma(T)] \tag{11.24a}$$

$$V_4 = V_{TO}[1 - sin2\Delta\gamma(T)] \tag{11.24b}$$

$$\Delta\gamma(T) = \alpha(T - T_0) \tag{11.24c}$$

where $\Delta\gamma(T)$ is polarization rotation angle variation induced by the temperature variation. Hence a temperature dependence of the photo-voltage difference can be obtained and be used to calculate the temperature. The parameters α and T_0 in $\Delta\gamma(T)$ of a temperature sensor can be obtained for each temperature sensor by putting the sensor in a temperature chamber with different temperature settings. Therefore, the temperature can be obtained using

$$T = \frac{1}{2\alpha} sin^{-1}\left(\frac{V_3 - V_4}{V_3 + V_4}\right) + T_0 \tag{11.24d}$$

The temperature sensor can be co-located with the current sensor, as shown in Figure 11.25c, to detect the temperature at the current sensor for correcting any errors caused by the temperature sensitivity of the Faraday material in the current sensor. Specifically, with temperature T determined, the corresponding $\theta_0(T)$ and $V_d(T)$ in Eq. (11.23) can be obtained, and the magnetic field H can be precisely determined using Eqs. (11.21c) and (11.23).

Figure 11.26 shows the experimental results of the current sensor of Figure 11.25a, in which a MO thick film of planar anisotropy described in Figure 6.23 is used. Figure 11.26a is the actual optical sensor head developed by the authors of this book for measuring an AC current flowing in the copper conductor, while Figure 11.26b shows the measured 50 Hz AC current as a function of time. Clearly, current variations as small as 2.07 mA can be clearly measured. Figure 11.26c shows the measured Faraday rotation angle (amplitude) as a function of rms current. A sensitivity of 0.029°/mA can be clearly achieved. Finally, the measurement sensitivity measurement is shown in Figure 11.26d, in which an rms current step of 1 mA can be clearly resolved.

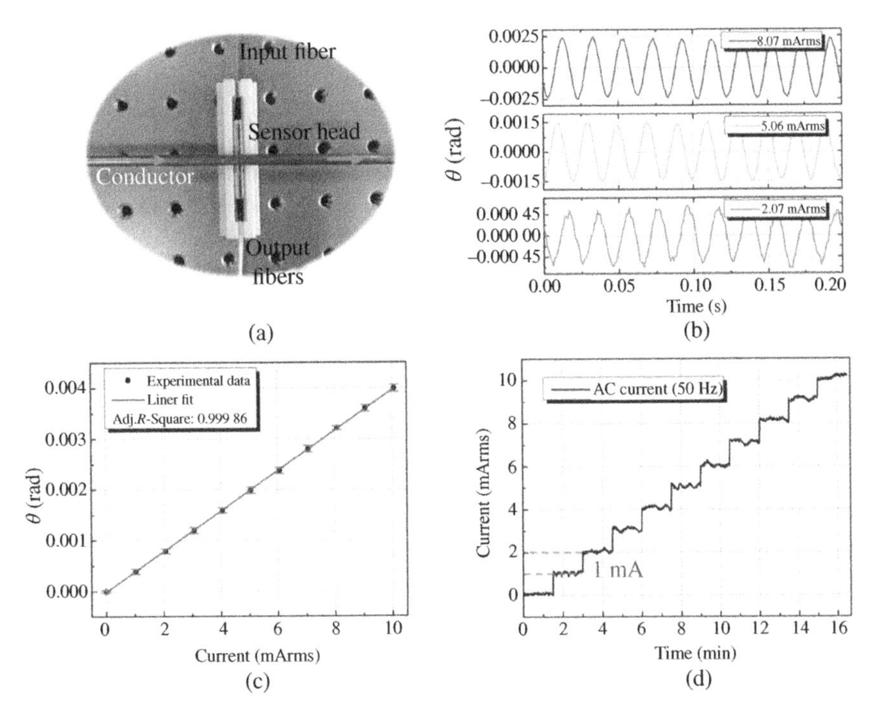

Figure 11.26 Experimental data of a transmissive current sensor using MO thick film of planar anisotropy. (a) Optical sensor head for measuring 50 Hz AC electrical current flowing in the conductor. (b) Measured Faraday rotation angle induced by the 50 Hz AC current as a function of time. (c) Measured peak Faraday rotation angle as a function of the rms current. (d) Measured current as a function of time when the input ramping up at 1 mA per step to show the measurement resolution and sensitivity.

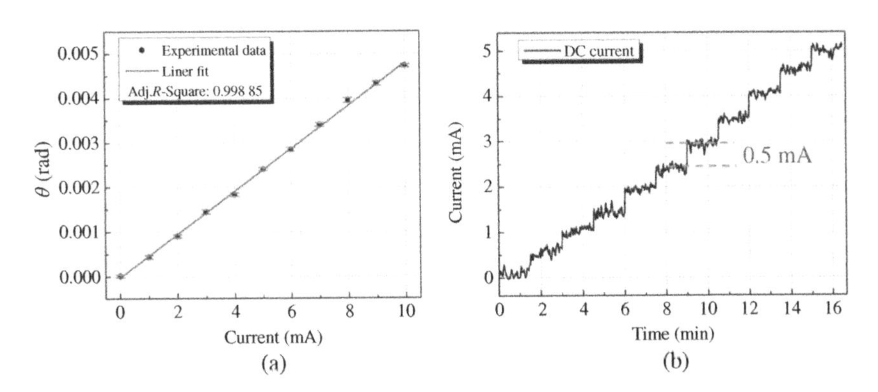

Figure 11.27 Experimental data of DC current measurement using the sensing head of Figure 11.26a. (a) Measured Faraday rotation angle as a function of current. (b) Result showing the measurement resolution of the DC current.

Figure 11.27 shows the experimental results of using the sensor head in Figure 11.26a for measuring the DC current. The measured Faraday rotation angle as a function of current is shown in Figure 11.27a, with a measurement sensitivity of 0.029°/mA obtained. The measurement resolution of 0.5 mA is obtained by increasing the input current 0.5 mA per step.

The transmissive configurations described in Figure 11.25 are input polarization sensitive and therefore either a low DOP light source or a PM fiber input is required. As mentioned earlier, the low DOP option may increase the cost of the light source, in addition to suffering a 3 dB loss due to polarizing optics at the input of the sensor head, while the PM fiber input option adds extra cost, especially when a long PM fiber lead is required if the interrogation unit is far away from the sensing point where the sensor head is located. To overcome these issues, a polarization insensitive design can be used, as described in Yao (2017b). With such a polarization insensitive sensor head, the input fiber lead can be the low cost standard SM fiber, with no restriction on the DOP of the light source. Interested readers are encouraged to read Yao (2017b).

11.3.2 Reflective Magnetic and Current Sensors Using MO Thick Film as the Sensing Medium

The transmissive magnetic and current sensors described in Section 11.3.1 require separate input and output fibers on both ends of the device, which makes the device bulky and difficult to install for applications requiring tight spaces. It is therefore desirable to design a reflective sensor system in which the input and output share the same fiber. In addition, because light passing through the sensing material twice, the measurement sensitivity is doubled. Finally, the reflective design allows the sensing head to be inserted in tight spaces, without having to manage the output fiber as in the transmissive sensor head.

Figure 11.28a shows the configuration of a reflective magnetic field and current sensor (Yao 2017c), in which either a broad bandwidth light source with a low DOP or a polarized light source can be used as the light source. Similar to Figure 11.25a, the low DOP source is for using low cost SM fiber to deliver light into the sensor head located remotely at the current sensing site. If a polarized or high DOP light source is used, a more expensive PM fiber must also be used, with its SOP aligned with the slow (or fast) axis of the PM fiber. In the sensor head, light first passes through a Wollaston PBS to define the input polarization. For the case of using a low DOP source, there is no requirement of the orientation of the Wollaston PBS with respect to the input light. On the other hand, for the case of PM fiber, the Wollaston PBS' passing axis should be aligned to the slow (or fast) axis of the PM fiber. After the Wollaston PBS, the linearly polarized light then passes through a permanent Faraday rotator with a fixed 22.5° polarization rotation made with a thick MO film of either perpendicular anisotropy (Figure 6.19) or planar anisotropy (Figure 6.23) under a saturation magnetic field provided by a permanent ring magnet. The light beam then passes through a sensing Faraday material, such as a length of Faraday crystal, or piece of Faraday film capable of rotating the polarization of light in response to a magnetic field parallel to the Faraday material. Again, a planar anisotropy Faraday thick film of Figure 6.23 is preferred due to its large rotation sensitivity and low domain scattering. A planar mirror is then used to reflect the light beam back to pass the sensing Faraday material and the 22.5° Faraday rotator twice before being split by the Wollaston PBS and finally coupled into the two fibers by the dual fiber collimator.

In the absence of the external magnetic field, 0° polarization rotation is produced by the sensing Faraday material and therefore the reflected light at the Wollaston PBS is 45° from the principal axis of the PBS (by passing the permanent 22.5° Faraday rotator twice) such that the light beam will split 50% into each of the two fibers. When the external magnetic field is non-zero, the SOP at the Wollaston will deviate from 45°, causing the optical powers into the two fibers unbalanced, with the power difference proportional to the external magnetic field. The optical power in each fiber is then detected by a PD before being converted into a photo-voltage. Therefore, by measuring the difference in detected photo-voltages by the two PDs, the magnetic field strength and direction can be

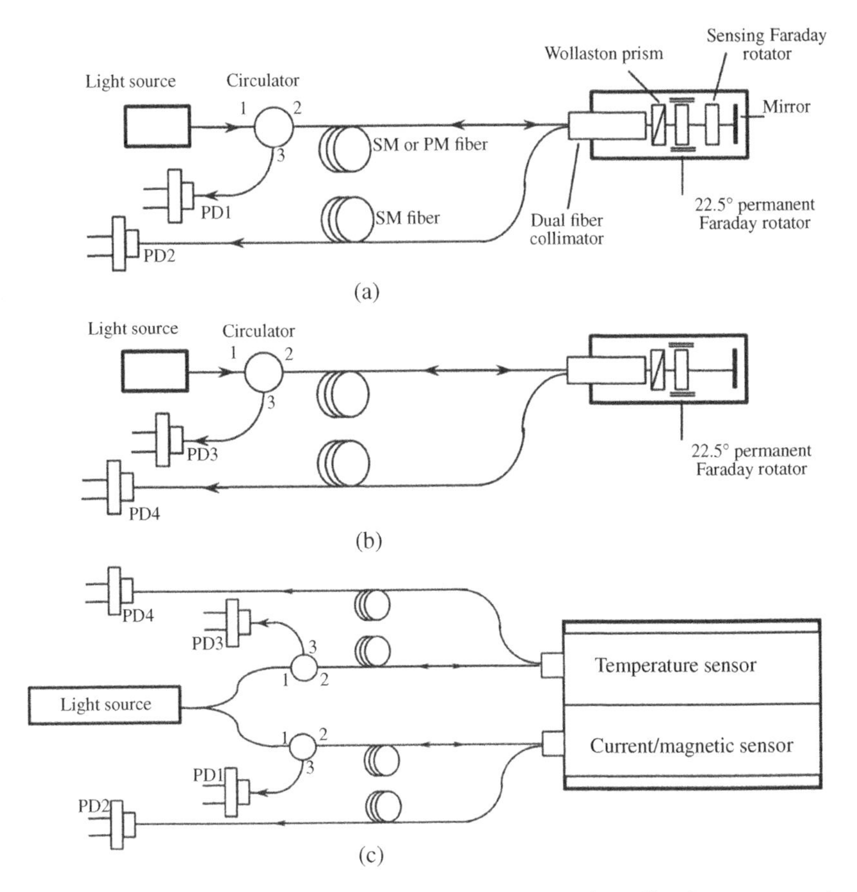

Figure 11.28 (a) Reflective magnetic and current sensor using a Faraday rotator as the sensing material. (b) A matching temperature sensor by removing the magnetic field sensing material in (a). (c) A pair of temperature and current/magnetic sensors co-located for accurate magnetic field and current sensing, where the temperature sensing is for compensating the temperature sensitivity of the Faraday rotator (Yao 2017c).

obtained. Similar to the derivations in Section 11.3.1, when the electrical gains for the detection circuits are properly adjusted, the photo-voltages from the two PDs, PD1 and PD2, can be expressed as

$$V_1 = V_0(1 + sin4\theta) \tag{11.25a}$$

$$V_2 = V_0(1 - sin4\theta) \tag{11.25b}$$

By comparing with Eq. (11.21), one may notice a difference of factor of 2 (2θ vs. 4θ), which is caused by the double pass of the light beam through the sensing Faraday material. The polarization rotation angle θ caused by the magnetic field and by the temperature can be expressed by the following equation:

$$sin4\theta = \frac{(V_1 - V_2)}{(V_1 + V_2)} \tag{11.25c}$$

$$\theta = \frac{1}{4}sin^{-1}\left[\frac{(V_1 - V_2)}{(V_1 + V_2)}\right] \tag{11.25d}$$

In general, the rotation angle θ is a function of both the magnetic field H and the temperature T, which can be expressed by Eq. (11.23). Therefore, in order to minimize the measurement inaccuracy caused by the temperature dependence of the sensing Faraday material and the bias 22.5° Faraday rotator, the temperature T must be precisely measured so that the SOP rotation caused by the temperature variation can be subtracted.

Figure 11.28b shows a temperature sensor head in which the Faraday sensing material in Figure 11.28a is removed. Similar to Eq. (11.24), the photo-voltages detected by PD3 and PD4 can be written as

$$V_3 = V_{T0}[1 + sin4\Delta\gamma(T)] \tag{11.26a}$$

$$V_4 = V_{T0}[1 - sin4\Delta\gamma(T)] \tag{11.26b}$$

$$\Delta\gamma(T) = \alpha(T - T_0) \tag{11.26c}$$

Again, due to the double pass, the signal is twice more sensitive to the temperature variation as that in Eq. (11.24). From Eq. (11.26), one obtains

$$T = \frac{1}{4\alpha} sin^{-1}\left(\frac{V_3 - V_4}{V_3 + V_4}\right) + T_0 \tag{11.26d}$$

When the T is obtained, the temperature induced rotation angle can be abstracted out and the true magnetic field induced rotation can be obtained. In addition, the Verdet constant $V_d(T)$ at temperature T can be used to calculate the final magnetic field or the corresponding electrical current.

Finally, Figure 11.28c shows a pair of magnetic and temperature sensors packaged together to sense both the magnetic field (or electrical current) and the temperature simultaneously. The temperature measured by the temperature sensor can be used to compensate the errors in the magnetic field measurement induced by temperature variations.

11.3.3 Reflective Current Sensor Using Optical Fiber as the Sensing Medium

The sensor systems based on the MO thick films described earlier are more suited for the magnetic field sensing, less suited for the electrical current sensing, because they are sensitive not only to the magnetic field produced by the electrical current in the intended conductor, but also to the interferences of other magnetic fields produced by the currents in the nearly conductors. In addition, they are also sensitive to the distance and the orientation with respect to the conductor to be sensed.

In order to minimize or eliminate such interferences, it is preferable to use a length of optical fiber as the sensing medium, in which the fiber is coiled around the conductor to be sensed, as shown in Figure 11.29a (Yao 2017c). This configuration is almost exactly the same as that of Figure 11.28a, except that the sensing Faraday medium is replaced by the fiber coil and the mirror is replaced by a 90° Faraday mirror. From the Ampere circuital law, $\oint H \cdot dl = I$, the fiber is only sensitive to the magnetic field produced by the electrical current enclosed by the fiber coil and is independent of the position and orientation of the electrical current inside the coil.

As described in Figure 6.24, the 90° Faraday mirror for eliminating the degrading effect of residual linear birefringence in the sensing fiber (Sections 9.4.3 and 9.4.4) so that only the Faraday effect induced circular birefringence is present to induce polarization rotation to be detected by the system. The SOP fluctuations associated with fiber motion and temperature variations can be eliminated. Consequently, the photo-voltages in PD1 and PD2 of Figure 11.29a can be expressed by Eqs. (11.25a) and (11.25b). The associated Faraday rotation angle θ can then be obtained by

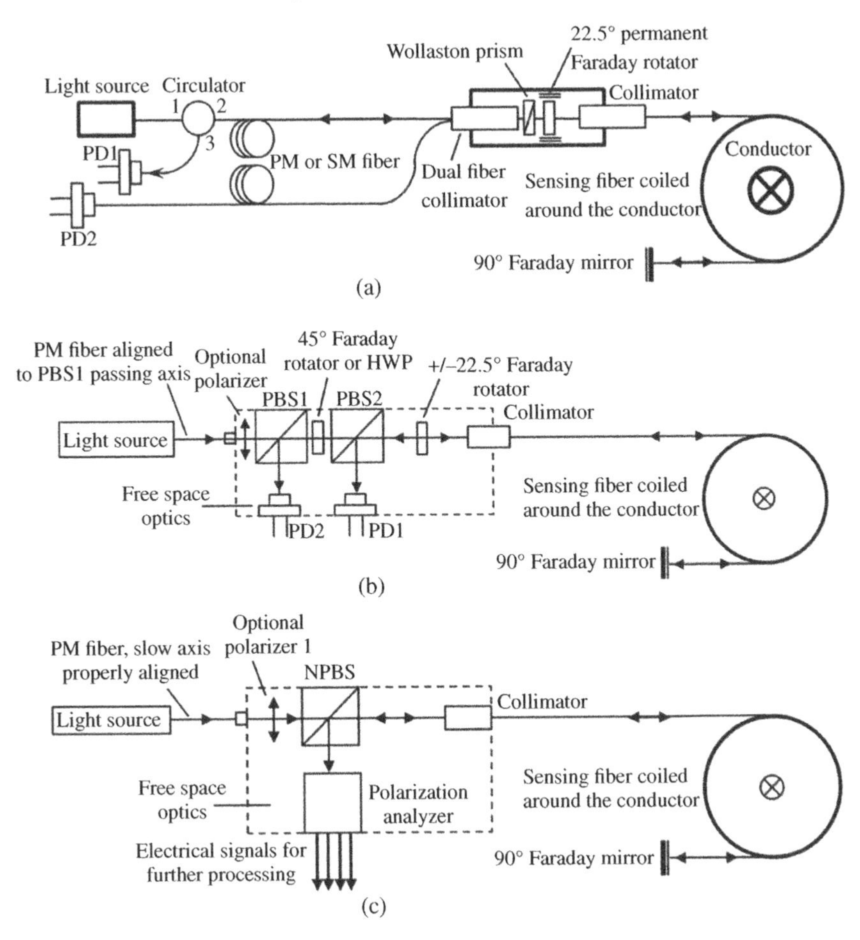

Figure 11.29 (a) A polarimetric electric current sensor using a length of optical fiber as the sensing medium. Hi-Bi spun optical fibers described in Chapters 3 and 9 are preferred in practice. (b) A configuration enabling infinite dynamic range for AC sensing. (c) A full polarimetric fiber optic current sensor with infinite dynamic range for both AC and DC sensing (Yao 2017c).

Eq. (11.25d). To eliminate the temperature dependence of the Verdet constant of the optical fiber, the temperature sensor of Figure 11.28b can be used to monitor the temperature at the sensing fiber so that the polarization deviation caused by the temperature dependence of the Verdet constant can be compensated digitally.

As can be seen from Eqs. (11.25a) and (11.25b), the received photo-voltage in PD1 and PD2 are sinusoidal functions of the Faraday rotation angle, whose range is limited to the first half of the sinusoidal period. When the Faraday rotation angle is larger than 90°, the detected photo-voltages are no longer monotonic, which makes it impossible to determine the direction of the Faraday rotation which is associated with the direction of the magnetic field or the electrical current flow. To overcome this difficulty, the configuration of Figure 11.29b is proposed.

As shown in Figure 11.29b, light from the light source is coupled to the optical head with a PM fiber or a SM fiber, and passes through an optional polarizer to clean up the polarization to be totally linear. The SOP of the light is aligned with the passing axis of PBS1 so that the light passes the PBS1 with a minimum insertion loss. After PBS1, the SOP is rotated 45° with a Faraday rotator or a HWP oriented 22.5° from the passing axis of PBS1 (the 45° SOP rotator). The PBS2 is orientated such that its passing axis is aligned to SOP of the light beam from the 45° SOP rotator to allow it passing with a minimum loss. After PBS2, light passes through a 22.5° Faraday rotator and then is coupled into a sensing fiber with a collimator. At the other end of the fiber, the light is reflected by a Faraday mirror to pass through the 22.5° Faraday rotator the second time with a total 45° SOP rotation when the magnetic field to be sensed is zero. The *s* polarization component of the light is reflected out by PBS2 into PD1 and the *p* component continues through PBS2, which is further rotated another 45° by the 45° SOP rotator so that it is totally reflected by PBS1 into PD2. Assuming the electrical gains are properly adjusted, the photo-voltages from the detected photocurrents in PD1 and PD2 can be expressed as

$$V_1 = V_0(1 + sin4\theta) \tag{11.27a}$$

$$V_2 = V_0(1 + cos4\theta) \tag{11.27b}$$

For AC sensing, one may separate the DC and AC components in Eqs. (11.27a) and (11.27b) with a bias-T, which can be separated as

$$V_{DC} = V_0 \tag{11.28a}$$

$$V_{1AC} = V_0 sin4\theta \tag{11.28b}$$

$$V_{2AC} = V_0 cos4\theta \tag{11.28c}$$

$$\theta = \frac{1}{4} tan^{-1}\left(\frac{V_{1AC}}{V_{2AC}}\right) \tag{11.28d}$$

Similar to the discussions on the sine–cosine OFD and the P-FOG, because both sine and cosine of the polarization rotation angle are present, any arbitrary large rotation angle can be obtained without ambiguity. Therefore, one may use a long sensing fiber to increase the detection sensitivity without causing detection ambiguity.

For DC sensing, one may obtain from Eqs. (11.27a) and (11.27b)

$$sin4\theta = \frac{(V_1 - V_0)}{V_0} \tag{11.29a}$$

$$cos4\theta = \frac{(V_2 - V_0)}{V_0} \tag{11.29b}$$

$$\theta = \frac{1}{4} tan^{-1} \frac{(V_1 - V_0)}{(V_2 - V_0)} \tag{11.29c}$$

A major difficulty with Eq. (11.29c) for DC sensing is the precise determination of V_0 at all times because it varies with the optical power from the light source. In order to overcome this difficulty, the full polarimetric configuration is proposed by the authors of this book, as described next.

As shown in Figure 11.29c, light from the light source is coupled to the optical head with a PM fiber or a SM fiber, and passes through a polarizer to make it totally linear before passing through a polarization insensitive beam splitter (NPBS) to be coupled into the sensing fiber. At the end of the sensing fiber, the light is reflected back by a 90° Faraday mirror toward the NPBS. About half of the light is directed by the NPBS toward an amplitude division or wave-front division polarimeter described in Chapter 8 to be analyzed.

Let $\mathbf{E}_{in} = E_0\hat{\mathbf{y}}$ be the optical field inputting to the sensing fiber; the returned optical beam from the 90° Faraday mirror and the sensing fiber can be expressed as

$$\mathbf{E}_{out} = E_0(cos2\theta\hat{\mathbf{x}} + sin2\theta\hat{\mathbf{y}}) \tag{11.30}$$

where θ is the single-pass Faraday rotation angle induced by the current in the sensing fiber.

Assuming the wave-front division polarimeter of Figure 11.20 is used to analyze the SOP of the return beam after it is reflected by the NPBS, we have

$$V_1 = \frac{\sigma}{2}|\mathbf{E}_{out}\cdot\hat{x}|^2 = \frac{1}{2}\sigma E_0^2 cos^2 2\theta \tag{11.31a}$$

$$V_2 = \frac{\sigma}{2}|\mathbf{E}_{out}\cdot\hat{y}|^2 = \frac{1}{2}\sigma E_0^2 sin^2 2\theta \tag{11.31b}$$

$$V_3 = \frac{\sigma}{4}|\mathbf{E}_{out}\cdot(\hat{x}+\hat{y})|^2 = \frac{1}{4}\sigma E_0^2(1 + sin4\theta) \tag{11.31c}$$

$$V_4 = \frac{\sigma}{4}|\mathbf{E}_{out}\cdot(\hat{x}-i\hat{y})|^2 = \frac{1}{4}\sigma E_0^2 \tag{11.31d}$$

where σ is a parameter to account for the overall optical loss and electrical gain of each channel, assuming the electrical gain of each PD is properly adjusted.

The Stokes parameters can be obtained as

$$S_0 = V_1 + V_2 = \frac{1}{2}\sigma E_0^2 \tag{11.32a}$$

$$s_1 = \frac{(V_1 - V_2)}{S_0} = cos4\theta \tag{11.32b}$$

$$s_2 = \frac{(2V_3 - S_0)}{S_0} = sin4\theta \tag{11.32c}$$

$$s_3 = \frac{(2V_4 - S_0)}{S_0} = 0 \tag{11.32d}$$

$$\theta = \frac{1}{4}tan^{-1}\left(\frac{S_2}{S_1}\right) \tag{11.32e}$$

Similar to the P-FOG, the SOP rotation angle induced by the electrical current traces out a circle on the Poincaré Sphere; however, unlike the P-FOG where the SOP circle is on the s_2 and s_3 planes, here the SOP circle is on the s_1 and s_2 planes. With this complete polarimetric configuration, unlimited dynamic range can be achieved for both AC and DC current sensing.

Figure 11.30 shows the experimental results of using a polarimetric fiber optic current sensor (P-FOCS) described in Figure 11.29a to measure AC electrical current of different rms values. In the experiment, a 10 m spun fiber from Fibercore is used as the sensing fiber and a SLED at 1310 nm is used as the light source. A slight deviation can be observed between the data from the P-FOCS and a standard current transformer (CT) at an rms current of 30 A, as shown in Figure 11.30a. No observable difference between the data from P-FOCS and the standard CT can be seen at an rms current above 120 A, as shown in Figure 11.30b–d, indicating excellent measurement accuracy.

Figure 11.31 shows the ratio error the P-FOCS at different rms current levels compared with a standard CT. It can be seen from Figure 11.31a, at a current of 120 A rms, the ratio error is about

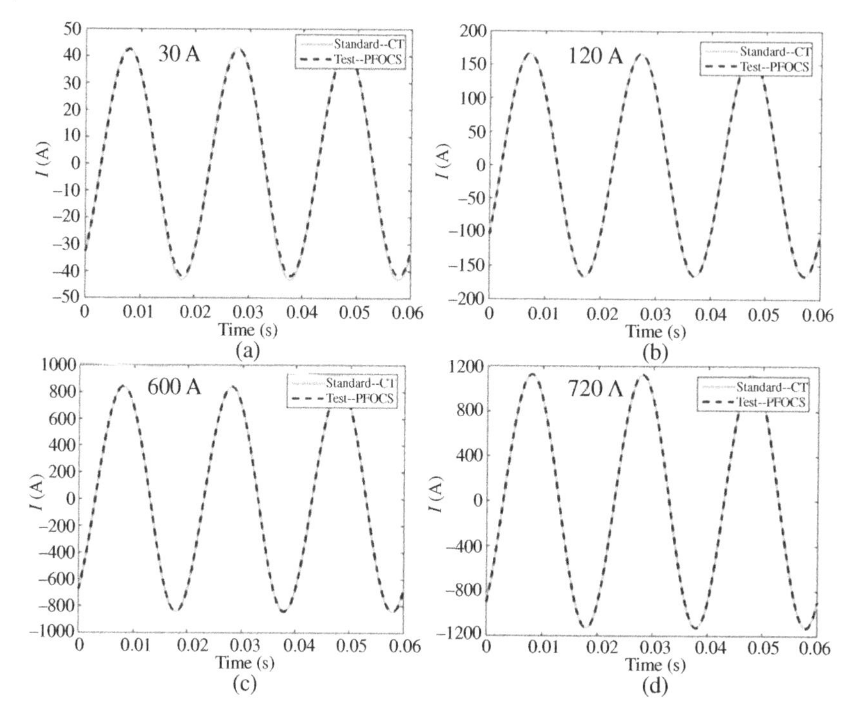

Figure 11.30 Experimental results of the fiber optic current sensor (FOCS) of Figure 11.29a. (a) 30 A rms 50 Hz AC current as a function of time. (b) 120 A rms 50 Hz AC current. (c) 600 A rms 50 Hz AC current. (d) 720 A rms 50 Hz AC current. Gray line: standard current transformer. Dashed line: measured result with FOCS. A slight deviation between the results of FOCS and the standard current transformer at 30 A rms can be seen.

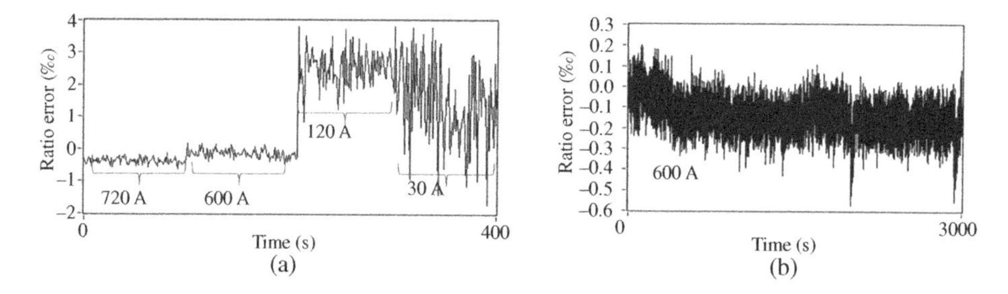

Figure 11.31 Ratio error of the P-FOCS at different current levels. (a) Short term ratio error at 30, 120, 600, and 720 A rms. (b) Long term ratio error measured at 600 A, with the largest ratio error less than 0.06%.

0.3%, while at a current at 600 A, the ratio error is less than 0.02% short term. From Figure 11.31b one can see that the maximum long term ratio error for 600 A rms current monitoring using the P-FOCS is less than 0.06%. The experimental results show the excellent performance of the P-FOCS suitable for high demanding current monitoring applications.

References

Baney, D. and Sorin, W. (1998). High resolution optical frequency analysis, Chapter 5. In: *Fiber Optic Test and Measurement* (ed. D. Derickson), 169–219. New Jersey: Prentice Hall.

Barczak, K., Pustelny, T., Dorosz, D., and Dorosz, J. (2009). New optical glasses with high refractive indices for applications in optical current sensors. *Acta Phys. Pol. A* 116 (3): 247–249.

Bergh, R., Lefèvre, H., and Shaw, H. (1981). All-single-mode fiber-optic gyroscope. *Opt. Lett.* 6: 198–200.

Born, B. and Wolf, E. (1999). *Principles of Optics: Electromagnetic Theory of Propagation, Interference and Diffraction of Light*, 7e. Cambridge University Press.

Brigida, A., Nascimento, I., Mendonca, S. et al. (2013). Experimental and theoretical analysis of an optical current sensor for high power systems. *Photonic Sensors* 3 (1): 26–34.

Cahill, R. and Udd, E. (1979). Phase-nulling fiber-optic laser gyro. *Opt. Lett.* 4: 93–95.

Ciminelli, C., Dell'Olio, F., Armenise, M. et al. (2013). High performance InP ring resonator for new generation monolithically integrated optical gyroscopes. *Opt. Express* 21 (1): 556–564.

Derickson, D. and Stokes, L. (1998). Wavelength meters, Chapter 4. In: *Fiber Optic Test and Measurement* (ed. D. Derickson), 131–168. New Jersey: Prentice Hall.

Doerr, C., Tamura, K., Shirasaki, M. et al. (1994). Orthogonal polarization fiber gyroscope with increased stability and resolution. *Appl. Opt.* 33: 8062–8068.

Ezekiel, S. and Balsamo, S. (1997). Passive ring resonator laser gyroscope. *Appl. Phys. Lett.* 30: 478–480.

Gardner, J. (1985). Compact Fizeau wavemeter. *Appl. Opt.* 24 (21): 3570–3573.

Gray, D., Smith, K., and Dunning, F. (1986). Simple compact Fizeau wavemeter. *Appl. Opt.* 25 (8): 1339–1343.

Gu, X., Müller, G.M., Frank, A., and Bohnert, K. (2013). Temperature-compensated fiber-optic current sensor. Optical Society of America. Frontiers in Optics 2013/Laser Science XXIX: paper FTh2B.4.

Hale, P. and Day, G. (1988). Stability of birefringent linear retarders (waveplates). *Appl. Opt.* 27 (24): 5146–5153.

Hall, J. and Carlsen, J. (ed.) (1997). *Laser Spectroscopy III*. Berlin: Springer-Verlag.

Huang, D., Swanson, E., Lin, C. et al. (1991). Optical coherence tomography. *Science* 254: 1178–1181.

Huber, R., Wojtkowski, M., and Fujimoto, J. (2006). Fourier Domain Mode Locking (FDML): a new laser operating regime and applications for optical coherence tomography. *Opt. Express* 14: 3225–3237.

Jenkins, S. and Hilkert, J. (2008). Sin/cosine encoder interpolation methods: encoder to digital tracking converters for rate and position loop controllers. *Proc. SPIE* 6971: 6971F.

Jia, Q., Han, Q., Liang, Z. et al. (2022). Temperature compensation of optical fiber current sensors with a static bias. *IEEE Sens. J.* 22 (1): 352–356.

Kim, J.C., Kim, J., Kim, C., and Choi, C. (2006). Ultra precision position estimation of servomotor using analog quadrature encoder. *J. Power Electron.* 6: 139–145.

Kurosawa, K. (1997). Optical current transducers using flint glass fiber as the Faraday sensor element. *Opt. Rev.* 4 (1A): 38–44.

Kurosawa, K., Yamashita, K., Sowa, T., and Yamada, Y. (2000). Flexible fiber Faraday effect current sensor using flint glass fiber and reflection scheme. *IEICE Trans. Electron.* E83-C (3): 326–330.

LamdaQuest. (2022). High speed tunable optical filter data sheet. http://www.lambdaquest.com/products.htm (last accessed 28 March 2022).

Lefèvre, H.C. (1993). *The Fiber Optic Gyroscope*. Boston: Artech House.

Lefèvre, H.C. (1997). Fundamentals of the interferometric fiber-optic gyroscope. *Opt. Rev.* 4: A20.

Li, Z., Meng, Z., Liu, T., and Yao, X.S. (2013). A novel method for determining and improving the quality of a quadrupolar fiber gyro coil under temperature variations. *Opt. Express* 21: 2521–2530.

Li, Z., Meng, Z., Wang, L. et al. (2015). Tomographic inspection of fiber coils using optical coherence tomography. *IEEE Photonics Technol. Lett.* 27: 549–552.

Li, J., Suh, M., and Vahala, K. (2017). Microresonator Brillouin gyroscope. *Optica* 4 (3): 346–348.

Liang, W., Ilchenko, V., Savchenkov, A. et al. (2017). Resonant microphotonics gyroscope. *Optica* 4 (1): 114–117.

Luna Innovations. (2022). Polarimeter POD-201 data sheet. POD-101D is obsolete, replaced by POD-201. https://lunainc.com/product/pod-201 (last accessed 28 March 2022).

Lynch, M. (1999). Orthogonal polarization fiber optic gyroscope with improved bias drift. Master of Science thesis. Massachusetts Institute of Technology.

Merlo, S., Norgia, M., and Donati, S. (2000). Fiber gyrocope principles, Chapter 16. In: *Handbook of Fiber Optic Sensing Technology* (ed. J.M. Lopez-Higuera). New Jersey: Wiley.

Nguyen, T, Ely, J., Szatkowski, G., Mata, C., and Mata, A. (2012). Fiber-optic sensor for aircraft lightning current measurement. *2012 International Conference on Lightning Protection (ICLP)*, Vienna, Austria, 2–7 September 2012 IEEE.

Papp, A. and Harms, H. (1980). Magnetooptical current transformer. 1: Principles. *Appl. Opt.* 19 (22): 3729–3734.

Sanders, G., Szafraniec, B., Liu, R. et al. (1995). Fiber-optic gyro development for a broad range of applications. In: *Proceedings of SPIE, Fiber Optic and Laser Sensors XIII, European Symposium on Optics for Environmental and Public Safety, 1995, Munich, Germany*, vol. 2510, 2–11.

Sheem, S. (1980). Fiber-optic gyroscope with [3 × 3] directional coupler. *Appl. Phys. Lett.* 37: 869–871.

Shupe, D. (1980). Thermally induced nonreciprocity in the fiber-optic interferometer. *Appl. Opt.* 19: 654–655.

Smith, A.M. (1980). Optical fibres for current measurement applications. *Opt. Laser Technol.* 12 (1): 25–29.

Snyder, J. (1982). Laser wavelength meters. *Laser Focus* 18: 55.

Tanaka, T., Igarashi, Y., Nara, M., and Yoshino, T. (1994). Automatic north sensor using a fiber-optic gyroscope. *Appl. Opt.* 33: 120–123.

Tran, M., Komljenovic, T., Hulme, J. et al. (2017). Integrated optical driver for interfermetric optical gyroscope. *Opt. Express* 25 (4): 3826–3840.

Vali, V. and Shorthill, R.W. (1976). Fiber ring interferometer. *Appl. Opt.* 15: 1099–1100.

Vobis, J. and Derickson, D. (1998). Optical spectrum analysis, Chapter 3. In: *Fiber Optic Test and Measurement* (ed. D. Derickson), 87–130. New Jersey: Prentice Hall.

Yao, X.S. (2006). Low-cost polarimetric detector. US Patent 7,372,568, filed 22 June 2006 and issued 13 May 2008

Yao, X.S. (2013). Non-interferometric fiber optic gyroscope based on polarization sensing. US Patent 9,823,075, filed 10 January 2013 and issued 21 November 2017.

Yao, X.S. (2014). Faraday current and temperature sensors. US Patent 9,733,133, filed 7 October 2014 and issued 15 August 2017.

Yao, X.S. (2017a). Faraday current and temperature sensors. US Patent 10,281,342, filed 15 August 2017 and issued 7 May 2019.

Yao, X.S. (2017b). Polarization insensitive current and magnetic sensors with active temperature compensation. US Patent 10,969,411, filed 16 February 2017 and issued 6 April 2021.

Yao, X.S. (2017c). Reflective current and magnetic sensors based on optical sensing with integrated temperature sensing. US Patent publication # US 2017/0234912, filed 16 February 2017 and published 17 August 2017.

Yao, X.S. and Chen, X. (2018). Sine–cosine optical frequency encoder devices based on optical polarization properties. US Patent 10,895,477, filed 9 May 2018 and issued 19 January 2021.

Yao, X.S., Zhang, B., Chen, X., and Willner, A. (2008). Real-time optical spectrum analysis of a light source using a polarimeter. *Opt. Express* 16 (22): 17854–17863.

Yao, X.S., Xuan, H., Chen, X. et al. (2019). Polarimetry fiber optic gyroscope. *Opt. Express* 27 (14): 19984–19995.

Yao, X.S., Ma, X., and Feng, T. (2022). Fast optical frequency detection techniques for coherent distributed sensing and communication systems. *Optical Fiber Communications Conference: Invited Presentation Paper M1D.5 in Advanced Coherent Technology Session*, San Diego, CA, 6–10 March 2022.

Zarinetchi, F., Smith, S., and Ezekiel, S. (1991). Stimulated Brillouin fiber-optic laser gyroscope. *Opt. Lett.* 16: 229–231.

Index

Polarization Measurement and Control in Optical Fiber Communication and Sensor Systems, First Edition.
X. Steve Yao and Xiaojun (James) Chen.

t

Printed and bound by CPI Group (UK) Ltd, Croydon, CR0 4YY

16/04/2025

14658352-0004